Springer Collected Works in M

For further volumes:
http://www.springer.com/series/11104

Joseph L. Walsh

Joseph L. Walsh

Selected Papers

Editors

Theodore J. Rivlin · Edward B. Saff

Reprint of the 2000 Edition

Springer

Author
Joseph L. Walsh (1895–1973)
University of Maryland
College Park
USA

Editors
Theodore J. Rivlin (1926–2006)
Chappaqua, NY
USA

Edward B. Saff
Department of Mathematics
Vanderbilt University
Nashville, TN
USA

ISSN 2194-9875
ISBN 978-1-4614-6132-6 (Softcover)
978-0-387-98782-8 (Hardcover)
DOI 10.1007/978-1-4614-6301-6
Springer New York Heidelberg Dordrecht London

Library of Congress Control Number: 2012954628

Mathematics Subject Classification (1991): 26Cxx, 30C15, 30C35, 41A20, 42C10, 65D07

Printed on acid-free paper

Springer is part of Springer Science+Business Media (www.springer.com)

Preface

In this volume we present a selection from the 281 published papers of Joseph Leonard Walsh (1895–1973), a complete list of which appears after this preface. In this list papers that appear in this volume have an asterisk next to their year of publication. We also include a list of Walsh's books, although no excerpts from them appear in this volume. This omission is due to limitation of the size of the volume. However, we wish to emphasize that many important contributions of Walsh appear solely in his books, the most influential of which are "Interpolation and Approximation by Rational Functions in the Complex Domain" and "The Location of Critical Points of Analytic and Harmonic Functions". Following the listing of papers and books is a biography of Walsh and summary of his work by Morris Marden, his first Ph.D. student, and additional material by D.V. Widder and W.E. Sewell.

The selected papers have been divided into seven broad sections. The sections are ordered following the evolution of Walsh's work. Appended to these sections are Commentaries on Walsh's work and a discussion of subsequent developments influenced by the work. We wish to express our gratitude to the following expert commentators for their contributions and their help in choosing the papers presented in this volume. They are Dieter Gaier, P.M. Gauthier, Q.I. Rahman, Walter Schempp and Ferenc Schipp. We wish also to express our thanks to Mrs. Elizabeth Walsh for help and encouragement in this project.

One of Walsh's papers has attained an unpredictably remarkable afterlife. It had not made a large impact for almost half a century and then, unexpectedly, sparked enormous activity which continues to the present day. The work referred to is "A closed set of normal orthogonal functions", Amer. J. Math., vol 45, 1923, pp. 5-24, which introduced what are now known as "Walsh Functions". There exists an immense literature about the theory and applications of these orthogonal functions. Further details will be found in the Commentary following the paper in Section 2.

J.L. Walsh was an Officer in the U.S. Navy in both the first and second World Wars, and is buried in Arlington National Cemetery. He had the ramrod appearance of a naval officer but was kind, helpful and supportive of his students and post-doctoral visitors. The editors of this volume, Ph.D. students of Professor Walsh at Harvard University and the University of Maryland, respectively, were motivated to prepare it by our affection for our teacher and respect for his achievements. We wish also to thank the Harvard Archives which allowed us to examine its ample holdings of Walsh material. A list of all of Walsh's Ph.D. students follows this Preface.

Finally, we wish also to express our thanks to Springer-Verlag, and especially Dr. Rüdiger Gebauer, for encouraging our efforts and being patient with us.

Chappaqua, New York — *Theodore J. Rivlin*
Tampa, Florida — *Edward B. Saff*

Contents

Bibliography of the Papers of Joseph L. Walsh[1]

[16] *Note on Cauchy's integral formula*, Ann. of Math. **18**(1916), 79–80.

[18*] *On the location of the roots of the Jacobian of two binary forms and of the derivative of a rational function,* Trans. Amer. Math. Soc. **19**(1918), 291–298.

[20-a] *On the proof of Cauchy's integral formula by means of Green's formula,* Bull. Amer. Math. Soc. **26**(1920), 155–157.

[20-b] *On the solution of linear equations in infinitely many variables by successive approximations*, Amer. J. Math. **42**(1920), 91–96.

[20-c*] *On the location of the roots of the derivative of a polynomial*, Ann. of Math. **22**(1920), 128–144.

[21-a*] *On the location of the roots of the derivative of a polynomial*, Comptes Rendus du Congrès International des Mathématiciens (Strasbourg, 1920), Publiés par Henri Villat, Toulouse, Édouard Privat, (1921), 339–342.

[21-b*] *On the location of the roots of the Jacobian of two binary forms and of the derivative of a rational function* Trans. Amer. Math. Soc. **22**(1921), 101–116.

[21-c] *A generalization of the Fourier cosine series*, Trans. Amer. Math. Soc. **22**(1921), 230–239.

[21-d] *On the transformation of convex point sets*, Ann. of Math. **22**(1921), 262–266.

[1]Reference numbers refer to the year in the twentieth century. An asterisk next to a reference number means that the article in question appears in this volume.

[21-e] *Sur la position des racines des dérivées d'un polynome*, C.R. Acad. Sci. Paris **172**(1921), 662–664.

[21-f] *Solution to proposed problem by Nathan Altschillercourt, "Find the surfaces all the plane sections of which are circles"*, Amer. Math. Monthly **28**(1921), 480.

[21-g] (With N. Wiener), *The equivalence of expansions in terms of orthogonal functions*, J. Math. Phys. **1**(1921), 103–122.

[21-h] *A theorem on cross-ratios in the geometry of inversion*, Ann. of Math. **23**(1921), 45–51.

[22-a] *A theorem on loci connected with cross-ratios*, Rend. Circ. Mat. Palermo **46**(1922), 236–248.

[22-b] *On the location of the roots of the derivative of a polynomial*, Proc. Nat. Acad. Sci. U.S.A. **8**(1922), 139–141.

[22-c] *A certain two-dimensional locus*, Amer. Math. Monthly **29**(1922), 112–114.

[22-d] *Some two-dimensional loci connected with cross ratios*, Trans. Amer. Math. Soc. **23**(1922), 67–88.

[22-e] *A generalization of normal congruences of circles*, Bull. Amer. Math. Soc. **28**(1922), 456–462.

[22-f] *On the convergence of the Sturm–Liouville series*, Ann. of Math. **24**(1922), 109–120.

[22-g*] *On the location of the roots of certain types of polynomials*, Trans. Amer. Math. Soc. **24**(1922), 163–180.

[22-h] *On the location of the roots of the Jacobian of two binary forms and of the derivative of a rational function*, Trans. Amer. Math. Soc. **24**(1922), 31–69.

[23-a] *Sur un théorème d'algèbre*, C.R. Acad. Sci. Paris **176**(1923), 1361–1364.

[23-b*] *A closed set of normal orthogonal functions*, Amer. J. Math. **45**(1923), 5–24.

[23-c] *A property of Haar's system of orthogonal functions*, Math. Ann. **90**(1923), 38–45.

[24-a] *Sur la détermination d'une analytique par ses valeurs sur un contour*, C.R. Acad. Sci. Paris **178**(1924), 58–60.

[24-b] *On the location of the roots of polynomials*, Bull. Amer. Math. Soc. **30**(1924), 51–62.

[24-c] *Some two-dimensional loci*, Quart. J. Pure Math. **1**(1924), 154–165.

[24-d] *On the expansion of analytic functions in series of polynomials*, Trans. Amer. Math. Soc. **26**(1924), 155–170.

[24-e] *On the location of the roots of Lamé's polynomials*, Tôhoku Math. J. **23**(1924), 312–317.

[24-f] *A generalization of evolutes*, Rend. Circ. Mat. Palermo **48**(1924), 23–27.

[24-g] (Book Review), *The Strasbourg Congress*, Bull. Amer. Math. Soc. **30**(1924), 461–464.

[24-h*] *An inequality for the roots of an algebraic equation*, Ann. of Math. **25**(1924), 285–286.

[24-i] *On Pellet's theorem concerning the roots of a polynomial*, Ann. of Math. **26**(1924), 59–64.

[25-a] *Sur la position des racines des fonctions entières de genre zero et un*, C.R. Acad. Sci. Paris **180**(1925), 2009–2011.

[26-a] *Note on the location of the roots of a polynomial*, Math. Z. **24**(1926), 733–742.

[26-b*] *Über die Entwicklung einer analytischen Funktion nach Polynomen*, Math. Ann. **96**(1926), 430–436.

[26-c*] *Über die Entwicklung einer Funktion einer komplexen Veränderlichen nach Polynomen*, Math. Ann. **96**(1926), 437–450.

[26-d] *Über den Grad der Approximation einer analytischen Funktion*, Sitzungsberichte Baverischen Akad. Wiss., (1926), 233–229.

[27-a] *A paradox resulting from integration by parts*, Amer. Math. Monthly **34**(1927), 88.

[27-b] *On the expansion of harmonic functions in terms of harmonic polynomials*, Proc. Nat. Acad. Sci. U.S.A. **13**(1927), 175–180.

[27-c] *On the degree of approximation to a harmonic function*, Bull. Amer. Math. Soc. **33**(1927), 591–598.

[28-a*] *On the expansion of analytic functions in series of polynomials and in series of other analytic functions*, Trans. Amer. Math. Soc. **30**(1928), 307–332.

[28-b] (Book Review), *The logarithmic potential*, by G.C. Evans, Amer. Math. Monthly **35**(1928), 254–257.

[28-c] *On approximation to an arbitrary function of a complex variable by polynomials*, Trans. Amer. Math. Soc. **30**(1928), 472–482.

[28-d*] *Über die Entwicklung einer harmonischen Funktion nach harmonischen Polynomen*, J. Reine Angew. Math. **159**(1928), 197–209.

[28-e] *On the degree of approximation to an analytic function by means of rational functions*, Trans. Amer. Math. Soc. **30**(1928), 838–847.

[29-a] *Note on the expansion of analytic functions in series of polynomials and in series of other analytic functions*, Trans. Amer. Math. Soc. **31**(1929), 53–57.

[29-b*] *The approximation of harmonic functions by harmonic polynomials and by harmonic rational functions*, Bull. Amer. Math. Soc. **35**(1929), 499–544.

[29-c] *On approximation by rational functions to an arbitrary function of a complex variable*, Trans. Amer. Math. Soc. **31**(1929), 477–502.

[29-d] *Boundary values of an analytic function and the Tchebycheff method of approximation*, Proc. Nat. Acad. Sci. U.S.A. **15**(1929), 799–802.

[30-a] *On the overconvergence of sequences of polynomials of best approximation*, Proc. Nat. Acad. Sci. U.S.A. **16**(1930), 297.

[30-b] *Boundary values of an analytic function and the Tchebycheff method of approximation*, Trans. Amer. Math. Soc. **32**(1930), 335–390.

[30-c] *On the overconvergence of sequences of polynomials of best approximation*, Trans. Amer. Math. Soc. **32**(1930), 794–816.

[31-a] *Note on the overconvergence of sequences of polynomials of best approximation*, Trans. Amer. Math. Soc. **33**(1931), 370–388.

[31-b] *The existence of rational functions of best approximation*, Trans. Amer. Math. Soc. **33**(1931), 668–689.

[31-c*] *On the overconvergence of certain sequences of rational functions of best approximation*, Acta Math. **57**(1931), 411–435.

[32-a] *On interpolation and approximation by rational functions with preassigned poles*, Trans. Amer. Math. Soc. **34**(1932), 22–74.

[32-b] *An expansion of meromorphic functions*, Proc. Nat. Acad. Sci. U.S.A. **18**(1932), 165–171.

[32-c*] *On polynomial interpolation to analytic functions with singularities*, Bull. Amer. Math. Soc. **38**(1932), 289–294.

[32-d] *On the overconvergence of sequences of rational functions*, Amer. J. Math. **54**(1932), 559–570.

[32-e] *On interpolation to harmonic functions by harmonic polynomials*, Proc. Nat. Acad. Sci. U.S.A. **18**(1932), 514–517.

[32-f] *Interpolation and functions analytic interior to the unit circle*, Trans. Amer. Math. Soc. **34**(1932), 523–556.

[33-a] *Bateman on mathematical physics*, Bull. Amer. Math. Soc. **39**(1933), 178–180. (Review).

[33-b] *Interpolation and an analogue of the Laurent development*, Proc. Nat. Acad. Sci. U.S.A. **19**(1933), 203–207.

[33-c] *The Cauchy-Goursat theorem for rectifiable Jordan curves*, Proc. Nat. Acad. Sci. U.S.A. **19**(1933), 540–541.

[33-d] *An extremal problem in analytic functions*, Proc. Nat. Acad. Sci. U.S.A. **19**(1933), 900–902.

[33-e] *Note on polynomial interpolation to analytic functions*, Proc. Nat. Acad. Sci. U.S.A. **19**(1933), 959–963.

[33-f] *Note on the location of the critical points of Green's functions*, Bull. Amer. Math. Soc. U.S.A. **39**(1933), 775–782.

[33-g] *A duality in interpolation to analytic functions by rational functions*, Proc. Nat. Acad. Sci. U.S.A. **19**(1933), 1049–1053.

[33-h] *On series of interpolation and the degree of convergence of sequences of analytic functions*, Tôhoku Math. J. **38**(1933), 375–389.

[33-i*] *Note on the location of the roots of the derivative of a polynomial*, Mathematica (Cluj) **8**(1933), 185–190.

[34-a] (With Helen G. Russell), *On the convergence and overconvergence of sequences of polynomials of best simultaneous approximation to several functions analytic in distinct regions*, Trans. Amer. Math. Soc. **36**(1934), 13–28.

[34-b*] *On approximation to an analytic function by rational functions of best approximation*, Math. Z. **38**(1934), 163–176.

[34-c*] *Note on the orthogonality of Tchebycheff polynomials on confocal ellipses*, Bull. Amer. Math. Soc. **40**(1934), 84–88.

[34-d] *Some interpolation series*, Amer. Math. Monthly **41**(1934), 300–308.

[34-e] *Sur l'interpolation par fonctions rationnelles*, C.R. Acad. Sci. Paris **198**(1934), 1377–1378.

[34-f] *Note on the location of the critical points of harmonic functions*, Proc. Nat. Acad. Sci. U.S.A. **20**(1934), 551–554.

[35-a] *Lemniscates and equipotential curves of Green's function*, Amer. Math. Monthly **42**(1935), 1–17.

[36-a] *A necessary condition for approximation by rational functions*, Bull. Amer. Math. Soc. **42**(1936), 219–221.

[36-b] *The divergence of sequences of polynomials interpolating in roots of unity*, Bull. Amer. Math. Soc. **42**(1936), 715–719.

[36-c] *Note on the behavior of a polynomial at infinity*, Amer. Math. Monthly **43**(1936), 461–464.

[36-d] *A mean-value theorem for polynomials and harmonic polynomials*, Bull. Amer. Math. Soc. **42**(1936), 923–930.

[37-a] *Note on the curvature of level curves of Green's function*, Proc. Nat. Acad. Sci. U.S.A. **23**(1937), 84–89.

[37-b] (With W.E. Sewell), *Note on degree of approximation to an integral by Riemann sums*, Amer. Math. Monthly **44**(1937), 155–160.

[37-c] *Note on the curvature of orthogonal trajectories of level curves of Green's function*, Proc. Nat. Acad. Sci. U.S.A. **23**(1937), 166–169.

[37-d*] *On the shape of level curves of Green's function*, Amer. Math. Monthly **44**(1937), 202–213.

[37-e] *Maximal convergence of sequences of harmonic polynomials*, Ann. of Math. **38**(1937), 321–354.

[37-f] (With G.M. Merriman), *Note on the simultaneous orthogonality of harmonic polynomials on several curves*, Duke Math. J. **3**(1937), 279–288.

[37-g*] (With W.E. Sewell), *Note on the relation between continuity and degree of polynomial approximation in the complex domain*, Bull. Amer. Math. Soc. **43**(1937), 557–563.

[38-a*] *Note on the curvature of orthogonal trajectories of level curves of Green's functions*, Bull. Amer. Math. Soc. **44**(1938), 520–523.

[38-b] (With W. Seidel), *On the derivatives of functions analytic in the unit circle*, Proc. Nat. Acad. Sci. U.S.A. bf 24(1938), 337–340.

[38-c] *On interpolation and approximation by functions analytic and bounded in a given region*, Proc. Nat. Acad. Sci. U.S.A. **24**(1938), 477–486.

[38-d*] (With W.E. Sewell), *Note on degree of trigonometric and polynomial approximation to an analytic function*, Bull. Amer. Math. Soc. **44**(1938), 865–873.

[39-a] (Book Review), *Convergence*, by W.L. Ferrar, Science **89**(1939), 59–60.

[39-b] *Note on the location of zeros of the derivative of a rational function whose zeros and poles are symmetric in a circle*, Bull Amer. Math. Soc. **45**(1939), 462–470.

[39-c] *On interpolation by functions analytic and bounded in a given region*, Trans. Amer. Math. Soc. **46**(1939), 46–65. MR **1**. 10.

[39-d*] *On the circles of curvature of the images of circles under a conformal map*, Amer. Math. Monthly **46**(1939), 472–485. MR **1**, 111.

[40-a*] *Note on the curvature of orthogonal trajectories of level curves of Green's function.* III, Bull. Amer. Math. Soc. **46**(1940), 101–108. MR **1**, 210.

[40-b*] *On the degree of convergence of sequences of rational functions*, Trans. Amer. Math. Soc. **47**(1940), 254–292. MR **1**, 309.

[40-c] *Note on the degree of convergence of sequences of analytic functions*, Trans. Amer. Math. Soc. **47**(1940), 293–304. MR **1**, 310.

[40-d] (With W.E. Sewell), *Note on degree of trigonometric and polynomial approximation to an analytic function, in the sense of least pth powers*, Bull. Amer. Math. Soc. **46**(1940), 312–319. MR **1**, 309.

[40-e] (With W.E. Sewell), *Sufficient conditions for various degrees of approximation by polynomials*, Duke Math. J. **6**(1940), 658–705. MR **2**, 80.

[41-a] (Book Review), *Sur les valeurs exceptionelles des fonctions méromorphes et de leurs dérivées*, By Georges Valiron, Actualités Sci. Indust., no. 570, Hermann, Paris, 1937; Bull. Amer. Math. Soc. **47**(1941), 7–8.

[41-b] (With W.E. Sewell), *On the degree of polynomial approximation to analytic functions: problem β*, Trans. Amer. Math. Soc. **49**(1941), 229–257. MR **2**, 276.

[41-c] (With W.E. Seidel), *On approximation by euclidean and noneuclidean translations of an analytic function*, Bull. Amer. Math. Soc. **47**(1941), 916–920. MR **4**, 10.

[42-a*] *Note on the coefficients of overconvergent power series*, Bull. Amer. Math. Soc. **48**(1942), 163–166. MR **3**, 201.

[42-b] (With W. Seidel), *On the derivatives of funtions analytic in the unit circle and their radii of univalence and of p-valence*, Trans. Amer. Math. Soc. **52**(1942), 128–216. MR **4**, 215.

[43-a] *Local civil time and date from time diagram*, Proc. U.S. Naval Institute **69**(Whole No. 479)(1943), 23–24.

[43-b] *A new diagram for a universal small-area plotting sheet*, Proc. U.S. Naval Institute **69**(Whole No. 487)(1943), 1221–1222.

[44-a] (With E.N. Nilson), *Interpolation and approximation by functions analytic and bounded in a given region*, Trans. Amer. Math. Soc. **55**(1944), 53–67. MR **5**, 115.

[46-a] *The running fix as used at sea*, (Some navigational wrinkles on "Pilot Chart of the North Atlantic Ocean," No. **1400**)(1946), Hydrographic Office, Washington, D.C.

[46-b] *Note on the location of the critical points of harmonic functions*, Bull. Amer. Math. Soc. **52**(1946), 346–347. MR **7**, 382.

[46-c*] *Overconvergence, degree of convergence, and zeros of sequences of analytic functions*, Duke Math. J. **13**(1946), 195–234. MR **8**, 201.

[46-d] *On degree of approximation on a Jordan curve to a function analytic interior to the curve by functions not necessarily analytic interior to the curve*, Bull. Amer. Math. Soc. **52**(1946), 449–453. MR **7**, 514.

[46-e] *Taylor's series and approximation to analytic functions*, Bull. Amer. Math. Soc. **52**(1946), 572–579. MR **8**, 114.

[46-f] (Book Review), *Table of* arc sin *x and Tables of associated Legendre functions*, by Lyman J. Briggs, et. al., Science **104**(1946), 41.

[46-g] *Note on the location of the zeros of the derivative of a rational function having prescribed symmetry*, Proc. Nat. Acad. Sci. U.S.A. **32**(1946), 235–237. MR **8**, 144.

[47-a] *A rigorous treatment of the first maximum problem in the calculus*, Amer. Math. Monthly **54**(1947), 35–36.

[47-b] *On the location of the critical points of harmonic measure*, Proc. Nat. Acad. Sci. U.S.A. **33**(1947), 18–20. MR **8**, 461.

[47-c] (With E.N. Nilson), *Note on the degree of convergence of sequences of polynomials*, Bull. Amer. Math. Soc. **53**(1947), 116–117. MR **9**, 23.

[47-d] *Note on the derivatives of functions analytic in the unit circle*, Bull. Amer. Math. Soc. **53**(1947), 515–523. MR **9**, 23.

[47-e] *Note on the critical points harmonic functions*, Proc. Nat. Acad. Sci. U.S.A. **33**(1947), 54–59. MR **8**, 513.

[47-f] *The location of the critical points of simply and doubly periodic functions*, Duke Math. J. **14**(1947), 575–586. MR **9**, 180.

[48-a] *On the critical points of functions possessing central symmetry on the sphere*, Amer. J. Math. **70**(1948), 11–21. MR **9**, 428.

[48-b] *Note on the location of the critical points of harmonic functions*, Bull. Amer. Math. Soc. **54**(1948), 191–195. MR **9**, 432.

[48-c] *The critical points of linear combinations of harmonic functions*, Bull. Amer. Math. Soc. **54**(1948), 196–205. MR **9**, 432.

[48-d] *Critical points of harmonic functions as positions of equilibrium in a field of force*, Proc. Nat. Acad. Sci. U.S.A. **34**(1948), 111–119. MR **9**, 432.

[48-e] *Methods of symmetry and critical points of harmonic functions*, Proc. Nat. Acad. Sci. U.S.A. **34**(1948), 257–271. MR **9**, 585.

[48-f] *On the location of the zeros of the derivatives of a polynomial symmetric in the origin*, Bull. Amer. Math. Soc. **54**(1948), 942–945. MR **10**, 250.

[49-a] (With W.E. Sewell and H.M. Elliott), *On the degree of convergence of harmonic polynomials to harmonic functions*, Proc. Nat. Acad. Sci. U.S.A. **35**(1949), 59–62. MR **10**, 374.

[49-b] (With E.N. Nilson), *On functions analytic in a region: Approximation in the sense of least pth powers*, Trans. Amer. Math. Soc. **65**(1949), 239–258. MR **10**, 524.

[49-c] (With A.S. Galbraith and W. Seidel), *On the growth of derivatives of functions omitting two values*, Trans. Amer. Math. Soc. **67**(1949), 320–326. MR **11**, 344.

[49-d] (With W.E. Sewell and H.M. Elliott), *On the degree of polynomial approximation to harmonic and analytic functions*, Trans. Amer. Math. Soc. **67**(1949), 381–420. MR **11**, 515.

[49-e] (With W.E. Sewell), *On interpolation to an analytic function in equidistant points: Problem β*, Bull. Amer. Math. Soc. **55**(1949), 1177–1180. MR **11**, 344.

[49-f] (Book Review), *Human behavior and the principle of least effort: an introduction to human ecology*, by G.K. Zipf, Scientific American, 56–58.

[50-a] *On distortion at the boundary of a conformal map*, Proc. Nat. Acad. Sci. U.S.A. **36**(1950), 152–156. MR **11**, 507.

[50-b] (With H. Margaret Elliott), *Polynomial approximation to harmonic and analytic functions: generalized continuity conditions*, Trans. Amer. Math. Soc. **68**(1950), 183–203. MR **11**, 515.

[50-c] *The location of critical points of harmonic functions*, Leopoldo Fejér et Frederico Riesz LXX Annos Natis Dedicatus, Pars B. Acta Sci. Math. Szeged **12**(1950). 61–65. MR **12**, 26.

[50-d] (With H.G. Russell), *On simultaneous interpolation and approximation by functions analytic in a given region*, Trans. Amer. Math. Soc. **69**(1950), 416–439. MR **12**, 813.

[51-a] *On Rouché's theorem and the integral-square measure of approximation*, Proc. Amer. Math. Soc. **2**(1951), 671–681. MR **13**, 335.

[51-b] *Note on the location of the critical points of a real rational function*, Proc. Amer. Math. Soc. **2**(1951), 682–685. MR **13**, 451.

[51-c*] *Note on approximation by bounded analytic functions*, Proc. Nat. Acad. Sci. U.S.A. **37**(1951), 821–826. MR **13**, 545.

[52-a] *On Rouché's theorem and the integral-square measure of approximation*, Proc. Internat. Congress Math., vol. **1**(1952), Amer. Math. Soc., Providence, R.I., 405–406.

[52-b] (With E.N. Nilson), *Note on overconvergence in sequences of analytic functions*, Proc. Amer. Math. Soc. **3**(1952), 442–443. MR **13**, 927.

[52-c] *Polynomial expansions of functions defined by Cauchy's integral*, J. Math. Pures Appl. (9) **31**(1952), 221–244, MR **14**, 547.

[52-d] *Note on the location of zeros of extremal polynomials in the non-euclidean plane*, Acad. Serbe Sci. Publ. Inst. Math. **4**(1952), 157–160. MR **14**, 164.

[52-e] (With Philip Davis), *Interpolation and orthonormal systems*, J. Analyse Math. **2**(1952), 1–28. MR **16**, 580.

[52-f] *Degree of approximation to functions on a Jordan curve*, Trans. Amer. Math. Soc. **73**(1952), 447–458. MR **14**, 630.

[52-g] (With H. Margaret Elliott), *Degree of approximation on a Jordan curve*, Proc. Nat. Acad. Sci. U.S.A. **38**(1952), 1058–1066. MR **14**, 741.

[53-a] *An interpolation series expansion for a meromorphic function*, Trans. Amer. Math. Soc. **74**(1953), 1–9. MR **14**, 741.

[53-b] *On continuity properties of derivatives of sequences of functions*, Proc. Amer. Math. Soc. **4**(1953), 69–75. MR **14**, 736; 1278.

[53-c*] (With T.S. Motzkin), *On the derivative of a polynomial and Chebyshev approximation*, Proc. Amer. Math. Soc. **4**(1953), 76–87. MR **15**, 701.

[53-d] *Note on the shape of level curves of Green's function*, Duke Math. J. **20**(1953), 611–615. MR **15**, 310.

[53-e] *Note on the shape of level curves of Green's function*, Amer. Math. Monthly, **60**(1953), 671–674. MR **15**, 424.

[53-f] (With David Young), *On the accuracy of the numerical solution of the Dirichlet problem by finite differences*, J. Res. Nat. Bur. Standards **51**(1953), 343–363. MR **15**, 562.

[54-a] *An interpolation problem for harmonic functions*, Amer. J. Math. **76**(1954), 259–272. MR **16**, 588.

[54-b] (With David Young), *On the degree of convergence of solutions of difference equations to the solution of the Dirichlet problem*, J. Math. Phys. **33**(1954), 80–93. MR **15**, 746.

[54-c] (With Philip Davis), *On representations and extensions of bounded linear functionals defined on classes of analytic functions*, Trans. Amer. Math. Soc. **76**(1954), 190–206. MR **15**, 803.

[54-d] (With J.P. Evans), *Note on the distribution of zeros of extremal polynomials*, Proc. Nat. Acad. Sci. U.S.A. **40**(1954), 332–337. MR **15**, 954.

[54-e] (With J.P. Evans), *On approximation by bounded analytic functions*, Arch. Math. **5**(1954), 191–196. MR **15**, 947.

[54-f] *Sur l'approximation par fonctions analytiques bornées*, C.R. Acad. Sci. Paris **239**(1954), 1339–1341. MR **16**, 811.

[54-g] Sur la représentation conforme des aires multiplement connexes, C.R. Acad. Sci. Paris **239**(1954), 1572–1574. MR **16**, 581.

[54-h] *Sur la représentation conforme des aires multiplement connexes*, C.R. Acad. Sci. Paris **239**(1954), 1756–1758. MR **16**, 811.

[54-i] (With M. Fekete), *On the asymptotic behavior of polynomials with extremal properties, and of their zeros*, J. Analyse Math. **4**(1954), 49–87. MR **17**, 354.

[54-j] *Détermination d'une fonction analytique par ses valeurs données dans une infinité dénombrable de points*, Bull. Soc. Math. Belg., (1954), 52–70. MR **17**, 601.

[55-a*] (With D. Gaier), *Zur Methode der variablen Gebiete bei der Randverzerrung*, Arch. Math. **6**(1955), 77–86. MR **16**, 348.

[55-b] (With T.S. Motzkin), *Least pth power polynomials on a real finite point set*, Trans. Amer. Math. Soc. **78**(1955), 67–81. MR **16**, 585.

[55-c] *A generalization of Jensen's theorem on the zeros of the derivative of a polynomial*, Amer. Math. Monthly **62**(1955), 91–93. MR **16**, 818.

[55-d] (With J.P. Evans), *On interpolation to a given analytic function by analytic functions of minimum norm*, Trans. Amer. Math. Soc. **79**(1955), 158–172. MR **16**, 1011.

[55-e] *Sur l'approximation par fonctions rationnelles et par fonctions holomorphes bornées*, Ann. Mat. Pura Appl. (4) **39**(1955), 267–277. MR **17**, 1077.

[56-a] (With Mishael Zedek), *On generalized Tchebycheff polynomials*, Proc. Nat. Acad. Sci. U.S.A. **42**(1956), 99–104. MR **17**, 730.

[56-b*] (With L. Rosenfeld), *On the boundary behavior of a conformal map*, Trans. Amer. Math. Soc. **81**(1956), 49–73. MR **17**, 836.

[56-c] *Best-approximation polynomials of given degree*, Proc. Sympos. Appl. Math., Vol. VI, Numerical Analysis, Santa Monica, Aug. 1953, Amer. Math. Soc., Providence, R.I.(1956), 213–218. MR **18**, 32.

[56-d*] *On the conformal mapping of multiply connected regions*, Trans. Amer. Math. Soc. **82**(1956), 128–146. MR **18**, 290.

[56-e] (With T.S. Motzkin), *Least pth power polynomials on a finite point set*, Trans. Amer. Math. Soc. **83**(1956), 371–396. MR **18**, 479.

[56-f] (With J.P. Evans), *On the location of the zeros of certain orthogonal functions*, Proc. Amer. Math. Soc. **7**(1956), 1085–1090. MR **18**, 725.

[56-g] *Note on degree of approximation to analytic functions by rational functions with preassigned poles*, Proc. Nat. Acad. Sci. U.S.A. **42**(1956), 927–930. MR **18**, 569.

[56-h] *"Birkhoff, George David"*, article in Encyclopedia Britannica. (1956).

[56-i] (With M. Fekete), *On restricted infrapolynomials*, J. Analyse Math. **5**(1956), 47–76. MR **19**, 263.

[57-a] (With T.S. Motzkin), *Underpolynomials and infrapolynomials*, Illinois J. Math. **1**(1957), 406–426. MR **19**, 643.

[57-b] (With T.S. Motzkin), *Polynomials of best approximation on a real finite point set*, Proc. Nat. Acad. Sci. U.S.A. **43**(1957), 845–846. MR **19**, 852.

[57-c] (With M. Fekete), *Asymptotic behavior of restricted extremal polynomials and of their zeros*, Pacific J. Math. **7**(1957), 1037–1064. MR **19**, 1045.

[57-d] (With David Young), *Lipschitz conditions for harmonic and discrete harmonic functions*, J. Math. Phys. **36**(1957), 138–150. MR **20**, #2532.

[58-a] *On approximation by bounded analytic functions*, Trans. Amer. Math. Soc. **87**(1958), 467–484. MR **20**, #3298.

[58-b] *A generalization of Faber's polynomials*, Math. Ann. **136**(1958), 23–33. MR **21**, #725.

[58-c] *Complex numbers and complex variables, Laplace's differential equation, and Conformal mapping*, three articles in the McGraw-Hill Encyclopedia of Science and Technology (1958).

[58-d] *Approximation by bounded analytic functions*, Seminars on Analytic Functions, Vol. II, published by U.S. Air Force, Office of Scientific Research, Washington, D.C.(1958), 73–87.

[58-e] *On infrapolynomials with prescribed constant term*, J. Math. Pures Appl. (9) **37**(1958), 295–316. MR **20**, #7098.

[59-a] *On extremal approximations*, On Numerical Approximation, Univ. of Wisconsin Press, Madison, Wis. (1959), 209–216. MR **21**, #421.

[59-b] (With T.S. Motzkin), *Location of zeros of infrapolynomials*, Compositio Math. **14**(1959), 50–70. MR **21**, #3539.

[59-c] *Approximation on a line segment by bounded analytic functions: Problem* β, Proc. Amer. Soc. **10**(1959), 270–272. MR **21**. #6443a.

[59-d] *Note on least-square approximation to an analytic function by polynomials, as measured by a surface integral*, Proc. Amer. Math. Soc. **10**(1959), 273–279. MR **23**, #A1047.

[59-e] *Approximation by bounded analytic functions: General configurations*, Proc. Amer. Math. Soc. **10**(1959), 280–285. MR **21**, #6443b.

[59-f] (With T.S. Motzkin), *Polynomials of best approximation on a real finite point set.* I, Trans. Amer. Math. Soc. **91**(1959), 231–245. MR **21**, #7388.

[59-g] (With H.G. Russell), *Integrated continuity conditions and degree of approximation by polynomials or by bounded analytic functions*, Trans. Amer. Math. Soc. **92**(1959), 355–370. MR **21**, #7311.

[59-h] *Note on approximation by bounded analytic functions* (Problem α), Math. Z. **72**(1959), 47–52. MR **22**, #776.

[59-i] (With T.S. Motzkin), *Polynomials of best approximation on an interval*, Proc. Nat. Acad. Sci. U.S.A. **45**(1959), 1523–1528. MR **22**, #9773.

[59-j] *Note on invariance of degree of polynomial and trigonometric approximation under change of independent variable*, Proc. Nat. Acad. Sci. U.S.A. **45**(1959), 1528–1533. MR **23**, #A1191.

[59-k] (With H.J. Landau), *On canonical conformal maps of multiply connected regions*, Trans. Amer. Math. Soc. **93**(1959), 81–96. MR **28**, #4093.

[59-l] *The analogue for maximally convergent polynomials of Jentzsch's theorem*, Duke Math. J. **26**(1959), 605–616.

[60-a] *Solution of the Dirichlet problem for the ellipse by interpolating harmonic polynomials*, J. Math. Mech. **9**(1960), 193–196. MR **22**, #4891.

[60-b] *On the asymptotic properties of extremal polynomials with prescribed constant term*, Math. Z. **73**(1960), 339–345. MR **22**, #2715.

[60-c] *Note on polynomial approximation on a Jordan arc*, Proc. Nat. Acad. Sci. U.S.A. **46**(1960), 981–983. MR **22**, #12338.

[60-d] *On degree of approximation by bounded harmonic functions*, J. Math. Pures Appl. (9) **39**(1960), 201–220. MR **27**, #1609.

[60-e] (With T.S. Motzkin), *Best approximators within a linear family on an interval*, Proc. Nat. Acad. Sci. U.S.A. **46**(1960), 1225–1233. MR **28**, #5279.

[60-f] *Degree of approximation by bounded harmonic functions*, Proc. Nat. Acad. Sci. U.S.A. **46**(1960), 1390–1393. MR **22**, #12338b.

[60-g] *Note on degree of approximation by bounded analytic functions: Problem* β, Trans. Amer. Math. Soc. **96**(1960), 246–258. MR **22**, #11140.

[61-a] *The circles of curvature of the curves of steepest descent of Green's function*, Amer. Math. Monthly **68**(1961), 323–329. MR **23**, #A3272.

[61-b] (With T.S. Motzkin), *Conformal maps of small disks*, Proc. Nat. Acad. Sci. U.S.A. **47**(1961), 1838–1843. MR **26**, #307.

[61-c] (With O. Shisha), *The zeros of infrapolynomials with some prescribed coefficients*, J. Analyse Math. **9**(1961), 111–160. MR **25**, #174.

[61-d] (With J.P. Evans), *Approximation by bounded analytic functions to functions represented by Dirichlet series*, Proc. Amer. Math. Soc. **12**(1961), 875–879. MR **25**, #4109.

[61-e] *A new generalization of Jensen's threorem on the zeros of the derivative of a polynomial*, Amer. Math. Monthly **68**(1961), 978–983. MR **24**, #A2009.

[62-a] *Degree of polynomial approximation to an analytic function as measured by a surface integral*, Proc. Nat. Acad. Sci. U.S.A. **48**(1962), 26–32. MR **24**, #A2176.

[62-b] (With J.H. Ahlberg and E.N. Nilson), *Best approximation properties of the spline fit*, J. Math. Mech. **11**(1962), 225–234. MR **25**, #738.

[62-c] *Asymptotic properties of polynomials with auxiliary conditions of interpolation*, Ann. Polon. Math. **12**(1962), 17–24. MR **27**, #1605.

[62-d] (With T.S. Motzkin), *Polynomials of best approximation on an interval* II, Proc. Nat. Acad. Sci. U.S.A. **48**(1962), 1533–1537. MR **26**, #2788.

[62-e] *On the convexity of the ovals of lemniscates*, Studies in Mathematical Analysis and Related Topics, Stanford University Press, Stanford, Calif.(1962), 419–423. MR **27**, #1606.

[62-f] *Approximation par les fonctions holomorphes bornées. Problème β'*, J. Math. Pures Appl. (9) **41**(1962), 213–232. MR **27**, #5914.

[63-a] (With T.S. Motzkin), *Zeros of the error function for Tchebycheff approximation in a small region*, Proc. London Math. Soc. (3) **13**(1963), 90–98. MR **26**, #1667.

[63-b] *Restricted infrapolynomials and trigonometric infrapolynomials*, Proc. Nat. Acad. Sci. U.S.A. **49**(1963), 302–304. MR **27**, #2613.

[63-c] *A generalization of Fejér's principle concerning the zeros of extremal polynomials*, Proc. Amer. Math. Soc. **14**(1963), 44–51. MR **27**, #271.

[63-d] *A sequence of rational functions with application to approximation by bounded analytic functions*, Duke Math. J. **30**(1963), 177–189. MR **30**, #2155.

[63-e] (With O. Shisha), *The zeros of infrapolynomials with prescribed values at given points*, Proc. Amer. Math. Soc. **14**(1963), 839–844. MR **27**, #3785.

[63-f] *Note on the convergence of approximating rational functions of prescribed type*, Proc. Nat. Acad. Sci. U.S.A. **50**(1963), 791–794. MR **28**, #400; **28** #1247.

[63-g] (Book Review), *Analytic function theory*, by E. Hille, SIAM Rev. **5**(1963), 377–378.

[64-a*] *Padé approximants as limits of rational functions of best approximation*, J. Math. Mech. **13**(1964), 305–312. MR **28**, #4283.

[64-b] (With O. Shisha). *On the location of the zeros of some infrapolynomials with prescribed coefficients*, Pacific J. Math. **14**(1964), 1103–1109. MR**30**, #232.

[64-c] (With O. Shisha), *Extremal polynomials and the zeros of the derivative of a rational function*, Proc. Amer. Math. Soc. **15**(1964), 753–758. MR **29**, #4876.

[64-d] *The convergence of sequences of rational functions of best approximation*, Math. Ann. **155**(1964), 252–264. MR **29**, #1484.

[64-e*] *A theorem of Grace on the zeros of polynomials, revisited*, Proc. Amer. Math. Soc. **15**(1964), 354–360. MR **28**, #4092.

[64-f] (With Z. Rubinstein), *On the location of the zeros of a polynomial whose center of gravity is given*, J. Analyse Math. **12**(1964), 129–142. MR **29**, #4877.

[64-g] (With A. Sharma), *Least squares and interpolation in roots of unity*, Pacific J. Math. **14**(1964), 727–730. MR **28**, #5278.

[64-h] *Surplus free poles of approximating rational functions*, Proc. Nat. Acad. Sci. U.S.A. **52**(1964), 896–901. MR **30**, #3983.

[64-i] (With Maynard Thompson), *Approximation with auxiliary conditions*, J. Math. Mech. **13**(1964), 1015–1019. MR **30**, #1253.

[64-j*] *The location of the zeros of the derivative of a rational function, revisited*, J. Math Pures Appl. (9) **43**(1964), 353–370. MR **31**, #3582.

[65-a] *Geometry of the zeros of the sums of linear fractions*, Trans. Amer. Math. Soc. **114**(1965), 30–39. MR **31**, #3579.

[65-b*] (With J.H. Ahlberg and E.N. Nilson), *Fundamental properties of generalized splines*, Proc. Nat. Acad. Sci. U.S.A. **52**(1965), 1412–1419. MR **36**, #6846.

[65-c] (With J.H. Ahlberg and E.N. Nilson), *Best approximation and convergence properties of higher-order spline approximations*, J. Math. Mech. **14**(1965), 231–243. MR **35**, #5823.

[65-d] (With A. Sinclair), *On the degree of convergence of extremal polynomials and other extremal functions*, Trans. Amer. Math. Soc. **115**(1965), 145–160. MR **33**, #7564.

[65-e] *The convergence of sequences of rational functions of best approximation* II, Trans. Amer. Math. Soc. **116**(1965), 227–237. MR **32**, #6120.

[65-f] (With J.H. Ahlberg and E.N. Nilson), *Extremal, orthogonality, and convergence properties of multidimensional splines*, J. Math. Anal. Appl. **12**(1965), 27–48. MR **37**, #661.

[65-g] (With J.H. Ahlberg and E.N. Nilson), *Convergence properties of generalized splines*, Proc. Nat. Acad. Sci. U.S.A. **54**(1965), 344–350. MR **36**, #6847.

[65-h] *Hyperbolic capacity and interpolating rational functions*, Duke Math. J. **32**(1965), 369–379. MR **31** #6081.

[65-i*] *The convergence of sequences of rational functions of best approximation with some free poles*, Approximation of Functions (Proc. Sympos. General Motors Res. Lab., 1964), Henry L. Garabedian, Ed., Elsevier, Amsterdam (1965), 1–16. MR **32**, #4441.

[66-a] (With T.S. Motzkin), *Mean approximation on an interval for an exponent less than one*, Trans. Amer. Math. Soc. **122**(1966), 443–460. MR **34**, #1769.

[66-b] *Approximation by polynomials: Uniform convergence as implied by mean convergence*, Proc. Nat. Acad. Sci. U.S.A. **55**(1966), 20–25. MR **32**, #5891.

[66-c] *Approximation by polynomials: Uniform convergence as implied by mean convergence* II, Proc. Nat. Acad. Sci. U.S.A. **55**(1966), 1405–1407. MR **35**, #4443.

[66-d] (With H.G. Russell), *Hyperbolic capacity and interpolating rational functions* II, Duke Math. J. **33**(1966), 275–279. MR **33**, #1624.

[66-e] *The convergence of approximating rational functions of prescribed type*, Contemporary Problems in the Theory of Analytic Functions (M.A. Lavrent'ev, Ed.) Proc. Internat. Conference on the Theory of Analytic Functions (Erevan, 1965), "Nauka", Moscow (1966), 304–308. (Russian). MR **35**, #3069.

[66-f] *Approximation by polynomials: Uniform convergence as implied by mean convergence* III, Proc. Nat. Acad. Sci. U.S.A. **56**(1966), 1406–1408. MR **35**, #4444.

[67-a] *Best approximation by rational functions and by meromorphic functions with some free poles*, J. Analyse Math. **18**(1967), 359–375. MR **36**, #1673.

[67-b] *On the convergence of sequences of rational functions*, SIAM J. Numer. Anal. **4**(1967), 211–221. MR **36**, #1675.

[67-c*] *An extension of the generalized Bernstein lemma*, Colloq. Math. **16**(1967), 91–92. MR **35**, #6844.

[67-d*] (With J.H. Ahlberg and E.N. Nilson). *Complex cubic splines*, Trans. Amer. Math. Soc. **129**(1967), 391–413. MR **36**, #573.

[68-a*] *Degree of approximation by rational functions and polynomials*, Michigan Math. J. **15**(1968), 109–110. MR **36**, #6845.

[68-b] *Note on classes of functions defined by integrated Lipschitz conditions*, Bull. Amer. Math. Soc. **74**(1968), 344–346. MR **36**, #2807.

[68-c] (With T.S. Motzkin), *A persistent local maximum of the pth power deviation on an interval*, $p < 1$, Pacific J. Math. **24**(1968), 133–142. MR **38**, #4868.

[68-d] *The convergence of sequences of rational functions of best approximation* III, Trans. Amer. Math. Soc. **130**(1968), 167–183. MR **36**, #1674.

[68-e*] *Approximation by bounded analytic functions: Uniform convergence as implied by mean convergence*, Trans. Amer. Math. Soc. **130**(1968), 406–413. MR **36**, #3997.

[68-f*] (With J.H. Ahlberg and E.N. Nilson), *Cubic splines on the real line*, J. Approx. Theory **1**, no. 1(1968), 5–10. MR **37**, #6650.

[69-a] (With E.B. Saff), *Extensions of D. Jackson's theorem on best complex polynomial mean approximation*, Trans. Amer. Math. Soc. **138**(1969), 61–69. MR **39**, #3001.

[69-b] *Inequalities expressing degree of convergence of rational functions*, J. Approx. Theory **2**(1969), 160–166. MR **39**, #7115.

[69-c] (With J.H. Ahlberg and E.N. Nilson), *Properties of analytic splines. I. Complex polynomial splines*, J. Math. Anal. Appl. **27**(1969), 262–278. MR **42**, #8136.

[69-d] *Note on approximation by bounded analytic functions, problem α: General configurations*, Aequationes Math. **3**(1969), 160–164. MR **41**, #5632.

[69-e] *Approximations to a function by a polynomial in a given function*, Amer. Math. Monthly **76**(1969), 1049–1050.

[69-f] (With Z. Rubinstein), *Extensions and some applications of the coincidence theorems*, Trans. Amer. Math. Soc. **146**(1969), 413–427. MR **40**, #4428.

[70-a*] (With W.J. Schneider), *On the shape of the level loci of harmonic measure*, J. Analyse Math. **23**(1970), 441–460. MR **42**, #6205.

[70-b] *Approximation by rational functions: Open problems*, J. Approx. Theory **3**(1970), 236–242. MR **43**, #538.

[70-c] *Note on degree of convergence of sequences of rational functions of prescribed type*, Proc. Nat. Acad. Sci. U.S.A. **67**(1970), 1188–1191. MR **42**, #4748.

[71-a] (With J.H. Ahlberg and E.N. Nilson), *Complex polynomial splines on the unit circle*, J. Math. Anal. Appl. **33**(1971), 234–257. MR **43**, #3696.

[71-b*] (With Dov. Aharonov), *Some examples in degree of approximation by rational functions*, Trans. Amer. Math. Soc. **159**(1971), 427–444. MR **44**, #6974.

[71-c] *Mean approximation by polynomials on a Jordan curve*, J. Approximation Theory **4**(1971), 263–268. MR **45**, #3726.

[72-a] (With Dov Aharonov), *On the convergence of rational functions of best approximation to a meromorphic function*, J. Math. Anal. Appl. **40**(1972), 418–426. MR **47**, #5263.

[72-b] (With Myron Goldstein), *Approximation by rational functions on Riemann surfaces*, Proc. Amer. Math. Soc. **36**(1972), 464–466. MR **47**, #2027.

[72-c] *Note on the convergence of sequences of rational functions of best approximation to a meromorphic function*, Proc. Nat. Acad. Sci. U.S.A. **69**(1972), 2963–2964. MR **46**, #7522.

[73-a*] (With T.S. Motzkin), *Equilibrium of inverse distance forces in three dimensions*, Pacific J. Math. **44**(1973), 241–250. MR **47**, #5234.

[73-b] (With E.B. Saff), *On the convergence of rational functions which interpolate in the roots of unity*, Pacific J. Math. **45**(1973), 639–641. MR **53**, #13583.

[73-c*] *History of the Riemann mapping theorem*, Amer. Math. Monthly **80**(1973), 270–276. MR **48** #2348.

[74-a] *The role of the pole in rational approximation*, Rocky Mountain J. Math. **4**(1974), 283–286. MR **49**, #7455.

[74-b] *Padé approximants as limits of rational functions of best approximation, real domain*, J. Approx. Theory **11**(1974), 225–230. MR **50**, #5287.

[76] (With P.M. Gauthier and Alice Roth), *Possibility of uniform approximation in the spherical metric*, Canad. J. Math. **28**(1976), 112–115. MR **57**, #9979.

Books by Joseph L. Walsh

[1] *Approximation by Polynomials in the Complex Domain*, Mémorial des Sciences Mathématiques, Gauthier-Villars, Paris, 1935, ii+72pp.

[2] *Interpolation and Approximation by Rational Functions in the Complex Domain*, Colloquium Publications, **20**, American Mathematical Society, Providence, R.I., 1935, ix+382pp.; 2nd edition, 1956; 3rd edition, 1960; 4th edition, 1965; 5th edition, 1969; Russian translation, IL, Moscow, 1961. MR **36**, #1671; 1672b, c.

[3] *A Bibliography on Orthogonal Polynomials* (with J.A. Shohat and Einar Hille), National Research Council, Bulletin No. 103, Washington, D.C., 1940, ix+204 pp.

[4] *The Location of Critical Points of Analytic and Harmonic Functions*, Colloquium Publications, **34**, American Mathematical Society, Providence, R.I., 1950, viii+384pp. MR **12**, 249.

[5] *Approximation by Bounded Analytic Functions*, Mémorial des Sciences Mathématiques, Fasc. 144, Gauthier-Villars, Paris, 1960, 66 pp. MR **22**, #9770.

[6] *A Rigorous Treatment of Maximum-Minimum Problems in the Calculus*, D.C. Heath, Boston, Mass., 1962, 22pp.

[7] *The Theory of Splines and Their Applications*, (with J.H. Ahlberg and E.N. Nilson), Academic Press, New York and London, 1967, xi+284 pp. MR **39**, #684.

Books by Joseph L. Walsh

Joseph L. Walsh's Ph.D. Students

1928 Morris Marden
1930 Orin J. Farrell
1931 Cecil T. Holmes
1932 Joseph L. Doob
1932 Helen G. Russell
1935 John H. Curtiss
1935 Yu-Cheng Shen
1936 Walter E. Sewell
1937 Zehman I. Mosesson
1938 Floyd E. Ulrich
1940 Maurice H. Heins
1940 Abraham Spitzbart
1941 Edwin N. Nilson
1942 Lynn H. Loomis
1947 Ivan R. Hershner, Jr.
1948 H. Margaret Elliott
1949 Helen Kelsall Nickerson
1952 Alan F. Kay
1952 Isaac E. Block
1953 Theodore J. Rivlin
1953 Lawrence Rosenfeld
1953 Jacqueline P. Evans
1954 Richard S. Varga
1956 Mishael Zedek
1957 Henry J. Landau
1961 Vincent C. Williams
1962 Dorothy B. Shaffer
1963 Victor M. Manjarrez
1964 Jerry L. Fields
1968 Edward B. Saff
1973 Marvin E. Ortel

Remarks on the Mathematical Researches of Joseph L. Walsh[1]

W.E. Sewell[2]

Polynomial approximation was neither discovered nor invented by J.L. Walsh (which may come as a surprise to some mathematicians). He is the one individual, however, who took a few scattered results on the subject and extended them, added mightily to them, and knit the whole together into a comprehensive, coherent theory. His book *Interpolation and Approximation by Rational Functions in the Complex Domain* has been, and still is (the fourth edition will shortly be published), the standard reference on the subject both here and abroad.

Walsh's study essentially begins with a proof, based on Runge's classical theorem on the approximation of the Cauchy integral, of the fundamental and basic result that a function $f(z)$ which is analytic on a finite closed Jordan region C can be uniformly approximated on C in the sense of Chebyshev by a polynomial $P(z)$. From here he attacked almost singlehandedly and simultaneously in all directions.

He utilized the theory of conformal mapping and the corresponding Green's function to extend the regions of analyticity and convergence from a finite closed Jordan set to a closed limited set C of the $z(= x + iy)$-plane whose complement K is connected and regular in the sense that K possesses a Green's function $G(x, y)$

[1]Received by the editors October 28, 1965. Dedicated to Professor J.L. Walsh on the occasion of his seventieth birthday.

[2]Special Research in Numerial Analysis, Duke University, Durham, North Carolina. This research was done at the Special research was done at the Special Research in Numerical Analysis, which is sponsored by the United States Army Reseach Office–Durham, under Contract DA 31-124-ARO-D-13.

with pole at infinity. The set C_ρ denotes the equipotential locus in K for which $G(x, y) = \log \rho > 0$. (It should be mentioned here that to Walsh conformal mapping is not only a tool but an interest in itself and his contribution, particularly to Green's theorem and the mapping of multiply connected regions, is noteworthy.) And from the possibility of approximation by polynomials and rational functions he directed his attention to the relation between the ultimate region of analyticity through analytic extension and the degree of convergence of various sequences of polynomials. A typical theorem is the following.

Let C be a closed limited point set whose complement is connected and regular. If the function $f(z)$ is single-valued and analytic on C, there exists a greatest number ρ (finite or infinite) such that $f(z)$ is single-valued and analytic at every point interior to C_ρ. If $R < \rho$ is arbitrary, there exist polynomials $p_n(z)$ of respective degrees $n = 0, 1, 2, \cdots$ such that

$$|f(z) - p_n(z)| \leq M/R^n \tag{1}$$

where M depends on R but not on n or z, is valid for z on C; but there exist no polynomials $p_n(z)$ such that (1) *is valid for z on C for $R > \rho$.*

These polynomials $p_n(z)$ depend on R, but there exist polynomials $P_n(z)$ independent of R of respective degrees n such that (1) is valid for z on C for every $R < \rho$, where M depends on R but not on n or z, and this sequence $P_n(z)$ was said by Walsh to converge to $f(z)$ on C *maximally, or with the greatest geometric degree of convergence.* This concept (and its natural extension to other degrees of convergence and to continuity as well as analyticity) has had a profound effect on the direction of effort in the field and has stimulated a great many subsequent results by other mathematicians.

From simple Chebyshev approximation he proceeded to consider approximation in the mean both for line and surface integrals, and in connection with the former he introduced his own brand of orthogonal functions. And in his study of polynomials, interpolation receives its fair share of attention both basically and as an auxiliary condition to approximation.

Much of the material on interpolation and approximation by rational functions was originally published in his book and is a model of methods, results, and exposition which has rarely been equalled in mathematical literature. A general problem in this theory can be stated in terms of the "Walsh array" (analogous to the Padé table):

$$\begin{matrix} R_{00}(z), & R_{01}(z), & R_{02}(z), & \cdots, \\ R_{10}(z), & R_{11}(z), & R_{12}(z), & \cdots, \\ \cdots, & \cdots, & \cdots, & \cdots, \end{matrix}$$

where $R_{mn}(z)$ is the rational function of the form

$$\frac{a_0z^m + a_1z^{m-1} + \cdots + a_m}{b_0z^n + b_1z^{n-1} + \cdots + b_n}, \qquad b_0z^n + b_1z^{n-1} + \cdots + b_n \not\equiv 0,$$

of best approximation to a function $f(z)$ defined on a set C of the z-plane. The problem is to investigate the convergence of various sequences formed from this array; some results are known, clearly much remains to be done.

After developing the theory on approximation to functions *analytic* on a given closed set and the *regions* of uniform convergence of the corresponding expansions, he devoted considerable attention to the problem of convergence and degree of convergence where the functions are *not* analytic on the closed set considered, but, for example, are analytic in the interior of the region with merely certain continuity properties on the boundary. He had proved the fundamental theorem of this theory in 1926—that *a function $f(z)$, analytic in a finite Jordan region C and continuous in the closed region $\bar{C}$ can be uniformly approximated in $\bar{C}$ by a polynomial in z.* The results, in general, may be described by the inequality

$$|f(z) - p_n(z)| \leq \omega(n), \quad z \text{ in } \bar{C}, \tag{2}$$

where $\omega(n)$ approaches zero as n becomes infinite, and the rate of approach depends upon the continuity properties of $f(z)$ on the boundary itself. Estimates of $\omega(n)$, where $f(z)$ and the boundary C have specific continuity properties, have been determined, but an interesting open problem is to find better estimates even for these specific cases and to investigate $\omega(n)$ for more general cases.

As stated above Walsh did not invent approximation by polynomials but he did originate the theory of approximation by bounded analytic functions. And he presented an excellent exposition of this theory in volume 144 of the Memorial des Sciences Mathématiques, entitled *Approximation by Bounded Analytic Functions*. A typical result is the following.

Let R be a Jordan region containing the closed Jordan region S, and let $U(z)$ denote the function harmonic in $R - S$, continuous in the corresponding closed region, equal to zero and unity on the respective boundaries of R and S. Let C_ρ denote generically the Jordan curve $U(z) = \rho, 0 < \rho < 1$, in $R - S$, and let $R_\rho, 0 < \rho < 1$, denote the point set S plus the points of $R - S$ at which we have $\rho < U(z) < 1$.

Let $f(z)$ be analytic in R_ρ, but not coincide with any function analytic throughout a region $R_{\rho'}, 0 < \rho' < \rho$. Let $f_M(z)$ be the (or a) function analytic and of modulus not greater than M in R, such that

$$\max[|f(z) - f_M(z)|, z \text{ on } S]$$

is least. Then we have

$$\limsup_{M\to\infty}[\max|f(z) - f_M(z)|, z \text{ on } S]^{1/\log M} = e^{\rho/(\rho-1)}.$$

The theory is extended to much more general regions, to interpolation plus approximation, and to functions with prescribed continuity conditions on the boundary of the region of analyticity. There are many open problems pointed out by Walsh in his essay (pp. 57–61) and for which his results and methods form a profitable starting point.

The above discussion is concerned with the complex domain, but Walsh's contribution to the approximation of harmonic functions by harmonic polynomials and by harmonic rational functions is voluminous and impressive.

At the beginning of this mathematical career Walsh wrote a paper, *On the location of the roots of the Jacobian of two binary forms, and of the derivative of a rational function*. His interest in critical points, defined both as the zeros of the derivative of an analytic function and also as the points where the first two partial derivatives of a harmonic function vanish, continued through the years and resulted in many significant papers and a second book by him in the Colloquium Publications of the American Mathematical Society, *The Location of Critical Points of Analytic and Harmonic Functions*. A large part of the material in this book had not been previously published, and it is a superb source of new results, novel methods, and interesting open problems. Although much of the book is concerned with harmonic functions, one of the most elegant theorems concerns the roots of the derivative of a polynomial:

Let the closed interior of the circles $C_1 : |z - \alpha_1| = r_1$ *and* $C_2 : |z - \alpha_2| = r_2$ *be the respective loci of* m_1 *and* m_2 *independent zeros of a variable polynomial* $p(z)$ *of degree* $m_1 + m_2$. *Then the locus of the zeros of the derivative* $p'(z)$ *consists of the closed interior of* C_1 *(if* $m_1 > 1$*), the closed interior of* C_2 *(if* $m_2 > 1$*), and the closed interior of the circle*

$$C : \left| z - \frac{m_2\alpha_1 + m_1\alpha_2}{m_1 + m_2} \right| = \frac{m_2 r_1 + m_1 r_2}{m_1 + m_2}.$$

Any of these circles C_1, C_2, *or* C *which is exterior to the other two circles contains respectively the following number of zeros of* $p'(z) : m_1 - 1, m_2 - 1, 1$.

A related area in which Walsh has found the answers to some of the long-standing and classical problems is the structure and behavior of extremal polynomials. An extremal polynomial, as its name implies, exhibits an extreme of some type in its behavior. For example, if $f(z)$ is defined on a set C of the complex plane then the (or a) polynomial $p_n(z)$ of given degree n which minimizes $|f(z) - p_n(z)|$, z on C, is an extremal polynomial and is called a *juxtapolynomial* to $f(z)$. If we put $f(z) = 0$ and require the leading coefficient of $p_n(z)$ to be unity then we have the classical Chebyshev or *infrapolynomial* of the set C. Infrapolynomials include as special cases many of the important extremal polynomials of analysis, and a study of them provides a uniform approach to all of these special polynomials.

Walsh has devoted considerable attention to juxtapolynomials especially where the function $f(z)$ is real and the set C lies on the real axis, and, in particular, to the relation between these polynomials and those of best approximation to $f(z)$ on C in the sense of least pth powers. His results on infrapolynomials with auxiliary conditions on the coefficients or of interpolation are numerous and outstanding, especially on the asymptotic behavior of such polynomials and of their zeros. For example, since Fekete proved that for C a closed bounded point set and for $p_n(z)$ the infrapolynomial of degree n on C,

$$\lim_{n\to\infty} [\max |p_n(z)|, z \text{ on } C]^{1/n} = \tau(C), \tag{3}$$

where $\tau(C)$ is the capacity (transfinite diameter) of C, the natural question has been as to whether (3) can be satisfied by a sequence of polynomials $p_n(z) = z^n + \cdots$ which are required to satisfy suitable auxiliary conditions of interpolation in a finite number of points. Walsh answered this question in the general case with the following theorem.

Let C be a closed bounded set on the z-plane of positive capacity whose complement K is a region (necessarily containing $z = \infty$), let $g(z)$ be the generalized Green's function for K with pole at infinity, and let $h(z)$ be the harmonic function conjugate to $g(z)$ on K. Let $\phi(z) = \exp[g(z) + ih(z)]$ in K. Let there be given a finite number of distinct points $z_1, z_2, \cdots, z_v$, and assigned values $A_{n1}, A_{n2}, \cdots, A_{nv}$, $n = v, v+1, v+2, \cdots$. A necessary and sufficient condition that there exist polynomials $p_n(z) = z^n + a_{n1}z^{n-1} + \cdots$, $n = v, v+1, v+2, \cdots$, which satisfy (3) *and*

$$p_n(z_k) = A_{nk}, \quad k = 1, 2, \cdots, v,$$

is that we have

$$\limsup_{n\to\infty} |A_{nk}|^{1/n} \le \tau(C) \cdot |\phi(z_k)|, \quad z_k \text{ in } K,$$

$$\limsup_{n\to\infty} |A_{nk}|^{1/n} \le \tau(C), \quad z_k \text{ in } C.$$

This area is still ripe with open problems.

Let me close with a word concerning my statement that Walsh attacked the problem "single-handedly"; this is not meant to imply that he at any time endeavored to establish a monopoly. He has collaborated widely, as his bibliographies indicate, and many important results are joint. The individuals have not been mentioned by name here, due to lack of space; this is my omission and one of which Walsh has never been guilty—he has without exception been extremely careful to give credit where credit was due. In fact, an outstanding and continuing characteristic of his professional career is his unselfish promotion of research. He has always encouraged others to attack problems which he has proposed or is pursuing. Throughout his books and articles you will find open problems, and he has worked with his students, former students, and colleagues frequently more to encourage them than to produce another paper under his own name. Even today papers are being written on problems which he proposed years ago, and which he left to others out of generosity, rather than perplexity.

Joseph Leonard Walsh[1]

D.V. Widder[2]

Few teachers can claim a record of 50 years of teaching at a single university. Professor Joseph Leonard Walsh has that distinction, for he was first an assistant in the Department of Mathematics at Harvard in 1915 and continued uninterrupted association therewith until his retirement in 1965. But these 50 years do not account for Walsh's total association with Harvard, for he took both his B.S. and Ph.D. degrees there. He advanced normally up the teaching ladder to the rank of full professor in 1935. He was chairman of the department from 1937 to 1942 and again briefly in the fall of 1950.

Surely the outstanding attribute of this man, that would occur first to any of his colleagues, was his devotion to research. He neglected none of the duties connected with his job, but he always arranged that they left him time for his writing. Even as department chairman or as president of the American Mathematical Society he remained faithful first and foremost to research. He posted no office hours, but all knew where he could be found—in Widener 474 deep in some problem of approximation. To that office he walked daily, rain or shine, and there he remained for most of the day. Four tremendous volumes of his papers (231 of them) stand in the department's treasure bookcase, attesting to the success of his labors.

[1]Received by the editors September 17, 1965. Dedicated to Professor J.L. Walsh on the occasion of his seventieth birthday.

[2]Department of Mathematics, Harvard University, Cambridge, Massachusetts.

Professor Walsh believed strongly in the principle of rotation of courses, and he activated the principle by teaching a great variety of subjects from year to year, including algebra, mechanics, differential equations, probability, number theory, potential theory, and of course approximation and function theory. Perhaps Math. 1 (calculus) and Math. 213 (complex variable) were his favorites. He also was firm in his belief that seasoned staff members should teach elementary courses, and more often than not he had a section of freshman calculus.

During his teaching years at Harvard Joe exhibited a happy admixture of informality and formality. In his early bachelor days he used to eat in Memorial Hall with a group of graduate students and teaching fellows. There he could hold his own in the rowdy tossing and receiving of butter and ice that sometimes developed. Once, on a dare, he lectured for an hour in his advanced course on the front steps of Jefferson Physical Laboratory, using the pavement as his blackboard. On the other hand the uncompromising precision apparent in his published works could show in his daily life. He could be a strict disciplinarian, even to the point of chiding a colleague for omitting from the minutes the number of a meeting or the precise time of adjournment. To whom can the department now turn to settle some fine point of parliamentary procedure?

His life as a scholar did not keep Joe from attending to important civic duties. During the Boston police strike of September, 1919, he enrolled in the volunteer police force that was temporarily entrusted with preserving the peace of the community. In two world wars he served in the Navy advancing from the rank of Ensign in 1918 to that of Lieutenant Commander in 1942 and to Captain in the early fifties. It was during his tour of service in World War Two that he met Elizabeth Cheney Strayhorn, an officer in the WAAC, whom he later married. Affectionately known to the department as "Liz", she has been ever the gracious hostess and sound adviser.

There was one period when the department accepted an annual challenge at golf from a team of mathematicians at Brown University, captained by the late R. G. D. Richardson. Walsh and Graustein were usually mainstays of Harvard's team. It must be recorded that Harvard was seldom the winner and that the quality of skill exhibited (at least by the undersigned) was more amusing than impressive. But the good fellowship between the teams was paramount.

Retirement from Harvard will not end Joe's career of teaching and research. For he is to join the staff of the University of Maryland in the spring term of 1966. No one will doubt that valuable contributions to research will continue to flow from his pen. The site of his home in Maryland will be within a few miles from the place of his birth. But the primary consideration in its choice was not that. It had to be at the right walking distance from the University.

Joseph L. Walsh

Joseph L. Walsh in Memoriam

Morris Marden

Threescore years and ten is the biblical measure of a man's normal life span. Yet in modern times his surpassing of this bound is not unusual. However, it becomes noteworthy when he is blessed with undiminished physical and mental vigor. This was true of Joe Walsh. "His spirit was marvelous until the end, and he spoke with gratitude for his many productive years".[1]

Joseph Leonard Walsh died on December 10, 1973 at the age of seventy-eight years in his home at University Park, Maryland. This site is not far from where he was born on September 21, 1895, as son of Reverend and Mrs. John Leonard Walsh. Most of his academic life as student, teacher and scholar was spent at Harvard University. In 1916 Harvard awarded him the S.B. degree, summa cum laude, and at the same time a Sheldon Traveling Fellowship for study at the Universities of Chicago and Wisconsin. On his return to Harvard in 1917, Walsh began some studies under Maxime Bôcher, but their progress was interrupted by World War I and his enlistment as an ensign in the U.S. Navy. In 1920 Harvard granted him a Ph.D. and also a second Sheldon Traveling Fellowship, this time for study in Paris under Paul Montel. Back from Europe in 1921 he joined the Harvard faculty, but in 1925 took a leave-of-absence for a year's research at Munich under Carathéodory. Returning again to Harvard, he was promoted through the ranks to a full professorship in 1935 and served as department chairman from 1937 to 1942. In the latter year he was recalled to active duty in the U.S. Navy as a

[1] Letter from Mrs. Joseph L. Walsh.

lieutenant commander. When he returned to Harvard in 1946, he was appointed to the prestigous Perkins Professorship, which he held until his retirement in 1966. A semester earlier he began a research professorship at the University of Maryland, in which position he remained fully active, working with Ph.D. and post doctoral students, until a few months before his death.

During Walsh's lifetime he received many academic and nonacademic honors. Among them was his election in 1936 to the National Academy of Science, and in 1937 to the vice-presidency of the American Mathematical Society. He was elected for a two-year term as president of the Society in 1949. This was a crucial period for the Society when it was experiencing growing pains due to a rapid increase in membership and in research publications. It was the period in which the Society created the post of Executive Director and moved its headquarters to Providence. Also during this period Walsh served as chairman of the organizing committee for the International Mathematical Congress, held in Cambridge during August 1950, the first since World War II. At about this time Walsh was recipient of a nonmathematical honor—promotion to the rank of captain in the U.S. Naval Reserve. In his later years Walsh was twice honored by the dedication of volumes of mathematical journals: SIAM J. (2) **3** (1966) on his seventieth birthday in 1965, and J. Approximation Theory **5** (1972) on his seventy-fifth birthday in 1970.

These mathematical honors were well deserved in view of the quantity and quality of his original research. Starting with his first publication in 1916, while still an undergraduate, he wrote, singly or jointly with students and others, a total of 279 research, expository, and book review articles as well as seven books. Though these papers covered a wide range of topics, they were, broadly speaking, concerned with four general areas:

(I) The relative location of the zeros of pairs of rational functions such as a polynomial and its derivative. (II) Zeros and topology of extremal polynomials. (III) The critical points and level lines of Green's function and other harmonic functions. (IV) Interpolation and approximation of continuous, analytic, or harmonic functions.

Regarding the general area (I), this was Walsh's first main research interest. His doctoral thesis was entitled *On the roots of the jacobian of two binary forms*. It was written under the guidance of Maxime Bôcher who had proved that if F and G are binary forms of the same degree and if all the zeros a_j of F lie in a circular region A and all the zeros b_j of G lie in a circular region B with $B \cap A = \emptyset$, then all the zeros c_k of the jacobian $J(F, G)$ lie in $A \cup B$. Like Bôcher, Walsh used geometric and physical methods, interpreting the c_k as equilibrium points in the field due to positive masses at the points a_j and negative masses at the points b_j with an inverse distance force law. Walsh's results are generalizations of the Lucas theorem that the convex hull of the zeros of a polynomial f contains all the critical points of f. These results are described in Walsh's papers and in M. Marden, *Geometry of polynomials,* 2nd ed., Math Surveys, no. 3, Amer. Math. Soc., Providence, R.I. 1966, MR **37** #1562. The most striking of these results are the following three:

(1) If an nth degree polynomial f has n_1 zeros in a disk $|z - c_1| \leq r_1$ and the remaining $n_2 = n - n_1$ zeros in the disk $|z - c_2| \leq r_2$, then any critical point of

f not in one of these disks lies in a third, "average" disk

$$|z - (n_2c_1 + n_1c_2)/n| \leq (n_2r_1 + n_1r_2)/n.$$

(2) If C_1, C_2 and C_3 are disjoint circular regions and if a rational function f has in the extended plane all its zeros in $C_1 \cup C_2$ and all its poles in C_3, then any critical point of f not in $C_1 \cup C_2 \cup C_3$ lies in a circular region C_4. The boundary ∂C_4 of C_4 is the locus of the point z_4 defined by the cross ratio $(z_1, z_2, z_3, z_4) =$ const. when z_1, z_2, z_3 vary independently on the circles ∂C_1, ∂C_2, ∂C_3 respectively.

(3) Let the form $\Phi(z_1, z_2, \cdots, z_n)$ be of degree one in each z_j, and of total degree n and symmetric in the set $z_1, z_2, \cdots, z_n$. Let C be a circular region containing the n points $z_j = z_{j0}$, $j = 1, 2, \cdots, n$. Then in C there exists at least one point ζ such that $\Phi(\zeta, \zeta, \cdots, \zeta) = \Phi(z_{10}, z_{20}, \cdots, z_{n0})$.

Regarding the general area (II), the methods and results were suggested in part by those in area (I). Given a closed bounded set E containing at least $n+1$ points and the class P_n of all polynomials $z_n + a_1z^{n-1} + \cdots + a_n$, an infrapolynomial p on E means a polynomial $p \in P_n$ with the property

$$\max_{z \in E} |p(z)| = \min_{q \in P_n} \max_{z \in E} |q(z)|.$$

The zeros of p play a role vis-à-vis set E similar to that of the critical points of a polynomial f vis-à-vis the zeros of f. For instance, Fejér proved that the zeros of p lie in the convex hull of E, and Fekete showed that p satisfies a form involving the points of E that is similar to the form for the logarithmic derivative of f in terms of the zeros of p. Walsh explored the subject of infrapolynomials intensively in papers which he published singly or jointly with Fekete, Motzkin, Shisha and Zedek.

Likewise the general area (III) was partly an offshoot of area (I). For example, Walsh proved that if G is the Green's function (with pole at infinity) for an unbounded region R with bounded boundary B, then all the critical points of G in R lie in the convex hull of B. Thus, the critical points of G play a role similar to the critical points of a polynomial whose zeros lie on B. In this connection Walsh also examined in detail the curvature and other characteristics of the level lines of Green's function. These theorems together with their generalizations to harmonic measures and other harmonic functions are developed in Walsh's papers and described in his book, *Critical points of analytical and harmonic functions*.

As for general area (IV), the subjects of interpolation and approximation encompass about half of Walsh's published articles as well as his now classical treatise entitled *Interpolation and approximation*. Among his many original results in this area, probably the most important are the following two:

(1) Every function continuous on a bounded Jordan arc J can be approximated on J uniformly by a polynomial in z. (2) Every function f analytic in a Jordan region B and continuous on its closure $\bar{B}$ can be uniformly approximated on $\bar{B}$ by a polynomial in z.

The first is a generalization of Weierstrass' theorem, which requires arc J to be a closed interval of the real axis. The second is a generalization of Runge's theorem, which requires f to be analytic in $\bar{B}$. To prove this second theorem, Walsh approximated B by a sequence of Jordan regions B_n with $\bar{B} \subset B_{n+1} \subset B_n$ for all n, and then applied Runge's theorem in $\bar{B}$ to the function $F_n(z) = f(\chi_n(z))$, where $w = \chi_n(z)$ maps B_n one-to-one conformally onto B. Walsh later extended this second theorem to sets B which are the union of a finite number of disjoint Jordan regions. Thus he paved the way for the more comprehensive theorem proved later by Mergelyan: if f is a function continuous on any closed bounded set S [that does not separate the plane] and analytic at all interior points of S, then it can be approximated on S uniformly by a polynomial in z.

Walsh maintained an active interest in interpolation, approximation and related topics over a period of about fifty years. Most recently he helped develop some of the fundamental theorems concerning spline interpolation and approximation both on the real line and in the complex plane. His contributions may be found in his published papers as well as in the monograph *The theory of splines and their applications*, which he wrote jointly with J. H. Ahlberg and E. N. Nilson.

In the related area of orthogonal expansions, special mention should be made of the so-called Walsh functions $\varphi_n^{(k)}(x)$. These are defined on the interval $0 \le x \le 1$ by the relations $\varphi_0(x) \equiv 1$;

$$\varphi_1^{(1)}(x) = 1, \quad 0 \le x < \frac{1}{2}; \quad \varphi_1^{(1)}(x) = -1, \quad \frac{1}{2} < x \le 1;$$

$$\varphi_{n+1}^{(2k-1)}(x) = \varphi_{n+1}^{(2k)}(x) = \varphi_n^{(k)}(2x), \qquad 0 \le x < \frac{1}{2};$$

$$\varphi_{n+1}^{(2k-1)}(x) = -\varphi_{n+1}^{(2k)}(x) = (-1)^{k+1}\varphi_n^{(k)}(2x-1), \quad \frac{1}{2} < x \le 1;$$

$$k = 1, 2, 3, \cdots, 2^{n-1}; \qquad n = 1, 2, 3, \cdots.$$

These functions, having some similarity to the Haar functions, were invented by Walsh in 1923 (see his paper 1923b). Now among the most widely used complete orthonormal systems, these functions serve as a valuable mathematical tool in communications engineering and other applied sciences.[2] They are treated in Harmuth's book on *Transmission of information by orthogonal functions*, 2nd ed., Springer-Verlag, New York, 1972; in a survey article by Balashov and Rubinshtein, *Series with respect to Walsh system and their generalization*, J. Soviet Math. **1** (1973), 727–763 which lists 140 references; in an article by N. J. Fine. *Encyclopaedic Dictionary of Physics*, suppl. vol. 4, Pergamon, New York, 1971; and in *Proc. Sympos. on Applications of Walsh Functions*, 1970, 1971, 1972 and 1973. A fifth such symposium was held March 18–20, 1974 at the Catholic University of America in Washington, D.C..

Not only did Walsh himself contribute to the above described areas of research, but also he introduced many of his students to these areas. This is evident from the titles of the doctoral theses written under his direction.

---

[2]Letter from T. J. Rivlin.

Walsh had altogether thirty-one Ph.D. students. The present writer succeeded in contacting nearly all of them to ask for their recollections of him as a research advisor, teacher and friend. In what follows are recorded some of this memorabilia.

Nearly all his former students agreed that, as a thesis advisor, Walsh was patient, generous and considerate, and freely gave help and encouragement. This was true, not only during the developmental stages of each thesis, but even during the years afterwards. It must have given him a great deal of satisfaction that over the years so many of his students continued to contribute to mathematical literature, and that they wrote special papers in his honor on the occasion of his seventieth and seventy-fifth birthdays.

Their remembrance of Walsh as a teacher almost always includes certain rituals accompaning each of his lectures. The opening ritual was to fling the classroom windows wide open regardless of outside temperatures and to deliver his lecture pacing back and forth on the platform while the students might be freezing in their seats. The closing ritual seemed to have been to toss his chalk into the waste basket from whatever position he ended the lecture. Besides, during the run of a lecture he was occasionally known to have used an inattentive student as a target for the chalk. Elemenatry calculus and introductory complex function theory were his favorite courses, but whatever the subject he prepared his lectures with meticulous care and presented them in a deep, musical voice.

Nearly all his former students were impressed by his love of walking. Regardless of the weather he would hike the mile and a half (or so) from his home near Fresh Pond to Widener Library or Seaver Hall. When he accepted the appointment to the University of Maryland, his first prerequisite for a new home was that it be within walking distance of the University.

Some of his former students recollect his sense of humor and his enjoyment of practical jokes. For example, at the 1948 summer meetings of the Society in New Haven, a group photograph was being taken with the mathematicians seated on wide circular bleachers. As the camera rotated, Walsh got his companion to run with him from one end of the bleachers to the other and thus they appeared twice in the same composite photograph.[3] As another instance, a former student who felt himself a novice in teaching sought pedagogic advice from Walsh, to which the latter gave the terse reply: "Always start writing in the upper left-hand corner of the blackboard".[4,5]

In summary, Walsh may be characterized as having had a strong sense of duty to religion, country and chosen work, as well as a love of art and music and a Thoreau-like love of nature.[6] Above all, he was a resolute, hard worker, known to have spent long hours daily in his Widener study. He explored his problems thoroughly along

---

[3]Letter from E. N. Nilson.

[4]Letter from T.J. Rivlin.

[5]Other anecdotes may be found in the two articles: D. V. Widder, *Joseph Leonard Walsh*, J. SIAM Numer. Anal. **3** (1966), 171–172; M. Marden, *Homage to Walsh*, J. Approximation Theory **5** (1972), ix-xiii.

[6]Letter from Maurice Heins.

all possible paths and byways, usually reporting his discoveries in a succession of papers. Often years later he revisited the same problems, searching for nuggets that he may have missed on his earlier explorations. When asked on one occasion how he managed to keep up the terrific pace of publication, Walsh pointed to the self-portrait of an artist showing Death standing nearby and the artist working to complete as much as possible in the little time that remained.[7] The verdict of history will surely be that Walsh did win this race against time. He did leave a permanent imprint upon the mathematics of this century and especially upon the men and women whose mathematical careers he helped to launch.

In the words of the poet, Henry Wadsworth Longfellow, [*Charles Sumner*, stanza 9]

"So when a great man dies
For years beyond our ken
The light he leaves behind him lies
Upon the paths of men."

Addendum to Joseph L. Walsh in Memoriam by Morris Marden

(i) The Chairman of the Organizing Committee for the International Mathematical Congress in Cambridge, August 1950 was Garrett Birkhoff not J. L. Walsh.
(ii) Walsh is listed as a Ph.D. student of G. D. Birkhoff.

The Editors

[7]Letter from E. N. Nilson.

Chapter 1

Zeros and Critical Points of Polynomials and Rational Functions

ON THE LOCATION OF THE ROOTS OF THE JACOBIAN OF TWO BINARY FORMS, AND OF THE DERIVATIVE OF A RATIONAL FUNCTION*

BY

J. L. WALSH

Introduction

Professor Bôcher has shown how the roots of certain algebraic invariants can be determined as the positions of equilibrium in the field of force due to properly situated repelling and attracting particles.† He considers a number of fixed particles either in a plane or on the surface of a sphere (the stereographic projection of the plane) and each of these particles is supposed to repel with a force equal to its mass divided by the distance. If a particle has negative mass, it attracts instead of repelling. The plane of the particles can be considered as the Gauss plane, and with this convention Bôcher proves the following theorem:‡

THEOREM I. *The vanishing of the jacobian of two binary forms f_1 and f_2 of degrees p_1 and p_2 respectively determines the points of equilibrium in the field of force due to p_1 particles of mass p_2 situated at the roots of f_1, and p_2 particles of mass $-p_1$ situated at the roots of f_2.*

Perhaps it is desirable briefly to indicate the proof of this theorem. We give the proof merely for the plane field. Let fixed particles of masses m_1, $m_2, \cdots, m_n$ be placed at the points represented by the complex quantities $e_1, e_2, \cdots, e_n$ respectively. Then at any point x of the plane, the force due to these fixed particles is in magnitude, direction, and sense

$$K\left(\frac{m_1}{x-e_1}+\frac{m_2}{x-e_2}+\cdots+\frac{m_n}{x-e_n}\right),$$

where the symbol K indicates the conjugate of the complex quantity following.

* Presented to the Society, February 23, 1918.

† Maxime Bôcher, *A problem in statics and its relation to certain algebraic invariants*, Proceedings of the American Academy of Arts and Sciences, vol. 40 (1904), p. 469. I am indebted to Professor Bôcher for a number of suggestions concerning the present paper.

‡ L. c., p. 476.

If homogeneous variables are introduced by the formulas

$$x = \frac{x_1}{x_2}, \qquad e_i = \frac{e_i'}{e_i''},$$

the plane field described in Theorem I becomes

$$K\left[x_2\left(p_2\sum_{i=1}^{i=p_1}\frac{e_i''}{e_i''x_1 - e_i'x_2} - p_1\sum_{i=p_1+1}^{i=p_1+p_2}\frac{e_i''}{e_i''x_1 - e_i'x_2}\right)\right]$$

$$(1) \qquad = K\left[x_2\left[p_2\frac{\frac{\partial f_1}{\partial x_1}}{f_1} - p_1\frac{\frac{\partial f_2}{\partial x_1}}{f_2}\right]\right],$$

where

$$f_1 = (e_1''x_1 - e_1'x_2)\cdots(e_{p_1}''x_1 - e_{p_1}'x_2),$$

$$f_2 = (e_{p_1+1}''x_1 - e_{p_1+1}'x_2)\cdots(e_{p_1+p_2}''x_1 - e_{p_1+p_2}'x_2).$$

The quantity in the brackets in (1) reduces to the quotient of the jacobian of f_1 and f_2 by f_1f_2, when Euler's theorem for homogeneous functions is applied. This completes the proof of Theorem I. It is to be noted that the jacobian vanishes not only at the points of no force, but also at the multiple roots of either form or a common root of the two forms; such a point is called a point of *pseudo-equilibrium*.

From the mechanical interpretation of Theorem I, Bôcher derives a number of results concerning the location of the roots of the jacobian with reference to the location of the roots of the ground forms.*

When we consider the mechanical system, it is intuitively obvious that there can be no position of equilibrium very near any of the fixed particles. In the Corollary to Theorem II of the present paper there is determined explicitly (and in an infinite variety of ways) a circle which can be drawn separating any one of these particles from the roots of the jacobian. If we have not one fixed particle but k particles, all attracting or all repelling, and if the remaining particles in the plane (or on the sphere) are sufficiently removed from those, then the mechanical system would lead us to expect that there could be no roots of the jacobian outside of and very near to a circle surrounding the k particles. This is a rough indication of the considerations that lead to Theorem II.

In the latter part of the paper some applications of these results are made to the roots of the derivative of a rational function.

* See, e. g., Theorem III below.

PART I

Let us consider the statical system in the plane due to fixed particles of the kind described. We shall make use of two lemmas, which, indeed, are more general than is necessary for our use.

LEMMA I. *If Q is a point exterior to the circle C whose center is O, then of all possible positions for a unit (repelling or attracting) particle on or within C, that position nearest to Q causes the particle to exert the greatest force at Q,—greatest not only in magnitude but also in component along QO. That position farthest from Q causes the particle to exert the least force at Q,—least not only in magnitude but also in component along QO.*

LEMMA II. *If Q is a point interior to the circle C whose center is O, then of all possible positions for a unit (repelling or attracting) particle on or outside of C, that position nearest to Q causes the particle to exert the greatest force at Q,—greatest not only in magnitude but also in component along QO. Of all possible positions for an attracting particle on or outside of C, that position on C which is farthest from Q causes the particle to exert the force at Q which has the greatest component in the direction and sense QO.*

The truth of each of these lemmas becomes evident upon inverting C in the circle of unit radius and center Q, noting that the force exerted at Q by a unit particle at R is in direction and magnitude $R'Q$, where R' is the inverse of R in the unit circle whose center is Q.

We shall now apply these lemmas to Theorem I. Suppose there is in the plane a circle C_1 which contains on or within its circumference k roots of f_1. Suppose there is a circle C_2—larger than C_1 and concentric with it—outside of which lie all the remaining $p_1 - k$ roots of f_1. Suppose further that there is a circle C_3—also larger than C_1 and concentric with it—outside of which lie all the roots of f_2. Then we shall try to determine a circle C_0 larger than C_1 and concentric with it and such that there is no root of the jacobian of f_1 and f_2 within the annular region between C_0 and C_1.

We denote by O the common center of C_1, C_2, and C_3, and the radii of these circles by a, b, and c, respectively. We have supposed that $a < b$, $a < c$. Set up the statical system of Theorem I and consider the force at a point Q between C_1 and the smaller of C_2 and C_3. The component in the direction and sense OQ of the force due to the k positive particles (each of mass p_2) on or within C is not less than $p_2 k/(a + r)$, where r is the distance OQ. The component in the direction and sense QO of the force due to the positive particles outside of C_2 (whose mass is $(p_1 - k)p_2$) is not greater than $(p_1 - k)p_2/(b - r)$. The component in the direction and sense QO of the negative particles outside of C_3 (whose mass is $-p_1p_2$) is not greater than

$p_1 p_2/(c+r)$. If Q is a point of equilibrium, we must have

$$\frac{p_2 k}{a+r} \leqq \frac{(p_1-k)p_2}{b-r} + \frac{p_1 p_2}{c+r}, \tag{2}$$

$$\frac{kbc - p_1 ab - (p_1-k)ac}{(p_1-k)b + p_1 c - ka} \leqq r. \tag{3}$$

If the left-hand member of (3) is positive, we construct the circle with that radius and center O, and denote this circle by C_0. Then *it is readily seen that C_0 always lies within C_2 (unless $k = p_1$, when C_0 and C_2 coincide), and C_0 may or may not lie within C_1 and may or may not lie within C_3.* If C_0 lies outside of C_1 but within C_3, then we have shown that the annular region between C_0 and C_1 contains no point of equilibrium. This region contains no root of either form and therefore no possible point of pseudo-equilibrium. Hence the annular region contains no root of the jacobian of the forms. If on the other hand, C_0 lies outside of C_3, then between C_1 and C_3 there is no root of the jacobian of the forms.

If C_0 lies outside of C_1, it is readily shown that there are precisely $k-1$ roots of the jacobian on or within C_1. Let the k roots of f_1 that are on or within C_1 move continuously so as to coincide at the point O, while the other roots of f_1 and all the roots of f_2 remain fixed. If Q is a position of equilibrium, inequality (2) obtains whenever Q is anywhere within C_1. Hence by (3) there is no root of the jacobian within C_1 except at O; and O is a $(k-1)$-fold root. During the change of the k roots of f_1 the roots of the jacobian change continuously (at least when we refer to the sphere instead of the plane) and are never in the annular region between C_1 and the nearer of C_0 and C_3. Hence at the start there were just $k-1$ roots of the jacobian on or within C_1.

Let us determine the circle C_0 by invariant elements. Suppose a line through O cuts the circles C_i in the points C_i' and C_i'' $(i = 0, 1, 2, 3)$ where the notation is such that O separates no pair of points C_i', C_j'. We find that*

$$(C_1'', C_2', C_3'', C_0') = p_1/k.$$

Hence, for the special case that C_1, C_2, and C_3 are concentric, with C_1 in the interior of C_2 and C_3, we have proved:

THEOREM II. *Suppose that f_1 and f_2 are two binary forms, the degree of f_1 being p_1, and suppose there are k roots of f_1 which lie in a closed region T_1 bounded by a circle C_1. Suppose there is a second closed region T_2 bounded by a circle C_2,*

* We are using the following definition for the cross-ratio:

$$(z_1, z_2, z_3, z_4) = \frac{(z_1-z_2)(z_3-z_4)}{(z_2-z_3)(z_4-z_1)}.$$

that T_2 has no point in common with T_1, and that T_2 contains the remaining $p_1 - k$ roots of f_1. Suppose further that there is a third closed region T_3 bounded by a circle C_3 coaxial with C_1 and C_2, that T_3 has no point in common with T_1, and that T_3 contains all the roots of f_2.

1. *If the circle C_0 described below lies in the region between C_1 and C_3, then there are no roots of the jacobian of f_1 and f_2 in the region included between C_1 and C_0; furthermore, there are just $k - 1$ roots of the jacobian in T_1.*

2. *If the circle C_0 lies in the region between C_2 and C_3, then there are no roots of the jacobian of f_1 and f_2 in the region included between C_1 and C_3; moreover, there are just $k - 1$ roots of the jacobian in T_1.*

In this theorem, C_0 denotes that circle of the coaxial family to which C_1, C_2, and C_3 belong which is the locus of points C_0' such that

$$(C_1'', C_2', C_3'', C_0') = p_1/k.$$

*C_i' and C_j'' denote the points in which any circle T orthogonal to the circles of the family cuts the circle C_i, and the notation is such that on T, neither null circle of the family shall separate any of the pairs of points C_i', C_j' $(i, j = 0, 1, 2, 3)$.**

This theorem is proved for the case that C_1, C_2, and C_3 are not concentric by making a linear transformation that transforms them into concentric circles, with C_1 in the interior of C_2 and C_3. (Such a transformation always exists.) Since the theorem is true for this particular case, and since everything used in the theorem is invariant under linear transformation, the theorem is true as stated.†

It is also true that Theorem II refers to the sphere as well as the plane, for everything essential in the theorem is invariant under stereographic projection.

* Reference to the italicized sentence immediately below (3) will show that 1 and 2 cannot occur at the same time. It may occur that C_0 lies in neither of these positions, which in the case of concentric circles means that C_0 lies within C_1; if this is true, the theorem makes no statement about the roots of the jacobian. Also, the circle C_0 may not exist, which means that the left-hand side of (3) is negative.

If no root of f_1 lies on C_1, if no root of f_1 lies on C_2 $(k \neq p_1)$, or if no root of f_2 lies on C_3, then in case (1) no root of the jacobian can lie on C_0. This is immediately seen by omitting the equality sign in (2) and hence in (3).

Of course, a theorem similar to II can be proved for Bôcher's covariant ϕ (l. c., p. 474).

Theorem II can be applied to the roots of special types of polynomials, but as Professor Curtiss pointed out to me, the following more general theorem can be proved by means of Lemma II. This more general theorem is a special case of the theorem just suggested concerning the covariant φ.

If $f(z)$ is a polynomial of degree n all of whose roots lie outside of a circle whose center is the origin and radius b, and if k_1 and k_2 are any positive numbers, then all the roots of $k_1 z f'(z) - k_2 f(z)$ lie outside the smaller of the two circles whose common center is the origin and whose radii are b and $k_2 b / (nk_1 - k_2)$ respectively. See Laguerre, *Œuvres*, vol. I, pp. 56, 133; see also the reference to Gonggrÿp below.

† We consider the exterior of a circle, including the boundary and the point at infinity, to be a closed region.

It is readily shown that Theorem II gives in general the largest region which will be free from roots of the jacobians of all pairs of forms which satisfy the hypothesis. Let us take the circles C_1, C_2, and C_3 in their original (concentric) positions, and first suppose C_0 to lie between C_1 and C_3. Then reference to inequality (2) shows that if $k \neq p_1$, the position of the particles which determine the field of force can be chosen so that (2) becomes an *equality*, and there will be a position of equilibrium on C_0. If $k = p$, C_0 and C_2 coincide, and we can consider C_2 to coincide with C_3. In this case, or if on the other hand C_0 lies outside of C_3, there can be chosen on C_3 a multiple root of f_2, which will be a root of the jacobian.

If we take C_1 a null circle P, and if we let C_2 and C_3 coincide and denote this circle by C, we have the following result:

COROLLARY. *Suppose that f_1 and f_2 are two binary forms, the degree of f_1 being p_1, and suppose that the circle C separates P (a k-fold root of f_1) from those roots of f_2 and f_1 (other than P) which do not lie on C itself. Then the circle C_0 separates P from those roots of the jacobian of f_1 and f_2 (other than P) which do not lie on C_0 itself, where C_0 is that circle of the coaxial family determined by C and P which is the locus of points C_0' such that*

$$(P, C', C'', C_0') = p_1/k;$$

C' and C'' denote the intersections of C with the circle through P and C_0' orthogonal to C.

In the corollary, it is of course true that C_0 lies between P and C unless $k = p_1$, when C_0 and C coincide.

If we take $k = p_1$, and if C_2 and C_3 are chosen coincident, Theorem II gives the following theorem, which is due to Bôcher:*

THEOREM III. *If the roots of a binary form f_1 of degree p_1 lie in a closed region T_1 and if the roots of a second binary form f_2 of degree p_2 lie in a second closed region T_2 which has no point in common with T_1, and if these two regions are bounded by arcs of circles each one of which circles separates the interior of T_1 from the interior of T_2, then the jacobian of f_1 and f_2 has just $p_1 - 1$ roots in T_1 and $p_2 - 1$ roots in T_2.*†

* As Professor Curtiss pointed out to me, the statement given by Bôcher (l. c., p. 478) is not quite accurate. This inaccuracy has been here corrected.

† There have recently been published two results which are special cases of Theorem III, although the authors were apparently not aware of the fact.

See L. R. Ford, *On the roots of a derivative of a rational function*, Proceedings of the Edinburgh Mathematical Society, vol. 33 (1915). Several of Ford's results are generalized in the present paper.

See also B. Gonggrÿp, *Quelques théorèmes*, etc., Liouville's Journal, ser. 7, vol. 1 (1915), p. 360. Compare the former reference to Laguerre.

PART II

All the theorems concerning jacobians which were proved by Bôcher, as well as the theorems of the present paper can be immediately applied to the roots of the derivative of a rational function.

Let us take any rational function not a constant, $f(z) = u(z)/v(z)$, and suppose (as we can do with no loss of generality) that u and v have no common factor containing z. Introduce homogeneous coördinates, setting $z = z_1/z_2$, and multiply the numerator and denominator of f by z_2^n, where n is the degree of f:*

$$f(z) = \frac{z_2^n u(z_1/z_2)}{z_2^n v(z_1/z_2)} = \frac{f_1(z_1, z_2)}{f_2(z_1, z_2)}.$$

If we express $f'(z)$, the derivative of $f(z)$, in terms of J, the jacobian of f_1 and f_2, we find

$$f'(z) = \frac{J}{n}\left(\frac{z_2}{f_2}\right)^2.$$

From this relation it follows that *the roots of f' are the roots of J and a double root at infinity, except that when one of these points is also a pole of f it cannot be a root of f'.*

We shall not attempt to carry over all the results concerning the roots of the jacobian to the corresponding results for the derivative of a rational function. We merely give a few examples by way of illustration.† The following theorem is a direct application of Theorem III.

If $f(z)$ is a rational function of degree n whose roots lie in a closed region T_1 and whose poles lie in a second closed region T_2 which has no point in common with T_1, and if these two regions are bounded by arcs of circles each one of which circles separates the interior of T_1 from the interior of T_2; then all the roots of the derivative of $f(z)$ lie in T_1 and T_2, except that there are two additional roots at infinity if $f(z)$ has no pole there. Except for these two possible roots, there are just $n - 1$ roots of $f'(z)$ in T_1, and if $f(z)$ has no multiple pole there are just $n - 1$ roots of $f'(z)$ in T_2.‡

* The degree of a rational function is the greater of the degrees of its numerator and denominator, or the common degree if the numerator and denominator have the same degree.

† Essentially the following theorem is given by Bôcher (l. c., p. 479): " If f_1 and f_2 are two forms and if all the roots of each form either lie on a circle C or are situated in pairs of points inverse with respect to C, then all the roots of the jacobian of f_1 and f_2 also lie on C or are situated in pairs of points inverse with respect to C. On any arc of C bounded by roots of f_1 (or of f_2) and containing no root of either form there is at least one root of the jacobian."

It is true that the force at any point of C (when the statical system is set up) is in direction tangent to C. Hence, if there are *two* circles C of the theorem stated in this footnote, their intersection must be a root of the jacobian or a root of one of the ground forms.

Both of these theorems can evidently be extended to the derivative of a rational function.

‡ This is a generalization of the well-known theorem of Lucas that " the roots of the deriva-

We can immediately obtain an upper bound for the moduli of the finite roots of the derivative of a rational function. Suppose f to have m_1 finite roots (or poles) and m_2 finite poles (or roots), $m_1 > m_2$. It follows from the corollary to Theorem II that if a circle whose radius is a includes all the finite roots and poles of f, then a concentric circle of radius $a(m_1 + m_2)/(m_1 - m_2)$ includes all the finite roots of f'.

CAMBRIDGE, MASS.
January, 1918

tive of a polynomial lie within or on the boundary of the smallest convex polygon enclosing the roots of the original polynomial."

As Bôcher points out, Theorem III is also a generalization of Lucas's theorem.

The following theorem is a corollary of Theorem II: *If $f(z)$ is a polynomial of degree n which has a k-fold root at P, and if a circle whose center is P and radius a includes no root of f other than P, then the circle whose center is P and radius ak/n includes no root of $f'(z)$ other than P.*

ON THE

LOCATION OF THE ROOTS OF THE DERIVATIVE OF A POLYNOMIAL

By J.-L. WALSH

If there are plotted in the plane of the complex variable the roots of a polynomial $f(z)$ and the roots of the derived polynomial $f'(z)$, there are interesting geometric relations between the two sets of points. It was shown by Gauss that the roots of $f'(z)$ are the positions of equilibrium in the field of force due to equal particles situated at each root of $f(z)$, if each particle repels with a force equal to the inverse distance. The derivative vanishes not only at the positions of equilibrium but also at the multiple roots of $f(z)$.

From Gauss's theorem follows immediately the theorem of Lucas that the roots of $f'(z)$ lie in any convex polygon in which lie the roots of $f(z)$. In the present note I wish to prove the following theorem, the connection of which with Lucas's theorem will be pointed out later.

THEOREM I. — *If m_1 roots of a polynomial $f(z)$ lie in or on a circle C_1 whose center is α_1 and radius r_1, and if all the remaining roots of $f(z)$, m_2 in number, lie in or on a circle C_2 whose center is α_2 and radius r_2, then all the roots of $f'(z)$ lie in or on C_1, C_2, and a third circle C_3 whose center is*

$$\frac{m_1\alpha_2 + m_2\alpha_1}{m_1 + m_2} \text{ and radius } \frac{m_1 r_2 + m_2 r_1}{m_1 + m_2}.$$

If the circles C_1, C_2, C_3 are mutually external they contain respectively the following numbers of roots of $f'(z)$: $m_1 - 1$, $m_2 - 1$, 1.

The centers of the circles C_1, C_2, C_3 are collinear. If $r_1 = r_2$, the three circles have the same radius. If $r_1 \neq r_2$, the point $\frac{r_1\alpha_2 - r_2\alpha_1}{r_1 - r_2}$ is a center of similitude for any pair of the circles C_1, C_2, C_3.

Theorem I is a generalization of the trivial theorem where $f(z)$ has but two roots, α_1 and α_2, of respective multiplicities m_1 and m_2. Then α_1 and α_2 are roots of $f'(z)$ of respective multiplicities $m_1 - 1$ and $m_2 - 1$, and there is a root of $f'(z)$ at the point $\frac{m_1\alpha_2 + m_2\alpha_1}{m_1 + m_2}$ which divides the segment (α_1, α_2) in the ratio $m_1 : m_2$.

We shall prove Theorem I from Gauss's theorem, making use of several lemmas.

Lemma I. — *If m particles lie on or within a circle C, the corresponding resultant force at any point P exterior to C is equivalent to the force at P due to m coincident particles on or within C.*

The force at P due to a particle at any point Q is in magnitude, direction, and sense Q'P, where Q' denotes the inverse of Q in the unit circle whose center is P. To replace m points Q by m coincident points we replace the m vectors Q'P by m coincident vectors, and hence replace their terminals Q' by the center of gravity of the m points Q'. Denote by C' the inverse of the circle C in the unit circle whose center is P. Then under the conditions of the lemma all the points Q lie in or on C, all the points Q' lie in or on C', their center of gravity lies in or on C', and hence the inverse of the center of gravity lies in or on C. This completes the proof of the lemma.

Lemma II. — *Let the points z_1 and z_2 be allowed to assume independently all positions on or within the respective circles C_1 and C_2 whose centers are α_1 and α_2 and radii r_1 and r_2 respectively. Then the locus of the point $\frac{m_1 z_2 + m_2 z_1}{m_1 + m_2}$ which divides the segment (z_1, z_2) in the constant ratio $m_1 : m_2 (m_1, m_2 > 0)$ is the interior (including the boundary) of the circle C_3 whose center is*

$$\frac{m_1\alpha_2 + m_2\alpha_1}{m_1 + m_2} \text{ and radius } \frac{m_1 r_2 + m_2 r_1}{m_1 + m_2}.$$

Under our hypothesis, the point $\frac{m_1 z_2 + m_2 z_1}{m_1 + m_2}$ lies on or within C_3. For we have

$$|z_1 - \alpha_1| \leqq r_1,$$
$$|z_2 - \alpha_2| \leqq r_2,$$

and hence

$$\left|\frac{m_1 z_2 + m_2 z_1}{m_1 + m_2} - \frac{m_1\alpha_2 + m_2\alpha_1}{m_1 + m_2}\right| = \left|\frac{m_1(z_2 - \alpha_2)}{m_1 + m_2} + \frac{m_2(z_1 - \alpha_1)}{m_1 + m_2}\right| \leqq \frac{m_1 r_2 + m_2 r_1}{m_1 + m_2}.$$

Any point z on or within C_3 corresponds to some pair of points z_1 and z_2 on or within C_1 and C_2 respectively. For let us set

$$z_1 - \alpha_1 = \left(z - \frac{m_1\alpha_2 + m_2\alpha_1}{m_1 + m_2}\right)\frac{r_1}{\dfrac{m_1 r_2 + m_2 r_1}{m_1 + m_2}},$$

$$z_2 - \alpha_2 = \left(z - \frac{m_1\alpha_2 + m_2\alpha_1}{m_1 + m_2}\right)\frac{r_2}{\dfrac{m_1 r_2 + m_2 r_1}{m_1 + m_2}}.$$

We are assuming that

$$\left|z - \frac{m_1\alpha_2 + m_2\alpha_1}{m_1 + m_2}\right| \leqq \frac{m_1 r_2 + m_2 r_1}{m_1 + m_2},$$

and hence

$$|z_1 - \alpha_1| \leqq r_1,$$
$$|z_2 - \alpha_2| \leqq r_2;$$

of course we have

$$z = \frac{m_1 z_2 + m_2 z_1}{m_1 + m_2}.$$

This completes the proof of Lemma II.

We are now in a position to prove Theorem I. Suppose a point z external to C_1 and C_2 to be a position of equilibrium in the field of force. The force at z due to the m_1 particles in or on C_1 is equivalent to the force at z due to m_1 particles coinciding at some point z_1 in or on C_1, and the force at z due to the m_2 particles in or on C_2 is equivalent to the force at z due to m_2 particles coinciding at some point z_2 in C_2. Then z divides the segment (z_1, z_2) in the ratio $m_1 : m_2$ and hence lies in or on C_3. Therefore the three circles C_1, C_2, C_3 contain all positions of equilibrium. They contain also all multiple roots of $f(z)$ and hence they contain all roots of $f'(z)$.

If the circles C_1, C_2, and C_3 are mutually external, we allow the m_1 roots of $f(z)$ in C_1 to move continuously in C_1 and to coalesce at a point in C_1, and similarly allow the m_2 roots of $f(z)$ in C_2 to move continuously in C_2 and to coalesce in C_2. In the final position the numbers of roots of $f'(z)$ in C_1, C_2, and C_3 are $m_1 - 1$, $m_2 - 1$, and 1 respectively. Throughout the motion of the roots of $f(z)$, the roots of $f'(z)$ move continuously, none enters or leaves any of the three circles, and hence the final number of roots of $f'(z)$ in each of those circles is the same as the initial number. The proof of Theorem I is now complete.

If no point of the circle C_2 is exterior to C_1, no point of C_3 is exterior to C_1, so we are led to the following theorem which is equivalent to the theorem of Lucas quoted above :

If all the roots of a polynomial lie on or within a circle, then all the roots of the derived polynomial lie on or within that circle.

For polynomials all of whose roots are real the following theorem is easily proved either from Theorem I or directly by a proof similar to the proof of that theorem :

THEOREM II. — *If m_1 roots of a polynomial $f(z)$ lie in a closed interval I_1 of the axis of reals whose center is α_1 and length l_1, and if all the remaining roots of $f(z)$, m_2 in number, lie in a closed interval I_2 of the axis of reals whose center is α_2 and length l_2, then all the roots of $f'(z)$ lie in I_1, I_2, and a third interval I_3 of the axis of reals whose center is*

$$\frac{m_1\alpha_2 + m_2\alpha_1}{m_1 + m_2} \text{ and length } \frac{m_1 l_2 + m_2 l_1}{m_1 + m_2}.$$

If the intervals I_1, I_2, I_3 are mutually external, they contain respectively the following numbers of roots of $f'(z)$: $m_1 - 1$, $m_2 - 1$, 1.

There are various generalizations of Theorems I and II. It is entirely incidental in Theorem I that we considered the *interiors* of two circles. By slight changes in the formulas used we obtain a similar theorem for the interior of one circle and the exterior of the other. The theorems can easily be extended to give results concerning the roots of the derivative of a rational function, or the roots of the jacobian of two binary forms. The last generalization is particularly interesting because the results are invariant under linear transformation of the complex variable. For these and other results the reader is referred to a paper by the present writer which has appeared and several others which are expected to appear in the *Transactions of the American Mathematical Society*.

Toulouse. - Imp. et Lib. Edouard Privat. — 3786

ON THE LOCATION OF THE ROOTS OF THE JACOBIAN OF TWO BINARY FORMS, AND OF THE DERIVATIVE OF A RATIONAL FUNCTION*

BY

J. L. WALSH

The present paper is an extension and in some respects a simplification of a recent paper published under the same title.† Both papers are based on a theorem (Theorem I, below) due to Professor Bôcher.‡ By means of the statical problem of determining the positions of equilibrium in a certain field of force, there are obtained some new results concerning the location of the roots of the jacobian of two binary forms relative to the location of the roots of the ground forms. Application is made to the roots of the derivative of a polynomial and to the roots of the derivative of a rational function. The present paper gives a proof and an application of a geometrical theorem (Theorem II) which may be not uninteresting.

Bôcher considers a number of fixed particles in a plane or by stereographic projection on the surface of a sphere, and supposes each particle to repel with a force equal to its mass (which may be positive or negative) divided by the distance. If the plane is taken as the Gauss plane, the following result is proved:§

THEOREM I. *The vanishing of the jacobian of two binary forms* f_1 *and* f_2 *of degrees* p_1 *and* p_2 *respectively determines the points of equilibrium in the field of force due to* p_1 *particles of mass* p_2 *situated at the roots of* f_1, *and* p_2 *particles of mass* $-p_1$ *situated at the roots of* f_2.

The jacobian vanishes not only at the points of no force, but also at the multiple roots of either form or a common root of the two forms; such a point is called a position of *pseudo-equilibrium*.

* Presented to the Society, Dec. 31, 1919.

† Walsh, these Transactions, vol. 19 (1918), pp. 291–298. This paper will be referred to as I.

‡ Maxime Bôcher, *A problem in statics and its relation to certain algebraic invariants*, Proceedings of the American Academy of Arts and Sciences, vol. 40 (1904), p. 469.

§ Bôcher's proof (l. c., p. 476) is reproduced in I, p. 291.

It is intuitively obvious that there can be no position of equilibrium very near any of the fixed particles, or very near and outside of a circle containing a number of fixed particles, all attracting or all repelling, if the other particles are sufficiently remote. We consider, then, a number of particles in a circle or more generally in a circular region. First we adjoin to the plane the point at infinity, and use the term *circle* to include point and straight line; then we define a *circular region* to be a closed region of the plane bounded by a circle, namely, the interior of a circle, the exterior of a circle including the point at infinity, a half plane, a point, or the entire plane. There will be no confusion in having the same notation for a circular region as for its boundary.

In the following development we shall use several lemmas.

LEMMA I. *The force at a point P due to k particles each of unit mass situated in a circular region C not containing P is equivalent to the force at P due to k coincident particles each of unit mass also in C.*

Denote by C' the inverse of C in the circle of unit radius and center P and by Q' the inverse of any point Q with regard to that circle. The force at P due to a particle at Q is in direction and magnitude PQ'. We replace k vectors PQ' by k coincident vectors having one terminal at P and the other at the center of gravity of the points Q'; these two sets of vectors have the same resultant. If any point Q is in the region C, its inverse Q' is in C', and the center of gravity of a number of such points Q' is also in C'. The inverse of this center of gravity is then in C.

LEMMA II. *In the field of force due to k positive particles at z_1, l positive particles at z_2, and $k+l$ negative particles at z_3, the only position of equilibrium is z_4 as determined by the cross-ratio*

$$\frac{(z_1-z_2)(z_3-z_4)}{(z_2-z_3)(z_4-z_1)} \equiv (z_1, z_2, z_3, z_4) = \frac{k+l}{k}.$$

The lemma is evidently true when one of the points z_1, z_2, z_3 is at infinity. The invariance of the positions of equilibrium under linear transformation follows from Theorem I and hence completes the proof.

We shall next prove a preliminary theorem, the proof of which is given in part by several succeeding lemmas.

THEOREM II. *If the envelopes of points z_1, z_2, z_3 are circular regions C_1, C_2, C_3 respectively, then the envelope of z_4, defined by the real constant cross-ratio*

$$\lambda = (z_1, z_2, z_3, z_4)$$

*is also a circular region.**

* The term *envelope* is used to denote the set of points which is the totality of positions assumed by each of the points z_1, z_2, z_3, z_4; the points z_1, z_2, z_3 are supposed to vary independently.

The proof of Theorem II which is presented in detail has some advantages and some dis-

We denote the envelope of z_4 by C_4, and we must show that C_4 is a region bounded by a single circle. First we consider several special cases of the theorem. If C_1, C_2, and C_3 are distinct points, C_4 is a point. If any of the regions C_1, C_2, C_3 is the entire plane, C_4 is also the entire plane. If $\lambda = 0$ and if C_1 and C_2 have a point in common, C_4 is the entire plane. If $\lambda = 0$ and C_1 and C_2 have no point in common, $z_3 = z_4$ and so C_4 coincides with C_3. If $\lambda = \infty$ and C_2 and C_3 have a common point, C_4 is the entire plane. If $\lambda = \infty$ and C_2 and C_3 have no common point, C_4 and C_1 are identical. If $\lambda = 1$ and C_1 and C_3 have a common point, C_4 is the entire plane. If $\lambda = 1$ yet C_1 and C_3 have no common point, C_4 is identical with C_2. In the sequel, unless it is explicitly stated to the contrary, we suppose λ to have none of the values $0, 1, \infty$. It follows that no two of the points z_1, z_2, z_3, z_4 coincide unless three of them coincide.

Except in the trivial case that C_1, C_2, C_3 are points, C_4 is evidently a two-dimensional continuum and is not necessarily the entire plane. The envelope C_4 is connected, for to join any pair of points z_4', z_4'' in C_4 by a curve in C_4, we need merely to choose any set of points corresponding to each, z_1', z_2', z_3'; z_1'', z_2'', z_3'', in the proper regions. Join z_1' and z_1'' by a continuous curve which lies in C_1, and similarly join z_2' and z_2'', and z_3' and z_3'', by continuous curves in C_2 and C_3 respectively. Allow z_1, z_2, z_3 to move from z_1', z_2', z_3' to z_1'', z_2'', z_3'' along these respective curves. The point z_4 corresponding moves from z_4' to z_4'' in C_4 and along a curve which is continuous because z_4 is a linear function of z_1, z_2, z_3.

Our next remark is stated explicitly as a lemma. It is readily stated and established for regions whose boundaries are curves much more general than circles, but we consider here merely the form under the hypothesis of Theorem II and for application to the proof of that theorem.

---

advantages over the following suggested method of proof. The theorem is evidently true when C_1, C_2, and C_3 are points. The theorem is easily proved when C_1 and C_2 are points but C_3 is not a point. By taking the envelope of the circular region C_4 in the preceding degenerate case, the theorem can be proved when C_1 is a point but neither C_2 nor C_3 is a point. The envelope of the region C_4 in this last degenerate case, as z_1 is allowed to vary over a region C_1 not a point, gives the envelope of z_4 for the theorem in its generality. I have not been able to carry through the actual analytic determination of the envelope by this method because the algebraic work is too laborious.

This suggested method of proof, however, shows at once that the boundary of the region C_4 in the general case is an algebraic curve or at least part of an algebraic curve.

It seems to me likely that Theorem II is true also when λ is imaginary, but I have not carried through the proof in detail.

In general the relation of the regions C_1, C_2, C_3, C_4 is not reciprocal. For example if C_1 is a point but neither C_2 nor C_3 is a point and if these regions lead to the fourth region C_4, then if we choose the circular regions C_2, C_3, C_4 as the original circular regions of the lemma, we cannot for any choice of λ be led to the region C_1. This lack of reciprocality does not depend on the degeneracy of one of the regions C_1, C_2, C_3, C_4.

LEMMA III. *If the point* z_4 *is on but not at a vertex of the boundary of* C_4,* *then any set of points* z_1, z_2, z_3 *corresponding lie on the boundaries of the respective regions* C_1, C_2, C_3; *the circle* C *through the points* z_1, z_2, z_3, z_4 *cuts the circles* C_1, C_2, C_3 *all at angles of the same magnitude; and if* C *is transformed into a straight line. the lines tangent to the circles* C_1, C_2, *and* C_3 *at the points* z_1, z_2, z_3 *respectively are parallel.*

The following proof is formulated only for the general case that none of the circles C_1, C_2, C_3 is a null circle, but no essential modification is necessary to include the degenerate cases.

When z_2 and z_3, and also the circle C are kept fixed, a continuous motion of z_1 along C also causes z_4 to move continuously along C. If the direction of motion of z_1 is reversed, the direction of motion of z_4 is also reversed. Hence z_4 is not on the boundary of C_4 unless z_1 is on the boundary of C_1, and as can be shown in an analogous manner, not unless z_2 and z_3 are on the boundaries of C_2 and C_3 respectively. The region C_4 is closed since the regions C_1, C_2, and C_3 are closed.

Let P be any point of the boundary of C_4. Transform P to infinity, so that the corresponding points z_1, z_2, z_3 lie on the same line L. We assume at first that L is not tangent to any of the circles C_1, C_2, C_3 nor to the boundary of C_4. The relative positions of the points z_1, z_2, z_3 on L together with the sense along L in which the region C_1 extends from z_1 determine uniquely the sense along L in which the regions C_2, C_3, C_4 must extend from z_2, z_3, P respectively. There is evidently a segment of L terminated by P composed entirely of points in C_4. If the lines tangent to the circles C_2 and C_3 at the points z_2 and z_3 are not parallel, it is possible slightly to rotate L about z_1 in one direction or the other into a new position L' and to determine a point z_2'' on L' and on the circle C_2 and a point z_3'' on L' and *interior to the region* C_3 such that the triangles $z_1 z_2 z_2''$ and $z_1 z_3 z_3''$ are similar and hence we have the relation

$$(z_1, z_2'', z_3'', P) = \lambda.$$

Then z_3'' can be moved in either sense along the line L' and still remain in its proper envelope, so there are corresponding points z_4'' on L' in either sense from P. Moreover, this is true for every position of L' if the angle from L to L' is in the proper sense and is sufficiently small, so if we transform P to the finite part of the plane and z_1 to infinity and notice that the lines L' are lines through the point P, it becomes evident that there are points z_4 in the neighborhood of P on any line L' through P which lies within a certain sector whose vertex is P, and there are points z_4 on L' in both directions

* It is of course true that the boundary of C_4 has no vertices, but that fact has not yet been proved.

from P. Hence if P is actually on the boundary of C_4, it must lie at a vertex of that boundary.*

The proof thus far has been formulated to prove that when P is at infinity the lines tangent to the circles C_2 and C_3 at z_2 and z_3 are parallel. The notation of the proof can easily be modified to show that the lines tangent to the circles C_1 and C_2 at z_1 and z_2 are parallel, and hence the lines tangent to C_1, C_2, C_3 at z_1, z_2, z_3 are parallel.

This same method of reasoning is readily used to prove that if the circle C of the lemma is tangent to one or two of the circles C_1, C_2, C_3 at the respective points z_1, z_2, z_3 but is not tangent to all these circles, the boundary of C_4 has a vertex at z_4. The circle C is not tangent to the boundary of C_4 unless C is tangent to C_1, C_2, and C_3. This consideration completes the proof of Lemma III.

It is desirable to make a revision in our use of the term *angle between two circles*. With Coolidge,† we consider circles to be described by a point moving in a counter-clockwise sense, and define the angle between two circles to be the angle between the half-tangents drawn at the intersection in the sense of description of the circles. When we are concerned with a single straight line, either sense may be given to it. We shall use this convention in proving the following lemma, which is a result purely of circle geometry which has not necessarily any connection with Theorem II. As stated and proved, it is slightly more general than is necessary for its application in the proof of that theorem.

LEMMA IV. *Suppose a variable circle C either to cut three distinct fixed non-coaxial circles C_1, C_2, C_3 all at the same angle or to cut a definite one of those circles at an angle supplementary to the angle cut on the other two. If the points z_1, z_2, z_3 are chosen as intersections of C with C_1, C_2, C_3 respectively such that when C is transformed into a straight line the lines tangent to C_1, C_2, C_3 at z_1, z_2, z_3 are all parallel, then the locus of the point z_4 defined by the real constant cross-ratio*

$$\lambda = (z_1, z_2, z_3, z_4)$$

is a circle C_4 which is also cut by C at an angle equal or supplementary to the angles cut on C_1, C_2, C_3.‡

This lemma is not true if the circles C_1, C_2, C_3 are coaxial circles having no point in common. For transform these circles into concentric circles. Then

* The method of proof used in this paragraph was suggested to me by Professor Birkhoff.

† *A treatise on the circle and the sphere*, p. 108.

‡ We remark that the circle C_4 can be constructed by ruler and compass whenever λ is rational or in fact whenever λ is given geometrically. For the circle C can be constructed by ruler and compass in any position; cf. Coolidge, l. c., p. 173. Hence we can determine any number of sets of points z_1, z_2, z_3 and therefore construct any number of points z_4, which enables us to construct C_4.

the circle C is a straight line orthogonal to these circles, C has two intersections with each, and on any particular circle C the points z_1, z_2, z_3 may be chosen on their proper circles so as to lead to four circles of type C_4, in general distinct, and concentric with C_1, C_2, C_3. All these four circles of type C_4 form the locus of points z_4. The situation is essentially the same if C_1, C_2, C_3 are coaxial circles having two common points; we are led to four circles C_4 which are in general distinct. But if we suppose C to vary continuously and also the points z_1, z_2, z_3, z_4 each to vary in one sense continuously, although of course we allow these points to go to infinity but not to occupy any position more than once, the lemma is true even for coaxial circles having no point or two points in common. These situations are included in the detailed treatments given under Cases I and II below.

This lemma breaks down also if the circles C_1, C_2, C_3 are coaxial circles all tangent at a single point, for we can consider the three points z_1, z_2, z_3 to coincide at that point; any circle C through that point satisfies the conditions of the lemma, any point of C can be chosen as z_4, whence it appears that the locus of z_4 is then the entire plane. But if we make not only our previous convention but in addition the convention that not all of the points z_1, z_2, z_3 shall lie at a point common to the three circles unless the fourth point coincides with them, then the lemma remains true. This situation is treated in detail under Case IV below.

The lemma is true but trivial in the degenerate cases $\lambda = 0, 1$, or ∞, for in these cases z_4 coincides with z_3, z_2, or z_1 respectively. The case that C_1, C_2, and C_3 are all null circles is likewise trivial. In the consideration of other cases we shall use the following theorem:

THEOREM. *If three circles be given not all tangent at one point, the circles cutting them at equal angles form a coaxial system, as do those cutting one at angles supplementary to the angles cut on the other two.**

Then as the circle C of Lemma IV varies, it always belongs to a definite coaxial system, unless C_1, C_2, C_3 are all tangent at a single point. This system may consist of (Case I) circles through two points, (Case II) non-intersecting circles, or (Case III) circles tangent to a line at a single point. Under Case IV will be treated the situation when C_1, C_2, C_3 are all tangent at a point. We consider these cases in order.

In Case I, transform to infinity one of the two points through which the coaxial family C passes, so that this family becomes the straight lines through a finite point q of the plane. In general q will be a center of similitude for each pair of the circles C_1, C_2, and C_3. These circles may or may not surround q.

* This statement differs from that of Coolidge, l. c., p. 111, Theorem 219, for we have adjoined to the plane the point at infinity. Theorem 220 seems to be erroneous; compare the four circles C_1, C_2, C_3, C_4 of Lemma IV.

Let z_4 be any point corresponding to the points z_1, z_2, z_3 on C_1, C_2, C_3 respectively. These four points lie on the line qz_4, and we have supposed that the lines tangent to C_1, C_2, C_3 at the points z_1, z_2, z_3 are parallel. Then when the line qz_4 (that is, the circle C) rotates about q, it will be seen that the point z_4 as determined by its constant cross-ratio with z_1, z_2, z_3 will trace a circle C_4 such that q is a center of similitude for any of the pairs of circles C_1, C_2, C_3, C_4. If these circles do not surround q, they have two common tangents belonging to the family C, and the properly chosen cross-ratio of the points of tangency is λ. If C_1, C_2, and C_3 are coaxial, C_4 is coaxial with them. Perhaps it is worth noticing that any circle C_4 such that q is a center of similitude for any pair of the circles C_1, C_2, C_3, C_4 is the circle C_4 of the lemma for a proper choice of λ; in particular C_4 may be the point q or the point at infinity.

Under Case I there are some special situations to be included. If one or more of the circles C_1, C_2, C_3 passes through q, then each of the other circles if not a null circle either is tangent to that circle at q or is a line parallel to the line tangent to that circle at q. If two of the original circles, for definiteness C_1 and C_2, are tangent at q and the other circle C_3 is a line parallel to their common tangent at q, then either z_4 coincides with z_1 and z_2 at q, or z_3 remains at infinity during the motion of C while z_4 traces a circle coaxial with C_1 and C_2; in particular this circle C_4 may be the null circle q. The four circles C_1, C_2, C_3, C_4 have a common tangent circle, namely the line tangent to C_1, C_2, C_4 at q. In the case just considered, one of the circles which passes through q, for definiteness C_1, may be tangent at q to the second circle C_2 which is a straight line. The circle C_3 is a line parallel to C_2. When the circle C varies, z_4 coincides with z_1 and z_2 at q, z_4 coincides with z_2 and z_3 at infinity, or the circle C coincides with C_2, z_1 with q, and z_3 with the point at infinity, while z_2 traces the line C_2 and hence z_4 also traces C_2. The circles C_1, C_2, C_3, C_4 have a common tangent circle C_2. If one of the original circles, for definiteness C_1, passes through q and the circles C_2 and C_3 are lines parallel to the tangent to C_1 and q, then the circle C_4 is a circle coaxial with C_2 and C_3 which may be the point at infinity. The four circles C_1, C_2, C_3, C_4 have as common tangent circle the line tangent to C_1 at q.

The general situation of Case I is not essentially changed and requires no further discussion if one of the circles C_1, C_2, C_3 is a point (q or the point at infinity) or if two of them are points (q and the point at infinity), except when at least one of the null circles lies on one of the non-null circles. In particular, if two circles, for example C_1 and C_2, are null circles and one of them (say C_2) lies on the non-null circle C_3, the locus of z_4 is a circle C_4 tangent to the circle C_3 at the point C_2. If the two null circles C_1 and C_2 both lie on the non-null circle C_3, the circle C is effectually the circle C_3, and C_4 coincides with C_3.

The special situations which we have considered under Case I may similarly degenerate by having one of the original circles a null circle. We shall discuss merely some typical examples. If C_1 and C_2 are tangent at q and C_3 is a null circle at infinity, C_4 is a circle tangent to C_1 and C_2 at q and may be the point q itself. If C_1 is a null circle at q, if C_2 is a circle passing through q, and if C_3 is a line parallel to the tangent to C_2 at q, C_4 is a circle tangent to C_2 at q. If C_1 is a null circle at q, if C_2 is a line passing through q, and C_3 is a line parallel to C_2, then C is essentially the single circle C_2, and C_4 coincides with C_2.

In Case II, the coaxial family C is composed of circles having no point in common, and hence there are two null circles of the family. Transform one of these null circles to infinity, so that the family C becomes a family of circles with a common center p. In the general case, the circles C_1, C_2, and C_3 are all of equal radii and any of them can be brought into coincidence with any other of them by a rotation about p. The point p is outside, on, or within all three circles according as it is outside, on, or within any one of them. Choose any point z_4 of the lemma; then z_1, z_2, z_3, z_4 lie on the circle C whose center is p. As C varies, its radius simply increases or decreases, and z_1, z_2, z_3 rotate about p so that the angles z_2pz_3, z_3pz_1, z_1pz_2 remain constant. Hence z_4 traces a circle C_4 whose radius is equal to the common radius of C_1, C_2, and C_3; moreover any two of the four circles C_1, C_2, C_3, C_4 can be brought into coincidence by a rotation about p. The four circles have two common tangent circles which belong to the family C, one of which may be the point p. The properly chosen cross-ratio of the points of tangency of a tangent circle is λ. Any circle is the circle C_4 of the lemma for a proper choice of λ provided it can be brought into coincidence with any of the circles C_1, C_2, C_3 by a rotation about p.

Another situation that may arise under Case II is that C_1, C_2, and C_3 are straight lines (that is, coaxial circles) through p and the point at infinity; then the locus of z_4 is a circle C_4 coaxial with them. There remains also the possibility that C_1, C_2, C_3 are straight lines all at the same distance from p. Then the circle C_4 is a line also at this same distance from p. There is a circle belonging to the family C which is tangent to C_1, C_2, C_3, C_4, and as before the cross-ratio of the points of contact is λ.

In Case III, the circles C belong to a coaxial family of circles all tangent at a point n, which point we transform to infinity. The circles C become parallel lines and in general C_1, C_2, C_3 become equal circles whose centers are collinear. As C moves parallel to itself, the points z_1, z_2, z_3 remain at equal distances from each other. The locus of z_4 either is a circle C_4 equal to C_1, C_2, and C_3 whose center is collinear with their centers or is the point at infinity. The four circles have two common tangent circles which belong to the family C, and the cross-ratio of the points of tangency of each of these circles is λ.

A degenerate case that should be mentioned is that the point n itself is one of the circles C_1, C_2, C_3. The results are essentially the same as in the general situation. In both the degenerate and the general situations any circle C_4 equal to C_1, C_2, C_3 and whose center is collinear with their centers is the circle C_4 of the lemma if λ is properly chosen.

A special case also occurs if one of the original circles, for definiteness C_1, is a straight line and the other two circles are straight lines parallel to the reflection of C_1 in any of the circles C. When C varies, either z_4 coincides with z_2 and z_3 at infinity, or z_1 is at infinity and z_4 traces a line parallel to C_2 and C_3.

A degenerate case occurs if one of the original circles, say C_3, is the point at infinity, while C_1 and C_2 are the reflections of each other in one of the circles C. Under the conditions of the lemma z_4 must coincide with z_3 at infinity, so C_4 coincides with C_3.

In Case IV, the circles C_1, C_2, C_3 are all tangent at a point m. Transform m to infinity, so that in any non-degenerate case C_1, C_2, C_3 become parallel lines. Under our convention that not all of the points z_1, z_2, z_3 shall lie at m unless z_4 coincides with them, we are led to four circles (in general distinct) according as we allow any one of the points z_1, z_2, z_3 or none of them constantly to lie at infinity. The additional convention already made that z_1, z_2, z_3, z_4 shall vary continuously in one sense and never coincide with any previous position enables us to choose simply one of these circles. The circle C is any straight line, and z_4 is either the intersection of C with a straight line C_4 parallel to C_1, C_2, C_3 or if none of the points z_1, z_2, z_3 is at infinity, z_4 may be constantly the point at infinity. The circles C_1, C_2, C_3, C_4 are all tangent at m.

Under Case IV should be mentioned the degenerate case that one of the circles C_1, C_2, C_3 is a null circle lying at the point of tangency of the other two circles. Our conventions enable us to choose a circle C_4 coaxial with C_1, C_2, C_3.

The proof of Lemma IV is now complete. It will be noticed that except in the special and degenerate cases, the result is entirely symmetric with respect to the four circles C_1, C_2, C_3, C_4. If we commence by choosing any three of those four circles and choose λ properly we shall be led to the other circle. If the last clause in the statement of the lemma is omitted, the lemma is true even if λ is not real.

There is a lemma corresponding to Lemma IV if we suppose two of the original circles, for example C_1 and C_2, to coincide, but suppose C_3 not to coincide with them. If we leave aside the easily treated cases $\lambda = 0, 1$, or ∞, we find either that the points z_1 and z_2 coincide on C_1, in which case z_4 coincides with them and traces the circle C_1, or that if C_1 is a non-null circle z_1 and z_2 do not coincide. In the latter case we are supposing the tangents to C at z_1 and z_2 to be parallel if C is transformed into a straight line and hence C must be orthogonal to C_1 and therefore by the conditions of the lemma also orthogonal to C_3. As before, when the circle C varies it constantly belongs to a definite coaxial system. The reader will easily treat the cases corresponding to Cases I, II, and III above, and also the degenerate case that C_3 is a null circle lying on C_1 and C_2. The results in the general case are quite analogous to the previous results if we notice that C_1, C_2, and C_3 are coaxial. For if C_3 is not a null circle, C cuts C_3 in two distinct points, and by their cross-ratio with z_1 and z_2 these lead to *two distinct circles* C_4 in addition to the circle C_1. Both of these new circles C_4 belong to the coaxial family determined by C_1 and C_3; as C moves it is constantly orthogonal to C_4 as well as to C_1, C_2, C_3. In general, then, the locus of z_4 when C_1 and C_2 coincide is C_1 and two other circles of the coaxial family determined by C_1 and C_3. These two other circles may in a degenerate case coincide, as the reader can easily determine. The convention formerly made, that the points z_1, z_2, z_3, z_4 vary in one sense continuously will of course restrict the locus of z_4 simply to one circle.

When the three circles C_1, C_2, C_3 coincide, we must consider C to coincide with them, or else at least two of the points z_1, z_2, z_3 to coincide and hence z_4 to coincide with them. That is, the circle C_4 corresponding to the circle C_4 of the lemma is the circle C_1.

Lemmas III and IV with the discussion supplementary to the latter do not give us immediately all the material necessary for the proof of Theorem II. For if C_1, C_2, C_3 are coaxial there are four circles, not necessarily all distinct, of the type C_4 of the lemma. If C_1, C_2, C_3 are not coaxial there are also four circles, not necessarily all distinct, of the type C_4 of the lemma, according as C cuts all the circles C_1, C_2, C_3 at equal angles or cuts one at an angle supplementary to the angle cut on the other two. It is conceivable that the boundary of the region C_4 of Theorem II should consist of arcs of more than one distinct circle; we proceed to show that this is in fact never the case.* The following lemma is essential in our proof.

* Whether the boundary of the region C_4 corresponds to motion of C cutting the three original circles at the same angle or a definite one of those circles at an angle supplementary to the angle cut on the other two depends on the relative positions of those circles, on whether the various regions are interior or exterior to their bounding circles, and on the value of λ—in short, on the order of the points z_1, z_2, z_3, z_4 on the circle C. When the regions C_1, C_2, C_3 are mutually external it is easy to prove by reasoning similar to that used in the proof of Lemma III that an arc of only one of the circles of type C_4 can be a part of the boundary of the region C_4. This fact can also be proved in the general case by that same method of reasoning, but the proof given in detail below is perhaps more satisfactory. It is desirable

LEMMA V. *In Theorem II, whenever the envelope of z_4 is not the entire plane, there is a circle S orthogonal to the four circles C_1, C_2, C_3, C_4.*

Whenever the regions C_1, C_2, C_3 have a common point, we may consider z_1, z_2, z_3 to coincide at that point, and consider the cross-ratio of any point z_4 in the plane with those three points to have the value λ, so the envelope of z_4 is the entire plane. In any other case there is a circle S orthogonal to the circles C_1, C_2, C_3. If not every pair of these three original circles intersect, choose two of them which do not intersect, and there will be two points inverse respecting both circles (these points are the null circles of the coaxial family determined by the two circles). Take the inverse of one of those points in the third of the original circles and pass a new circle S through all three points. Then S is orthogonal to the three original circles. If each of the circles C_1, C_2, C_3 has a point in common with the other two, we can transform two of the circles into straight lines (if one of the circles is a null circle the other two circles pass through that null circle and hence the region C_4 is the entire plane). If these two lines are not parallel, the third circle cannot be a straight line nor can it surround the intersection of the other two lines. Hence there is a circle orthogonal to all three circles. If the two lines are parallel the third circle cannot be a straight line. Then there is a circle, in this case a straight line, orthogonal to all three circles. This completes the proof of Lemma V.

Let us transform into a straight line any particular circle S orthogonal to the three original circles and let us suppose not every point of S to be a point of the region C_4; for definiteness assume the point at infinity not to belong to C_4. The positions which each of the three points z_1, z_2, z_3 of Theorem II may occupy fill an entire segment of S, and hence the points z_4 on S which correspond to points z_1, z_2, z_3 on S fill an entire segment of S; we denote this segment by σ. The terminal points of the segment σ are the intersections of S with one of the circles of type C_4 of Lemma IV; we denote that circle by C_4' and the other three circles of that type by C_4'', C_4''', C_4''''. The entire configuration is symmetric with respect to S, so the centers of all the circles C_4', C_4'', C_4''', C_4'''' lie on S. Moreover, S belongs to all four types of circles C of Lemma IV, since it is orthogonal to C_1, C_2, C_3. Hence the intersections of all the circles C_4'', C_4''', C_4'''' are points z_4 which correspond to points z_1, z_2, z_3 lying on S, and hence all those intersections lie on the segment σ. Then of the circles C_4'', C_4''', C_4'''' each is interior to or coincident with C_4'.

Either the entire interior or the entire exterior of each of the circles C_4', C_4'', C_4''', C_4'''' belongs to the region C_4. For the points z_4 which correspond to

that most of the material making up that proof should be given anyway, as a test whether the region C_4 is the entire plane, as giving a ruler-and-compass construction for the circle C_4, and as describing more in detail the entire situation with which we are concerned.

points z_1, z_2, z_3 in the proper regions and on the circle C of Lemma IV fill an entire arc of C, extending from one intersection of C with the circle C_4 to the other intersection. The entire exterior of our circle C_4' does not belong to the region C_4, for the point at infinity does not belong to that region. Hence the entire interior of C_4' does belong to the region C_4. No point external to C_4' can be a point of the boundary of C_4, for none of the circles C_4'', C_4''', C_4'''' has a point exterior to C_4'. Hence the region C_4 is the interior of C_4', under our assumption that not every point of S belongs to the region C_4.

Let us notice that we can allow any or all of the circles C_1, C_2, C_3 to move continuously so as to remain orthogonal to S, so as never to intersect any former position, and so as always to enlarge the regions C_1, C_2, C_3. Then the circle C_4' grows larger and larger, never intersecting its former position, until it becomes the point at infinity, in which case the region C_4 is the entire plane. If the regions C_1, C_2, C_3 are enlarged still further, the region C_4 still remains the entire plane.

Whether or not we assume that not every point of S belongs to the region C_4, we can start with a situation in which not every point of S is a point of C_4 and enlarge the regions C_1, C_2, C_3 in the manner described so as to attain any situation desired in which the region C_4 is not the entire plane. At every stage the region C_4 is a circular region. This completes the proof of Theorem II. We have also obtained a test whether or not the region C_4 is the entire plane. *A necessary and sufficient condition that the region C_4 of Theorem II be the entire plane is that the point z_4 may occupy any position on S and still correspond to points z_1, z_2, z_3 in their proper envelopes and also on S.*

The preceding developments give a comparatively simple ruler-and-compass construction for the circle C_4, whenever λ is rational or is given geometrically. The circle S can be constructed by ruler and compass.* The two points of intersection of S and C_4 can be determined by means of their cross-ratio with properly chosen intersections of S and C_1, C_2, C_3. Since S and C_4 are orthogonal, C_4 can then be constructed.

We shall apply Theorem II in proving our principal theorem.

THEOREM III. *Let f_1 and f_2 be binary forms of degrees p_1 and p_2 respectively, and let the circular regions C_1, C_2, C_3 be the respective envelopes of m roots of f_1, the remaining $p_1 - m$ roots of f_1, and all the roots of f_2. Denote by C_4 the circular region which is the envelope of points z_4 such that*

$$(z_1, z_2, z_3, z_4) = \frac{p_1}{m},$$

when z_1, z_2, z_3 have the respective envelopes C_1, C_2, C_3. Then the envelope of

* Coolidge, l. c., p. 173.

the roots of the jacobian of f_1 and f_2 is the region C_4, together with the regions C_1, C_2, C_3 except that among the latter the corresponding region is to be omitted if any of the numbers m, $p_1 - m$, p_2 is unity. If a region C_i $(i = 1, 2, 3, 4)$ has no point in common with any other of those regions which is a part of the envelope of the roots of the jacobian, it contains of those roots precisely $m - 1$, $p_1 - m - 1$, $p_2 - 1$, or 1 according as $i = 1, 2, 3$, or 4.

We shall first show by the aid of Lemmas I and II and of Theorems I and II that no point not in C_1, C_2, C_3, or C_4 can be a root of the jacobian. For if a point z_4 is not in C_1, C_2, or C_3 and is a root of the jacobian, it is a position of equilibrium and not of pseudo-equilibrium. The force at z_4 will not be changed if we replace the particles in each of the regions C_1, C_2, C_3 by the same number of coincident particles at points z_1, z_2, z_3 in the respective regions. Then z_4 is a position of equilibrium in the new field of force and hence by Lemma II we have

$$(z_1, z_2, z_3, z_4) = \frac{p_1}{m},$$

and therefore z_4 lies in C_4.

Any point in C_4 can be a root of the jacobian, for we need merely find points z_1, z_2, z_3 in the regions C_1, C_2, C_3 such that

$$(z_1, z_2, z_3, z_4) = \frac{p_1}{m}$$

and allow all the roots of the ground forms in each of those regions to coincide at those points. Any point of a region C_1, C_2, C_3 which is the envelope of more than one root of a ground form can be a position of pseudo-equilibrium and hence a root of the jacobian. If any of the regions C_1, C_2, C_3 is the envelope of merely one root of a ground form, then no point in that region but not in any other of the regions C_1, C_2, C_3, C_4 can be a position of equilibrium or of pseudo-equilibrium and hence no such point can be a root of the jacobian. If a point is common to two of the regions C_1, C_2, C_3, C_4 it is a point of C_4 and hence is a point of the envelope of the roots of the jacobian.

We have now proved the theorem except for its last sentence, to the demonstration of which we now proceed. When the roots of the ground forms in the regions C_1, C_2, C_3 coincide, the regions C_1, C_2, C_3, C_4 contain respectively the following numbers of roots of the jacobian: $m - 1$, $p_1 - m - 1$, $p_2 - 1$, 1. The roots of the jacobian vary continuously when the roots of the ground forms vary continuously; no root of the jacobian can enter or leave any of the regions C_1, C_2, C_3, C_4 which has no point in common with any other of those regions which is a part of the envelope of the roots of the jacobian.

The proof of Theorem III is now complete.* It applies to the sphere as well as the plane, since everything essential in the theorem is invariant under stereographic projection.

Instead of considering primarily the jacobian of two binary forms as heretofore, we may consider a rational function $f(z)$, introduce homogeneous coördinates, and compute the value of the derivative $f'(z)$ in terms of J, the jacobian of the binary forms which are the numerator and denominator of $f(z)$. We find that *the roots of $f'(z)$ are the roots of J and a double root at infinity, except that when one of these points is also a pole of $f(z)$ it cannot be a root of $f'(z)$.*† Application of Theorem III gives a theorem analogous to Theorem III, but which we state in a form slightly different from the statement of that theorem.

THEOREM. *If the circular regions C_1, C_2, C_3 contain respectively m roots (or poles) of a rational function $f(z)$ of degree p, all the remaining roots (or poles) of $f(z)$, and all the poles (or roots) of $f(z)$, then all the roots of $f'(z)$ lie in the regions C_1, C_2, C_3, and a fourth circular region C_4 determined as the envelope of points z_4 such that*

$$(z_1, z_2, z_3, z_4) = \frac{p}{m},$$

while the envelopes of z_1, z_2, z_3 are respectively C_1, C_2, C_3,—except that there are two roots at infinity if $f(z)$ has no pole there. Except for these two additional roots, if any of the regions C_i $(i = 1, 2, 3, 4)$ has no point in common with any other of those regions which contains a root of $f'(z)$, then that region contains the following number of roots of $f'(z)$ for $i = 1, 2, 3, 4$ respectively:

$$m - 1, \qquad p - m - 1, \qquad q_3 - 1, \qquad 1;$$

or

$$q_1 - 1, \qquad q_2 - 1, \qquad p - 1, \qquad 1,$$

according as C_1 contains m roots or m poles of $f(z)$; here q_i indicates the number of distinct poles of $f(z)$ in C_i.

Perhaps the following special cases of this theorem are worth stating explicitly.

If $f(z)$ is a rational function whose m_1 finite roots (or poles) lie on or within a circle C_1 with center α_1 and radius r_1 and whose m_2 finite poles (or roots) lie on or within a circle C_2 with center α_2 and radius r_2, and if $m_1 > m_2 > 0$, then

* It may be noticed that this proof does not explicitly use the fact that C_4 is a circular region.

If C_1, C_2, C_3 are coaxial circles with no point in common, Theorem III reduces essentially to Theorem II (I, p. 294). If $m = 0$ or $p_1 - m = 0$, the regions C_1, C_2, and C_4 can be considered to coincide: this gives Theorem III (I, p. 296), which is due to Bôcher.

† See I, p. 297.

all the finite roots of $f'(z)$ *lie in* C_1, C_2, *and a third circle* C_3 *whose center is*

$$\frac{m_1 \alpha_2 - m_2 \alpha_1}{m_1 - m_2}$$

and radius

$$\frac{m_1 r_2 + m_2 r_1}{m_1 - m_2}.$$

If $f(z)$ *has no finite multiple poles, and if* C_1, C_2, C_3 *are mutually external, they contain respectively the following numbers of roots of* $f'(z)$: $m_1 - 1$, $m_2 - 1$, 1. *Under the given hypothesis, if* $m_1 = m_2$ *and if* C_1 *and* C_2 *are mutually external, these circles contain all the finite roots of* $f'(z)$.*

If $f(z)$ *is a polynomial* m_1 *of whose roots lie on or within a circle* C_1 *whose center is* α_1 *and radius* r_1, *and if the remaining* m_2 *roots lie on or within a circle* C_2 *whose center is* α_2 *and radius* r_2, *then all the roots of* $f'(z)$ *lie on or within* C_1, C_2, *and a third circle* C_3 *whose center is*

$$\frac{m_1 \alpha_2 + m_2 \alpha_1}{m_1 + m_2}$$

and radius

$$\frac{m_1 r_2 + m_2 r_1}{m_1 + m_2}.$$

If these circles are mutually external, they contain respectively the following number of roots of $f'(z)$: $m_1 - 1$, $m_2 - 1$, 1.

If $f(z)$ *is a polynomial of degree* n *with a* k*-fold root at* P, *and with the remaining* $n - k$ *roots in a circular region* C, *then all the roots of* $f'(z)$ *lie at* P, *in* C, *and in a circular region* C' *obtained by shrinking* C *toward* P *as center of similitude in the ratio* $1 : k/n$. *If* C *and* C' *have no point in common they contain respectively* $n - k - 1$ *roots and* 1 *root of* $f'(z)$.†

A special case of this last theorem is the following

THEOREM. *If a circle includes all the roots of a polynomial* $f(z)$, *it also includes all the roots of* $f'(z)$.

* A more restricted theorem than this has been proved not merely for rational functions but also for the quotient of two entire functions. See M. B. Porter, *Proceedings of the National Academy of Sciences*, vol. 2 (1916), pp. 247, 335.

There is no theorem analogous to the theorem of the present paper if $m_1 = m_2$ and if C_1 and C_2 are not mutually external. For we may consider all the roots and all the poles of $f(z)$ to coincide, so that $f(z)$ reduces to a constant and every point of the plane is a root of $f'(z)$.

† This theorem is true whether the circle C surrounds, passes through, or does not surround P, and whether the region C is interior or exterior to the circle C. The special case where P is the center of the circle C and the region C is external to that circle was pointed out in a footnote, I, p. 298. The special case where C does not surround P and the region C is interior to the circle C was pointed out to me by Professor D. R. Curtiss.

The latter theorem is equivalent to the well-known theorem of Lucas:

If all the roots of a polynomial $f(z)$ *lie on or within any convex polygon, then all the roots of* $f'(z)$ *lie on or within that polygon.*

HARVARD UNIVERSITY,
CAMBRIDGE, MASS.,
May, 1920.

ON THE LOCATION OF THE ROOTS OF THE DERIVATIVE OF A POLYNOMIAL.*

By J. L. Walsh.

1. Introduction: Jensen's Theorem. This paper contains some geometric results concerning the relative positions of the roots of a polynomial and those of its derivative. Although not entirely restricted to real polynomials, and although the cubic is especially treated in detail, most of the results here presented are naturally connected with the following theorem of Jensen's:

If circles are described whose diameters are the segments joining pairs of conjugate imaginary roots of a real polynomial $f(z)$, then every non-real root of the derivative $f'(z)$ lies on or within those circles.†

For brevity we shall call the circles with which this theorem is concerned *Jensen circles.*

The succeeding developments follow largely from Gauss's theorem that the roots of the derivative are the positions of equilibrium in the field of force due to particles one situated at each root of the original polynomial, each particle repelling as the inverse distance. The derived polynomial has roots not only at the positions of equilibrium but also at the multiple roots of the original polynomial. When we are concerned with real polynomials especially it seems natural to study the field of force due to two particles.

2. The Field of Force due to Two Particles. In the field of force due to particles of the kind described, the force at a point P due to a particle at Q is in direction, magnitude, and sense $Q'P$, where Q' is the inverse of Q in the unit circle whose center is P. The force at P due to k particles is equivalent to k coincident vectors with one terminal at P and the other at the center of gravity of the inverses of the positions of those k particles. In the sequel we shall have frequent occasion to use this fact.

---

* Presented to the American Mathematical Society, December 31, 1919.

† This theorem was stated without proof by Jensen, Acta Mathematica, vol. XXXVI (1912), p. 190. Attention was called to it by Professor D. R. Curtiss in an abstract published in the Bulletin of the American Mathematical Society, vol. XXVI, p. 62. No proof of Jensen's theorem has previously been published.

Two recent papers by J. S. Nagy, Jahresbericht der Deutschen Mathematiker-Vereinigung, Bd. 27 (1918), pp. 37, 44, contain some results concerning the roots of the derivative of a polynomial, more particularly if all the roots of the original polynomial are real.

According to the previous notation, the force at P due to unit particles at Q and R is in direction, magnitude, and sense $S'P$, where S' is the mid-point of the segment $Q'R'$. Since S' is the harmonic conjugate of the point at infinity with respect to Q' and R', and since cross-ratios are invariant under inversion, it follows that *the force at P due to unit particles at Q and R respectively is equivalent to the force at P due to two coincident particles situated at S, the harmonic conjugate of P with respect to Q and R.* The point S may of course be constructed by ruler and compass; we shall describe the case where Q and R are the points $+ i$ and $- i$. At P construct the tangent to the circle through P, $+ i$, and $- i$. Using as center the intersection of this tangent with the axis of imaginaries describe a circle through P. This circle intersects at the point S the circle through P, $+ i$, and $- i$. An alternate construction is found by noticing that the lines joining the origin with P and S are symmetric respecting the coördinate axes. If P is on either coördinate axis a construction is used which differs slightly from either of these but which is easily devised.

We obtain immediately some results concerning the field of force. It is symmetric respecting each coördinate axis; at a point on either axis the force is directed along that axis. At any point on the unit circle whose center is the origin, the force is horizontal. Inside that circle but above the axis of reals, the force has a component vertically downward. Outside that circle but above the axis of reals the force has a component vertically upward. The line of action of a force always cuts the axis of imaginaries between the points $+ i$ and $- i$.

On any circular arc bounded by the points $+ i$ and $- i$, the force has a minimum on the axis of reals. On the unit circle whose center is the origin, the minima occur at $+ 1$ and $- 1$, where the force is of magnitude 1.

The relation between P and S is reciprocal; when expressed in terms of complex variables as coördinates that relation is linear. Hence when one of the points P, S moves in a circle so does the other.

We shall now proceed to determine the lines of force. The field of force is given by

$$\frac{f'(\bar{z})}{f(\bar{z})} = \frac{2(x - iy)}{(x - iy)^2 + 1},$$

and this leads to the differential equation

$$\frac{dy}{dx} = \frac{y}{x} \cdot \frac{x^2 + y^2 - 1}{x^2 + y^2 + 1}.$$

The solution of this gives us the lines of force

$$x^2 - y^2 + 1 = Cxy,$$

(together with $xy = 0$) which are equilateral hyperbolas having the origin as center and passing through the points $+ i$ and $- i$.

What is the locus of points such that the lines of action of the forces there all pass through a fixed point $(a, 0)$*? The line of action of the force at (x_1, y_1) is given by

$$x_1 x - y_1 y + 1 = \frac{C}{2}(x_1 y + y_1 x).$$

We have also the equations

$$a x_1 + 1 = \frac{C}{2} a y_1, \qquad x_1^2 - y_1^2 + 1 = C x_1 y_1,$$

from which we obtain

$$\left(x_1 + \frac{1}{a}\right)^2 + y_1^2 = 1 + \frac{1}{a^2} \qquad (a \neq 0),$$

which is a circle easily constructed by ruler and compass and which passes through the points $+ i$ and $- i$.

The results of this paragraph have been deduced on the assumption that we have merely two single particles, whose distance apart is 2. It is obvious what are the results for k particles at each of two points with any distance between them.

3. Some Immediate Results. A proof of Jensen's theorem is now evident. At a point not on the axis of reals nor on or within any Jensen circle, the force has a vertical component in direction away from the axis of reals—this is true of the force due to any pair of conjugate imaginary roots of $f(z)$, also true of the force due to any real root of $f(z)$, so it is true of the resultant. Hence such a point cannot be a position of equilibrium. Such a point cannot be a multiple root of $f(z)$ and hence cannot be a root of $f'(z)$.† We may add that a point of a Jensen circle not on the axis of reals cannot be a root of $f'(z)$ unless it is on or within another Jensen circle or is a multiple root of $f(z)$.

Jensen's theorem can be generalized as follows:

If all the roots of a polynomial $f(z)$ not on a line L nor one of a pair situated symmetrically with respect to L lie on one side of L, then on the opposite side of L there are no roots of $f'(z)$ except on or within circles whose diameters are the segments joining those symmetric roots.

This theorem and likewise Jensen's theorem are limiting cases of the following:

* For a point not on the axis of reals we are in general led to a cubic equation.

† A more immediate but less elegant proof can be given by the method of inversion indicated at the beginning of § 2.

If $f(z)$ is a polynomial such that every root not in nor on a circle C is one of a pair of roots symmetric with respect to C, then all the roots of $f'(z)$ lie in or on C and in or on circles whose diameters are the segments joining those pairs of symmetric roots.

The proof here is similar to the proof of Jensen's theorem—the force at a point P external to C has an outward component along the line from the center of C to P. Similarly, it is readily shown by the geometric construction given that the force at P (external to the other circles considered) due to any pair of roots symmetric respecting C has also an outward component along that line.

There are an unlimited number of theorems that can be written down immediately. We give a few more examples.

*If a circle C contains all but P and Q, two equal roots of a polynomial $f(z)$, contains neither of them, but has its center on the line PQ and not on the segment PQ, then no roots of $f'(z)$ lie in that semicircular region bounded by the perpendicular bisector of PQ and that half of the circle on PQ as a diameter which is nearer C.** *There are also no roots of $f'(z)$ outside of the circle on PQ as a diameter and in that half plane bounded by the perpendicular bisector of PQ which does not contain C.*

The limiting case of this theorem is also true—C is replaced by a straight line.

If $f(z)$ is a real polynomial having equal roots at $+ i$ and $- i$, if there is no other root of $f(z)$ whose abscissa is less than $\alpha > 1$, and if the point $z = 1$ is interior to no Jensen circle, then there is no non-real root of $f'(z)$ in the circle whose center is $- 1/\alpha$ and which passes through the points $+ i$ and $- i$.

In this proof we need consider only that part of the last circle lying to the right of the axis of imaginaries. The line of action of the force at any point inside that circle due to the particles at $+ i$ and $- i$ cuts the axis of reals to the left of the point $x = \alpha$. The line of action of the force due to the remaining particles (whether these be real or in conjugate imaginary pairs) cuts that axis in a point not to the left of the point $x = \alpha$.

All of the theorems stated in this paragraph may give a closer idea of the location of the roots of the derived polynomial than does Lucas's theorem that the roots of $f'(z)$ lie in any convex polygon enclosing the roots of $f(z)$. Since the location of the roots of $f'(z)$ does not in general depend on the solution of a quadratic equation, it is not to be expected that the ruler-and-compass constructions used here would determine

* Varying the roots of $f(z)$ which are on or in C and finally allowing them to coalesce at a point in C on PQ (this method is used frequently later) shows that the other semicircular region with these boundaries has in its interior precisely one root of $f'(z)$.

precisely the location of the roots. In exceptional cases, however, that location can be found.*

If a polynomial $f(z)$ has only four roots, and if these are equal in multiplicity and located at the vertices of a rectangle $ABCD$, then the circles whose diameters are the two longer sides of the rectangle, AB and CD, intersect in roots of $f'(z)$. For the force at each of those points due to each of the pairs of particles (A, B) and (C, D) is in direction parallel to the shorter sides of the rectangle, and the two forces at each point are equal in magnitude and opposite in direction. Symmetry shows that a third root of $f'(z)$ lies at the center of the rectangle. If the rectangle becomes a square, the center is a three-fold root of $f'(z)$.

Another example of finding the explicit location of the roots of $f'(z)$ is given in the last paragraph of this paper.

4. A Theorem Complementary to Jensen's Theorem. The question of how many roots of $f'(z)$ are situated in a Jensen circle, and how many in the intervals of the axis of reals readily suggests itself. An answer is given in Grace's theorem, which indeed is true for both real and non-real polynomials:

If a polynomial $f(z)$ of degree n has roots at $+ i$ and $- i$, there is at least one root of $f'(z)$ on or in the circle whose center is the origin and radius $\cot \pi/n$.†

This maximum value for the modulus of the root of $f'(z)$ which is nearest the origin is actually assumed when the roots of $f(z)$ are the vertices of a regular polygon of n sides whose center is $\cot \pi/n$.

A theorem other than this can also be proved for real polynomials, and which may give more definite results than Grace's theorem; first we prove:

In the interior of any interval of the axis of reals containing no root of $f(z)$ and exterior to Jensen's circles, there is at most one root of $f'(z)$.

The theorem refers merely to the *interior* of an interval, so it is sufficient to prove the theorem assuming that neither extremity is a root of either $f(z)$ or $f'(z)$. Consider the interval $\alpha \leqq x \leqq \beta$, and suppose first that the forces at α and β are in the same direction—toward the right for definiteness. Move all the roots of $f(z)$ whose abscissas are greater than β horizontally and continuously to the right, and allow them to become infinite. The roots of $f'(z)$ also move continuously (at least when we consider the stereographic projection of the plane) and one or more may become infinite. The forces at α and β continually increase in magnitude

* There is of course the trivial case where M and N are respectively m- and n-fold roots of $f(z)$. There being no other roots, the point dividing the segment MN in the ratio $m : n$ is a root of $f'(z)$.

† Proceedings of the Cambridge Philosophical Society, vol. XI (1901–02), p. 352. Proved later but independently by Heawood, Quarterly Journal of Mathematics, vol. XXXVIII (1907), p. 84.

and are never zero. Hence throughout the process there are a fixed number of roots in the interval—a number which at the end of the process is evidently zero.

Secondly we consider the case that the force at α is directed toward the right, and the force at β toward the left. Here there is evidently at least one root in the interval. Let μ_1 and μ_2 be any pair of conjugate imaginary roots of $f(z)$ whose common abscissa is greater than β. Allow them to move continuously toward the right, always remaining conjugate imaginary, along the circle through μ_1, μ_2, and α, and finally to coincide on the axis of reals. This motion keeps constant the force at α and continually increases the magnitude of the force at β. Treat in this manner all the pairs of conjugate imaginary roots of $f(z)$ whose abscissas are greater than β, and correspondingly treat all the conjugate imaginary roots of $f(z)$ whose abscissas are less than α, moving them so that the force at β is kept constant. During the whole motion the roots of $f'(z)$ vary continuously, none can enter or leave the interval, and hence at the beginning there was precisely one root of $f'(z)$ in the interval.

Thirdly, it is conceivable that the force at α should be directed toward the left and that at β toward the right. This means that the force at α due to the particles whose abscissas are greater than β is greater than the force at β due to those particles, which is impossible.

From the theorem just proved and by similar methods we shall deduce:

If a Jensen circle has on or within it k roots of $f(z)$ and is not interior to nor has a point in common with any exterior Jensen circle, then it has on or within it not more than $k + 1$ nor less than $k - 1$ roots of $f'(z)$.

Denote by μ and ν ($\mu < \nu$) the intercepts of this circle C with the axis of reals. If the forces at μ and ν are respectively directed toward the right and left, and if the particles whose abscissas are less than μ are moved by translation to the left and to infinity, one and only one root of $f'(z)$ will issue from C, and that toward the left. If the particles whose abscissas are greater than ν are translated horizontally to the right and to infinity, one and only one root of $f'(z)$ will issue from C, and that toward the right. Finally $k - 1$ roots of $f'(z)$ remain in or on C, and therefore the original number was $k + 1$. If the forces at μ and ν are respectively toward the left and right, and if the particles exterior to C are translated horizontally to infinity, although during the motion one root of $f'(z)$ may enter C, it will eventually issue from C. The final number of roots on or within C is the same as the original number, $k - 1$.* In a similar manner it is

* Immediately we obtain the theorem: *If $f(z)$ is a real polynomial with two simple roots at $+i$ and $-i$, with m roots whose abscissas are greater than $m + 1$, n roots whose abscissas are less than $-n - 1$, and no other roots, then $f'(z)$ has one real root and no other root in the unit circle whose center is the origin.* Cf. the theorem of § 6.

shown that if the forces at μ and ν are in the same sense, there are on or within C precisely k roots of $f'(z)$. The cases where μ or ν or both are roots of $f'(z)$ are easily treated.

5. A Theorem Related to Jensen's Theorem. Jensen's theorem gives a configuration—the Jensen circles and the axis of reals—in which all roots of $f'(z)$ must lie, and it is easy to see that no more restricted locality will satisfy the conditions of the theorem. First, any point of the axis of reals may be a multiple root of $f(z)$ and hence a root of $f'(z)$. Second, we may have a root of $f'(z)$ as near as desired to a point ρ interior to a Jensen circle in any preassigned configuration. Let the line of action of the force at ρ due to the particles μ and ν determining the Jensen circle intersect the axis of reals at a point σ. If the distance from ρ to ρ', its harmonic conjugate respecting μ and ν, is commensurable with the distance from ρ to σ, then neglecting the other particles in the field, by a proper choice of the multiplicities of the roots of $f(z)$ at μ, ν, and σ, we can make ρ a root of $f'(z)$. If the two distances are not commensurable, and taking account of the other particles in the plane, we can make the multiplicities of the roots of $f(z)$ at μ, ν, and σ very large in comparison with the other roots of $f(z)$, and in such ratio that there is a root of $f'(z)$ as near to ρ as desired.

This reasoning refers to a preassigned configuration rather than a preassigned polynomial; it is of course impossible if the degree of $f(z)$ is limited; by considering polynomials of a fixed degree we may expect to obtain some results concerning a region more restricted.

Considering the polynomial

$$f(z) = (z^2 + 1)(z - \alpha)^{n-2},$$

where α is real, we shall prove that the non-real roots of $f'(z)$ lie on the circle whose center is the origin and radius $\sqrt{(n-2)/n}$. We eliminate α from the equations obtained from the real and pure imaginary parts of the equation

$$(z - \alpha)^{-n+3} f'(z) = n(x + iy)^2 - 2\alpha(x + iy) + n - 2 = 0, \qquad y \neq 0;$$

$$x^2 + y^2 = \frac{n-2}{n}.$$

When α is large and positive, one of the roots of $f'(z)$ different from α is near the origin and the other near the point $2\alpha/n$. As α decreases, the former root moves to the right, while the latter moves to the left. The two roots coalesce at $\sqrt{(n-2)/n}$ when $\alpha = \sqrt{n(n-2)}$. As α continues to decrease, the roots move on the circle already determined, remaining conjugate imaginary, and when $\alpha = 0$ those two roots are $\pm i\sqrt{n(-2)}/n$. The path of the roots when α further decreases is found from symmetry.

We shall extend this result and prove:

If $f(z)$ is a real polynomial of degree n whose roots are all real except simple roots at $+i$ and $-i$, then all the non-real roots of $f'(z)$ lie in or on the circle whose center is the origin and radius $\sqrt{(n-2)/n}$.

In the proof, we first notice—this is in the nature of a lemma—that if the force at a point P is in direction along a line l and due to k particles on a line λ, then the force is not greater than the force at P due to k coincident particles situated at the intersection of l and λ. The lemma is proved by the method of inversion previously described (§ 2).

We next consider the force at points along the arc of a circle bounded by the points $+i$ and $-i$, the arc intersecting the axis of reals at β, between the points $\sqrt{(n-2)/n}$ and 1. The lines of action of the force at points of the arc due to the particles at $+i$ and $-i$ all pass through the point $2\beta/(1-\beta^2)$. The force at β due to $n-2$ particles at $2\beta/(1-\beta^2)$ is less than the force at β due to the two particles at $+i$ and $-i$, for if we have

$$\frac{n-2}{\dfrac{2\beta}{1-\beta^2}-\beta} \geqq \frac{2}{2\,\dfrac{1+\beta^2}{2\beta}},$$

are led to

$$\frac{n-2}{n} \geqq \beta^2,$$

which is contrary to our assumption. The force at any point on the arc considered due to the particles at $+i$ and $-i$ increases in magnitude as we move from the axis of reals toward $+i$ or $-i$. Moreover the force due to $n-2$ particles at $2\beta/(1-\beta^2)$ decreases, so no point on the arc can be a position of equilibrium when the $n-2$ particles coalesce. From the lemma, then, no such point can be a position of equilibrium in any other case.

The treatment of the arcs of circles which have a point in common with the circle whose center is the origin and radius $\sqrt{(n-2)/n}$ can readily be made, noting as before that the force due to the particles at $+i$ and $-i$ increases as we move away from the axis of reals, while the force due to the $n-2$ coincident particles decreases. This completes the proof.

6. Sufficient Conditions for the Reality of the Roots of $f'(z)$. When all the roots of $f(z)$ except two non-real roots are sufficiently removed from the latter, the roots of $f'(z)$ in the corresponding Jensen circle are real. This paragraph gives a theorem containing sufficient conditions for the reality of the roots of $f'(z)$, which theorem is stated simply to concern one Jensen circle, but may of course be applied to several in order.

If $f(z)$ is a real polynomial with simple roots at $+i$ and $-i$, m roots *whose abscissas are greater than* $\sqrt{m(m+2)}$, *n roots whose abscissas are less than* $-\sqrt{n(n+2)}$, *and with no other roots, then* $f(z)$ *has precisely one root in the interval* $(-\sqrt{n/(n+2)}, \sqrt{m/(m+2)})$ *and no non-real root in the Jensen circle whose center is the origin.*

The degenerate cases here are first $m = 0$, $n \neq 0$, and all the n roots concentrated at $-\sqrt{n(n+2)}$, in which case $f'(z)$ has a double root at $-\sqrt{n/(n+2)}$; and second $m \neq 0$, $n = 0$, all the m roots concentrated at $\sqrt{m(m+2)}$, in which case $f'(z)$ has a double root at $\sqrt{m/(m+2)}$. In either of these cases we make the convention that simply one of those roots belongs to the interval mentioned.

In any non-degenerate case, the force at $\sqrt{m/(m+2)}$ is directed toward the right, for otherwise we have the force at that point due to the particles at $+i$ and $-i$ less than the force at that point due to the m particles:

$$\frac{2}{\sqrt{\frac{m+2}{m}}+\sqrt{\frac{m}{m+2}}} < \frac{m}{\sqrt{m(m+2)}-\sqrt{\frac{m}{m+2}}}, \qquad m < m,$$

which is absurd. Similarly, the force at $-\sqrt{n/(n+2)}$ is directed toward the left, so the interval of the theorem contains at least one root of $f'(z)$.

Suppose neither of the points $+1$ and -1 to lie on or within any Jensen circle except of course the unit circle C whose center is the origin. Then C contains at least one root and not more than three roots of $f'(z)$. In fact, by considering the forces at the points $+1$ and -1 we immediately determine from the results of § 4 whether C contains one, two, or three roots, and it is then evident that all those roots are real.

We shall now prove that under no circumstances consistent with our hypothesis can any point of C except $+1$ and -1 be a root of $f'(z)$. First suppose a point of C in the first quadrant to lie in or on one of the Jensen circles pertaining to the m roots. At such a point (x, y), the horizontal component of the force due to the two particles at $+1$ and -1 is greater than the horizontal component of the force due to the m particles. Assuming the contrary, we must have

$$\frac{2}{2x} \leqq \frac{2}{2(\sqrt{m(m+2)}-1)} + \frac{m-2}{\sqrt{m(m+2)}-1},$$

$$x \geqq \frac{\sqrt{m(m+2)}-1}{m-1} > 1,$$

which is impossible. The proof just given is also valid for the point $+1$,—if that point is on or within one of the Jensen circles belonging to the m roots, it is not a root of $f'(z)$.

It is conceivable, secondly, that a point $P : (x, y)$ in the first quadrant and on C should lie exterior to all the Jensen circles pertaining to the m roots, and yet should lie interior to one or more of the Jensen circles pertaining to the n roots, and should be a root of $f'(z)$. We shall prove the impossibility of this, roughly, as follows: such a root of $f'(z)$ must be near the point $+ 1$, for as we move upward from that point along C the horizontal force at P due to the particles at $+ i$ and $- i$ becomes greater and eventually exceeds the force at P due to the m particles. The force due to the m particles is such that the vector representing the total force at P due to the m particles and the particles at $+ i$ and $- i$ is inclined to the horizontal at a comparatively steep angle. In order for the force at P due to two or more of the n particles to be inclined at that same angle, P must be quite near the center of the corresponding Jensen circle, which proves to be impossible.

Consider the slope of the line of action of the force due to the two particles at $+ i$ and $- i$ and to the m particles (the force always with a component toward the left),—this slope is numerically least when the m roots are all concentrated at $\sqrt{m(m + 2)}$. For invert the configuration in the unit circle whose center is P, except that the point $(- x, y)$ is to be inverted into a point Q by means of a circle whose center is P and radius $\sqrt{-1}$. When we replace the m particles of the theorem by m coincident particles, the terminal of the vector corresponding is seen to lie in or on the boundary of the sector of a circle, which sector is bounded by the line through P and $\sqrt{m(m + 2)}$, and by the circle which is the inverse (regarding the unit circle whose center is P) of the reflection of that line in the axis of reals. The point of contact of a tangent from Q to that circle cannot lie between the point which is the inverse of $\sqrt{m(m + 2)}$ and the intersection of the line PQ with the circle. This fact is proved most easily, perhaps, by inverting the inverse figure (including Q) again in the unit circle whose center is P. The details are omitted here, but this completes the proof that the slope is numerically least when the m roots are concentrated at $\sqrt{m(m + 2)}$.

If all the m roots are located at $\sqrt{m(m + 2)}$, the total force at P due to the m particles and to the particles at $+ i$ and $- i$ has a slope numerically equal to

$$\frac{mxy}{x\sqrt{m(m + 2)^3} - mx^2 - m(m + 2) - 1},$$

assuming that the force has a component toward the left. If this quantity is less than

$$\frac{my}{\sqrt{m(m + 2)^3} - m - m(m + 2) - 1},$$

we shall have

$$-mx - m(m+2)x - x < -mx^2 - m(m+2) - 1,$$

$$mx(1-x) + m(m+2)x + x > m(m+2) + 1,$$

which inequality is false for $x < 1$.

We turn now to consideration of the force due to one pair of the n particles, using running coördinates (ξ, η). The slope of the line of action of the force at (ξ, η) is

$$\frac{\eta}{\xi} \cdot \frac{\xi^2 + \eta^2 - 1}{\xi^2 + \eta^2 + 1},$$

and the locus of points at which the force has the slope $-\mu$ is given by

$$(\eta + \mu\xi)(\xi^2 + \eta^2) - (\eta - \mu\xi) = 0.$$

All points in the first quadrant and interior to the corresponding Jensen circle at which the slope of the force is numerically greater than μ lie above the line $\eta - \mu\xi = 0$. For the point P considered above, we have

$$\mu \leqq \frac{y}{\sqrt{n(n+2)}} < y.$$

Then if P is a position of equilibrium we must have

$$y > \frac{my}{\sqrt{m(m+2)^3} - m - m(m+2) - 1}, \qquad \sqrt{m(m+2)} > m + 1,$$

which is impossible.

We have therefore proved that no point of the circle C except $+1$ or -1 can be a root of $f'(z)$, and that $+1$ is not a root if it lies on or within one of the Jensen circles pertaining to the m roots, nor -1 if it lies on or within one of the Jensen circles pertaining to the n roots.

If the Jensen circle of any pair of the m roots incloses or intersects the unit circle whose center is the origin, continuously move those roots toward the right along the circle joining them with the point -1, and move them until the Jensen circle cuts the axis of reals slightly to the right of the point $+1$, and so that there is no root of $f'(z)$ on the axis of reals between the point of intersection and the point $+1$. During this motion the force at the point -1 is constant, so there is no change in the number of roots inside C. Similarly move any pair of the n roots whose Jensen circle incloses or cuts the unit circle, keeping the force constant at the point $+1$. In this final position, the forces at the points $+1$ and -1 are in the same direction as were the forces in the initial position, never having changed sense. In the final position—which is of course the initial position so far as concerns Jensen circles not inclosing nor having a point in common with C—the circle C contains one, two, or three roots of $f'(z)$ according as the forces at $+1$ and -1 are both, one, or neither directed away from the origin. The forces at the points $\sqrt{m/(m+2)}$, $-\sqrt{n/(n+2)}$ are initially and finally directed away from the origin, so it is clear where the one, two, or three roots of $f'(z)$ lie inside the circle C,—and in the same intervals in the initial and final positions. The reader will readily take up the possibility that one or both of the points $+1$ and -1 may be a root of $f'(z)$, and this will complete the proof of the theorem.

It is to be noticed that the intervals given in the hypothesis of the theorem are the smallest which will insure the reality of the roots of $f'(z)$. For if we allow, for example, an abscissa smaller than $\sqrt{m(m+2)}$ for the m roots, we can concentrate them at the point nearest the origin and remove the n roots so far (by changing either their abscissas or ordinates) that their influence in the field of force is as small as desired. Hence $f'(z)$ will have two non-real roots in the circle C.

7. The Reality and Non-Reality of the Roots of $f'(z)$. A General Theorem. Having derived a sufficient condition for the reality of roots of $f'(z)$, we shall now derive a sufficient condition for the non-reality of roots. A number of results will then be collected into a general theorem.

If $f(z)$ is a real polynomial of degree $n > 2$ with simple roots at $+i$ and $-i$, and if all the other roots of $f(z)$ are interior to the interval $(0, \sqrt{n(n-2)})$, then $f'(z)$ has precisely two non-real roots.

First, $f'(z)$ can have no root interior to a finite interval of the axis of reals bounded by the origin and a root of $f(z)$ but containing no root of $f(z)$. For consideration of the forces at such a point x due respectively to the particles at $+i$ and $-i$ and the particles on the axis of reals would lead to the inequalities

$$\frac{2}{x+\frac{1}{x}} > \frac{n-2}{\sqrt{n(n-2)}-x},$$

$$0 > \left(\sqrt{nx} - \sqrt{\frac{n-2}{x}}\right)^2.$$

Second, there can be no more than one root of $f'(z)$ interior to an interval of the axis of reals bounded by two roots of $f(z)$. If there were, by moving to the left the root of $f(z)$ which is the right-hand boundary of that interval, eventually at least two roots of $f'(z)$ must become imaginary. For when a k-fold root and an l-fold root of $f(z)$ coalesce, the point is a $(k+l-1)$-fold root of $f'(z)$.

It will now be shown that no point (x, y) interior to the circle whose center is the origin and radius unity and whose ordinate is positive and less than $\frac{1}{2}$ can be a root of $f'(z)$, assuming that there is at least one root of $f(z)$ in the interval of the theorem and whose abscissa is less than x. The vertical component of the force at (x, y) due to the particle in that position is not less than the component for a particle at the origin:

$$\frac{y}{x^2 + y^2}.$$

The vertical component of the force at (x, y) due to the two particles at $+ i$ and $- i$ is numerically

$$\frac{2y(1 - x^2 - y^2)}{(x^2 + y^2)^2 + 2(x^2 - y^2) + 1}.$$

Assuming that (x, y) is a root of $f'(z)$, we have

$$\frac{2y(1 - x^2 - y^2)}{(x^2 + y^2)^2 + 2(x^2 - y^2) + 1} \geqq \frac{y}{x^2 + y^2}, \qquad 4y^2 - 1 \geqq 3(x^2 + y^2)^2,$$

which is impossible if $y < \frac{1}{2}$.

This completes the proof that there is not more than one root of $f'(z)$ interior to any interval of the axis of reals bounded by two roots of $f(z)$, and hence there are precisely two non-real roots of $f'(z)$. It may be added that the entire argument remains valid if in the plane there are k particles of positive abscissas none of whose Jensen circles includes nor has a point in common with the unit circle whose center is the origin—this last-named circle contains precisely two non-real roots of $f'(z)$.

We shall summarize a number of the previous results in the theorem:

If $f(z)$ is a polynomial of degree n with simple roots at $+ i$ and $- i$, and if all the remaining $n - 2$ roots are real and—

(1) *concentrated at $\sqrt{n(n - 2)}$, $f'(z)$ has a double root at $\sqrt{(n - 2)/n}$*

(2) *with abscissas not less than $\sqrt{n(n - 2)}$, $f'(z)$ has all its roots real, precisely one of which lies in the interval $(0, \sqrt{(n - 2)/n})$.*

(3) *with abscissas non-negative but less than $\sqrt{n(n - 2)}$, $f'(z)$ has precisely two non-real roots.*

(4) *with abscissas unrestricted, the non-real roots of $f'(z)$ lie on or within the circle whose center is the origin and radius $\sqrt{(n - 2)/n}$.*

(5) *with abscissas unrestricted but coincident, the non-real roots of $f'(z)$ lie on that circle.*

8. Variation of the Roots of a Real Cubic. This paragraph considers how the roots of the derivative of a real cubic vary when one of the roots of the cubic varies, and also how the roots of a real cubic may vary so that the roots of the derivative are fixed. We shall have frequent occasion to use the well-known theorem that the roots of a polynomial and those of its derivative have a common center of gravity.

There has already been described (§ 5, $n = 3$) the variation of the roots of the derivative of a real cubic with two fixed non-real roots when the third (real) root of the cubic varies. Of course the variation of the roots of the derivative when the real root is fixed and the two non-real roots move horizontally is essentially identical with that. When the real root is fixed and the two non-real roots move in another manner, the motion of the roots of the derivative is easily determined. For example, if the two non-real roots of the cubic move in a vertical line, and if the roots of the derivative are not real they also move in a vertical line.

For the sake of completeness we consider also a cubic whose roots 1, -1, α are all real. When α is very large and positive, the two roots of the derivative are approximately at the origin and the point $2\alpha/3$. When α decreases, both roots move to the left, and when $\alpha = 1$ these roots have reached the points $-\frac{1}{3}$ and 1 respectively. When α continues to decrease, the roots continue their motion to the left, and when $\alpha = 0$ these roots are at $\pm \frac{1}{3}\sqrt{3}$. For negative values of α the location of the roots is obtained from symmetry.

When the cubic has two coincident roots at 0 and a third root at α, the roots of the derivative are at 0 and $2\alpha/3$.*

We shall now consider what real cubics have given fixed points as the roots of their derivatives, first choosing those points at $+i$ and $-i$. The cubic itself must be of the form

$$f(z) = z^3 + 3z + C,$$

and therefore we have to study the (C, z) transformation. We shall describe the result rather in terms of the variation of α, the one real root of $f(z)$. When $\alpha = 0$, the other two roots of $f(z)$ are at the points $\pm \sqrt{3}i$. When α moves to the right or left, these other roots move toward the left or right, one on each branch of the hyperbola

$$x^2 - \frac{y^2}{3} = -1.$$

The common abscissa of the two non-real roots of $f(z)$ is always $-\alpha/2$. This completely determines the motion.

Suppose two fixed points $+1$ and -1 are the roots of the derivative

* If $f(z)$ is a non-real cubic two of whose roots are fixed while the third traces a line bisecting their segment, the roots of $f'(z)$ trace a cubic curve having that line as asymptote. Only the degenerate cases of the cubic curve have been considered in detail here.

of a real cubic. How do the roots of the cubic vary? If those roots are all real, there is one of them in each of the intervals $(-\infty, -1)$, $(-1, +1)$, $(+1, +\infty)$, with the obvious convention regarding roots at the ends of these intervals. When one real root of the cubic is $\alpha = 2$, the other roots coalesce at -1. When α moves to the left, the other roots move along the axis of reals, one to the right and one to the left. The former coincides with α at the point 1, when the latter has reached the point -2. As α further moves to the left, the former root moves from 1 toward the right, while the latter moves from -2 also to the right. When $\alpha = 0$, these roots are at $\pm\sqrt{3}$. When α reaches -1, the root moving from the left coincides with it, whereas the other root is at the point 2. As α moves to the left from the point -1, the other two roots move toward each other, and coalesce at $+1$ when $\alpha = -2$. As α continues its motion to the left, those roots move along the right-hand branch of the hyperbola

$$x^2 - \frac{y^2}{3} = 1.$$

The common abscissa of those two imaginary roots is always $-2\alpha/3$. Symmetry now gives us a complete discussion of the situation.

When the real cubic has two real coincident roots at the origin, the roots of the cubic lie one on each of the lines

$$y = 0, \qquad y = \sqrt{3}x, \qquad y = -\sqrt{3}x.$$

The three roots lie always at the vertices of an equilateral triangle whose center is the origin.

9. Ruler-and-Compass Construction for the Roots of the Derivative of a Cubic. It is to be expected that the roots of the derivative of a cubic have some interesting properties relative to the triangle whose vertices are the roots of the original cubic. In fact, it has been proved that these two points are the foci of the maximum ellipse which can be inscribed in that triangle, which ellipse touches the sides of the triangle at their mid-points.* We shall use this property to give a ruler-and-compass construction for the roots of the derivative. Of course those roots depend on the solution of a quadratic, so it is known a priori that they can be located by ruler and compass.

Let A, B, and C be the roots of the original cubic, let F be the mid-point of AB, and let the intersection of the medians of the triangle be M.

* This seems first to have been proved by F. J. van den Berg, Nieuw Archief von Wiskunde, 1882, 1884, 1888. That reference is not available to the present writer, but is given indirectly by E. Cesàro, Periodico di Mat., vol. XVI (1900–01), p. 81. See also M. Bôcher, these Annals, vol. VII (1892), p. 70; Grace, l. c.; Heawood, l. c.

Let a line through M parallel to AC intersect AB in D. Then MD and MB are in direction conjugate diameters of the ellipse. Determine the length MG such that

$$\overline{MG}^2 = DF \cdot FB,$$

which construction is readily made. Lay off this length from M on a line through M parallel to AB. Then MF and MG are in direction and length conjugate diameters of the ellipse. For if any tangent meets two conjugate semidiameters of an ellipse, the rectangle under its segments is equal to the square of the parallel semidiameter.*

Knowing in direction and magnitude two conjugate semidiameters of the ellipse, we can find the foci.† From F draw FN perpendicular to MG and produce FN its own length to H. Join MH, and on MH as diameter describe a circle whose center is denoted by K. Join FK, cutting the circle in P and Q. Lay off on MP, $MX = FQ$ and on MP lay off $MY = FP$. Then MX and MY are the axes of the ellipse; the foci may be found as the intersection with MX of a circle whose center is Y and radius MX.*

This construction can be greatly simplified in some special cases, notably if the polynomial is real or more generally if the triangle ABC is isosceles. If ABC is an equilateral triangle, the intersection of the medians is a double root of $f'(z)$. If $AB < BC = CA$, the circle with M as center and MF as radius is the major auxiliary circle of the ellipse. Let this circle cut AC in the points S and T. If R is the mid-point of AC, lines through R parallel to MS and MT respectively cut CM in the foci. For the length of a line through the center parallel to either focal radius vector and terminated by the tangent is the semi-major axis.§ If $AB > BC = CA$, denote by V the intersection of RM with AB. Then FV is the semi-major axis, so we can complete the construction as before.

There seems to be no obvious construction applicable when the points A, B, C are collinear, but we can get a rather simple procedure with the aid of the equations involved. The center of gravity of the three roots is easily found by ruler and compass, so we can choose that point as the

* See, e.g., Casey, Analytical Geometry (1893), p. 231. In fact we may simply lay off $MG : AB = 1 : \sqrt{12}$. The corresponding result given by Grace, l. c., p. 356, contains a numerical error.

† The construction which follows is due to Mannheim and given by Casey, l. c., p. 210.

‡ We indicate briefly another construction. Lucas has shown that in the sense of least squares, the line passing nearest to the three points A, B, C is the major axis of the ellipse. That line can be constructed by ruler and compass. See Coolidge, American Mathematical Monthly, vol. XX (1912–13), p. 187. Knowing the major axis of the ellipse in position, from any of the sides of the triangle and its mid-point (a tangent to the ellipse and its point of contact) we can construct the major auxiliary circle and hence find the foci.

§ Salmon, Conic Sections, p. 175, Ex. 2.

origin of coördinates. If two of the roots of the polynomial are denoted by α and β, we have to deal with the polynomials

$$f(z) = (z - \alpha)(z - \beta)(z + \alpha + \beta),$$

$$f'(z) = 3z^2 - (\alpha^2 + \alpha\beta + \beta^2).$$

By means of a succession of right triangles, we readily construct

$$\sqrt{(\alpha + \beta)^2 + \alpha^2 + \beta^2},$$

and we easily divide it in the ratio $1 : \sqrt{6}$.

HARVARD UNIVERSITY,
February, 1920.

BULETINUL SOCIETĂŢII DE ŞTIINŢE DIN CLUJ (ROMÂNIA)
BULLETIN DE LA SOCIÉTÉ DES SCIENCES DE CLUJ (ROUMANIE)
Tome VII, p. 521—526
25 janvier 1934.

NOTE ON THE LOCATION OF THE ROOTS OF THE DERIVATIVE OF A POLYNOMIAL

By **J. L. Walsh**
Cambridge (Mass.) U. S. A.

Received 1st nov. 1933.

§ 1. INTRODUCTION. GAUSS's theorem is familiar:

If particles are placed at the roots of a polynomial $f(z)$, and if each particle repels with a force equal to the inverse distance, then the positions of equilibrium in the corresponding field of force are roots of the derivative $f'(z)$. The derivative $f'(z)$ has no other roots except the multiple roots of $f(z)$, all of which are roots of $f'(z)$.

From GAUSS's theorem can be proved immediately the well known theorem of LUCAS:

Any convex region which contains the roots of $f(z)$ also contains the roots of $f'(z)$.

The term *contains* can be interpreted here and below either in the sense of *containing on or within the boundary* or of *containing in the interior.*

If restricted polynomials are considered, sharper results than LUCAS's theorem can be obtained. The most interesting of such results is JENSEN's theorem: (1)

If the polynomial $f(z)$ is real, then all the non-real roots of the derivative $f'(z)$ lie on or within the circles whose diameters are the segments joining the pairs of conjugate imaginary roots of $f(z)$.

In the field of force due to two equal particles of the kind introduced in GAUSS's theorem, the lines of force are equilateral hyperbolas which pass through those particles and whose common center is the point midway between the particles (2). This suggests that JENSEN's theorem is closely related to equilateral hyberbolas. Indeed, CURTISS expresses JENSEN's theorem essentially as follows: (3).

(1) JENSEN, *Acta Math.* vol. 36 (1912), p. 190. WALSH, *Annals of Math.*, vol. 22 (1920), pp. 128—144.

(2) The proof is not difficult. The details are given by WALSH, loc. cit.

(3) *Bull. Amer. Math. Soc.* vol. 26 (1919-20), pp. 53, 61—62.

Let $f(z)$ be a real polynomial and let P be a point not on the axis of reals. If the rectangular hyperbola whose center is on the axis of reals, of which P is a vertex, contains on or within it no root of $f(z)$, then P cannot be a root of $f'(z)$.

By the *interior* of a hyperbola we mean all points not on the curve which are separated from the center of the hyberbola by the curve. By the *exterior* of a hyperbola we mean all points not on the curve and not so separated.

JENSEN's theorem is concerned with a polynomial whose roots are symmetric with respect to a line. It is interesting also to study polynomials whose roots are symmetric with respect to a point. Here too the roots (other than those at the point of symmetry) occur in pairs, and from the field of force one might expect equilateral hyperbolas to be of signifiance. This is in fact the case, and the primary object of the present note is to study such polynomials. Our main result is

Theorem I. *Let $f(z)$ be a polynomial whose roots are symmetric with respect to a point O.*

1. If all the roots of $f(z)$ lie in a double sector whose vertex is O, and whose angle is not greater than $\pi/2$, then all roots of the derivative $f'(z)$ also lie in that sector.

2. If all the roots of $f(z)$ lie on or interior to an equilateral hyperbola whose center is O, then all roots of $f'(z)$ also lie on or interior to that hyberbola, except for a simple root at O itself.

3. If all the roots of $f(z)$ lie on or exterior to an equilateral hyperbola whose center is O, then all the roots of $f'(z)$ also lie on or exterior to that hyperbola.

4. If all the roots of $f(z)$ lie on an equilateral hyperbola whose center is O, then all the roots of $f'(z)$ also lie on that hyperbola except for a simple root at O. Any finite arc of the hyperbola bounded by roots of $f(z)$ contains in its interior precisely one (a simple) root of $f'(z)$.

The given regions containing the roots of $f(z)$ are all considered as closed.

§ 2. PROOF OF THEOREM I. It is an interesting exercise to prove Theorem I by study of the field of force involved. The hyperbolas enter directly, and one needs to take into account only the direction of the forces involved. If according to Theorem I the point P cannot be a root of $f'(z)$, then in each case all the forces at P due to pairs of particles at the roots of $f(z)$ — all of these forces lie in a certain sector at P of angle less than π, and hence those forces cannot be in equilibrium. We do not carry out this proof in detail, however, for another proof is also instructive.

In our proof of Theorem I we choose O as the origin; this choice involves no loss of generality. The given polynomial is also polynomial in z^2, except for a possible power of z: $f(z)=z^k\varphi(z^2)$.

In order to study the roots of $f'(z)$, we make the substitution $w=z^2$, and study the polynomial in w defined by the equations

$F(w)=f(w^{1/2})$, if $f(0)\neq 0$,

$F(w)=[f(w^{1/2})]^2$, if $f(0)=0$.

In the first case $[f(0)\neq 0]$, the roots of $F'(w)$ and of $f'(z)$ correspond exactly, except that $z=0$ is a root of $f'(z)$ whereas $w=0$ need not be a root of $F'(w)$. In the second case $[f(0)=0]$, the roots of $F'(w)$ correspond both to the roots of $f(z)$ and the roots of $f'(z)$.

Under the transformation $w=z^2$, an arbitrary straight line in the w-plane corresponds in the z-plane to an equilateral hyberbola whose center is the origin and reciprocally, except that a straight line through the origin $w=0$ corresponds to two perpendicular straight lines through the origin $z=0$. Angles at the origin in the w-plane are twice the corresponding angles at the origin in the z-plane. The transformation may be studied by setting $w=u+iv$, $z=x+iy$, so the transformation can be expressed $u=x^2-y^2$, $v=2xy$. The line $Au+Bv=C$ corresponds to $A(x^2-y^2)+2Bxy=C$, and the properties mentioned may be read of directly.

Proof of Theorem I is now immediate. In case 1, all roots of $F(w)$ lie in a (single) sector of the w-plane whose vertex is the origin and whose angle is not greater than π. By Lucas's theorem, all roots of $F'(w)$ also lie in that sector. Hence the corresponding roots of $f'(z)$ lie in the given double sector of the z-plane, as does also the possible extraneous root $z=0$. The proof is complete.

The remaining parts of Theorem I are proved in precisely the same manner. In case 1, if the angle of the given sector is $\pi/2$, the given regions of the z-plane correspond to a half-plane of the w-plane bounded by a line through the origin $w=0$; in case 2, the given regions of the z-plane correspond to a half-plane of the w-plane for which the origin $w=0$ is an exterior point; in case 3, the given region of the z-plane corresponds to a half-plane for which the origin $w=0$ is an interior point; in case 4, the given curve of the z-plane corresponds to a straight line of the w-plane. In each case, Lucas's theorem yields the result. As a limiting form of case 4, it is also allowable that the roots of $f(z)$ should lie on two perpendicular lines, but here there need be no root of $f'(z)$ at $z=0$ if $f(0)=0$, and there may be a multiple root of $f'(z)$ at $z=0$ if $f(0)\neq 0$.

As a corollary to Theorem I, we remark that in case 1 no point (other than 0) of the lines bounding the sector can be a root of $f'(z)$ unless it is a multiple root of $f(z)$ or unless all the roots of $f(z)$ lie on that bounding line. In cases 2 and 3, no point of the equilateral hyperbola in question can be a root of $f'(z)$ unless it is a multiple root of $f(z)$ or unless all the roots of $f(z)$ lie on that hyperbola.

It follows from the method of proof of Theorem I that the various parts of that theorem are in a sense the „best" results obtainable without actually taking into account the multiplicities of the various roots of $f(z)$. In case 1, for instance, the conclusion is false without the restriction that the angle α of the sector be not greater than $\pi/2$. This can be illustrated by the example

$$f(z)=(z-a-ib)(z-a+ib)(z+a-ib)(z+a+ib),\quad 0<a<b,$$
$$f'(z)=4z(z^2-a^2+b^2).$$

The roots of $f'(z)$ other than 0 are pure imaginary, hence exterior to the closed double sector $-\alpha\leqq\theta\leqq\alpha, \pi-\alpha\leqq\theta\leqq\pi+\alpha$ (where $\alpha=\tan^{-1}b/a$) in which the roots of $f(z)$ lie.

Theorem I may obviously be more powerful than Lucas's theorem in application to a particular polynomial.

§ 3. Generalizations. It will be noticed that Theorem I can be applied repeatedly in determining regions free from zeros of $f'(z)$. Thus, if $f(z)$ is a polynomial whose roots are symmetric with respect to a point O and if all the roots of $f(z)$ lie on or within a hyperbola whose center is O and eccentricity less than $2^{1/2}$, then all roots of $f'(z)$ also lie on or within that hyperbola. Formal proofs of these two statements are by no means difficult.

For the direct application of Theorem I to a specific polynomial $f(z)$, it would be natural to construct a polygon whose sides are equilateral hyberbolas; thus we need to construct the unique equilateral hyperbola whose center is O which passes through a pair of roots of $f(z)$ which lie in a (single) sector whose vertex is O and angle less than $\pi/2$. This equilateral hyperbola can naturally be constructed by points by ruler and compass. Let the given points (roots of $f(z)$) be P and Q and their mid-point M. Let the line OM cut the circle constructed on PQ as diameter in the points S and T. The asymptotes of the desired hyperbola are parallel to PS and PT. If an arbitrary line through O cuts PS in S′ and PT in T′, then the point of intersection Q′ of a parallel to PT through S′ and of a parallel to PS through T′ is a point of the equilateral hyperbola whose center is O and which passes through P and Q. The proof is left to the reader.

It would seem that any theorem whatever concerning the roots of the derivative of a polynomial could be applied in the w-plane and thus used to yield a new result in the z-plane. Most such theorems involve circles or higher curves in the w plane and hence involve still higher curves in the z-plane. A circle in the w-plane corresponds to a bicircular quartic in the z-plane. By way of illustration, Jensen's theorem can be transformed from the w-plane to the z-plane, either by the transformation $w = z^2$ or by a more general transformation $w-\beta=(z-\alpha)^2$. The new result involves certain bicircular quartics and also involes the symmetry of points with respect to a hyperbola. The actual formulation of the new result can be done by the reader.

It is obvious too that many other results related to Theorem I are easily established, by the use of transformations other than $w=z^2$; the transformation $w=z^n$ is of particular value. We state a single example:

Theorem II. *Let $f(z)$ be a polynomial whose roots have n-fold symmetry about a point O, that is to say, a rotation of $2\pi/n$ about O leaves the totality of roots unchanged. If all the roots of $f(z)$ lie in the n (simple) sectors found from a fixed sector of angle not greater than π/n, by rotating it successively through angles $2\pi/n$, $4\pi/n$, $6\pi/n, \ldots, 2\pi$, then all the roots of $f'(z)$ lie in these n sectors.*

§ 4. Applications. One of the most interesting applications of the location of the roots of the derivative of a polynomial is to the study of the location of the critical points (i. e., points where both first partial derivatives vanish) of Green's function for a multiply connected region. We shall not into details in the present note(4), but merely remark that either one can apply Theorem I directly, or one can apply Lucas's theorem (or other theorems) in the w-plane (where $w=z^2$), and interpret the result in the original z $(=x+iy)$-plane. We state the following theorem by way of illustration:

Theorem III. *Let R be an infinite region whose boundary B is finite and symmetric in the origin O, and let $G(x, y)$ be Green's function (assumed to exist) for R with pole at infinity.*

1. If B lies in a (closed) double sector whose vertex is O and angle not greater than $\pi/2$, then all the critical points of $G(x, y)$ also lie in that sector.

2. If each point of B lies on or within an equilateral hyperbola whose center is O, then all critical points of $G(x, y)$ other than O also lie on or within that hyperbola.

3. If each point of B lies on or exterior to an equilateral hyperbola whose center is O, then all critical points of $G(x, y)$ also lie on or exterior to that hyperbola.

4. If B consists wholly of points of an equilateral hyperbola whose center is O, then all critical points of $G(x, y)$ other than O lie on that hyperbola. Each finite arc of that hyperbola bounded by points of B and containing in its interior no points of B contains in its interior precisely one critical point of $G(x, y)$.

(4) Compare Walsh, *Bull. Amer. Math. Soc.*, vol. 39 (1933), pp. 775—782.

ON THE LOCATION OF THE ROOTS OF CERTAIN TYPES OF POLYNOMIALS*

BY

J. L. WALSH

When we study the dependence of a variable on k other variables which vary independently, our problem may be very much simplified if we can consider all or some of these independent variables to coincide and thus study the dependence of our original dependent variable on one new variable or at least on a number of new independent variables less than k. The present writer has recently published a theorem (Theorem II, below) which enables us to make a reduction of this sort in the study of the relations between the roots of certain types of polynomials. The present paper aims to prove Theorem I (below), which is a much more general result of the same nature, and to indicate various applications of that theorem. The applications given are extremely simple and follow from Theorem I with practically no further machinery.† The most interesting application is Theorem VI.

Our problem is, more explicitly, to study the geometric relationship of the roots of a polynomial

$$f(z) = (z - a_1)(z - a_2) \cdots (z - a_m)$$

to the roots of a related polynomial

$$\phi(z) = (z - b_1)(z - b_2) \cdots (z - b_n)$$

* Presented to the Society, December 29, 1920, and April 23, 1921.

† When this paper was first offered for publication, the writer believed Theorem I to be new. Professor D. R. Curtiss has kindly pointed out its connection with a theorem due to Grace and has indicated an entirely new proof of Theorem I; the reader will refer to Professor Curtiss's note which immediately follows this paper.

The *point of view* of the present paper in the proof and application of Theorem I seems to be new, and also the results obtained except where otherwise stated.

This paper is the development of a short note published in the Paris Comptes Rendus, March, 1921, to which explicit reference is made later, and which contained in outline the proof of Theorem I. In the interval between the publication of that note and the publication of the present paper, there have appeared a number of other papers dealing with Grace's Theorem. See Szegö, Mathematische Zeitschrift, vol. 13 (1922), pp. 28–55; Cohn, Mathematische Zeitschrift, vol. 14 (1922), pp. 110–148; Egerváry, Acta Litterarum ac Scientiarum, Regiae Universitatis Hungaricae Francisco-Josephinae, vol. 1 (1922), pp. 39–45; Fekete, same Journal, vol. 1 (1923), pp. 98–100.

The present paper has very little in common with these other papers, but with Szegö's Theorem 9 the reader should compare our Theorem X, and with Szegö's Theorems 13–15 compare our Theorem IV and also Walsh, American Mathematical Monthly, vol. 29 (1922), pp. 112–114.

whose roots are supposed to be known.* When we study the functional dependence of a root of f, say a_1 for definiteness, on the b_i, it may occur that without changing a_1 we can replace the n roots b_i by n roots of a polynomial ϕ which coincide at some point b which bears a simple relation to the roots b_i. Then we can study the dependence of a_1 on b instead of on $b_1, b_2, \cdots, b_n$, which is frequently a simplification of our problem.

To the plane of the complex variable we adjoin the point at infinity; infinity is to be considered simply as an ordinary value of the variable. As a geometrical consequence we shall consider the term *circle* to include the possibility of a straight line. We shall have occasion to deal with *circular regions*, by which we mean a closed portion of the plane bounded by a single circle, that is, the interior of a circle, the exterior of a circle (including the point of infinity), a half-plane, a point, or the entire plane; the points of the boundary are always to be included in the region.

I. A GENERAL THEOREM

We proceed to the proof of our main result:

THEOREM I. *Let $f(z)$ be a polynomial in z whose coefficients are polynomials linear in and symmetric in each of the sets of variables*

$$\{\alpha_1, \alpha_2, \cdots, \alpha_k\}, \quad \{\beta_1, \beta_2, \cdots, \beta_l\}, \quad \cdots, \quad \{\lambda_1, \lambda_2, \cdots, \lambda_q\}.\dagger$$

Let these points $\{\alpha_i\}$, $\{\beta_i\}$, $\cdots$, $\{\lambda_i\}$ lie in circular regions C_α, C_β, $\cdots$, C_λ. Then for any fixed values of these variables and of z we can always make all the $\{\alpha_i\}$ coincide in C_α, all the $\{\beta_i\}$ coincide in C_β, etc., without altering the value of $f(z)$.

The theorem also obtains if we replace $f(z)$ by the quotient of two polynomials of the type described, except that we are to consider the conclusion of the theorem satisfied if these two polynomials vanish simultaneously.‡ Of course if we have two polynomials of the type required, their sum is also of that type.

We shall prove the theorem considering the $\{\beta_i\}$, $\{\gamma_i\}$, $\cdots$, $\{\lambda_i\}$ always as fixed and showing that we can make the $\{\alpha_i\}$ coincide in C_α as stated. Then we can consider the $\{\alpha_i\}$ fixed and coincident, and the $\{\gamma_i\}$, $\cdots$, $\{\lambda_i\}$ fixed; our former reasoning will show that the $\{\beta_i\}$ can be made to coincide. Continued reasoning in this manner will evidently complete the proof of the theorem. It will be convenient to assume that the value of $f(z)$ considered

* Professor Curtiss has recently published a very interesting report on this general field, Science, vol. 55 (1922), pp. 189–194.

† The coefficients of $f(z)$ need not be homogeneous in each of these sets of variables, but each coefficient must be a linear combination of the elementary symmetric functions of each of these sets with coefficients linear combinations of the elementary symmetric functions of the other sets. These linear combinations may, moreover, contain constant terms.

‡ This is what actually occurs in the situation of Theorem II if we choose P inside C.

in the hypothesis is zero; this involves no loss of generality, for the addition of a constant term to $f(z)$ does not alter the properties required.

We now wish to show that the α_i can all be made to coincide. If the region C_α is a point, the statement is trivial. If we prove the theorem where C_α is a circular region not the whole plane we have proved it where C_α is the whole plane.

Consider the α_i to vary independently and to have the region C_α as their common locus. Then the relation

$$f(z) = 0 \tag{1}$$

defines z as an analytic function of the α_i and hence z will have a certain locus Z; this locus Z will be a closed point set since C_α is closed. If (1) degenerates and does not effectively contain z at all, we introduce an auxiliary variable ζ by placing

$$F(\zeta) \equiv f(z) - \zeta = 0;$$

the new function $F(\zeta)$ surely contains ζ and we may reason with it as before for $f(z)$.

Let α be any point interior to C_α. There is a certain locus Z of points z corresponding to the null-circle α as the locus of the $\{\alpha_i\}$. Make use of the auxiliary circle C of the coaxial family determined by α and C_α. Let C commence with the position α and gradually enlarge and coincide with C_α, the region bounded by C and containing α always considered as the locus of the $\{\alpha_i\}$. The locus Z also grows larger and varies continuously with C (except as noted below), for the roots z of (1) are continuous functions of the $\{\alpha_i\}$. When the region C continues to enlarge, beyond C_α if necessary, the locus Z eventually becomes the entire plane; this happens ordinarily before C has swept through the entire plane. The point z with which we start is a point of the locus Z surely when C coincides with C_α and possibly before.

The case that the locus Z does not vary continuously with C occurs only if, for some choice of the $\{\alpha_i\}$, (1) vanishes identically in z. When this occurs, the roots $\{z\}$ of (1) do not vary continuously with the $\{\alpha_i\}$. The locus Z may enlarge suddenly and become the entire plane while its boundary does not necessarily sweep over the whole plane.

Whether or not this phenomenon takes place, corresponding to our original point z there is some circle C such that for no circle C', the region C' smaller than and entirely contained in the region C, can z be a point of the locus Z of the roots of (1). This statement and in fact Theorem I as well are true if z is a point of the locus corresponding to α. The statement is true in any other case. For if there is a sequence of sets of points $\{\alpha_i\}$ in a sequence of regions each contained in the preceding, there is at least one limiting set of points $\{\alpha_i\}$ contained in all the regions and hence in the limit region. The relation (1) obtains for this limit set.

Our entire proof of Theorem I rests on the remark that when we fix z and all but two of the points $\{\alpha_i\}$, equation (1) becomes a homographic and involutory relation between the other two of these points and hence when one of these points traces a circle the other also traces a circle. The proof is complicated, however, by the possibility that this relation may degenerate and may effectively contain but one or neither of these variables.*

We prove now a very special case of Theorem I, namely, that if a circular region C contains two points α_1 and α_2 connected by a relation of the form

$$a_1\alpha_1\alpha_2 + a_2(\alpha_1 + \alpha_2) + a_3 = 0, \tag{2}$$

then C contains a root α' of the equation

$$a_1\alpha'^2 + 2a_2\alpha' + a_3 = 0. \tag{3}$$

Transform (2) by an auxiliary linear transformation so that the point at infinity is a double point of the transformation (α_1, α_2) defined by (2). The line through α_1 and α_2 is transformed into itself by the transformation (α_1, α_2), for the three points $(\alpha_1, \alpha_2, \infty)$ go into $(\alpha_2, \alpha_1, \infty)$. When α_1 is moved on this line, α_2 moves on the line in the opposite sense, for if the two points move always in the same sense, when α_1 is moved along the finite segment from its original position to the original position of α_2, α_2 has moved in the same sense, has at no time coincided with the point at infinity and hence has not reached the original position of α_1. The points α_1 and α_2 moving on the line joining them in opposite senses must have a double point D between them. It is thus evident that the region C which contains α_1 and α_2 must contain D or the point at infinity and hence a solution of (3).

The reasoning just used supposes implicitly that the relation (2) does not degenerate. If (2) does not contain α_2 for a particular value of α_1, that value of α_1 is a solution of (3). If (2) is satisfied identically, any values α_1 and α_2 suffice.

We now return to the proof of Theorem I in its general form. Let C be the

* Thus we may start with the auxiliary polynomial

$$F(z) = (z - \alpha_1)(z - \alpha_2)\cdots(z - \alpha_k) \qquad k > 3$$

and consider for the polynomial of Theorem I

$$f(z) \equiv F'(z).$$

The relation

$$f(z) = 0 \tag{1'}$$

if $\alpha_{k-1} = \alpha_k = z$ contains effectively none of the variables $\alpha_1, \alpha_2, \cdots, \alpha_{k-2}$. If $\alpha_k = z$, we have (1′) reducing to

$$(z - \alpha_1)(z - \alpha_2)\cdots(z - \alpha_{k-1}) = 0,$$

and hence we have no effective homographic relationship between α_{k-1} and α_k.

This example is not an unnatural one; indeed it is one of the first to which we should think of applying Theorem I; compare Theorem II and its applications to which a reference is given.

smallest region of the kind already described for which z is a point of the locus Z. There is at least one of the points α_i on C. We shall show that all those points can be made to coincide on C.

Two points α_1 and α_2 which are on C can be made to coincide either on C or within C. For we fix z and the remainder of the points $\{\alpha_i\}$ and hence have a relation of type (2) between α_1 and α_2. A double point of the transformation is either on the circle C or interior to the region C.

Let us combine in this manner as many of the points $\{\alpha_i\}$ on C as possible, so as to leave the smallest number of points $\{\alpha_i\}$ actually on C. This number of points we denote by n; it is greater than zero and cannot be reduced.

We shall prove by induction that all these n points on C can be made to coalesce on C. This fact is evidently true for two points, for no double point of the corresponding transformation can be interior to C. We assume explicitly that whenever there are given $n - 1$ points of this sort on an arc A of C, they can be made to coincide on A without changing any other point α_i, or the relation (1); we shall prove this fact for n points. It involves no loss of generality to suppose $n - 1$ of these points at an end point X of A and the nth point at the other end point Y.

We consider one of the points α_1 at X and the point α_n at Y as connected by (1), while z and the other $\{\alpha_i\}$ remain fixed. When α_1 moves on C, α_n moves on C in the opposite sense, and the transformation (α_1, α_n) has a double point D interior to A. Consider α_1 and α_n to coincide at D. The new arc bounded by X and D contains the n points. We can make coincide, at a point X', $n - 1$ points α_i, X' on the arc XD, and the other point α_n will be at D. Then we have the n points on an arc A' bounded by X' and Y', the arc A' shorter than A and every point of A' a point of A. Moreover, there are $n - 1$ points at X', the end point nearest X, and the other is at $Y' = D$, the end point nearest Y.

We can continue in this manner successively to shorten our arc A, and we can shorten it indefinitely. For let us suppose we have a sequence of arcs (using the natural extension of our former notation), XY, $X'Y'$, $X''Y''$, $\cdots$, and suppose the points X, X', X'', $\cdots$ have a limit point x and the points Y, Y', Y'', $\cdots$ a limit point y different from x and of such a nature that we cannot make our arc of type A shorter than xy. We have $n - 1$ coincident points α_i, successively approaching x, and one point α_n simultaneously approaching y. From the continuity of the left-hand member of (1) we can therefore suppose these $n - 1$ points α_i to coincide at x and the other point α_n to lie at y. Then the procedure formerly used shows that we can bring the n points into an arc shorter than xy all of whose points are points of xy. This contradiction completes the proof of our statement that the arcs A, A',

A'', $\cdots$ shorten indefinitely, and hence the points $\alpha_1, \alpha_2, \cdots, \alpha_n$ can all be made to coincide at the limit point P of those arcs, with (1) still satisfied.

If $k = n$, Theorem I is proved. If $k > n$, consider the transformation (α_1, α_k) defined as before. We have supposed that none of the points $\alpha_1, \alpha_2, \cdots, \alpha_n$ can be moved from the circle C, so the transformation cannot contain α_k and hence α_k can be moved as near P as desired. Thus any other of the points $\alpha_{n+1}, \alpha_{n+2}, \cdots, \alpha_k$ can be moved as near P as desired without changing the value of $f(z)$. Hence all these points can be made to approach P and so the value of $f(z)$ is unchanged if all the $\alpha_1, \alpha_2, \cdots, \alpha_k$ coincide at P. Theorem I is now completely proved.

II. Special cases of Theorem I, with applications

There are two results, essentially special cases of Theorem I, which are particularly interesting in their applications.

Theorem II. *If the points $\alpha_1, \alpha_2, \cdots, \alpha_k$ lie in a circular region C and if z is exterior to C, the root of the equation in α*

$$\frac{1}{z-\alpha_1} + \frac{1}{z-\alpha_2} + \cdots + \frac{1}{z-\alpha_k} = \frac{k}{z-\alpha}$$

*lies in C.**

Theorem II is in reality a special case of the extension of Theorem I where we consider the quotient of two polynomials of the kind described. The denominator polynomial cannot vanish, since z is exterior to C.

For applications of Theorem II we refer to the citation already made and that made in the next following footnote. We proceed to another special case of Theorem I:

Theorem III. *If the points $\alpha_1, \alpha_2, \cdots, \alpha_k$ lie in a circular region C, the equation in α*

$$(z-\alpha_1)(z-\alpha_2)\cdots(z-\alpha_k) = (z-\alpha)^k \tag{4}$$

has at least one root in C.

Theorem III can be proved independently of Theorem I in a manner precisely analogous to the proof of Theorem II (loc. cit.). This proof of Theorem III involves a transcendental transformation of the α- (or (x, y)-) plane:

$$\alpha - z = x + iy = e^{u+iv},$$

and a study of the transform of C and certain centers of gravity in the (u, v)-plane. This proof gives in certain cases more detailed information than does Theorem I regarding the root α of (4). Thus if z is exterior to C, which is

* Walsh, these Transactions, vol. 22 (1921), p. 102; Lemma I.
Theorem II is closely connected with another more simple corollary of Theorem I, namely, that if k equal particles lie in a circle their center of gravity also lies in that circle.

the interior of a circle, we can write

$$k \log (z - \alpha) = \log (z - \alpha_1) + \log (z - \alpha_2) + \cdots + \log (z - \alpha_k),$$

where all the logarithms involved have the coefficients of $\sqrt{-1}$ in their pure imaginary parts lying between two numbers whose difference is less than π. We add the remark, without proof at this time, that there is no other finite region bounded by a regular curve which possesses the property of the finite circular region indicated by Theorem III.*

Theorem III can easily be extended to give some information concerning the location of *all* the roots of equation (4) in α. In fact, if α' is any root, the other roots are given by

$$(z - \alpha)^k = (z - \alpha')^k,$$
$$z - \alpha = \omega (z - \alpha'),$$

where ω is a kth root of unity. If we suppose that α' is in C, all the roots of (4) must lie in the k circular regions obtained by revolving C about z as center of rotation through the angles $2j\pi/k$, $j = 1, 2, \cdots, k$. In particular suppose one of these k circular regions, say C_1, is external to all the other $k - 1$ circular regions. Then we can prove that C_1 has on or within it precisely one root of (4). For consider the points $\alpha_1, \alpha_2, \cdots, \alpha_k$ to move continuously always remaining in C and to coincide at the center of C. In this situation C_1 contains precisely one root of (4). The roots of (4) vary continuously with the points $\alpha_1, \alpha_2, \cdots, \alpha_k$; none can enter or leave C_1 during motion of the kind indicated; so in the original situation C_1 contained precisely one root of (4).

However, our purpose is not primarily to study the roots of (4), but rather to use Theorems I and III as tools in proving more interesting relations. In preparation for these results we now prove the

Lemma. *If the interiors and boundaries of the two circles C_1 and C_2, whose centers are α_1 and α_2 and radii r_1 and r_2 respectively, are the loci of two points z_1 and z_2, then as z_1 and z_2 vary independently, the locus of the point z which divides the segment (z_1, z_2) in the constant ratio $(m_1 : m_2)$,*

$$z = \frac{m_2 z_1 + m_1 z_2}{m_1 + m_2} \qquad (m_1 \neq -m_2, \quad m_1 m_2 \neq 0), \tag{5}$$

is the interior and boundary of a circle C whose center is

$$\frac{m_2 \alpha_1 + m_1 \alpha_2}{m_1 + m_2}$$

and whose radius is

$$\left| \frac{m_2 r_1}{m_1 + m_2} \right| + \left| \frac{m_1 r_2}{m_1 + m_2} \right| .\dagger$$

* For the corresponding fact for Theorem II, compare Walsh, these Transactions, vol. 24 (1922), pp. 31–69; Theorem III.

† The terminology that z divides the segment (z_1, z_2) in the ratio $(m_1 : m_2)$ is usual when

If z_1 and z_2 are on or within C_1 and C_2 respectively,

$$|z_1 - \alpha_1| \leqq r_1, \qquad |z_2 - \alpha_2| \leqq r_2,$$

we have

$$z - \frac{m_2 \alpha_1 + m_1 \alpha_2}{m_1 + m_2} = \frac{m_2}{m_1 + m_2}(z_1 - \alpha_1) + \frac{m_1}{m_1 + m_2}(z_2 - \alpha_2),$$

which is in absolute value less than or equal to

$$\left|\frac{m_2 r_1}{m_1 + m_2}\right| + \left|\frac{m_1 r_2}{m_1 + m_2}\right|.$$

It remains to be shown that given any point z on or within C, we can properly determine z_1 and z_2. In order to do this, we merely place

$$\frac{m_2(z_1 - \alpha_1)}{m_1 + m_2} = \left(z - \frac{m_2 \alpha_1 + m_1 \alpha_2}{m_1 + m_2}\right) \frac{\left|\frac{m_2 r_1}{m_1 + m_2}\right|}{\left|\frac{m_2 r_1}{m_1 + m_2}\right| + \left|\frac{m_1 r_2}{m_1 + m_2}\right|},$$

$$\frac{m_1(z_2 - \alpha_2)}{m_1 + m_2} = \left(z - \frac{m_2 \alpha_1 + m_1 \alpha_2}{m_1 + m_2}\right) \frac{\left|\frac{m_1 r_2}{m_1 + m_2}\right|}{\left|\frac{m_2 r_1}{m_1 + m_2}\right| + \left|\frac{m_1 r_2}{m_1 + m_2}\right|}.$$

Then z_1 and z_2 lie in the proper regions and satisfy (5), so the proof of the lemma is complete.

This lemma gives no result if $m_1 + m_2 = 0$; to treat this case we take (5) in the form

$$(6) \qquad m_2(z - z_1) = -m_1(z - z_2),$$

which is equivalent for our present purpose. If C_1 and C_2 are mutually external there is no point z different from the point at infinity which satisfies (6). If C_1 and C_2 are not mutually external we may choose $z_1 = z_2$, and every point of the plane satisfies (6). In the theorems below we give the general formulas for the determination of C with the understanding that when $m_1 + m_2 = 0$, C is considered to contain no finite point of the plane or every finite point of the plane according as C_1 and C_2 are or are not mutually external.*

---

m_1 and m_2 are real; whether m_1 and m_2 are real or not, we simply understand that statement to mean that z is given by (5).

This lemma was proved by the present writer and by the present method for m_1 and m_2 real and positive, in a note in the *Comptes Rendus du Congrès international*, Strasbourg, 1920. See also Theorem II of the reference given in connection with the present Theorem II.

* Thus, a careful statement of Theorem IV for the case $A = 1$ is

THEOREM IVa. *If the points $a_1, a_2, \cdots, a_k$ have as their locus a circle C_1 whose center is α and radius r_1, and if the points $b_1, b_2, \cdots, b_k$ have as their locus a circle C_2 whose center is β*

We now apply Theorem III and the Lemma to prove

THEOREM IV. *If the points* $a_1, a_2, \cdots, a_k$ *have as their locus (the interior and boundary of) a circle* C_1 *whose center is* α *and radius* r_1, *and if the points* $b_1, b_2, \cdots, b_k$ *have as their locus a circle* C_2 *whose center is* β *and radius* r_2, *the roots of the equation*

(7) $(z - a_1)(z - a_2)\cdots(z - a_k) - A(z - b_1)(z - b_2)\cdots(z - b_k) = 0$

have as their locus the k *circles with the respective radii*

$$\frac{r_1 + |A^{1/k}|r_2}{|1 - A^{1/k}|}$$

and centers

$$\frac{\alpha - A^{1/k}\beta}{1 - A^{1/k}} \tag{8}$$

where $A^{1/k}$ *takes all the* k *values possible. Any one of these* k *circles which is external to all the others contains precisely one root of* (7).

It is evident that any polynomial equation can be written in the form (7), and in an infinite variety of ways.

We prove Theorem IV as follows. If a point z is a root of (7), an application of Theorem III shows that

$$(z - a)^k - A(z - b)^k = 0,$$

where a and b are some points on or within C_1 and C_2 respectively. Hence we have

$$z - a = A^{1/k}(z - b),$$

$$z = \frac{a - A^{1/k}b}{1 - A^{1/k}},$$

where $A^{1/k}$ is some kth root of A, so z is on or within one of the k circles of the theorem. Variation of a and b independently and over C_1 and C_2 as loci shows that the k circles are the actual loci of the roots of (7). Continuous motion of the points $a_1, a_2, \cdots, a_k, b_1, b_2, \cdots, b_k$ so as to remain in their proper regions

and radius r_2, *and if* C_1 *and* C_2 *are mutually external, then the roots of the equation*

(7') $(z - a_1)(z - a_2)\cdots(z - a_k) = (z - b_1)(z - b_2)\cdots(z - b_k)$

have as their locus the $k - 1$ *circles with the respective radii*

$$\frac{r_1 + r_2}{|1 - \omega|}$$

and centers

$$\frac{\alpha - \omega\beta}{1 - \omega}$$

where ω *takes the* $k - 1$ *values of the* k*th roots of unity which differ from unity itself. Any one of these* $k - 1$ *circles which is external to all the others contains precisely one root of* (7').

If C_1 *and* C_2 *are not mutually external, the locus of the roots of* (7') *is the entire plane.*

The details of the proof of Theorem IV*a* are similar to those of the proof of Theorem IV and are left to the reader.

and finally to coincide shows in precisely the manner used in considering the roots of (4) that any one of the k circles which is exterior to all the others contains precisely one root of (7).

Theorem IV reduces to the lemma when $k = 1$.

A few remarks on the geometry of the situation are not out of place. The k points (8) all lie on the circle which is the locus of points z such that

$$\frac{|z-\alpha|}{|z-\beta|} = |A^{1/k}|,$$

which circle is of course a circle of the coaxial family determined by α and β as null circles. Each of the k points also lies on a circle which is the locus of points z such that

$$\frac{\dfrac{z-\alpha}{|z-\alpha|}}{\dfrac{z-\beta}{|z-\beta|}} = \frac{A^{1/k}}{|A^{1/k}|};$$

that is, the argument of the ratio $(z-\alpha)/(z-\beta)$ is constant. These k circles belong to the coaxial family of all circles passing through α and β, the family conjugate to the former coaxial family mentioned. These same k circles can be arranged in order so that at α and β each cuts its predecessor at an angle of $2\pi/k$.

It is quite easy for us to obtain results concerning the roots of the derivatives of equation (7). The mth derivative $(k > m \geqq 0)$, except for a constant factor, can be written

$$(9)\quad (z-a_1')(z-a_2')\cdots(z-a_{k-m}') - A(z-b_1')(z-b_2')\cdots(z-b_{k-m}') = 0.$$

The points a_j' all lie on or within C_1 and the points b_j' all lie on or within C_2, by the theorem of Lucas concerning the roots of the derivative of a polynomial. Hence all the roots of (9) lie on or within the $k-m$ circles with the respective radii

$$\frac{r_1 + |A^{1/(k-m)}|r_2}{|1-A^{1/(k-m)}|}$$

and centers

$$\frac{\alpha - A^{1/(k-m)}\beta}{1-A^{1/(k-m)}},$$

where $A^{1/(k-m)}$ takes all the $k-m$ values possible. These circles form the locus of the roots of (9). Any one of these $k-m$ circles which is external to the remaining $k-m-1$ circles contains precisely one root of (9). This new result includes Theorem IV as the case $m = 0$.

We now state another theorem which results from Theorem III and which like Theorem IV refers to the sum of two polynomials.*

THEOREM V. *If the points* $a_1, a_2, \cdots, a_k$ *lie on or within a circle* C *whose center is* α *and radius* r, *all the roots of the equation*

$$(z - a_1)(z - a_2) \cdots (z - a_k) - A = 0 \tag{10}$$

lie on or within one of the k *circles which have the common radius* r *and whose centers are the points* $\alpha + A^{1/k}$, *where* $A^{1/k}$ *takes all the* k *possible values. If these* k *circles are mutually external, each contains precisely one root of* (10).

Of course any polynomial equation can be written in the form (10) and in an infinite variety of ways. The proof of the theorem follows the proof of Theorem IV. If C is the locus of $a_1, a_2, \cdots, a_k$, the k circles form the locus of the roots of (10).

Equation (10) is particularly interesting because the derived equation is independent of A. All the roots of the mth derived equation, $m < k$, lie on or within C. Theorem V can also be extended to a polynomial of the form

$$\frac{d}{dz}[(z - a_1)(z - a_2) \cdots (z - a_k)] - A = 0,$$

but this generalization as well as Theorem V itself is included in a more general theorem to be proved from Theorem I.

III. SOME DIRECT APPLICATIONS OF THEOREM I

We now proceed to derive a number of results directly from Theorem I instead of from Theorem II or Theorem III.

If in Theorem I the polynomial $f(z)$ contains merely one set of variables $\{\alpha_i\}$ referring to a single circular region, the locus Z can have as its boundary no point other than a point of the curve traced by z when $\alpha = \alpha_1 = \alpha_2 = \cdots = \alpha_k$ traces the circle C_α. This leads to

THEOREM VI. *Let the common locus of all the roots of a polynomial*

$$f(z) = (z - \alpha_1)(z - \alpha_2) \cdots (z - \alpha_k)$$

be the interior and boundary of a circle C *whose center is* α *and radius* r. *Then the locus of the roots of the polynomial*

$$A_0 f(z) + A_1 f'(z) + \cdots + A_{k-1} f^{(k-1)}(z) + A_k f^{(k)}(z) \tag{11}$$

is composed of the interiors and boundaries of the circles whose common radius is r *and whose centers are the roots of*

$$A_0 (z - \alpha)^k + kA_1 (z - \alpha)^{k-1} + \cdots + k(k-1) \cdots 2 \cdot 1 A_k. \tag{12}$$

* If in Theorem IV we replace (7) by the sum of two polynomials not of the same degree and neither of degree zero, we are led to the determination of a locus which is not generally bounded by circles. Compare 4 of the next to the last paragraph of this paper.

Any of these circles having no point in common with any other of these circles contains a number of roots of (11) *equal to the multiplicity of the center of that circle as a root of* (12).

A detailed proof of this theorem is quite simple. For any particular root z of (11) we consider the points $\{\alpha_i\}$ to coincide at $\bar{\alpha}$ on or in C. We have then

$$A_0(z-\bar{\alpha})^k + kA_1(z-\bar{\alpha})^{k-1} + \cdots + k(k-1)\cdots 2\cdot 1A_k = 0.$$

Since $|\bar{\alpha}-\alpha| \leqq r$, we know that z must lie in or on one of the circles determined. Conversely, any point z in or on one of these circles is a root of (11) for some choice of the $\{\alpha_i\}$; denote the center of the circle by β. We need merely choose $\alpha_1 = \alpha_2 = \cdots = \alpha_k = \alpha + z - \beta$, which is a point on or within C.

If one of the circles C_1 has no point in common with any other of those circles, we may vary the α_i at will and no root of (11) can enter or leave C_1; these roots vary continuously if the α_i vary continuously. In particular if

$$\alpha_1 = \alpha_2 = \cdots = \alpha_k = \alpha,$$

we have the number of roots of (11) proper to C_1. It is similarly true that a number of these circles $C_1, \cdots, C_l$ having no point in common with any other of the circles contain a number of roots of (11) equal to the sum of the multiplicities of their centers considered as roots of (12).

Theorem VI enables us to give a result concerning the location of the roots of the sum of any number of arbitrary polynomials; $f(z)$ is to be chosen one of those polynomials of highest degree.

If in Theorem VI we place $A_0 = A_2 = A_3 = \cdots = A_k = 0,\ A_1 \neq 0$, we have

A circle which contains all the roots of a polynomial contains also all the roots of its derivative.

This is essentially the theorem of Lucas:

*A convex polygon which contains all the roots of a polynomial contains also all the roots of its derivative.**

The similar results obtained from Theorem VI by placing $A_l \neq 0$ but all the other A_i zero $(l > 0)$ give the theorem of Lucas for the other derivatives of $f(z)$.

Theorem I applies just as well to the integral of a polynomial as to its derivative, but for a result of the nature of those just given it is necessary to choose a *particular* integral. Thus we shall prove

THEOREM VII. *Let the common locus of all the roots of the polynomial*

$$f(z) = (z-\alpha_1)(z-\alpha_2)\cdots(z-\alpha_k)$$

* In both of these statements as well as many other similar results, the term *contains* may be interpreted to include the possibility of points on the boundary or to exclude that possibility. Whichever interpretation is considered in the hypothesis, the conclusion will be true under that same interpretation.

be the interior and boundary of the circle C whose center is α and radius r. Then the locus of the roots of the polynomial

$$F(z) = \int_0^z f(z)\,dz$$

is composed of the interiors and boundaries of the $k+1$ circles whose centers are $(1-\omega)\alpha$ and radii $|1-\omega|r$, where ω takes all the values of the $(k+1)$st roots of unity. Any of these circles which is entirely exterior to all the others contains precisely one root of $F(z)$.

If z is a root of $F(z)$ for particular values of the α_i, we must have for some choice of $\bar{\alpha} = \alpha_1 = \alpha_2 = \cdots = \alpha_k$ in or on C,

$$F(z) = \frac{(z-\bar{\alpha})^{k+1}}{k+1} - \frac{(-\bar{\alpha})^{k+1}}{k+1} = 0,$$
$$z = (1-\omega)\bar{\alpha},$$

where $\omega^{k+1} = 1$. From this equation we have at once the desired inequality:

$$z - (1-\omega)\alpha = (1-\omega)(\bar{\alpha} - \alpha),$$
$$|z - (1-\omega)\alpha| \leqq |1-\omega|r,$$

which shows that z is on or within one of the $k+1$ circles. On the other hand, any point z on or within one of these $k+1$ circles is a root of $F(z)$ for proper choice of the $\{\alpha_i\}$. In fact, if ω is the particular root of unity corresponding to that circle we have merely to place $\bar{\alpha} = \alpha_1 = \alpha_2 = \cdots = \alpha_k$, where

$$\bar{\alpha} - \alpha = \frac{z - (1-\omega)\alpha}{1-\omega};$$

the exceptional case $\omega = 1$ is trivial.

The proof of the theorem is now complete except for the last sentence; this is proved in precisely the same manner as the corresponding statement in Theorem VI. Like Theorem VI, Theorem VII illustrates the remark made just previous to the statement of Theorem VI.

We shall next prove further results concerning the roots of the derivative of a polynomial, making continued use of the Lemma. The special case of the Lemma where m_1 and m_2 are positive is particularly simple; this special case leads to

THEOREM VIII. *Denote by $g(z)$ the polynomial*

$$(z - z_1)^{m_1}(z - z_2)^{m_2}$$

*and by $m_1^{(n)} : m_2^{(n)}$ $(n = 1, 2, \cdots, m)$ the ratios in which the m distinct roots $z^{(n)}$ of $g^{(k)}(z)$ (the kth derivative of $g(z)$) divide the segment (z_1, z_2).** *Let the interiors and boundaries of circles C_1 and C_2 whose centers are α_1 and α_2 and radii r_1 and r_2 be the loci respectively of m_1 and m_2 roots of a polynomial $f(z)$ which*

* The reader will easily prove from Rolle's Theorem that no point $z^{(n)}$ distinct from z_1 and z_2 can be a multiple root of $g^{(k)}(z)$.

has no other roots. Then the locus of the roots of $f^{(k)}(z)$, *the kth derivative of* $f(z)$, *is composed of the m circles* $C^{(n)}$ *whose centers are*

$$\frac{m_2^{(n)}\alpha_1 + m_1^{(n)}\alpha_2}{m_1^{(n)} + m_2^{(n)}}$$

and whose radii are

$$\frac{m_2^{(n)} r_1 + m_1^{(n)} r_2}{m_1^{(n)} + m_2^{(n)}}.$$

If one of these circles $C^{(n)}$ *is exterior to all the others, it contains a number of roots of* $f(z)$ *equal to the multiplicity of* $z^{(n)}$ *as a root of* $g^{(k)}(z)$.

The proof of this theorem is quite similar to the proofs of Theorems VI and VII and will therefore be omitted.*

If a number of circles $C^{(n)}$ of Theorem VIII overlap and are entirely exterior to all the other circles $C^{(n)}$, they contain together a number of roots of $f^{(k)}(z)$ equal to the sum of the multiplicities of the corresponding roots $z^{(n)}$ of $g^{(k)}(z)$. A similar situation for one as for several circles arises in connection with Theorems VI and VII. In the future we shall indicate this general fact by saying that the number of roots of $f^{(k)}(z)$ *proper* to a circle $C^{(n)}$ is the multiplicity of the point $z^{(n)}$ as a root of $g^{(k)}(z)$.

In Theorem VIII we are in reality considering a polynomial

$$f(z) = \phi(z) \cdot \psi(z),$$

where the m_1 roots of $\phi(z)$ lie on or within C_1 and the m_2 roots of $\psi(z)$ lie on or within C_2. Then our conclusion refers to the polynomial

$$f^{(k)}(z) = \phi \cdot \psi^{(k)} + \frac{k}{1!}\phi' \cdot \psi^{(k-1)} + \frac{k(k-1)}{2!}\phi'' \cdot \psi^{(k-2)} + \cdots + \phi^{(k)} \cdot \psi.$$

When the result is expressed in this form it can be given a large extension. Under the same hypothesis with respect to ϕ and ψ we consider the polynomial

$$A_0\phi \cdot \psi^{(k)} + A_1\phi' \cdot \psi^{(k-1)} + A_2\phi'' \cdot \psi^{(k-2)} + \cdots + A_k\phi^{(k)} \cdot \psi. \tag{13}$$

If z is a root of (13) for a particular choice of the roots of ϕ and ψ, there exist α and β in or on C_1 and C_2 respectively and such that

$$\begin{aligned} & m_2(m_2-1)\cdots(m_2-k+1)A_0(z-\alpha)^{m_1}(z-\beta)^{m_2-k} \\ & \quad + m_1 \cdot m_2(m_2-1)\cdots(m_2-k+2)A_1(z-\alpha)^{m_1-1}(z-\beta)^{m_2-k+1} \\ & \quad + m_1 \cdot (m_1-1) \cdot m_2 \cdot (m_2-1) \\ & \qquad \cdots (m_2-k+3)A_2(z-\alpha)^{m_1-2}(z-\beta)^{m_2-k+2} \\ & \quad \cdot \quad \cdot \quad \cdot \quad \cdot \quad \cdot \quad \cdot \quad \cdot \quad \cdot \\ & \quad + m_1(m_1-1)\cdots(m_1-k+1)A_k(z-\alpha)^{m_1-k}(z-\beta)^{m_2} = 0. \end{aligned} \tag{14}$$

* This theorem was published in a short note by the present writer, Paris Comptes Rendus, vol. 172 (1921), pp. 662–664, and its proof as there indicated contains the germ of the proof of Theorem I. The special case of Theorem VIII for the case $k = 1$ had been previously proved by means of Theorem II; see the reference given in connection with that theorem.

Equation (14) is essentially an equation in homogeneous coördinates μ_1, μ_2:

$$
\begin{aligned}
& m_2(m_2-1)\cdots(m_2-k+1)A_0\,\mu_1^{m_1}\mu_2^{m_2-k} \\
& \quad + m_1\cdot m_2\cdot(m_2-1)\cdots(m_2-k+2)A_1\,\mu_1^{m_1-1}\mu_2^{m_2-k+1} \\
(15)\quad & \quad + m_1(m_1-1)\cdot m_2\cdot(m_2-1)\cdots(m_2-k+3)A_2\,\mu_1^{m_1-2}\mu_2^{m_2-k+2} \\
& \quad \cdots\cdots\cdots\cdots\cdots\cdots\cdots\cdots \\
& \qquad\qquad + m_1(m_1-1)\cdots(m_1-k+1)A_k\,\mu_1^{m_1-k}\mu_2^{m_2} = 0,
\end{aligned}
$$

whose distinct solutions, m in number, we denote by $\mu_1^{(n)} : \mu_2^{(n)}$, $n = 1, 2, \cdots, m$. From (14) and (15) we see that for some n we have

$$\frac{z-\alpha}{z-\beta} = \frac{\mu_1^{(n)}}{\mu_2^{(n)}}, \qquad z = \frac{\mu_2^{(n)}\alpha + \mu_1^{(n)}\beta}{\mu_1^{(n)} + \mu_2^{(n)}}.$$

The Lemma immediately gives us certain circles in at least one of which must be located the point z. It is easily seen that these circles give the exact locus of z, and we have

THEOREM IX. *Let the interiors and boundaries of circles C_1 and C_2, whose centers are α_1 and α_2 and radii r_1 and r_2, be respectively the loci of the m_1 roots of a polynomial ϕ and the m_2 roots of a polynomial ψ. Then the locus of the roots of* (13) *is composed of the interiors and boundaries of the m circles $C^{(n)}$ whose centers are the points*

$$\frac{\mu_2^{(n)}\alpha_1 + \mu_1^{(n)}\alpha_2}{\mu_1^{(n)} + \mu_2^{(n)}}$$

and radii

$$\frac{|\mu_2^{(n)}|\,r_1 + |\mu_1^{(n)}|\,r_2}{|\mu_1^{(n)} + \mu_2^{(n)}|} \qquad (n = 1, 2, \cdots, m),$$

where $\mu_1^{(n)} : \mu_2^{(n)}$ are the m distinct roots of (15). *The number of roots of* (13) *proper to a circle $C^{(n)}$ is the multiplicity of the ratio $\mu_1^{(n)} : \mu_2^{(n)}$ as a root of* (15).

The methods that we have been using enable us also to obtain a result for the roots of the derivative of an entire transcendental function of the simple type

$$f(z) = e^{P(z)}\cdot Q(z)$$

where

$$
\begin{aligned}
P(z) &= a(z-\alpha_1)(z-\alpha_2)\cdots(z-\alpha_p), \qquad a \neq 0, \\
Q(z) &= b(z-\beta_1)(z-\beta_2)\cdots(z-\beta_q), \qquad b \neq 0.
\end{aligned}
$$

The roots of $f'(z)$ are given by the equation

$$Q(z)P'(z) + Q'(z) = 0. \tag{16}$$

If all the α_i and β_i lie in or on a circle C whose center is α and radius r, and if z is a root of (16), we shall have for some $\bar{\alpha}$ and $\bar{\beta}$ in or on C

$$ap(z-\bar{\beta})^q(z-\bar{\alpha})^{p-1} + q(z-\bar{\beta})^{q-1} = 0.$$

Then we must have either

$$(z-\bar\beta)^{q-1}=0$$

or

$$ap(z-\bar\beta)(z-\bar\alpha)^{p-1}+q=0.$$

If this latter equation is verified we find from Theorem III that there exists a point $\bar\gamma$ in or on C such that

$$ap(z-\bar\gamma)^p+q=0,$$

$$z-\bar\gamma=\omega\sqrt[p]{\frac{-q}{ap}},\qquad \text{where}\qquad \omega^p=1.$$

It therefore follows that

$$z-\left(\alpha+\omega\sqrt[p]{\frac{-q}{ap}}\right)=\bar\gamma-\alpha,$$

which is less than or equal to r in absolute value. Then all the roots of $f'(z)$ lie in C and the p circles of common radius r whose centers are the points

$$\alpha+\omega\sqrt[p]{\frac{-q}{ap}}.$$

We state this fact and its obvious converse in

THEOREM X. *Let the interior and boundary of the circle C whose center is α and radius r be the common locus of all the roots of the two polynomials*

$$P(z)=a(z-\alpha_1)(z-\alpha_2)\cdots(z-\alpha_p),\qquad a\neq 0,$$
$$Q(z)=b(z-\beta_1)(z-\beta_2)\cdots(z-\beta_q),\qquad b\neq 0.$$

Then the locus of the roots of the derivative of the function

$$e^{P(z)}\cdot Q(z)$$

is composed of C (unless $q=1$) and the p circles whose common radius is r and whose centers are the points

$$\alpha+\omega\sqrt[p]{\frac{-q}{ap}},$$

where ω takes the values of the pth roots of unity. There are $q-1$ roots of the derivative proper to C and one proper to each of the other p circles.

In the proof of this theorem we have used not the fact that the interior and boundary of C is the locus of the roots of $P(z)$ but that it is the locus of the roots of $P'(z)$, which gives in reality a more general result. Thus if $p=1$, there is no restriction whatever on the root of $P(z)$.

We now prove a theorem which is a generalization of our result concerning the roots of (16) corresponding to our former generalization of Theorem VIII.

THEOREM XI. *Let the interior and boundary of a circle C whose center is* α *and radius is r be the common locus of the roots of the polynomials* $\phi_1, \phi_2, \cdots, \phi_q$ *of respective degrees* $k_1, k_2, \cdots, k_q$. *Then the locus of the roots of the polynomial*

$$(17) \qquad \phi_1 \cdot \phi_2 \cdot \phi_3 \cdot \cdots \cdot \phi_q - A\phi_1^{(n_1)} \cdot \phi_2^{(n_2)} \cdot \cdots \cdot \phi_q^{(n_q)}, \qquad n_i \leqq k_i,$$

is composed of the circle C (unless $n_i = k_i$ *for* $i = 1, 2, \cdots, q$*) and the circles,* $n_1 + n_2 + \cdots + n_q$ *in number, whose centers are the points*

$$\alpha + \omega[k_1 \cdot (k_1 - 1) \cdots (k_1 - n_1 + 1) \cdot k_2 \cdot (k_2 - 1) \cdots (k_2 - n_2 + 1) \cdots k_q \cdot (k_q - 1) \cdots (k_q - n_q + 1) \cdot A]^{1/(n_1+n_2+\cdots+n_q)}$$

where ω *takes all possible values of an* $(n_1 + n_2 + \cdots + n_q)$*th root of unity, and whose common radius is* r. *There are*

$$(k_1 + k_2 + \cdots + k_q) - (n_1 + n_2 + \cdots + n_q)$$

roots of (17) *proper to C and one root proper to each of the* $n_1 + n_2 + \cdots + n_q$ *other circles.*

Under the hypothesis of the theorem, if z is a root of (17) there exist points $\alpha', \alpha'', \cdots, \alpha^{(q)}$ on or within C and such that we have

$$(z - \alpha')^{k_1}(z - \alpha'')^{k_2} \cdots (z - \alpha^{(q)})^{k_q} - [k_1 \cdot (k_1 - 1) \cdots (k_1 - n_1 + 1) \cdot k_2 \cdot (k_2 - 1) \cdots (k_2 - n_2 + 1) \cdot k_q \cdot (k_q - 1) \cdots (k_q - n_q + 1) \cdot A \cdot (z - \alpha')^{k_1 - n_1}(z - \alpha'')^{k_2 - n_2} \cdots (z - \alpha^{(q)})^{k_q - n_q}] = 0.$$

Then z must satisfy one of the equations

$$(z - \alpha')^{k_1 - n_1}(z - \alpha'')^{k_2 - n_2} \cdots (z - \alpha^{(q)})^{k_q - n_q} = 0,$$

$$(z - \alpha')^{n_1}(z - \alpha'')^{n_2} \cdots (z - \alpha^{(q)})^{n_q} - k_1(k_1 - 1) \cdots (k_q - n_q + 1)A = 0.$$

If z satisfies this latter equation we know from Theorem III that there exists a point $\bar{\alpha}$ on or within C and such that

$$(z - \bar{\alpha})^{n_1+n_2+\cdots+n_q} = k_1(k_1 - 1) \cdots (k_q - n_q + 1)A.$$

That is, we have the equation

$$z - \{\alpha + \omega[k_1(k_1 - 1) \cdots (k_q - n_q + 1)A]^{1/(n_1+n_2+\cdots+n_q)}\} = \bar{\alpha} - \alpha,$$

where $\omega^{n_1+n_2+\cdots+n_q} = 1$. The absolute value of the right-hand member is not greater than r, so z must lie in or on one of the $n_1 + n_2 + \cdots + n_q$ circles of the theorem.

The detailed proof of the remainder of Theorem XI now requires no further analytical work and is left to the reader.

The present paper aims merely to give some of the more immediate results that can be obtained from Theorem I. Further results can be obtained by extending in various directions:

1. Application of these methods to other and more general types of polynomials.

2. Extension of the results to other circular regions, namely, half-planes and the exteriors of circles.

3. Expression of the results in a manner which shall be independent of linear transformation of the complex variable.

4. Detailed determination and study of loci which naturally arise and which are of the same general nature as the locus determined in the lemma. For example, if the loci of α and β are two circular regions, determine the locus of the points z defined by

$$(z-\alpha)^m = A(z-\beta)^n.$$

For only the simplest polynomials $f(z, \alpha_i, \beta_i)$ do we get a number of circular regions as the locus of the roots of f when the loci of the roots of the α_i and β_i are circular regions. All the problems of the present paper have been characterized by a certain *linearity*, either the original problems themselves or those problems after a simple transformation. In more general problems, where the locus of the roots does not ordinarily consist of circular regions, the exact locus should be determined and also a simple even if rough approximation.

5. Application of these results and methods to the case of real polynomials.*

The present writer hopes to return later to a consideration of these questions.

---

* Thus the following theorem results from Theorem VI:

Let all the roots of a polynomial $f(z)$ of degree k lie in an interval of the axis of reals whose center is α and length l. Denote the roots of the polynomial

$$A_0 z^k + kA_1 z^{k-1} + \cdots + k(k-1)\cdots 2\cdot 1\cdot A_k \tag{1'}$$

by $z_1, z_2, \cdots, z_k$, which roots are supposed real. Then all the roots of the polynomial

$$A_0 f + A_1 f' + \cdots + A_k f^{(k)} \tag{2'}$$

lie in the intervals of the axis of reals whose centers are the points $\alpha + z_i$ $(i = 1, 2, \cdots, k)$ and whose common length is l. The number of roots of (2′) proper to each of these intervals is the multiplicity of the corresponding z_i as a root of (1′).

HARVARD UNIVERSITY,
CAMBRIDGE, MASS. (JUNE, 1921.)

A THEOREM OF GRACE ON THE ZEROS OF POLYNOMIALS, REVISITED

J. L. WALSH[1]

A theorem of Grace [1] has shown itself highly useful (e.g. [2]) in the study of the geometry of the zeros of polynomials. We shall treat not that theorem in full generality, but the following special case:

LEMMA 1. *If we have* $m_k > 0$, $\sum m_k = 1$, $|\alpha_k| \leqq 1$, $k = 1, 2, \cdots, n$, $|z| > 1$, *then the equation in* α

$$\prod_{k=1}^{n} (z - \alpha_k)^{m_k} = z - \alpha \tag{1}$$

has a solution α *which satisfies* $|\alpha| \leqq 1$. *Indeed there exists such a solution* α *satisfying*

$$\min_k \arg[(z - \alpha_k)/z] \leqq \arg[(z - \alpha)/z] \leqq \max_k \arg[(z - \alpha_k)/z], \tag{2}$$

where these three arguments are values of any $\arg[(z-\beta)/z]$ *chosen continuous for fixed* z *and for all* β *with* $|\beta| \leqq 1$.

The latter part of Lemma 1 is unusual, but is to be established below. Lemma 1 is valid [2, Theorem III] without the hypothesis $|z| > 1$ if (2) is omitted.

If the m_k and α_k are fixed in Lemma 1 and $|z|$ is large, the point α (which depends on z, with $|\alpha| \leqq 1$) lies near the center of gravity of the α_k, as we see by writing (1) in the form

$$\begin{aligned} &\sum m_k \log\left(1 - \frac{\alpha_k}{z}\right) \\ &\equiv \sum m_k \left[-\frac{\alpha_k}{z} - \frac{1}{2}\left(\frac{\alpha_k}{z}\right)^2 - \frac{1}{3}\left(\frac{\alpha_k}{z}\right)^3 - \cdots \right] \\ &\equiv \log\left(1 - \frac{\alpha}{z}\right) \equiv \left[-\frac{\alpha}{z} - \frac{1}{2}\left(\frac{\alpha}{z}\right)^2 - \frac{1}{3}\left(\frac{\alpha}{z}\right)^3 - \cdots \right]; \end{aligned} \tag{3}$$

yet this important fact is not mentioned in Lemma 1. The purpose of this paper is to prove, and to apply in illustrative cases, a revision of Lemma 1:

Received by the editors October 1, 1962 and, in revised form, February 2, 1963.

[1] This research was supported (in part) by the Air Force Office of Scientific Research. Abstract in Notices Amer. Math. Soc. 9 (1962), 484.

LEMMA 2. *If we have* $m_k>0$, $\sum m_k=1$, $|\alpha_k|\leqq 1$, $\sum m_k\alpha_k=0$, $|z|>1$ *(where* $k=1, 2, 3, \cdots, n$*) then there exists an* α *such that* $|\alpha|\leqq 1/|z|$, *with*

$$\sum m_k \log\left(1-\frac{\alpha_k}{z}\right)=\log\left(1-\frac{\alpha}{z}\right), \tag{4}$$

where $\arg(1-\alpha/z)$ *may be chosen as in* (2).

Precisely the remark already made, that for large $|z|$ the point α lies near the center of gravity of the α_k, applies also to another lemma which is of frequent use in the study of the geometry of zeros of polynomials, namely

LEMMA 3. *If we have* $m_k>0$. $\sum m_k=1$, $|\alpha_k|\leqq 1$, $k=1, 2, \cdots, n$, $|z|>1$, *then* α *as defined by the equation*

$$\sum_{k=1}^{n}\frac{m_k}{z-\alpha_k}=\frac{1}{z-\alpha} \tag{5}$$

satisfies $|\alpha|\leqq 1$.

The analogue of (3) is here the equivalent of (5):

$$\sum\frac{m_k}{z-\alpha_k}\equiv\sum\frac{m_k}{z}\left[1+\frac{\alpha_k}{z}+\left(\frac{\alpha_k}{z}\right)^2+\cdots\right]$$

$$\equiv\frac{1}{z-\alpha}\equiv\frac{1}{z}\left[1+\frac{\alpha}{z}+\left(\frac{\alpha}{z}\right)^2+\cdots\right].$$

As the author has recently indicated [4], there exists a modification of Lemma 3 which bears precisely the same relation to Lemma 3 that Lemma 2 bears to Lemma 1:

LEMMA 4. *Under the conditions of Lemma* 3 *and with* $\sum m_k\alpha_k=0$, *we have* $|\alpha|\leqq 1/|z|$.

Before proceeding with the proof of Lemma 2 we remark that the unit disk $|\zeta|<1$ is mapped conformally onto a convex region R of the w-plane by the transformation $w=\log(1-\zeta)$. It follows that the image R_r in the w-plane of the disk $|\zeta|\leqq r$ (<1) is also convex.

Equation (4), which defines α, expresses $\log(1-\alpha/z)$ in the w-plane as the center of gravity of the weighted points $w=\log(1-\alpha_k/z)$ of R. To prove Lemma 1 it is sufficient to choose $z=1/r$, where r is less than unity. Here we have $|\alpha_k/z|\leqq r$, so all points $\log(1-\alpha_k/z)$ lie in R_r as does their center of gravity. That is, $\log(1-\alpha/z)$ lies in R_r and (to return to the ζ-plane) $\zeta=\alpha/z$ satisfies $|\alpha/z|\leqq r$, whence $|\alpha|\leqq 1$.

The last part of Lemma 1 follows at once, so Lemma 1 is established.

The proof of Lemma 1 as just given has not been previously published in detail; this proof was outlined in [2, see Theorem III]. We proceed with the proof of Lemma 2. For $|z|>1$ we set $z'=1/z$ with $|z'|<1$, and define $\alpha=\alpha(z')$ by equation (4) in the form

$$\sum m_k \log(1-\alpha_k z') = \log(1-\alpha z'). \tag{6}$$

We choose $\alpha(z')$ as in the proof of Lemma 1, and it follows from Lemma 1 that we have $|\alpha(z')|\leqq 1$. By the proof of Lemma 1 we may write

$$\begin{aligned}\alpha(z') &\equiv [1-\prod(1-\alpha_k z')^{m_k}]/z' \\ &\equiv \left[1-\prod\left(1-m_k\alpha_k z' + \frac{m_k(m_k-1)}{2!}\alpha_k^2 z'^2 - \cdots\right)\right] \Big/ z', \\ \alpha(0) &= 0.\end{aligned}$$

Schwarz's lemma applied to $\alpha(z')$ now yields

$$|\alpha| \leqq |z'| = 1/|z|, \tag{7}$$

so Lemma 2 is established.

We remark that if also $\sum m_k\alpha_k^2=0$, then we have $\alpha'(0)-0$, and there follows $|\alpha|\leqq 1/|z|^2$. A similar remark follows if we have $\alpha''(0)=0$, etc.

It is to be mentioned that there exists no number λ (<1) such that for all z with $|z|>1$ inequality (7) can be replaced by

$$|\alpha| \leqq \lambda/|z| \tag{8}$$

with the hypothesis of Lemma 2. Indeed we choose $z>1$, $n=2$, $m_1=m_2$, $\alpha_1=1$, $\alpha_2=-1$, so there follows

$$(z-1)(z+1) = z^2-1 = (z-\alpha)^2$$
$$\alpha = z - \sqrt{(z^2-1)}.$$

When $z\to 1$, we have $\alpha\to 1$, which contradicts (8).

In the special case just considered except that $\alpha_1=-\alpha_2$, $|\alpha_1|=1$, and $|z|>1$, we have

$$(z-\alpha_1)(z+\alpha_1) = (z-\alpha)^2, \qquad \alpha_1^2 = \alpha(2z-\alpha),$$
$$|\alpha(2z-\alpha)| = 1. \tag{9}$$

It follows that the locus of α is one oval of lemniscate (9), and we have $\max|\alpha|=|z|-(|z|^2-1)^{1/2}$, $\min|\alpha|=(|z|^2+1)^{1/2}-|z|$. If we require merely $\alpha_1=-\alpha_2$, $|\alpha_1|\leqq 1$, the locus of α is the closed interior of the same oval of (9).

The proof of Lemma 4, announced in [3] and hitherto unpublished, is primarily geometric in character and relatively involved. The comparatively simple proof published in [4] results from later suggestions made to the present writer by Mr. S. Jacobs, Professor L. Carleson, and Dr. Z. Rubinstein. The present proof of Lemma 2 is quite similar to the latter proof [4] of Lemma 4.

Lemmas 1 and 3 are the special cases $f(\zeta)\equiv\log(1-\zeta)$ and $f(\zeta)\equiv 1/(1-\zeta)$ of the following proposition: *Let* $w=f(\zeta)$ *be schlicht-convex in* $|\zeta|<1$, *hence also in* $|\zeta|<r$ (<1). *Let* z *be fixed,* $|z|>1$, *and suppose* $|\alpha_k|\leq 1$, $m_k>0$, $\sum_1^n m_k=1$. *Then the equation in* α

$$\sum_{k=1}^{n} m_k f\left(\frac{\alpha_k}{z}\right)=f\left(\frac{\alpha}{z}\right)$$

has a solution $\alpha=\alpha(z)$ *with* $|\alpha|\leq 1$. *This solution can be restricted so as to be unique.* Lemmas 2 and 4 are special cases of the corollary: *If* $\sum m_k\alpha_k=0$, *then* $\alpha(\infty)=0$, *and we have* $|\alpha(z)|\leq 1/|z|$ *for* $|z|>1$.

Lemmas 2 and 4 have important differences. The latter determines in fact for each even n and for every z with $|z|>1$ the actual locus of the point α, but Lemma 2 does not. Moreover each holds for $|z|>1$ and not for $|z|<1$. This is not serious in Lemma 4, for if $n>3$ each point of $|z|<1$ is a possible multiple point α_k and is ordinarily a point of the locus of zeros of the polynomial studied, but not so in the applications of Lemma 2, where the point set $|z|\leq 1$ requires separate and independent study, perhaps with the aid of the (valid) first part of Lemma 1, but involving now $|z|\leq 1$ and without the conclusion concerning arg $(1-\alpha/z)$.

Inequality (7) is not sharp, and we indicate rapidly a slight improvement. Let us set $\phi(z')\equiv\sum m_k\log(1-\alpha_k-z')$, $\psi(z')\equiv 1-\exp[\phi(z')]$, $z'=1/z$, $\psi''(0)=-\phi''(0)=\sum m_k\alpha_k^2=b$, whence $|b|\leq 1$. We set also $\Psi(z')\equiv\psi(z')-bz'^2/2$, whence $\Psi(0)=\Psi'(0)=\Psi''(0)=0$, and for $|z'|<1$ we have $|\Psi(z')|<1+|b|/2$. Schwarz's lemma then yields, for $|z'|<1$,

$$\left|\psi(z')-\frac{bz'^2}{2}\right|\leq\left(1+\frac{|b|}{2}\right)|z'|^3,$$

$$|\psi(z')|=|\alpha z'|\leq\frac{|bz'^2|}{2}+\left(1+\frac{|b|}{2}\right)|z'|^3,$$

$$|\alpha|\leq|z'|^2+\frac{|b|}{2}(|z'|+|z'|^2);$$

when $z'=0$ this inequality still holds, with $\alpha=0$, so we have

$$(10) \qquad |\alpha| \leqq \frac{1}{|z|^2} + \frac{|b|}{2}\left(\frac{1}{|z|} + \frac{1}{|z|^2}\right) \leqq \frac{1}{2|z|} + \frac{3}{2|z|^2}.$$

The extreme members of (10) give a stronger inequality than (7) whenever we have $|z| > 3$.

We turn now to some relatively immediate applications of Lemma 2.

THEOREM 1. *Suppose* $|\alpha_k| \leqq 1$ *for* $k = 1, 2, \cdots, n$, *with* $\sum \alpha_k = 0$; *we set* $p(z) \equiv \prod (z - \alpha_k) - C$, *where the constant* C *is arbitrary. Then for* $|z| \leqq 1$ *all zeros of* $p(z)$ *lie in the* n *circles* $|z - C^{1/n}| \leqq 1$ *and for* $|z| > 1$ *in the* n *lemniscate regions*

$$(11) \qquad |z(z - C^{1/n})| \leqq 1,$$

where $C^{1/n}$ *takes all* n *values.*

For $|z| \leqq 1$ we apply the analogue [2, Theorem III] of Lemma 1 involving merely $|\alpha| \leqq 1$; if $p(z) = 0$ we have

$$(z - \alpha)^n - C = 0,$$

whence $\alpha = z - C^{1/n}$, so the first part of the conclusion follows. For $|z| > 1$ we apply Lemma 2, to obtain $|\alpha| \leqq 1/|z|$,

$$|z - C^{1/n}| \leqq 1/|z|,$$

and (11) follows. Thus all zeros of $p(z)$ lie in the parts (if any) of the n discs $|z - C^{1/n}| \leqq 1$ contained in the closed unit disc plus the parts (if any) of the regions (11) contained in the exterior of the unit disc. Any intersection of the set (11) with the unit circumference lies in one of the n discs, and conversely, any intersection of one of the n discs with the unit circumference lies in one of the sets (11), so the boundaries of the n discs join continuously with the boundaries of the respective n lemniscate regions. However, if we have $|C| > 2^n$, the n discs lie exterior to the unit disc; all zeros of $p(z)$ lie in the n regions consisting each of the closed interior of that oval of a lemniscate (11) containing $C^{1/n}$. It is readily shown by the method of continuity that if these latter n regions are mutually disjoint, each contains precisely one zero of $p(z)$. Even if we have $1 < |C| < 2^n$, it may occur that the point set already described as containing all zeros of $p(z)$ falls into n mutually disjoint closed regions, each bounded in part by an arc in $|z| \leqq 1$ of a circle $|z - C^{1/n}|$ and in part by an arc in $|z| > 1$ of the corresponding lemniscate; in this case, too, each such closed region contains precisely one zero of $p(z)$.

It may be noted that if $|C| < 2^n$ the lemniscate $|z(z - C^{1/n})| = 1$ consists of a single Jordan curve, whereas if $|C| > 2^n$ it consists of two mutually exterior ovals contained in the respective closed discs whose centers are zero and $C^{1/n}$, having the common radius $[C^{1/n}-(C^{2/n}-4)^{1/2}]/2$, a radius less than unity. This radius approaches zero as $|C^{1/n}|$ becomes infinite, in great contrast to the corresponding locus [2, Theorem V] yielded by application of the first part of Lemma 1 but for $|z|>1$, which consists of discs whose centers are the points $C^{1/n}$ and which have the common radius unity.

We state another application of Lemma 2, related to Theorem IV of [2].

THEOREM 2. *Suppose we have* $|\alpha_k-a|\leqq r_1$ *and* $|\beta_k-b|\leqq r_2$, $k=1, 2, \cdots, n$, *with* $\sum\alpha_k = na$, $\sum\beta_k = nb$. *We set* $p(z) \equiv \prod(z-\alpha_k) - A\prod(z-\beta_k)$, *where* A *is an arbitrary constant. Then if* $A\neq 1$ *all zeros of* $p(z)$ *lie in the* n *loci*

$$\left|z-\frac{a-bA^{1/n}}{1-A^{1/n}}\right| \leqq |1-A^{1/n}|^{-1}\left[r_1\cdot\min\left(1,\frac{r_1}{|z-a|}\right) + r_2|A|^{1/n}\cdot\min\left(1,\frac{r_2}{|z-b|}\right)\right], \tag{12}$$

where $A^{1/n}$ *is in turn each* n*th root of* A.

If $A=1$ *and we have*

$$r_1\cdot\min\left(1,\frac{r_1}{|z-a|}\right) + r_2\cdot\min\left(1,\frac{r_2}{|z-b|}\right) > |a-b|, \tag{13}$$

then all zeros of $p(z)$ *lie in the* $n-1$ *loci* (12) *where* $A^{1/n}$ *is in turn each* n*th root of unity except unity. If* $A=1$ *and* (13) *is false, we draw no conclusion concerning the location of* z.

The boundary of the locus (12) consists formally of four algebraic curves. If $r_1=0$ these curves are either circles or lemniscates, perhaps degenerate. If $a=b$ each of these curves is either obviously a circle or (12) has the form

$$|z-a| \leqq B_1 + \frac{B_2}{|z-a|}, \qquad B_1\geqq 0,\ B_2\geqq 0, \tag{14}$$

which can be written

$$\left[|z-a| - \frac{B_1}{2} + \frac{1}{2}(B_1^2+4B_2)^{1/2}\right] \left[|z-a| - \frac{B_1}{2} - \frac{1}{2}(B_1^2+4B_2)^{1/2}\right] \leqq 0. \tag{15}$$

If $B_2=0$, (14) defines a disc; if $B_2>0$, the first factor of (15) is automatically positive for all z; thus (14) defines a circular disc.

The proof of Theorem 2 is similar to that of Theorem 1 and is left to the reader. However, we remark that the proof continually involves equations of the form

$$z-\alpha=A^{1/n}(z-\beta),\quad A^{1/n}\neq 1;\qquad z-\alpha=z-\beta. \tag{16}$$

When α and β are known to lie in given discs, the first of equations (16) implies that z lies in a related disc as in (12); but the second equation implies no conclusion concerning the location of z if the given discs containing α and β are not disjoint, and is an impossibility (z does not exist) if those discs are disjoint.

REFERENCES

1. J. H. Grace, *The zeros of a polynomial*, Proc. Cambridge Philos. Soc. **11** (1902), 352–357.

2. J. L. Walsh, *On the location of the roots of certain types of polynomials*, Trans. Amer. Math. Soc. **24** (1922), 163–180.

3. ———, *The location of the zeros of the derivative of a rational function, revisited*, Abstract 617-244, Notices Amer. Math. Soc. **8** (1961), 447.

4. ———, *The location of the zeros of the derivative of a rational function, revisited*, J. Math. Pures Appl. (to appear).

HARVARD UNIVERSITY

The location of the zeros of the derivative of a rational function, revisited;

By J. L. WALSH (1).

There are in the literature ([1], [2], [3], [4]) a number of theorems relating to the location in the plane of the complex variable of the zeros of the derivative of a rational function, or more generally (the technique is the same) of the zeros of a function of the form

$$(1) \qquad \sum_{k=1}^{m} \frac{m_k}{z-\alpha_k} - \sum_{k=1}^{n} \frac{n_k}{z-\beta_k} \qquad (m_k > 0,\ n_k > 0),$$

where the α_k and the β_k are known to lie in given circular regions; the conclusion then refers to certain new regions shown to contain respectively specified numbers of zeros of the function (1). These theorems have in common the property that they are sharp when applied to the totality of zeros of all possible polynomials satisfying the respective hypotheses, but are not sharp when applied to an individual fixed polynomial.

The principal tool in the proof of such theorems is essentially due to E. Laguerre, but the following formulation ([2], § 1.5.1)

---

(1) This work was supported (in part) by the Air Force Office of Scientific Research.

is more convenient and more intuitively suggestive for our present purposes :

LEMMA 1. — *If the α_k lie in a circular region* D (*i. e., the closed interior or exterior of a circle, or half-plane*) *not containing z, then α as defined by*

$$\sum_{k=1}^{m} \frac{m_k}{z - \alpha_k} = \sum_{k=1}^{m} \frac{m_k}{z - \alpha} \qquad (m_k > 0), \tag{2}$$

also lies in D.

If all the quantities in (2) are replaced by their complex conjugates, the kth term in the first member represents m_k times the vector to z from the inverse of α_k in the unit circle whose center is z; the center of gravity of these m initial points (with weights m_k) is the initial point of a vector to z which is the sum of the m original vectors divided by $\sum m_k$. The inverses of the points α_k lie in the inverse of D, as does their center of gravity, so the inverse α of the center of gravity lies in D.

Since the conjugate of the vectors $\frac{m_k}{z - \alpha_k}$ represents a vector from α_k toward z of magnitude $\frac{m_k}{z - \alpha_k}$, it represents the force at z due to a particle of mass m_k at α_k which repels according to the law of inverse distance. The language of this mechanical analogy (first considered by Gauss) is both convenient and intuitively suggestive, so we shall frequently employ it. The particle α of mass $\sum m_k$ in Lemma 1 is then *the particle equivalent to the m particles α_k*, in the sense that it exerts the same force at z and has the same total mass as the original particles; α lies in D.

The advantage of Lemma 1 is that for z exterior to D the first member of (2) depends no longer on the m parameters α_k but on the single parameter α in D; thus the location of the zeros of (1) can be much more readily studied, often as a relatively simple geometric problem. Clearly only the ratios of the m_k are significant.

When the α_k are allowed to vary independently in D, the locus of α defined by (2) for fixed z is indeed D; but if the α_k are fixed and z is suitably restricted, the locus of α is also restricted; for instance as $|z| \to \infty$ the point α approaches the center of gravity of the α_k

$$\alpha \sum \frac{m_k}{1 - \frac{\alpha_k}{z}} = \sum \frac{m_k \alpha_k}{1 - \frac{\alpha_k}{z}}.$$

1. The object of the present paper is to establish and apply a new lemma analogous to Lemma 1 but where *now the weighted* α_k *are required to have their center of gravity in the center of* D. The new lemma then replaces the old one in application to functions of form (1), but the principle is identical, namely that of replacing a number of weighted points α_k or β_k in D by a single point in D, in the sense of (2), and thereby materially simplifying (1). Whereas the conclusions of the older theorems deal largely with circles, the conclusions of the new theorems deal largely with lemniscates and curves of higher degree. The new results are considerably sharper than the old ones when applied to an individual rational function rather than an entire class of rational functions.

2. We are now in a position to consider our principal lemma :

LEMMA 2. — *If we have* $m_k > 0$, $|\alpha_k| \leqq r$, $|\alpha| > r$, $\sum_1^m m_k \alpha_k = 0$, *then* α *as defined by* (2) *with* $z = a$ *satisfies* $|\alpha| \leqq \frac{r^2}{|a|}$.

In the proof we take $r = 1$ and $\sum_1^m m_k = 1$, which involves no loss of generality; our conclusion becomes

$$|\alpha| \leqq \frac{1}{|a|}. \tag{3}$$

If we choose $m = 2$, $m_1 = m_2$, $\alpha_1 = -\alpha_2 = b e^{i\theta}$, $0 < b \leqq 1$, we have by (2)

$$\frac{1}{a - b e^{i\theta}} + \frac{1}{a + b e^{i\theta}} = \frac{2}{a - \alpha}, \qquad \alpha = \frac{b^2 e^{2i\theta}}{a},$$

so (3) is satisfied. This establishes Lemma 2 for the case $m = 2$, and also shows that for fixed a the locus of α in Lemma 2 must contain the entire disc $|z| \leq \frac{1}{a}$.

We set $z' = \frac{1}{z}$ with $|z'| < 1$, and define $\alpha = \alpha(z')$ as in Lemma 1

$$\sum \frac{m_k}{1 - \alpha_k z'} = \frac{1}{1 - \alpha z'};$$

it follows from Lemma 1 that we have $|\alpha(z')| \leq 1$. Then we have

$$\alpha = \alpha(z') = \frac{\sum \frac{m_k}{1 - \alpha_k z'} - 1}{\sum \frac{m_k z'}{1 - \alpha_k z'}} = \frac{\sum \frac{m_k \alpha_k}{1 - \alpha_k z'}}{\sum \frac{m_k}{1 - \alpha_k z'}}, \qquad \alpha(0) = 0.$$

By Schwarz's Lemma there now follows $|\alpha(z')| \leq |z'|$, which for $z' = \frac{1}{a}$ is (3).

We remark that if also $\alpha'(0) = \sum m_k \alpha_k^2 = 0$, then we have $|\alpha| \leq \frac{1}{a^2}$, and similarly for higher derivatives.

The proof of Lemma 2 as just given is due to suggestions made by Mr. S. Jacobs, Professor L. Carleson, Dr. Z. Rubinstein, and the present writer. This proof is much shorter than the writer's original proof, which is geometric and based on a detailed study for fixed a of the possible positions of the various α_k when the point α lies on the boundary of its locus.

Lemma 2 has been formulated for the case of a finite number of particles in $|z| \leq r$ with center of gravity 0, but holds also for an infinite sum and even for any positive distribution of matter in $|z| \leq r$ with center of gravity 0 which is represented by a Stieltjes line or surface integral. Such a distribution can be represented as the limit of a sum of positive particles, always with 0 as center of gravity; the conclusion of Lemma 2 holds for the particle equivalent to the sum, and hence for the particle equivalent to the limit of the sum.

3. We are now in a position to use Lemma 2 to establish a number of new theorems, analogues of known results which are themselves applications of Lemma 1.

THEOREM 1. — *Suppose we have*

$$|\alpha_k - \alpha_0| \leq r, \qquad m_k > 0, \qquad \sum_1^m m_k = M, \qquad \sum_1^m m_k \alpha_k = M\alpha_0, \qquad |z - \alpha_0| > r.$$

Then if z is a zero of the equation

$$\sum_1^m \frac{m_k}{z - \alpha_k} = A \qquad (A \neq 0), \tag{4}$$

we have also

$$\left| (z - \alpha_0)\left[z - \left(\alpha_0 + \frac{M}{A} \right) \right] \right| \leq r^2. \tag{5}$$

In particular, if

$$p(z) \equiv \prod_1^m (z - \alpha_k),$$

equation (4) *with each* $m_k = 1$ *is* $(z \neq \alpha_k)$

$$\frac{p'(z)}{p(z)} = A.$$

By Lemma 2, equation (4) can be written

$$\frac{M}{z - \alpha} = A, \qquad |\alpha - \alpha_0| \leq \frac{r^2}{|z - \alpha_0|}, \tag{6}$$

so we may write

$$\left| z - \frac{M}{A} - \alpha_0 \right| \leq \frac{r^2}{|z - \alpha_0|},$$

which is (5).

Without the hypothesis $\sum m_k \alpha_k = M\alpha_0$, we can deduce by Lemma 1, equation (6) but now with $|\alpha - \alpha_0| \leq r$, and the conclusion ([3], Theorem VI) is

$$\left| z - \left(\alpha_0 + \frac{M}{A} \right) \right| \leq r. \tag{7}$$

When $\frac{M}{A}$ is large in comparison with α_0, inequality (7) restricts z (with $|z-\alpha_0|>r$) to a certain disc of radius r, whereas (5) restricts z much more severely.

Indeed, let us consider in the (x, y)-plane the lemniscate

$$(8) \qquad |z(z-2x_0)|=s, \qquad 0<s<x_0^2.$$

The intercepts on the x-axis are given by $z^2-2x_0z=\pm s$, whence the four intercepts are $x=x_0\pm(x_0^2\pm s)^{\frac{1}{2}}$. The lemniscate consists of two mutually disjoint ovals, of which one contains the origin and the other the point $2x_0$. The reader will easily prove :

REMARK. — *The closed interior of the lemniscate* (8) *lies in the two closed discs with respective centers*

$$x_0\pm\frac{1}{2}\left[(x_0^2+s)^{\frac{1}{2}}+(x_0^2-s)^{\frac{1}{2}}\right]$$

and common radius

$$\frac{1}{2}\left[(x_0^2+s)^{\frac{1}{2}}-(x_0^2-s)^{\frac{1}{2}}\right].$$

The lemniscate consists of two ovals, each doubly tangent to the boundary of one of these discs. The respective discs and ovals lie also interior to the discs with centers 0 *and* $2x_0$ *and common radius* $s^{\frac{1}{2}}$.

Suppose in Theorem 1 we have $r<\frac{M}{2|A|}$ Then the lemniscate (5) lies in the discs with respective centers

$$\alpha_0+\frac{M}{2A}\pm\frac{|A|}{2A}\left[\left(\frac{M^2}{4|A|^2}+r^2\right)^{\frac{1}{2}}+\left(\frac{M^2}{4|A|^2}-r^2\right)^{\frac{1}{2}}\right]$$

and the common radius

$$\frac{1}{2}\left[\left(\frac{M^2}{4|A|^2}+r^2\right)^{\frac{1}{2}}-\left(\frac{M^2}{4|A|^2}-r^2\right)^{\frac{1}{2}}\right];$$

the second of these discs lies in the given disc $|z-\alpha_0|\leq r$ and has at most a minor role in the conclusion of the theorem.

When A is small $\left(\text{that is to say, when } \frac{M}{A} \text{ is large}\right)$ the first of these discs is remote from α_0 and has a radius which is not only small but approaches zero as $\frac{M}{A} \to \infty$, if r is fixed. This fact stands in great contrast to the disc (7) whose radius does not depend on A and M.

In the sequel we shall limit ourselves primarily to the zeros of the derivatives of various types of rational functions, hence to particles of integral mass. The reader will notice that this is merely a matter of convenience, and that more general results involving particles of mass not necessarily integral follow without change of method.

4. A second application of Lemma 2 will yield

THEOREM 2. — *Let m zeros α_k of a polynomial $p(z)$ of degree $m+n$ lie in the disc* C_1 :

$$|z-\alpha_0| \leq r_1 \qquad \textit{with} \quad \sum_1^m \alpha_k = m\alpha_0$$

and the remaining n zeros β_k lie in the disc C_2 :

$$|z-\beta_0| \leq r_2 \qquad \textit{with} \quad \sum_1^n \beta_k = n\beta_0.$$

If we have

$$|z-\alpha_0| > r_1, \qquad |z-\beta_0| > r_2,$$

and if z is a zero of the derivative $p'(z)$, then we have

$$\left|z - \frac{n\alpha_0 + m\beta_0}{m+n}\right| \leq \frac{1}{m+n}\left[\frac{nr_1^2}{|z-\alpha_0|} + \frac{mr_2^2}{|z-\beta_0|}\right]. \tag{9}$$

The classical equation

$$\frac{p'(z)}{p(z)} \equiv \sum_1^m \frac{1}{z-\alpha_k} + \sum_1^n \frac{1}{z-\beta_k}$$

gives by Lemma 2

$$\frac{m}{z-\alpha} + \frac{n}{z-\beta} = 0,$$

where α and β satisfy respectively

$$|\alpha - \alpha_0| \leqq \frac{r_1^2}{|z - \alpha_0|} \quad \text{and} \quad |\beta - \beta_0| \leqq \frac{r_2^2}{|z - \beta_0|};$$

we have

$$z = \frac{n\alpha + m\beta}{m + n},$$

$$z - \frac{n\alpha_0 + m\beta_0}{m + n} = \frac{n(\alpha - \alpha_0) + m(\beta - \beta_0)}{m + n},$$

which implies (9).

If Lemma 1 is used instead of Lemma 2 we have merely

$$|\alpha - \alpha_0| \leqq r_1, \quad |\beta - \beta_0| \leqq r_2,$$

and (9) is replaced ([2], § 1.5) by

$$(10) \qquad \left| z - \frac{n\alpha_0 + m\beta_0}{m + n} \right| \leqq \frac{nr_1 + mr_2}{m + n}.$$

If the disc (10) is disjoint from C_1 and C_2, and if z satisfies the hypothesis of the theorem, we have by (10)

$$|z - \alpha_0| \geqq \frac{m|\alpha_0 - \beta_0|}{m + n} - \frac{nr_1 + mr_2}{m + n}$$

and a similar inequality for $|z - \beta_0|$. Consequently (9) may be replaced by

$$(11) \quad \left| z - \frac{n\alpha_0 + m\beta_0}{m + n} \right| \leqq \frac{nr_1^2}{m|\alpha_0 - \beta_0| - nr_1 - mr_2} + \frac{mr_2^2}{n|\alpha_0 - \beta_0| - mr_2 - nr_1}.$$

The second member of (11) is small if $|\alpha_0 - \beta_0|$ is relatively large, and indeed approaches zero as $|\alpha_0 - \beta_0| \to \infty$ while m, n, r_1, r_2 remain fixed, in contrast to the second member of (10). Of course the method of proof of (11) may be iterated.

In the special case $r_1 = 0$ of Theorem 2, inequality (9) defines the closed interior of a lemniscate, to which the above Remark may be applied.

The method of continuity due to Bôcher ([2], § 4.2.3) enables us to draw some conclusions regarding the number of zeros in a region. Under the conditions of the first part of Theorem 1, with $r < \frac{M}{2|A|}$, if the α_k are all distinct, equation (4) has precisely m roots, which vary continuously with the α_k. When two α_k coalesce, a root of (4) coalesces with them. If all the α_k remain in their assigned region and vary continuously with center of gravity α_0 so as to coincide at α_0 [say replace each α_k by $\alpha_0 + (\alpha_k - \alpha_0)\rho$ where ρ is real and varies continuously from 1 to 0], precisely $m - 1$ roots of (4) have disappeared ond one root lies in $z = \alpha_0 + \frac{M}{A}$. The locus (5) falls into the closed interiors of two ovals, one of which lies interior to C : $|z - \alpha_0| = r$ and the other R is exterior to C and contains $z = \alpha_0 + \frac{M}{A}$. During this variation, no root of (4) can cross C, so *in the original configuration there were precisely* $m - 1$ *zeros of* (4) *in* C *and one in* R. Thus in the second part of Theorem 1, if $r < \frac{m}{2|A|}$, the oval R contains precisely one zero of $Ap(z) - p'(z)$ and C contains $m - 1$ zeros. Incidentally, it is obvious by Lemma 1 or Lemma 2 that if $A = 0$ in Theorem 1 all zeros of (4) and of $p'(z)$ lie in C; this is essentially the theorem of Lucas concerning the zeros of the derivative of a polynomial.

By the method of continuity it follows similarly that in Theorem 2 if the locus (9) plus C_1 and C_2 falls into three mutually disjoint point sets containing respectively the points α_0, β_0, and $\frac{n\alpha_0 + m\beta_0}{m+n}$, these sets contain the following numbers of zeros of $p'(z)$: $m-1$, $n-1$, 1.

5. Theorem 2 admits interesting applications in certain cases of symmetry, of which we give an illustration :

Theorem 3. — *Let the real polynomial* $p(z)$ *of degree* $2m$ *have* m *zeros in* C_1 : $|z - bi| < r$ *with their center of gravity in* bi *and* m *zeros in* C_2 : $|z + bi| \leq r$ *with their center of gravity in* $-bi$, *where b is positive,* $r < (2 - 2^{\frac{1}{2}})b$. *Then there is only one real zero of* $p'(z)$, *which lies on the interval*

$$(12) \qquad |z| \leq \frac{\left[(4r^4 + b^4)^{\frac{1}{2}} - b^2\right]^{\frac{1}{2}}}{2^{\frac{1}{2}}};$$

all other zeros of $p'(z)$ *lie in* C_1 *and* C_2.

Here (9) takes the form

$$(9') \qquad |z| \leq \frac{r^2}{2}\left[\frac{1}{|z - bi|} + \frac{1}{|z + bi|}\right],$$

a locus which is symmetric in both axes and contains the points $\pm bi$ and the origin in its interior. A half line $y =$ Cte, $x \geq 0$, can intersect the boundary of the locus in at most one point, for as x increases the first member increases monotonically and the second member decreases monotonically. To find the intersection y_0 of the boundary with the positive y-axis, $0 < y_0 < b - r$, we have

$$y_0 = \frac{r^2}{2}\left[\frac{1}{b - y_0} + \frac{1}{b + y_0}\right], \qquad y_0^3 - b^2 y_0 + br^2 = 0,$$

and there follows $y_0 < b - r$ provided $r < (2 - 2^{\frac{1}{2}})\,b$. The closed region belonging to (9′) and containing the origin is disjoint from the remainder of (9′), and by virtue of this last inequality on r, the point $z = (b - r)\,i$ does not lie in the locus (9′); any other point z on the circumference $|z - bi| = r$ has the same value for $|z - bi|$ and larger values for $|z|$ and $|z + bi|$, hence does not lie in the locus; no point of the locus lies exterior to that circumference with an ordinate greater than $b + r$ or greater than y_0.

The closed point set determined by (9′) and containing the origin is disjoint from the remainder of (9′), which lies wholly in C_1 and C_2, so by the method of continuity that closed point set is connected and contains precisely one zero x_0 of $p'(z)$, necessarily real. The respective particles in C_1 and C_2 equivalent to the given particles there lie in the circles

$$|z \pm bi| \leq \frac{r^2}{|x_0 - bi|} = \frac{r^2}{(b^2 + x_0^2)^{\frac{1}{2}}},$$

so if x_0 is a position of equilibrium we must have

$$|x_0| \leq \frac{r^2}{(b^2 + x_0^2)^{\frac{1}{2}}},$$

which implies (12). Any configuration which satisfies the hypothesis of Theorem 3 except that $r = \left(2 - 2^{\frac{1}{2}}\right) b$, is the limit of configurations with $r < \left(2 - 2^{\frac{1}{2}}\right) b$, and (12) persists, so in the hypothesis of Theorem 3 we may also write $r \leq \left(2 - 2^{\frac{1}{2}}\right) b$. For the theorem without conditions on the center of gravity of the zeros ([2], § 3.7) one requires $2r < b$, and (12) is replaced by $|x| \leq r$.

THEOREM 4. — *Let* $R(z)\,(\not\equiv \mathrm{Cte})$ *be a rational function whose* m *finite zeros* α_k *satisfy* $|\alpha_k - \alpha_0| \leq r_1$, $\sum_1^m \alpha_k = m\alpha_0$, *and whose* n *finite poles* β_k *satisfy* $|\beta_k - \beta_0| \leq r_2$ *with* $\sum_1^n \beta_k = n\beta_0$. *Suppose* $R'(z) = 0$ *with* $|z - \alpha_0| > r_1$, $|z - \beta_0| > r_2$. *Then* (i) *if* $m \neq n$ *we have*

$$(13) \qquad \left| z - \frac{n\alpha_0 - m\beta_0}{n - m} \right| \leq \frac{1}{|n - m|}\left[\frac{nr_1^2}{|z - \alpha_0|} + \frac{mr_2^2}{|z - \beta_0|}\right]$$

and (ii) *if* $m = n$ *we have*

$$(14) \qquad |\alpha_0 - \beta_0| \leq \frac{r_1^2}{|z - \alpha_0|} + \frac{r_2^2}{|z - \beta_0|}.$$

Since z is a zero of $R'(z)$ we have

$$\sum_1^m \frac{1}{z - \alpha_k} - \sum_1^n \frac{1}{z - \beta_k} = 0,$$

$$(15) \qquad \frac{m}{z - \alpha} - \frac{n}{z - \beta} = 0, \qquad |\alpha - \alpha_0| \leq \frac{r_1^2}{|z - \alpha_0|}, \qquad |\beta - \beta_0| \leq \frac{r_2^2}{|z - \beta_0|},$$

and in case (i)

$$z = \frac{n\alpha - m\beta}{n - m}, \qquad z - \frac{n\alpha_0 - m\beta_0}{n - m} = \frac{n(\alpha - \alpha_0) - m(\beta - \beta_0)}{n - m},$$

so (13) follows. In case (ii) we have from (15) the equation $\alpha = \beta$, whence (14) follows. Neither (13) nor (14) can be satisfied if $|z|$ is sufficiently large, except that if $\alpha_0 = \beta_0$, inequality (14) is satisfied for all $z\,(\neq \alpha_0)$ and the conclusion places no restriction on z and is vacuous. If $\alpha_0 = \beta_0$, inequality (13) becomes

$$|z - \alpha_0|^2 \leq \frac{nr_1^2 + mr_2^2}{|n - m|}.$$

Without the restriction on the centers of gravity of the α_k and β_k, the second member of (13) is replaced ([2], § 4.2.4) by

$$\frac{nr_1 + mr_2}{|n - m|},$$

and (14) is replaced by

$$(16) \qquad |\alpha_0 - \beta_0| \leq r_1 + r_2.$$

Inequality (16) asserts merely that the two discs of the hypothesis intersect; the conclusion that otherwise no zero of $R'(z)$ lies exterior to those discs is well known ([2], § 4.2), as Bôcher's theorem.

If one of the numbers r_1, r_2 is zero, inequality (13) represents the closed interior of a lemniscate and (14) a closed disc. In Theorem 4 the roles of zeros and poles may be reversed.

6. From the general standpoint of circle geometry and linear transformations of the plane, one may formulate the following theorem. *Let there be given a (closed) circular region* C *and two fixed points* Z *and* z *exterior to* C. *Let a number of particles have* C *as their locus with the requirement that their center of gravity with respect to* Z *shall be the inverse of* Z *in the boundary of* C. *Then the locus of the center of gravity of these particles with respect to* z *is the closed region not containing* Z *bounded by that circle of the coaxal family determined by the boundary of* C *and the null-circle* Z *which passes through the harmonic conjugate of* z *with*

respect to the intersections with the boundary of C *of the circle through* z *of the conjugate coaxal family.* The center of gravity of weighted particles α_k with respect to Z is defined as the point α such that the linear transformation of the complex variable which carries Z to infinity carries α into the center of gravity of the images of the α_k. It is sufficient to establish this theorem in the special case $Z = \infty$, which is Lemma 2.

So far as concerns applications to the zeros of the derivative of a rational function, Lemma 2 is by far the most important special case, for if a set of points (weighted particles) in the plane is given, their center of gravity is easily found, as is a disc with the center of gravity as center containing the given points. Nevertheless we shall devote some attention to the special case of this theorem when Z is the center of a disc whose closed exterior is C. This case arises naturally in numerous situations where the given points are symmetrically located with respect to Z, and also where those points are obtained by a reciprocal transformation from the case of Lemma 2. Such a transformation yields

Lemma 3. — *Suppose we have*

$$|\alpha_k| \geqq r(>0), \qquad \sum_1^m \frac{m_k}{\alpha_k} = 0, \qquad m_k > 0, \qquad r > |a| > 0;$$

then α *as defined by* (2) *with* $|z| = |a|$ *satisfies* $|\alpha| \geqq \dfrac{r^2}{|a|}$.

The condition $\sum \frac{m_k}{\alpha_k} = 0$ seems somewhat artificial as relating to a center of gravity, although if $|\alpha_k|$ is independent of k that condition is equivalent to $\sum m_k \alpha_k = 0$. But the ordinary center of gravity of variable particles that become infinite has no meaning whatever in the sense of limit unless their behavior is more precisely defined. We proceed to consider some applications of Lemma 3, and shall find companion pieces to a number of results already established.

Theorem 5. — *Let m zeros α_k of a polynomial $p(z)$ satisfy*

$$|\alpha_k - \alpha_0| \leqq r_1, \qquad \sum_1^m \alpha_k = m\alpha_0,$$

and the remaining n zeros β_k satisfy

$$|\beta_k - \beta_0| \geqq r_2, \qquad \sum_1^n \frac{1}{\beta_k - \beta_0} = 0.$$

If $p'(z) = 0$, $|z - \alpha_0| > r_1$, $|z - \beta_0| < r_2$, then we have

$$\left| z - \frac{n\alpha_0 + m\beta_0}{m+n} \right| \geqq \frac{1}{m+n} \left[\frac{mr_2^2}{|z - \beta_0|} - \frac{nr_1^2}{|z - \alpha_0|} \right]. \tag{17}$$

We may write similarly to the proof of Theorem 2

$$z = \frac{n\alpha + m\beta}{m+n}, \qquad |\alpha - \alpha_0| \leqq \frac{r_1^2}{|z - \alpha_0|}, \qquad |\beta - \beta_0| \geqq \frac{r_2^2}{|z - \beta_0|},$$

$$z - \frac{n\alpha_0 + m\beta_0}{m+n} = \frac{n(\alpha - \alpha_0) + m(\beta - \beta_0)}{m+n},$$

which implies (17). If $\alpha_0 = \beta_0$, inequality (17) reduces to

$$|z - \alpha_0|^2 \geqq \frac{mr_2^2 - nr_1^2}{m+n}. \tag{18}$$

Inequality (18) defines no useful locus unless

$$r_1 < \frac{mr_2^2 - nr_1^2}{m+n} < r_2.$$

Theorem 5 is a companion piece to Theorem 2, and is sharper than the result ([2], § 3.1.1) found by omitting the conditions on the centers of gravity, in which the second member of (17) is replaced by

$$\frac{mr_2 - nr_1}{m+n}.$$

If $r_1 = 0$ inequality (17) represents the exterior of a lemniscate. The following elementary remark is perhaps of interest. Let P_1 and P_2 be distinct fixed points and P a variable point, and let $PP_1 \cdot PP_2 = h$ be a lemniscate of two ovals, of which one contains P_2 in its interior and of which the points A and B are respectively the nearest to P_1 and the farthest from P_1. Then the circle whose center is P_2 and radius P_2A contains in its interior the closed interior of the oval, and the circle whose center is P_2 and radius P_2B contains in its closed exterior the closed exterior of the oval; for if P lies on the oval we have

$$P_1A \leqq PP_1 \leqq P_1B, \quad P_2B = \frac{h}{P_1B} \leqq PP_2 = \frac{h}{PP_1} \leqq \frac{h}{P_1A} = P_2A.$$

This remark is more easily applied than the previous remark, and pertains to the exterior as well as the interior of a lemniscate.

A companion piece to Theorem 4, proved by this same method, is

THEOREM 6. — *Let* $R(z)$ ($\not\equiv$ Cte) *be a rational function whose m finite zeros* α_k *with center of gravity* α_0 *satisfy*

$$|\alpha_k - \alpha_0| \leqq r_1$$

and whose n finite poles β_k *satisfy*

$$\sum_1^n \frac{1}{\beta_k - \beta_0} = 0, \qquad |\beta_k - \beta_0| \geqq r_2.$$

If $R'(z) = 0$ *with* $|z - \alpha_0| > r_1$, $|z - \beta_0| < r_2$, *then* (i) *if* $m \neq n$ *we have*

$$\left| z - \frac{n\alpha_0 - m\beta_0}{n - m} \right| \geqq \frac{1}{|n - m|} \left[\frac{mr_2^2}{|z - \beta_0|} - \frac{nr_1^2}{|z - \alpha_0|} \right], \tag{19}$$

and (ii) *if* $m = n$ *we have*

$$|\alpha_0 - \beta_0| \geqq \frac{r_2^2}{|z - \beta_0|} - \frac{r_1^2}{|z - \alpha_0|}. \tag{20}$$

Compare here ([2], § 4.4).

Case (ii) when applicable is more precise than Bôcher's Theorem, which without any requirement on centers of gravity asserts that if

$$|z - \alpha_0| \leqq r_1 \quad \text{and} \quad |z - \beta_0| \geqq r_2$$

are mutually disjoint, then no zero of $R'(z)$ satisfies both $|z - \alpha_0| > r_1$ and $|z - \beta_0| < r_2$. When $\alpha_0 = \beta_0$, (19) simplifies materially; and (20) reduces to the inequality $r_1^2 \geq r_2^2$. If this inequality is satisfied, it places no restriction on z and the conclusion of the theorem is vacuous; every point of the plane belongs to the locus of zeros of $R'(z)$. If this inequality is not satisfied, the conclusion asserts that if $R'(z) = 0$ with $|z - \alpha_0| > r_1$ and $|z - \alpha_0| < r_2$ an impossibility occurs, so the conclusion is equivalent to the statement that $R'(z) = 0$ is impossible with $r_1 < |z - \alpha_0| < r_2$, a statement that follows also from Bôcher's theorem.

7. In each of the Theorems 2-6 we have essentially determined the actual *locus* of the finite zeros of $p'(z)$ or $R'(z)$, taken for all possible values of m and n. For instance in Theorem 1 the locus of z under the given hypothesis consists of the closed set (5), and if all values of m are admitted also the open disc $|z - \alpha_0| < r$; see below, and compare also the references to [2], especially concerning geometric loci. This is the significance of such an inequality as (14) with $\alpha_0 = \beta_0$ in Theorem 4, which is comparable to (20); the inequality (14) is satisfied for all z for which it has a meaning, the inequality places no restriction on z except that the inequality is not defined when $z = \alpha_0$, and the locus of z is the entire plane with $z = \alpha_0$ deleted. However, in considering the locus of the finite zeros of $R'(z)$, we should note that a point of the boundary of one of the given circular regions required to contain poles of $R(z)$ is not part of the locus of zeros of $R'(z)$ unless it occurs in a supplementary point set derived by the theorem itself; on this topic the reader may compare ([2], § 4.2.2). For instance, in Theorem 2 any point of C_1 or C_2 can be a multiple zero of $p(z)$ and hence a zero of $p'(z)$. Moreover, if z satisfies (9) there exist α and β satisfying all the conditions involved in the proof of Theorem 2; it will be recalled that Lemma 2 determines the actual *locus* of α under all conditions of its hypothesis.

However, Lemma 2 does not determine the locus of α if all the masses are unity and their number is odd, and the applications of Lemma 2 are not sharp if an odd number of unit particles are prescribed to lie in a given circular region. It would be worth-while to sharpen Lemma 2 in this respect.

Each of the Theorems 2, 4, 5, 6 is analogous to an older theorem involving certain circular regions required to contain all or some of the zeros or poles of a rational function $R(z)$, without any restriction on a center of gravity, a theorem asserting that all zeros of $R'(z)$ lie on a specified point set, and whose proof depends essentially on Lemma 1. All such older theorems are contained in a very general theorem due to Marden [1], which is especially emphasized in ([4], § 23). In a similar way, each of the Theorems 2, 4, 5, 6 is also a special case of a new very general theorem, where in the hypotheses we now adjoin the conditions of Lemmas 2 and 3 on the centers of gravity — or still more generally where we impose those conditions on some sets of zeros and poles of $R(z)$ but not on others; for instance, we may use those conditions only where finite circular regions are involved, and otherwise be content with use of Lemma 1. In each case the result is proved by the replacement of each of various sets of particles by a single equivalent particle, which materially simplifies the study of the locus of the zeros of $R'(z)$. Every such new locus is defined by the same formulas as in Marden's theorem in their original (i. e., unsimplified) form, except that now each circle bounding a region to which Lemma 2 or Lemma 3 is applied has in effect a radius that depends on z. Certain higher plane curves are involved in this locus, curves that also deserve further study both in detail and by consideration of approximating curves as we have done for lemniscates in the remark following Theorem 1. Special cases, especially those involving symmetry, are also of interest.

For definiteness consider now the situation of Theorems 2 and 5, which are more or less typical. It is interesting to compare the locus L_1 of zeros of $p'(z)$ in the analogue of Theorem 2 or 5 but by use of Lemma 1, with the locus L_2 found by use of Lemmas 2 and 3. It is clear that L_2 is a subset of L_1, and it follows by the method of continuity that if L_1 consists of N mutually disjoint closed sets then L_2 consists of a least N such subsets, at least one in each of the subsets of L_1. The method shows also that the subsets of L_2 as well as those of L_1 contain fixed characteristic numbers of the zeros of $p'(z)$, numbers independent of $p(z)$ satisfying the hypothesis of the theorem. Each locus L_1 or L_2 contains all the zeros of $p'(z)$ when the given zeros of $p(z)$ lie at the centers of their assigned discs (at infinity if the assigned regions are infinite), and the numbers of the zeros of this special $p'(z)$ (including those that may tend to infinity) in the various subsets of L_1 and L_2 are characteristic of the subsets.

By virtue of the fact that Lemma 2 extends to more general distributions of mass, as does Lemma 3, those lemmas in extended form apply directly to the study of the location of critical points of harmonic and analytic functions, precisely as an extension of Lemma 1 applies ([2], chap. VI-IX). The harmonic functions may be for instance harmonic measures, or functions with logarithmic poles such as Green's functions and their linear combinations; the analytic functions may be those corresponding to the harmonic functions just mentioned, or entire or meromorphic functions. The results are then analogues and broad generalizations of the present Theorems 1-6, and of the results in [2].

REFERENCES.

[1] M. Marden, *The geometry of the zeros of a polynomial in a complex variable* [*Math. Surveys* (*Amer. Math. Soc.*), vol. 3, 1949].

[2] J. L. Walsh, *The Location of Critical Points*, Colloquium Publications of the American Mathematical Society, vol. 34, 1950.

[3] J. L. Walsh, *On the location of the roots of certain types of polynomials* *Trans. Amer. Math. Soc.*, vol. 24, 1922, p. 163-180).

[4] W. Specht, *Algebraische Gleichungen mit reellen oder komplexen Koeffizienten*, Enzyklopädie der mathematischen Wissenschaften, Stuttgart, 1958.

(Manuscrit reçu le 11 août 1962.)

AN INEQUALITY FOR THE ROOTS OF AN ALGEBRAIC EQUATION.

BY J. L. WALSH.

This note gives an inequality for the roots of an algebraic equation which can be proved from the following theorem:

If the roots of the equation

$$x^m + c_1 x^{m-1} + \cdots + c_m = 0$$

lie on or within a circle C whose radius is r and whose center is the point $x = \alpha$, then all the roots of the equation

$$x^m + c_1 x^{m-1} + \cdots + c_m = M \tag{1}$$

lie on or within the m circles whose centers are the m points $\alpha + M^{\frac{1}{m}}$ and *whose common radius is r.**

In the application of this theorem we shall use merely the fact that the roots of (1) lie on or within the circle whose center is α and radius $r + |M^{\frac{1}{m}}|$.

All roots of the equation

$$x^2 + a_1 x = 0$$

lie on or within the circle whose center is the origin and radius $|a_1|$. Then all roots of

$$x^2 + a_1 x + a_2 = 0$$

lie on or within the circle whose center is the origin and radius $|a_1| + \sqrt{|a_2|}$. Hence all roots of

$$x^3 + a_1 x^2 + a_2 x = 0$$

lie on or within the same circle, so all roots of

$$x^3 + a_1 x^2 + a_2 x + a_3 = 0$$

* Walsh, Trans. Amer. Math. Soc., vol. 24 (1922).

lie on or within the circle whose center is the origin and radius $|a_1| + \sqrt{|a_2|} + \sqrt[3]{|a_3|}$.

Continuation of this reasoning shows that all roots of the equation

$$x^n + a_1 x^{n-1} + a_2 x^{n-2} + \cdots + a_n = 0 \tag{3}$$

lie on or within the circle whose center is the origin and whose radius is

$$|a_1| + \sqrt{|a_2|} + \sqrt[3]{|a_3|} + \cdots + \sqrt[n]{|a_n|}. \tag{4}$$

This upper limit (4) for the moduli of the roots of (3) is actually attained when all but one of the coefficients a_i vanish, and is thus attained not merely for a single type of equation (3) but for n types of such equations. The upper limit (4) is particularly useful when all but one of the coefficients a_i are small when compared with the remaining coefficient.

If we use the fact that all the roots of equation (2) lie on or within the circle whose center is $-a_1/2$ and radius $|a_1/2|$, we find that all the roots of (3) lie on or within the circle whose center is $-a_1/2$ and whose radius is

$$\left|\frac{a_1}{2}\right| + \sqrt{|a_2|} + \sqrt[3]{|a_3|} + \cdots + \sqrt[n]{|a_n|}.$$

Commentary by Q.I. Rahman

Comments on [18*], [21-a*], and [21-b*]

[18*] and [21-b*] are amongst the earliest papers dealing with extensions of the well-known Gauss-Lucas theorem to rational functions.

Professor Walsh wrote several other papers on this topic and the large number of results obtained by him constitute a comprehensive study of the critical points of the quotient, product, linear combinations, etc. of two and more polynomials: [22-d], [22-h], [39b], [46-g], [64-j*], [65-a].

THEOREM I of [21-a*] and THEOREM III of [21-b*] or rather the special case of the latter mentioned on pages 114–115 [equivalently, pages 28–29 of this selected works] are referred to as Walsh's two-circle theorems. They have been found useful in the study of a variety of questions. As example, we mention Kakeya's problem [1] seeking bounds for $k-1$ critical points of a polynomial in terms of a bound for k of its zeros. For an elegant proof of Walsh's two-circle theorem for polynomials and a contribution to Kakeya's problem see [6]. Extended versions of the two-circle theorem have been used in the study of the moduli of the zeros of polynomials [5]. For extensions to the case of a rational function whose zeros and poles are distributed over any finite number of prescribed circular regions see [2, 3].

LEMMA I of [21-b*] is equivalent to a well-known theorem of Laguerre (see [4, pp. 49–50]). For remarks on this result see the comments on [22-g*] where it appears as THEOREM II.

THEOREM II in [21-b*] is called Walsh's cross-ratio theorem. For related results involving an arbitrary finite number of circular regions see [4, §21].

The ideas contained in the earlier papers of Walsh on the location of critical points of polynomials and rational functions apply to the critical points of an arbitrary analytic function, of Green's function and of other harmonic functions. His book [7] is an excellent reference for developments in these directions.

References

[1] S. Kakeya, *On zeros of a polynomial and its derivatives*, Tôhoku Math. J. **11** (1917), 5–16.

[2] M. Marden, *On the zeros of linear partial fractions*, Trans. Amer. Math. Soc. **32** (1930), 81–109.

[3] M. Marden, *On the zeros of the derivative of a rational function*, Bull. Amer. Math. Soc. **42** (1936), 400–405.

[4] M. Marden, *Geometry of Polynomials*, Math. Surveys No. 3, Amer. Math. Soc. R. I., 1966.

[5] P. Montel, *Sur les modules des zeros des polynomes*, Ann. École Norm. Sup. (3) **40** (1923), 1–34.

[6] A. M. Ostrowski, *On the moduli of zeros of derivatives of polynomials*, J. Reine Angew. Math. **230** (1968), 40–50.

[7] J. L. Walsh, *The Location of Critical Points of Analytic and Harmonic Functions*, Amer. Math. Soc. Colloq. Publ. vol. 34, Providence, R. I., 1950.

Comments on [20-c*] and [33-i*]

About the same time that Walsh's paper [20-c*] appeared Echols [3] came up with another proof of Jensen's theorem. A bit later Sz.-Nagy [7] proved that *if* g *is a polynomial with real coefficients or a canonical product of genus* 0 *or* 1 *such that* $g(x) \in \mathbf{R}$ *for* $x \in \mathbf{R}$, *then the non-real critical points of*

$$f(z) := e^{-az^2+bz}g(z), \quad a, b \in \mathbf{R}, a \geq 0$$

lie on or inside the Jensen circles of f. For analogous results about the imaginary zeros of the n-th derivative of a real "entire function of genus 1*" and some interesting applications see [5, 6].

It was shown by de Bruijn [2] that *if for each non-real zero* $z_\mu = x_\mu + iy_\mu$ *of a real polynomial* f *and* $h \in \mathbf{R}$ *we define*

$$D_\mu(h) := \{z \in \mathbf{C} : |z - x_\mu|^2 \leq y_\mu^2 - h^2\}$$

($D_\mu(h) = \emptyset$ *if* $|h| > |y_\mu|$), *then for any* $h \in \mathbf{R}\backslash\{0\}$ *the non-real zeros of*

$$f(z + ih) - \gamma f(z - ih), \text{ where } \gamma \in \mathbf{C}, |\gamma| = 1,$$

lie in the union of the disks $D_\mu(h)$. Jensen's theorem can be easily deduced from this result. Some further extensions and applications are also given in [2].

In addition to [20-c*] and [33-i*] several other papers of Walsh deal with analogues, refinements and extensions of Jensen's theorem. As illustration we quote the following result from [33-f].

Let R be an infinite region whose boundary B is a finite Jordan configuration, and let $G(x, y)$ *be the Green's function for R with pole at infinity. Let B be symmetric in the axis of reals and let Jensen circles be constructed with diameters the segments joining all possible pairs of points of B symmetric in the axis of reals. Then every non-real critical point of* $G(x, y)$ *in R lies on or interior to at least one of these Jensen circles.*

According to one of the results proved in [46-g] *if f is a real rational function which has no poles interior to the unit circle C and whose zeros are inverse to its poles with respect to C, then all non-real zeros of* f' *interior to C lie on or within the non-Euclidean circles constructed using non-Euclidean segments joining pairs of conjugate imaginary zeros of f as non-Euclidean diameters.*

The main result proved in [55-c] is the following:

Let all the zeros of the real polynomial f lie in the strip $S : \alpha < \operatorname{Re} z < \beta$. *Let* γ *be an arbitrary real point not in* (α, β) *and through the pair* $z_\mu, \overline{z_\mu}$ *of conjugate imaginary zeros of f draw the circle* $\Gamma_{\mu,\gamma}$ *tangent to the line* γz_μ *at* z_μ. *If* D_μ *denotes the intersection of the closed disks bounded by* $\Gamma_{\mu,\gamma}$, *where* γ *takes all real values not belonging to* (α, β), *then the non-real zeros of* f' *must lie in the union of the sets* D_μ.

The following result appears in [61-e].

Let $r(z)$ *be a real rational function not identically constant all of whose finite zeros lie in the half-plane* $x > 0$ *and all of whose finite poles lie in the half-plane* $x < 0$, *except that* $z = 0$ *may be a zero or a pole. Let a circle* $\Gamma(z_k)$ *be drawn with center on the axis of reals passing through each conjugate pair* $(z_k, \overline{z_k})$ *of zeros and of poles, where* $\Gamma(z_k)$ *is tangent at* z_k *to the line* Oz_k. *Then all non-real zeros of* $r'(z)$ *lie in the closed interiors of the* $\Gamma(z_k)$.

Amongst other papers inspired by Jensen's theorem are [1] and [4].

References

[1] G. Anchochea, *Sur les polynômes dont les zéros sont symmétriques par rapport á un contour circulaire*, C. R. Acad. Sci., Paris **221** (1945) 13–15.

[2] N. G. Bruijn, *The roots of trigonometric integrals*, Duke Math. J. **17** (1950), 197–226.

[3] W. H. Echols, *Note on the roots of the derivative of a polynomial*, Amer. Math. Monthly, **27** (1920), 299–300.

[4] P. Montel, *Remarque sur la note précédente*, C. R. Acad. Sci. Paris **221** (1945), 15.

[5] G. Pólya, *Some problems connected with Fourier's work on transcendental equations*, Quart. J. Math., Oxford Ser., **1** (1930), 21–34.

[6] G. Pólya, *Über die Realität der Nullstellen fast aller Ableitungen gewisser ganzer Funktionen*, Math. Ann. **114** (1937), 622–634.

[7] J. von Sz.-Nagy, *Zur Theorie der algebraischen Gleichungen*, Jber. Deutsch. Math.-Verein. **31** (1922), 238–251.

Comments on [22-g*]

1. THEOREM I is a result of fundamental importance. It is referred to as *Walsh's coincidence theorem*. Without loss of generality, it may be stated as follows (see [11, pp. 62–63]):

Let Φ be a symmetric n-linear form of total degree n in $z_1, z_2, \ldots, z_n$ and let C be a circular domain containing the n points $z_1^{(0)}, z_2^{(0)}, \ldots, z_n^{(0)}$. Then in C there exists at least one point ζ such that

$$\Phi(\zeta, \zeta, \ldots, \zeta) = \Phi(z_1^{(0)}, z_2^{(0)}, \ldots, z_n^{(0)}).$$

As shown by Curtiss [4] it is equivalent to the following theorem due to Grace [8] and we shall therefore call it the Grace-Walsh (or simply G-W) theorem.

If

$$f(z) = \sum_{k=0}^{n} C(n,k)A_k z^k, \qquad g(z) = \sum_{k=0}^{n} C(n,k)B_k z^k, \qquad A_n B_n \neq 0$$

are apolar, i.e. their coefficients satisfy the equation

$$A_0B_0 - C(n,1)A_1B_{n-1} + C(n,2)A_2B_{n-2} - \cdots + (-1)^n A_n B_0 = 0,$$

then each circular domain containing all the zeros of one contains at least one zero of the other.

Various other formulations of the G-W theorem are known [6, pp. 11–12]. It has a variety of applications as amply illustrated in [18] and [12, pp. 16–27]. If f is a *real-valued* differentiable function on $[-1, 1]$ such that $f(-1) = f(1)$, then according to Rolle's theorem f' must have a zero in $(-1, 1)$. As the example $e^{i\pi z}$ shows, this may not hold if f is allowed to take complex values on $[-1, 1]$. However, from the G-W theorem it follows [8, 18, Satz 5] that *if f is an arbitrary polynomial assuming the same value at the points $-1, +1$ then f' must have a zero in each of the two half-planes* $\{z : \text{Re } z \leq 0\}$, $\{z : \text{Re } z \geq 0\}$. Szegő [18, Section 23] also used it to give an alternative proof of the fact [9] that *if f is a polynomial of degree at most n such that $f'(z) \neq 0$ in $|z| < 1$, then f is univalent in* $|z| < \sin(\pi/n)$. The two results just mentioned were used in [3] to show that *if $f(z) := c\prod_{\nu=1}^{n}(z - z_\nu)$ has all its zeros in $|z| \leq 1$, then each of the disks* $|z - z_\nu| \leq 2^{1/n} \leq 1 + 1/n, (\nu = 1, \ldots, n)$ *contains at least one zero of f'*. This is the closest anybody has come to proving Sendov's conjecture for polynomials of arbitrary degree n. According to the conjecture, each disk $D_\nu := \{z : |z - z_\nu| \leq 1\}, (\nu = 1, \ldots, n)$ contains at least one zero of f'. Although the conjecture remains open for polynomials of degree ≥ 7 it was shown by Gacs

[7] that a zero of f' can be found even in

$$\Delta_\nu := \left\{ z : \left| z - \frac{z_\nu}{2} \right| \le 1 - \frac{|z_\nu|}{2} \right\} \subset D_\nu, \nu = 1, \ldots, n \qquad \text{if } n \le 5.$$

The G-W Theorem (in the form of coincidence theorem) played an important part in his reasoning.

2. Let $z \in \mathbf{C}$ be an arbitrary point different from all the zeros of a polynomial f of degree n. According to a famous theorem of Laguerre [10], *if α is a root of the equation*

$$\frac{f'(z)}{f(z)} = \frac{n}{z - \alpha}$$

and γ a circle (or a straight line) passing through z and α, then each of the (two) open circular domains bounded by γ contains a zero of f unless all the zeros lie on γ. Although THEOREM II can be easily deduced from Laguerre's, it remains a result of fundamental importance. It has numerous applications. The theorem says, in particular, that if all the zeros of a polynomial f lie in $|z| \ge 1$, then to each z in the open unit disk U there corresponds a number $\phi(z)$ of $\overline{U}$ such that

$$\frac{f'(z)}{f(z)} = \frac{n}{z - \frac{1}{\phi(z)}}.$$

Dieudonné [5] made the simple but important observation that as a function of z, $\phi(z)$ is holomorphic in U. With this it becomes possible to apply results from the theory of bounded analytic functions to the study of polynomials (see for example [5, 17]).

It was shown by Ostrowski [13] that Walsh's two-circle theorem [21-a*, Theorem I] can be easily deduced from THEOREM II.

The significance of the circular regions in the statement of THEOREM II is underlined by the following result of Walsh proved in [22-h, Theorem III].

If a circular region C has the property that the force at any external point P due to every set of k unit particles in C is equivalent to the force at P due to k unit particles coinciding at some point of C, then C is a circular region.

3. THEOREM III is also a very important result. It says, in particular, that if $f(z) := \prod_{\nu=1}^{n}(1 - z\zeta_\nu)$ is a polynomial of degree n having all its zeros in $|z| \ge 1$, then for each z in the open unit disk U there exists a number ζ belonging to $\overline{U}$ such that $f(z) = (1 - z\zeta)^n$. Again, it was Dieudonné [5] who made the observation that ζ may be chosen so as to define a function $\zeta(z)$ holomorphic in U. THEOREM III and its variants have been used in the study of the zeros of linear combinations of two or more polynomials. The case of two polynomials is completely settled by THEOREM IV; for a result concerning several polynomials see [15]. Some other applications of THEOREM III can be found in [17].

4. THEOREM VII may be seen as a converse of the Gauss-Lucas theorem. The result was generalized by Biernacki [2] who proved that *if all the zeros of a*

polynomial f lie in a convex region K containing point a, then all the zeros of $F(z) = \int_a^z f(t)dt$ lie in the domain bounded by the envelope of all the circles passing through a and having centers on the boundary of K.

5. THEOREM IX is a very general result about the location of the zeros of the linear combinations of the product $f_1^{(j)}(z) \cdot f_2^{(n-j)}(z)$ of the derivatives of two polynomials f_1 and f_2. It is capable of many applications. THEOREM IX can be specialized to conclude that if all the zeros of an n-th degree polynomial f lie in $|z| \leq r$ and if all the zeros of the polynomial

$$\Phi(z) := \lambda_0 + C(n,1)\lambda_1 z + \cdots + C(n,n)\lambda_n z^n$$

lie in the circular domain

$$|z| \leq s|z - \tau|, \ s > 0,$$

then all the zeros of the polynomial

$$\sum_{\nu=0}^{n} \frac{1}{\nu!}(\tau z)^{\nu} \lambda_{\nu} f^{(\nu)}(z)$$

lie in the disk $|z| \leq r \max\{1, s\}$. This latter result has been applied [14, 16] to obtain extensions of S. Bernstein's inequality for the derivatives of a trigonometric polynomial.

Editors' Note: Walsh's theorem on the location of the zeros of linear combinations of derivatives of a polynomial (which is reproduced in [11, Theorem 18.1]) was the essential ingredient used by Baratchart, Saff and Wielonsky [1], in proving the convergence of sequences of rational functions that are determined by interpolation to the exponential function. Indeed, Walsh's result enables one to extend the known convergence of Padé approximants (which interpolate the exponential function in the origin) to rational interpolation of the exponential in arbitrary bounded schemes of real points.

References

[1] L. Baratchart, E. B. Saff and F. Wielonsky, *Rational interpolation of the exponential function*, Canad. J. Math., Vol. **XX**, (1995), 1–27.

[2] M. Biernacki, *Sur les zéros des polynômes*, Ann. Univ. Mariae Curie-Sklodowwska Sect. A **9** (1955), 81–98.

[3] B. D. Bojanov, Q. I. Rahman and J. Szynal, *On a conjecture of Sendov about the critical points of a polynomial*, Math. Z. **190** (1985), 281–285.

[4] D. R. Curtiss, *A note on the preceding paper*, Trans. Amer. Math. Soc. **24** (1922), 181–184.

[5] J. Dieudonné, *Sur quelques propriétés des polynômes*, Actualités Sci. Indust. No 114, Hermann, Paris, 1934.

[6] J. Dieudonné, *La théorie analytique des polynômes d'une variable*, Mémor. Sci. Math. No. 93. (1938).

[7] F. Gacs, *On polynomials whose zeros are in the unit disk*, J. Math. Anal. Appl. **36** (1971), 627–637.

[8] J. H. Grace, *The zero of a polynomial*, Proc. Cambridge Philos. Soc. **11** (1902), 352–357.

[9] S. Kakeya, *On zeros of a polynomial and its derivatives*, Tôhoku Math. J., **11** (1917), 5–11.

[10] E. Laguerre, Œuvres, t.1, pp. 48–66, 133–143.

[11] M. Marden, *Geometry of Polynomials*, Math. Surveys No. 3, Amer. Math. Soc., Providence, R. I., 1966.

[12] P. Montel, *Leçons sur les fonctions univalentes ou multivalentes*, Gautier - Villars, Paris, 1933.

[13] A. M. Ostrowski, *On the moduli of zeros of derivatives of polynomials*, J. Reine Angew. Math. **230** (1968), 40–50.

[14] Q. I. Rahman, *Functions of exponential type*, Trans. Amer. Math. Soc. **135** (1969), 295–309.

[15] Q. I. Rahman, *Zeros of linear combinations of polynomials*, Canad. Math. Bull. **15** (1972), 139–142.

[16] Q. I. Rahman and G. Schmeisser, *Les inégalités de Markoff et de Bernstein*, Séminaire de Mathématiques Supérieures. No. 86 (Eté, 1981) Les Presses de l'Université de Montréal, Qué. 1983.

[17] Q. I. Rahman and J. Stankiewicz, *Differential inequalities and local valency*, Pacific J. Math. **54** (1974), 165–181.

[18] G. Szegő, *Bemerkungen zu einem Satz von J. H. Grace über die Wurzeln algebraischer Gleichungen*, Math. Z. **13** (1922), 28–55.

Comments on [64-e*], [64-j*]

Let us recall that if all zeros on a polynomial $f(z) = \prod_{k=1}^{n}(z - a_k)$ lie in $|z| \leq \rho_0$, then [20-c*, THEOREMS II, III] for $|z| > \rho_0$ we have

$$\frac{f'(z)}{f(z)} = \frac{n}{z - \alpha(z)}, \qquad f(z) = (z - \beta(z))^n,$$

where

$$|\alpha(z)| \leq \rho_0, \qquad |\beta(z)| \leq \rho_0. \tag{1}$$

This is all that can be said about $|\alpha(z)|$, $|\beta(z)|$ if z is an arbitrary point outside the disk $|z| \leq \rho_0$. The main conclusion coming out of [64-e*], [64-j*] is that both $\alpha(z)$, $\beta(z)$ approach the center of gravity $c := (\alpha_1 + \cdots + \alpha_n)/n$ of the zeros $\alpha_1, \ldots, \alpha_n$ of f as $|z| \to \infty$. It is easily seen that in order to find out how fast $|\alpha(z) - c|$, $|\beta(z) - c|$ tend to zero as $|z|$ tends to infinity it is desirable to take c as center and consider the smallest disk $|z - c| \leq \rho_c$ containing all the zeros of f. The above mentioned results [20-c*, THEOREMS II, III] imply that if the zeros of f lie in $|z - a| \leq \rho_a$ for any $a \in \mathbf{C}$, then for $|z - a| > \rho_a$ the numbers $\alpha(z)$, $\beta(z)$

satisfy

$$|\alpha(z) - a| \le \rho_a, \qquad |\beta(z) - a| \le \rho_a.$$

However, if a is chosen to be c, then for $|z - c| > \rho_c$

$$|\alpha(z) - c| \le \frac{(\rho_c)^2}{|z - c|}, \qquad |\beta(z) - c| \le \frac{(\rho_c)^2}{|z - c|}. \tag{2}$$

Even more can be said if $|z - c|$ is large. This was shown by Walsh himself in [64-e*] but some refinements are given in [2].

To understand better the significance of choosing c as the center of a circle containing all the zeros of a polynomial f we observe first that c may be taken to be zero. A simple translation does it. Assuming the center of gravity of the zeros of $f(z) := \prod_{k=1}^{n}(z - \alpha_k)$ to be zero amounts to requiring that $nf(z) - zf'(z)$ is a polynomial of degree at most $n - 2$. It also means that if $g(z) := z^n f(1/z) = \prod_{k=1}^{n}(1 - \alpha_k z)$, then $g'(0) = 0$. Given that $|\alpha_k| \le \rho_0$ for $1 \le k \le n$, if

$$\{g(z)\}^{\varepsilon} = 1 + \sum_{k=2}^{\infty} b_{k,\varepsilon} z^k, \ (\varepsilon = 1 \text{ or } -1)$$

then [1], [2]

$$|b_{k,\varepsilon}| \le \left(\frac{n}{2}\right)^{k/2} \rho^k, \qquad k \ge 2. \tag{3}$$

This is somewhat surprising since $|b_{k,1}|$ can be as large as $\binom{n}{k}\rho^k$ if $g'(0)$ is not assumed to be zero. Inequality (3) which is closely related to (2) plays an important role in the proof [1] of the fact that the derivative f' of an entire function has bounded index only if f has bounded value distribution. This illustrates further the importance of choosing the center of gravity $(\alpha_1 + \cdots + \alpha_n)/n$ as the center of a disk containing all the zeros $\alpha_1, \ldots, \alpha_n$ of a polynomial.

For applications of (2), other than those contained in [64-e*], [2], see [64-f] where some extensions have also been indicated.

References

[1] W. K. Hayman, *Differential inequalities and local valency*, Pacific J. Math. **44** (1973), 117–137.

[2] Q. I. Rahman and J. Stankiewicz, *Differential inequalities and local valency*, Pacific J. Math. **54** (1974), 165–181.

Comments on [24-h*]

As an addendum to [24-h*] we mention:

Q.I. Rahman, *A bound for the moduli of the zeros of polynomials*, Canad. Math. Bull. **13** (1970), 541–542.

Chapter 2

Walsh Functions

Reprinted from THE AMERICAN JOURNAL OF MATHEMATICS, Vol. XLV, No. 1, Jan., 1923.

A CLOSED SET OF NORMAL ORTHOGONAL FUNCTIONS.*

BY J. L. WALSH.

Introduction.

A set of normal orthogonal functions $\{\chi\}$ for the interval $0 \leqq x \leqq 1$ has been constructed by Haar,† each function taking merely one constant value in each of a finite number of sub-intervals into which the entire interval (0, 1) is divided. Haar's set is, however, merely one of an infinity of sets which can be constructed of functions of this same character. It is the object of the present paper to study a certain new closed set of functions $\{\varphi\}$ normal and orthogonal on the interval (0, 1); each function φ has this same property of being constant over each of a finite number of sub-intervals into which the interval (0, 1) is divided. In fact each function φ takes only the values $+1$ and -1, except at a finite number of points of discontinuity, where it takes the value zero.

The chief interest of the set φ lies in its similarity to the usual (e.g., sine, cosine, Sturm-Liouville, Legendre) sets of orthogonal functions, while the chief interest of the set χ lies in its *dissimilarity* to these ordinary sets. The set φ shares with the familiar sets the following properties, none of which is possessed by the set χ: the nth function has $n-1$ zeroes (or better, sign-changes) interior to the interval considered, each function is either odd or even with respect to the mid-point of the interval, no function vanishes identically on any sub-interval of the original interval, and the entire set is uniformly bounded.

Each function χ can be expressed as a linear combination of a finite number of functions φ, so the paper illustrates the changes in properties which may arise from a simple orthogonal transformation of a set of functions.

In § 1 we define the set χ and give some of its principal properties. In § 2 we define the set φ and compare it with the set χ. In § 3 and § 4 we develop some of the properties of the set φ, and prove in particular that every continuous function of bounded variation can be expanded in terms of the φ's and that every continuous function can be so developed in the sense not of convergence of the series but of summability by the first Cesàro mean. In § 5 it is proved that there exists a continuous function which

* Presented to the American Mathematical Society, Feb. 25, 1922.

† *Mathematische Annalen*, Vol. 69 (1910), pp. 331–371; especially pp. 361–371.

cannot be expanded in a convergent series of the functions φ. In § 6 there is studied the nature of the approach of the approximating functions to the sum function at a point of discontinuity, and in § 7 there is considered the uniqueness of the development of a function.

§ 1. Haar's Set χ.

Consider the following set of functions:

$$f_0(x) \equiv 1, \qquad 0 \leqq x \leqq 1,$$

$$f_1^{(1)}(x) \equiv \begin{cases} 1, & 0 \leqq x < \frac{1}{2}, \\ 0, & \frac{1}{2} < x \leqq 1, \end{cases} \qquad f_1^{(2)}(x) \equiv \begin{cases} 1, & \frac{1}{2} < x \leqq 1, \\ 0, & 0 \leqq x < \frac{1}{2}, \end{cases}$$

. . . .

$$f_k^{(i)}(x) \equiv \begin{cases} 1, & \frac{i-1}{2^k} < x < \frac{i}{2^k}, \\ 0, & 0 \leqq x < \frac{i-1}{2^k}, \text{ or } \frac{i}{2^k} < x \leqq 1, \end{cases} \qquad \begin{matrix} i = 1, 2, 3, \cdots, 2^k, \\ k = 1, 2, 3, \cdots, \infty; \end{matrix}$$

these functions may be defined at a point of discontinuity to have the average of the limits approached on the two sides of the discontinuity.

If we have at our disposal all the functions $f_k^{(i)}$, it is clear that we can approximate to any continuous function in the interval $0 \leqq x \leqq 1$ as closely as desired and hence that we can expand any continuous function in a uniformly convergent series of functions $f_k^{(i)}$. For a continuous function $F(x)$ is uniformly continuous in the interval $(0, 1)$, and thus uniformly in that entire interval can be approximated as closely as desired by a linear combination of the functions $f_k^{(i)}$ where k is chosen sufficiently large but fixed. The approximation can be made better and better and thus will lead to a uniformly convergent series of functions $f_k^{(i)}$.

Haar's set χ may be found by normalizing and orthogonalizing the set $f_k^{(i)}$, those functions to be ordered with increasing k, and for each k with increasing i. The set χ consists of the following functions:*

$$\chi_0(x) \equiv 1, \qquad 0 \leqq x \leqq 1, \qquad \chi_1(x) \equiv \begin{cases} 1, & 0 \leqq x < \frac{1}{2}, \\ -1, & \frac{1}{2} < x \leqq 1, \end{cases}$$

$$\begin{aligned} \chi_2^{(1)}(x) &= \sqrt{2}, & \chi_2^{(2)} &= 0, & 0 &\leqq x < \tfrac{1}{4}, \\ &= -\sqrt{2}, & &= 0, & \tfrac{1}{4} &< x < \tfrac{1}{2}, \\ &= 0, & &= \sqrt{2}, & \tfrac{1}{2} &< x < \tfrac{3}{4}, \\ &= 0, & &= -\sqrt{2}, & \tfrac{3}{4} &< x \leqq 1, \end{aligned}$$

. . . .

* L. c., p. 361.

$$\begin{aligned}
\chi_n^{(k)} &= \sqrt{2^{n-1}}, & \frac{k-1}{2^{n-1}} < x < \frac{2k-1}{2^n}, & \qquad k = 1, 2, 3, \cdots, 2^{n-1}, \\
&= -\sqrt{2^{n-1}}, & \frac{2k-1}{2^n} < x < \frac{k}{2^{n-1}}, & \qquad n = 1, 2, 3, \cdots, \infty, \\
&= 0, & 0 < x < \frac{k-1}{2^{n-1}} \text{ or } \frac{k}{2^{n-1}} < x < 1. &
\end{aligned}$$

The same convention as to the value of $\chi_n^{(k)}$ at a point of discontinuity is made as for the $f_n^{(k)}$, and $\chi_n^{(k)}(0)$ and $\chi_n^{(k)}(1)$ are defined as the limits of $\chi_n^{(k)}$ as x approaches 0 and 1.

For any particular value of N, all the functions $f_n^{(k)}$, $n < N$, can be expressed linearly in terms of the functions $\chi_n^{(k)}$, $n < N$, and conversely.

Let $F(x)$ be any function integrable and with an integrable square in the interval (0, 1); its formal development in terms of the functions χ is

$$\begin{aligned}
F(x) \sim \chi_0(x) \int_0^1 F(y)\chi_0(y)dy + \chi_1(x) \int_0^1 F(y)\chi_1(y)dy + \cdots \\
+ \chi_n^{(k)}(x) \int_0^1 F(y)\chi_n^{(k)}(y)dy + \cdots .
\end{aligned} \tag{1}$$

This series (1) is formed with coefficients determined formally as for the Fourier expansions, and it is well known that $S_m(x)$, the sum of the first m terms of this series, is that linear combination $F_m(x)$ of the first m of the functions χ which renders a minimum the integral

$$\int_0^1 (F(x) - F_m(x))^2 dx.$$

That is, $S_m(x)$ is in the sense of least squares the best approximation to $F(x)$ which can be formed from a linear combination of the first m functions χ; it is likewise true that $S_m(x)$ is the best approximation to $F(x)$ which can be formed from a linear combination of those functions $f_n^{(k)}$ that are dependent on the first m functions χ.

Let $F(x)$ be continuous in the closed interval (0, 1). If ϵ is any positive number, there exists a corresponding number n such that

$$|F(x') - F(x'')| < \epsilon \qquad \text{whenever} \qquad |x' - x''| < \frac{1}{2^n}.$$

We interpret $S_{2^n}(x)$ as a linear combination of the functions $f_n^{(k)}$. The multiplier of the function $f_n^{(k)}$ which appears in $S_{2^n}(x)$ is chosen so as to furnish the best approximation in the interval $\left(\frac{k-1}{2^n}, \frac{k}{2^n}\right)$ to the function $F(x)$, so it is evident that $S_{2^n}(x)$ approximates to $F(x)$ uniformly in the entire interval (0, 1) with an approximation better than ϵ. The function

$S_{2^{n+1}}(x)$ cannot differ from $F(x)$ by more than ϵ at any point of the interval (0, 1), and so for all the functions $S_{2^{m+1}}(x)$. Thus we have

THEOREM I. *If $F(x)$ is continuous in the interval* (0, 1), *series* (1) *converges uniformly to the value $F(x)$ if the terms are grouped so that each group contains all the 2^{n-1} terms of a set $\chi_n^{(k)}$, $k = 1, 2, 3, \cdots, 2^{n-1}$.*

Haar proves that the series actually converges uniformly to $F(x)$ without the grouping of terms,* and establishes many other results for expansions in terms of the set χ; to some of these results we shall return later.

§ 2. The Set φ.

The set φ, which it is the main purpose of this paper to study, consists of the following functions:

$$\varphi_0(x) \equiv 1, \qquad 0 \leqq x \leqq 1, \qquad \varphi_1(x) \equiv \begin{cases} 1, & 0 \leqq x < \frac{1}{2}, \\ -1, & \frac{1}{2} < x \leqq 1, \end{cases}$$

$$\varphi_2^{(1)}(x) \equiv \begin{cases} 1, & 0 \leqq x < \frac{1}{4},\ \frac{3}{4} < x \leqq 1, \\ -1, & \frac{1}{4} < x < \frac{3}{4}, \end{cases}$$

$$\varphi_2^{(2)}(x) \equiv \begin{cases} 1, & 0 \leqq x < \frac{1}{4},\ \frac{1}{2} < x < \frac{3}{4}, \\ -1, & \frac{1}{4} < x < \frac{1}{2},\ \frac{3}{4} < x \leqq 1, \end{cases}$$

. . . .

$$\varphi_{n+1}^{(2k-1)}(x) \equiv \begin{cases} \varphi_n^{(k)}(2x), & 0 \leqq x < \frac{1}{2}, \\ (-1)^{k+1}\varphi_n^{(k)}(2x-1), & \frac{1}{2} < x \leqq 1, \end{cases} \tag{2}$$

$$\varphi_{n+1}^{(2k)}(x) \equiv \begin{cases} \varphi_n^{(k)}(2x), & 0 \leqq x < \frac{1}{2}, \\ (-1)^{k}\varphi_n^{(k)}(2x-1), & \frac{1}{2} < x \leqq 1, \end{cases}$$

$$k = 1, 2, 3, \cdots, 2^{n-1}, \qquad n = 1, 2, 3, \cdots, \infty.$$

In general, the function $\varphi_n^{(1)}$, $n > 0$, is to be used, with the horizontal scale reduced one half and the vertical scale unchanged, to form the functions $\varphi_{n+1}^{(1)}$ and $\varphi_{n+1}^{(2)}$ in each of the halves $(0, \frac{1}{2})$, $(\frac{1}{2}, 1)$ of the original interval; the function $\varphi_{n+1}^{(1)}(x)$ is to be even and the function $\varphi_{n+1}^{(2)}$ odd with respect to the point $x = \frac{1}{2}$. Similarly the function $\varphi_n^{(k)}$ is to be used to form the functions $\varphi_{n+1}^{(2k-1)}$ and $\varphi_{n+1}^{(2k)}$, the former of which is even and the latter odd with respect to the point $x = \frac{1}{2}$. All the functions $\varphi_n^{(k)}$ are to be taken positive in the interval $\left(0, \frac{1}{2^n}\right)$. The function $\varphi_n^{(k)}$ is to be defined at points of discontinuity as were the functions f and χ, and at $x = 0$ to have the value 1, and at $x = 1$ to have the value $(-1)^{k+1}$.† The function

* L. c., p. 368.

† If it is desired to develop periodic functions by means of the set φ [or the similar sets f and χ] simultaneously in all the intervals $\cdots$, $(-2, -1)$, $(-1, 0)$, $(0, 1)$, $(1, 2)$, $\cdots$, it will be wise to change these definitions at $x = 0$ and $x = 1$ so that always the value of $\varphi_n^{(k)}(x)$ is the arithmetic mean of the limits approached at these points to the right and to the left.

$\varphi_n^{(k)}$ is odd or even with respect to the point $x = \frac{1}{2}$ according as k is even or odd.

The functions φ_0, φ_1, $\varphi_2^{(1)}$, $\varphi_2^{(2)}$ have 0, 1, 2, 3 zeroes (i.e., sign-changes) respectively interior to the interval (0, 1). The function $\varphi_{n+1}^{(2k-1)}(x)$ has twice as many zeroes as the function $\varphi_n^{(k)}$; and $\varphi_{n+1}^{(2k)}(x)$ has one more zero, namely at $x = \frac{1}{2}$, than has $\varphi_{n+1}^{(2k-1)}(x)$. Thus the function $\varphi_n^{(k)}$ has $2^{n-1} + k - 1$ zeroes; this formula holds for $n = 2$ and follows for the general case by induction. Hence each function $\varphi_n^{(k)}$ has one more zero than the preceding; the zeroes of these functions increase in number precisely as do the zeroes of the classical sets of functions—sine, cosine, Sturm-Liouville, Legendre, etc. We shall at times find it convenient to use the notation $\varphi_0, \varphi_1, \varphi_2, \cdots$ for the functions $\varphi_n^{(k)}$; the subscript denotes the number of zeroes.

The orthogonality of the system φ is easily established. Any two functions $\varphi_n^{(k)}$ are orthogonal if $n < 3$, as may be found by actually testing the various pairs of functions. Let us assume this fact to hold for $n = 1, 2, 3, \cdots, N - 1$; we shall prove that it holds for $n = N$. By the method of construction of the functions φ, each of the integrals

$$\int_0^{1/2} \varphi_N^{(k)}(x)\varphi_m^{(l)}(x)dx, \qquad \int_{1/2}^1 \varphi_N^{(k)}(x)\varphi_m^{(l)}(x)dx, \qquad m \leqq N,$$

is the same except possibly for sign as an integral

$$\int_0^1 \varphi_{N-1}^{(j)}(y)\varphi_{m-1}^{(i)}(y)dy$$

after the change of variable $y = 2x$ or $y = 2x - 1$. Each of these two integrals [in fact, they are the same integral] whose variable is y has the value zero, so we have the orthogonality of $\varphi_N^{(k)}(x)$ and $\varphi_m^{(l)}(x)$:

$$\int_0^1 \varphi_N^{(k)}(x)\varphi_m^{(l)}(x)dx = 0.$$

This proof breaks down if the two functions $\varphi_{N-1}^{(j)}(y)$, $\varphi_{m-1}^{(i)}(y)$ are the same, but in that case either $\varphi_N^{(k)}(x)$ and $\varphi_m^{(l)}(x)$ are the same and we do not wish to prove their orthogonality, or one of the functions $\varphi_N^{(k)}(x)$, $\varphi_m^{(l)}(x)$ is odd and the other even, so the two are orthogonal.

Each of the functions $\varphi_n^{(k)}(x)$ is normal, for we have

$$|\varphi_n^{(k)}(x)| \equiv 1$$

except at a finite number of points.

Each of the functions $\chi_0, \chi_1, \chi_2^{(1)}, \chi_2^{(2)}, \cdots, \chi_{n+1}^{(2^n)}$ can be expressed linearly in terms of the functions $\varphi_0, \varphi_1, \varphi_2^{(1)}, \varphi_2^{(2)}, \cdots, \varphi_{n+1}^{(2^n)}$. Thus for $n = 1$ we have

$$\chi_0 = \varphi_0, \quad \chi_1 = \varphi_1, \quad \chi_2^{(1)} = \tfrac{1}{2}\sqrt{2}(\varphi_2^{(1)} + \varphi_2^{(2)}), \quad \chi_2^{(2)} = \tfrac{1}{2}\sqrt{2}(-\varphi_2^{(1)} + \varphi_2^{(2)}).$$

It is true generally that except for a constant normalizing factor $\sqrt{2}$, the function $\chi_{n+1}^{(k)}$, $k \leqq 2^{n-1}$, is the same linear combination of the functions $\frac{1}{2}[\varphi_{n+1}^{(2k-1)} + \varphi_{n+1}^{(2k)}]$ as is $\chi_n^{(k)}$ of the functions $\varphi_n^{(k)}$, and the function $\chi_{n+1}^{(k)}$, $k > 2^{n-1}$, is the same linear combination of the functions $\frac{1}{2}(-1)^{k+1}[\varphi_{n+1}^{(2k-1)} - \varphi_{n+1}^{(k)}]$ as is $\chi_n^{(k-2^{n-1})}$ of the functions $\varphi_n^{(k)}$.

It is similarly true that all the functions $\varphi_0, \varphi_1, \cdots, \varphi_{n+1}^{(2^n)}$ can be expressed linearly in terms of the functions $\chi_0, \chi_1, \cdots, \chi_{n+1}^{(2^n)}$. Thus we have for $n = 2$,

$$\varphi_0 = \chi_0, \quad \varphi_1 = \chi_1, \quad \varphi_2^{(1)} = \tfrac{1}{2}\sqrt{2}(\chi_2^{(1)} - \chi_2^{(2)}), \quad \varphi_2^{(2)} = \tfrac{1}{2}\sqrt{2}(\chi_2^{(1)} + \chi_2^{(2)}).$$

The general fact appears by induction from the very definition of the functions φ.

The set χ is known to be closed;* it follows from the expression of the χ in terms of the φ that the set φ is also closed.

The definition of the functions $\varphi_n^{(k)}$ enables us to give a formula for $\varphi_n^{(k)}(x)$. Let us set, in binary notation,

$$x = \frac{a_1}{2^1} + \frac{a_2}{2^2} + \frac{a_3}{2^3} + \cdots, \qquad a_i = 0 \text{ or } 1.$$

If x is a binary irrational or if in the binary expansion of x there exists $a_i \neq 0$, $i > n$, the following formulas hold for $\varphi_n^{(k)}$:

$$\begin{aligned}
\varphi_0 &= 1, & \varphi_1 &= (-1)^{a_1}, \\
\varphi_2^{(1)} &= (-1)^{a_1+a_2}, & \varphi_2^{(2)} &= (-1)^{a_2}, \\
\varphi_3^{(1)} &= (-1)^{a_2+a_3}, & \varphi_3^{(2)} &= (-1)^{a_1+a_2+a_3}, \\
\varphi_3^{(3)} &= (-1)^{a_1+a_3}, & \varphi_3^{(4)} &= (-1)^{a_3}, \\
\varphi_4^{(1)} &= (-1)^{a_3+a_4}, & \varphi_4^{(2)} &= (-1)^{a_1+a_3+a_4}, \\
\varphi_4^{(3)} &= (-1)^{a_1+a_2+a_3+a_4}, & \varphi_4^{(4)} &= (-1)^{a_2+a_3+a_4}, \\
\varphi_4^{(5)} &= (-1)^{a_2+a_4}, & \varphi_4^{(6)} &= (-1)^{a_1+a_2+a_4}, \\
\varphi_4^{(7)} &= (-1)^{a_1+a_4}, & \varphi_4^{(8)} &= (-1)^{a_4}, \\
&\cdots\cdots\cdots\cdots\cdots\cdots, \\
&\cdots\cdots\cdots\cdots\cdots\cdots.
\end{aligned} \tag{3}$$

The general law appears from these relations; always we have

$$\begin{aligned}
\varphi_n^{(1)} &= (-1)^{a_{n-1}+a_n}, \\
\varphi_n^{(k)} &= \varphi_{k-1}\varphi_n^{(1)}.
\end{aligned} \tag{4}$$

A general expression for $\varphi_n^{(k)}(x)$ when x is a binary rational can readily be computed from formulas (3), for we have expressions for the values of $\varphi_n^{(k)}$ for neighboring larger and smaller values of the argument than x.

* That is, there exists no non-null Lebesgue-integrable function on the interval (0, 1) which is orthogonal to all functions of the set; l. c., p. 362.

§ 3. Expansions in Terms of the Set $\{\varphi\}$.

The following theorem results from Theorem I by virtue of the remark that all the functions $\varphi_n^{(k)}$ can be expressed in terms of the functions $\chi_n^{(i)}$ and conversely, and from the least squares interpretation of a partial sum of a series of orthogonal functions:

THEOREM II. *If* $F(x)$ *is continuous in the interval* $(0, 1)$, *the series*

$$F(x) \sim \varphi_0(x)\int_0^1 F(y)\varphi_0(y)dy + \varphi_1(x)\int_0^1 F(y)\varphi_1(y)dy + \cdots \varphi_i^{(j)}(x)\int_0^1 F(y)\varphi_i^{(j)}(y)dy + \cdots, \tag{5}$$

converges uniformly to the value $F(x)$ *if the terms are grouped so that each group contains all the* 2^{n-1} *terms of a set* $\varphi_n^{(k)}$, $k = 1, 2, 3, \cdots, 2^{n-1}$.

Series (5) after the grouping of terms is precisely the same as series (1) after the grouping of terms.

Theorem II can be extended to include even discontinuous functions $F(x)$; we suppose $F(x)$ to be integrable in the sense of Lebesgue. Let us introduce the notation

$$F(a+0) = \lim_{\epsilon=0} F(a+\epsilon), \qquad F(a-0) = \lim_{\epsilon=0} F(a-\epsilon), \qquad \epsilon > 0,$$

and suppose that these limits exist for a particular point $x = a$. We introduce the functions

$$F_1(x) = \begin{cases} F(x), & x < a, \\ F(a-0), & x \geqq a, \end{cases} \qquad F_2(x) = \begin{cases} F(a+0), & x \leqq a, \\ F(x), & x > a, \end{cases} \tag{6}$$

The least squares interpretation of the partial sums $S_{2^n}(x)$ of the series (1) or (5) as expressed in terms of the $f_i^{(j)}$ gives the result that if $h_1 < F(x) < h_2$ in any interval, then also $h_1 < S_{2^n}(x) < h_2$ in any completely interior interval if n is sufficiently large. It follows that $F_1(x)$ is closely approximated at $x = a$ by its partial sum S_{2^n} if n is sufficiently large, and that this approximation is uniform in any interval about the point $x = a$ in which $F_1(x)$ is continuous. A similar result holds for $F_2(x)$.

The function $F_1(x) + F_2(x)$ differs from the original function $F(x)$ merely by the function

$$G(x) = \begin{cases} F(a+0), & x < a, \\ F(a-0), & x > a. \end{cases}$$

The representation of such functions by sequences of the kind we are considering will be studied in more detail later (§ 6), but it is fairly obvious that such a function is represented uniformly except in the neighborhood of the point a. If $F(x)$ is continuous at and in the neighborhood of a, or if a is dyadically rational, the approximation to $G(x)$ is uniform at the point a as well. Thus we have

THEOREM III. *If $F(x)$ is any integrable function and if $\lim_{x=a} F(x)$ exists for a point a, then when the terms of the series* (5) *are grouped as described in Theorem II, the series so obtained converges for $x = a$ to the value $\lim_{x=a} F(x)$. If $F(x)$ is continuous at and in the neighborhood of a, then this convergence is uniform in a neighborhood of a.*

If $F(x)$ is any integrable function and if the limits $F(a-0)$ and $F(a+0)$ exist for a dyadically rational point $x = a$, then the series with the terms grouped converges for $x = a$ to the value $\frac{1}{2}[F(a+0) + F(a-0)]$; this convergence is uniform in the neighborhood of the point $x = a$ if $F(x)$ is continuous on two intervals extending from a, one in each direction.

It is now time to study the convergence of series (5) when the terms are not grouped as in Theorems II and III. We shall establish

THEOREM IV. *Let the function $F(x)$ be of limited variation in the interval $0 \leqq x \leqq 1$. Then the series* (5) *converges to the value $F(x)$ at every point at which $F(a+0) = F(a-0)$ and at every point at which $x = a$ is dyadically rational. This convergence is uniform in the neighborhood of $x = a$ in each of these cases if $F(x)$ is continuous in two intervals extending from a, one in each direction.*

Since $F(x)$ is of limited variation, $F(a+0)$ and $F(a-0)$ exist at every point a. Theorem IV tacitly assumes $F(x)$ to be defined at every point of discontinuity a so that $F(a) = \frac{1}{2}[F(a+0) + F(a-0)]$.

Any such function $F(x)$ can be considered as the difference of two monotonically increasing functions, so the theorem will be proved if it is proved merely for a monotonically increasing function. We shall assume that $F(x)$ is such a function, and positive. We are to evaluate the limit of

$$\int_0^1 F(y) K_n^{(k)}(x, y) dy,$$

$$K_n^{(k)}(x, y) = \varphi_0(x)\varphi_0(y) + \varphi_1(x)\varphi_1(y) + \cdots + \varphi_n^{(k)}(x)\varphi_n^{(k)}(y).$$

We have already evaluated this limit for the sequence $k = 2^{n-1}$, so it remains merely to prove that

$$\lim_{n=\infty} \int_0^1 F(y) Q_n^{(k)}(x, y) dy = 0, \tag{7}$$

$$Q_n^{(k)}(x, y) = \varphi_n^{(1)}(x)\varphi_n^{(1)}(y) + \varphi_n^{(2)}(x)\varphi_n^{(2)}(y) + \cdots + \varphi_n^{(k)}(x)\varphi_n^{(k)}(y),$$

whatever may be the value of k.

We shall consider the function $F(x)$ merely at a point $x = a$ of continuity; that is, we study essentially the new functions F_1 and F_2 defined by equations (6). In the sequel we suppose a to be dyadically irrational; the necessary modifications for a rational can be made by the reader.

The following formulas are easily found by the definition of the $Q_n^{(k)}$; both x and y are supposed dyadically irrational:

$$Q_2^{(1)}(x, y) = \pm 1,$$

$$Q_2^{(2)}(x, y) = \begin{cases} 0 \text{ if } x < \tfrac12,\ y > \tfrac12 \text{ or if } x > \tfrac12,\ y < \tfrac12, \\ \pm 2 \text{ if } x < \tfrac12,\ y < \tfrac12 \text{ or if } x > \tfrac12,\ y > \tfrac12, \end{cases}$$

$$\cdots\cdots\cdots\cdots$$

$$Q_n^{(1)}(x, y) = \pm 1,$$

$$Q_n^{(2)}(x, y) = \begin{cases} 0 \text{ if } x < \tfrac12,\ y > \tfrac12 \text{ or if } x > \tfrac12,\ y < \tfrac12, \\ 2Q_{n-1}^{(1)}(2x, 2y) \text{ if } x < \tfrac12,\ y < \tfrac12, \\ 2Q_{n-1}^{(1)}(2x-1, 2y-1) \text{ if } x > \tfrac12,\ y > \tfrac12, \end{cases}$$

$$\cdots\cdots\cdots\cdots$$

$$Q_n^{(2k)}(x, y) = \begin{cases} 0 \text{ if } x < \tfrac12,\ y > \tfrac12 \text{ or if } x > \tfrac12,\ y < \tfrac12, \\ 2Q_{n-1}^{(k)}(2x, 2y) \text{ if } x < \tfrac12,\ y < \tfrac12, \\ 2Q_{n-1}^{(k)}(2x-1, 2y-1) \text{ if } x > \tfrac12,\ y > \tfrac12, \end{cases}$$

$$Q_n^{(2k+1)}(x, y) = \begin{cases} \pm 1 \text{ if } x < \tfrac12,\ y > \tfrac12 \text{ or if } x > \tfrac12,\ y < \tfrac12, \\ \dfrac{Q_n^{(2k)} + Q_n^{(2k+2)}}{2} \text{ if } x < \tfrac12,\ y < \tfrac12 \text{ or if } x > \tfrac12,\ y > \tfrac12. \end{cases}$$

The integral in (7) for $x = a$ is to be divided into three parts. Consider an interval bounded by two points of the form $x = \frac{\rho}{2^\nu}$, $x = \frac{\rho+1}{2^\nu}$, where ρ and ν are integers and such that

$$\frac{\rho}{2^\nu} < a < \frac{\rho+1}{2^\nu}.$$

Then we have

$$\int_0^1 F_1(y)Q_n^{(k)}(a, y)dy = \int_0^{\rho/2^\nu} F_1(y)Q_n^{(k)}(a, y)dy$$
$$+ \int_{\rho/2^\nu}^{(\rho+1)/2^\nu} F_1(y)Q_n^{(k)}(a, y)dy + \int_{(\rho+1)/2^\nu}^1 F_1(y)Q_n^{(k)}(a, y)dy. \tag{8}$$

These integrals on the right need separate consideration.

Let us set

$$\frac{\rho}{2^\nu} = \frac{\mu_1}{2^1} + \frac{\mu_2}{2^2} + \frac{\mu_3}{2^3} + \cdots + \frac{\mu_\nu}{2^\nu}, \qquad \mu_i = 0 \text{ or } 1.$$

The first integral in the right-hand member of (8) can be written

$$\int_0^{\mu_1/2^1} + \int_{\mu_1/2^1}^{(\mu_1/2^1)+(\mu_2/2^2)} + \cdots \int_{(\rho/2^\nu)-(\mu_\nu/2^\nu)}^{\rho/2^\nu} F_1(y)Q_n^{(k)}(a, y)dy. \tag{9}$$

Each of these integrals is readily treated. Thus, on the interval $0 \leqq y \leqq \frac{\mu_1}{2^1}$, $Q_n^{(k)}(a, y)$ takes only the values ± 1 or 0, is 0 if k is even and has the value $\pm \varphi_n^{(k)}(y)$ if k is odd. It is of course true that

$$\lim_{n=\infty} \int_0^1 \Phi(y)\varphi_n^{(k)}(y)dy = 0 \tag{10}$$

no matter what may be the function $\Phi(y)$ integrable in the sense of Lebesgue and with an integrable square.* Hence we have

$$\lim_{n=\infty} \int_0^{\mu_1/2^1} F_1(y)Q_n^{(k)}(a, y)dy = 0.$$

On the interval $\frac{\mu_1}{2^1} \leqq y \leqq \frac{\mu_1}{2^1} + \frac{\mu_2}{2^2}$, the function $Q_n^{(k)}(a, y)$ takes only the values $0, \pm 1, \pm 2$, and except for one of these numbers as constant factor, has the value $\varphi_n^{(k)}(y)$. It is thus true that

$$\lim_{n=\infty} \int_{\mu_1/2^1}^{(\mu_1/2^1)+(\mu_2/2^2)} F_1(y)Q_n^{(k)}(a, y)dy = 0.$$

From the corresponding result for each of the integrals in (9) and a similar treatment of the last integral in the right-hand member of (8), we have

$$\begin{aligned} &\lim_{n=\infty} \int_0^{p/2^\nu} F_1(y)Q_n^{(k)}(a, y)dy = 0, \\ &\lim_{n=\infty} \int_{(p+1)/2^\nu}^1 F_1(y)Q_n^{(k)}(a, y)dy = 0. \end{aligned} \tag{11}$$

We shall obtain an upper limit for the second integral in (8) by the second law of the mean. We notice that

$$\left| \int_\xi^{(p+1)/2^\nu} Q_n^{(k)}(a, y)dy \right| \leqq \tfrac{1}{2},$$

whatever may be the value of ξ. In fact, this relation is immediate if n

* This well-known fact follows from the convergence of the series

$$\Sigma\, (a_n^{(k)})^2,$$

proved from the inequality

$$\int_0^1 (\Phi(x) - a_0\varphi_0 - a_1\varphi_1 - a_2^{(1)}\varphi_2^{(1)} - \cdots - a_n^{(k)}\varphi_n^{(k)})^2 dx \geqq 0,$$

where $a_n^{(k)} = \int_0^1 \Phi(y)\varphi_n^{(k)}(y)dy$.

s small and it follows for the larger values of n by virtue of the method of construction of the $Q_n^{(k)}$. Moreover, if $n \geqq \nu$ and if $\xi = \frac{\rho}{2^\nu}$, this integral has the value zero. We therefore have from the second law of the mean, $n \geqq \nu$,

$$\int_{\rho/2^\nu}^{(\rho+1)/2^\nu} F_1(y) Q_n^{(k)}(a, y)dy = F_1\left(\frac{\rho}{2^\nu}\right) \int_{\rho/2^\nu}^{\xi} Q_n^{(k)}(a, y)dy$$
$$+ F_1\left(\frac{\rho+1}{2^\nu}\right) \int_{\xi}^{(\rho+1)/2^\nu} Q_n^{(k)}(a, y)dy$$
$$= \left[F_1(a) - F_1\left(\frac{\rho}{2^\nu}\right)\right] \int_{\xi}^{(\rho+1)/2^\nu} Q_n^{(k)}(a, y)dy.$$

By a proper choice of the point $\frac{\rho}{2^\nu}$ we can make the factor of this last integral as small as desired; the entire expression will be as small as desired for sufficiently large n. The relations (11) are independent of the choice of $\frac{\rho}{2^\nu}$, so (7) is completely proved for the function F_1. A similar proof applies to F_2, so (7) can be considered as completely proved for the original function $F(x)$.

The uniform convergence of (5) as stated in Theorem IV follows from the uniform continuity of $F(x)$ and will be readily established by the reader.

§ 4. Further Expansion Properties of the Set φ.

The least square interpretation already given for the partial sums and the expression of the φ's in terms of the f's show that if the terms of (5) are grouped as in Theorems II and III, the question of convergence or divergence of the series at a point depends merely on that point and the nature of the function $F(x)$ in the neighborhood of that point. This same fact for series (5) when the terms are not grouped follows from (8) and (10) if $F(x)$ is integrable and with an integrable square. We shall further extend this result and prove:

THEOREM V. *If $F(x)$ is any integrable function, then the convergence or divergence of the series (5) at a point depends merely on that point and on the behavior of the function in the neighborhood of that point. If in particular $F(x)$ is of limited variation in the neighborhood of a point $x = a$, and if a is dyadically rational or if $F(a-0) = F(a+0)$, then series (5) converges for $x = a$ to the value $\frac{1}{2}[F(a-0) + F(a+0)]$. If $F(x)$ is not only of limited variation but is also continuous in two neighborhoods one on each side of a, and if a is dyadically rational or if $F(a-0) = F(a+0)$, the convergence of (5) is uniform in the neighborhood of the point a.*

Theorem V follows immediately from the reasoning already given and from (10) proved without restriction on Φ; we state the theorem for any bounded normal orthogonal set of functions ψ_n:

Theorem VI. *If* $\{\psi_n(x)\}$ *is a uniformly bounded set of normal orthogonal functions on the interval* (0, 1), *and if* $\Phi(x)$ *is any integrable function, then*

$$\lim_{n=\infty} \int_0^1 \Phi(x)\psi_n(x)dx = 0. \tag{12}$$

Denote by E the point set which contains all points of the interval for which $|\Phi(x)| > N$; we choose N so large that

$$\int_E |\Phi(x)|dx < \epsilon,$$

where ϵ is arbitrary. Denote by E_1 the point set complementary to E; then we have

$$\int_0^1 \Phi(x)\psi_n(x)dx = \int_E \Phi(x)\psi_n(x)dx + \int_{E_1} \Phi(x)\psi_n(x)dx.$$

It follows from the proof of (10) already indicated that the second integral on the right approaches zero as n becomes infinite. The first integral is in absolute value less than $M\epsilon$ whatever may be the value of n, where M is the uniform bound of the ψ_n. It therefore follows that these two integrals can be made as small as desired, first by choosing ϵ sufficiently small and then by choosing n sufficiently large.*

It is interesting to note that Theorem VI breaks down if we omit the hypothesis that the set ψ_n is uniformly bounded. In fact Theorem VI does not hold for Haar's set χ. Thus consider the function

$$\Phi(x) = (x - \tfrac{1}{2})^{-\nu}, \qquad \nu < 1.$$

We have

$$\int_0^1 \Phi(x)\chi_n^{(2^n-2+1)}(x)dx = \sqrt{2^{n-1}} \int_{1/2}^{1/2+1/2^n} (x - \tfrac{1}{2})^{-\nu}dx$$

$$- \sqrt{2^{n-1}} \int_{1/2+1/2^n}^{1/2+1/2^{n-1}} (x - \tfrac{1}{2})^{-\nu}dx = \frac{(2^{n-1})^{\nu-(1/2)}}{1-\nu}[2^\nu - 1].$$

Whenever $\nu \geqq \tfrac{1}{2}$, it is clear that (12) cannot hold, and if $\nu > \tfrac{1}{2}$, there is a sub-sequence of the sequence in (12) which actually becomes infinite.

* Theorem VI is proved by essentially this method for the set $\psi_n(x) = \sqrt{2}\sin n\pi x$ by Lebesgue, *Annales scientifiques de l'école normale supérieure*, ser. 3, Vol. XX, 1903. See also Hobson, *Functions of a Real Variable* (1907), p. 675, and Lebesque, *Annales de la Faculté des Science de Toulouse*, ser 3, Vol. I (1909), pp. 25–117, especially p. 52.

We turn now from the study of the convergence of such a series expansion as (5) to the study of the summability of such expansions, and are to prove

THEOREM VII. *If $F(x)$ is continuous in the closed interval* (0, 1), *the series* (5) *is summable uniformly in the entire interval to the sum $F(x)$.*

If $F(x)$ is integrable in the interval (0, 1), *and if $F(a-0)$ and $F(a+0)$ exist, and if either $F(a-0) = F(a+0)$ or a is dyadically rational, then the series* (5) *is summable for $x = a$ to the value $\frac{1}{2}[F(a-0) + F(a+0)]$. If $F(x)$ is continuous in the neighborhood of the point $x = a$, or if a is dyadically rational and $F(x)$ continuous in the neighborhood of a except for a finite jump at a, the summability is uniform throughout a neighborhood of that point.*

In this theorem and below, the term *summability* indicates summability by the first Cesàro mean.

We shall find it convenient to have for reference the following

LEMMA. *Suppose that the series*

$$\begin{aligned}(b_1 + b_2 + \cdots + b_{n_1}) + (b_{n_1+1} + b_{n_1+2} + \cdots + b_{n_2}) + \cdots \\ + (b_{n_k+1} + b_{n_k+2} + \cdots + b_{n_{k+1}}) + \cdots\end{aligned} \tag{13}$$

converges to the sum B and that the sequence

$$\begin{gathered}b_1,\quad \frac{2b_1 + b_2}{2},\quad \frac{3b_1 + 2b_2 + b_3}{3},\quad \cdots \\ \frac{(n_1-1)b_1 + (n_1-2)b_2 + \cdots + b_{n_1-1}}{n_1 - 1}, \\ \frac{(n_1-1)b_1 + \cdots + b_{n_1-1}}{n}, \\ \frac{(n_1-1)b_1 + (n_1-2)b_2 + \cdots + b_{n_1-1} + b_{n_1+1}}{n_1+1}, \\ \frac{(n_1-1)b_1 + \cdots + b_{n_1-1} + 2b_{n_1+1} + b_{n_1+2}}{n_1+2},\quad \cdots \\ \frac{(n_1-1)b_1 + \cdots + b_{n_1-1} + (n_2-n_1-1)b_{n_1+1} + (n_2-n_1-2)b_{n_1+2} + \cdots + b_{n_2-1}}{n_2-1}, \\ \cdots,\end{gathered} \tag{14}$$

converges to zero. Then the series

$$b_1 + b_2 + b_3 + \cdots \tag{15}$$

is summable to the sum B.

This lemma involves merely a transformation of the formulas involving the limit notions. Insert zeroes in series (13) so that the parentheses are respectively the n_1-th, n_2-th, n_3-th terms of the new series; this new series converges to the sum B and hence is summable to the sum B. The term-by-term difference of the new series and (15) is the series

$$\begin{aligned} b_1 + b_2 + \cdots + b_{n_1-1} - (b_1 + b_2 + \cdots + b_{n_1-1}) + b_{n_1+1} + b_{n_1+2} \\ + \cdots + b_{n_2-1} - (b_{n_1+1} + b_{n_1+2} + \cdots + b_{n_2-1}) + \cdots, \end{aligned} \tag{16}$$

which is to be shown to be summable to the sum zero. The sequence corresponding to the summation of (16) is precisely (14).

A sufficient condition for the convergence to zero of (14) is that we have, independently of m,

$$\lim_{k=\infty} \frac{mb_{n_k+1} + (m-1)b_{n_k+2} + \cdots + b_{n_k+m}}{m} = 0, \qquad m \leqq n_{k+1} - n_k, \tag{17}$$

for from a geometric point of view each term of the sequence (14) is the center of gravity of a number of terms such as occur in (17), each term weighted according to the number of b_i that appear in it. An (ϵ, δ)-proof can be supplied with no difficulty.

For the case of Theorem VII let us assume $F(x)$ integrable and that $F(a-0)$ and $F(a+0)$ exist. The series (15) is to be identified with the series (5), and (13) with (5) after the terms are grouped as in Theorem III. The sum that appears in (17) is, then, for $x = a$,

$$\begin{aligned} \frac{1}{m}\int_0^1 [m\varphi_n^{(1)}(a)\varphi_n^{(1)}(y) + (m-1)\varphi_n^{(2)}(a)\varphi_n^{(2)}(y) + \cdots \\ + \varphi_n^{(m)}(a)\varphi_n^{(m)}(y)]F(y)dy, \quad m \leqq 2^{n-1}. \end{aligned} \tag{18}$$

We shall prove that (18) formed for the function $F_1(y)$ defined in (6) and for a dyadically irrational has the limit zero as n becomes infinite.

Let us notice that

$$\begin{aligned} \frac{1}{m}\int_0^1 |m\varphi_n^{(1)}(a)\varphi_n^{(1)}(y) + (m-1)\varphi_n^{(2)}(a)\varphi_n^{(2)}(y) + \cdots \\ + \varphi_n^{(m)}(a)\varphi_n^{(m)}(y)|dy = 1. \end{aligned} \tag{19}$$

This follows directly from (3) and (4). The value of the integral in (19) is unchanged if we replace a by any dyadic irrational b. Choose $0 < b < 2^{-n}$, so that all the functions $\varphi_0, \varphi_1, \varphi_2, \cdots, \varphi_{m-1}$ are positive for $x = b$. Then the integrand in (19) can be reduced merely to $m\varphi_0(y)$, so (19) is proved.

Let us consider the integral (18) formed for the function $F_1(y)$ to be divided as in (8), where as before

$$\frac{\rho}{2^\nu} < a < \frac{\rho+1}{2^\nu},$$

and let us denote by (20), (21), (22), (23) respectively the entire integral and its three parts. Then (22) can be made as small as desired simply by proper choice of the point $\frac{\rho}{2^\nu}$, for in the interval $\left(\frac{\rho}{2^\nu}, \frac{\rho+1}{2^\nu}\right)$ we can make $|F_1(y) - F_1(a)|$ uniformly small, we have established (19), and we have also

$$\int_{\rho/2^\nu}^{(\rho+1)/2^\nu} [m\varphi_n^{(1)}(a)\varphi_n^{(1)}(y) + (m-1)\varphi_n^{(2)}(a)\varphi_n^{(2)}(y) + \cdots + \varphi_n^{(m)}(a)\varphi_n^{(m)}(y)]F_1(a)dy = 0$$

if merely $n > \nu$.

The integral (21) is the average of m integrals of the type that appear in (8):

$$\int_0^{\rho/2^\nu} F_1(y)Q_n^{(k)}(a, y)dy, \qquad k = 1, 2, \cdots, m.$$

Thus the entire integral (21) approaches zero as n becomes infinite. Treatment in a similar way of the integral (23) proves that (20) approaches zero. It is likewise true that (18) formed for the function $F_2(y)$ also approaches zero as n becomes infinite. This completes the proof of the second sentence in Theorem VII for a dyadic irrational; we omit the proof for a dyadic rational. The uniformity of the continuity of $F(x)$ gives us readily the remaining parts of Theorem VII.

§ 5. Not Every Continuous Function Can Be Expanded in Terms of the φ.

The summability of the expansions of continuous functions in terms of the functions φ is another point of resemblance of those functions to the Fourier sine and cosine functions. Still another point of resemblance which we shall now establish is that there exists a continuous function whose expansion in terms of the φ's does not converge at every point of the interval.

Our proof rests on a beautiful theorem due to Haar,* by virtue of which the existence of such a continuous function will be shown if we prove merely that

$$\int_0^1 |K_n^{(k)}(a, y)|dy \tag{24}$$

is not bounded uniformly for all n and k. The point a is a point of divergence of the expansion of the continuous function and for our particular case may be chosen any point of the interval (0, 1). We shall study (24) in detail merely for a dyadically irrational; the integral (24) is independent of the point a chosen if a is dyadically irrational.

* L. c., p. 335. This condition holds for any set of normal orthogonal functions and is necessary as well as sufficient, if a slight restriction is added.

The integral (24) is bounded uniformly for all the values n if $k = 2^{n-1}$, so it will be sufficient to consider the integral

$$c_n^{(k)} = \int_0^1 |Q_n^{(k)}(a, y)|\,dy.$$

The following table shows the value of $c_n^{(k)}$ for small values of n and for each value of k:

$n = 2$	1								1							
$n = 3$	1				1				$1\frac{1}{2}$				1			
$n = 4$	1		1		$1\frac{1}{2}$		1		$1\frac{3}{4}$		$1\frac{1}{2}$		$1\frac{3}{4}$		1	
$n = 5$	1,	1,	$1\frac{1}{2}$,	1,	$1\frac{3}{4}$,	$1\frac{1}{2}$,	$1\frac{3}{4}$,	1,	$1\frac{7}{8}$,	$1\frac{3}{4}$,	$2\frac{1}{8}$,	$1\frac{1}{2}$,	$2\frac{1}{8}$,	$1\frac{3}{4}$,	$1\frac{7}{8}$,	1,
	.	.	.	.	.	.	.	.	.	.	.	.	.	.	.	.

We have the general formulas

$$c_n^{(1)} = c_n^{(2n+1)} = 1,$$
$$c_n^{(k)} = c_{n+1}^{(2k)},$$
$$c_{n+1}^{(2k+1)} = \tfrac{1}{2}\left[c_n^{(k)} + c_n^{(k+1)}\right] + \tfrac{1}{2},$$

so the $c_n^{(k)}$ are not uniformly bounded.

THEOREM VIII. *If a point a is arbitrarily chosen, there will exist a continuous function whose φ-development does not converge at a.*

§ 6. The Approximation to a Function at a Discontinuity.

We have considered in § 3 and § 4 with a fair degree of completeness the nature of the approach to $F(x)$ of the formal development of an arbitrary function $F(x)$ in the neighborhood of a point of continuity of $F(x)$. We shall now consider the approach to $F(x)$ of this formal development in the neighborhood of a point of discontinuity of $F(x)$. We study this problem merely for a function which is constant except for a single discontinuity, a finite jump, but this leads directly to similar results for any function $F(x)$ at an isolated discontinuity which is a finite jump, if $F(x)$ is of such a nature that the expansion of $F(x)$ would converge uniformly in the neighborhood of the point of discontinuity were that discontinuity removed by the addition of a function constant except for a finite jump.

Let us consider the function

$$f(x) = \begin{cases} 1, & 0 \leqq x < a, \\ 0, & a < x \leqq 1. \end{cases}$$

If a is dyadically rational, $f(x)$ can be expressed as a finite sum of functions φ,* and thus is represented uniformly, if we make the definition $f(a)$

* A discontinuity at $x = 0$ or $x = 1$ is slightly different [compare the first footnote of § 2]. Under the present definition of the φ's it acts like an artificial discontinuity in the interior of the interval and has no effect on the sequence representing the function.

$= \frac{1}{2}[f(a-0)+f(a+0)]$; this follows from the evident possibility of expanding $f(x)$ in terms of the functions $f_0, f_1, f_2^{(1)}, \cdots$.

If the point a is dyadically irrational, $f(x)$ *cannot be expanded in terms of the* φ. The formal development of $f(x)$ converges in fact for every value of x other than a and diverges for $x = a$.* The convergence for $x \neq a$ follows, indeed, from Theorem IV. We proceed to demonstrate the divergence.

Use the dyadic notation

$$a = \frac{a_1}{2^1} + \frac{a_2}{2^2} + \frac{a_3}{2^3} + \cdots, \qquad a_n = 0 \text{ or } 1.$$

The partial sum

$$S_n^{(k)}(x) = \varphi_0(x)\int_0^1 f(y)\varphi_0(y)dy + \varphi_1(x)\int_0^1 f(y)\varphi_1(y)dy + \cdots + \varphi_n^{(k)}(x)\int_0^1 f(y)\varphi_n^{(k)}(y)dy$$

is in the sense of least squares the best approximation to $f(x)$ that can be formed from the functions $\varphi_0, \varphi_1, \cdots, \varphi_n^{(k)}$. It is therefore true that when $k = 2^{n-1}$, on every subinterval $\left(\frac{r}{2^n}, \frac{r+1}{2^n}\right)$ on which $f(x)$ is constant, $S_n^{(k)}(x)$ is also constant and equal to $f(x)$. On that subinterval $\left(\frac{m}{2^n}, \frac{m+1}{2^n}\right)$ which contains the point a, $S_n^{(k)}$ has the value

$$2^n a - m = \frac{a_{n+1}}{2^1} + \frac{a_{n+2}}{2^2} + \frac{a_{n+3}}{2^3} + \cdots, \tag{25}$$

which lies between zero and unity. Thus $S_n^{(k)}(x)$ $[n > 1]$ is a function with two points of discontinuity and which takes on three distinct values at its totality of points of continuity.

The infinite series corresponding to the sequence (25) is

$$\left(\frac{a_2}{2^1} + \frac{a_3}{2^2} + \frac{a_4}{2^3} + \cdots\right) + \left(\frac{a_3}{2^2} + \frac{a_4}{2^3} + \cdots - \frac{a_2}{2}\right) + \left(\frac{a_4}{2^2} + \frac{a_5}{2^3} + \frac{a_6}{2^4} + \cdots - \frac{a_3}{2}\right) + \left(\frac{a_5}{2^2} + \frac{a_6}{2^3} + \frac{a_7}{2^4} + \cdots - \frac{a_4}{2}\right) + \cdots. \tag{26}$$

Not all the numbers a_n after a certain point can be zero and not all of them

* This was pointed out for the set χ by Faber, *Jahresbericht der deutschen Mathematiker-Vereinigung*, Vol. 19 (1910), pp. 104–112.

can be unity, so the general term of the series (26) cannot approach zero and the sequence (25) cannot converge.

It is likewise true that the sequence (25) is not always summable and if summable may not be summable to the value $\frac{1}{2}$. Thus if we choose

$$a = \frac{1}{2} + \frac{1}{2^2} + \frac{0}{2^3} + \frac{1}{2^4} + \frac{1}{2^5} + \frac{0}{2^6} + \frac{1}{2^7} + \cdots,$$

the sequence (25) is summable to the sum $\frac{2}{3}$. Likewise the sequence $S_n^{(k)}(x)$ for $x = a$ and where we consider all values of n and k, is summable to the value $\frac{2}{3}$.

The general behavior of $S_n^{(k)}(x)$ for $f(x)$ where we do not make the restriction $k = 2^{n-1}$ is quite easily found from the behavior for $k = 2^{n-1}$ and the relation

$$\varphi_n^{(i)}(a) \int_0^1 f(y)\varphi_n^{(i)}(y)dy = \varphi_n^{(k)}(a) \int_0^1 f(y)\varphi_n^{(k)}(y)dy,$$

which holds for all values of i, k, and n.

In fact there occurs a phenomenon quite analogous to Gibbs's phenomenon for Fornier's series. For the set φ, the approximating functions are uniformly bounded. The peaks of the approximating function $S_n^{(k)}$ disappear entirely for $k = 2^{n-1}$ but reappear (usually altered in height) for for larger values of n.

It is clear that the facts concerning the approximating curves for $f(x)$ hold without essential modification for a function of limited variation at a simple finite discontinuity, and that the facts for the summation of the approximating sequence hold without essential modification for a function continuous except at a simple finite discontinuity.

§ 7. The Uniqueness of Expansions.

We now study the possibility of a series of the form

$$a_0\varphi_0(x) + a_1\varphi_1(x) + \cdots + a_n\varphi_n(x) + \cdots \tag{27}$$

which converges on $0 \leqq x \leqq 1$ to the sum zero, with the possible exception of a certain number of points x. Faber has pointed out* that there exists a series of the functions $\chi_n^{(k)}(x)$ which converges to zero except at one single point, and the convergence is uniform except in the neighborhood of that point.

We state for reference the easily proved

LEMMA. *If the series* (27) *converges for even one dyadically irrational value of x, then* $\lim_{n=\infty} a_n = 0$.

* L. c., p. 111.

This lemma results immediately from the fact that $\varphi_n^{(k)}(x) = \pm 1$ if x is dyadically irrational.*

We shall now use this lemma to establish

THEOREM IX. *If the series* (27) *converges to the sum zero uniformly except in the neighborhood of a single value of* x, *then* $a_n = 0$ *for every* n.

We phrase the argument to apply when this exceptional value x_1 is dyadically irrational. If $x_1 > \frac{1}{2}$, we have for $0 \leqq x \leqq \frac{1}{2}$,

$$a_0\varphi_0(x) + a_1\varphi_1(x) + \cdots + a_n\varphi_n(x) + \cdots = 0,$$
$$(a_0 + a_1)\varphi_0(y) + (a_2 + a_3)\varphi_1(y) + (a_4 + a_5)\varphi_2(y) + \cdots = 0,$$

for every value of $y = 2x$. Then we have from the uniformity of the convergence,

$$a_0 + a_1 = 0, \qquad a_2 + a_3 = 0, \qquad a_4 + a_5 = 0, \qquad \cdots. \tag{28}$$

If $x_1 < \frac{3}{4}$, we have for $\frac{3}{4} \leqq x \leqq 1$,

$$a_0\varphi_0(x) + a_1\varphi_1(x) + \cdots + a_n\varphi_n(x) + \cdots = 0,$$

or for $0 \leqq y \leqq 1$, $y = 4x - 3$,

$$(a_0 - a_1 + a_2 - a_3)\varphi_0(y) + (a_4 - a_5 + a_6 - a_7)\varphi_1(y) \\ + (a_{4n} - a_{4n+1} + a_{4n+2} - a_{4n+3})\varphi_n(y) + \cdots = 0.$$

From the uniformity of the convergence we have

$$a_0 - a_1 + a_2 - a_3 = 0,$$
$$a_4 - a_5 + a_6 - a_7 = 0,$$
$$\cdot \quad \cdot \quad \cdot \quad \cdot \quad \cdot \quad \cdot,$$

or from (28),

$$a_0 = - a_1 = - a_2 = a_3,$$
$$a_4 = - a_5 = - a_6 = a_7,$$
$$\cdot \quad \cdot \quad \cdot \quad \cdot \quad \cdot \quad \cdot.$$

If $x_1 > \frac{5}{8}$, we have for $\frac{5}{8} \leqq x \leqq \frac{3}{4}$,

$$a_0\varphi_0(x) + a_1\varphi_1(x) + \cdots = 0,$$

or for $0 \leqq y \leqq 1$, $y = 8x - 5$,

$$(a_0 - a_1 - a_2 + a_3 - a_4 + a_5 + a_6 - a_7)\varphi_0(y) \\ + (a_8 - a_9 - a_{10} + a_{11} - a_{12} + a_{13} + a_4 - a_{15})\varphi_1(y) + \cdots = 0.$$

Then each of these coefficients must vanish, and hence

$$a_0 = - a_1 = - a_2 = a_3 = a_4 = - a_5 = - a_6 = a_7.$$

* This lemma is closely connected with a general theorem due to Osgood, *Transactions of the American Mathematical Society*, Vol. 10 (1909), pp. 337–346.

See also Plancherel, *Mathematische Annalen*, Vol. 68 (1909–1910), pp. 270–278.

Continuation in this way together with the Lemma shows that every a_n must vanish. This reasoning is typical and does not essentially depend on our numerical assumptions about x_1. Then Theorem IX is proved.

The reasoning is precisely similar if instead of the hypothesis of Theorem IX we admit the possibility of a finite number of points in the neighborhood of each of which the convergence is not assumed uniform:

THEOREM X. *If the series*

$$a_0\varphi_0(x) + a_1\varphi_1(x) + \cdots + a_n\varphi_n(x) + \cdots$$

converges to the sum zero uniformly, $0 \leqq x \leqq 1$, *except in the neighborhood of a finite number of points, then* $0 = a_1 = a_2 = \cdots = a_n = \cdots$.

HARVARD UNIVERSITY,
May, 1922.

Commentary by F. Schipp

Introduction

A number of major fields of modern mathematics developed in the first decades of this century. Among them are measure theory, functional analysis, and real analysis. That was the time for the mathematical foundation of probability theory and a new age began in the theory of Fourier series. There are three orthogonal systems which played a fundamental role in this process. One of them was introduced by the Hungarian mathematician Haar [19] in 1910, the other was published by the German Rademacher [31] in 1922, and the third one is due to Walsh [37] (=[23-b*]) in 1923. All these systems consist of piecewise constant functions. This is why they may have seemed to be rather artificial compared to the classical trigonometric, Legendre and Sturm-Liouville systems.

Haar proved in his doctoral dissertation in 1909 that, with respect to the system named after him, the Fourier series of every continuous function is uniformly convergent. This result was especially interesting since it was known by Du Bois Reymond that the trigonometric system fails to have this property. Moreover, as we know it now there is no uniformly bounded orthonormal system with the above property [27]. Walsh knew the results of Haar and he used them in his paper. On the other hand he didn't, maybe couldn't know about Rademacher's paper of 1922 when he handed his work to the American Mathematical Society in February 25, 1922. Since the system introduced by Walsh contains the one of Rademacher as a subsystem, we may say that Walsh rediscovered it. However there was no hint about the connection between the two systems. That connection and its importance was observed by Paley [28] in 1932.

The Walsh System

The Walsh functions take on only the values $+1$ and -1 except for the points of Q which is the set of the so called dyadic rationals, i.e. $Q := \{p2^{-q} : 0 \le p \le 2^n, p, q, \in N\}$ $(N := \{0, 1, 2, \ldots\})$. This feature makes them essentially different from the trigonometric functions. However, as it is already mentioned in the introductory part of Walsh's paper, they imitate some properties of the *sine* and *cosine* systems. For instance, the n-th Walsh function changes its sign n times in $[0, 1)$, it is either odd or even with respect to the midpoint of $[0, 1)$, and they form a uniformly bounded orthonormal system. Walsh was probably motivated by these properties in creating his system. Later it turned out that from several points of view it is more convenient to change the original definition so that the Walsh functions will be continuous from the right at the points of Q. It is quite interesting that some of the above properties characterize the Walsh system. Indeed, an orthonormal system defined on $[0, 1)$ of which the functions are right-continuous, take on only the values $+1$ and -1, and the n-th one changes its sign n times on $[0, 1)$ is necessarily equal to the Walsh system [9], [25]. Walsh observed and used in the investigation of Fourier series with respect to his system that the first 2^n Walsh and Haar functions span the same linear subspace. Proving the analogue to the Dirichlet-Jordan and the Fejér theorems he showed that the Walsh-Fourier series of a function of bounded variation is convergent at every point of continuity, and it is uniformly $(C, 1)$ summable in case of a continuous function. He also proved that to any point of $[0, 1)$ there exists a continuous function for which the Walsh-Fourier series diverges at that point. The paper also contains a unicity theorem.

Paley's paper of 1932 is fundamental in Walsh-Fourier analysis, in which he gave a decomposition formula for Walsh functions w_n as products of Rademacher functions r_n. Indeed, if $n = \sum_{k=0}^{\infty} n_k 2^k (n_k = 0 \text{ or } 1)$ is the binary expansion of $n \in N$ then $w_n = \prod_{k=0}^{\infty} r_k^{n_k}$. This relation between the two systems is expressed in modern terminology by saying the Walsh system $W = (w_n, n \in N)$ is the product system of the Rademacher system $R = (r_n, n \in N)$. W differs from the original Walsh system $\Phi = (\phi_n, n \in N)$ only in the order of its elements. That is why it is called the Walsh system ordered in Paley's sense, or briefly the Walsh-Paley system. It is interesting that this order of Walsh functions was earlier used by Kaczmarcz [23] in a paper published in 1929. His method was based not on the product decomposition but on a recursive definition later used also in the book of Kaczmarcz and Steinhaus [24]. Most mathematicians prefer the Walsh-Paley system, while in technical applications the original Walsh system is generally used. The reason behind this is probably the product decomposition on the one side and the greater similarity to the trigonometric system on the other side. Concerning the latter we mention that the Walsh system is often divided into *sal* and *cal* functions similarly to the *sine* and the *cosine* functions (see e.g. [22]).

As it turned out, the Walsh and the Walsh-Paley systems behave similarly in most problems. It can be shown that the Φ system itself can be considered as a product system based on the system $(r_{n-1}r_n, n \in N)(r_{-1} = 1)$. As a consequence of this, there exists a one-to-one measure preserving map T of $[0, 1)$ onto itself that transforms the two systems into each other: namely, for any $x \in [0, 1)$ and $n \in N$ we have $\phi_n(x) = w_n(T(x))$. T can be easily defined by the binary coefficients of x (see [32]). It is worth mentioning that the relation between the values of Walsh functions at a point and its binary coefficients has also been described by Walsh.

Another rearrangement of the Walsh system is closely connected with the matrices introduced by Hadamard [20] in 1893. These matrices are defined by means of the Kronecker product as follows

$$H_0 = (1), \qquad H_1 = \begin{pmatrix} +1 & +1 \\ +1 & -1 \end{pmatrix}, \qquad H_{n+1} = H_1 \otimes H_n \qquad (n \in N).$$

The Hadamard matrices of dimension $2^n \times 2^n$ can be expressed by Walsh functions. Indeed, $h^n_{kj} = w_{\hat{k}}(j2^{-n})$, where $\hat{k}$ arises from k by reversing its binary digits, i.e. $\hat{k} = k_{n-1} + k_{n-2}2 + \cdots + k_0 2^{n-1}$. Šneider [33] used the permutation $k \to \hat{k}$ to define another rearrangement of the Walsh system in 1948, which he named after Kaczmarcz.

The Impact of Walsh Functions on Modern Mathematics

It is not easy to list even the most significant results of modern mathematics that are related to the Walsh system. The Walsh system is discussed in almost every recently published book on harmonic analysis or Fourier series. We may say that it is considered to be the most important orthogonal system besides the trigonometric one from both theoretical and practical points of view. The Walsh functions are investigated or used in many papers and there are many more publications concerning their application to technology (see [5], [20], [21]). For the period before 1970, a good survey is given in the paper of Balašov and Rubinštein [3]. There is a well written summary about the results from the decade 1970-1980 due to Wade [35]. Since then, two comprehensive monographs on dyadic harmonic analysis (which has become an independent field of mathematics) were published [18], [32]. Here we only mention some examples because of its connection with other fields of mathematics.

Probability Theory

Many concepts and results in probability theory are related to the Walsh and the Rademacher series. The Rademacher system provides one of the simplest examples of a system of independent functions. The 2^n-th partial sums of the Walsh series form a martingale, and the full sequences of the partial sums are closely connected with tree-like martingales. The three series theorem of Kolmogorov is the generalization of a result of Kolmogorov and a result of Rademacher with respect to the convergence of Rademacher series. The famous Paley inequality for Walsh series inspired important investigations in martingale theory, which resulted in inequalities for the quadratic variation, for the maximal function and for the martingale transform. These inequalities turned out to be fundamental not only in martingale theory but also in the theory of Hardy spaces and the consequences spread to many other fields. Here we only mention the research activities of Burkholder and his collaborators [7] and the books [12], [16], [29]. The extension of the martingale inequalities for the tree-like case has been given in [32]. It led to a theorem analogous to the one of Carleson with respect to the almost everywhere convergence of the trigonometric Fourier series of functions belonging to L^2.

Harmonic Analysis

Vilenkin [34] and Fine [14] independently observed that the Walsh system can be identified as the character system of the so called dyadic or Cantor group. In other words, the Walsh system is the "trigonometric system" of the dyadic group. This observation gave new momentum to the investigations and made it possible to use the methods and concepts of harmonic analysis, such as invariant measure, convolution, moduli of continuity, etc. In many problems it is more appropos to consider the Walsh functions to be defined on the dyadic group, which is similar to $[0, 1)$ in several respects, but for instance its topological structure is essentially different from that of $[0, 1)$. In this way the Walsh functions become continuous, and differentiable with respect to the concept of the dyadic derivative introduced by Gibbs [17], and Butzer and Wagner [8]. Moreover, they are eigenfunctions of the operator of dyadic differentiation. From this perspective the Walsh system, considered earlier as quite pathological, could be integrated with harmonic analysis, and it became one of the most important models, simpler than the classical one in several respects.

Functional Analysis

The Walsh system played an important role in the solution of many problems related to bases. For instance, using Paley's inequality, Marcinkiewicz [26] proved that the Haar system is an unconditional basis in L^p for $p > 1$. The inequality analogous to Paley's inequality holds for the Franklin system too [6]. Consequently, it is also an unconditional basis in $L^p (p > 1)$ spaces. The Walsh functions appeared in the solution of a famous problem of Banach. Enflo [13] constructed a separable Banach space not having a Schauder basis by using Walsh functions.

Generalizations

There are a number of orthonormal systems used in modern mathematics which can from one aspect or another be considered as generalizations of the Walsh system. Some of them are based on the algebraic property that the dyadic group can be decomposed into the direct product of countably many cyclic groups of order two. Replacing them by cyclic groups of order p or by arbitrary cyclic groups of finite order and taking the corresponding character system we obtain the systems investigated in [10], [30]. They are special cases of periodic multiplicative systems introduced and investigated by Vilenkin in 1947. They are now called Vilenkin systems. Concerning these we refer the reader to the survey paper [36] and the book [1].

The transform based on the Walsh functions and analogous to the trigonometric Fourier transform was defined by Fine [15] in 1950. In connection with the results and generalizations of the Walsh transform we refer to the monographs [18], [32].

It was the Walsh system again which inspired concept building and investigations concerning the various types of multiplicative systems. This program which is closely connected with probability theory was started by Alexits [2] in the seventies. Based on the relation between the Walsh and the Haar systems and replacing the Haar system by the Franklin system, Ciesielski [11] introduced a uniformly bounded orthonormal system of piecewise linear functions. Many of the convergence properties of the Walsh system are inherited by this system and by various types of multiplicative systems (see [32]).

Technical Applications

The fact that the Walsh functions take on only the values $+1$ and -1 makes it easy to compute with them. It also makes their computerization and technical adaptations simple. As examples we mention the generators of Walsh signals and the fast algorithm denoted by FWT for the fast computation of the Walsh-Fourier coefficients. It is generally accepted that, while the trigonometric system is the adequate mathematical tool for electronic systems containing classical circuit elements, the corresponding system for the electric systems built from integrated circuits is the Walsh system. Due to this, the Walsh system is used in more and more areas of technology [4], [5], [21], [22].

References

[1] G. N. Agaev, N. Ja. Vilenkin, G. M. Dzafarli and A. I. Rubinštein, *Multiplicative systems and harmonic analysis on zero-dimensional groups*, Baku "ELM", 1981 (in Russian).

[2] G. Alexits, *Approximation Theory (Selected papers)*, Ed. K. Tandori, Akadémiai Kiadó, Budapest, 1983.

[3] L. A. Balašov and A. I. Rubinštein, *Series with respect to the Walsh system and their generalizations*, J. Soviet Math. **1** (1973), 727–763.

[4] C. A. Bass, *Applications of Walsh functions*, Symposium and Workshop, Naval Research Lab., Washington D.C., 1970.

[5] K. G. Beauchamp, *Walsh functions and their applications*, Academic Press, New York, 1975.

[6] S. V. Bočkarev, *A method of averaging in the theory of orthogonal series and some problems in the theory of bases*, Trudy Mat. Inst. Steklov. No. 146, 1978.

[7] D. L. Burkholder, *Distribution function inequalities for martingales*, Ann. Prob. **1** (1973), 19–42.

[8] P. L. Butzer and H. J. Wagner, *Walsh series and the concept of a derivative*, Appl. Anal. **3** (1973), 29–46.

[9] J. S. Byrnes and D. A. Swick, *Instant Walsh functions*, SIAM Review **12** (1970), 131.

[10] H. E. Chrestenson, *A class of generalized Walsh functions*, Pacific J. Math. **5** (1955), 17–31.

[11] Z. Ciesielski, *A bounded orthonormal system of polygonals*, Studia Math, **31** (1968), 339–346.

[12] R. Durrett, *Brownian motion and martingales in analysis*, Wadsworth Math. Series, Belmont, California, 1984.

[13] P. Enflo, *A counterexample to the approximation property in Banach spaces*, Acta Math. **139** (1973), 309–317.

[14] N. J. Fine, *On the Walsh functions*, Trans. Amer. Math. Soc. **65** (1949), 372–414.

[15] N. J. Fine, *The generalized Walsh functions*, Trans. Amer. Math. Soc. **69** (1950), 66–77.

[16] A. M. Garsia, *Martingale inequalities*, W. A. Benjamin, Inc., Reading, Mass., 1973.

[17] J. E. Gibbs and M. S. Millard, *Walsh functions as solutions of a logical differential equation*, National Physical Lab. (1969), Middlesex, England, DES Report No. 1.

[18] B. I. Golubov, A. V. Efimov and V. A. Skvorcov, *Walsh series and transforms*, Nauka, Moskow, 1987.

[19] A. Haar, *Zur Theorie der orthogonalen Funktionensysteme*, Mat. Ann., **69** (1910), 331–371.

[20] J. Hadamard, *Résolution d'une question relative aux déterminants*, Bull. Sci. Math. **17** (1893), 240–246.

[21] H. F. Harmuth, *Transmission of information by orthogonal functions*, Springer-Verlag, Berlin, 1972.

[22] H. F. Harmuth, *Sequency theory*, Academic Press, New York, N.Y., 1977.

[23] S. Kaczmarcz, *Über ein Orthogonalsystem*, Comt. Rendu Congres Math. Warszawa (1929), 189–192.

[24] S. Kaczmarcz, and H. Steinhaus, *Theorie der Orthogonalreihen*, Monogr. Mat. Vol. 6, Warszawa-Lwow, 1935.

[25] S. V. Levizov, *Some properties of the Walsh system*, Mat. Zametki, **27** (1980), 715-720.

[26] J. Marcinkiewicz, *Quelques théorèmes sur les séries orthogonales*, Ann. Soc. Polon. Math. **16** (1937), 85–96.

[27] A. M. Olevskiĭ, *Fourier series with respect to general orthogonal systems*, Springer-Verlag, Berlin, 1975.

[28] R. E. A. C. Paley, *A remarkable system of orthogonal functions*, Proc. London Math. Soc., **34** (1932), 241–279.

[29] K. E. Petersen *Brownian motion, Hardy spaces and bounded mean oscillation*, LMS Lecture Notes. Cambridge Univ. Press, Cambridge, 1977.

[30] J. J. Price, *Certain groups of orthonormal step functions*, Canad. J. Math. **9** (1957), 413–425.

[31] H. A. Rademacher, *Einige Sätze über Reihen von allgemeinen Orthogonalen-funktionen*, Math. Ann., **87** (1922), 112–138.

[32] F. Schipp, W. R. Wade, P. Simon and J. Pál, *Walsh series: an introduction to dyadic harmonic analysis*, Adam Hilger, Bristol, New York, 1990.

[33] A. A. Šneider, *On series with respect to Walsh functions with monotone coefficients*, Izv. Acad. Nauk SSSR, Ser. Mat. **12** (1948), 179–192.

[34] N. Ya. Vilenkin, *A class of complete orthonormal series*, Izv. Akad. Nauk SSSR, Ser. Mat. **11** (1947), 363–400.

[35] W. R. Wade, *Recent developments in the theory of Walsh series*, Internat. J. Math., **5** (1982), 625–673.

[36] W. R. Wade, *Vilenkin-Fourier series and approximation*, Coll. Math. Soc. J. Bolyai, **58** Approx. Theory, 1990, 699–734.

[37] J. L. Walsh, *A closed set of normal orthogonal functions*, Amer. J. Math., **45** (1923), 5–24.

Commentary by T. J. Rivlin

1. The question of the discovery of the Walsh Functions has been discussed often. In the "Collected Papers of Hans Rademacher", M.I.T. 1974, the Editor, Emil Grosswald writes the following:

"This phase of Rademacher's work culminated with paper 13, "Einige Sätze über Reihen von allgemeinen Orthogonalenfunktionen," published in 1922. Here he introduced the systems of orthogonal functions now generally known as the Rademacher functions. Fortuitous circumstances delayed the publication of a second paper, containing the completion of the system of Rademacher functions. In the meantime, Walsh also obtained and published the completion of the system of Rademacher functions, where-upon Rademacher decided not publish his own manuscript. The few persons who had the opportunity to read it considered it to be extremely beautiful, so that it is most unfortunate that it appears to have been lost."

2. Henning F. Harmuth, in a letter to the Editors of this volume, wrote the following:

"The Influence of Walsh Functions on Electrical Communications"

"Before about 1960 electrical communications was solidly based on the system of sine and cosine functions. Communications engineers were trained to see electromagnetic signals not as functions of times but as functions of frequency in the

Fourier transform domain. This theoretical development produced impressive practical results but it also led to the unnecessarily restrictive small-relative-bandwidth technology of radio transmission.

Walsh functions have many features that are similar to those of sinusoidal functions. One can define a sequency in analogy to frequency, one can build sequency rather than frequency filters, there is a Walsh-Fourier transform, etc. But the Walsh functions are two-valued and thus were better suited than sinusoidal functions for the emerging semiconductor technology.

Walsh functions helped us to overcome our habitual thinking in terms of sinusoidal functions and opened the way to what eventually became the large-relative-bandwidth technology or - less precise - the ultra-wideband technology of radio transmission."

Professor Harmuth was a pioneer of applications of Walsh Functions in communications and his books on this subject (see [21] and [22] in the commentary of F. Schipp) examined the possibilities in detail.

In each year from 1969 to 1974 a Symposium on "Applications of Walsh Functions" was held in Washington, D.C. The Symposiums of 1969 (April 1), 1970 (March 31, April 1, 2, 3) and 1971 (April 13, 14, 15) took place at the Naval Research Laboratory. The Symposiums of 1972 (March 27, 28, 29), 1973 (April 16, 17, 18) and 1974 (March 18, 19, 20) were held at The Catholic University of America. Proceedings of the Symposiums of 1970-73 may be purchased from the National Technical Information Service, U.S. Department of Commerce, Springfield, VA, 22151. The copyright of the 1974 Symposium, "Applications of Walsh Functions and Sequency Theory" is owned by the Institute of Electrical and Electronics Engineers, Inc., New York, NY 10017 (IEEE Cat. no. 74CH0861-5EMC).

At the 1969 Symposium the opening remarks by Professor Walsh are as follows: "Let us look at the historical situation regarding orthogonal functions at the beginning of the twentieth century. There were then well known and useful kinds of such functions: the trigonometric functions which occur in Fourier series; orthogonal polynomials such as those of Legendre, Hermite, and Laguerre; Bessel's functions, the Sturm-Liouville series, and other special functions. But, there was no general theory embracing all such systems of functions.

At that time the great mathematician David Hilbert was at Göttingen. His Hilbert space consists of all points $A : \{a_1, a_2, \ldots\}$ such that $|a_1|^2 + |a_2|^2 + \cdots$ converges, and thus is an extension of ordinary Euclidean space to a countable number of dimensions. In the Hilbert space there exists a *distance* between two arbitrary points, and hence the study of analytic geometry and analysis can be carried out.

Hilbert is especially important here for his work on integral equations, and for his influence on students.

Erhard Schmidt received his Ph.D. in Göttingen in 1905, with a thesis on integral equations. However, he showed that a more or less arbitrary sequence of functions can be orthogonalized. This fact had been proved several decades earlier by Gram, but Schmidt gave emphasis and significance to the result, and thus indicated that systems of orthogonal functions are by no means exceptional but can be produced at will.

The Riesz-Fischer theorem first appeared in 1907, proved simultaneously by Fischer and F. Riesz. The theorem asserts that if a set of normal orthogonal functions $f_n(z)$ is given on an interval [0, 1] together with a set of coefficients a_n such that $|a_1|^2 + |a_2|^2 + \cdots$ converges, then there exists a function $F(x)$ integrable (Lebesgue) together with its square, such that

$$a_n = \int_0^1 F(x) f_n(x)\,dx \quad \text{for all } n.$$

This converse, that if $F(z)$ is a function integrable together with its square and the coefficients a_n are as just defined, then $|a_1|^2 + |a_2|^2 + \cdots$ converges, follows by

$$|a_1|^2 + |a_2|^2 + \cdots + |a_n|^2 \leq \int_0^1 F(x)^2\,dx,$$

and had been known previously. The beauty and simplicity of these theorems was, and still is, almost overwhelming.

The Hungarian mathematician, Alfred Haar, received his Ph.D. in Göttingen, in 1909, for a thesis concerned with convergence properties of series of orthogonal functions, and also containing a new set (now called the Haar system) of such functions. Here was a set of orthogonal functions each taking essentially only two values such that the formal expansion of an arbitrary continuous function in those functions converges uniformly to the given function, a property not possessed by any orthogonal set known up to that time.

Little progress on the subject of orthogonal functions was made during the decade containing World War I.

In 1922, H. Rademacher of Hamburg exhibited a new set of orthogonal functions taking on only the values ± 1, in the interval [0, 1], which was of use in his study of the general theory of orthogonal functions. However, the Rademacher functions are not complete; there exist other non-trivial functions orthogonal to all of them.

In 1923, J. L. Walsh of Cambridge, Massachusetts published a set of orthogonal functions which are complete on the interval [0, 1]; they take only the values ± 1, and are similar in oscillation and many other properties to the trigonometric functions. They have turned out to have important practical applications in calculation and in numerous special branches of science.

The functions of Rademacher and of Walsh were discovered quite independently. Rademacher's manuscript was received by the editors on October 8, 1921; Walsh presented his functions to the American Mathematical Society at a meeting on February 25, 1922, and published an abstract of his paper in Bull. Amer. Math. Soc. Vol. 28 (1922) p. 241. His manuscript was dated May 1922.

P. Franklin of Cambridge, Massachusetts published in 1928 a result related to the functions of Haar. He obtained orthogonal functions by orthogonalizing the integrals of the Haar functions, and thus obtained a set of continuous orthogonal functions in terms of which an arbitrary continuous function can be uniformly expanded.

The usefulness of all these functions both in theoretical work and in engineering applications still seems to be in its infancy. In your use of them and in your efforts in general, I wish you every success."

References

[1] E. Schmidt, *Zur Theorie der linearen und nichtlinearen Integralgleichungen*, I. Teil. *Entwicklung willkürlicher Funktionen nach Systemen vorgeschriebener.* Math. Ann. **63** (1905), 433–476.

[2] A. Haar, *Zur Theorie der orthogonalen Funktionensysteme*, Erste Mitteilung. Math. Ann. **69** (1909), 331–371.

[3] H. Rademacher, *Einige Sätze über Reihen von allgemeinen Orthogonalenfunktionen*, Math. Ann. **87** (1922), 112–138.

[4] J. L. Walsh, *A closed set of normal orthogonal functions*, Amer. Jour. Math. **45** (1923), 5–24.

[5] P. Franklin, *A set of continuous orthogonal functions*, Math. Ann. **100** (1928), 522–529.

The 1969 Symposium is described in detail in the volume of the 1974 Symposium. This 1974 volume also contains a "Bibliography on Walsh and Walsh Related Functions" which consists of 45 pages! This Bibliography includes a reference to a short article by N. J. Fine, entitled "Walsh Functions", which appeared in the "Encyclopaedic Dictionary of Physics", J. Thewlis (ed.), Pergamon Press, Oxford, Supplementary Volume 4, 1971. This article gives an excellent exposition of the purely mathematical aspects of the Walsh Functions, with a Bibliography of 70 publications. Fine was a major contributor to the mathematical side of the Walsh Functions.

Interest in Walsh Functions has not flagged in recent years. A volume with the title "Walsh Functions in Signal and Systems Analysis and Design", edited by Spyros G. Tzafestas was published in 1985 by Van Nostrand Reinhold Company Inc. in the series "Benchmark Papers in Electrical Engineering and Computer Science", Volume 31. On the mathematical side, the book of F. Schipp, et al, (see [32] in the Commentary of F. Schipp) appeared in 1990.

Chapter 3

Qualitative Approximation

Über die Entwicklung einer analytischen Funktion nach Polynomen.

Von

J. L. Walsh[1]) in München.

Viele Resultate über die Entwicklung einer analytischen Funktion nach Polynomen sind wohlbekannt[2]), darunter das Theorem von Runge:

Es sei die Funktion $f(z)$ eine analytische Funktion von z in einem einfach zusammenhängenden Bereiche R der z-Ebene. Dann läßt sich $f(z)$ in eine Reihe nach Polynomen von z entwickeln, welche in jedem ganz innerhalb R gelegenen abgeschlossenen Bereiche gleichmäßig konvergiert.

Der Zweck dieses Artikels ist, ein Resultat anzugeben, das in bezug auf bestimmte Funktionen noch allgemeiner ist, als das Theorem von Runge, und das außerdem erlaubt, das Theorem von Runge sehr leicht zu beweisen:

Satz. *Es sei $f(z)$ eine analytische Funktion von z im Inneren einer Jordanschen Kurve C, und es sei $f(z)$ stetig im abgeschlossenen Bereiche, welcher aus der Kurve C und ihrem Inneren besteht. Dann läßt sich $f(z)$ im ganzen abgeschlossenen Bereiche in eine Reihe nach Polynomen von z entwickeln; diese Reihe konvergiert gleichmäßig in demselben abgeschlossenen Bereiche.*

Dieser Satz kann, wenn die Kurve analytisch ist, leicht durch den Gebrauch konformer Abbildung bewiesen werden[3]). Auch im allgemeineren Falle werden die gewünschten Resultate durch einen Courantschen Satz[4])

---

[1]) Fellow, International Education Board.

[2]) Siehe z. B. Montel, „Leçons sur les Séries à une Variable Complexe" (Paris 1910), wo die Literatur zitiert ist.

[3]) Walsh, „On the Expansion of Analytic Functions in terms of Polynomials", Trans. Amer. Math. Soc. 26 (1924), S. 155–170, Theorem III.

[4]) „Über eine Eigenschaft der Abbildungsfunktionen bei konformer Abbildung", Göttinger Nachrichten, Math.-phys. Klasse, 1914, S. 101–109; 1922, S. 69–70.

über konforme Abbildung erreicht. Prof. Carathéodory hat mich angeregt, die Möglichkeit einer Anwendung dieses Courantschen Satzes auf das vorliegende Problem nachzuprüfen.

Der Einfachheit halber nehmen wir an, daß der Punkt $z = 0$ im Inneren von C liegt; diese Annahme schränkt die Allgemeinheit nicht ein. Wir betrachten eine Folge $\{C_n\}$ von ineinander eingeschachtelten Jordanschen Kurven in der z-Ebene, die im Äußern von C liegen, und so, daß die Bedingungen des Satzes von Courant erfüllt sind[a]). Wir bilden das Innere von C, C_n, resp. auf das Innere des Einheitskreises Γ in der u-Ebene ab, so daß die Punkte $z = 0$ und $u = 0$, und in diesen Punkten auch die positiven Richtungen der Achsen des Reellen einander entsprechen. Wir bezeichnen mit $u = \varphi(z)$, $\varphi_n(z)$ die Funktionen, welche diese Abbildungen liefern, und mit $z = \psi(u)$, $\psi_n(u)$ resp. die Umkehrfunktionen. Durch die Abbildung $u = \varphi_n(z)$ wird die Kurve C in die im Inneren von Γ liegende Jordansche Kurve γ_n transformiert.

Die Funktionen $f[\psi(u)]$ und $f\{\psi[\varphi_n(z)]\}$ sind im Inneren von Γ und C_n resp. analytisch und in den entsprechenden abgeschlossenen Bereichen stetig. Um unseren Satz zu beweisen, genügt es zu zeigen, daß es zu jedem $\varepsilon > 0$ ein n gibt, so daß für irgendeinen Punkt z auf oder im Inneren von C die Ungleichung

$$(1) \qquad |f\{\psi[\varphi_n(z)]\} - f(z)| < \frac{\varepsilon}{2}$$

gilt. Denn nach dem Rungeschen Theorem können wir die Funktion $f\{\psi[\varphi_n(z)]\}$ auf C und im Inneren von C beliebig genau durch ein Polynom $p(z)$ annähern:

$$|f\{\psi[\varphi_n(z)]\} - p(z)| < \frac{\varepsilon}{2}.$$

Hieraus erfolgt unmittelbar die Schlußfolgerung unseres Satzes:

$$|f(z) - p(z)| < \varepsilon$$

für alle betreffenden Punkte, da gleichmäßige Entwicklung und Annäherung mit beliebiger Genauigkeit vollständig äquivalent sind.

Wir beweisen die Ungleichung (1) durch die Transformation $z = \psi_n(u)$ auf die u-Ebene, d. h. wir wollen zeigen, daß für jedes u auf oder im Inneren von γ_n die Ungleichung

$$(2) \qquad |f[\psi(u)] - f[\psi_n(u)]| < \frac{\varepsilon}{2}$$

[a]) Eine solche Folge läßt sich mit Hilfe eines Quadratnetzes im Äußern von C konstruieren. Vgl. Osgood, Funktionentheorie (Leipzig 1912), S. 156.

befriedigt ist. Es ist hinreichend, die Ungleichung (2) für alle auf γ_n liegenden Punkte u zu beweisen. Wenn der Punkt u auf γ_n liegt, so liegen beide Punkte $z = \psi(u)$ und $z = \psi_n(u)$ im abgeschlossenen Bereich, welcher aus der Kurve C und ihrem Inneren besteht; in diesem abgeschlossenen Bereiche ist $f(z)$ stetig und damit gleichmäßig stetig. Das heißt, die Ungleichung (2) gilt gleichmäßig für alle betreffenden Werte von u, wenn wir nur n so auswählen können, daß $|\psi(u) - \psi_n(u)|$ beliebig klein ist, gleichmäßig für alle auf γ_n liegenden Punkte u. Diese letzte Bedingung läßt sich erfüllen infolge der in Γ gleichmäßigen Konvergenz der Folge $\{\psi_n\}$, welche aus dem Courantschen Satz erfolgt[6]). Unser Satz ist hiermit bewiesen.

Wenn jeder der Bereiche $C_1, C_2, \ldots, C_m$ durch eine Jordansche Kurve begrenzt ist und wenn der kürzeste Abstand der abgeschlossenen Bereiche C_i, C_j $(i, j = 1, 2, \ldots, m,\ i \neq j)$ voneinander nicht null ist, so ist jede innerhalb $C_1 + C_2 + \ldots + C_m$ analytische, in der entsprechenden abgeschlossenen Punktmenge stetige Funktion in eine gleichmäßig konvergierende Reihe von Polynomen entwickelbar. Wir führen den Beweis nicht aus, weil derselbe wegen einer ähnlichen Erörterung von Montel[7]) sehr einleuchtend ist.

Die Frage, ob eine innerhalb eines Bereiches analytische und im entsprechenden abgeschlossenen Bereiche stetige Funktion in eine im abgeschlossenen Bereiche gleichmäßig konvergente Reihe von Polynomen entwickelbar ist, gilt in der Enzyklopädie als ungelöst[8]). In dieser Beziehung machen wir vier Bemerkungen, deren Tragweite in Wirklichkeit noch größer ist, als wir hier zu erörtern brauchen.

1. *Unser Satz erstreckt sich nicht bis zum allgemeinsten einfach zusammenhängenden Bereiche der z-Ebene.* Zum Beispiel betrachten wir als Bereich einen außerhalb eines Kreises K liegenden Streifen A, der an einem Ende geschlossen ist und sich spiralförmig gegen K unendlich oft

[6]) A. a. O. S. 108, Satz IV. Unser Satz kann auch durch den Satz IIIa von Courant bewiesen werden.

[7]) A. a. O. Kap. IV. Die Methode stammt von Runge her.

[8]) Hilb und Szász, Bd. II$_2$, Heft 8, S. 1276 (Sept. 1924). Siehe auch Montel, a. a. O. S. 66–71, wo die Frage im Zusammenhang mit der Annäherungsmethode von Tschebyscheff betrachtet ist. Wir haben also bewiesen, daß für eine im Innern einer Jordanschen Kurve analytische und im entsprechenden abgeschlossenen Bereich stetige Funktion die Methode von Tschebyscheff eine gleichmäßig konvergente Folge von Polynomen liefert, deren Grenze die ursprüngliche Funktion ist.

Hilb und Szász bemerken auch, daß Konvexität des Bereiches genügt; ich habe das unabhängig davon auch bemerkt (a. a. O. S. 168, Fußnote), April 1924.

windet und sich beliebig nahe dem Kreis nähert. Die Funktion

$$f(z) = \frac{1}{z-a},$$

wo $z = a$ der Mittelpunkt von K ist, ist überall im abgeschlossenen Bereiche A analytisch, doch nicht in eine im abgeschlossenen Bereiche gleichmäßig konvergente Reihe von Polynomen entwickelbar. Denn die Summe $F(z)$ einer solchen Reihe ist innerhalb K analytisch und in der abgeschlossenen Kreisfläche stetig. Auf der Kreislinie kann aber nicht $F(z) = f(z)$ sein, wegen der Werte der Integrale:

$$\int_K F(z)\,dz = 0, \qquad \int_K f(z)\,dz = 2\pi i.$$

2. *Gegeben ein einfach zusammenhängender Bereich C. Es ist nicht wahr, daß die Möglichkeit, eine beliebige im Inneren von C analytische, im abgeschlossenen Bereiche stetige Funktion in eine im abgeschlossenen Bereiche gleichmäßig konvergente Reihe von Polynomen zu entwickeln, ergibt, daß C durch eine Jordansche Kurve begrenzt ist.* Zum Beispiel, betrachten wir den Bereich C in der $z(= re^{i\theta})$-Ebene:

$$C\colon\ 1 > r \geqq 0, \quad -\pi < \theta < \pi.$$

Jede Funktion $f(z)$, die im Inneren von C analytisch und im entsprechenden abgeschlossenen Bereiche stetig ist, ist auch im ganzen Inneren des Bereiches $1 > r \geqq 0$ analytisch*), und deshalb läßt sich dieselbe auf die beschriebene Weise entwickeln. Trotzdem ist der Bereich C nicht durch eine Jordansche Kurve begrenzt.

Ein Bereich, der die besagte Eigenschaft hat, braucht nicht einfach zusammenhängend zu sein; das Beispiel

$$C'\colon 1 > r > 0$$

ist dem soeben angegebenen Beispiel ähnlich.

3. Wir nennen einen Randpunkt Q eines Bereiches *hebbar*, wenn eine Umgebung von Q existiert, in der kein Punkt liegt, der sich nicht im abgeschlossenen Bereiche befindet. Wenn wir hebbare Randpunkte ausschließen, so gilt der folgende Satz, dessen Beweis wir skizzieren:

Es sei ein zusammenhängender Bereich C in der z-Ebene gegeben. Eine notwendige und hinreichende Bedingung dafür, daß jede im Bereiche C mit Einschluß des Randes reguläre analytische Funktion in eine im abgeschlossenen Bereiche C gleichmäßig konvergente Reihe von Polynomen entwickelbar sei, ist, daß C entweder mit der ganzen Ebene mit Einschluß des Punktes $z = \infty$ zusammenfalle oder mit einem endlichen einfach zu-

*) Siehe z. B. Osgood, a. a. O. S. 315.

sammenhängenden Bereich, dessen Rand die Ebene in genau zwei zusammenhängende Bereiche zerteile[10]).

Wenn der Bereich die ganze Ebene ist, so muß die betreffende Funktion eine Konstante sein, die natürlich entwickelbar ist. Sonst muß der Bereich C beschränkt sein, weil jedes nicht konstante Polynom im Punkte ∞ den Wert ∞ hat und eine Reihe von solchen Polynomen in der Umgebung des Punktes ∞ nicht gleichmäßig gegen eine stetige Funktion konvergieren kann. Von nun an betrachten wir nur endliche Bereiche. Wir nennen $\bar{C}$ den C entsprechenden abgeschlossenen Bereich.

Die besagte Bedingung ist notwendig. Sonst existiert ein in C nicht enthaltener Punkt P: $z = z_1$, der sich nicht mit dem Punkt ∞ durch einen Streckenzug, welcher keinen Punkt von $\bar{C}$ enthält, verbinden läßt. Sämtliche Punkte, die sich mit P durch einen Streckenzug verbinden lassen, der keinen Punkt von $\bar{C}$ enthält, bilden einen einfach zusammenhängenden Bereich B, dessen Rand aus lauter Punkten von C besteht. Die Funktion

$$f(z) = \frac{1}{z - z_1}$$

ist in jedem Punkt von C regulär; wäre sie in C entwickelbar, so wäre die Summe $S(z)$ der Reihe im Inneren von B regulär analytisch, was nach Voraussetzung ausgeschlossen ist.

Wir benutzen hier nämlich den folgenden Satz: *Stimmen die Werte von zwei in einem einfach zusammenhängenden Bereiche B definierten monogenen analytischen Funktionen $f_1(z)$ und $f_2(z)$ auf dem Rande von B überein, und sind die Funktionen in der in B liegenden Umgebung des Randes von B regulär, dann sind die Funktionen identisch.* Wenn nämlich die Funktion $z = \psi(u)$ den Einheitskreis der u-Ebene auf den Bereich B abbildet, so ist die Funktion

$$F(u) = \begin{cases} 0, & |u| = 1, \\ f_1[\psi(u)] - f_2[\psi(u)], & |u| < 1, \\ f_1\left[\psi\left(\frac{1}{\bar{u}}\right)\right] - f_2\left[\psi\left(\frac{1}{\bar{u}}\right)\right], & |u| > 1, \end{cases}$$

für $|u| = 1$ regulär und Null, und daher identisch Null.

[10]) Diese Bedingung lautet auch so, *daß C entweder mit der ganzen Ebene zusammenfalle oder ein endlicher Bereich sei, dessen Komplementärmenge* (d. h. Komplementärmenge des abgeschlossenen Bereiches, in bezug auf die ganze Ebene) *zusammenhänge.*

Diese Bereiche finden sich in einer Klasse von Bereichen, deren schöne Eigenschaften Prof. Carathéodory studiert hat, Math. Annalen 72 (1912), S. 107–144, Kap. III.

Die Bedingung ist hinreichend. Denn nach der Regularität von $f(z)$ im *abgeschlossenen* Bereiche gibt es eine von z_0 unabhängige positive Größe ε, so daß $f(z)$ im Kreise $|z - z_0| < \varepsilon$ regulär ist, wenn nur z_0 in $\bar{C}$ liegt. Es existiert also eine Jordansche Kurve J, welche $\bar{C}$ enthält und deren Inneres aus lauter Regulärpunkten von $f(z)$ besteht; das Resultat folgt hiernach aus dem Theorem von Runge.

Wir beweisen die Existenz der Kurve J genauer. Es gibt in der Tat eine endliche Anzahl von Kreisen $K_1, K_2, \ldots, K_N$, jeder vom Halbmesser ε und mit dem Mittelpunkt auf dem Rande von C, die den Rand von C überdecken. Die Jordansche Kurve, welche aus den zu äußerst liegenden Bogen dieser Kreise besteht, enthält etwa M einfach zusammenhängende Bereiche $B_1, B_2, \ldots, B_M$, deren Punkte weder zu C noch zu $K_1, K_2, \ldots, K_N$ gehören. Ein beliebiger Punkt A_l von B_l läßt sich mit dem Punkt ∞ durch einen Streckenzug verbinden, welcher keinen Punkt von $\bar{C}$ enthält. Wir können diesen Streckenzug in einen Bereich einschließen, der gleichfalls keinen Punkt von $\bar{C}$ enthält und der durch einen anderen Streckenzug S_l begrenzt ist. Die oben gebrauchte Jordansche Kurve J besteht aus Bögen von $K_1, K_2, \ldots, K_N$ und aus Strecken von $S_1, S_2, \ldots, S_M$, die in den Kreisen $K_1, K_2, \ldots, K_N$ liegen.

4. *Es sei die Funktion $f(z)$ im Inneren einer Jordanschen Kurve C analytisch und im entsprechenden abgeschlossenen Bereiche $\bar{C}$ stetig. Eine notwendige und hinreichende Bedingung dafür, daß jede im Inneren von C analytische und im abgeschlossenen Bereiche $\bar{C}$ stetige Funktion $F(z)$ in eine im abgeschlossenen Bereiche $\bar{C}$ gleichmäßig konvergierende Reihe von Polynomen von $f(z)$ entwickelbar sei:*

$$F(z) = \sum_{i=0}^{\infty} \sum_{n=0}^{i} c_{in} [f(z)]^n, \tag{3}$$

besteht darin, daß $f(z)$ im abgeschlossenen Bereiche schlicht sei.

Eine Funktion $f(z)$ heißt im abgeschlossenen Bereiche $\bar{C}$ *schlicht*, wenn die Gleichung

$$f(z_1) = f(z_2),$$

wo z_1 und z_2 zu $\bar{C}$ gehören, die Gleichung $z_1 = z_2$ stets ergibt.

Die Bedingung ist notwendig. Sonst haben wir

$$f(z_1) = f(z_2),$$

wo $z_1 \neq z_2$ ist und z_1 und z_2 in $\bar{C}$ sind. Die Funktion $F(z) = z$ ist dann in $\bar{C}$ nicht entwickelbar, weil die Reihen (3) für $z = z_1$ und für $z = z_2$ identisch sind, jedoch $F(z_1) \neq F(z_2)$ ist.

Die Bedingung ist hinreichend. Der Bereich $\bar{C}$ entspricht einem Bereiche $\bar{B}$ in der u-Ebene mittels der ein-eindeutigen Transformation $u = f(z)$. Der Bereich $\bar{B}$ ist durch eine Jordansche Kurve begrenzt. Wir haben also nach unserem Hauptsatze die im abgeschlossenen Bereiche $\bar{B}$ gleichmäßig konvergierende Entwicklung

$$F[\varphi(u)] = \sum_{i=0}^{\infty} \sum_{n=0}^{i} c_{in} u^n, \tag{4}$$

wo $F(z)$ eine beliebige in C analytische, in $\bar{C}$ stetige Funktion ist, und wo $\varphi(u)$ die Umkehrfunktion von $f(z)$ ist. Die Reihen (3) und (4) sind äquivalent.

Dieser Satz gilt auch für einen unendlichen Bereich $\bar{C}$, der durch eine Jordansche Kurve begrenzt ist.

(Eingegangen am 24. 10. 1925.)

Über die Entwicklung einer Funktion einer komplexen Veränderlichen nach Polynomen.

Von

J. L. Walsh in München[1]).

In dem vorliegenden Artikel soll in der Hauptsache ein Beweis des folgenden Satzes entwickelt werden:

Satz I. *Ist die komplexe Funktion $F(z)$ der komplexen Veränderlichen $z = x + iy$ stetig auf einer Jordanschen Kurve C, die den Punkt $z = 0$ einschließt, so läßt sich $F(z)$ in eine auf C gleichmäßig konvergierende Reihe von Polynomen in z und $1/z$ entwickeln.*

Dieser Satz ist eine Verallgemeinerung des klassischen Satzes von Weierstraß, daß jede stetige periodische Funktion von $\Phi(\theta)$ von Periode 2π sich gleichmäßig entwickeln läßt in eine Reihe von trigonometrischen Polynomen:

$$\Phi(\theta) = \sum_{i=0}^{\infty} \sum_{n=0}^{i} (a_{in} \cos n\theta + b_{in} \sin n\theta).$$

Satz I ist mit dem Satze von Weierstraß identisch, wenn die Kurve C der Einheitskreis ist, denn die Relationen

$$\sin n\theta = \frac{z^n - z^{-n}}{2i}, \qquad \cos n\theta = \frac{z^n + z^{-n}}{2};$$

$$z^n = \cos n\theta + i \sin n\theta, \qquad z^{-n} = \cos n\theta - i \sin n\theta$$

gelten dann auf C. Jedes trigonometrische Polynom ist ein Polynom von z und $1/z$, und umgekehrt.

Satz I findet sich schon bewiesen für den Fall, daß die Kurve C eine analytische Kurve ist[2]).

[1]) National Research Fellow.

[2]) Walsh, Transactions of the American Mathematical Society 26 (1924), S. 168, Fußnote.

Satz I ist dem folgenden neulich bewiesenen Satz[3]) ähnlich, welchen wir im Beweis gebrauchen:

Satz II. *Ist die Funktion $F(z)$ im Inneren einer Jordanschen Kurve C analytisch und im entsprechenden abgeschlossenen Bereiche stetig, so läßt sich $F(z)$ in eine im abgeschlossenen Bereiche gleichmäßig konvergierende Reihe von Polynomen in z entwickeln.*

Satz II wird durch einen Satz von Courant über konforme Abbildung[4]) gewonnen, dessen Benutzung mir Prof. Carathéodory geraten hat.

Zum Beweise von Satz I benutzen wir den folgenden Hilfssatz:

Satz III. *Es sei B ein ringförmiger Bereich, der durch zwei Jordansche Kurven C_1 und C_2, die keinen gemeinsamen Punkt besitzen, begrenzt wird, und es sei die Funktion $F(z)$ im Inneren von B analytisch, im entsprechenden abgeschlossenen Bereiche stetig. Wenn der Nullpunkt im Inneren von C_1 und C_2 liegt, so läßt sich $F(z)$ gleichmäßig im abgeschlossenen Bereiche B nach Polynomen von z und $1/z$ entwickeln. Wenn C_2 im Inneren von C_1 liegt, so ist diese Entwicklung die Summe einer auf und innerhalb C_1 gleichmäßig konvergierenden Reihe von Polynomen von z und einer auf und außerhalb C_2 gleichmäßig konvergierenden Reihe von Polynomen von $1/z$.*

Es seien K_1 und K_2 zwei im Bereiche B liegende analytische Jordansche Kurven, so daß jede die Kurve C_2 einschließt, und es liege K_2 im Inneren von K_1. Die Funktionen

$$(1) \qquad F_1(z) = \frac{1}{2\pi i}\int_{K_1} \frac{F(t)\,dt}{t-z}, \qquad F_2(z) = \frac{1}{2\pi i}\int_{K_2} \frac{F(t)\,dt}{t-z},$$

wo die Integrale im positiven Sinne in bezug auf den durch K_1 und K_2 begrenzten Bereich erstreckt sind, verhalten sich im Inneren von K_1 bzw. im Äußeren von K_2 analytisch. Wenn z zwischen K_1 und K_2 liegt, so gilt die Gleichung

$$(2) \qquad F(z) = F_1(z) + F_2(z).$$

Die Integrale (1) sind von der besonderen Wahl der Kurven K_1 und K_2

[3]) Walsh, Mathematische Annalen 96 (1926), S. 430—436.

(Bemerkung bei der Korrektur.) Ich habe erst neulich (am 23. Juli 1926) durch eine freundliche mündliche Mitteilung von Herrn Marcel Riesz erfahren, daß dieser Satz durch die von Carleman benutzten Methoden sich beweisen läßt. Carleman hat einen ähnlichen aber weniger allgemeinen Satz bewiesen. Vgl. „Über die Approximation analytischer Funktionen", Arkiv för Matematik, Astronomi och Fysik 17 (1922–23).

Herr Riesz hat diese Tatsache vor drei Jahren bemerkt, ohne sie zu publizieren.

[4]) Göttinger Nachrichten 1914, S. 101—109, Satz IV.

unabhängig; wir lassen diese Kurven die Kurven C_1 bzw. C_2 annähern. Die Funktion $F_1(z)$ ist im ganzen Inneren von K_1 regulär-analytisch. Wenn z zwischen K_1 und K_2 liegt, aber sich einem Punkt von C_2 nähert, so nähern sich $F(z)$ und $F_1(z)$ — und deshalb auch $F_2(z)$ — stetigen Grenzwerten. Die Funktion $F_1(z)$ beträgt sich ebenso, wenn z gegen einen Punkt von C_1 geht. D. h. die Funktionen $F_1(z)$ und $F_2(z)$ sind im Inneren von C_1 analytisch, im entsprechenden abgeschlossenen Bereiche stetig, bzw. im Äußeren von C_2 (inklusive des Punktes ∞) analytisch, im entsprechenden abgeschlossenen Bereiche stetig, wenn passende Definitionen dieser Funktionen auf den Kurven gegeben werden. Die Gleichung (2) gilt für alle im abgeschlossenen Bereiche B liegenden Punkte.

Die Funktion $F_1(z)$ läßt sich nach Satz II auf und im Inneren von C_1 nach Polynomen von z gleichmäßig entwickeln. Man sieht durch die Transformation $z = 1/z'$, daß die Funktion $F_2(z)$ sich auf und im Äußeren von C_2 nach Polynomen von z' gleichmäßig entwickeln läßt. Die gliedweise Summe dieser beiden Entwicklungen befriedigt die Behauptungen unseres Satzes.

Jetzt können wir den Satz I leicht beweisen: es sei $u = f(z)$ eine Funktion, die das Innere von C auf das Innere des Einheitskreises Γ in der u-Ebene abbildet, so daß $f(0) = 0$ ist, und es sei $z = \varphi(u)$ die Umkehrfunktion von $f(z)$. Die Funktion $F[\varphi(u)]$ ist auf Γ definiert und stetig und läßt sich dort nach Polynomen von u und $1/u$ nach dem Weierstraßschen Satze gleichmäßig entwickeln. D. h. wir haben auf C die gleichmäßige Entwicklung

$$F(z) = \sum_{i=0}^{\infty} \sum_{n=-i}^{i} c_{in} [f(z)]^n. \tag{3}$$

Die Funktion $f(z)$ läßt sich nach Satz II auf C durch Polynome von z gleichmäßig approximieren. Die Funktion $1/f(z)$ ist im Inneren von C überall regulär-analytisch, außer im einzelnen Punkte $z = 0$. Nach Satz III läßt sich $1/f(z)$ auf C durch Polynome von z und $1/z$ gleichmäßig approximieren[5]). Der Satz I folgt nun mit Hilfe der Gleichung (3).

[5]) Ich verdanke Prof. Hartogs die Idee der folgenden Bemerkung.

Satz III ist an und für sich vielleicht nicht ohne Interesse. Man kann aber beweisen, daß die Funktion $1/f(z)$ sich auf C nach Polynomen von z und $1/z$ gleichmäßig approximieren läßt; daher folgt Satz I ohne Gebrauch von Satz III. Die Funktion $f(z)$ hat nämlich eine einfache Nullstelle in $z = 0$, und die Funktion $1/f(z)$ dort einen einfachen Pol:

$$\frac{1}{f(z)} = \frac{a}{z} + f_1(z),$$

wo $f_1(z)$ im Inneren von C analytisch ist und im entsprechenden abgeschlossenen Bereich stetig. Die Gleichung gilt im abgeschlossenen Bereich. In diesem abgeschlossenen Bereich läßt sich $f_1(z)$ durch Polynome von z gleichmäßig approximieren, also $1/f(z)$ durch Polynome von z und $1/z$.

Einige weitere Tatsachen sind noch bemerkenswert.

1°. Satz I kann auch folgendermaßen ausgesprochen werden:

Ist die Funktion $F(z)$ auf der Jordanschen Kurve C stetig, so läßt sich $F(z)$ in eine auf C gleichmäßig konvergierende Reihe von rationalen Funktionen von z entwickeln.

Die Entwicklung, die im Satze I betrachtet wird, ist schon eine Entwicklung nach rationalen Funktionen, und behält diese Eigenschaft nach einer ganzen oder gebrochenen linearen Transformation der Ebene. Wir brauchen also, um die Äquivalenz dieser Sätze nachzuweisen, nur das Umgekehrte zu betrachten.

Wenn eine Funktion $F(z)$ nach rationalen Funktionen entwickelbar ist, so haben höchstens endlichviele dieser Funktionen Pole auf C. Jede rationale Funktion, die keinen Pol auf C hat, läßt sich auf C nach Satz III durch Polynome in z und $1/z$ gleichmäßig approximieren, wenn die Lage der Kurve C die verlangte ist. Die Funktion $F(z)$ läßt sich also auf C nach Polynomen von z und $1/z$ entwickeln.

Die jetzige Formulierung des Satzes I ist aber auch gültig, wenn die Kurve C den Nullpunkt nicht einschließt, und auch wenn die Kurve C sich ins Unendliche erstreckt.

2°. *Ist die Funktion $f(z)$ auf einer Jordanschen Kurve C stetig, so ist eine notwendige und hinreichende Bedingung dafür, daß eine beliebige auf C stetige Funktion $F(z)$ in eine auf C gleichmäßig konvergierende Reihe von Polynomen von $f(z)$ und $1/f(z)$ entwickelbar sei, daß die Transformation $u = f(z)$ die Kurve C ein-eindeutig auf eine den Punkt $u = 0$ einschließende Jordansche Kurve in der u-Ebene abbilde.*

Die Bedingung ist notwendig. Zuerst bilden die Punkte $u = f(z)$ eine Jordansche Kurve K, weil die Abbildung $u = f(z)$ ein-eindeutig ist. Die Transformation $u = f(z)$ ist nämlich ebenso wie $f(z)$ selbst eindeutig. Die Umkehrfunktion $z = \varphi(u)$ ist gleichfalls eindeutig. In der Tat, wäre

$$f(z_1) = f(z_2), \quad z_1 \neq z_2, \quad z_1, z_2 \text{ auf } C,$$

so wäre die Funktion $F(z) = z$ nicht entwickelbar, denn die Reihen für $z = z_1$ und $z = z_2$ wären dieselben, mit $F(z_1) \neq F(z_2)$.

Liegt zweitens der Punkt $u = 0$ auf K, so ist $1/f(z)$ in einem Punkt unendlich und es können nur endlich viele Glieder der Entwicklung von $F(z)$ negative Potenzen von $f(z)$ enthalten; also ist jede Funktion $F(z)$ nur nach Polynomen von $f(z)$ gleichmäßig entwickelbar. D. h. jede auf K stetige Funktion $F[\varphi(u)]$ wäre nach Polynomen von u gleichmäßig entwickelbar. Liegt anderseits $u = 0$ außerhalb K, so ist $1/u$ auf und innerhalb K regulär-analytisch und dort nach Polynomen von u gleichmäßig entwickelbar; infolgedessen ist ebenfalls jede auf K stetige Funktion $F[\varphi(u)]$ nach Polynomen von u gleichmäßig entwickelbar. Hier haben wir einen Widerspruch, weil die Funktion $F[\varphi(u)] = \frac{1}{u-u_0}$, wo u_0 innerhalb K liegt, stetig ist und doch nicht nach Polynomen von u gleichmäßig entwickelbar*).

Die Bedingung ist hinreichend. Die beliebige stetige Funktion $F(z) = F[\varphi(u)]$ ist auf K' nach Polynomen von u und $1/u$ gleichmäßig entwickelbar, d. h. nach Polynomen von $f(z)$ und $1/f(z)$.

Es ist nach dieser Erörterung klar, daß es keine auf C stetige Funktion $f(z)$ gibt, derartig, daß jede auf C stetige Funktion $F(z)$ nach Polynomen nur von $f(z)$ entwickelbar ist.

Für die Gültigkeit der Bemerkung 2° braucht nicht die Kurve C endlich zu sein.

3°. Es sei B ein Bereich, der durch zwei Jordansche Kurven C_1 und C_2 begrenzt ist, und möge C_1 außerhalb der Kurve C_2 liegen. Es existiert dann keine Funktion $f(z)$, die im Inneren von B analytisch, im entsprechenden abgeschlossenen Bereiche stetig ist, so daß eine beliebige, im Inneren von B analytische, im entsprechenden abgeschlossenen Bereiche stetige Funktion $F(z)$ nach Polynomen von $f(z)$ gleichmäßig entwickelbar ist. Eine notwendige und hinreichende Bedingung dafür, daß jede solche Funktion $F(z)$ nach Polynomen von $f(z)$ und $1/f(z)$ gleichmäßig entwickelbar sei, ist, daß durch die Transformation $u = f(z)$ der Bereich B einem Bereiche B' entspreche, welcher durch zwei Jordansche Kurven ohne gemeinsamen Punkt begrenzt ist, und daß jede dieser Kurven den Nullpunkt einschließe.

*) Eine solche Entwicklung würde eine im ganzen Inneren von C analytische Funktion darstellen, was wegen des folgenden Satzes unmöglich ist:

Stimmen die Werte von zwei in einem einfach zusammenhängenden Bereiche B definierten monogenen analytischen Funktionen auf dem Rande von B überein, und sind die Funktionen in der in B liegenden Umgebung des Randes von B regulär, in der entsprechenden abgeschlossenen Umgebung stetig, dann sind die Funktionen identisch.

Ein Beweis dieses Satzes findet sich skizziert bei Walsh, Math. Annalen loc. cit. Es ist auch nicht notwendig für die Gültigkeit dieses Satzes bzw. seines dort gegebenen Beweises, daß die Werte der fraglichen Funktionen auf dem *ganzen* Rande von B übereinstimmen.

Wir haben in der Tat durch die dort gegebene Skizze den folgenden Satz:

Es sei die Funktion $F(z)$ auf dem Rande eines beschränkten Bereiches B regulär-analytisch. Eine notwendige und hinreichende Bedingung dafür, daß $F(z)$ auf dem Rande von B (bzw. im ganzen abgeschlossenen Bereich B) nach Polynomen von z gleichmäßig entwickelbar sei, besteht darin, daß $F(z)$ in jedem Punkt regulär-analytisch sei, der sich nicht mit dem Punkt ∞ durch einen Streckenzug, der keinen Punkt des Randes von B enthält, verbinden läßt.

Dieser Satz gilt auch, wenn der Bereich B mehrfach zusammenhängend ist.

Der Beweis ist einfach und wird dem Leser überlassen.

4°. Man kann ein weitergehendes Resultat als den Satz I herleiten, wenn die Funktion $F(z)$ auf C eine Bedingung von Lipschitz befriedigt und wenn die Kurve C im Sinne von Osgood[7]) regulär ist.

Die Funktion $F(z)$ läßt sich nämlich als die Summe

(4) $$F(z) = F_1(z) + F_2(z)$$

schreiben, wo $F_1(z)$ im Inneren von C analytisch ist und im entsprechenden abgeschlossenen Bereich stetig, und wo $F_2(z)$ im Äußeren von C analytisch ist und im entsprechenden abgeschlossenen Bereich stetig[8]).

Entwickelt man die Funktionen $F_1(z)$ und $F_2(z)$, so bekommt man nach (4) die Entwicklung

(5) $$F(z) = \sum_{i=1}^{\infty} \sum_{n=0}^{i} a_{i_n} z^n + \sum_{i=1}^{\infty} \sum_{n=0}^{-i} a_{i_n} z^n$$

wo die Reihen auf und im Inneren von C, bzw. auf und im Äußeren von C gleichmäßig konvergieren. Die Entwicklung (5) ist natürlich nicht möglich für alle auf C bloß *stetigen* Funktionen, wenn auch C regulär ist.

5°. *Es sei der Bereich B durch die Jordanschen Kurven $C_0, C_1, \ldots, C_n$ begrenzt, wo B innerhalb C_0 und außerhalb $C_1, C_2, \ldots, C_n$ liegt, und wo die Kurve C_i keinen gemeinsamen Punkt mit der Kurve C_j $(i, j = 0, 1, \ldots, n,\ i \neq j)$ hat. Es seien $z_1, z_2, \ldots, z_n$ beliebige Punkte innerhalb $C_1, C_2, \ldots, C_n$ bzw. Ist die Funktion $F(z)$ innerhalb B analytisch und im entsprechenden abgeschlossenen Bereiche stetig, so ist sie in demselben abgeschlossenen Bereiche die Summe von $(n+1)$ Funktionen:*

$$F(z) = F_0(z) + F_1(z) + F_2(z) + \ldots + F_n(z),$$

wo $F_0(z)$ im Inneren von C_0 regulär und im entsprechenden abgeschlossenen Bereiche stetig ist und wo $F_k(z)$, $k = 1, 2, \ldots, n$, im Äußeren von C_k regulär und im entsprechenden abgeschlossenen Bereiche stetig ist. Die Funktion $F(z)$ ist also im abgeschlossenen Bereiche B als die Summe von $(n+1)$ Reihen entwickelbar:

$$F(z) = \sum_{i=0}^{\infty} \sum_{m=0}^{i} a_{im} z^m + \sum_{i=0}^{\infty} \sum_{m=0}^{i} a'_{im} (z - z_1)^{-m}$$

$$+ \sum_{i=0}^{\infty} \sum_{m=0}^{i} a''_{im} (z - z_2)^{-m} + \ldots + \sum_{i=0}^{\infty} \sum_{m=0}^{i} a^{(n)}_{im} (z - z_n)^{-m}.$$

Jede dieser Reihen konvergiert im abgeschlossenen Bereiche B gleichmäßig; die erste konvergiert auf und im Inneren von C_0 gleichmäßig, die

[7]) Funktionentheorie I (Leipzig 1912), S. 51.

[8]) Plemelj, Monatshefte für Math. und Phys. 19 (1909), S. 205–210.

$(k+1)$-te, $k=1,2,\ldots,n$, konvergiert auf und im Äußeren von C_k gleichmäßig.

Der Beweis dieses Satzes ist fast genau derselbe wie der Beweis des Satzes III.

6°. *Jede Funktion $F(z)$, die auf einem Jordanschen Kurvenstück C' stetig ist, läßt sich auf C' nach Polynomen von z gleichmäßig entwickeln.*

Man kann eine Jordansche Kurve C konstruieren, von welcher das Stück C' ein Bogen ist[9]). Wir nehmen an, daß der Nullpunkt innerhalb C liegt. Die Funktion $F(z)$ läßt sich auf C fortsetzen, so daß die erweiterte Funktion $F(z)$ auf der ganzen Kurve C stetig ist. Nach Satz I läßt sich die Funktion $F(z)$ auf C nach Polynomen von z und $1/z$ gleichmäßig entwickeln.

Die Funktion $1/z$ läßt sich auf C' nach Polynomen von z gleichmäßig approximieren. In der Tat, sei K ein Jordansches Kurvenstück, welches den Nullpunkt mit dem Punkt ∞ verbindet, und welches keinen Punkt von C' enthält[10]). Die in solcher Weise aufgeschnittene Ebene ist ein einfach zusammenhängender Bereich B, in dessen Inneren die Funktion $1/z$ regulär-analytisch ist, und dessen Inneres den Punkt ∞ nicht enthält. Nach dem wohlbekannten Satze von Runge ist also die Funktion $1/z$ im Inneren von B nach Polynomen von z entwickelbar und die Reihe konvergiert gleichmäßig in jedem ganz im Inneren von B liegenden abgeschlossenen Bereich, daher auf C'.

Da $F(z)$ auf C' nach Polynomen von z und $1/z$ gleichmäßig entwickelbar ist, und da die Funktion $1/z$ sich auf C' durch Polynome von z gleichmäßig approximieren läßt, so ist $F(z)$ auf C' nach Polynomen von z gleichmäßig entwickelbar.

Die Bemerkung 6° ist eine direkte Verallgemeinerung des klassischen Weierstraßschen Satzes, daß jede in einem abgeschlossenen Intervall $a \leqq x \leqq b$ stetige Funktion $F(x)$ sich in diesem Intervall nach Polynomen in x gleichmäßig entwickeln läßt.

Diese Bemerkung 6° hat der Verfasser erst nach einer Unterredung mit Professor Hartogs gemacht. Professor Hartogs hatte schon diese Bemerkung für ein analytisches Kurvenstück C' gebraucht, um weitere Anwendungen auf Entwicklungen nach Polynomen zu machen. Die allgemeinere Bemerkung 6° ist ebenfalls weiterer Anwendungen fähig[11]).

---

[9]) Vgl. von Kerékjártó, Topologie I, S. 69 (Berlin 1923).

[10]) Ein einfacher Bogen zerlegt die Ebene nicht. von Kerékjártó, l. c. S. 67.

[11]) Hartogs und Rosenthal, eine Arbeit, die in den Mathem. Annalen erscheint. (Bemerkung bei der Korrektur.) Noch eine weitere Anwendung ist folgende: *Ist die Funktion $F(z)$ auf einer beschränkten abzählbaren Punktmenge M, die*

(Fortsetzung auf der nächsten Seite.)

Für die ursprüngliche Bemerkung 6° muß die Kurve C' ganz im Endlichen liegen, aber für die folgende Bemerkung — die sehr leicht zu beweisen ist — kann die Kurve C' sich ins Unendliche erstrecken.

Es sei die Funktion $f(z)$ auf einem Jordanschen Kurvenstück C' stetig. Eine notwendige und hinreichende Bedingung dafür, daß jede auf C' stetige Funktion nach Polynomen von $f(z)$ gleichmäßig entwickelbar sei, besteht darin, daß $f(z_1) \neq f(z_2)$ sei, wenn $z_1 \neq z_2$ ist und z_1 und z_2 auf C' liegen.

Diese Bedingung ist nämlich die, daß die Transformation $u = f(z)$ das Kurvenstück C' eineindeutig auf ein beschränktes Jordansches Kurvenstück abbildet.

Wenn die Funktion $f(z)$ reell ist, so ist also notwendig und hinreichend, daß $f(z)$ monoton im engeren Sinne sei.

Sind die Punkte der Kurve C' als eine eindeutige stetige Funktion $z(t)$ des reellen Parameters t gegeben, wo $z(t_1) \neq z(t_2)$ ist, wenn $t_1 \neq t_2$ ist, so heißt *monoton im engeren Sinne*, daß $t_1 < t_2$ stets $f[z(t_1)] < f[z(t_2)]$ ergibt, oder daß $t_1 < t_2$ stets $f[z(t_1)] > f[z(t_2)]$ ergibt. Wenn die Kurve C' ein Intervall der reellen Axe ist, so ist diese Bedingung, daß $f(z)$ streng monoton sei[19]).

Es gibt also keine stetige reelle oder komplexe Funktion $f(z)$ von Periode p, derartig, daß eine beliebige stetige Funktion $F(z)$ der reellen Veränderlichen z von Periode p nach Polynomen von $f(z)$ gleichmäßig approximierbar ist. Nach Bemerkung 2° gibt es auch keine reelle stetige Funktion $f(z)$ von Periode p, so daß eine beliebige stetige Funktion $F(z)$ der reellen Veränderlichen z von Periode p nach Polynomen von $f(z)$ und $1/f(z)$ gleichmäßig entwickelbar ist. Eine notwendige und hinreichende Bedingung für eine solche *komplexe* Funktion $f(z)$ ist natürlich in Bemerkung 2° enthalten.

7°. Wenn man eine stetige Funktion einer reellen Veränderlichen x durch Polynome von $f(x)$ gleichmäßig approximieren will, so ist es keines-

---

nur endlich viele Häufungspunkte besitzt, stetig, dann ist $F(z)$ auf M nach Polynomen in z gleichmäßig entwickelbar. Man konstruiert in der Tat ein Jordansches Kurvenstück C', welches die Menge M enthält, und man erweitert die Definition der Funktion $F(z)$, so daß sie überall auf C' definiert und stetig ist. Die Funktion $F(z)$ ist auf C' nach Polynomen gleichmäßig entwickelbar, also auch auf M. Diese Aussage wurde von Herrn Szegö und mir zusammen formuliert.

[19]) Eine solche Transformation $u = f(z)$ wird oft von Lebesgue, S. Bernstein, Jackson, de la Vallée-Poussin und anderen in der Theorie der Approximation durch Polynome einer reellen Veränderlichen gebraucht, um Resultate über rationale Polynome bei der Annäherung durch trigonometrische Polynome anzuwenden, und umgekehrt. Vgl. de la Vallée-Poussin, Approximation des fonctions d'une variable réelle (Paris 1919).

wegs notwendig, daß $f(x)$ selbst stetig sei[18]). Wir beweisen in der Tat das folgende für reelle Funktionen:

Es sei die reelle Funktion $f(x)$ definiert auf der beschränkten Punktmenge C der reellen Achse. Eine notwendige und hinreichende Bedingung dafür, daß eine willkürliche auf C im engeren Sinne stetige Funktion $F(x)$ nach Polynomen von $f(x)$ gleichmäßig entwickelbar sei, besteht darin, daß $f(x)$ eine beschränkte Funktion sei, deren Umkehrfunktion eindeutig und im engeren Sinne stetig ist.

Die Funktion $g(z)$ heißt *im engeren Sinne stetig*, wenn $\lim_{n\to\infty} z_n = z_0$ nur dann in sich schließt, daß $\lim_{n\to\infty} g(z_n)$ existiert, wenn $g(z_n)$ definiert ist, und daß diese Grenze gleich $g(z_0)$ ist, wenn auch $g(z_0)$ definiert ist.

Diese Bedingungen sind notwendig. Die Funktion $f(x)$ muß beschränkt sein, sonst ist keine beschränkte Funktion $F(x)$ gleichmäßig entwickelbar, außer einer Konstanten. Die Umkehrfunktion $x = \varphi(u)$ von $u = f(x)$ muß eine eindeutige Funktion von u sein. Sonst hätten wir

$$f(x_1) = f(x_2), \qquad x_1 \neq x_2.$$

In diesem Falle wäre die Funktion $F(x) \equiv x$ nicht gleichmäßig entwickelbar, weil die Reihen für $x = x_1$ und $x = x_2$ dieselben wären, mit $F(x_1) \neq F(x_2)$.

Die Funktion $\varphi(u)$ muß im engeren Sinne stetig sein, sonst hätten wir

$$\lim_{n\to\infty} u_n = u_0, \qquad \lim_{n\to\infty} \varphi(u_n) = \varphi_0,$$
$$\lim_{n\to\infty} u'_n = u_0, \qquad \lim_{n\to\infty} \varphi(u'_n) = \varphi'_0 \neq \varphi_0,$$

wo die Werte u'_n alle gleich u_0 sein dürfen. Wir hätten auch anderseits für $F(x) \equiv x$ die Reihen

$$F(x) = \sum_{i=0}^{\infty} \sum_{n=0}^{i} c_{in} [f(x)]^n, \quad c_{in} \text{ konstant,} \tag{6}$$

$$\varphi(u_k) = \sum_{i=0}^{\infty} \sum_{n=0}^{i} c_{in} u_k^n, \qquad \varphi(u'_k) = \sum_{i=0}^{\infty} \sum_{n=0}^{i} c_{in} (u'_k)^n.$$

---

[18]) Die Funktion $f(x)$ kann ja in jedem Punkt unstetig sein. Es sei zum Beispiel das Intervall $0 \leqq x \leqq 1$; wir setzen

$$f(x) = \begin{cases} x, & x \text{ rational}, \\ 2+x, & x \text{ irrational}. \end{cases}$$

Man kann jede für $0 \leqq x \leqq 1$ stetige Funktion durch Polynome von $f(x)$ gleichmäßig approximieren.

Die entsprechenden Gleichungen für die Grenzwerte:

$$\varphi_0 = \sum_{i=0}^{\infty} \sum_{n=0}^{i} c_{in} u_0^n, \qquad \varphi_0' = \sum_{i=0}^{\infty} \sum_{n=0}^{i} c_{in} u_0^n,$$

enthalten also den Widerspruch $\varphi_0 = \varphi_0'$.

Die Bedingung ist hinreichend. Die Funktion $F[\varphi(u)]$ ist vielleicht nicht auf einer abgeschlossenen Punktmenge definiert, aber läßt sich erweitern, so daß sie auf einer abgeschlossenen Punktmenge definiert und stetig ist und zwar so, daß sie für alle Werte von u definiert und stetig ist. Wenn M so gewählt ist, daß $|f(x)| \leqq M$ ist, so ist $F[\varphi(u)]$ für $-M \leqq u \leqq M$ nach Polynomen von u gleichmäßig entwickelbar:

$$F[\varphi(u)] = \sum_{i=0}^{\infty} \sum_{n=0}^{i} c_{in} u^n$$

ist, welches gleichwertig mit (6) ist. Hiermit ist unser Beweis vollendet.

Dieser Satz gilt auch für eine unbeschränkte Punktmenge C, wenn in der Definition der Stetigkeit im engeren Sinne auch der Wert ∞ als Grenzwert für abhängige und unabhängige Veränderliche zulässig ist, und wenn wir nur beschränkte Funktionen $F(x)$ entwickeln.

8°. Man kann auch analytische Funktionen einer komplexen Veränderlichen durch Polynome einer *unstetigen* Funktion gleichmäßig approximieren. Wir geben nur ein sehr einfaches Beispiel; die Erörterung läßt sich leicht verallgemeinern und der Leser wird dann einen allgemeinen Satz formulieren können.

In der $z(=x+iy)$-Ebene sei $u = f_1(z)$ die Funktion, die das Innere des Halbkreises $y > 0$, $|z| < 1$ auf das Innere des Kreises $C_1 : |u| \leqq 2$ abbildet und sei $u = f_2(z)$ die Funktion, die das Innere des Halbkreises $y < 0$, $|z| < 1$ auf das Innere einer beliebigen Jordanschen Kurve C_2 abbildet, die im Kreise $|u - 3| \leqq \frac{1}{2}$ liegt. Die Funktionen $f_1(z)$ bzw. $f_2(z)$ sind in diesen abgeschlossenen Halbkreisen stetig, wenn passende Definitionen der Funktionen auf den Rändern gegeben werden. Es sei C_3 eine beliebige geschlossene Jordansche Kurve, die im Kreise $|u - 5i| \leqq 1$ liegt, und C_4 ein beliebiges Jordansches Kurvenstück: $u = \omega(t)$, $0 \leqq t \leqq 1$ [wobei $\omega(t_1) \neq \omega(t_2)$ ist, wenn $t_1 \neq t_2$ ist], das im Inneren des Kreises $|u - 9 + 6i| = 3$ liegt. Es sei $u = f_3(z)$ die Funktion, die das Innere des Kreises $C: |z| = 1$ auf das Innere von C_3 abbildet; die Funktion $f_3(z)$ ist im abgeschlossenen Bereich $|z| \leqq 1$ stetig.

Wir betrachten jetzt die Funktion:

$$f(z) = \begin{cases} f_1(z), & y > 0,\ |z| < 1, \\ f_2(z), & y < 0,\ |z| < 1, \\ \omega(z), & 0 < x < 1, \\ f_1(z), & -1 \leqq x < 0,\ x \text{ rational}, \\ f_2(z), & -1 < x < 0,\ x \text{ irrational}, \\ f_3(z), & |z| = 1,\ z^2 \neq 1, \\ 1492, & z = 0, \\ 1776, & z = 1, \end{cases} \quad \Bigg\} \ y = 0,$$

die im ganzen abgeschlossenen Bereich C: $|z| \leqq 1$ definiert ist. Wir behaupten: *Ist $F(z)$ im Inneren von C analytisch, im entsprechenden abgeschlossenen Bereich stetig, so läßt sich $F(z)$ nach Polynomen von $f(z)$ im abgeschlossenen Bereich gleichmäßig entwickeln.*

Es sei $z = \varphi(u)$ die Umkehrfunktion von $u = f(z)$. Die Funktion $F[\varphi(u)]$ ist, wenn die Definition passend erweitert wird, im Inneren von C_1, C_2, C_3 analytisch, im entsprechenden abgeschlossenen Bereich stetig. Nach Bemerkung 6° läßt sich $F[\varphi(u)]$ auf C_4 durch Polynome von u gleichmäßig approximieren. Wir setzen noch

$$F[\varphi(u)] = \begin{cases} F(0), & |u - 1492| \leqq 1, \\ F(1), & |u - 1776| \leqq 1. \end{cases}$$

Die in solcher Weise erweiterte Funktion $F[\varphi(u)]$ ist also nach Polynomen von u gleichmäßig entwickelbar[14]), und es bleibt nur u durch $f(z)$ zu ersetzen, um die Behauptung zu erweisen.

Eine solche Funktion $f(z)$, die die Eigenschaft hat, daß jede Funktion $F(z)$, die für $|z| < 1$ analytisch und im abgeschlossenen Bereich C: $|z| \leqq 1$ stetig ist, sich nach Polynomen von $f(z)$ gleichmäßig entwickeln läßt, braucht aber in keinem Punkt stetig zu sein. Die Funktion

$$f(z) = \begin{cases} z, & x \text{ und } y \text{ rational}, \\ f_3(z), & \text{in jedem anderen Punkte}, \end{cases}$$

wo $f_3(z)$ die obige Bedeutung hat, besitzt die besagte Eigenschaft.

9°. Wir fügen noch einen Satz über den *Grad der Approximation* hinzu:

Es sei die Funktion $F(z)$ auf einer Jordanschen Kurve C definiert. Eine notwendige und hinreichende Bedingung dafür, daß Polynome n-ten

[14]) Walsh, Math. Annalen, loc. cit.

Grades $V_n(z)$ *existieren,* $n = 0, 1, 2, \ldots$, *derartig, daß die Ungleichheit*

$$(7) \qquad |F(z) - V_n(z)| < \frac{B}{R^n},$$

$B, R > 1$, konstant und von n, z unabhängig,

für sämtliche z *auf* C *befriedigt sei, besteht darin, daß eine auf und im Inneren von* C *regulär-analytische Funktion* $F(z)$ *existiere, die auf* C *mit der gegebenen Funktion* $F(z)$ *übereinstimmt.*

Wenn die Ungleichheit (7) für sämtliche z auf C befriedigt ist, so existiert natürlich eine im Inneren von C reguläre, im entsprechenden abgeschlossenen Bereiche stetige Funktion $f(z)$, die auf C mit der gegebenen Funktion $F(z)$ übereinstimmt, so daß (7) für alle z im abgeschlossenen Bereiche befriedigt ist; die Funktion $F(z)$ ist bloß die Grenzfunktion der Folge $V_n(z)$.

Das Wesentliche dieses Satzes findet sich schon bei Szegö[15]), obgleich nicht ausdrücklich betont, aber nur für den Fall, daß die Kurve C analytisch ist.

Wir geben den Beweis dieses Satzes, wenn die Kurve C der Einheitskreis ist. Einerseits ist, wenn die Funktion $F(z)$ auf und im Inneren von C regulär-analytisch ist, $F(z)$ regulär-analytisch in einem mit C konzentrischen, aber größeren Kreis, und die Abschnitte $V_n(z)$ der Taylorschen Entwicklung um den Nullpunkt von $F(z)$ befriedigen die Bedingung (7). Anderseits bekommt man für die Taylorschen Koeffizienten c_n von $F(z)$, wenn $F(z)$ auf C gegeben ist, so daß (7) befriedigt ist,

$$c_n = \frac{1}{2\pi i}\int_C F(z)\, z^{-n-1}\, dz = \frac{1}{2\pi i}\int_C [F(z) - V_{n-1}(z)]\, z^{-n-1}\, dz,$$

$$(8) \qquad |c_n| \leq \frac{B}{R^{n-1}}.$$

Aus (8) folgt unmittelbar, daß die Taylorsche Entwicklung der vorher beschriebenen Funktion $f(z)$ in einem Kreis vom Radius ϱ $(1 < \varrho < R)$ gleichmäßig konvergiert. Die Funktion $F(z)$ ist daher auf C regulär-analytisch.

[15]) Math. Zeitschr. 9 (1921), S. 218–270, insbesondere S. 263–267; der Beweis dafür, daß die besagte Bedingung hinreichend ist, stammt im wesentlichen von Fejér her.

Dieses Resultat wurde, wenn C eine Ellipse ist, von S. Bernstein schon früher bewiesen: Mémoires Acad. Roy. de Belgique, Cl. des Sc. (2) 4 (1912), Sätze 24, 61.

Für den Kreis vgl. auch de la Vallée-Poussin, loc. cit. Ch. VIII, IX.

(Bemerkung bei der Korrektur.) Siehe auch eine Arbeit von Walsh, Münchner Berichte, 1926.

Dieser soeben gegebene Beweis stammt im wesentlichen von Szegö; er gilt fast unverändert für irgendeine analytische Jordansche Kurve C, wenn man nicht mehr die Taylorsche Entwicklung, sondern die Entwicklung nach den zur Kurve C gehörenden Polynomen $P_n(z)$ (von Szegö) gebraucht.

Der Satz erstreckt sich aber bis zur allgemeinsten Jordanschen Kurve C. Die besagte Bedingung ist hinreichend. Die Funktion $F(z)$ ist auf und im Inneren von C regulär-analytisch, darum in einem größeren abgeschlossenen Bereiche regulär-analytisch, der aus einer außerhalb C liegenden *analytischen* Jordanschen Kurve C_1 und ihrem Inneren besteht. Eine Folge von Polynomen existiert mit der Eigenschaft (7) für jedes z auf und im Inneren von C_1, infolgedessen für jedes z auf und im Inneren von C.

Die Bedingung ist notwendig. Es gibt nach einem Satz von Carathéodory[16]) außerhalb bzw. innerhalb C liegende *analytische* Jordansche Kurven C_1 und C_2, so daß das Äußere von C_1 bzw. C_2 auf das Äußere des Kreises $|u| = \varrho$ bzw. $|u| = 1$ $(1 < \varrho < R)$ abgebildet wird durch eine und dieselbe Transformation $u = \varphi(z)$, wobei $\varphi(\infty) = \infty$ ist. Die Entwicklung der schon definierten Funktion $f(z)$ nach den zu C_2 gehörenden Polynomen $P_n(z)$ von Szegö konvergiert gleichmäßig im abgeschlossenen Bereiche, welcher aus der Kurve C_1 und ihrem Inneren besteht, wegen der zu (8) entsprechenden Abschätzung. Der Satz ist also vollständig bewiesen.

Der Satz von Carathéodory und folglich auch unser Satz (mit dem obigen Beweis) gilt für allgemeinere Bereichgrenzen als Jordansche Kurven.

Wir greifen den folgenden Satz heraus:

Es sei die Punktmenge C die volle Grenze eines beschränkten (ev. mehrfach zusammenhängenden) Bereiches B, und es sei die Funktion $F(z)$ auf C definiert. Eine notwendige und hinreichende Bedingung dafür, daß Polynome n-ten Grades $V_n(z)$ existieren, $n = 0, 1, 2, \ldots$, derartig, daß die Ungleichheit (7) für sämtliche z auf C befriedigt sei, besteht darin, daß eine im abgeschlossenen Inneren von C regulär-analytische Funktion $F(z)$ existiere, die auf C mit der gegebenen Funktion $F(z)$ übereinstimmt. Hier heißt das a b g e s c h l o s s e n e I n n e r e *von C die Punktmenge $\bar{C}$ aller Punkte, deren keiner sich mit dem Punkt ∞ durch einen Streckenzug verbinden läßt, der keinen Punkt von C enthält. Diese Punktmenge $\bar{C}$ ist also abgeschlossen; jeder Punkt von C selbst gehört dazu.*

---

[16]) Math. Annalen 72 (1912), S. 107–144, Kap. III.

Die Bedingung ist hinreichend. Die Funktion $F(z)$ ist auf und innerhalb einer Jordanschen Kurve analytisch, welche jeden Punkt des abgeschlossenen Inneren von C enthält[17]). Polynome existieren also, welche die Eigenschaft (7) für alle zu $\bar{C}$ gehörenden Punkte z besitzen.

Die Bedingung ist auch notwendig. Die zu $\bar{C}$ komplementäre Menge (in bezug auf die ganze Ebene) ist ein Bereich, dessen Grenze zur Punktmenge C gehört. Der Satz von Carathéodory und seine Anwendung gelten also wie vorher, wenn B einfach zusammenhängend ist. Wenn der Bereich B nicht einfach zusammenhängend ist, so ist noch eine kurze Erörterung nötig. Die Punktmenge B', die aus den Punkten von B und den Punkten des Inneren jedes ganz in B liegenden Polygons besteht, ist ein einfach zusammenhängender Bereich, von welchem jeder Grenzpunkt auch ein Grenzpunkt von B ist. Das abgeschlossene Innere der Grenze von B' fällt mit dem abgeschlossenen Inneren $\bar{C}$ der Grenze von B zusammen und enthält natürlich jeden Punkt der Bereiche B und B'. Die Grenzfunktion $f(z)$ der Folge $V_n(z)$ ist auf der ganzen Punktmenge $\bar{C}$ definiert, und die Ungleichheit (7) gilt für $f(z)$ statt $F(z)$ für jeden Punkt z von $\bar{C}$. Wir gebrauchen den Bereich B' statt B in der Anwendung des Satzes von Carathéodory.

---

[17]) Vgl. Walsh, Math. Annalen, loc. cit., Bemerkung 3.

(Eingegangen am 26. 1. 1926.)

ON THE EXPANSION OF ANALYTIC FUNCTIONS IN SERIES OF POLYNOMIALS AND IN SERIES OF OTHER ANALYTIC FUNCTIONS*

BY

J. L. WALSH

1. **Introduction.** The present paper is substantially a continuation of a previous paper† in which polynomial developments of an arbitrary analytic function were considered, culminating in three theorems. The first of these theorems is in essence a modification and completion of a result due to Birkhoff, a generalization of Taylor's development about the origin in the plane of the complex variable x:‡

THEOREM I. *Let the functions*

$$p_0(x),\ p_1(x),\ p_2(x),\ \cdots$$

be analytic for $|x| \leqq 1+\epsilon$, *and such that on and within the circle* γ': $|x| = 1+\epsilon$, *we have*

$$| p_k(x) - x^k | \leqq \epsilon_k \qquad (k = 0,1,2,\cdots), \tag{1}$$

where the series $\sum \epsilon_k^2$ *converges to a sum less than unity, and where the series* $\sum \epsilon_k$ *converges. Then there exists a set of functions* $P_k(x)$ *continuous for* $|x| \geqq 1$, *analytic for* $|x| > 1$,§ *zero at infinity, and such that*

$$\int_\gamma P_k(x)p_i(x)dx = \delta_{ki} = \begin{cases} 0, & i \neq k, \\ 1, & i = k, \end{cases} \qquad \gamma: |x| = 1. \tag{2}$$

If $F(x)$ *is any function integrable and with an integrable square (in the sense of Lebesgue), then the two series*

$$\sum_{k=0}^{\infty} a_k x^k, \qquad a_k = \frac{1}{2\pi i}\int_\gamma F(x)x^{-k-1}dx, \tag{3}$$

$$\sum_{k=0}^{\infty} c_k p_k(x), \qquad c_k = \int_\gamma F(x)P_k(x)dx, \tag{4}$$

have on γ *(and hence*‖ *in the closed region* $|x| \leqq 1$*) essentially the same con-*

* Presented to the Society, September 9, 1927; received by the editors in May, 1927.

† These Transactions, vol. 26 (1924), pp. 155–170. We shall refer to this paper as I.

‡ I, p. 159.

§ See below, § 2.

‖ A convergent series of constant terms dominates the term-by-term difference of series (3) and (4) for $|x| = 1$ and hence for $|x| \leqq 1$.

vergence properties, in the sense that their term-by-term difference approaches uniformly and absolutely the sum zero. In particular if $F(x)$ is continuous for $|x|\leqq 1$, analytic for $|x|<1$, and satisfies a Lipschitz condition on γ, then the series (4) *converges uniformly to the sum $F(x)$ in the closed region $|x|\leqq 1$.*

In I this theorem was applied, after conformal transformation, to obtain the two other theorems mentioned, the first on the expansion of an analytic function in terms of polynomials, the second including the analogue of the Laurent series. In the present paper we treat (Part A) more in detail the analogy between the two series (3) and (4), considering arbitrary series of type (4), the analogue of Abel's theorem and its converse, convergence properties on circles other than γ, and the uniqueness of expansions. In Part B we apply these results to the case of polynomials belonging to a given region, and collect the main results of the paper in Theorem IX. We consider in particular the expansion of a discontinuous function, in Theorem XI. It is found that under certain conditions Gibbs's phenomenon occurs, precisely as for Fourier's series. In Part C we study the use of polynomial expansions in connection with multiply-connected regions, obtaining certain results on the boundary values of analytic functions.

A. Series of analytic functions

2. **Modification of proof of Theorem I.** The proof of Theorem I given in I is needlessly complicated. It is perhaps worth while to present in some detail a modification, for we shall need later certain inequalities obtained.

We apply the Lemma used in I, choosing the interval $0\leqq\phi\leqq 2\pi$ as the circle γ: $|x|=1$, using $x=e^{i\phi}$ on γ. The functions $\{u_n(\phi)\}$ and $\{U_n(\phi)\}$ are taken (modifying the argument of I, pp. 162–3) simply as

$$(5)\qquad u_n(\phi)=\frac{x^n}{(2\pi)^{1/2}},\qquad U_n(\phi)=\frac{p_n(x)}{(2\pi)^{1/2}}\qquad (n=0,1,2,\cdots).$$

Thus we have immediately

$$(6)\qquad c_{nk}=\int_0^{2\pi}(U_n-u_n)\bar{u}_k d\phi,$$

$$(7)\qquad \sum_{k=0}^{\infty} c_{nk}\bar{c}_{nk}\leqq\int_0^{2\pi}(U_n-u_n)(\bar{U}_n-\bar{u}_k)d\phi\leqq\epsilon_n^2,$$

$$c_{nk}=\int_0^{2\pi}(U_n-u_n)\bar{u}_k d\phi=\frac{1}{2\pi i}\int_\gamma (p_n(x)-x^n)\frac{dx}{x^{k+1}}$$

$$=\frac{1}{2\pi i}\int_{\gamma'}(p_n(x)-x^n)\frac{dx}{x^{k+1}},$$

$$(8) \qquad | c_{nk} | \leqq \frac{\epsilon_n}{(1+\epsilon)^k} \cdot$$

The function $V_k(\phi)$ of I is therefore given by the equation*

$$(9) \qquad V_k(\phi) = \sum_{i=0}^{\infty} (d_{ki} + \delta_{ki}) u_i .$$

We have, however, the inequalities

$$(10) \qquad | d_{ki} + c_{ik} | \leqq \frac{p_i p_k}{1-p}, \qquad p_i^2 = \sum_{j=0}^{\infty} | c_{ij} |^2, \qquad p^2 = \sum_{i,j=0}^{\infty} | c_{ij} |^2,$$

$$(11) \qquad \sum_{i=0}^{\infty} | d_{ki} | \ll \sum_{i=0}^{\infty} | d_{ki} + c_{ik} | + \sum_{i=0}^{\infty} | c_{ik} | .$$

We define the functions $P_k(x)$ so as to make the two following series identical:

$$\sum_{k=0}^{\infty} c_k p_k(x), \qquad c_k = \int_\gamma F(x) P_k(x) dx,$$

$$(12) \qquad \sum_{k=0}^{\infty} b_k U_k(\phi), \qquad b_k = \int_\gamma F(x) \bar{V}_k(\phi) d\phi .$$

That is, we set

$$c_k = \frac{b_k}{(2\pi)^{1/2}}, \qquad P_k(x) = \frac{1}{(2\pi)^{1/2}} \bar{V}_k(\phi) \frac{d\phi}{dx} \cdot$$

We have of course

$$x = e^{i\phi}, \qquad dx = ie^{i\phi} d\phi, \qquad \frac{d\phi}{dx} = \frac{1}{ix} \cdot$$

It follows, then, directly from (9) and (5) that the functions $P_k(x)$ are continuous for $|x| \geqq 1$, analytic for $|x| > 1$, and zero at infinity. Moreover, if the series $\sum \epsilon_k$ is dominated by a convergent geometric series, then the functions $P_k(x)$ are analytic likewise for $|x| = 1$.† In fact, the series (11) is also dominated by a convergent geometric series, by virtue of (7):

$$p_i \leqq \epsilon_i ,$$

* See Walsh, these Transactions, vol. 22 (1921), p. 234, where the Lemma used in I is proved, and inequalities (10) and (11) likewise derived.

† The writer withdraws the statement in I, pp. 159, 163, that the functions $P_k(x)$ are analytic *on* γ, when the ϵ_k are not further restricted. Thus in the proof of I, Theorem I, we choose the ϵ_k so that the series $\sum \epsilon_k$ is dominated by a convergent geometric series.

and by virtue of (8). Then the series (9), when conjugate complex quantities are taken, is a Laurent series whose coefficients are dominated by a convergent geometric series, so $\overline{V}_k(x)$, and hence also $P_k(x)$, is analytic for $|x|=1$.

3. **Development of continuous functions on γ.** If the function $F(x)$ is continuous for $|x|\leqq 1$ and analytic for $|x|<1$, then the Taylor development of $F(x)$ about the origin converges, when summed by the method of Cesàro, uniformly for $|x|\leqq 1$ to the value $F(x)$. In fact the Taylor development is on γ precisely the Fourier development of $F(x)$, which when summed as described converges uniformly on γ to the value $F(x)$, hence uniformly on and within γ to the value $F(x)$. The Taylor series itself converges for $|x|<1$, by the usual inequalities for the coefficients of a power series, and hence converges to the value $F(x)$, because in case of a convergent series the sum assigned by the Cesàro summation process is the sum of the series.

Application of this remark yields, if we remember that series (3) and (4) have essentially the same convergence properties in the entire closed region $|x|\leqq 1$,

THEOREM II. *If $F(x)$ is continuous for $|x|\leqq 1$ and analytic for $|x|<1$, then the series* (4) *converges uniformly to the sum $F(x)$ in any closed region $|x|\leqq|x_0|<1$, and the sequence formed from* (4) *by the Cesàro summation method converges uniformly for $|x|\leqq 1$, to the sum $F(x)$.*

We turn now from the consideration of series (4) arising from functions $F(x)$ given on γ, to the consideration of series of the form

$$\sum_{k=0}^{\infty} g_k p_k(x) \tag{13}$$

with arbitrary coefficients g_k.

4. **Convergence of arbitrary series** (13). If no further restriction is placed on the quantities ϵ_k than in Theorem I, it is not true that the convergence of (13) for $x=x_0$ enables us to conclude the convergence of (13) for all values of x such that $|x|<|x_0|$. Let us set, in fact,

$$p_0(x) = 1, \quad p_k(x) = x^k - \delta^k, \quad k > 0,$$

where δ is positive and so small that for $\epsilon_0=0$, $\epsilon_k=\delta^k$, $k>0$, the required conditions on ϵ_k are fulfilled. Then every series (13) converges for $x=\delta$, yet need not converge for every x such that $|x|<\delta$. Indeed, under this same definition for $p_k(x)$, every series

$$\sum_{k=0}^{\infty} g_{2^k} p_{2^k}(x)$$

converges whenever $x=\omega\delta$, $\omega^{2^n}=1$, n being integral. This series converges therefore on a point set everywhere dense on the circle $|x|=\delta$, yet does not necessarily converge for $x=0$.

Under suitable restrictions on the ϵ_k we can prove the result for series (13) which is analogous to the well known result for Taylor's series:

THEOREM III. *If the series* (13) *converges for* $x=x_0$, *where* $|x_0|\leqq 1+\epsilon$, *and if the series* $\sum_{k=0}^{\infty}\epsilon_k t^k$ *converges for every (finite) value of* t, *then the series* (13) *converges for all values of* x *such that* $|x|<|x_0|$, *and the convergence is uniform for all values of* x *such that* $|x|\leqq|x_1|<|x_0|$.

We naturally assume $x_0\neq 0$; the contrary case is without content. We prove actually a stronger theorem than that stated, for we use not the convergence of (13) for $x=x_0$ but merely the boundedness of the terms of the series.

The inequality

$$|p_k(x_0)-x_0^k|\leqq\epsilon_k$$

gives at once the double inequality

$$1-\frac{\epsilon_k}{|x_0^k|}\leqq\frac{|p_k(x_0)|}{|x_0^k|}\leqq 1+\frac{\epsilon_k}{|x_0^k|}\cdot$$

But we have $\lim_{k\to\infty}\epsilon_k/|x_0{}^k|=0$, and hence $\lim_{k\to\infty}|p_k(x_0)|/|x_0{}^k|=1$. Therefore if the quantities $g_kp_k(x_0)$ are uniformly bounded, so also are the quantities $g_kx_0{}^k$, and conversely. From the boundedness of the $g_kx_0{}^k$:

$$|g_kx_0^k|<M,$$

follows the inequality

$$|g_k|<\frac{M}{|x_0^k|}\cdot$$

We now make use of the inequality

$$|p_k(x)|\leqq|x|^k+\epsilon_k,$$

or

$$\sum_{k=0}^{\infty}|g_kp_k(x)|\ll\sum_{k=0}^{\infty}|g_k|\cdot|x|^k+\sum_{k=0}^{\infty}|g_k|\epsilon_k.$$

The first series on the right converges uniformly for $|x|\leqq|x_1|$, since the individual terms of that series are bounded for $x=x_0$. The second series on the right, which does not contain x, converges. Hence the series on the left converges uniformly for $|x|\leqq|x_1|$, and Theorem III is established.

The argument just given includes practically a proof of the fact that for points x such that $|x|\leqq 1+\epsilon$, the two series

$$\sum_{k=0}^{\infty} g_k x^k \text{ and } \sum_{k=0}^{\infty} g_k p_k(x)$$

have the same points of convergence, of absolute convergence, of divergence, of summability, and the same regions (or point sets) of uniform convergence. The only exception here occurs if the Taylor series converges only at the point $x=0$. For if there exists a single value of x, say $x_0 \neq 0$, of the kind considered so that either of these two series converges, we have

$$|g_k| < \frac{M}{|x_0^k|}.$$

It follows that for all values of x, $|x| \leqq 1+\epsilon$, we have uniformly

$$|g_k p_k(x) - g_k x^k| \leqq \frac{M\epsilon_k}{|x_0{}^k|}.$$

The series $\sum_{k=0}^{\infty} M\epsilon_k / |x_0{}^k|$ converges and does not contain x, so the statement is proved.

There is no exceptional case here, even if the Taylor series converges only for $x=0$, provided $p_k(0)=0$, $k=1, 2, \cdots$; compare Theorems VII and IX below.

We can state now two interesting results of this discussion; the first result gives the radius of convergence in terms of the coefficients. Here and in the remainder of the paper we assume, unless otherwise stated, the series $\sum \epsilon_k t^k$ to converge for all values of t.

THEOREM IV. *If we set*

$$\overline{\lim_{k\to\infty}} |g_k|^{1/k} = \frac{1}{\rho},$$

then if $\rho \leqq 1+\epsilon$, *series* (13) *converges for* $|x| < \rho$ *and diverges for* $1+\epsilon > |x| > \rho$; *if* $\rho > 1+\epsilon$, *series* (13) *converges for* $|x| \leqq 1+\epsilon$.

The generalized theorem of Abel yields its analogue for the series (13):

THEOREM V. *If the series* (13) *converges for the value* $x=x_1$, *where* $|x_1| \leqq 1+\epsilon$, *then this series converges uniformly in the closed region bounded by two arbitrary line segments terminating at the point* x_1 *and by an arc of the circle* $|x| = |x_2| < |x_1|$.

If (13) *converges uniformly on an arc* x_1x_2 *of the circle* $|x| = |x_1|$, *then this series converges uniformly in the closed region bounded by this circular arc, by two arbitrary line segments lying in the region* $|x| \leqq |x_1|$ *and terminated respectively by* x_1 *and* x_2, *and by an arc of a circle* $|x| = |x_3| < |x_1|$.

5. **Analogue of Tauber's theorem.** For series (13) we can give likewise a converse (not exact) of Abel's Theorem:

THEOREM VI. *If the series* (13) *is such that* $\lim_{k\to\infty} g_k/k = 0$, *and if for radial approach* to the point* x_1 *on* γ *we have*

$$\lim_{x\to x_1} f(x) = g,$$

where $f(x)$ *denotes the value of the* (*convergent for* $|x|<1$) *series* (13), *then we have also*

$$\sum_{k=0}^{\infty} g_k p_k(x_1) = g.$$

If we define the numbers b_k by the relations

$$b_k = (2\pi)^{1/2} g_k,$$

and then set

$$a_k = b_k + c_{0k}b_0 + c_{1k}b_1 + c_{2k}b_2 + \cdots \qquad (k = 0,1,2,\cdots), \tag{14}$$

we find by the use of the Schwarz inequality in conjunction with (8),

$$|a_k - b_k| \leq \frac{\left(\sum_{i=0}^{\infty} |b_i|^2\right)^{1/2}}{(1+\epsilon)^k}. \tag{15}$$

By our hypothesis on the g_k, the series $\sum_{i=0}^{\infty} |b_i|^2$ converges. The a_k defined by (14) are such that $\lim_{k\to\infty} a_k/k = 0$, by (15). These numbers a_k are such that $\sum_{k=0}^{\infty} |a_k|^2$ converges, hence, by the Riesz-Fischer theorem, there exists a function $F(x)$ defined on γ, integrable and with an integrable square, whose coefficients with respect to the normal orthogonal system $\{u_k\}$ used in §2 are the numbers a_k. The numbers b_k, subjected to the condition that $\sum_{k=0}^{\infty} |b_k|^2$ should converge, are uniquely determined by (14),† and hence the numbers $g_k = b_k/(2\pi)^{1/2}$ are the coefficients of $F(x)$ in its expansion (4). Thus (3) and (13) have essentially the same convergence properties in and on γ.

By Tauber's theorem,‡ we have, since $\lim_{k\to\infty} a_k/k = 0$,

$$\sum_{k=0}^{\infty} a_k x_1^k = g,$$

* The result holds also for approach in various other ways. See for example Landau, *Ergebnisse der Funktionentheorie*, Berlin, 1916, Kap. III. Compare our application of Theorem VI in Theorem IX.

† See the reference in § 2 to the proof of the Lemma used in I.

‡ See Landau, loc. cit.

and thus we have as well

$$\sum_{k=0}^{\infty} g_k p_k(x_1) = g,$$

and the theorem is established.

Many other results similar to Theorem VI, analogues of results for Taylor's series, might be established. We choose, however, to treat the equivalence of series (3) and (4) on circles other than γ.

6. **Properties of series on circles other than γ.** We prove the following theorem:

THEOREM VII. *If $p_k(x)$ has at least a k-fold root at the origin, then an arbitrary function $F(x)$ integrable and with an integrable square on the circle Γ: $|x| = \mu < 1+\epsilon$ can be formally expanded in a series of type (4), where the coefficients are found by integration over Γ. This series (4) and the Taylor development (formal) of $F(x)$ have on and within the circle Γ the same convergence properties, in the sense that their term-by-term difference converges absolutely and uniformly on Γ and in its interior to the sum zero.*

We perform the substitution $z = x/\mu$, $x = \mu z$, so as to apply Theorem I directly to the unit circle in the z-plane. We require for application of Theorem I the inequality

$$(16) \qquad \left| \frac{p_k(\mu z)}{\mu^k} - z^k \right| \leqq \epsilon_k \text{ for all } |z| \leqq 1 + \epsilon', \quad \epsilon' > 0.$$

Expansion of the function $F(\mu z)$ on the circle $|z| = 1$ in terms of the functions $p_k(\mu z)/\mu^k$, which approximate to the functions z^k, will yield of course a formal expansion of $F(x)$ on the circle Γ in terms of the functions $p_k(x)$. The Taylor expansion of $F(\mu z)$, a power series in z, transforms into a Taylor expansion of $F(x)$, a power series in x.

Our original inequality

$$| p_k(x) - x^k | \leqq \epsilon_k, \quad |x| \leqq 1 + \epsilon,$$

may be written

$$\left| \frac{p_k(x)}{x^k} - 1 \right| \leqq \frac{\epsilon_k}{|x^k|}, \quad x \neq 0.$$

But the function $(p_k(x)/x^k) - 1$ is analytic without exception for $|x| \leqq 1+\epsilon$, when properly defined for $x = 0$, and its greatest absolute value in that closed region is taken on for $|x| = 1+\epsilon$. Thus we have

$$\left| \frac{p_k(x)}{x^k} - 1 \right| \leqq \frac{\epsilon_k}{(1+\epsilon)^k}, \quad |x| \leqq 1 + \epsilon.$$

Transformation to the z-plane gives the equivalent inequalities

$$\left| \frac{p_k(\mu z)}{\mu^k z^k} - 1 \right| \leqq \frac{\epsilon_k}{(1+\epsilon)^k}, \quad \mu |z| \leqq 1 + \epsilon,$$

$$\left| \frac{p_k(\mu z)}{\mu^k} - z^k \right| \leqq \frac{\epsilon_k |z^k|}{(1+\epsilon)^k}, \quad \mu |z| \leqq 1 + \epsilon. \tag{17}$$

The right-hand member of (17) is not greater than ϵ_k provided we restrict z as follows:

$$|z| \leqq \frac{1+\epsilon}{\mu} \text{ if } \mu > 1, \quad |z| \leqq 1 + \epsilon \text{ if } \mu \leqq 1. \tag{18}$$

The upper limits of z in (18) are both greater than unity, so (17) yields directly (16), and Theorem VII is completely established.

The proof of Theorem VII has not assumed any restriction on the quantities ϵ_k beyond that of Theorem I. In fact, the condition that $p_k(x)$ should have at least a k-fold root at the origin is a very favorable one with reference to successive approximations and equivalence of expansions. With the conditions imposed, the requirements on the ϵ_k of Theorem I can be considerably lightened; we do not, however, carry out the details here.

We remark, too, that a result similar to Theorem VII is readily proved under the assumption that a convergent geometric series dominates the series $\sum \epsilon_k$, without the assumption that $p_k(x)$ has at least a k-fold root at the origin; but here there is in general a lower limit *greater than zero* on the radius μ of the circle Γ. We omit the proof of this remark.

The following theorem is by no means the most general result of its kind that can be easily established:

THEOREM VIII. *If $p_k(x)$ has at least a k-fold root at the origin, and if the series $\sum \epsilon_k t^k$ converges for every t, then the expansion of type* (4) *of any function $F(x)$ analytic at the origin is unique. The functions $P_k(x)$ of Theorem* I *are analytic over the entire plane except at the origin.*

If $F(x)$ is analytic on and within γ, there cannot exist two distinct expansions of $F(x)$ of the form

$$F(x) = \sum_{k=0}^{\infty} c_k p_k(x), \quad F(x) = \sum_{k=0}^{\infty} g_k p_k(x) \tag{19}$$

both of which converge uniformly on γ. For multiplication of these series through by $P_k(x)\,dx$ and integration term by term over γ gives by (2) the equality of c_k and g_k.

We return to the more general situation of Theorem VIII. If two series of the form (19) both converge at even a single point for which $|x| \leqq 1+\epsilon$, they converge uniformly on and within some circle Γ' whose center is the origin. By the remark just made concerning uniqueness of expansions, and by the proof of Theorem VII, the two expansions are identical if they represent the same function on any circle whatever whose center is the origin. If the two series represent $F(x)$ in a region lying interior to the circle $|x| = 1+\epsilon$, they both represent that function throughout their entire regions of convergence interior to the circle $|x| = 1+\epsilon$.

The analyticity of the functions $P_k(x)$ of Theorem I over the entire plane except at the origin follows from (9) and (11) as used in §2, with the new properties of the series $\sum \epsilon_k$. Compare also Theorem IXa, which does not use those new properties. Under the present hypothesis, then, the integrals which appear in (4) can be taken over any rectifiable Jordan curve which lies interior to γ and in whose interior the origin lies, provided the function $F(x)$ is analytic for $|x| \leqq 1$. The functions $P_k(x)$ which arise in Theorem I for the circle γ, and the functions $P_k(x)$ which arise in the proof of Theorem VII by application of Theorem I to the transform in the z-plane of the circle Γ are identical; this can be verified by making the change of variable in all the formulas involved.

B. Series of polynomials

7. **Application of results of A.** We now apply Theorem I and the theorems which have just been proved in connection with it, deriving results as in I (p. 163 et seq.) for expansions of arbitrary functions in terms of polynomials. We choose the quantities ϵ_k to satisfy the requirements of Theorem I and also so that $\sum \epsilon_k t^k$ converges for every t.* The polynomial $p_k(z)$ is to be chosen so as to have a k-fold root at the origin; this choice is possible; compare I, p. 164, or Theorem X below. Then we have

Theorem IX. *In the plane of the complex variable z let C be a simple closed finite analytic curve*† *which includes in its interior the origin. Then*

* See also the condition of I, p. 164, and its application in the proof of Theorem IXa.

† That is to say, a curve whose points can be put into one-to-one (regular-) analytic correspondence with the points of a circle. It is then a classical theorem in the study of conformal mapping that the region interior to C can be mapped on the interior of a circle so that the mapping is one-to-one and conformal in the *closed* regions considered, therefore one-to-one and conformal in larger regions including those closed regions in their interiors. See Picard, *Traité d'Analyse*, II, Paris, 1893, pp. 272, 276, or Bieberbach, *Einführung in die konforme Abbildung*, Berlin, 1913, p. 120.

From this theorem it follows at once, in the notation of I or of Theorem IX, that for points z_1 and z_2 on C, the quotients $(z_1-z_2)/(\phi(z_1)-\phi(z_2))$ and $(\phi(z_1)-\phi(z_2))/(z_1-z_2)$ are uniformly bounded. Hence a function which satisfies a Lipschitz condition on C corresponds to a function which satisfies a Lipschitz condition on the circle γ, and conversely.

the interior of C can be mapped one-to-one and conformally on the interior of the unit circle γ *in the* x*-plane by some transformation* $x=\phi(z)$, $z=\psi(x)$, *where* $\phi(0)=0$, *and the transformation will be one-to-one and conformal for the mapping of the closed interior of* $C_{1+\epsilon}$, *an analytic Jordan curve in whose interior C lies, onto the closed interior of the circle* $|x|=1+\epsilon$, $\epsilon>0$. *In general we denote by* C_ρ *the transform of the circle* $|x|=\rho$, *where* $0<\rho<1+\epsilon$.

Then there exists a set $\{p_k(z)\}$ *of polynomials in z and a set of functions* $\{s_k(z)\}$ *analytic at every point of the extended plane except the origin, zero at infinity, and such that*

$$\int_{C_\rho} s_k(z)p_i(z)dz = \begin{cases} 0, & i \neq k, \\ 1, & i = k. \end{cases}$$

If the function $F(z)$ *is analytic interior to* C_ρ, *continuous in the corresponding closed region, and satisfies a Lipschitz condition on* C_ρ, *then the series*

$$\sum_{k=0}^{\infty} a_k p_k(z), \quad a_k = \int_{C_\rho} F(z)s_k(z)dz, \tag{20}$$

converges uniformly in this closed region to the value $F(z)$. *If* $F(z)$ *is required merely to be analytic interior to* C_ρ *and continuous in the corresponding closed region, then* (20) *converges uniformly to the value* $F(z)$ *interior to an arbitrary curve* $C_{\rho'}$, $\rho'<\rho$, *and when summed by the method of Cesàro,* (20) *converges uniformly to the value* $F(z)$ *in the closed region bounded by* C_ρ.

If $F(z)$ *is an arbitrary function defined on* C_ρ, *integrable and with an integrable square, and if the condition**

$$\int_{C_\rho} F(z)z^k dz = 0 \qquad (k = 0,1,2,\cdots),$$

is satisfied, then the two series

$$\sum_{k=0}^{\infty} a_k' x^k, \qquad a_k' = \frac{1}{2\pi i}\int_{|x|=\rho} F[\psi(x)]x^{-k-1}dx,$$

and (20) *transformed by* $z=\psi(x)$ *have essentially the same convergence properties on and within the circle* $|x|=\rho$, *in the sense that their term-by-term difference converges absolutely and uniformly to the sum zero for* $|x|\leq\rho$.

---

* No condition is necessary here if we use

$$a = \int_{|x|=\rho} F[\psi(x)]P_k(x)dx$$

instead of (20).

An arbitrary series of the form

$$\sum_{k=0}^{\infty} g_k p_k(x) \tag{21}$$

which converges for a single point z *on* C_ρ, *converges uniformly interior to* $C_{\rho'}$, *if* $\rho' < \rho$. *If* (21) *diverges for a point* z *on* C_ρ *that series diverges for all points* z *exterior to* C_ρ *and interior to* $C_{1+\epsilon}$. *If in general we set*

$$\limsup_{k\to\infty} |g_k|^{1/k} = \frac{1}{\rho}$$

then if $\rho \leqq 1+\epsilon$, *series* (21) *converges for* z *interior to* C_ρ *and diverges for* z *exterior to* C_ρ *but interior to* $C_{1+\epsilon}$; *if* $\rho > 1+\epsilon$, *series* (21) *converges for* z *on or interior to* $C_{1+\epsilon}$. *If* $0 < \rho < 1+\epsilon$, *some singular point of the function represented by the series lies on the curve* C_ρ.

If (21) *converges for the value* $z = z_1$ *on* C_ρ, *then this series converges uniformly in the closed region bounded by two arbitrary line segments terminating in* z_1, *and by an arc of a curve* $C_{\rho'}$, *where* $\rho' < \rho$. *If* (21) *converges uniformly on an arc* z_1z_2 *of the curve* C_ρ *then this series converges uniformly in the closed region bounded by that arc, by two arbitrary line segments whose interiors are interior to* C_ρ *and which are terminated respectively by* z_1 *and* z_2, *and by an arc of the curve* $C_{\rho'}$, *where* $\rho' < \rho$.

If (21) *is such that* $\lim_{k\to\infty} g_k(k\rho^k)^{-1} = 0$, *and if for approach along the normal* to* C_ρ *to the point* z_1 *on* C_ρ *we have*

$$\lim_{z\to z_1} f(z) = g,$$

where $f(z)$ *denotes the value of the (convergent for* z *interior to* C_ρ*) series* (21), *then we have also*

$$\sum_{k=0}^{\infty} g_k p_k(z_1) = g.$$

An arbitrary series (21), *convergent for a single value of* z *interior to* $C_{1+\epsilon}$ *and not the origin, is the unique expansion of form* (20) *of some function* $F(z)$ *analytic on and within some curve* C_ρ.

The only part of this theorem not a direct result of our previous theorems is the fact that $s_k(z)$ is analytic over the entire z-plane except at the origin. This should give the reader no difficulty, using Theorem VIII and the method of I, p. 165; compare also §9.

* A more general theorem might easily be announced; see the footnote to Theorem VI. Here we do not apply Theorem VI directly, but the more general theorem suggested in connection with Theorem VI.

It will be noticed that Theorem IX does not mention convergence of the series (20) or (21) outside of $C_{1+\epsilon}$. The reason for this omission will become clearer after we have proved a general theorem on approximation.

8. **A general theorem on approximation.** We prove a much more general theorem than necessary for our immediate purposes:*

THEOREM X. *If the function $f(z)$, defined on the bounded point set S, can be approximated on that point set as closely as desired by a polynomial in z, and if there be given any p points $z_1, z_2, \cdots, z_p$ of S together with an arbitrary $\epsilon > 0$, then there exists a polynomial $p(z)$ such that*

$$|p(z) - f(z)| \leqq \epsilon, \quad z \text{ on } S,$$

and

$$p(z_i) = f(z_i) \qquad (i = 1, 2, \cdots, p).$$

We prove Theorem X by means of Lagrange's Interpolation Formula, and find it convenient first to prove the following

LEMMA. *If $z_1, z_2, \cdots, z_p$, R are considered fixed, if we have*

$$|G_k| \leqq \eta \qquad (k = 1, 2, \cdots, p),$$

and if $G(z)$ denotes the polynomial defined by Lagrange's Interpolation Formula which takes on the values G_k in the points z_k, $k = 1, 2, \cdots, p$, then there exists a constant M independent of η so that we have

$$|G(z)| \leqq M\eta \text{ for all } z, \ |z| \leqq R.$$

For simplicity we take R so large that $|z_k| \leqq R$, $k = 1, 2, \cdots, p$. The Lagrange Formula is

$$G(z) = \sum_{\nu=1}^{p} G_\nu \frac{(z - z_1) \cdots (z - z_{\nu-1})(z - z_{\nu+1}) \cdots (z - z_p)}{(z_\nu - z_1) \cdots (z_\nu - z_{\nu-1})(z_\nu - z_{\nu+1}) \cdots (z_\nu - z_p)},$$

so the Lemma is obvious if we merely set

$$M = \sum_{\nu=1}^{p} \frac{(2R)^{p-1}}{|z_\nu - z_1| \cdots |z_\nu - z_{\nu-1}| \cdot |z_\nu - z_{\nu+1}| \cdots |z_\nu - z_p|}.$$

* It seems inconceivable that this theorem is not already in the literature, but the writer has been unable to find a reference to it. The corresponding theorem for approximation by trigonometric polynomials is given by D. Jackson, Bulletin of the American Mathematical Society, vol. 32 (1926), pp. 259–262.

Theorem X holds of course for approximation of real functions by means of real polynomials, and can be extended (1) by requiring the agreement of certain derivatives of the approximating polynomial with the corresponding derivatives of the given function, (2) by assigning as the values of the polynomial (and derivatives) not the exact but values near to the values of the function (and derivatives), (3) by noticing that for points z_k off S arbitrary values $f(z_k)$ may be assigned.

Theorem X follows easily now. Choose R so large that all points of S lie in the circle $|z| \leqq R$. Choose a polynomial $q(z)$ (which exists by hypothesis) so that we have

$$|q(z) - f(z)| \leqq \frac{\epsilon}{1+M}, \quad z \text{ on } S.$$

For the polynomial $G(z)$ we assign the values

$$G(z_k) = q(z_k) - (fz_k) \qquad (k = 1, 2, \cdots, p),$$

so that we have by the Lemma

$$|G(z)| \leqq \frac{M\epsilon}{1+M}, \quad z \text{ on } S.$$

Then the polynomial

$$p(z) = q(z) - G(z)$$

satisfies all the requirements of Theorem X.

The polynomials $p_k(z)$ of Theorem IX are polynomials which uniformly approximate to the functions $[\phi(z)]^k$ respectively on and within $C_{1+\epsilon}$. By a classical theorem due to Runge, these polynomials may be subjected to the auxiliary condition of uniformly approximating other analytic functions —let us say constants—in arbitrary non-intersecting closed Jordan regions outside of $C_{1+\epsilon}$. In particular we may by Theorem X require that the polynomials $p_k(z)$ shall actually take on arbitrarily preassigned values at an arbitrary number of points exterior to $C_{1+\epsilon}$. Thus we may choose (1) the value zero for the points $z_1, z_2, \cdots, z_p$ (independent of k), in which case all series of the form (21) converge at those points, or we may choose (2)

$$p_k(z_i) = k! \qquad (i = 1, 2, \cdots, p; \quad k = 1, 2, 3, \cdots),$$

in which case no series (21) not convergent throughout the interior of $C_{1+\epsilon}$ converges at the points z_i. It is because of this difference in behavior that Theorem IX omits mention of the convergence or divergence of (21) outside of $C_{1+\epsilon}$.

9. **Further properties of expansions.** One interesting property of series (20) has not yet been mentioned, which brings out still more clearly the analogy with Taylor's series:

THEOREM IXa. *The coefficients a_k in* (20) *can be written in the form*

$$a_k = A_0^{(k)}F(0) + A_1^{(k)}F'(0) + \cdots + A_k^{(k)}F^{(k)}(0),$$

where $A_i^{(k)}$ is a constant independent of $F(z)$, and where $F^{(i)}(0)$ indicates the ith derivative of $F(z)$ at the origin.

Differentiation of (20), with insertion of the value $z=0$, yields

$$\begin{aligned} F(0) &= a_0 p_0(0), \\ F'(0) &= a_1 p_1'(0), \\ F''(0) &= a_1 p_1''(0) + a_2 p_2''(0), \\ F'''(0) &= a_1 p_1'''(0) + a_2 p_2'''(0) + a_3 p_3'''(0), \\ & \cdots\cdots\cdots\cdots \end{aligned}$$

Here we use the fact that $p_0(z)$ is constant, and that $p_k(z)$ has at least a k-fold root at the origin and hence (I, p. 164), having precisely k roots interior to $C_{1+\epsilon}$, has precisely a k-fold root at the origin. This system of equations is therefore such that $p_k^{(k)}(0)$ is always different from zero, $k=0, 1, 2, \cdots$, and hence the system can be solved for the coefficients a_k linearly in terms of the $F^{(i)}(0)$.

As an application of Theorem IXa, it may be noticed that $s_k(z)$ can be written in the form

$$S_0(z) = \frac{B_1^{(0)}}{z}, \quad S_k(z) = \frac{B_2^{(k)}}{z^2} + \frac{B_3^{(k)}}{z^3} + \cdots + \frac{B_{k+1}^{(k)}}{z^{k+1}}, \quad k > 0;$$

this follows directly from the integral formula for the derivatives of $F(z)$.

One may consider in some detail the expansion of an arbitrary function $\Phi(z)$, analytic on and interior to C, in terms not of the polynomials $p_k(z)$ but in terms of their derivatives $p_k'(z)$. Let $F(z)$ be any integral of $\Phi(z)$, so that we have for z on and interior to C,

$$F(z) = a_0 p_0(z) + a_1 p_1(z) + a_2 p_2(z) + \cdots, \qquad a_k = \int_C F(z) s_k(z) dz,$$

$$F'(z) = \Phi(z) = a_1 p_1'(z) + a_2 p_2'(z) + \cdots.$$

The term $a_0 p_0'(z)$ is here omitted, for $p_0(z)$ is a constant.

The integral

$$\int_C s_k(z) dz, \quad k > 0,$$

is equal to zero, for this integral may be written

$$\frac{1}{p_0(z)} \int_C s_k(z) p_0(z) dz,$$

known to vanish by Theorem IX. Hence the indefinite integral $\sigma_k(z)$ of $s_k(z)$ is single-valued in and on C.

Let us integrate

$$a_k = \int_C F(z)s_k(z)dz, \quad k > 0,$$

by parts, $\int u dv = uv - \int v du$, setting $u = F(z)$, $dv = s_k(z)dz$. We find

$$a_k = -\int_C \Phi(z)\sigma_k(z)dz.$$

As is to be expected, we find also by partial integration

$$\int_C \sigma_k(z)p_i'(z)dz = -\int_C s_k(z)p_i(z)dz = -\delta_{ik}.$$

That is, there exists a set of functions $\{-\sigma_k(z)\}$ such that the two sets $\{p_k'(z)\}$ and $\{-\sigma_k(z)\}$ are biorthogonal. An arbitrary function $\Phi(z)$ analytic on and within C can be expanded in the series

$$\Phi(z) = a_1p_1'(z) + a_2p_2'(z) + \cdots, \quad a_k = -\int_C \Phi(z)\sigma_k(z)dz,$$

which converges uniformly on and within C.

Both this remark on the derived functions and Theorem IXa can be applied in the x-plane to the series (4) under the hypothesis of Theorem VIII.

10. **Expansion of discontinuous functions.** There are considered in I not merely series such as (20), but likewise series in polynomials $q_k(z)$ in the reciprocal of z.* These series are used in I to expand arbitrary functions satisfying a Lipschitz condition on C. In order to study the expansion of *discontinuous* functions in such series, we investigate the function or functions represented by Cauchy's Integral

$$F(z) = \frac{1}{2\pi i}\int_C \frac{f(t)dt}{t - z},$$

where the given function $f(t)$ is discontinuous. We shall suppose C to be the same curve previously considered, although the discussion holds under much broader conditions.

A particularly simple kind of discontinuity, that of a finite jump, is typified by the function $f(t) = \log t$, where we choose as that branch of the

* We notice that the argument used in I, p. 166, to prove $b_0 = 0$ can be somewhat shortened. We choose, in fact, $q_0(z) = 1$, $q_k(\infty) = 0$ for $k > 0$. Then in the expansion of $f_2(z)$:

$$f_2(z) = \sum_{k=0}^{\infty} b_k q_k(z),$$

it is obvious that $b_0 = 0$ when $f_2(z) = 0$ for $z = \infty$.

function $\log t$ the branch which is real and positive for the smallest real positive value of t on C, say t_0. Consider in general the plane cut along the line $0t_0$, and from t_0 to infinity along a curve exterior to C. The determination of the branch of $\log t$ considered is then made by the use of continuity in the cut plane. The function $f(t)$ is continuous on C except for a finite jump at t_0 of magnitude $2\pi i$.

We evaluate the integral

$$F(z) = \frac{1}{2\pi i}\int_C \frac{\log t}{t-z}\,dt$$

by partial integration, $\int u\,dv = uv - \int v\,du$, setting $u = \log t$, $dv = dt/(t-z)$. We find

$$F(z) = \frac{1}{2\pi i}\log t\cdot\log(t-z)\Big|_C - \frac{1}{2\pi i}\int_C \frac{\log(t-z)}{t}\,dt.$$

The first term in the right-hand member has the value $2\pi i + \log t_0 + \log(t_0 - z)$, or $\log(t_0 - z)$, according as z lies interior or exterior to C. The proper determination of $\log(t-z)$ is to be found by continuity, moving t along C until it coincides with t_0, then by moving z, not crossing the cut, until z coincides with the origin. The second term in the right-hand member has a zero derivative with respect to z, if z lies interior to C, as is seen by direct computation. The value of the integral, z interior to C, is therefore a constant equal to the value for $z=0$:

$$-\frac{1}{2\pi i}\int_C \frac{\log t}{t}\,dt = -\frac{1}{4\pi i}\log^2 t\Big|_C = -\pi i - \log t_0.$$

The value of this same integral

$$-\frac{1}{2\pi i}\int_C \frac{\log(t-z)}{t}\,dt$$

when z lies exterior to C is, by Cauchy's Formula, $-\log(-z)$. We have finally, therefore,

$$f_1(z) \equiv F(z) = \pi i + \log(t_0 - z), \quad z \text{ interior to } C,$$

$$f_2(z) \equiv -F(z) = -\pi i - \log(t_0 - z) + \log z, \quad z \text{ exterior to } C.$$

As a check we have $f(z) = f_1(z) + f_2(z)$ when z lies on C (except in case $z = t_0$, where the functions $f_1(z)$ and $f_2(z)$ are, strictly speaking, not defined), as we should have by the results of Plemelj.* The function $f_1(z)$ is analytic on

* See I, p. 167. The validity of the equation $f(z) = f_1(z) + f_2(z)$ is dependent, provided $f(z)$ satisfies certain large conditions of integrability, merely on the behavior of the function $f(z)$ in the neighborhood of the point z considered; the satisfaction of a Lipschitz condition in such a neighborhood is sufficient for the validity of the equation.

and interior to C except at the single point t_0; the function $f_2(z)$ is analytic on and exterior to C except at t_0 and vanishes at infinity; both functions are integrable and have an integrable square on C. We notice too by direct computation

$$\int_C f_1(t)t^n dt = 0 \qquad (n = 0, 1, 2, \cdots),$$

$$\int_C f_2(t)t^n dt = 0 \qquad (n = -1, -2, -3, \cdots),$$

from which follow the formulas (notation of I) for the coefficients in the expansion of $f_1(t)$ and $f_2(t)$:

$$(22) \qquad \begin{aligned} &\int_C f_1(t)t_k(t)dt = 0 \qquad (k = 1, 2, 3, \cdots), \\ &\int_C f_2(t)s_k(t)dt = 0 \qquad (k = 0, 1, 2, \cdots). \end{aligned}$$

We use here in proving (22) the fact that $t_k(t)$ is analytic on and within C, hence on C can be expressed as a uniformly convergent series of polynomials in t; likewise $s_k(t)$ is analytic on and exterior to C, hence on C can be expressed as a uniformly convergent series of polynomials in $1/t$ each without constant term. Such series may be integrated term by term, even after multiplication by $f_1(z)$ or $f_2(z)$.

The development of $f_1(z)$ on C in terms of the polynomials $p_k(z)$ has essentially the same convergence properties as the development of the function

$$\pi i + \log [t_0 - \psi(x)]$$

in a Fourier series (which is precisely the same as the development of the function in a Taylor or Laurent series) on the unit circle γ in the x-plane. The development of $f_2(z)$ on C in the polynomials $q_k(z)$ has essentially the same convergence properties as the development of

$$-\pi i - \log [t_0 - \psi_1(x)] + \log \psi_1(x)$$

in a Fourier (or Laurent) series on γ, where we may take the solutions $x = x_0$ of the two equations

$$\psi(x) = t_0, \quad \psi_1(x) = t_0$$

equal,* $\psi_1(x)$ being a mapping function for the exterior of γ onto the exterior of C, with correspondence of the points at infinity. These two functions

* Rotation of axes does not alter the convergence properties of a Taylor or Fourier development.

just considered in the x-plane are both integrable with an integrable square and on γ possess continuous derivatives except at the point x_0. The developments of the two functions converge therefore to the values of the respective functions except at x_0, and uniformly except in the neighborhood of x_0. In the neighborhood of the point x_0 the term-by-term sum of the two developments converges like the Fourier development of

$$\log \frac{t_0 - \psi(x)}{t_0 - \psi_1(x)} + \log \psi_1(x).$$

The latter term contains the only discontinuity, a finite jump of magnitude $2\pi i$. Gibbs's phenomenon therefore occurs in its characteristic form at this point x_0; the series converges to the value which is the arithmetic mean of the limits approached in the two directions on γ at x_0 by the function developed.

The Fourier development of $f_2(t)$ transformed by either $t=\psi(x)$ or $t=\psi_1(x)$ but interpreted for the same values of t has essentially the same convergence properties in the two cases.

The discussion we have given is not essentially dependent on the particular choice of t_0 made originally. We may therefore state

THEOREM XI. *If the function $F(z)$ satisfies a Lipschitz condition on C, and if the function*

$$f(z) = F(z) + k_1 \log z + k_2 \log z + \cdots + k_m \log z,$$

where each term $k_i \log z$ is continuous on C except at a single point z_i of C, $z_i \neq z_k$ $(i \neq k)$, $k_k \neq 0$,—be developed in a series (2) *as in* I (*p.* 156), *then the Fourier development of $f(z)$ on the unit circle $|x|=1$, where $z=\psi(x)$, and the series*

$$f(z) = a_0 p_0(z) + [a_1 p_1(z) + b_1 q_1(z)] + [a_2 p_2(z) + b_2 q_2(z)] + \cdots \tag{23}$$

have essentially the same convergence properties on C. In particular* (23) *exhibits Gibbs's phenomenon at the points z_k precisely as does a Fourier series. On any closed arc of C containing no point z_k, the series*

$$a_0 p_0(z) + a_1 p_1(z) + a_2 p_2(z) + \cdots \tag{24}$$

* The statement that two series *have the same convergence properties* is used in two senses in the literature, to indicate (1) that their term-by-term difference converges uniformly to the sum zero, or (2) that their term-by-term difference converges absolutely and uniformly to the sum zero. The present writer has hitherto consistently used the second of these two meanings, but in the present case implies (1) instead of (2). The treatment given here considers uniform convergence but not absolute convergence.

converges to the value $f_1(z)$ *and the series*

$$b_1q_1(z) + b_2q_2(z) + \cdots \tag{25}$$

converges uniformly to the value $f_2(z)$, *where*

$$f_1(z) \equiv \frac{1}{2\pi i}\int_C \frac{f(t)dt}{t-z}$$

is analytic interior to C and continuous in the corresponding closed region except at the points z_k, *and*

$$f_2(z) \equiv \frac{1}{2\pi i}\int_C \frac{f(t)dt}{t-z}$$

is analytic exterior to C, vanishes at infinity, and is continuous in the corresponding closed region except at the points z_k. *For* $z=z_k$ *the two series* (24) *and* (25) *diverge with infinite sum, whereas the series* (23) *converges and its sum is the arithmetic mean of the two limits approached by* $f(z)$ *as z moves in opposite senses on C and approaches* z_k. *If an arbitrary neighborhood of each of the points* z_k *is cut out of the closed interior of C, the series* (24) *converges uniformly to the value* $f_1(z)$ *in the remaining closed region. If an arbitrary neighborhood of each of the points* z_k *is cut out of the closed exterior of C, the series* (25) *converges uniformly to the value* $f_2(z)$ *in the remaining closed region.*

Theorem XI is proved under the hypothesis on the ϵ_k that $\sum \epsilon_k t^k$ converges for every t.

Actual formulas for $f_1(z)$ and $f_2(z)$ in terms of logarithms and the functions represented by the integral

$$\frac{1}{2\pi i}\int_C \frac{F(t)dt}{t-z}$$

can easily be written down. Of course, any function which is smooth except for a finite number of finite jumps can be put into the form of $f(z)$ of this theorem.

The conclusion of Theorem XI naturally holds for the formal Laurent development of a discontinuous function of the kind considered, if the curve C is a circle. In particular it is the divergence of the Taylor series for log $(x-a)$ for $x=a$ that enables us to conclude the divergence of (24) and (25) for $z=z_k$.

C. Boundary values of an analytic function

11. A condition for analyticity. We now take up the study of the boundary values of an analytic function, later for a multiply-connected region but first for a simply-connected region:*

THEOREM XII. *If the function $f(z)$ is continuous on the analytic Jordan curve C, and if we have*

$$\int_C f(z)z^n dz = 0 \qquad (n = 0,1,2,\cdots), \tag{26}$$

then there exists a function $F(z)$ analytic interior to C, continuous in the closed region which consists of C and its interior, and which coincides with $f(z)$ on C.

If the curve C is the unit circle $|z|=1$, the theorem is surely true. In fact the formal Laurent development of $f(z)$ is of the form of a Taylor series, since by (26) the coefficients of the negative powers of z vanish:

$$f(z) \sim a_0 + a_1 z + a_2 z^2 + \cdots, \quad a_n = \frac{1}{2\pi i}\int_C \frac{f(z)}{z^{n+1}}dz.$$

This development is precisely the formal Fourier development of $f(z)$ for $0 \leqq \phi \leqq 2\pi$ if we set $z = e^{i\phi}$. The Fourier development, when summed by the method of Cesàro, converges uniformly on C, since $f(z)$ is continuous. Each term of the corresponding sequence is analytic on and interior to C, hence the sum of the series is analytic interior to C, continuous in the corresponding closed region, and is equal to $f(z)$ on C. This completes the proof of the theorem when C is the unit circle.

If C is not the unit circle, we map the interior of C onto the interior of the unit circle γ in the x-plane, the transformation being as usual $x = \phi(z)$, $z = \psi(x)$. The function $[\phi(z)]^n\phi'(z)$ is analytic in and on C, hence on C can be (by Runge's theorem) uniformly expanded in a series of polynomials in z:

$$[\phi(z)]^n\phi'(z) = \pi_0(z) + \pi_1(z) + \pi_2(z) + \cdots, \; n \geqq 0.$$

This series converges uniformly on C even after multiplication term by term by the continuous function $f(z)$. Term-by-term integration of the new series thus formed yields, by virtue of (26),

$$\int_C f(z)[\phi(z)]^n\phi'(z)dz = 0 \qquad (n = 0,1,2,\cdots).$$

---

* In connection with this problem and the conditions derived, see F. and M. Riesz, *Comptes Rendus du Congrès* (1916) *des Mathématiciens Scandinaves*, Uppsala, 1920, pp. 27–44; Privaloff, *L'Intégrale de Cauchy*, Saratow, 1919; Kakeya, Tôhoku Mathematical Journal, vol. 5 (1914), pp. 40–44, as well as the references given in I, p. 167.

We have, then,

$$\int_\gamma f[\psi(x)]x^n dx = 0 \qquad (n = 0, 1, 2, \cdots),$$

so by the special case of the theorem already proved there exists a function analytic interior to γ, continuous in the corresponding closed region, and coinciding on γ with the function $f[\psi(x)]$. Transformation by the formula $x = \phi(z)$ gives the required function in the z-plane.

Conditions (26), it may be remarked, are all independent of each other and none of them may be omitted. If all of those conditions except a finite number are satisfied, then there exists a function $F(z)$ and a polynomial $P(z)$ in $1/z$ such that $F(z)$ is analytic interior to C, continuous in the corresponding closed region, and such that

$$F(z) + P(z) = f(z), \quad z \text{ on } C.$$

The polynomial $P(z)$ is uniquely determined if we require that it shall vanish at infinity; otherwise is uniquely determined only to within an additive constant.

An alternate statement for Theorem XII is

THEOREM XIIa. *If the function $f(z)$ is continuous on the analytic Jordan curve C, and if we have*

$$\int_C f(z)\omega(z)dz = 0, \tag{27}$$

for every function $\omega(z)$ analytic in the closed region interior to C, then there exists a function $F(z)$ analytic interior to C, continuous in the closed region which consists of C and its interior, and which coincides with $f(z)$ on C.*

The equivalence of (26) and (27) is easy to show. If (27) holds, (26) is surely satisfied. If (26) holds, then an arbitrary function $\omega(z)$ of the kind considered in Theorem XIIa can be uniformly expanded, in the closed region consisting of C and its interior, in a series of polynomials:

$$\omega(z) = \pi_0(z) + \pi_1(z) + \pi_2(z) + \cdots.$$

Term-by-term multiplication of this series by $f(z)$ and term-by-term integration yield, by means of (26), equation (27). Thus (27) is both a necessary and a sufficient condition for the existence of $F(z)$.

There is a similar statement for functions representing the boundary values of a function analytic at infinity:

* We have here an equivalent condition if $\omega(z)$ is required to be analytic merely interior to C and continuous in the corresponding closed region.

THEOREM XIII. *Let the function $f(z)$ be continuous on the analytic Jordan curve C, in whose interior the origin lies; then the two equivalent conditions*

$$\text{(A)} \qquad \int_C f(z)z^k dz = 0 \qquad (k = -1, -2, -3, \cdots);$$

$$\text{(B)} \qquad \int_C f(z)\omega(z)dz = 0,$$

for every function $\omega(z)$ analytic exterior to C (also at infinity), continuous in the corresponding closed region, and zero at infinity—these two conditions are each necessary and sufficient that there should exist a function $F(z)$ zero at infinity, analytic exterior to C (including the point at infinity), continuous in the corresponding closed region, and equal to $f(z)$ on C.

The proof of this theorem is easy and will be omitted.

If in condition (A) we omit $k=-1$, and in (B) require that $\omega(z)$ should have a double root at infinity, those two conditions remain equivalent. The conditions are then necessary and sufficient for the existence of $F(z)$, analytic exterior to C (including the point at infinity), continuous in the corresponding closed region, and equal to $f(z)$ on C. We cannot say, however, that $F(z)$ vanishes at infinity.

It will be noticed that if the continuous function $f(z)$ satisfies (26) as well as (A) of Theorem XIII, the two functions defined interior and exterior to C respectively are analytic in the neighborhood of C, hence analytic also on C. The function analytic exterior to C vanishes at infinity, so $f(z)$ is identically zero.

12. **Extension to multiply-connected regions.** In extending Theorems XII and XIII to the case of regions bounded by several contours, we shall mention merely the analogue of condition (A) although the analogue of condition (B) is easily included.

THEOREM XIV. *If the analytic Jordan curve C' lies interior to the analytic Jordan curve C, if the origin lies interior to C', and if the functions $f_1(z)$ and $f_2(z)$ continuous on C and C' respectively satisfy the conditions*

$$(28) \qquad \int_C f_1(z)z^k dz = \int_{C'} f_2(z)z^k dz \qquad (k = \cdots -2, -1, 0, 1, 2, \cdots),$$

then there exists a function $F(z)$ analytic in the annular region bounded by C and C', continuous in the corresponding closed region, and which on C and C' coincides with $f_1(z)$ and $f_2(z)$ respectively.

It is sufficient to establish Theorem XIV in the case that C is a circle. For if the theorem is true in that case, we shall prove it to be true in the general case. Let $z=\psi(x)$, $x=\phi(z)$ denote as usual the functions which map the interior of C onto the interior of the unit circle γ in the x-plane. Since $\phi(0)=0$, the curve C' corresponds to an analytic Jordan curve γ' in whose interior the origin $x=0$ lies.

Conditions (28) lead to the equations

$$\int_C f_1(z)\,[\phi(z)]^k\phi'(z)dz = \int_{C'} f_2(z)\,[\phi(z)]^k\phi'(z)dz \tag{29}$$
$$(k=\cdots -2,\ -1,0,1,2,\cdots).$$

For the function $[\phi(z)]^k\ \phi'(z)$ is analytic in the closed region bounded by C and C', hence in that closed region can be uniformly expanded in a series of polynomials in z and $1/z$.* This expansion can be integrated term by term on C or C', after multiplication through by $f_1(z)$ or $f_2(z)$. Computation of the two members of (29) by the use of this series makes their equality evident in the light of (28).

Equations (29) are precisely the equations

$$\int_\gamma f_1[\psi(x)]x^kdx = \int_{\gamma'} f_2[\psi(x)]x^kdx \qquad (k=\cdots -2,\ -1,0,1,2,\cdots),$$

sufficient for the existence of a function $f(x)$ analytic in the annular region bounded by γ and γ', continuous in the corresponding closed region, and equal to $f_1[\psi(x)]$ and $f_2[\psi(x)]$ on γ and γ'; this is sufficient for the existence of the function $F(z)$ of the theorem.

It remains, then, to prove Theorem XIV when C is a circle. Consider the formal Taylor development of the function $f_1(z)$:

$$f_1(z) \sim a_0 + a_1z + a_2z^2 + \cdots, \qquad a_k = \frac{1}{2\pi i}\int_C \frac{f_1(z)}{z^{k+1}}dz.$$

This series converges interior to C, defining a function $F_1(z)$ analytic interior to C, and the series converges uniformly on any curve Γ interior to C. Thus if Γ is an arbitrary rectifiable Jordan curve interior to C and includes in its interior the origin, we have

$$a_{k-1} = \int_C f_1(z)z^{-k}dz = \int_\Gamma F_1(z)z^{-k}dz \quad (k=1,2,3,\cdots).$$

Hence the function $f_2(z)=F_1(z)$ is continuous on C' and satisfies the conditions

* This is very easy to prove by writing the function involved as the sum of two functions, given by Cauchy's integral taken over C and C' respectively. The one function is analytic on and interior to C, the other on and exterior to C'.

$$\int_{C'}[f_2(z)-F_1(z)]z^{-k}dz=\int_{C'}f_2(z)z^{-k}dz-\int_{C}f_1(z)z^{-k}dz=0$$
$$(k=1,\ 2,\ 3,\ \cdots).$$

By Theorem XIII there exists a function $F_2(z)$ which is analytic exterior to C', continuous in the corresponding closed region, and which on C' coincides with $f_2(z)-F_1(z)$. The function $F_2(z)-f_1(z)$ is continuous on C. From the relations

$$\int_{C}F_2(z)z^kdz=\int_{C'}F_2(z)z^kdz=\int_{C'}f_2(z)z^kdz-\int_{C'}F_1(z)z^kdz$$
$$=\int_{C'}f_2(z)z^kdz=\int_{C}f_1(z)z^kdz \qquad (k=0,1,2,\cdots),$$

we deduce by Theorem XII the existence of a function $\Phi(z)$ analytic interior to C, continuous in the corresponding closed region, and coinciding on C with $F_2(z)-f_1(z)$:

$$\Phi(z)=F_2(z)-f_1(z),\qquad z \text{ on } C. \tag{30}$$

We have, however,

$$-\int_{C}\Phi(z)z^kdz=\int_{C}[F_2(z)-\Phi(z)]z^kdz=\int_{C}f_1(z)z^kdz$$
$$(k=-1,\ -2,\ -3,\ \cdots),$$

so that the two functions $-\Phi(z)$ and $F_1(z)$ have the same coefficients in their Taylor development about the origin and are therefore identical.

The function $F_2(z)-\Phi(z)$ is then analytic interior to the annular region bounded by C and C', continuous in the corresponding closed region, by (30) equals $f_1(z)$ on C, and by the definition of $F_2(z)$ equals $f_2(z)$ on C'.

Theorem XIV, whose proof is now complete, can be extended to regions of higher connectivity:

THEOREM XV. *Let R be the region bounded by an analytic Jordan curve C_0 and by non-intersecting analytic Jordan curves $C_1, C_2, \cdots, C_n$ lying interior to C_0. If the function $f(z)$ is continuous on C, the complete boundary of R, then a necessary and sufficient condition that there exist a function $F(z)$ analytic in R, continuous in the corresponding closed region, and equal to $f(z)$ on C, is*

$$\int_C f(z)z^k dz = 0 \qquad (k = 0,1,2,\cdots),$$

(31)

$$\int_C f(z)(z - z_i)^k dz = 0 \qquad (i = 1,2,\cdots,n;\ k = 1,2,3,\cdots),$$

where z_i *is an arbitrary fixed point interior to* C_i. *The integrals in* (31) *are to be taken over the complete boundary of* R, *in the positive sense on that boundary.*

The proof of Theorem XV is similar to the proof of Theorem XIV and is omitted. Theorem XV remains true if the Jordan curves C_i are no longer required to be analytic, provided they are regular in the sense of Osgood,* and provided the function $f(z)$ satisfies a Lipschitz condition on C. The proof of this new theorem is likewise fairly simple, as an application of the theorem of Plemelj used in I, p. 167. It will be noticed too that in the proofs of the theorems given we need not require that the curves used be analytic; it is sufficient if the derivative $\phi'(z)$ of the mapping function used in each case is continuous in the closed region which we map.

The existence of other theorems which lie not far away is obvious; we give a single example related to Theorem XIV:

THEOREM XVI. *Let the analytic Jordan curve* C' *lie interior to the analytic Jordan curve* C, *let the origin lie interior to* C', *and let the functions* $f_1(z)$ *and* $f_2(z)$ *be defined and continuous on* C *and* C' *respectively. Then a necessary and sufficient condition for the existence of a function* $F(z)$ *analytic interior to* C, *continuous in the corresponding closed region, and coinciding on* C *and* C' *with* $f_1(z)$ *and* $f_2(z)$ *respectively, is*

$$\int_C f_1(z)z^k dz = \int_{C'} f_2(z)z^k dz = 0 \qquad (k = 0,1,2,\cdots),$$

$$\int_C f_1(z)z^k dz = \int_{C'} f_2(z)z^k dz \qquad (k = -1,-2,-3,\cdots).$$

Theorems XII–XVI have obvious application to expansions in terms of polynomials, particularly in connection with such theorems as IX, which do not demand analyticity for the development of a given function.

---

* *Funktionentheorie*, Leipzig, 1912, p. 51.

HARVARD UNIVERSITY,
CAMBRIDGE, MASS.

Journal für die reine und angewandte Mathematik.
Herausgegeben von K. Hensel.
Druck und Verlag Walter de Gruyter & Co. Berlin W 10.

Sonderabdruck aus Bd. 159. Heft 4. 1928.

Über die Entwicklung einer harmonischen Funktion nach harmonischen Polynomen.

Von *J. L. Walsh*[1]) in Cambridge (Mass.).

Die Theorie der Entwicklung einer harmonischen Funktion in der Ebene nach harmonischen Polynomen wurde bisher gar nicht tief untersucht, obwohl die verwandte Theorie der Entwicklung einer analytischen Funktion nach Polynomen schon ziemlich weit entwickelt ist[2]). Im vorliegenden Artikel werden das Analogon des *Runge*schen Satzes für analytische Funktionen, und noch allgemeinere Sätze erörtert; hauptsächlich wird aber das allgemeinste beschränkte, einfach zusammenhängende Gebiet G festgestellt, von der Beschaffenheit, daß jede in G harmonische, im entsprechenden abgeschlossenen Bereich stetige Funktion sich nach harmonischen Polynomen gleichmäßig in diesem Bereich entwickeln läßt. Dazu brauchen wir in erster Linie die schönen Resultate von *Lebesgue*[3]) über harmonische Funktionen.

I.

Ein *Gebiet* C ist eine zusammenhängende Punktmenge, die aus lauter inneren Punkten besteht; der entsprechende *Bereich* $\overline{C}$ ist die abgeschlossene Punktmenge, die aus dem Gebiet und seinen Randpunkten besteht.

Eine Funktion $u(x, y)$ heißt *harmonisch* in einem Punkte P, wenn in jedem Punkte einer Umgebung von P die Funktion $u(x, y)$ und ihre partiellen Ableitungen der ersten und zweiten Ordnung stetig sind, und die Laplacesche Differentialgleichung

$$\frac{\partial^2 u}{\partial x^2} + \frac{\partial^2 u}{\partial y^2} = 0$$

befriedigt wird.

Der Satz von *Runge* für analytische Funktionen lautet:

Es sei $f(z)$ eine Funktion von z, analytisch in einem einfach zusammenhängenden Gebiet G der z-Ebene, das den Punkt ∞ in seinem Inneren nicht enthält. Dann läßt sich $f(z)$ in eine Reihe von Polynomen in z entwickeln, welche in jedem innerhalb G liegenden Bereich gleichmäßig konvergiert.

---

[1]) National Research Fellow.

[2]) Vgl. aber *Bergmann*, Mathematische Annalen Bd. 86 (1922), S. 238—271, der verschiedene Entwicklungen harmonischer Funktionen betrachtet, namentlich die, die durch Trennung von Real- und Imaginärteilen spezieller Entwicklungen einer analytischen Funktion hervorgehen. Insbesondere enthält seine Arbeit den Sätzen I und II nahe liegende Resultate.

[3]) „Sur le problème de Dirichlet", Rendiconti del Circolo Matematico di Palermo 24 (1907). S. 371—402.

Wir beweisen nun das Analogon:

Satz I. *Es sei $u(x, y)$ eine Funktion von (x, y), harmonisch in einem einfach zusammenhängenden Gebiet G der (x, y)-Ebene, das den Punkt ∞ in seinem Inneren nicht enthält. Dann läßt sich $u(x, y)$ in eine Reihe von harmonischen Polynomen in (x, y) entwickeln, welche in jedem innerhalb G liegenden Bereich gleichmäßig konvergiert.*

Es sei $v(x, y)$ eine in G zu $u(x, y)$ konjugierte Funktion; die Funktion $v(x, y)$ ist in G eindeutig, wenn man nur ein Element von $v(x, y)$ und seine harmonischen Fortsetzungen auf innerhalb G liegenden Kurven betrachtet. Wir wenden den Satz von *Runge* auf die Funktion $f(z) = u(x, y) + iv(x, y)$ an, und erhalten als ihre Entwicklung in G eine nach Polynomen in z fortschreitende Reihe. Die reelle Reihe, deren jedes Glied der reelle Teil des entsprechenden Gliedes jener Reihe ist, hat in G die Summe $u(x, y)$. Jedes Glied dieser reellen Reihe ist ein harmonisches Polynom in (x, y), und die Reihe konvergiert auf die besagte Weise.

Runge beweist auch den folgenden allgemeineren Satz:

Ist $f(z)$ in den außereinanderliegenden, einfach zusammenhängenden Gebieten $G_1, G_2, \ldots, G_n$ analytisch, wovon keins den Punkt ∞ in seinem Inneren enthält, dann läßt sich $f(z)$ in eine Reihe von Polynomen in z entwickeln, welche in beliebigen innerhalb $G_1, G_2, \ldots, G_n$ bzw. liegenden Bereichen $B_1, B_2, \ldots, B_n$ gleichmäßig konvergiert.

Die Übertragung dieses Satzes auf harmonische Funktionen geschieht wie bei Satz I und liefert ein entsprechendes Analogon für harmonische Funktionen, von welchem wir später Gebrauch machen wollen. Von nun an betrachten wir nur reelle Funktionen.

Der folgende Satz ist noch allgemeiner als Satz I, in bezug auf mancherlei Funktionen:

Satz II. *Es sei die Funktion $u(x, y)$ im Inneren einer Jordanschen Kurve harmonisch und im entsprechenden Bereich stetig. Dann läßt sich $u(x, y)$ in eine im ganzen Bereich gleichmäßig konvergierende Reihe von harmonischen Polynomen in (x, y) entwickeln.*

Man leitet ohne Mühe den Satz I aus dem Satz II her. Wir bilden nämlich (nach *Runge*) eine Folge von ineinandergeschachtelten ganz in G liegenden *Jordan*schen Kurven $K_1, K_2, \ldots$, wovon jede im Inneren der folgenden liegt, und so, daß jeder Punkt von G im Inneren einer Kurve K_n liegt. Wir bilden auch eine Folge $\varepsilon_1 > \varepsilon_2 > \varepsilon_3 \ldots$, deren Grenzwert Null ist. Nach Satz II existieren harmonische Polynome $P_n(x, y)$, so daß die Bedingungen

$$|u(x, y) - P_n(x, y)| < \varepsilon_n, \qquad (x, y) \text{ im Inneren von } K_n, \quad n = 1, 2, 3, \ldots,$$

befriedigt sind. Die Reihe

$$P_1(x, y) + \sum_{n=1}^{\infty} [P_{n+1}(x, y) - P_n(x, y)]$$

ist dann die im Satz I verlangte.

In einem Gebiet oder Bereich sind natürlich gleichmäßige Entwicklung und Annäherung mit beliebiger Genauigkeit vollständig äquivalent.

Satz II kann durch den Gebrauch konformer Abbildung, genau wie im entsprechenden Falle analytischer Funktionen, bewiesen werden [1]). Wir führen also diesen Beweis nicht an, schon aus dem Grund, weil dieser Satz in dem allgemeineren Satz V enthalten ist.

---

[1]) Vgl. *Walsh*, Mathematische Annalen Bd. 96 (1926), S. 430—436.

II.

Wir beweisen jetzt ein sehr allgemeines Resultat über Entwicklungen nach harmonischen Polynomen.

Satz III. *Ist die reelle Funktion $U(x, y)$ auf einer beschränkten Punktmenge M stetig, welche die volle Begrenzung eines (eventuell mehrfach zusammenhängenden) unendlichen Gebietes B ist, so läßt sich $U(x, y)$ auf M nach harmonischen Polynomen gleichmäßig entwickeln.*

Wir skizzieren den Beweis bloß, weil er den Beweisen von *Lebesgue* ähnlich ist. Die Punktmenge M ist natürlich abgeschlossen, weil sie der Rand eines Gebietes ist, und liegt wegen ihrer Beschränktheit im Inneren eines Kreises K_1. Es existiert nun eine reelle überall in K_1 (inklusive des Randes) stetige Funktion $U_0(x, y)$, die mit $U(x, y)$ auf M übereinstimmt [1]). Die Funktion $U_0(x, y)$ ist in K_1 durch Polynome in (x, y) gleichmäßig approximierbar. D. h., ist ein beliebiges $\varepsilon > 0$ gegeben, so existiert solch ein reelles Polynom $P(x, y)$, daß

$$|P(x, y) - U_0(x, y)| < \varepsilon$$

für alle (x, y) auf und innerhalb K_1 gilt. Um Satz III zu beweisen, genügt es also zu zeigen, daß die Funktion $P(x, y)$ auf M durch harmonische Polynome gleichmäßig entwickelbar oder (was äquivalent ist) gleichmäßig approximierbar ist.

Man kann — etwa durch Zerlegung der Ebene in Quadrate — eine Folge von abgeschlossenen Punktmengen $M_1, M_2, M_3, \ldots$ finden, so daß 1. die Menge M_k aus einer endlichen Anzahl von in B und in K_1 liegenden, voneinander getrennten *Jordan*schen Kurven mit Einschluß ihrer Inneren besteht; 2. die Menge M in jeder Menge M_k enthalten ist; 3. kein Punkt der Begrenzung von M_k der Menge M_{k+1} gehört; 4. jeder bestimmte Punkt von B nur einer endlichen Anzahl der Mengen M_k angehört.

Jedes der Integrale

$$\iint_{M_k} \left[\left(\frac{\partial P}{\partial x}\right)^2 + \left(\frac{\partial P}{\partial y}\right)^2\right] dx\, dy, \qquad k = 1, 2, 3, \ldots,$$

ist nicht größer als das Integral

$$\iint_{K_1} \left[\left(\frac{\partial P}{\partial x}\right)^2 + \left(\frac{\partial P}{\partial y}\right)^2\right] dx\, dy;$$

diese Integrale sind also beschränkt. Es existiert folglich eine Funktion $U_k(x, y)$, für $k = 1, 2, 3, \ldots$, die im Inneren von M_k harmonisch, auf M_k stetig ist, und die auf dem Rande von M_k mit $P(x, y)$ übereinstimmt. Wir erhalten also die Ungleichheiten [2])

$$\iint_{M_k} \left[\left(\frac{\partial P}{\partial x}\right)^2 + \left(\frac{\partial P}{\partial y}\right)^2\right] dx\, dy \geqq \iint_{M_k} \left[\left(\frac{\partial U_k}{\partial x}\right)^2 + \left(\frac{\partial U_k}{\partial y}\right)^2\right] dx\, dy,$$

so daß diese letzten Integrale ebenfalls beschränkt sind.

Wir beweisen nun:

$$\lim_{k \to \infty} U_k(x, y) = P(x, y) \quad \textit{gleichmäßig auf } M.$$

Wir nehmen das Gegenteil an und stoßen dabei auf einen Widerspruch. Dann existieren nämlich Punkte Q_k, $k = 1, 2, 3, \ldots$, von M, so daß $\lim_{k \to \infty} Q_k = Q$ (unvermeidlich ein Punkt von M) ist, und $\lim_{k \to \infty} U_k(Q_k) = L \neq P(Q)$ ist. Wir erhalten infolgedessen, eventuell nach Weglassung einer endlichen Anzahl der ursprünglichen Punkte Q_k, die Ungleichheit

[1]) *Lebesgue*, l. c., S. 379—380.

[2]) *Lebesgue*, S. 395—396.

$$|U_k(Q_k) - P(Q)| > 2\delta, \quad k = 1, 2, 3, \ldots, \quad \text{wo } \delta > 0 \text{ ist.}$$

Wir nehmen einen Kreis K_0, dessen Mittelpunkt Q ist, und dessen Radius so klein ist, daß die Ungleichheit

$$|P(x, y) - P(Q)| < \delta$$

für jedes (x, y) auf oder im Inneren von K_0 gilt.

Wenn k genügend groß ist, liegt der Punkt Q_k und auch Punkte der Grenze von M_k im Kreise K_0. Im folgenden betrachten wir nur solche Werte von k. Wir bezeichnen mit K irgendeinen Kreis (also veränderlich!) mit dem Mittelpunkt Q, dessen Halbmesser größer als QQ_k und kleiner als der Halbmesser von K_0 ist. Die Punkte, die sich mit Q_k durch einen Streckenzug verbinden lassen, der keinen Punkt von K oder von der Begrenzung von M_k enthält, bilden ein Gebiet $K^{(k)}$, welches Q_k enthält und dessen sämtliche Punkte zu M_k gehören; man vergleiche Satz IV. Sämtliche Randpunkte von $K^{(k)}$ können nicht Randpunkte von M_k sein; sonst hätten wir, da die Randpunkte von $K^{(k)}$ im Inneren von K_0 liegen und außerdem $U_k(x, y) = P(x, y)$ auf dem Rande von M_k ist,

$$|U_k(x, y) - P(Q)| < \delta,$$

worin (x, y) ein beliebiger Randpunkt von $K^{(k)}$ ist. Aber dann wäre die Ungleichheit

$$|U_k(Q_k) - P(Q)| > 2\delta$$

für eine *harmonische* Funktion $U_k(x, y)$, die ihr Maximum auf dem Rande des Bereiches annimmt, unmöglich. Wir erhalten also, wiederum durch diese bekannte Eigenschaft einer harmonischen Funktion, die Ungleichheit

$$|U_k(Q_K^{(k)}) - P(Q)| > 2\delta,$$

worin $Q_K^{(k)}$ ein auf K liegender Punkt von M_k ist.

Auf wenigstens einem in M_k liegenden Bogen von K hat die Funktion $U_k(x, y)$ eine Schwankung größer als δ, — auf einem Bogen nämlich, der $Q_K^{(k)}$ enthält, und der durch einen Punkt des Randes von M_k begrenzt ist. Wir haben also, mit Polarkoordinaten (ϱ, θ), deren Pol in Q liegt,

$$\int \left(\frac{\partial U_k}{d\theta}\right)^2 d\theta > \frac{\delta^2}{2\pi},$$

wo das Integral über den besagten Bogen von K erstreckt ist [1]). Man bekommt ferner

$$\iint \left[\left(\frac{\partial U_k}{\partial \varrho}\right)^2 + \left(\frac{1}{\varrho}\frac{\partial U_k}{\partial \theta}\right)^2\right] d\varrho \cdot \varrho d\theta$$

$$= \iint \left[\left(\frac{\partial U_k}{\partial x}\right)^2 + \left(\frac{\partial U_k}{\partial y}\right)^2\right] dx\, dy > \frac{\delta^2}{2\pi} \log \frac{[\text{Radius von } K_0]}{QQ_k},$$

worin die Integrale über den in M_k liegenden Teil des Rings erstreckt sind, der durch K_0 und den Kreis, dessen Mittelpunkt Q ist und der durch Q_k hindurchgeht, begrenzt ist.

Die Integrale

$$\iint_{M_k} \left[\left(\frac{\partial U_k}{\partial x}\right)^2 + \left(\frac{\partial U_k}{\partial y}\right)^2\right] dx\, dy$$

sind also nicht beschränkt, wir haben den besagten Widerspruch, und so haben wir bewiesen, daß

$$\lim_{k \to \infty} U_k(x, y) = P(x, y) \textit{ gleichmäßig auf } M \textit{ ist}$$ [2]).

[1]) *Lebesgue*, S. 388.

[2]) Vgl. auch *Lebesgue*, S. 398—399.

Nun ist die Funktion $U_k(x, y)$ auf M nach harmonischen Polynomen gleichmäßig approximierbar, laut des Analogons des allgemeineren Satzes von *Runge*. Die Funktion $P(x, y)$ und daher die Funktion $U(x, y)$ ist also auf M durch harmonische Polynome gleichmäßig mit beliebiger Genauigkeit approximierbar, und Satz III ist bewiesen.

Der Bequemlichkeit halber beweisen wir jetzt einen Hilfssatz:

Satz IV. *Es sei M irgendeine beschränkte abgeschlossene Punktmenge der Ebene. Dann bildet die Gesamtheit M' von Punkten, die sich mit einem beliebigen nicht zu M gehörenden Punkt A durch einen Streckenzug, der keinen Punkt von M enthält, verbinden lassen, ein Gebiet, dessen Rand aus lauter Punkten von M besteht.*

Der Punkt A ist offenbar ein innerer Punkt der Menge M', da eine ganze Umgebung von A keinen Punkt von M enthält. Wenn zwei Punkte von M' durch einen Streckenzug verbunden werden können, der keinen Punkt von M enthält, so sind auch alle Punkte dieses Streckenzuges innere Punkte von M'. Man kann zwei beliebige Punkte von M' mit A, also auch miteinander, durch Streckenzüge von der besagten Eigenschaft verbinden. Die Punktmenge M' hängt also zusammen, besteht aus lauter inneren Punkten und ist daher ein Gebiet.

Existieren in jeder Umgebung eines Punktes B Punkte sowohl von M' wie solche, die nicht zu M' gehören, so sind in jeder Umgebung von B auch Punkte von M vorhanden. Sonst könnte man auch B mit A verbinden, durch einen Streckenzug mit den verlangten Eigenschaften. Der Punkt B wäre dann ein (innerer) Punkt von M', was ein Widerspruch ist. In jeder Umgebung von B sind also Punkte von M, daher ist B auch ein Punkt von M. D. h. jeder Randpunkt von M' ist ebenfalls ein Randpunkt von M, und der Satz ist bewiesen.

Satz IV läßt sich unmittelbar anwenden, indem wir folgendes beweisen:

Ist N irgendeine beschränkte Punktmenge, die aus lauter inneren Punkten besteht und deren Begrenzung M auch die volle Begrenzung eines unendlichen Gebietes ist, so existiert eine auf N harmonische, auf $N + M$ stetige Funktion $u(x, y)$, die beliebige vorgeschriebene stetige Randwerte $U(x, y)$ auf M annimmt. Diese Funktion $u(x, y)$ ist nach harmonischen Polynomen in (x, y) auf $N + M$ gleichmäßig entwickelbar.

In der Tat bilden sämtliche Punkte, die sich mit einem beliebigen Punkt A von N durch einen Streckenzug, der keinen Punkt von M enthält, verbinden lassen, ein Gebiet, welches einfach zusammenhängt, und dessen Begrenzung aus lauter Punkten von M besteht. Die Punktmenge N besteht infolgedessen aus einer endlichen oder (abzählbar) unendlichen Anzahl von Gebieten, deren Begrenzungen aus lauter Punkten von M bestehen. Die in Satz III betrachtete Entwicklung von $U(x, y)$ konvergiert gleichmäßig auf M, daher gleichmäßig auf $N + M$, und die Summe ist auf $N + M$ stetig, auf N harmonisch.

Wir studieren insbesondere einfach zusammenhängende Gebiete und greifen die folgende Verallgemeinerung des Satzes II heraus:

Satz V. *Es sei C ein beschränktes Gebiet, dessen Rand auch der Rand eines unendlichen Gebietes sei. Es sei die Funktion $u(x, y)$ in C harmonisch und im entsprechenden Bereich $\overline{C}$ stetig. Dann ist $u(x, y)$ in eine im Bereich $\overline{C}$ gleichmäßig konvergierende Reihe von harmonischen Polynomen in (x, y) entwickelbar.*

Die Sätze II, III, V können als Verallgemeinerungen des klassischen *Weierstrass*schen Satzes über Entwicklungen nach trigonometrischen Polynomen angesehen werden. Dieser Satz behauptet, daß eine beliebige stetige Funktion $f(\theta)$ der Periode 2π in eine gleichmäßig konvergierende Reihe, deren jedes Glied ein trigonometrisches Polynom von der Gestalt

$$\sum_{n=0}^{n=k} (a_n \cos n\theta + b_n \sin n\theta)$$

ist, entwickelt werden kann. Wir führen Polarkoordinaten (r, θ) in der Ebene ein, und sprechen denselben Satz in einer anderen Form aus:

Eine beliebige auf dem Einheitskreise $K(r = 1)$ stetige Funktion $U(r, \theta)$ kann in eine auf K gleichmäßige konvergierende Reihe entwickelt werden, deren jedes Glied ein harmonisches Polynom der Gestalt

$$\sum_{n=0}^{n=k} r^n (a_n \cos n\theta + b_n \sin n\theta)$$

ist.

Man kann natürlich dieses Polynom als Polynom in (x, y) schreiben.

In diesem Satze ist es völlig gleichgültig, ob wir die Funktion $U(r, \theta)$ auf K oder die entsprechende Lösung $u(r, \theta)$ des *Dirichlet*schen Problems für den Kreis mit den Randwerten $U(r, \theta)$ entwickeln. Denn die Entwicklung von $U(r, \theta)$ konvergiert gleichmäßig auf K, daher gleichmäßig im abgeschlossenen Inneren von K, und stellt im Inneren von K die Funktion $u(r, \theta)$ dar. Dieser Satz von *Weierstrass*, wie auch Satz III, gibt also einen Beweis der Existenz der Lösung des *Dirichlet*schen Problems. Sätze II, III, V reduzieren sich auf den *Weierstrass*schen Satz wenn der fragliche Rand des Gebietes ein Kreis wird.

III.

Wir wenden uns jetzt zum umgekehrten Problem. Von welchen beschränkten Gebieten C ist es wahr, daß jede im Gebiete C harmonische, im entsprechenden abgeschlossenen Bereiche $\bar{C}$ stetige Funktion in eine im Bereiche $\bar{C}$ gleichmäßig konvergierende Reihe entwickelbar ist?

Ein Gebiet C, welches die besagte Eigenschaft hat, braucht nicht einfach zusammenhängend zu sein, wie das Gegenbeispiel

$$C: 0 < \sqrt{x^2 + y^2} < 1$$

zeigt. Denn jede im Gebiet C harmonische, im entsprechenden Bereich $\bar{C}$ stetige Funktion ist auch im Nullpunkt harmonisch, darum im Bereich $\bar{C}$ gleichmäßig entwickelbar.

Im folgenden schließen wir isolierte Randpunkte aus [1]), indem wir unseren Hauptsatz beweisen:

Satz VI. *Es sei C ein beschränktes, endlichfach zusammenhängendes Gebiet der (x, y)-Ebene ohne isolierte Randpunkte. Eine notwendige und hinreichende Bedingung dafür, daß jede in C harmonische, im entsprechenden Bereiche $\bar{C}$ stetige Funktion in eine im Bereiche $\bar{C}$ gleichmäßig konvergierende Reihe von harmonischen Polynomen in (x, y) entwickelbar sei, besteht darin, daß C ein (einfach zusammenhängendes) Gebiet sei, dessen Rand auch der Rand eines unendlichen Gebietes ist.*

Daß diese Bedingung hinreichend ist, ist schon in Satz V enthalten; wir brauchen also nur das Umgekehrte zu betrachten.

Es sei G die Punktmenge, die aus sämtlichen Punkten besteht, die sich mit dem Punkt ∞ durch einen Streckenzug verbinden lassen, der keinen Punkt vom Bereich $\bar{C}$ enthält; die Punktmenge G ist also ein Gebiet, dessen Randpunkte auch Randpunkte von C sind. Ist Q irgendein dem Bereich $\bar{C}$ nicht angehörender Punkt, der auch kein

[1]) Diese Annahme ist teilweise entbehrlich. Vgl. *Lebesgue*, S. 379, 390.

Punkt von G ist, so bilden sämtliche Punkte, die sich mit Q durch einen $\bar{C}$ nicht treffenden Streckenzug verbinden lassen, auch ein Gebiet G', dessen Rand aus lauter Punkten von $\bar{C}$ besteht.

Wir nehmen nun an, daß ein Randpunkt M von C kein solcher von G ist, und stoßen dabei auf einen Widerspruch. Denn es möge kein Punkt von G auf oder im Inneren des Kreises K mit dem Mittelpunkt M und dem Halbmesser R liegen. Die Funktion

$$U(r, \theta) = \begin{cases} 1 - \dfrac{r}{R}, & r \leqq R, \\ 0, & r \geqq R, \end{cases}$$

ist offenbar in jedem Randpunkt von C definiert und stetig, wobei (r, θ) Polarkoordinaten mit dem Pol M sind. Nach *Lebesgue*[1]) existiert eine im Gebiet C harmonische, im Bereich $\bar{C}$ stetige Funktion $u(x, y)$, deren Randwerte die Werte $U(r, \theta)$ sind. Nach Voraussetzung ist diese Funktion $u(x, y)$ in $\bar{C}$ gleichmäßig entwickelbar. Die Entwicklung konvergiert gleichmäßig auf dem Rande von C, also in jedem Punkt eines solchen wie oben betrachteten Gebietes G'. Jeder Punkt von K ist entweder ein Punkt von $\bar{C}$ oder ein Punkt eines solchen Gebietes G'. Die Entwicklung von $u(x, y)$ ist also eine Reihe, die auf K gleichmäßig konvergiert, ihre Summe ist darum im Inneren von K harmonisch, im entsprechenden Bereich stetig. Die Werte dieser Summe auf K sind aber alle kleiner als 1, da eine harmonische Funktion ihr Maximum auf dem Rande eines Bereichs erreicht. Nach dem *Gauß*schen Mittelwertsatz für harmonische Funktionen ist die Summe der Reihe auch in M kleiner als 1, also nicht $U(r, \theta)$, welches ein Widerspruch ist. Satz VI ist also bewiesen.

Satz VI ist umso merkwürdiger, weil das analoge Resultat für analytische Funktionen einer komplexen Veränderlichen nicht gilt. Man zeigt in der Tat folgendes[2]):

Es sei C ein beschränktes, einfach zusammenhängendes Gebiet der z-Ebene. Eine notwendige und hinreichende Bedingung dafür, daß jede in $\bar{C}$ regulär-analytische Funktion nach Polynomen von z in $\bar{C}$ gleichmäßig entwickelbar sei, besteht darin, daß die (in Bezug auf die ganze Ebene) zu $\bar{C}$ komplementäre Menge zusammenhänge.

Wir nehmen z. B. als Gebiet C einen, am einen Ende geschlossenen Streifen, der außerhalb eines Kreises K' liegt und sich um den Kreis K' unendlich oft herumwindet, indem er an ihn beliebig nahe herankommt. Die Kreislinie K' sowie der abgeschlossene Streifen gehört dem entsprechenden Bereiche $\bar{C}$ an.

Ist $u(x, y)$ in C harmonisch, in $\bar{C}$ stetig, so ist $u(x, y)$ in $\bar{C}$ nach harmonischen Polynomen in (x, y) gleichmäßig entwickelbar.

Die Funktion $1/z$, worin $z = x + iy$ ist und $z = 0$ der Mittelpunkt von K' ist, ist in $\bar{C}$ regulär-analytisch, also stetig, doch nicht in $\bar{C}$ nach Polynomen in z gleichmäßig entwickelbar. Zum Beweis nehmen wir an, daß eine solche Entwicklung existiert:

$$\frac{1}{z} = \sum_{k=0}^{\infty} p_k(z), \qquad p_k(z) \text{ ein Polynom in } z.$$

Wir integrieren gliedweise diese gleichmäßig konvergierende Reihe über den Kreis K':

$$\frac{1}{2\pi i}\int_{K'} \frac{dz}{z} = \sum_{k=0}^{\infty}\left[\frac{1}{2\pi i}\int_{K'} p_k(z)\,dz\right],$$

und erhalten also den Widerspruch $1 = 0$.

---

[1]) Loc. cit., S 398.

[2]) *Walsh*, loc. cit. Hier sind aber gewisse Randpunkte ausgeschlossen, genau wie im Satz VIa.

Wir beantworten noch eine dem Satz VI naheliegende Frage, nämlich, von welchen beschränkten Bereichen $\bar{C}$ ist es wahr, daß jede im *Bereich* $\bar{C}$ harmonische Funktion $u(x, y)$ nach harmonischen Polynomen in (x, y) gleichmäßig in $\bar{C}$ entwickelbar ist? Wir bemerken zuerst, daß ein Randpunkt R von $\bar{C}$, wovon eine Umgebung aus lauter Punkten von $\bar{C}$ besteht, gar keine Rolle spielt. In solch einem Punkt R ist die Funktion $u(x, y)$ auch harmonisch; irgendein innerer Punkt R von $\bar{C}$ kann als solch ein Randpunkt betrachtet werden, und umgekehrt, wenn man nur eine kleine Änderung in der Definition des entsprechenden Gebietes vornimmt. Wir nennen diese Randpunkte *hebbar*, und beweisen

Satz VIa. *Es sei $\bar{C}$ ein beschränkter Bereich der (x, y)-Ebene ohne hebbare Randpunkte. Eine notwendige und hinreichende Bedingung dafür, daß jede im (abgeschlossenen) Bereich $\bar{C}$ harmonische Funktion in eine im Bereich $\bar{C}$ gleichmäßig konvergierende Reihe von harmonischen Polynomen in (x, y) entwickelbar sei, besteht darin, daß $\bar{C}$ ein Bereich sei, dessen Rand auch der Rand eines unendlichen Bereiches ist.*

Es ist schon im Satz V enthalten, daß diese Bedingung hinreichend ist; wir betrachten jetzt das Umgekehrte.

Es sei G die Punktmenge, die aus sämtlichen Punkten besteht, die sich mit dem Punkt ∞ durch einen Streckenzug verbinden lassen, der keinen Punkt von $\bar{C}$ enthält; die Punktmenge G ist also ein Gebiet, dessen Randpunkte auch Randpunkte von $\bar{C}$ sind.

Angenommen der Satz sei nicht wahr, es existiere also ein Randpunkt R von $\bar{C}$, dessen jede Umgebung Punkte Q enthält, die weder zu $\bar{C}$ noch G gehören. Es liege kein Punkt von G innerhalb eines Kreises mit dem Mittelpunkt R und Radius r, und es liege der spezielle Punkt Q, der weder zu $\bar{C}$ noch G gehört, innerhalb eines Kreises mit dem Mittelpunkt R und Radius $r/3$. Die Funktion

$$u(x, y) = \log \overline{PQ}, \qquad P: (x, y),$$

ist in jedem Punkt von $\bar{C}$ harmonisch, doch nicht in $\bar{C}$ nach harmonischen Polynomen in (x, y) gleichmäßig entwickelbar. Denn die Werte dieser Funktion auf dem Rande von G sind alle größer als $\log 2r/3$. Die Summe einer solchen Entwicklung hätte auch Werte größer als $\log 2r/3$ auf dem Rande von G, also auch in jedem Punkt (insbesondere in jedem Punkt von $\bar{C}$), der sich nicht mit dem Punkt ∞ durch einen Streckenzug verbinden läßt, der keinen Randpunkt von G enthält. Der Wert der Funktion $u(x, y)$ im Punkt R, der zu $\bar{C}$ gehört, ist aber kleiner als $\log r/3$, und der Satz ist bewiesen.

Im Vorhergehenden ist das Wesentlichste des folgenden Satzes schon enthalten:

Satz VII. *Es sei die stetige Funktion $u(x, y)$ auf irgendeiner beschränkten abgeschlossenen Punktmenge C definiert. Es sei D die Punktmenge, die aus sämtlichen Punkten besteht, die sich mit dem Punkt ∞ durch einen Streckenzug verbinden lassen, der keinen Randpunkt von C enthält. Eine notwendige und hinreichende Bedingung dafür, daß $u(x, y)$ nach harmonischen Polynomen auf C gleichmäßig entwickelbar sei, besteht darin, daß $u(x, y)$ auf C mit der auf der zu $\bar{D}$ komplementären Menge E harmonischen Funktion $U(x, y)$ übereinstimme, die auf E durch die Randwerte $u(x, y)$ auf dem Rande von D definiert ist.*

Die Funktion $U(x, y)$ ist die Summe der im Satz III betrachteten Entwicklung, worin D die Rolle von B spielt. Die Funktion $U(x, y)$ ist in jedem Punkt der Menge C definiert. Jeder Punkt von C ist, in der Tat, entweder ein Randpunkt von D oder gehört zu einem Gebiet, welches aus sämtlichen Punkten besteht, die sich mit einem Punkt von

E durch einen Streckenzug verbinden lassen, der keinen Randpunkt von D enthält; ein solches Gebiet ist durch Randpunkte von D, also durch Randpunkte von C, begrenzt. Die im Satz III betrachtete Entwicklung konvergiert gleichmäßig auf dem Rande von D, also gleichmäßig auf C.

Die Bedingung dieses Satzes ist nach Satz III offenbar hinreichend. Sie ist auch notwendig. Wir haben ja zwei gleichmäßige Entwicklungen der Funktion $U(x, y)$ auf dem Rande von D, die eine nach der jetzigen Voraussetzung, die andere nach Satz III. Die gliedweise Differenz dieser Entwicklungen konvergiert auch gleichmäßig auf dem Rande von D. Dort hat sie die Summe Null, daher (nach dem Vorhergehenden) ist die Summe auch auf C Null, und die zwei Funktionen $U(x, y)$ und $u(x, y)$ stimmen auf C überein.

Es sei ausdrücklich bemerkt, in Zusammenhang mit Satz VII, daß die Funktionen $u(x, y)$ und $U(x, y)$ nicht auf der ganzen zu $\overline{D}$ komplementären Menge E übereinzustimmen brauchen. Z. B., die Punktmenge C möge der oben betrachtete Streifen sein, und die Funktion $u(x, y)$ möge $\log \sqrt{x^2 + y^2}$ sein. Diese Funktion ist im Bereich $\overline{C}$ gleichmäßig entwickelbar, aber natürlich nicht auf der ganzen Punktmenge E.

Es ist leicht, die Bedingung vom Satz VII auf irgendeine auf einer beliebigen beschränkten Punktmenge C_1 definierte Funktion $u_1(x, y)$ auszudehnen. Man betrachtet die Punktmenge C, die aus der Punktmenge C_1 und ihrer Ableitung besteht, und man definiert die neue Funktion $u(x, y)$ gleich der Funktion $u_1(x, y)$ auf C_1, und in jedem Punkt von $C - C_1$ gleich irgendeinem Limes von $u_1(x, y)$ in jenem Punkt. Eine notwendige und hinreichende Bedingung dafür, daß $u_1(x, y)$ auf C_1 nach harmonischen Polynomen gleichmäßig entwickelbar sei, besteht darin, daß $u(x, y)$ auf C (stetig sei, und auch dort) nach harmonischen Polynomen gleichmäßig entwickelbar sei.

Aus Satz VII ergibt sich, daß irgendeine Funktion $u(x, y)$, die stetig auf dem Rande (oder auf einer abgeschlossenen Teilmenge des Randes) eines im Satz V betrachteten Gebietes (also insbesondere auf irgendeiner *Jordan*schen Kurve oder auf einem *Jordan*schen Kurvenstück) ist, auch auf dieser Punktmenge gleichmäßig entwickelbar ist.

Dem Vorhergehenden ähnliche Überlegungen liefern drei ähnliche Sätze:

Es sei C irgendeine beschränkte abgeschlossene Punktmenge. Eine notwendige und hinreichende Bedingung dafür, daß jede auf C stetige Funktion nach harmonischen Polynomen auf C gleichmäßig entwickelbar sei, bsteht darin, daß C die volle Begrenzung eines unendlichen Bereiches sei.

Es sei C irgendeine beschränkte abgeschlossene Punktmenge, von der jeder Punkt innerer Punkt bzw. Häufungspunkt derselben ist. Eine notwendige und hinreichende Bedingung dafür, daß jede auf (der abgeschlossenen Punktmenge) C harmonische Funktion nach harmonischen Polynomen auf C gleichmäßig entwickelbar sei, besteht darin, daß der Rand von C auch der Rand eines unendlichen Bereiches sei.

Es sei C irgendeine beschränkte abgeschlossene Punktmenge ohne hebbare Randpunkte, von der jeder Punkt innerer Punkt bzw. Häufungspunkt derselben ist. Eine notwendige und hinreichende Bedingung dafür, daß jede im Inneren von C harmonische, auf C stetige Funktion nach harmonischen Polynomen auf C gleichmäßig entwickelbar sei, besteht darin, daß der Rand von C auch der Rand eines unendlichen Bereiches sei.

Wir betrachten nicht ausführlicher im Zusammenhang mit der Frage von Satz VI allgemeinere Gebiete C als endlichfach zusammenhängende. Wir bemerken hier nur eines: Wenn das Randwertproblem für C (mit C) immer eine Lösung hat, und wenn die Punktmenge C aus lauter inneren Punkten besteht, so ist die notwendige und hinreichende Bedingung von Satz VI die, daß der Rand von C auch der Rand eines unendlichen Gebietes sei.

Wir wenden uns jetzt zu Entwicklungen von in mehrfach zusammenhängenden Gebieten harmonischen Funktionen.

IV.

Die Erörterung von II läßt sich auf mehrfach zusammenhängende Bereiche anwenden. Wir beweisen:

Satz VIII. *Es sei die Funktion $u(x, y)$ im Gebiet C harmonisch, im entsprechenden Bereich stetig, wobei C durch die Jordanschen Kurven C_1 und C_2 begrenzt ist, derart, daß jeder Punkt von C_2 im Inneren von C_1 liegt und der Nullpunkt durch C_2 umschlossen wird. Wird dann die Konstante k passend gewählt, so läßt sich die Funktion $u(x, y) - k \log \sqrt{x^2 + y^2}$ im Bereich $\bar{C}$ in eine Reihe von harmonischen Polynomen in (x, y) plus eine Reihe von harmonischen Polynomen in $\left(\frac{x}{x^2 + y^2}, \frac{y}{x^2 + y^2}\right)$ gleichmäßig entwickeln.*

Man zeigt nach *Osgood* [1]) mit Hilfe der *Greens*chen Formel, daß, wenn k passend gewählt wird, die Gleichung

$$u(x, y) = u_1(x, y) + u_2(x, y) + k \log \sqrt{x^2 + y^2}$$

im Bereich $\bar{C}$ gilt, worin $u_1(x, y)$ innerhalb C_1 harmonisch und im entsprechenden Bereich stetig ist, und $u_2(x, y)$ außerhalb C_2 (inklusive des Punktes ∞) harmonisch und im entsprechenden Bereich stetig ist.

Satz V läßt sich unmittelbar auf die Funktion $u_1(x, y)$ anwenden, und auf die Funktion $u_2(x, y)$, nachdem man eine Transformation durch reziproke Radien mit dem Pol im Nullpunkt ausgeführt hat. Wir erhalten also die verlangte Entwicklung. Die Reihe der Polynome in (x, y) konvergiert gleichmäßig im abgeschlossenen Inneren von C_1, und die Reihe der Polynome in $\left(\frac{x}{x^2 + y^2}, \frac{y}{x^2 + y^2}\right)$ konvergiert gleichmäßig im abgeschlossenen Äußeren (inklusive des Punktes ∞) von C_2.

Es ist vielleicht nicht ohne Interesse, zu bemerken, daß die Funktion $\log r$ in keinem den Bedingungen von Satz VIII befriedigenden Bereich $\bar{C}$ in eine gleichmäßig konvergierende Reihe harmonischer Polynome in (x, y) und $\left(\frac{x}{x^2 + y^2}, \frac{y}{x^2 + y^2}\right)$ entwickelbar ist. Angenommen, daß eine solche Entwicklung existiert:

$$\log r = \sum_{k=1}^{\infty} \sum_{n=-k}^{k} r^n (a_n \cos n\theta + b_n \sin n\theta),$$

so sei J eine in C liegende den Nullpunkt umschließende analytische *Jordan*sche Kurve. Wir dürfen diese Reihe gliedweise in der Richtung der Normale von J differentieren, und nachher über J gliedweise integrieren [2]). Die Integrale dürfen auf irgendeinem Kreise, dessen Mittelpunkt im Nullpunkt liegt, berechnet werden. Wir bekommen also den Widerspruch

$$2\pi = 0.$$

---

[1]) Die Einzelheiten sind ausführlich von *Osgood* gegeben worden; man braucht im übrigen nur zu bemerken, daß die Stetigkeit von u_1 auf C_1 bzw. von u_2 auf C_2 sich aus der Stetigkeit von u_2 auf C_1 bzw. von u_1 auf C_2 ergibt. Siehe *Osgood*, Funktionentheorie I (Leipzig 1912), S. 642—644.

Die Konstante k ist unmittelbar bekannt, wenn die Funktion $u(x, y)$ gegeben ist; loc cit., S. 644.

[2]) Vgl. *Osgood*, loc. cit., S. 652—653. *Osgood* betrachtet hauptsächlich die partiellen Ableitungen, wir gebrauchen hier aber die Derivierten in der Richtung der Normale von J.

Nun zeigen wir durch ein Gegenbeispiel, daß Satz VIII nicht gilt, wenn man nur voraussetzt, daß die Kurven C_1 und C_2 ineinander liegen, ohne getrennt zu sein. Es seien C_1 der Kreis $x^2 + y^2 = 4$, und C_2 der Kreis $(x-1)^2 + y^2 = 1$. Wir bezeichnen mit $\theta_1 (0 \leqq \theta_1 < 2\pi)$ den Winkel XAP und mit $\theta_2 (0 \leqq \theta_2 < 2\pi)$ den Winkel XBP, wo A, B, bzw. X die Punkte $(2, 0)$, $(1, 0)$, $(4, 0)$ sind, und P ein beliebiger Punkt (x, y) ist. Die Funktion

$$u(x, y) = \theta_2 - 2\theta_1$$

ist also im Gebiet harmonisch, welches durch C_1 und C_2 begrenzt wird, und im entsprechenden Bereich stetig; man setzt $u(A) = -\pi$. Die Funktion $u(x, y)$ läßt sich aber nicht in eine Summe

$$(1) \qquad u(x, y) = u_1(x, y) + u_2(x, y) + k \log \overline{BP}, \qquad P : (x, y),$$

spalten, wo $u_1(x, y)$ innerhalb C_1 harmonisch, im entsprechenden Bereich stetig bleibt, und $u_2(x, y)$ sich außerhalb C_2 (inklusive des Punktes ∞) harmonisch, im entsprechenden Bereich stetig verhält. Die Funktion $u(x, y) - k \log \overline{BP}$ ist also für kein k als die Summe einer Reihe von harmonischen Polynomen in (x, y) und einer Reihe von harmonischen Polynomen in $\left(\frac{x-1}{(x-1)^2 + y^2}, \frac{y}{(x-1)^2 + y^2}\right)$ entwickelbar.

Um diese Behauptung zu beweisen, nehmen wir an, daß (1) erfüllt ist. Die Funktion $u(x, y) + \theta_1 = \theta_2 - \theta_1$ ist außerhalb C_2 eindeutig, harmonisch (inklusive des Punktes ∞), im entsprechenden Bereich stetig, außer dem einzelnen Punkt A. Nach (1) ist gleichfalls die Funktion $u_1(x, y) + \theta_1 + k \log \overline{BP}$ außerhalb C_2 (inklusive des Punktes ∞) harmonisch, im entsprechenden Bereich stetig, außer dem einzelnen Punkt A, wo sie beschränkt ist. Nach Voraussetzung ist aber $u_1(x, y)$ innerhalb C_1 harmonisch, im abgeschlossenen Bereich stetig. Daher ist die Funktion $u_1(x, y) + \theta_1$ auch im Punkt A harmonisch, kraft eines Satzes von *Osgood*[1]), da sie in der Umgebung von A harmonisch, eindeutig und beschränkt ist. Andrerseits ist $u_1(x, y)$ im abgeschlossenen Inneren von C_1 stetig, darum θ_1 *im abgeschlossenen Inneren von* C_1 stetig, welches unmöglich ist. Die Behauptung, daß $u(x, y)$ sich nicht auf die durch (1) angezeigte Weise spalten läßt, ist also bewiesen.

Wir können trotzdem eine einigermaßen komplizierte Entwicklung dieser Funktion $u(x, y)$ herleiten. Es sei C_1' ein veränderlicher Kreis, welcher im Äußeren von C_1 liegt und sich C_1 annähert, und C_2' ein veränderlicher Kreis, welcher im Inneren von C_2 liegt und sich C_2 annähert. Es existiert eine Funktion $u'(x, y)$ harmonisch im Gebiet, das durch C_1' und C_2' begrenzt ist, und welche die Funktion $u(x, y)$ im Bereich, der durch C_1 und C_2 begrenzt ist, gleichmäßig und beliebig nahe approximiert[2]). Die Funktion $u'(x, y) - k \log \sqrt{x^2 + y^2}$ läßt sich für passendes k in diesem Bereich durch harmonische Polynome in (x, y) und harmonische Polynome in $\left(\frac{x-1}{(x-1)^2 + y^2}, \frac{y}{(x-1)^2 + y^2}\right)$ gleichmäßig approximieren, also läßt sich $u(x, y)$ in diesem Bereich durch eine Reihe, deren Glieder von der Form

$$k \log \sqrt{x^2 + y^2} + \text{harm. Polyn. in } (x, y) + \text{harm. Polyn. in } \left(\frac{x-1}{(x-1)^2 + y^2}, \frac{y}{(x-1)^2 + y^2}\right)$$

sind, beliebig nahe gleichmäßig entwickeln.

---

[1]) Loc cit., S. 647; eine passende Definition der Funktion im Punkt A muß natürlich gegeben werden.

[2]) *Lebesgue*, S. 398—399; vgl. auch den Beweis des Satzes III.

27*

Wir formulieren später eine Verallgemeinerung dieser Bemerkung.

Satz VIII läßt sich auf verschiedene Weisen ausdehnen; wir beweisen z. B.:

Satz IX. *Es sei C ein beschränktes Gebiet, dessen Begrenzung aus abgeschlossenen Punktmengen $C_0, C_1, \ldots, C_n$ besteht, worin C_0, bzw. $C_1, \ldots, C_n$ in der Begrenzung eines Gebietes $G_0, G_1, \ldots, G_n$ enthalten ist, welches den zu C nicht gehörenden Punkt ∞, bzw. $(x_1, y_1), \ldots, (x_n, y_n)$ enthält. Besitzen die Punktmengen $C_i, C_j (i, j = 0, 1, \ldots, n,\ i \neq j)$ keinen gemeinsamen Punkt, und ist die Funktion $u(x, y)$ im Gebiet C harmonisch, im entsprechenden Bereich $\overline{C}$ stetig, so läßt sich bei passender Wahl der Konstanten $k_1, k_2, \ldots, k_n$ die Funktion*

$$(2)\quad u(x, y) + k_1 \log \sqrt{(x - x_1)^2 + (y - y_1)^2} + \cdots + k_n \log \sqrt{(x - x_n)^2 + (y - y_n)^2}$$

als die Summe von $n + 1$ in $\overline{C}$ gleichmäßig konvergierenden Reihen von harmonischen Polynomen in

$$(3)\quad (x, y),\ \left(\frac{x - x_1}{(x - x_1)^2 + (y - y_1)^2},\ \frac{y - y_1}{(x - x_1)^2 + (y - y_1)^2}\right), \ldots,$$

$$\left(\frac{x - x_n}{(x - x_n)^2 + (y - y_n)^2},\ \frac{y - y_n}{(x - x_n)^2 + (y - y_n)^2}\right)$$

bzw. entwickeln. Diese Reihen konvergieren gleichmäßig auf der zum Gebiet G_i komplementären Menge (in bezug auf die ganze Ebene), $i = 0, 1, \ldots, n$ bzw.

Dieser Satz ergibt sich unmittelbar (laut Satz V) aus der von *Osgood* betrachteten Spaltung von $u(x, y)$ [1]) in eine Summe von $n + 1$ Funktionen, wovon jede in einem das Gebiet C enthaltenden, durch $C_0, C_1, \ldots, C_n$ bzw. begrenzten einfach zusammenhängenden Gebiet harmonisch und im entsprechenden Bereich stetig ist.

Wenn man die Annahme nicht macht, daß die Punktmengen $C_i, C_j, i, j = 0, 1, \ldots, n, i \neq j$, keinen gemeinsamen Punkt besitzen, so fährt man fort wie im obigen betrachteten speziellen Falle. Man bekommt folgendermaßen eine ähnliche in $\overline{C}$ gültige gleichmäßige Entwicklung in eine Reihe bzw. mehrere Reihen, je nachdem keine Gruppe aus den Punktmengen $C_0, C_1, \ldots, C_n$ von den übrigen getrennt ist oder nicht. Spalten sich diese Punktmengen in etwa $k + 1$ Gruppen, wovon jede Gruppe keinen gemeinsamen Punkt mit einer anderen hat, so spaltet sich die Funktion $u(x, y)$ in k logarithmische Glieder wie die Glieder von (2), plus $k + 1$ Funktionen $u_i(x, y), i = 0, 1, \ldots, k$, deren jede in einem das Gebiet C enthaltenden *einfach zusammenhängenden* Gebiet G_i' harmonisch und im entsprechenden Bereich stetig ist. Die Begrenzung jedes Gebietes G_i' ist eine der besagten Gruppen (etwa Γ_i) aus den Punktmengen $C_0, C_1, \ldots, C_n$. Nach den Ergebnissen von *Lebesgue* [2]) läßt sich die Funktion $u_i(x, y)$ in G_i' durch eine Funktion $u_i'(x, y)$ beliebig nahe approximieren, und die Funktion $u_i'(x, y)$ ist in einem den Bereich $\overline{G_i'}$ enthaltenden veränderlichen Gebiet G_i'' harmonisch. Das Gebiet G_i'' ist durch getrennte *Jordan*sche Kurven begrenzt; jedes Gebiet G_j, dessen Begrenzung zu Γ_i gehört, enthält eine und nur eine dieser Kurven. Das Gebiet G_i'' enthält genau diejenigen Punkte unter $\infty, (x_1, y_1), \ldots, (x_n, y_n)$, die in G_i' liegen.

Satz IX liefert jetzt eine Approximation zu $u_i'(x, y)$ in $\overline{G_i'}$, also zu $u_i(x, y)$ in $\overline{G_i'}$ (daher auch in $\overline{C}$), durch logarithmische Glieder plus harmonische Polynome in den zu den unter $\infty, (x_1, y_1), \ldots, (x_n, y_n)$ außerhalb $\overline{G_i'}$ liegenden Punkten entsprechenden Funktionen (3).

---

[1]) Loc. cit., S. 642—643. Bei *Osgood* kommt nur der Fall in Betracht, daß die Punktmengen $C_0, C_1, \ldots, C_n$ *Jordan*sche Kurven sind. Die Erörterung gilt aber in unserem allgemeineren Fall.

[2]) Loc cit., S. 398—399.

So bekommen wir endlich die gesuchte in $\overline{C}$ gleichmäßig konvergierende Entwicklung von $u(x, y)$. Sie besteht aus k logarithmischen Gliedern wie die Glieder von (2) plus $k+1$ Reihen, jede gleichmäßig konvergent nicht nur in $\overline{C}$ sondern in einem größeren einfach zusammenhängenden Bereich $\overline{G}_i'$. Jedes Glied einer solchen Reihe ist eine Summe logarithmischer Glieder wie die Glieder von (2), plus harmonische Polynome in gewissen Funktionen (3); diese logarithmischen Glieder sowohl wie diese Funktionen (3) entsprechen Punkten ∞, $(x_1, y_1), \ldots, (x_n, y_n)$, die außerhalb $\overline{G}_i$ liegen.

Die Sätze VIII und IX, und natürlich auch das soeben betrachtete Resultat, lassen Ausdehnungen zu, die man durch eine beliebige Transformation durch reziproke Radien erreicht.

THE APPROXIMATION OF HARMONIC FUNCTIONS BY HARMONIC POLYNOMIALS AND BY HARMONIC RATIONAL FUNCTIONS*

BY J. L. WALSH

1. *Introduction.* The following theorem of Weierstrass is classical.

Let the function $f(\theta)$ be continuous for all values of the argument and periodic with period 2π. Then if an arbitrary positive ϵ be given, there exists a trigonometric polynomial which differs from $f(\theta)$ at most by ϵ; that is, the inequality

$$\left| f(\theta) - \sum_{k=0}^{n}(a_k \cos k\theta + b_k \sin k\theta) \right| \leqq \epsilon \tag{1}$$

holds for all values of θ.

An equivalent statement of the conclusion of this theorem is that $f(\theta)$ *can be expanded in a series of the form*

$$\sum_{n=0}^{\infty} \sum_{k=0}^{n}(a_{nk} \cos k\theta + b_{nk} \sin k\theta),$$

which converges uniformly for all values of θ. This is of course a general fact, that if a given function can be uniformly approximated as closely as desired by a linear combination of other functions, then that function can be expanded in a uniformly convergent series of which each term is a linear combination of those other functions, and conversely. This fact is easy to prove and will be frequently used in the sequel.

If in the (x, y)-plane we introduce polar coordinates (r, θ) and consider the function $f(\theta)$ defined on the unit circle C, Weierstrass's theorem refers to the approximation of a function $f(\theta)$ continuous on C by trigonometric polynomials, or what is the same thing, by polynomials of the form

* An address delivered by invitation of the program committee at the meeting of the Society in New York, February 23, 1929.

$$\sum_{k=0}^{n} r^k(a_k \cos k\theta + b_k \sin k\theta). \tag{2}$$

A polynomial of the form (2) is of course harmonic, for the expressions $r^k \cos k\theta$ and $r^k \sin k\theta$ are respectively the real parts of the analytic functions z^k, $-iz^k$, and can be written as harmonic polynomials in x and y.

If we interpret Weierstrass's theorem as yielding a uniform *expansion* rather than uniform approximation of $f(\theta)$, we have a sequence $\{p_n(r, \theta)\}$ of functions each of type (2) which converges uniformly to the function $f(\theta)$ on C and likewise converges uniformly in the *closed* interior of C. For a necessary and sufficient condition for uniform convergence on C is that if an arbitrary ϵ be given, there exist M so that

$$| p_m(r,\theta) - p_n(r,\theta) | \leqq \epsilon, \quad m,n > M,$$

holds on C. But this inequality holding for $r = 1$ holds also for $r \leqq 1$, since the functions $p_n(r, \theta)$ are harmonic for $r \leqq 1$ and a harmonic function has its maximum and minimum values on the boundary of the region considered. The limit $f(r, \theta)$ of this sequence $\{p_n(r, \theta)\}$ is thus harmonic* interior to C, continuous in the corresponding closed region, and on C coincides with the given continuous function $f(\theta)$. That is, by Weierstrass's theorem we have established the existence of the function $f(r, \theta)$, the solution of the Dirichlet problem for the region interior to C and for arbitrary continuous boundary values $f(\theta)$; the solution of the Dirichlet problem is known to be unique.

The problem of the uniform approximation to $f(\theta)$ *on* C by harmonic polynomials is precisely equivalent to the problem of the uniform approximation to $f(r, \theta)$ *in the closed interior of* C, for a function harmonic in a region has no maximum or minimum interior to that region. Inequality (1) is equivalent to the inequality

---

* A function $u(x, y)$ is harmonic in a region if it is continuous there, together with its first and second partial derivatives, and if $(\partial^2 u/\partial x^2) + (\partial^2 u/\partial y^2) = 0$. A function is harmonic at a point if it is harmonic throughout a neighborhood of that point.

$$\left| f(r,\theta) - \sum_{k=0}^{n} r^k(a_k \cos k\theta + b_k \sin k\theta) \right| \leq \epsilon, \qquad r \leq 1;$$

each of these inequalities implies the other.

The discussion just given suggests the general problem of studying the approximation to a given continuous function on an arbitrary point set C by means of harmonic polynomials or more generally by harmonic rational functions, particularly in the more interesting cases that the given function is harmonic in the interior points of C, or that C is itself the boundary of a region. The present paper is devoted to results on this general problem, chiefly a *report* so far as concerns approximation by polynomials, but a detailed exposition of results on approximation by more general rational functions, for these results are in the main new. As in the special case already considered to some extent, namely the theory of approximation to functions of a single real variable by trigonometric polynomials, there are particularly three special problems to be taken up here: first, the general problem of the possibility of approximation, corresponding to Weierstrass's theorem as already mentioned; second, the problem of approximation by *special types* of polynomials corresponding for example to expansion in Fourier's series; and third, the study of the *degree* of approximation, the study of the relation between the continuity properties of the functions approximated and the least maximum error for approximation by a polynomial of degree n. We shall examine these problems in some detail in the order given, first the problems for approximation by polynomials and later the corresponding problems for approximation by more general rational functions.

In the theory of functions of the complex variable $z=x+iy$, similar problems arise in connection with the approximating on a given point set of a given continuous or analytic function by means of polynomials in z or rational functions of z. This analogous theory frequently suggests results in the present theory of harmonic functions, and may even furnish methods of proving those results; but the connection between the two theories is not always simple, and some situations in each theory have no analogs in the other. The two theories both have interesting applications to and connections with harmonic and analytic continuation, conformal mapping, the study of the Dirichlet problem, and topology.

2. *General Approximation by Harmonic Polynomials.* The fundamental theorem in the approximation of analytic functions by polynomials is due to Runge [1]*:

Let $f(z)$ be a function of z, analytic in a simply-connected region† *C of the z-plane which does not contain the point at infinity in its interior. Then $f(z)$ can be developed in C in a series of polynomials in z, which converges uniformly on any closed point set interior to C.*

We shall use Runge's theorem to prove the following analog for harmonic functions.‡

Let $u(x, y)$ be a function of (x, y), harmonic in a simply-connected region C of the (x, y)-plane which does not contain the point at infinity in its interior. Then $u(x, y)$ can be developed in C in a series of harmonic polynomials in (x, y), which converges uniformly on any closed point set interior to C.

Let $v(x, y)$ be a function conjugate to $u(x, y)$ in C; the function $v(x, y)$ is single-valued in C if one considers only an element of $v(x, y)$ and its harmonic extensions along curves interior to C. Runge's theorem as applied to the function $f(z) = u(x, y) + iv(x, y)$ yields a development of $f(z)$ in polynomials in z. The series whose terms are the real parts of the respective terms of that development represents the function $u(x, y)$ and has the required properties with respect to convergence.

---

* The numbers in square brackets refer to the works of the authors indicated, in the bibliography at the close of this paper.

† A region is a connected set of interior points. A region plus its boundary points forms the corresponding *closed* region.

‡ Compare Walsh [1, p. 198]. Bergmann [1] uses similar reasoning, but does not bring out clearly the uniformity of the convergence of the series of harmonic polynomials.

The essential content for our present purposes of the theorem just proved may also be stated as follows.

If the function $u(x, y)$ is harmonic in a closed Jordan region, then $u(x, y)$ can be developed in that closed region in a uniformly convergent series of harmonic polynomials in (x, y).

The present writer has proved, however, results much more general than this [1], regarding the uniform development of arbitrary harmonic functions. We state a number of these results, simply mentioning at this time the fact that the proof is based on Lebesgue's important work [1] on harmonic functions. We shall go into more detail regarding methods in §5, when we take up the more general problem of development in terms of harmonic rational functions.

Let C be an arbitrary limited simply-connected region of the (x,y)-plane. A necessary and sufficient condition that an arbitrary function harmonic in C, continuous in the corresponding closed region, be uniformly developable in the closed region by harmonic polynomials in (x, y), is that the boundary of C be also the boundary of an infinite region.

Let C be an arbitrary limited closed point set. A necessary and sufficient condition that an arbitrary function continuous on C be uniformly developable on C by harmonic polynomials in (x, y), is that C be the boundary of an infinite region.

An arbitrary limited closed point set C divides the plane in general into a number of regions; in particular one of these, D, is infinite. A continuous function $u(x, y)$ defined on C is in particular defined on the boundary B of D, and on this boundary can, by the theorem just stated, be uniformly developed in harmonic polynomials. This expansion of $u(x, y)$ on B converges uniformly on B, hence uniformly on the entire point set E which is complementary (with respect to the entire plane) to D, and therefore defines a function $U(x, y)$ on the entire set E. The set C is a subset of E. The functions $u(x, y)$ and $U(x, y)$ coincide on B. *A necessary and sufficient condition that $u(x, y)$ be uniformly developable on C in harmonic polynomials is that $u(x, y)$ and $U(x, y)$ should coincide on C.*

These theorems obviously include Weierstrass's theorem on the approximation to an arbitrary continuous function by trigonometric polynomials. These theorems include also Weierstrass's classical theorem that an arbitrary function continuous on a closed interval of the axis of reals can be uniformly approximated on that closed interval as closely as desired by a polynomial in the real variable. For if the point set C is chosen as such an interval, the only harmonic polynomials $1, r \cos \theta, r \sin \theta, \cdots, r^n \cos n\theta, r^n \sin n\theta, \cdots$ not vanishing identically on C can be written on C in the form $1, x, x^2, \cdots$, so that on C a harmonic polynomial is a polynomial in x.

With these results, the most important questions regarding the expansion of arbitrary functions on *limited* point sets are answered. The corresponding questions for unlimited point sets have apparently not been treated previously. On some unlimited point sets C only the polynomial 1 of the set $\{r^n \cos n\theta, r^n \sin n\theta\}$ is uniformly bounded, so the only functions continuous on C which can be uniformly approximated on C by harmonic polynomials are constants. On cértain other unlimited point sets C only a finite number of the harmonic polynomials $\{r^n \cos n\theta, r^n \sin n\theta\}$ are continuous, and a necessary and sufficient condition that a function continuous on such a point set C be uniformly developable on C by harmonic polynomials is that the function should be a linear combination of those particular harmonic polynomials. On other unlimited point sets C an infinity of the harmonic polynomials $\{r^n \cos n\theta, r^n \sin n\theta\}$ are continuous, so necessary and sufficient conditions for the expansion of arbitrary functions on such point sets C are not obvious, nor can they be obtained without modification of the methods hitherto employed.

It is illuminating to contrast the present problem with the problem of approximating given analytic or continuous functions by means of polynomials in the complex variable. It is not yet known what is the most general region C such that an arbitrary function analytic interior to C and continuous in the corresponding closed region can be approximated uniformly in the closed region as closely as desired by polynomials in the complex variable. It can be shown by an example [Walsh, 1, p. 203] however, that this most general region C is not the most general region such that an arbitrary function harmonic interior to C and continuous in the corresponding closed region can be approximated uniformly in the closed region as closely as desired by harmonic polynomials. *Unlimited* point sets C are easy to treat for approximation by polynomials in the complex variable. Every such polynomial not a constant becomes infinite on C, so the only continuous functions which can be uniformly approximated on C are constants, and every such function can be so approximated.

3. *Harmonic Polynomials Belonging to a Region.* It is natural to inquire whether more precise results than those just considered are obtainable regarding ***expansions in a given region in terms of a particular set of harmonic polynomials belonging to that region.*** For instance, if the region is chosen as the interior of the unit circle C, then an arbitrary function $f(x, y)$ harmonic in the closed interior of C can be expanded on the circumference in a series of Fourier:

$$(3)\quad f(x,y) = \frac{a_0}{2} + \sum_{k=1}^{\infty}(a_k \cos k\theta + b_k \sin k\theta),$$

$$a_k = \frac{1}{\pi}\int_0^{2\pi} f \cos k\theta\, d\theta,\ b_k = \frac{1}{\pi}\int_0^{2\pi} f \sin k\theta\, d\theta,$$

where the integrals are computed over C. The series converges uniformly on C. The series

$$(4)\qquad \frac{a_0}{2} + \sum_{k=1}^{\infty} r^k(a_k \cos k\theta + b_k \sin k\theta)$$

is a series of harmonic polynomials which converges uniformly on C and hence converges uniformly throughout the closed interior of C. The sum of the series (4) coincides with $f(x, y)$ on C and hence throughout the interior of C as well.

If we start not with the unit circle but a more general Jordan curve C as the boundary of our region, can there be found harmonic polynomials $\{p_k(x, y)\}$ such that an arbitrary function harmonic on and within C can be expanded in a series

$$\sum_{k=0}^{\infty} a_k p_k(x, y) \tag{5}$$

which converges uniformly in the closed interior of C? The answer here is affirmative, if the curve C is analytic. In fact, Faber [1], Fejér [1], Szegö [1], Carleman [1], Bergmann [1], Bochner [1], and others have studied polynomials $\{s_k(z)\}$ in the complex variable z belonging to such a region and have shown that an arbitrary function $f(z)$ analytic in the closed region can be expanded in a series

$$\sum_{k=0}^{\infty} c_k s_k(z)$$

which converges uniformly in the closed region. A suitable set of harmonic polynomials of the kind we desire can be found by separating $s_k(z)$ into its real and pure imaginary parts. Thus, let the function $u(x, y)$ be given harmonic in the closed interior of C; there exists a function $v(x, y)$ conjugate to it, likewise harmonic in this closed region. Then by the properties of the polynomials $s_k(z)$, there exists an expansion

$$u(x, y) + iv(x, y) = \sum_{k=0}^{\infty} (c_k' + ic_k'')(s_k'(x, y) + is_k''(x, y)),$$

where $c_k = c_k' + ic_k''$, $s_k(z) = s_k'(x, y) + s_k''(x, y)$. From this we derive the series

$$u(x, y) = \sum_{k=0}^{\infty} [c_k' s_k'(x, y) - c_k'' s_k''(x, y)], \tag{6}$$

so we have an expansion of type (5)* by setting

$$p_{2k}(x,y) = s_k'(x,y), \qquad p_{2k+1}(x,y) = s_k''(x,y).$$

In particular if the derivatives of the polynomials $s_k(z)$ are orthogonal with respect to the area interior to C, the formulas for the coefficients in (6) are remarkably simple in form, as Bergmann has shown [1].

There are, however, various other ways of determining sets of harmonic polynomials belonging to a given region. Let us consider the plane of the auxiliary variables (x', y'), and map conformally the interior of C onto the interior of the unit circle Γ of the (x', y')-plane. A function harmonic in the closed interior of C corresponds to a function harmonic in the closed interior of Γ, and can be uniformly expanded in the closed interior of Γ in a series (4) of harmonic polynomials in (x', y'). The harmonic polynomials $r^n \cos n\theta$, $r^n \sin n\theta$ in the (x', y')-plane correspond in the (x, y)-plane to functions harmonic in the closed interior of C; moreover these functions in the (x, y)-plane can be replaced by harmonic *polynomials* in (x, y) which differ only slightly from them, without essentially altering the convergence properties of series expansions of arbitrary functions in terms of them. More explicitly, we state the following theorem [Walsh, 2].

Let C be a simple finite analytic curve in the (x, y)-plane. Then there exist harmonic polynomials $\{p_k(x, y)\}$ such that if $f(x, y)$ is defined and continuous on C and on C is of bounded variation, then $f(x, y)$ can be developed into a series

$$f(x,y) = \sum_{k=0}^{\infty} a_k p_k(x,y), \tag{7}$$

which converges uniformly in the closed interior of C. Series (7) thus represents a function harmonic interior to C, continuous in the corresponding closed region, and having the value $f(x, y)$ on C. There exist continuous functions $\{q_k(x, y)\}$ on C

* Removal of brackets in series (6) can be justified in the cases to which reference has been made.

with which the polynomials $\{p_k(x, y)\}$ *form a biorthogonal set:*

$$\int_C p_k(x,y)q_m(x,y)ds = \begin{cases} 0, & \text{if } k \neq m, \\ 1, & \text{if } k = m. \end{cases}$$

The coefficients of (7) *are given by the formulas*

$$a_k = \int_C f(x,y)q_k(x,y)ds\ ; \tag{8}$$

the functions $q_k(x, y)$ *depend on* C *but not on* $f(x, y)$.

If the function $f(x, y)$ *is known merely to be continuous on* C, *then the series* (7), *where the* a_k *are given by* (8), *converges throughout the interior of* C, *uniformly on any closed point set interior to* C; *if summed by the method of Cesàro this series converges uniformly on and within* C *and thus represents a solution of the Dirichlet problem for the region interior to* C *and the boundary values* $f(x, y)$.

If the function $f(x, y)$ *is of bounded variation on* C, *but not necessarily continuous, then the series* (7), *where the* a_k *are given by* (8), *converges at every point of the closed region, uniformly on any closed point set interior to* C. *The function represented is bounded in the closed interior of* C *and approaches the boundary values* $f(x,y)$ *continuously at every point of continuity of* $f(x, y)$.

This theorem as stated applies only to an analytic curve C, but the writer has some further results, as yet unpublished, which apply to much more general curves.

Still another method of defining a set of harmonic polynomials belonging to a given region bounded by a rectifiable Jordan curve is that of orthogonalization. We begin with the harmonic polynomials

$$1,\ r\cos\theta,\ r\sin\theta,\ r^2\cos 2\theta,\ r^2\sin 2\theta,\ \cdots$$

and orthogonalize them with respect to the given curve C. This yields a set of polynomials $\{p_k(x, y)\}$ such that

$$\int_C p_k(x,y)p_m(x,y)ds = \begin{cases} 0, & k \neq m, \\ 1, & k = m. \end{cases}$$

An arbitrary function $f(x, y)$ defined on C can be expanded formally in terms of these polynomials:

$$(9) \qquad f(x,y) \sim \sum_{k=0}^{\infty} a_k p_k(x,y), \quad a_k = \int_C f(x,y) p_k(x,y) ds.$$

This orthogonalization method was used by Szegö [1] in the corresponding case of polynomials in the complex variable. The results in the present case have been established by Merriman [1], who determines the asymptotic formulas for the polynomials $\{p_k(x, y)\}$, and shows, under suitable restrictions, that the series (9) converges interior to C or even uniformly in the closed interior of C. The sum of the series is of course the solution of the Dirichlet problem for the continuous boundary values $f(x, y)$.

It would be of interest to investigate the corresponding problems where the harmonic polynomials are orthogonalized not with respect to arc length on the boundary, but with respect to the area of the region interior to C [compare Bergmann 1, Carleman 1]. The former method has the theoretical advantage, however, of not requiring for expansion the knowledge of the values of the harmonic function except on the boundary of the region. Another general problem which deserves treatment is the detailed study of the convergence on C of expansions in terms of the special polynomials of Faber and others, either for the case of analytic or of harmonic functions, where the given function to be expanded is not known to be analytic or harmonic in the *closed* region.

4. *Degree of Approximation.* We have considered thus far the *possibility* of approximation by harmonic polynomials and the expansion in terms of a *particular set* of harmonic polynomials. We turn now to consideration of the *degree* of approximation, that is, the relation between the properties of the function approximated, with regard to existence of derivatives etc. on the one hand, and the asymptotic properties of the maximum error in the best approximation

by a harmonic polynomial of degree n,* on the other hand. A simple result [Walsh, 3] here is the analog and consequence of the corresponding result in approximation of analytic functions by means of polynomials in the complex variable.

Let C be an arbitrary closed Jordan region of the (x, y)-plane, and let $w=\phi(z)$, $z=x+iy$, be a function which maps conformally the exterior of C onto the exterior of the unit circle in the w-plane so that the points at infinity correspond to each other. Let C_R denote the curve $|\phi(z)|=R$, $R>1$, that is, the transform in the z-plane of the circle $|w|=R$.

A necessary and sufficient condition that an arbitrary function $u(x, y)$, defined in C, be harmonic in the (closed) region C is that there should exist harmonic polynomials $p_n(x, y)$ of degree n, $n=0, 1, 2, \cdots$, and numbers M, $R>1$, such that the inequalities

$$|u(x,y) - p_n(x,y)| \leqq \frac{M}{R^n}, \tag{10}$$

where M and R are independent of n and of (x, y), should be valid for every point (x, y) of C.

If the polynomials $p_n(x, y)$ are given so that (10) *is satisfied for every (x, y) of C, the sequence $\{p_n(x, y)\}$ converges everywhere interior to C_R and uniformly on any closed point set interior to C_R, so the function $u(x, y)$ is harmonic throughout the interior of C_R.*†

If $u(x, y)$ is given harmonic in the closed region interior to C_ρ, the polynomials $p_n(x, y)$ can be chosen to satisfy (10) *with $R=\rho$, for (x, y) in C.*

---

* The polynomial $a_0x^n+a_1x^{n-1}y+ \cdots +a_ny^n+b_0x^{n-1}+b_1x^{n-2}y+ \cdots$ is considered to be of degree n, but if the term degree is used *in the restricted sense*, is of degree n if and only if at least one of the coefficients a_i is different from zero.

† Here and below we tacitly assume that if $u(x, y)$ is not originally supposed to be defined on the entire point set considered, then the definition in the new points is to be made by harmonic extension—or what amounts to the same thing—by means of the convergent series of harmonic polynomials.

This theorem seems to be the only one in the literature concerning degree of approximation by harmonic polynomials, except in the case that C itself is a circle, when approximation by harmonic polynomials reduces to approximation on the circumference by trigonometric polynomials, for which well known results have been obtained by Bernstein, Jackson, Montel, de la Vallée-Poussin, and others [see de la Vallée-Poussin, 1]. There is obviously occasion here for further investigation. If $u(x, y)$ is given harmonic within, continuous on and within the Jordan curve C, what is the relation between the maximum error for the best approximation of $u(x, y)$ on C by a harmonic polynomial of degree n on the one hand, and the continuity properties of the curve C and of the function $u(x, y)$ on C on the other hand? If $u(x, y)$ is given harmonic within, continuous on and within C_R, what is the relation between the maximum error for the best approximation on C by a harmonic polynomial of degree n on the one hand, and the continuity properties of the curve C and of the function $u(x, y)$ on the curve C_R on the other hand?

Let us stop for a moment to consider the Tchebycheff harmonic polynomial for the function $u(x, y)$ on a point set C, that is, the harmonic polynomial $p_n(x, y)$ of degree n for which the maximum $|u(x, y)-p_n(x, y)|$, (x, y) on C, is least. It is convenient here to refer to a general theorem due to Haar [1], which deals with approximation on a given point set C of a given function $u(x, y)$ by linear combinations of given functions $\{u_i(x, y)\}$; the coefficients a_i are to be determined so that the maximum

$$(10') \qquad |u(x,y) - a_1u_1(x,y) - \cdots - a_mu_m(x,y)|$$

is least for (x, y) on C. If C is closed and the u's are continuous, such a determination of the coefficients is possible [Haar, loc. cit.] and the corresponding linear combination of the $u_i(x, y)$ may be called a Tchebycheff polynomial of order m. Haar's theorem asserts—except for incidental restrictions on the continuity of the functions and the closure of the point set—that *a necessary and sufficient condition for the uniqueness of the Tchebycheff polynomial of order m for approximation on C to an arbitrary function continuous on C, is that no function*

$$(10'') \qquad \begin{gathered} A_1u_1(x, y) + \cdots + A_mu_m(x, y), \\ |A_1| + |A_2| + \cdots + |A_m| \neq 0, \end{gathered}$$

shall vanish at more than $m-1$ points of C. It is immaterial whether one studies on the one hand the approximation in a closed Jordan region C by harmonic polynomials of a function harmonic interior to the region, continuous in the closed region, or on the other hand the approximation on the boundary of the region of the boundary values of the harmonic function. For the maximum in C of the function in (10′) must occur on the boundary of C. Fréchet has proved that the Tchebycheff trigonometric polynomial is unique for approximation to an arbitrary continuous function $f(\theta)$ with period 2π on the interval $0 \leqq \theta \leqq 2\pi$. That is, from our present standpoint (compare §1), *the Tchebycheff harmonic polynomial is unique for approximation to a continuous function on a circumference C, or for approximation on and within a circle to a function continuous there, harmonic in the interior.* This statement is true, however, only with a restriction. The Tchebycheff polynomial which is a linear combination of the functions 1, $r \cos \theta$, $r \sin \theta$, $\cdots$, $r^n \cos n\theta$, $r^n \sin n\theta$ is indeed unique, for an arbitrary linear combination (10″) of these functions vanishes on an algebraic curve of degree n, and this curve has at most $2n$ points in common with the circumference C.* In fact, this method proves that *the corresponding Tchebycheff harmonic polynomial is unique for approximation on an ellipse C to a function continuous on C, or for approximation on and within C of a function harmonic interior to C, continuous in the corresponding closed region.*

The Tchebycheff polynomial, which is a linear combination

* The circumference cannot be a branch of this curve, for the function (10″) cannot vanish everywhere on any (limited) Jordan curve.

of the functions 1, $r \cos \theta$, $r \sin \theta, \cdots, r^{n-1} \cos (n-1)\theta$, $r^{n-1} \sin (n-1)\theta$, $r^n \cos n\theta$, is not necessarily unique either when C is a circle or an ellipse. Indeed, if $n=1$, the function (10″) is A_1+A_2x, which obviously for suitable choice of A_1 and A_2 vanishes at more than a single point of the circle or ellipse C.

It would be an interesting investigation, to determine what algebraic curves C have this property, that the Tchebycheff harmonic polynomial for approximation to an arbitrary function continuous on C is unique. This is a subject dealing with the real intersections of plane algebraic curves, in which a single point of intersection counts merely as a single point, in spite of singularities and multiple points (in the usual sense of the term) of either curve.

Whether or not the Tchebycheff harmonic polynomial $t_n(x, y)$ of degree n is unique, a Tchebycheff polynomial $t_n(x, y)$ of degree n exists, and we can derive certain properties of the sequence $\{t_n(x, y)\}$. In the notation of the previous theorem, we shall prove the following facts.

Let the function $u(x, y)$ be harmonic in the closed interior of C, and have at least one singularity on the curve C_ρ but no singularity interior to C_ρ. Then a sequence $\{t_n(x, y)\}$ of Tchebycheff harmonic polynomials for the function $u(x, y)$ considered in the closed interior of C (or what is essentially the same, considered on C itself), whether or not the Tchebycheff polynomial is unique, converges throughout the interior of C_ρ, and uniformly on any closed point set interior to C_ρ. The limit of the sequence throughout the interior of C_ρ is $u(x, y)$. The sequence can converge uniformly in no region $C_{\rho'}$, with $\rho' > \rho$.

The proof is immediate. There exists, by the theorem already stated, some sequence $\{p_n(x, y)\}$, where $p_n(x, y)$ is of degree n, such that we have

$$(10) \qquad |u(x,y) - p_n(x,y)| \leqq \frac{M}{R^n}, \qquad (x,y) \text{ in } C,$$

provided merely that $R<\rho$. If (10) is valid for the polynomial $p_n(x, y)$, it is also valid if $p_n(x, y)$ is replaced by $t_n(x, y)$:

$$| u(x,y) - t_n(x,y) | \leqq \frac{M}{R^n}, \qquad (x,y) \text{ in } C,$$

provided still that $R<\rho$. Again by virtue of the theorem stated, the sequence $\{t_n(x, y)\}$ must converge interior to every C_R for which $R<\rho$, hence throughout the interior of C_ρ, uniformly on any closed point set interior to C_ρ. In particular, if the function $u(x, y)$ has no singularity except at infinity, the sequence converges at every point of the plane, uniformly on any limited closed point set. In the general case, the sequence $\{t_n(x, y)\}$ cannot converge uniformly on any curve $C_{\rho'}$ for which $\rho'>\rho$, for then it would likewise converge uniformly in a region containing C_ρ, which has on it a singularity of the function $u(x, y)$.

The theorem just established is the analog of a theorem due to Faber [2, p. 105] for the case of approximation to an analytic function by polynomials in the complex variable. The proofs are, however, different, and the present proof applies without essential change in that other situation, even if C is a Jordan arc or a certain more general point set, instead of a Jordan region.

It will be noticed that the sequence $\{t_n(x, y)\}$ may converge for certain points (x, y) exterior to C_ρ. Let us suppose for instance that the region C is symmetric on the x-axis and that the function $u(x, y)$ satisfies the equation $u(x, y) = -u(x, -y)$. Then a given $t_n(x, y)$ may be replaced by another Tchebycheff polynomial $t_n'(x, y)$ of degree n which satisfies the equation $t_n'(x, y) = -t_n'(x, -y)$ and which approximates $u(x, y)$ on C as closely as does $t_n(x, y)$. If we have

$$| u(x,y) - t_n(x,y) | \leqq \epsilon, \qquad (x,y) \text{ on } C,$$

we have likewise by symmetry

$$| u(x, -y) - t_n(x, -y) | \leqq \epsilon, \qquad (x,y) \text{ on } C,$$

that is,

$$|u(x,y)+t_n(x,-y)|\leqq\epsilon,\qquad (x,y) \text{ on } C.$$

If we set

$$t_n'(x,y)=\frac{t_n(x,y)-t_n(x,-y)}{2},$$

the harmonic polynomial $t_n'(x, y)$ obviously satisfies the functional equation considered, and we have the inequality

$$|u(x,y)-t_n'(x,y)|\leqq\epsilon,\qquad (x,y) \text{ on } C\,;$$

that is to say, the polynomial $t_n'(x, y)$ is as good an approximation to $u(x, y)$ on C as is $t_n(x, y)$.* Interpret $t_n'(x, y)$ as a linear combination of the functions $r^n \cos n\theta$, $r^n \sin n\theta$, hence as a linear combination of the functions $r^n \sin n\theta$. The sequence $\{t_n'(x, y)\}$ converges for $\theta=0$ for *all* values of r, no matter how small C may be or where the singularities of $u(x, y)$ may lie.

We have thus far restricted our entire discussion of the degree of approximation by harmonic polynomials to the consideration of point sets C which are regions. That is probably the simplest case; the study of the same problems where C is, let us say, a Jordan arc, is more complicated. Complication arises because a harmonic polynomial may vanish identically on C, so that convergence on C of a sequence of harmonic polynomials, even so that (10) is satisfied on C, does not imply convergence of the sequence for points not on C. Let us treat here in detail the simplest pos-

* We have essentially proved here that if the function $u(x, y)$ can be uniformly expanded by harmonic polynomials on a point set C which is symmetric in the x-axis and if we have $u(x, y)=-u(x, -y)$, then $u(x, y)$ can likewise be uniformly expanded on C by harmonic polynomials $p_n(x, y)$ which satisfy the equation $p_n(x, y)=-p_n(x,-y)$.

Another theorem of the same general nature easily proved by methods already used elsewhere for functions of a complex variable is that if an arbitrary function $u(x, y)$ can be uniformly expanded by harmonic polynomials on a point set C, then $u(x, y)$ is the uniform limit of a sequence of harmonic polynomials each of which is equal to $u(x, y)$ in n arbitrary preassigned points of C.

sible case, namely that C is an interval of the axis of reals; the result is also valid if C is any line segment.

Suppose that $u(x, y)$ is defined on C and that a set of polynomials $\{p_n(x, y)\}$ exists so that (10) holds on C for $R>1$. On C a harmonic polynomial of degree n is a polynomial in x of degree n. Inequality (10), valid for a polynomial $p_n(x, y)$ in x of degree n, implies the convergence of this sequence of polynomials *considered as polynomials in the complex variable* x throughout the interior of C_R (which is defined here as already indicated for a *region* C), uniformly on any closed point set interior to C_R, and hence represents interior to C_R an analytic function of the complex variable x. This is indeed a theorem due to Bernstein, and in the present case the curve C_R is a certain ellipse whose foci are the extremities of the interval C. It follows that *there exists a function* $U(x, y)$, *harmonic interior to* C_R, *coinciding on* C *with the given function* $u(x, y)$. There are two important differences between this result and the result established for the case that C is a Jordan region. First, we have not shown in the present case, nor is it necessarily true, that the original given sequence of harmonic polynomials $\{p_n(x, y)\}$ converges everywhere interior to C_R. Second, the function $U(x, y)$ is not uniquely determined by the requirements of being harmonic interior to C_R and coinciding on C with the given function $u(x, y)$; if any such function $U(x, y)$ is at hand, we may find another by adding to it an arbitrary function harmonic interior to C_R and vanishing on C.

Reciprocally, if it is desired to establish (10) when the function $u(x, y)$ is given harmonic on and within C_R, that can always be done if C is an arbitrary Jordan arc, or indeed a much more general point set. The result follows from the corresponding result for the development of analytic functions in terms of polynomials in the complex variable [compare Walsh, 3].

Before we leave the subject of approximation by harmonic polynomials and turn to harmonic rational functions, we mention another topic which seems not to have been treated in the literature and yet which deserves to be investigated, namely, interpolation by means of harmonic polynomials. When does a harmonic polynomial of degree n exist which takes on preassigned values at $2n+1$ points? When is it unique? What of the asymptotic character of the polynomial as the number of points becomes infinite? What application is there to approximation and expansion, if these points are chosen on the boundary of a region? These questions have been answered in special cases, corresponding to trigonometric interpolation [see Faber, 3, Jackson, 1, de la Vallée-Poussin, 1], and also for the analogous problem of interpolation by polynomials in the complex variable [Fejér, 1]. The former case yields satisfactory results (compare §1) for interpolation by harmonic polynomials either on a circumference or on and within a circle, but for more general situations the questions seem still to be untouched.*

5. *General Approximation by Harmonic Rational Functions.* The results thus far established for approximation by harmonic polynomials have precise analogs for approximation of arbitrary harmonic functions by harmonic rational functions; these analogs will now be treated in the same order, and the proofs follow, in the main, the proofs for the simpler case. The results in the present case, however, are mostly new to the literature and must therefore be treated here in some detail. This newness explains the apparent lack of balance between the treatment of the two cases of approximation, by harmonic polynomials and by more general harmonic rational functions.

---

* These results of Fejér do yield simple theorems immediately. Thus we may make the following statement.

Let C be an arbitrary Jordan curve, and let the function $u(x, y)$ be harmonic on and within C. Then there exists a sequence of harmonic polynomials $\{p_n(x, y)\}$ of respective degrees n coinciding with $u(x, y)$ at n points P_i of C, which converge uniformly on and within C to the function $u(x, y)$. These n points P_i can be chosen as points of C which correspond, under conformal mapping of the exterior of C on the exterior of a circle K so that the points at infinity correspond to the vertices of a regular polygon of n sides inscribed in K.

Let the function $u(x, y)$ be harmonic in the closed region C bounded by two analytic Jordan curves C_1 and C_2, with C_2 interior to C_1 and the origin interior to C_2. Green's formula

$$u(x,y) = \frac{1}{2\pi}\int_{C_1+C_2}\left[u\frac{\partial \log r}{\partial n} - \log r\frac{\partial u}{\partial n}\right]ds,$$

where the integrals are to be taken in the positive sense with respect to the region, n being the inner normal, breaks up the function $u(x, y)$ into a function harmonic on and interior to C_1, plus a function harmonic on and exterior to C_2, plus a multiple of $\log r$, where $r^2 = x^2 + y^2$ [see Osgood, 1, pp. 642–644, Walsh, 1, p. 206]. Consequently the function $u(x, y)$ can be approximated in the closed region C as closely as desired by a harmonic polynomial in (x, y) plus a harmonic polynomial in

$$\left(\frac{x}{x^2+y^2}, \frac{y}{x^2+y^2}\right)$$

plus a multiple of $\log r$. A similar situation obtains if C is bounded not by two Jordan curves but by n Jordan curves, and is not materially altered if these bounding curves are not analytic, nor if $u(x, y)$ is not harmonic in the closed region, but is continuous in the closed region and harmonic in the interior.

We are in a position however to establish a more general* theorem.

Let C be a closed point set which does not contain the point at infinity and which contains no region of infinite connectivity not included in a larger region of finite connectivity belonging to C. Let $f(x, y)$ be an arbitrary function continuous on C and

* More general, that is, so far as concerns the point sets and functions considered; actually less specific in one regard, for the older theorem splits up the given function into $2n-1$ functions, of which $n-1$ are logarithms and left unchanged in the approximation. The other n functions are harmonic in simply-connected regions in whose interiors C lies, and *each* of these n functions can be uniformly approximated or expanded in terms of harmonic rational functions in the corresponding simply-connected region.

harmonic in the interior points of C. Then if an arbitrary positive ϵ *be given, there exists a harmonic function* $\phi(x, y)$, *namely, a polynomial in* (x, y) *and in*

$$\left(\frac{x - x_i}{(x - x_i)^2 + (y - y_i)^2}, \frac{y - y_i}{(x - x_i)^2 + (y - y_i)^2}\right),$$

such that we have

$$(11) \qquad \begin{aligned} &|f(x, y) - \phi(x, y) - a_1 \log r_1 \\ &\qquad - a_2 \log r_2 - \cdots - a_m \log r_m| < \epsilon, \quad (x, y) \text{ in } C, \end{aligned}$$

where the a_i *are suitable constants and* $r_i^2 = (x - x_i)^2 + (y - y_i)^2$. *Here the points* (x_i, y_i), $i = 1, 2, \cdots$, *are exterior to C, and can be preassigned, one in each of the regions R into which C separates the plane, although all of such preassigned points do not necessarily appear in* (11).

In particular if C has no interior points, an arbitrary function $f(x, y)$ *continuous on C can be so approximated.*

Let us outline the proof; the method is essentially due to Lebesgue. Let K be a circle which contains C. There exists a function $F(x, y)$ continuous on and within K and which coincides with $f(x, y)$ on C. There exists a polynomial $p(x, y)$ in (x, y) which is not necessarily harmonic but which throughout K differs from $F(x, y)$ by less than $\epsilon/3$. Divide the plane into squares and continue subdivision indefinitely, by halving the sides of the squares already constructed, so that we determine a sequence of closed point sets S_i, each of which consists of a finite number of regions each bounded by a finite number of non-intersecting Jordan curves, of such a nature that each S_i contains C in its interior, so that each S_i contains its successors, but so that every point exterior to C is exterior to some S_i. By suitable modification of these point sets S_i if necessary, we can make sure that no two of the closed regions composing any particular S_i have a point in common. Let $h_i(x, y)$ be the function harmonic throughout the interior of S_i, continuous on S_i, and coinciding on the boundary of S_i with the function $p(x, y)$.

Then we have [Walsh, 1, p. 199] $\lim_{i\to\infty} h_i(x, y) = p(x, y)$ uniformly on the boundary points of C. That is, in particular we can choose k so that $|h_k(x, y) - p(x, y)| < \epsilon/3$ uniformly on the boundary of C. But on C the function $h_k(x, y)$ can be uniformly approximated by a function of the sort considered in the theorem:

$$(12)\quad \begin{aligned} | h_k(x, y) - \phi(x, y) - a_1 \log r_1 \\ - \cdots - a_m \log r_m | < \frac{\epsilon}{3}, \quad (x, y) \text{ on } C, \end{aligned}$$

and the points (x_i, y_i) lie exterior to C. In fact, if the points (x_i, y_i) are not preassigned, we can [loc. cit., p. 208] satisfy (12) on the entire point set S_k. If the points (x_i, y_i) are preassigned, we can make the approximation (12) not on the point set S_k but on a point set S_k' of the same connectivity as S_k, bounded by a finite number of non-intersecting Jordan curves, which is contained in S_k, which contains C, but which contains precisely those of the preassigned points (x_i, y_i) which lie in regions R lying entirely in S_k. In this new approximation (12), which holds in S_k' and hence in C, we use only the preassigned points (x_i, y_i).

Combination of the inequalities obtained yields (11) uniformly for all points on the boundary of C. But all the functions in the left-hand member of (11) are harmonic in the interior points of C, continuous on the corresponding closed point set, and such a function has no maximum or minimum in an interior point. Hence inequality (11) holds for *all* points of C, and the theorem is established.

We mention explicitly that it is not true that an arbitrary function continuous in a limited closed region and harmonic interior to the region can be uniformly approximated in that region as closely as desired by a harmonic rational function, either with or without logarithmic terms as considered in (11). Consider for example the region C formed from the circle $x^2+y^2<1$ by cutting out the line segment $y=0$, $-1/2 \leqq x \leqq 1/2$; let $f(x, y)$ be the function harmonic interior to this region, continuous in the closed region, zero on the circumference, and unity on the line segment. The function $f(x, y)$ cannot be uniformly approximated in the closed region. For the approximating functions have only isolated singularities [Osgood 1, p. 680], are continuous without exception in the closed region (else are not uniformly bounded), hence are harmonic in the closed region. These approximating functions can be uniformly approximated as closely as desired in the closed region C by harmonic polynomials in (x, y), so $f(x, y)$ can also be uniformly approximated in C as closely as desired by a harmonic polynomial in (x, y). But if such a polynomial differs from $f(x, y)$ by less than ϵ for points on $x^2+y^2=1$, that polynomial differs from zero by less than ϵ for all points $x^2+y^2<1$, which is a contradiction for $x=y=0$ if $\epsilon<1/2$.

The general theorem we have proved, culminating in inequality (11), is obviously not an exhaustive discussion of its subject-matter. If C is a closed region of infinite connectivity, is it true that an arbitrary function $f(x, y)$ harmonic interior to C and continuous in the closed region, can in the closed region be uniformly approximated as closely as desired, as in (11)? Is our general theorem true without any restriction as to regions of infinite connectivity? Is it true that if C is an arbitrary closed point set without interior points, then an arbitrary function $f(x, y)$ continuous on C can be uniformly approximated on C as closely as desired by a rational harmonic function *without logarithmic terms*?

Even though we are not in a position to answer this last question, there are specific closed point sets C on which an arbitrary continuous function can certainly be approximated as closely as desired by a harmonic rational function. In fact, *an arbitrary closed point set C which consists of a finite number of Jordan arcs which do not divide the plane into an infinite number of regions has this property.* More generally, [compare Walsh, 5] *if C is such a point set that an arbitrary function continuous on C can be approximated on C as closely as desired by a rational function of the complex variable, then*

C also has the property considered. For let $f(x, y)$ be the given (real) function; a rational function $p(z)$ of z exists so that we have $|f(x, y)-p(z)|<\epsilon$, (x, y) on C, where ϵ is preassigned. Hence we have also $|f(x, y)-r(x, y)|<\epsilon$, (x, y) on C, where $r(x, y)$ is a rational harmonic function of (x, y), the real part of $f(z)$.

We do not attempt to approximate the most general function harmonic in a multiply-connected region C by harmonic rational functions without the use of logarithmic terms, for that is impossible, as we shall now prove, if the closed region cannot be considered as a closed simply-connected region.* We choose C limited, so there exists an analytic Jordan curve J interior to C, whose interior contains points not belonging to the closed region C; let us assume the origin to be such a point, so that the function $\log r$, where $r^2=x^2+y^2$, is harmonic in the closed region C. Assume the approximation possible, so that $\log r$ can be expanded in C in a uniformly convergent series of harmonic rational functions. These rational functions have only isolated singularities and are continuous in C, hence harmonic throughout the interior of C. Differentiate this series term by term in the direction of the normal to J, and integrate the resulting series over J term by term [Osgood 1, pp. 652–653]. For each term of the series, the result of this process is zero [Osgood 1, p. 680], but for the function $\log r$ the result is 2π. We thus have the contradiction $2\pi=0$, and the statement is established.

The entire discussion we have given enables us to answer the question as to whether a function given on an arbitrary closed point set C can be uniformly approximated on that point set by harmonic rational functions plus logarithmic terms, with singularities in assigned points not belonging to C, provided that no region of infinite connectivity is involved. The facts and proofs are so similar to the corresponding facts and proofs for the case of approximation by harmonic polynomials [Walsh, 1] that they are omitted.

* Already proved [Walsh 1, p. 206] in a special case.

It is not our purpose to study non-uniform expansion of harmonic functions, but one result is now interesting and yet so obvious that we mention it here; it is still an open problem to determine in general what functions can be expanded not necessarily uniformly in terms of harmonic rational functions either with or without logarithmic terms.

Let C_0 be an arbitrary open point set, which may be empty. Let $C_1, C_2, \cdots$ be closed point sets (any or all of which may be empty) mutually exclusive and having no common point with C_0, and on each of the point sets $C_0, C_1, \cdots$ let the function $u(x, y)$ be expansible by harmonic rational functions plus logarithmic terms. Then the function $u(x, y)$ can be expanded on $C_0+C_1+C_2+\cdots$ in a series of rational harmonic functions plus logarithmic terms. The series converges uniformly on each of the point sets $C_1, C_2, \cdots$ on which $u(x, y)$ is uniformly expansible, and if $u(x, y)$ is harmonic on C_0, except possibly for logarithmic singularities or singularities corresponding to rational harmonic functions, then the series converges uniformly on every closed point set contained in C_0.

We use $\psi(x, y)$ generically to denote a rational harmonic function with logarithmic terms, such as occurs in (11). Let $S_1, S_2, \cdots$ be closed point sets, each consisting of a finite number of mutually exclusive regions, each bounded by a finite number of non-intersecting Jordan curves, and such that S_k lies interior to C_0, that S_k lies interior to S_{k+1}, and that every point of C_0 lies in some S_k. Let us suppose

$$u(x,y) = \lim_{n\to\infty} \psi_{kn}(x,y), \quad [(x,y) \text{ on } C_k, \ k = 0,1,2, \cdots],$$

where this limit is approached uniformly if $u(x, y)$ is uniformly expansible on C_k, $k\geqq 1$, and uniformly on every closed point set in C_0 for $k=0$ if that is possible. Then we can determine a closed point set B_{11}, consisting of a finite number of mutually exclusive closed regions each bounded by a finite number of non-intersecting Jordan curves, which contains C_1 but has no point in common with S_1. We can also determine $\psi_1(x, y)$ (see the lemma below) such that

$$|\psi_1(x,y) - \psi_{02}(x,y)| < \tfrac{1}{2}, \qquad (x,y) \text{ on } S_1,$$
$$|\psi_1(x,y) - \psi_{11}(x,y)| < \tfrac{1}{2}, \qquad (x,y) \text{ on } B_{11}.$$

Let B_{12} be a closed point set of the topological simplicity of B_{11}, which contains C_1 but has no point in common with S_2 or C_2. Let B_{21} be a closed point set likewise of the topological simplicity of B_{11}, which contains C_2 but has no point in common with S_2 or B_{12}. Choose $\psi_2(x, y)$ so that we have

$$|\psi_2(x,y) - \psi_{03}(x,y)| < \tfrac{1}{3}, \qquad (x,y) \text{ on } S_2,$$
$$|\psi_2(x,y) - \psi_{12}(x,y)| < \tfrac{1}{3}, \qquad (x,y) \text{ on } B_{12},$$
$$|\psi_2(x,y) - \psi_{21}(x,y)| < \tfrac{1}{3}, \qquad (x,y) \text{ on } B_{21}.$$

We continue this process; in general B_{1n} shall contain C_1 but shall have no point in common with S_n or $C_2, C_3, \cdots, C_n$. The point set $B_{2,n-1}$ shall contain C_2 but shall have no point in common with $C_3, \cdots, C_n, S_n$ or B_{1n}. The point set B_{n1} shall contain C_n but shall have no point in common with S_n or $B_{1n}, B_{2,n-1}, \cdots, B_{n-1,2}$. The function $\psi_n(x, y)$ is then to be chosen so that we have

$$|\psi_n(x,y) - \psi_{0,n+1}(x,y)| < \frac{1}{n+1}, \quad (x,y) \text{ on } S_n,$$

$$|\psi_n(x,y) - \psi_{1n}(x,y)| < \frac{1}{n+1}, \quad (x,y) \text{ on } B_{1n},$$

$$\cdots\cdots\cdots\cdots$$

$$|\psi_n(x,y) - \psi_{n1}(x,y)| < \frac{1}{n+1}, \quad (x,y) \text{ on } B_{n1},$$

and the sequence $\{\psi_n(x, y)\}$ has the property required in the theorem. Uniform convergence of the sequence $\{\psi_n(x, y)\}$ on a closed point set belonging to $C_0+C_1+\cdots$ but not necessarily a C_k depends merely on the uniform convergence of the corresponding sequence (or sequences) $\{\psi_{kn}(x, y)\}$ on that point set. The function $u(x, y)$, if harmonic on C_0 except possibly for logarithmic singularities or singularities corresponding to rational harmonic functions, can be chosen arbitrarily on C_0, and the sequence $\{\psi_n(x, y)\}$ converges to the value $u(x, y)$, uniformly on any closed point set contained in C_0.

If none of the point sets $C_0, C_1, C_2, \cdots$ separates any point of another of those point sets from the point at infinity the theorem is true if the words *harmonic rational functions plus logarithmic terms* are replaced by *harmonic polynomials*. For uniform convergence on any closed point set contained in C_0, however, we require that $u(x, y)$ should be harmonic in C_0 and that C_0 should be composed of mutually exclusive simply-connected regions. In the discussion just given we have had occasion to apply the following lemma.

LEMMA. *If S_1 and S_2 are mutually exclusive closed point sets each consisting of a finite number of mutually exclusive regions each bounded by a finite number of non-intersecting Jordan curves, and if the functions $\psi_1(x, y)$ and $\psi_2(x, y)$ are rational harmonic functions with logarithmic terms, then if $\epsilon>0$ be given there exists a rational harmonic function $\psi(x, y)$ with logarithmic terms so that we have*

$$|\psi(x,y) - \psi_1(x,y)| < \epsilon, \qquad (x,y) \text{ in } S_1,$$
$$|\psi(x,y) - \psi_2(x,y)| < \epsilon, \qquad (x,y) \text{ in } S_2.$$

For simplicity in the proof we assume that both S_1 and S_2 are limited. We write $\psi_1(x, y) = \psi_1'(x, y) + \psi_1''(x, y)$, where $\psi_1'(x, y)$ has no singularities in S_1 and $\psi_1''(x, y)$ has none in S_2. Similarly set $\psi_2(x, y) = \psi_2'(x, y) + \psi_2''(x, y)$, where $\psi_2'(x, y)$ has no singularities in S_2, and $\psi_2''(x, y)$ none in S_1; this splitting up is possible in the present case, but a slight modification may be necessary, due to the presence of logarithmic terms, if the point at infinity belongs to S_1 or S_2. It is possible [Walsh, 1, p. 208] to determine $\psi'(x, y)$ so that the inequalities

$$|\psi'(x,y) - \psi_1'(x,y) + \psi_2''(x,y)| < \epsilon, \qquad (x,y) \text{ in } S_1,$$
$$|\psi'(x,y) - \psi_2'(x,y) + \psi_1''(x,y)| < \epsilon, \qquad (x,y) \text{ in } S_2,$$

are satisfied. These inequalities, if we set

$$\psi(x,y) = \psi'(x,y) + \psi_1''(x,y) + \psi_2''(x,y),$$

are the inequalities it is desired to establish.

We mention explicitly the general question of the possibility of the expansion of given functions in terms of harmonic polynomials or in terms of harmonic rational functions when the restriction of uniformity of convergence is not made. This question seems not to be completely answered in the literature, although Lavrentieff [1] has recently announced some results without proof, and Hartogs and Rosenthal [1] have published an important paper on the corresponding subject for expansion in terms of polynomials in the complex variable. It is worth noting that Osgood's classical theorem [2] in this corresponding subject has the following analog in the present one.

If there converges in a region R a sequence of functions $\{u_n(x, y)\}$ harmonic in R, then this sequence converges uniformly in some sub-region of R.

From this theorem follows directly Osgood's better known result, that *if there converges in a region R a sequence of functions $\{f_n(z)\}$ analytic in that region, then this sequence converges uniformly in some sub-region of R.* For since the sequence $\{f_n(z)\}$ converges, the sequence $\{u_n(x, y)\}$ of the real parts of these functions converges in R, hence uniformly in some sub-region R' of R. Convergence of the sequence $\{iv_n(x, y)\}$ of the pure imaginary parts of the functions $f_n(z)$ at a single point of R' is now sufficient to ensure convergence of the sequence $\{iv_n\}$ and hence of the sequence $\{f_n(z)\}$ uniformly in any simply-connected closed proper sub-region of R'.

Let us indicate briefly the proof of the theorem for harmonic functions. We prove first, after Montel [1, p. 109], but this is only a modification of Osgood's proof, that the sequence $\{u_n(x, y)\}$ is uniformly bounded in some sub-region of R. Otherwise we should have some $|u_{n_1}(x, y)| > 1$ at some point of R, hence in some sub-region R_1 of R. If the sequence $\{u_n(x, y)\}$ is not uniformly bounded in R_1, we must have $|u_{n_2}(x, y)| > 2$ at some point of R_1, and therefore in some sub-region R_2 of R_1. Proceeding in this way we arrive at a sequence $\{u_{n_k}(x, y)\}$ greater in absolute value than $\{k\}$ in R_k. There is at least one point common to all the R_k, and at this point the original sequence cannot converge.

From the boundedness of the original sequence in some sub-region R' it follows [compare Osgood, 2] that the first partial derivatives of these functions are uniformly bounded in an arbitrary closed sub-region of R', hence that the functions $\{u_n(x, y)\}$ are equicontinuous in this sub-region and therefore converge uniformly there.

We shall not go into great detail on the subject of rational harmonic functions *belonging to a region*, for the discussion can be made to depend upon the discussion for harmonic polynomials belonging to a region. Let a region C be bounded by the Jordan curves $C_0, C_1, \cdots, C_k$, of which no two have a common point, and so that the curves $C_1, C_2, \cdots, C_k$ lie interior to C_0. Let $u(x, y)$ be an arbitrary function harmonic interior to C and continuous in the closed region. As has already been suggested, Green's formula applied to $u(x, y)$ not for the region C but for a neighboring region C' of the same connectivity interior to C splits up the function $u(x, y)$ into k logarithmic terms with singularities at infinity and at points interior to $C_1, C_2, \cdots, C_k$ respectively, which may be preassigned, plus k functions harmonic exterior respectively to $C_1, C_2, \cdots, C_k$, including the point at infinity, and continuous in the corresponding closed regions, plus a function harmonic interior to C_0 and continuous in the corresponding closed region. If the curves $C_0, C_1, \cdots, C_k$ are suitably restricted, the regions interior to C_0 and exterior to $C_i (i=1, 2, \cdots, k)$ respectively have associated with them (compare §3) sets of rational functions, in fact polynomials in (x, y) and in

$$\left(\frac{x - x_i}{(x - x_i)^2 + (y - y_i)^2}, \frac{y - y_i}{(x - x_i)^2 + (y - y_i)^2}\right)$$

respectively, and the last mentioned $k+1$ functions can be respectively expanded in terms of these functions.

It would be an interesting problem, however, to construct

for the region C *a single set* of normal orthogonal harmonic rational functions (with also k logarithmic functions) in terms of which an arbitrary harmonic function could be expanded. Here orthogonal may mean (1) with respect to area (that is, surface integral over C) or (2) with respect to length (that is, line integral over $C_0, C_1, \cdots, C_k$). Such a set of functions can easily be constructed by the process of orthogonalization, starting with the logarithmic and rational functions we have been using. If interpretation (2) is used, it is true that *if the curves $C_0, C_1, \cdots, C_k$ bounding the region C are analytic, and if the function $u(x, y)$ is harmonic interior to C, continuous in the corresponding closed region, then the formal expansion of $u(x, y)$ in terms of these normal orthogonal functions belonging to C converges to the value $u(x, y)$ throughout the interior of C, uniformly on any closed point set interior to C.* This follows from the reasoning as given by Merriman [1]; but questions of asymptotic character of the normal orthogonal functions and of uniform convergence of the formal expansion in the closed region C, are there left unanswered under the present general hypothesis on the region C.

6. *Degree of Approximation by Rational Harmonic Functions.* Our results on the degree of approximation to a harmonic function by rational harmonic functions are to be obtained with the help of the corresponding results on the degree of approximation to an analytic function by rational functions of the complex variable, so as a preliminary study we need to consider the relation between the degree of a rational harmonic function and the degree of the corresponding analytic function. In this discussion we use the word *degree* to indicate degree *in the restricted sense.**

If $u(x, y)$ is a harmonic polynomal of degree n, then its conjugate function $v(x, y)$ is likewise a harmonic polynomial of degree n and hence the function $f(z) = u(x, y) + iv(x, y)$ is a

* Compare §4. The degree of a rational function is the greater of the degrees of numerator and denominator, or the common degree if the two are of the same degree.

polynomial in z of degree n. In fact, we may define $v(x, y)$ by

$$v(x,y) = \int_{(0,0)}^{(x,y)} \left(-\frac{\partial u}{\partial y}dx + \frac{\partial u}{\partial x}dy \right) + C,$$

where C is an arbitrary constant. The function $f(z) = u(x, y) + iv(x, y)$ can have no singular points other than those of $u(x, y)$, hence is an entire function. In the Taylor's development of $f(z)$ about the origin, the coefficients of powers of z higher than the nth vanish; otherwise $u(x, y)$ is not a polynomial of degree n.

The situation is somewhat more complicated if $u(x, y)$ is a harmonic rational function not a polynomial. Osgood [1, p. 680] has shown that if $u(x, y)$ is harmonic and rational, then $f(z) = u(x, y) + iv(x, y)$, where $v(x, y)$ is defined by the equation just given, is a rational function of z. If $f(z)$ is given, with no factor containing z common to numerator and denominator,

$$f(z) = \frac{a_0 z^m + a_1 z^{m-1} + \cdots + a_m}{b_0 z^n + b_1 z^{n-1} + \cdots + b_n}, \qquad a_0 b_0 \neq 0,$$

the real part is found by multiplying numerator and denominator by the conjugate complex quantity of the original denominator,

$$f(z) = \frac{a_0 \bar{b}_0 z^m \bar{z}^n + \cdots + a_m \bar{b}_n}{b_0 \bar{b}_0 z^n \bar{z}^n + \cdots + b_n \bar{b}_n},$$

and then separating into real and pure imaginary parts. The function $u(x, y)$ is then a rational function whose numerator is of total degree $m+n$ or less, whose denominator is of degree $2n$, and with no factor common to numerator and denominator; the actual degree of $u(x, y)$ is $m+n$ or less, or $2n$ according as $m \geqq n$ or $m < n$. Reciprocally, let $u(x, y)$ be given in its lowest terms; denote the degrees of numerator and denominator by p and $2q$ respectively, and those of the numerator and denominator of the corresponding analytic function $f(z)$, which is determined only to within an additive constant, by m and n. A necessary and sufficient condition that $u(x, y)$ be harmonic at infinity is that $f(z)$ be analytic at infinity, that is, $m \leqq n$; in this case we have $n=q$, $m \geqq p-q$; the actual degree of $u(x, y)$ is $2n$ and that of $f(z)$ is n. A necessary and sufficient condition that $u(x, y)$ be singular at infinity is that $f(z)$ be singular at infinity, that is, $m>n$. Here we may write $f(z)$ as a rational function whose numerator is of degree $n-1$ or less and denominator of degree n plus a polynomial of degree $m-n$. Then $u(x, y)$ is a rational function whose numerator is of degree $2n-1$ or less and denominator of degree $2n$ plus a polynomial of degree $m-n$, or a rational fraction whose numerator is of degree $p=m+n$ and denominator of degree $2q=2n$; the actual degree of $u(x, y)$ is $m+n$ and that of $f(z)$ is m.

We can prove the following result concerning approximation by harmonic rational functions; henceforth we use the word *degree* in its inclusive sense.

Let S be an arbitrary closed Jordan region of the (x, y)-plane. If there exist rational harmonic functions $r_n(x, y)$, real parts of rational functions of $z=x+iy$ of respective degrees n such that we have*

$$| u(x,y) - r_n(x,y) | \leqq \frac{M}{R^n}, \qquad R > 1,$$

for all points (x, y) of S and for all sufficiently large n, and if the singular points of the functions $r_n(x, y)-r_{n-1}(x, y)$ have no limit point on S, then $u(x, y)$ is the real part of a function $f(z)$ meromorphic on S. If in addition the functions $r_n(x, y)$ have no singularities on S, then $u(x, y)$ is harmonic on S.

Let $w=\Phi(z)$ denote a function which maps the complement of S onto the exterior of the unit circle in the w-plane, so that the points at infinity correspond to each other. Let S_R denote the Jordan curve $|\Phi(z)|=R$, where $R>1$. If the singularities of the functions $r_n(x, y)-r_{n-1}(x, y)$ have no limit point interior to

---

* The rational function of z is not determined uniquely by $r_n(x, y)$, but merely to within an additive constant. Nevertheless the degree, and nature and location of the singularities of this rational function are determined by $r_n(x, y)$ uniquely.

S_ρ, then the sequence $\{r_n(x, y)\}$ converges interior to S_ν, where $\nu = (1+\rho R^{1/2})/(\rho + R^{1/2})$, and the convergence is uniform on any closed point set interior to S_ν. Hence $f(z)$ is meromorphic interior to S_ν and if $r_n(x, y)$ has no singularities interior to S_ν, the function $u(x, y)$ is harmonic interior to S_ν.

If the function $r_n(x, y) - r_{n-1}(x, y)$ is the real part of a function which has at most n poles, for n sufficiently large, we may set $\nu = (1+\rho R)/(\rho + R)$; and in particular if the only singularities of $r_n(x, y)$ lie at infinity, we may set $\rho = \infty$, $\nu = R$, so that the sequence $\{r_n(x, y)\}$ converges interior to S_R.

The proof follows the proof of the corresponding theorem [Walsh 3, Theorem I] for approximation by means of polynomials. If the functions $f(z)$ and $t_n(z)$, of which $u(x, y)$ and $r_n(x, y)$ are respectively the real parts, are properly chosen, and if S' is an arbitrary closed region interior to S, then the inequality $|f(z) - t_n(z)| \leqq M'/R^n$ is valid for z in S'. If the region S' is a region bounded by a Jordan curve uniformly near the boundary of S, the function $w = \Phi'(z)$ which maps the complement of S' onto the exterior of the unit circle in the w-plane differs little from the function $w = \Phi(z)$, and the curve $S_R' : |\Phi'(z)| = R$ lies uniformly near the curve $S_R : |\Phi(z)| = R$. The theorem follows by virtue of the corresponding theorem for approximation to analytic functions by means of rational functions [Walsh 4, Theorem IV].

It is a positive simplification here not to mention the degree of $r_n(x, y)$ except in connection with $t_n(z)$, for such widely differing functions as

$$x, \quad \frac{x}{x^2+y^2}, \quad \frac{3x^2+3y^2+x}{x^2+y^2}$$

belong in the same category.

We turn now to the study of the approximation of functions by rational harmonic functions in multiply connected regions. Here logarithmic terms are essential for the approximation of the most general harmonic functions, as we have seen, so we prove the following theorem.

Let C be a closed region bounded by Jordan curves C_0, C_1, $\cdots$, C_k, such that no two of these curves have a common point and so that C_1, C_2, $\cdots$, C_k lie interior to C_0. If the sequence

$$
\begin{aligned}
&r_n(x,y) + A_{n1}\log\left[(x-x_{n1})^2+(y-y_{n1})^2\right] \\
(13)\qquad &+ A_{n2}\log\left[(x-x_{n2})^2+(y-y_{n2})^2\right] \\
&+\cdots + A_{nk}\log\left[(x-x_{nk})^2+(y-y_{nk})^2\right], \\
&\qquad\qquad (n=1,2,\cdots),
\end{aligned}
$$

converges uniformly in C, where $r_n(x, y)$ is a harmonic rational function of (x, y) with no singularities interior to C, and where the point (x_{ni}, y_{ni}) lies in or on C_i and approaches a limit (x_i, y_i) as n becomes infinite, then $\lim_{n\to\infty} A_{ni}$ exists, $i=1, 2, \cdots, k$, and each of the sequences*

$$(14)\quad r_n(x,y),\ A_{ni}\log\left[(x-x_{ni})^2+(y-y_{ni})^2\right],\ (i=1,2,\cdots,k),$$

converges uniformly in any closed region interior to C; if the functions in (14) have no singularity in C and no limit point of singularities in C, this convergence is uniform in the closed region C.

The proof is simple; choose an analytic Jordan curve C_i' interior to C and enclosing C_i but none of the curves C_1, C_2, $\cdots$, C_{i-1}, C_{i+1}, $\cdots$, C_k in its interior. The sequence of the derivatives of (13) in the direction of the normal ν to C_i' converges uniformly on C_i', and can be integrated on C_i' term by term [Osgood 1, pp. 652–653, Walsh 1, p. 206]. It follows from the theorem of Osgood already quoted that

$$\int_{C_i'} \frac{\partial r_n(x,y)}{\partial \nu} ds = 0;$$

for we have

$$\int_{C_i'} \frac{\partial r_n(x,y)}{\partial \nu} ds = \int_{C_i'} \frac{\partial s_n(x,y)}{\partial s} ds = 0,$$

* It is of course sufficient if the point (x_i, y_i) lies on or within C_i without the assumption that (x_{ni}, y_{ni}) lies interior to or on C_i; we integrate (as below) over C_i' when n is chosen so large that all points (x_{ni}, y_{ni}) but no points (x_{nj}, y_{nj}), $i\neq j$, lie interior to C_i'.

where $s_n(x, y)$ is a function conjugate to $r_n(x, y)$. The total result of the integration of the nth term of the sequence (13) is then precisely $4\pi A_{ni}$, so $\lim_{n\to\infty} A_{ni} = A_i$ exists. It follows that

$$\lim_{n\to\infty} A_{ni} \log [(x - x_{ni})^2 + (y - y_{ni})^2] = A_i \log [(x - x_i)^2 + (y - y_i)^2],$$

uniformly in any closed region interior to C, and hence it follows that the sequence $\{r_n(x, y)\}$ converges uniformly as asserted.

In the next theorem we shall not trouble to consider sequences so general as (13).

Let C be a closed region bounded by Jordan curves C_0, C_1, $\cdots$, C_k, such that no two of these curves have a common point and such that C_1, C_2, $\cdots$, C_k lie interior to C_0. A necessary and sufficient condition that the function $u(x, y)$ defined in C should be harmonic in (the closed region) C, is that there should exist rational harmonic functions $r_n(x, y)$, real parts of rational functions of z of degrees $(k+1)n$, $n = 0, 1, 2, \cdots$, so that we have

$$(15)\qquad \left| u(x,y) - \sum_{i=1}^{k} A_{in} \log [(x - x_i)^2 + (y - y_i)^2] - r_n(x,y) \right| \leqq \frac{M}{R^n}, \quad R > 1,$$

for all points (x, y) of C; here the A_{in} are constants and the points (x_i, y_i) are supposed to lie interior to C_i respectively. It follows that $\lim_{n\to\infty} A_{in} = A_i$ *exists if* (15) *is satisfied.*

The sufficiency of this condition follows, as in the proof already given, from Osgood's theorem and from the corresponding result for approximation of analytic functions of the complex variable [Walsh 4, Theorem V]. It is found by the integration that

$$|A_{in} - A_i| \leqq M'/R_n;$$

hence (15) holds if A_{in} is replaced by A_i, provided that M is replaced by a suitable M''. The necessity of the condition is likewise easy to establish. The points (x_i, y_i) may be chosen arbitrarily interior to the curves C_i respectively. Then [compare Osgood 1, pp. 642–644; Walsh 1, p. 206] in C the function $u(x, y)$ may be expressed as the sum of the k logarithmic terms which appear in (15) plus k functions harmonic respectively on and exterior to C_i, and at infinity, for $i=1, 2, \cdots, k$, plus a function harmonic in the closed interior to C_0. The results of §4 can now be applied, and yield the desired rational functions $r_n(x, y)$.

This theorem can be extended by considering (1) variable logarithmic terms, (2) more explicit regions or point sets for the location of the singularities of the functions $r_n(x, y)$, (3) more explicit regions (that is, regions containing C in their interiors) for the harmonic character of $u(x, y)$ and for the convergence of the sequence $\{r_n(x,y)\}$, (4) more general boundaries for regions than Jordan curves. There is comparatively little difficulty involved in making any of these generalizations however, and the essential reasoning involved already appears in the literature (either here or elsewhere), so these generalizations are left to the reader.

It is also possible to study sequences of harmonic rational functions which satisfy such a relation as (15) not in a region but on a Jordan arc or curve. Here the situation is not so simple, as we have indicated in the analogous situation in §4. If (15) holds merely on the unit circle C, for instance, the sequence of harmonic rational functions need not converge elsewhere; for the function $A_n \log r$ vanishes on C and a change in A_n does not alter (15), yet may alter the convergence off of C of the sequence in (15). The rational harmonic functions are not uniquely determined, moreover, by their values on C; the two functions $r^n \cos n\theta$ and $r^{-n} \cos n\theta$ are equal on C. Knowledge even of the *location* of the singularities of a harmonic rational function (such as $ar^n \cos n\theta + br^{-n} \cos n\theta$) and of the value on C does not determine the rational function uniquely. Nor is a function $u(x, y)$ harmonic in a region containing C in its interior determined by its values on C.

If (15) holds on the unit circle C for rational functions $\{r_n(x, y)\}$ with singularities only at the origin and at infinity and where (x_i, y_i), $i=1$, is the origin, there exists a function $U(x, y)$ harmonic on C, in fact harmonic for $r<R$, and coinciding on C with the values $u(x, y)$. Indeed, the logarithmic term of (15) vanishes on C and hence may be entirely omitted. The rational function $r_n(x, y)$ is a polynomial in

$$r^{-n}\cos n\theta,\ r^{-n}\sin n\theta,\ \cdots,\ 1,\ \cos\theta,\ \sin\theta,\ \cdots,\ r^n\cos n\theta,\ r^n\sin n\theta,$$

but the negative powers of r may be changed into positive powers without altering $r_n(x, y)$ or (15) on C, so we are dealing with a sequence of harmonic polynomials, converging like a geometric series on C. This situation has already been treated in §4, and the existence of the required function $U(x, y)$ is established there.

7. *Expansions in Three Dimensions*. The theory of the expansion of harmonic functions in three dimensions is not nearly so far developed as the corresponding theory in two dimensions, but we shall discuss a few results and a few open problems.

It is classic that Poisson's integral yields directly an expansion of functions which are given harmonic interior to the sphere, continuous in the corresponding closed region.

If the function $u(x, y, z)$ is harmonic interior to a sphere and continuous in the corresponding closed region, then interior to the sphere $u(x, y, z)$ can be expanded in a series of harmonic polynomials in (x, y, z), the series converging uniformly in any closed region interior to the sphere.

This theorem will be used in proving the analog of the corollary (§2) to Runge's theorem. This analog has already been proved by Bergmann [1], but only for the case of convex regions. The method we shall use is closely related to that of Bergmann, and both methods are intimately related to that of Runge.

If the function $u(x, y, z)$ is harmonic in a closed region S bounded by a simple closed (limited) surface, then in that closed region the function $u(x, y, z)$ can be approximated uni-

formly as closely as desired by a harmonic polynomial in (x, y, z).

We shall find it convenient to have for reference the following lemma.

LEMMA. *Let the two points* (x_0, y_0, z_0) *and* (x', y', z') *and the simple polygonal curve C joining them lie exterior to the closed point set S. If the function* $u(x, y, z)$ *can be uniformly approximated in S as closely as desired by a harmonic function whose only singularity lies in* (x_0, y_0, z_0), *then this function* $u(x, y, z)$ *can likewise be uniformly approximated in S as closely as desired by a harmonic function whose only singularity lies in* (x', y', z') *and which is rational except for the factor* $[(x-x')^2+(y-y')^2+(z-z')^2]^{-1/2}$.

We give the proof of this lemma for the case that both (x_0, y_0, z_0) and (x', y', z') are finite points, but that is simply a matter of convenience. The proof holds with only obvious changes even if (x', y', z') is the point at infinity,* and it is in this latter form that the lemma will be applied later. When (x', y', z') is the point at infinity, the approximating functions are polynomials in (x, y, z), and the factor $[(x-x')^2+(y-y')^2+(z-z')^2]^{-1/2}$ does not enter.

Construct a finite sequence of spheres $S_0, S_1, \cdots, S_n$ extending from (x_0, y_0, z_0) to (x', y', z'), so that no sphere contains in its interior a point of S, but so that successive spheres of the sequence have a region common to them. The point (x_0, y_0, z_0) shall lie interior to S_0 and so shall another particular point (x_1, y_1, z_1) of C. The sphere S_1 shall contain (x_1, y_1, z_1) in its interior and likewise another particular point (x_2, y_2, z_2) which lies on C between (x_1, y_1, z_1) and (x', y', z'). We proceed in this way to construct $S_2, S_3, \cdots, S_{n-1}$, finally arriving at the sphere S_n which contains in its interior (x_n, y_n, z_n) and (x', y', z'). The construction of these spheres is surely possible, for it is no loss of generality to assume that C is composed of merely a finite number of line

* It is to be remembered that vanishing at infinity is a necessary condition that a function should be harmonic there.

segments and has the length l. Assume likewise that the nearest distance from C to a point of S is δ. Choose the sequence of points (x_0, y_0, z_0), (x_1, y_1, z_1), $\cdots$, (x_n, y_n, z_n), (x', y', z') on C such that the greatest distance between two successive points is less than $\delta/2$. The sphere S_i may be taken as the sphere with center (x_i, y_i, z_i) and diameter δ.

Let an arbitrary positive ϵ be given. We choose the harmonic function $r_0(x, y, z)$, whose only singularity lies in (x_0, y_0, z_0), such that we have

$$| u(x,y,z) - r_0(x,y,z) | < \frac{\epsilon}{n+2}, \qquad (x,y,z) \text{ in } S.$$

This choice is possible by hypothesis. We next choose the harmonic function $r_1(x, y, z)$, whose only singularity lies in (x_1, y_1, z_1), rational except for the factor $[(x-x_1)^2+(y-y_1)^2+(z-z_1)^2]^{-1/2}$, such that we have

$$(16) \quad | r_0(x,y,z) - r_1(x,y,z) | < \frac{\epsilon}{n+2}, \qquad (x,y,z) \text{ in } S.$$

We prove by means of an inversion in the unit sphere whose center is (x_1, y_1, z_1) that such choice is possible:

$$(17) \qquad \xi = \frac{x - x_1}{r^2}, \qquad \eta = \frac{y - y_1}{r^2}, \qquad \zeta = \frac{z - z_1}{r^2},$$

$$r^2 = \frac{1}{\rho^2} = (x - x_1)^2 + (y - y_1)^2 + (z - z_1)^2.$$

The function $\rho_0(\xi, \eta, \zeta) = (1/\rho)\, r_0(x, y, z)$, where $\xi, \eta, \zeta, x, y, z$ are connected by the relations (17), is a harmonic function of (ξ, η, ζ) on and within S_0', the transform of the sphere S_0 under the inversion. Then by the theorem already stated, the function $r'(\xi, \eta, \zeta)$, a polynomial in (ξ, η, ζ), can be determined so that we have

$$\left| \frac{1}{\rho} r_0(x,y,z) - r'(\xi,\eta,\zeta) \right| < \frac{\epsilon}{(n+2)d}, \quad (\xi,\eta,\zeta) \text{ interior to } S_0',$$

where d is the distance from (x_1, y_1, z_1) to the farthest point of S_0'. This inequality implies the inequality

$$| r_0(x,y,z) - \rho r'(\xi,\eta,\zeta) | < \frac{\epsilon}{n+2}, \quad (x,y,z) \text{ exterior to } S_0.$$

But $(1/r)r'(\xi, \eta, \zeta)$ is a harmonic function of (x, y, z), rational except for the factor $1/r$ and with its only singularity in the point (x_1, y_1, z_1); hence its equal $\rho r'(\xi, \eta, \zeta)$ may be identified with the desired function $r_1(x, y, z)$ and yields (16).

This same process can now be continued, and yields the additional inequalities

$$| r_1(x,y,z) - r_2(x,y,z) | < \frac{\epsilon}{n+2}, \qquad (x,y,z) \text{ in } S,$$

$$\cdots\cdots\cdots\cdots$$

$$| r_n(x,y,z) - r_{n+1}(x,y,z) | < \frac{\epsilon}{n+2}, \qquad (x,y,z) \text{ in } S.$$

Hence we have finally

$$| u(x,y,z) - r_{n+1}(x,y,z) | < \epsilon, \qquad (x,y,z) \text{ in } S,$$

where $r_{n+1}(x, y, z)$ represents a harmonic function whose only singularity lies in (x', y', z') and is, except for the factor $[(x-x')^2+(y-y')^2+(z-z')^2]^{-1/2}$, rational in (x, y, z). That is, the lemma is established.

We need a second lemma in the proof. For the sake of simplicity we state and prove the lemma for integrals of a function of a single variable, but the proof holds without essential change for multiple integrals, and to double integrals we shall apply the lemma. Likewise in the proof we consider merely a single parameter α, but no difficulty is introduced by the appearance of several, and several parameters appear in the application we shall make.

LEMMA. *Let $f(x, \alpha)$ be a continuous real function of the arguments for $a \leqq x \leqq b$, $\alpha_1 \leqq \alpha \leqq \alpha_2$. If an arbitrary positive δ be given, then there exists δ' so that the inequality*

$$(18) \qquad \left| \int_a^b f(x,\alpha)dx - \sum_{i=1}^{n} f(\xi_i,\alpha)(x_i - x_{i-1}) \right| \leqq \delta,$$

$$a = x_0 \leqq x_1 \leqq \cdots \leqq x_n = b, \qquad x_{i-1} \leqq \xi_i \leqq x_i,$$

holds uniformly for all α, $\alpha_1 \leqq \alpha \leqq \alpha_2$, *provided merely that we have* $|x_i - x_{i-1}| \leqq \delta'$.

By the mean value theorem for integrals, we can write

$$\int_a^b f(x,\alpha)dx = \sum_{i=1}^n f(\eta_i,\alpha)(x_i - x_{i-1}),$$

where $x_{i-1} \leqq \eta_i \leqq x_i$, and where η_i depends on α. Let us choose δ' so small that the inequality $|\zeta_i - \zeta_i'| \leqq \delta'$ implies

$$|f(\zeta_i,\alpha) - f(\zeta_i',\alpha)| \leqq \delta/(b - a)$$

uniformly for all α, $\alpha_1 \leqq \alpha \leqq \alpha_2$; such a δ' exists by the uniform continuity of $f(x, \alpha)$. Then under the hypothesis $|x_i - x_{i-1}| \leqq \delta'$, we have also $|\xi_i - \eta_i| \leqq \delta'$, from which follows the inequality

$$\left| \sum_{i=1}^n f(\eta_i,\alpha)(x_i - x_{i-1}) - \sum_{i=1}^n f(\xi_i,\alpha)(x_i - x_{i-1}) \right| \leqq \delta,$$

which is precisely (18).

We are now in a position to prove the theorem. Denote by σ a simple closed surface consisting entirely of a finite number of portions of planes parallel to the coordinate planes, which contains S in its interior, and on and within which the given function $u(x, y, z)$ is harmonic. The value of $u(x, y, z)$ in S is given by Green's integral

$$(19)\qquad u(x,y,z) = \frac{1}{4\pi}\int_\sigma\int \left[\frac{1}{r}\frac{\partial u}{\partial n} - u\frac{\partial\left(\frac{1}{r}\right)}{\partial n}\right]d\sigma,$$

where n represents the exterior normal. Let an arbitrary positive ϵ be given. We divide σ into a finite number n of pieces σ_i, so that each piece consists only of a portion of a plane parallel to one of the coordinate planes, so that each piece σ_i lies interior to a sphere S_i which contains on or within it no point of S, and so that we have

$$(20)\qquad \left|\frac{1}{4\pi}\int\int_{\sigma}\frac{1}{r}\frac{\partial u}{\partial n}d\sigma-\frac{1}{4\pi}\sum_{i=1}^{n}\frac{1}{r_i}\left(\frac{\partial u}{\partial n}\right)_i\Delta_i\sigma\right|\leq\frac{\epsilon}{4},\quad (x,y,z)\text{ in }S,$$

$$\left|\frac{1}{4\pi}\int\int_{\sigma}u\frac{\partial(1/r)}{\partial n}d\sigma-\frac{1}{4\pi}\sum_{i=1}^{n}u_i\left[\frac{\partial(1/r)}{\partial n}\right]_i\Delta_i\sigma\right|\leq\frac{\epsilon}{4},\quad (x,y,z)\text{ in }S,$$

where r_i is the distance from (x, y, z) to a particular but arbitrary point P_i of σ_i, where $(\partial u/\partial n)_i$ is the normal derivative of $u(x, y, z)$ at P_i, where $\Delta_i\sigma$ is the area of σ_i, where u_i is the value of $u(x, y, z)$ at P_i, and where $[\partial(1/r)/\partial n]_i$ is the value of $\partial(1/r)/\partial n$ at P_i. The last expression may also be written as the value at P_i of a partial derivative of $1/r$ with respect not to (x, y, z) but to running coordinates on σ, and this latter form serves better to indicate that we are dealing with a harmonic function of (x, y, z).

Each term of the sums in (20) is a function of (x, y, z) harmonic not merely in the closed region S but throughout the exterior of the corresponding sphere S_i, even at infinity. It is immediately seen by inversion with P_i as center of inversion that in the (x, y, z)-space each of these terms can be approximated by a harmonic function of (x, y, z) whose only singularity lies in P_i and which is rational except for the factor $1/r_i$, where r_i indicates distance measured from P_i. The approximation can be made uniformly as closely as desired in the exterior of the sphere S_i, and hence uniformly as closely as desired in the closed region S. But each point P_i can be joined to the point at infinity by a Jordan curve which does not meet S. Then by the first lemma we can approximate each term of the sums in (20) as closely as desired uniformly in the closed region S by a harmonic polynomial in (x, y, z). That is, we can determine harmonic polynomials $p_1(x, y, z)$ and $p_2(x, y, z)$ such that we have

$$(21)\qquad \left|\frac{1}{4\pi}\sum_{i=1}^{n}\frac{1}{r_i}\left(\frac{\partial u}{\partial n}\right)_i\Delta_i\sigma-p_1(x,y,z)\right|\leq\frac{\epsilon}{4},\quad (x,y,z)\text{ in }S,$$

$$\left|\frac{1}{4\pi}\sum_{i=1}^{n}u_i\left(\frac{\partial(1/r)}{\partial n}\right)_i\Delta_i\sigma-p_2(x,y,z)\right|\leq\frac{\epsilon}{4},\quad (x,y,z)\text{ in }S.$$

It follows immediately that we have

$$(22)\quad | u(x,y,z) - [p_1(x,y,z) - p_2(x,y,z)] | \leqq \epsilon, \; (x,y,z) \text{ in } S,$$

and the theorem is proved.

The present proof has the advantage over the proof of §2 of holding (with only minor modifications) in two, three, or n dimensions, and since it does not involve the theory of functions of a complex variable, has the additional advantage of being more satisfactory from the standpoint of the purist. This remark applies also to the results which immediately follow.

Another remark is of interest. If the point (x, y, z) is exterior to the closed surface σ, the integral (19) has the value zero, where r still indicates distance from (x, y, z) to an arbitrary point of σ. Indeed it is a general theorem that if the functions u and v are both harmonic on and within σ, then we have

$$\int_\sigma\int \left[v\frac{\partial u}{\partial n} - u\frac{\partial v}{\partial n}\right] d\sigma = 0.$$

Thus if the function $u(x, y, z)$ is given harmonic on a point set S composed of the closed interiors of two non-intersecting simple closed surfaces, two new non-intersecting simple closed surfaces σ_1 and σ_2 can be constructed which consist entirely of a finite number of portions of planes parallel to the coordinate planes, such that each point of S lies interior either to σ_1 or σ_2, and such that $u(x, y, z)$ is harmonic on and within both σ_1 and σ_2. Then formulas (19), (20), (21), (22) are valid without any change whatever, if we set $\sigma = \sigma_1 + \sigma_2$. A similar fact holds if S falls into n distinct parts. In fact we can prove the following theorem.

Let S be an arbitrary closed limited point set whose complementary set (with respect to the entire space) is connected. Then if $u(x, y, z)$ is harmonic on S, the function $u(x, y, z)$ can be uniformly approximated on S as closely as desired by a harmonic polynomial in (x, y, z).

There exists a point set σ consisting of a finite number of mutually exclusive closed limited simply-connected regions

$\sigma_1, \sigma_2, \cdots, \sigma_n$ containing all points of S in their interiors, the individual regions bounded entirely by a finite number of portions of planes parallel to the coordinate planes, and such that $u(x, y, z)$ is harmonic on the entire set $\sigma_1+\sigma_2+\cdots+\sigma_n$. The previous method applies without change and the theorem follows.

The following application corresponds to the theorem of Runge; the proof is omitted.

Let C be a point set composed of a finite or infinite number of mutually exclusive simply-connected regions, none of which contains the point at infinity in its interior. If the function $u(x, y, z)$ is harmonic interior to each of these regions, then $u(x, y, z)$ can be expanded on C in a series of harmonic polynomials in (x, y, z), and the series converges uniformly on any closed point set contained in C.

We add one further result on general approximation. A closed region S is said to be *convex with respect to the interior point P* if S is bounded by a surface which is cut in a single point by each ray through P. A region which is convex in the usual sense of the word is convex with respect to each of its interior points.

If the region S is a limited closed region convex with respect to some interior point P, then an arbitrary function $u(x, y, z)$, harmonic interior to S and continuous in the closed region, can be approximated as closely as desired uniformly in the closed region S by a harmonic polynomial in (x, y, z).

Choose P as origin of coordinates, which involves no loss of generality, and consider the transformation

$$x' = \rho x, \qquad y' = \rho y, \qquad z' = \rho z, \qquad \rho > 1,$$

which transforms S into a region S' which contains S in its interior. The function $u(x, y, z)$ is transformed into a new function

$$u'(x, y, z) = u\left(\frac{x}{\rho}, \frac{y}{\rho}, \frac{z}{\rho}\right)$$

defined throughout S'. It follows from the uniform continuity of $u(x, y, z)$ in S that if $\epsilon > 0$ be given, then ρ can be chosen so near to unity that we have

$$|u'(x,y,z) - u(x,y,z)| < \frac{\epsilon}{2}, \qquad (x,y,z) \text{ in } S.$$

For $u'(x, y, z) - u(x, y, z)$ is the same as $u(x/\rho, y/\rho, z/\rho) - u(x, y, z)$, which becomes uniformly small with $\rho - 1$. The function $u'(x, y, z)$ is harmonic in the closed region S, so there exists a harmonic polynomial $p(x, y, z)$ such that we have

$$|u'(x,y,z) - p(x,y,z)| < \frac{\epsilon}{2}, \qquad (x,y,z) \text{ in } S,$$

whence follows

$$|u(x,y,z) - p(x,y,z)| < \epsilon, \qquad (x,y,z) \text{ in } S,$$

and the theorem is established.*

It will be noticed that the study of the approximation to an arbitrary function in the plane is fairly well developed, while for space the study is only begun. In particular the general question as to when a function harmonic interior to a three-dimensional region and continuous in the closed region (or more generally, defined on a given closed point set) can be uniformly approximated in that closed region (or on that point set) as closely as desired by a harmonic polynomial, seems to be well worth investigating. There is an essential difference in methods and results for two and for three dimensions, for in the latter case the Dirichlet problem does not always have a solution, even for simply-connected regions. Methods of solution of the problem of approximation in three dimensions would presumably depend on the study of the solution of the Dirichlet problem for variable regions, and the result would presumably depend to some extent on whether the region is normal [compare Kellogg, 1].

* The corresponding theorem for analytic functions of a complex variable was given independently by Hilb and Szász, Encyklopädie der mathematischen Wissenschaften, vol. 2, C, II, p. 1276, and by Walsh, Transactions of this Society, vol. 26 (1924), pp. 155–170; p. 168, footnote.

We mention other problems for three or more dimensions: the study of harmonic polynomials belonging to a general region*, the degree of approximation to functions by harmonic polynomials, the question of interpolation by harmonic polynomials, approximation by harmonic rational functions,—the solution of all of these problems still lies in the future.

REFERENCES AND BIBLIOGRAPHY

Bergmann, 1. Mathematische Annalen, vol. 86 (1922), pp. 238–271.
2. Mathematische Annalen, vol. 96(1927), pp. 248–263.
Bochner, 1. Mathematische Zeitschrift, vol. 14 (1922), pp. 180–207.
Carleman, 1. Arkiv för Matematik, Astronomi och Fysik, vol. 17 (1922–23).
Faber, 1. Mathematische Annalen, vol. 57 (1903), pp. 389–408.
2. Journal für Mathematik, vol. 150 (1920), pp. 79–106.
3. Mathematische Annalen, vol. 69 (1910), pp. 372–443.
Fejér, 1. Göttinger Nachrichten, 1918, pp. 319–331.
Haar, 1. Mathematische Annalen, vol. 78 (1918), pp. 294–311.
Hartogs and Rosenthal, 1. Mathematische Annalen, vol. 100 (1928), pp. 212–263.
Jackson, 1. Transactions of this Society, vol. 21 (1920), pp. 321–332.
Kellogg, 1. This Bulletin, vol. 32 (1926), pp. 601–625.
Lavrentieff, 1. Comptes Rendus, vol. 184 (1927), pp. 1634–35.
Lebesgue, 1. Rendiconti di Palermo, vol. 24 (1907), pp. 371–402.
Merriman, 1. *On the expansion of harmonic functions in terms of normal-orthogonal harmonic polynomials*, not yet published. Presented to the Society (New York), April 1928.
Montel, 1. *Séries de Polynomes*, Paris, 1910.
Osgood, 1. *Funktionentheorie*, vol. 1, Leipzig, 1912.
2. Annals of Mathematics, (2), vol. 3 (1901–02), pp. 25–34.
Runge, 1. Acta Mathematica, vol. 6 (1885), pp. 229–244.
Szegö, 1. Mathematische Zeitschrift, vol. 9 (1921), pp. 218–270.
de la Vallée-Poussin, 1. *L'Approximation des Fonctions*, Paris, 1919.
Walsh, 1. Journal für Mathematik, vol. 159 (1928), pp. 197–209.
2. Proceedings of the National Academy of Sciences, vol. 13 (1927), pp. 175–180.
3. This Bulletin, vol. 33 (1927), pp. 591–598.
4. Transactions of this Society, vol. 30 (1928), pp. 838–847.
5. *On approximation by rational functions to an arbitrary function of a complex variable*, shortly to appear in the Transactions of this Society.

HARVARD UNIVERSITY

* For convex regions a treatment is given by Bergmann, 2.

Commentary by P.M. Gauthier[1]

As the late Morris Marden [16][2] emphasized in his obituary for Joseph L. Walsh, the subject of interpolation and approximation of continuous, holomorphic, or harmonic functions encompasses about half of Walsh's published articles as well as his treatise [29] entitled "Interpolation and Approximation by Rational Functions in the Complex Domain". In this survey, we shall limit ourselves to Walsh's qualitative results in this area, that is, his density results. Such results focus on the possibility of approximation rather than on algorithms and speed of approximation. But it is the qualitative theory that contains what are probably the most important of Walsh's many original results in the general area of interpolation and approximation, namely his Jordan-arc theorem and his Jordan-domain theorem. We shall discuss these and other qualitative results of Walsh as well as some of their offshoots.

The qualitative theory of approximation started with a double big-bang in 1885, the year in which Karl Runge [21] and Karl Weierstrass [30] respectively published their fundamental approximation theorems.

As every analyst knows, the Weierstrass theorem states that every continuous function on a closed interval can be uniformly approximated by polynomials. Such an interval can be considered as lying on the real axis of the complex plane and such polynomials can be considered as complex polynomials whose coefficients happen to be real. Suppose now that K is any compact subset of the finite complex plane $\mathbf{C}$ and that f is a function defined on K. Let $P(K)$ denote the uniform closure

[1]Research supported in part by NSERC (Canada) and FGAR (Québec).
[2]See pages xxxix–xliv of this selected works.

of the polynomials on K and $R(K)$ the uniform closure of the rational functions whose poles lie outside K. The qualitative theory seeks to ascertain whether or not f lies in $P(K)$ or $R(K)$. Of course, any such f must lie in the class $A(K)$ of functions continuous on K and holomorphic on the interior of K.

Runge's theorem states that any function f, holomorphic on (a neighbourhood of) a compact subset K of $\mathbf{C}$, can be uniformly approximated by rational functions; moreover, if the complement $\mathbf{C}\backslash K$ is connected, then the approximating functions can, in fact, be taken to be polynomials. It follows that if $\mathbf{C}\backslash K$ is connected, the problem of approximating a given function f on K by polynomials reduces to the problem of approximating f by functions holomorphic on K.

A Jordan domain in the plane is, by definition, the homeomorphic image of a closed disk; a Jordan arc is the homeomorphic image of a closed interval. In the late 20's J. L. Walsh obtained the following two results (respectively, [24, 25] (=[26-b*, 26-c*])), which are among the most beautiful in the entire theory of qualitative approximation.

Theorem 1 (Jordan domain). *If K is a closed Jordan domain in $\mathbf{C}$, then any f in $A(K)$ can be uniformly approximated by polynomials.*

Theorem 2 (Jordan arc). *Any continuous function on a Jordan arc in $\mathbf{C}$ can be uniformly approximated by polynomials.*

Although significant progress was made by Lavrentiev [15], it was not until the 50's that the ultimate result in qualitative polynomial approximation was obtained by Mergelyan [17] who showed (constructively) that $A(K) = P(K)$ if and only if $\mathbf{C}\backslash K$ is connected.

To better appreciate the magnitude of the two above-mentioned results of Walsh, it is appropriate to recall how exceedingly complicated Jordan arcs and curves can be. The very depth of the Jordan-curve theorem strongly suggests that such complexity may be lurking. Indeed, Denjoy [8] asserts that through any compact totally-disconnected subset of $\mathbf{R}^n$, one can weave a Jordan arc. This remarkable fact follows from the earlier note of Denjoy [7] and also was proved differently by Antoine [1] in his thesis.

Stochastic analogs of the Walsh Jordan-domain and Jordan-arc theorems were obtained in [2]. However, we shall not discuss these; we refer the interested reader directly to [2] and also to [5] for other results on stochastic approximation.

The Walsh Jordan-arc theorem in its original form fails in $\mathbf{C}^n$, $n > 1$. In fact, Wermer [31] and Rudin [20] constructed Jordan arcs in $\mathbf{C}^3$ and $\mathbf{C}^2$ respectively on which not every continuous function can be approximated by polynomials. The problem of finding the appropriate extension to $\mathbf{C}^n$ of the Walsh Jordan-arc theorem, that is, of characterizing those Jordan arcs in $\mathbf{C}^n$ on which every continuous function can be approximated by polynomials, is one of the most notorious and long-standing open problems in several complex variables.

Attempts to find a satisfactory analog in $\mathbf{C}^n$ of the Walsh Jordan-domain theorem gave rise in the late 60's to striking developments in the theory of integral representations (generalizations of the Cauchy formula) in $\mathbf{C}^n$. Although the modern integral representation theory has been only partially successful in attaining its original goal of extending Walsh's Jordan-domain theorem, it has more than paid for itself by resolving a number of deep problems in several complex variables and renewing the interest of analysis in general for this area. Indeed, the integral representations approach led to a *perestroika* in several complex variables which dominated research in this area for many years (see, for example [13] and [19]).

The conditions for rational approximation proved more difficult to ferret out than those for polynomial approximation. In 1931, Hartogs and Rosenthal [11] published their beautiful theorem that on any compact set of planar measure zero, every continuous function can be uniformly approximated by rational functions. Throughout his lifetime Walsh, himself, had many partial results on the qualitative theory of rational approximation, many of which can be found in the various editions of his book (op. cit.). A complete solution, however, awaited Vitushkin [23], who employed constructive methods involving capacities. Although Mergelyan's polynomial approximation theorem has subsequently been proved by soft (non constructive) methods, no alternate proof has so far been found for the theorem of Vitushkin.

In the two years following the publication of his fundamental papers on approximation by complex polynomials on Jordan arcs and Jordan domains, Walsh [27, 28](=[28-d*, 29-b*]) published results on analogous questions for approximation by harmonic polynomials and (in the terminology of Walsh) harmonic rational functions. We now describe these harmonic results.

First of all, in these two papers (op. cit.), Walsh proved a harmonic analog of the Runge theorem, and in the case of approximation by harmonic polynomials, even obtained an analog of the Mergelyan theorem [17] on approximation by complex polynomials. Namely, Walsh showed that a necessary and sufficient condition in order that every function continuous on a compact set K in $\mathbf{C}$ and harmonic on the interior of K be approximated uniformly by harmonic polynomials is that the boundary of K coincide with the boundary of the unbounded complementary component of K. Let us refer to this as the harmonic Mergelyan theorem of Walsh. (May Walsh forgive me. It is perhaps the Mergelyan theorem which should be called the complex Walsh theorem of Mergelyan, since Walsh obtained results on such approximations, both complex and harmonic, earlier. However, since the present survey discusses several theorems of Walsh, such a nomenclature would be too ambiguous here.) From this, in particular, followed the harmonic analogs of Walsh's Jordan-arc and Jordan-domain theorems. A complete solution of the problem of harmonic rational approximation, harmonic analogs of the Vitushkin theorem [23] on complex rational approximation, however, was not found until the seminal work of Keldysh [12].

In the second of these two harmonic papers, Walsh [28](=[29-b*]) also obtained a now classical result known as the Walsh-Lebesgue theorem which asserts that if K is a compact subset of $\mathbf{C}$ having connected complement, then any continuous function on the boundary of K can be uniformly approximated by harmonic polynomials. In view of the maximum principle, this leads to a solution of the Dirichlet problem on K. This Walsh-Lebesgue theorem has also served as a catalyst for entire chapters in the theory of function algebras such as the theory of Dirichlet algebras and logmodular algebras. For a general reference on the theory of function algebras, we refer the reader to, for example [10].

Harmonic approximation in $\mathbf{R}^n$, for $n > 2$, has been more successful than holomorphic approximation in $\mathbf{C}^n$, for $n > 1$. Indeed, Walsh's harmonic Runge theorem and its proof hold in any dimension. The extensions of Walsh's harmonic Mergelyan theorem (see [4]) and the Walsh-Lebesgue theorem (see [6, p. 157]) involve the potential-theoretic notion of thinness. In $\mathbf{R}^2$, these conditions involving thinness have equivalent purely topological formulations; this is no longer so in $\mathbf{R}^n, n > 2$. Thus, the higher dimensional analogs of the Walsh Jordan-domain and Jordan-arc theorems are no longer corollaries of the harmonic Mergelyan theorem. Nevertheless, "Jordan-disc" and Jordan-arc theorems (see, respectively, [14, Theorem 5.20] and [3]) do hold in $\mathbf{R}^n$. The "Jordan-ball" theorem is, however, false (see, for example [4]). Here, by a Jordan-disc we mean a homeomorphic image of a closed disc and by a Jordan-ball in $\mathbf{R}^n$ we mean a homeomorphic image of an n-ball. In $\mathbf{R}^2$ both of these notions coincide with that of a closed Jordan domain. The Keldysh theorem also can be stated in terms of thinness and holds in $\mathbf{R}^n$ for all n (Keldysh stated it in $\mathbf{R}^3$).

At this point I would like to point out a certain anachronism in the way I have presented the results of Keldysh and Walsh on harmonic approximation. I have stated that the harmonic Mergelyan theorem of Walsh characterizes those compact sets on which each continuous function, harmonic on the interior, can be approximated uniformly by harmonic polynomials. In $\mathbf{R}^2$ the Jordan-arc and Jordan-domain theorems are corollaries. In fact, Walsh considered these two corollaries separately. Holomorphic and harmonic functions are, of course, defined on open sets. Walsh and Keldysh (and even many analysts to this day) considered functions continuous on a compact set K and harmonic on its interior only in the case when K is a "closed domain". Thus, they investigated what they considered to be two distinct problems: firstly, approximation of functions harmonic on a domain and continuous up to the boundary; secondly, approximation of continuous functions on a closed bounded set having no interior. I have preferred to present the theorems of Walsh and Keldysh in the more modern way which unifies these two problems.

The importance of the interplay between approximation and interpolation is underscored by the title of Walsh's earler mentioned book. This is a sort of love-hate dichotomy which is caricaturized by the following anecdote. Most operating watches (approximation) never have the right time. However, a broken watch (interpolation) is correct twice a day. But Walsh [26] (=[28-a*]) showed that *if* you can have your cake, then you can eat it too. Namely, he showed that if you can approximate, then you can approximate *and* interpolate. Precisely, Walsh proved that if a function f can be uniformly approximated on a compact subset K of $\mathbf{C}$ by polynomials in z, then given points $z_1, z_2, \ldots, z_n$ in K, f can be uniformly approximated by polynomials p which also satisfy the interpolating conditions $p(z_i) = f(z_i), i = 1, 2, \ldots, n$. Thus, Walsh showed that if there are watches that keep arbitrarily good time, then, for any finite number of preassigned moments, there are such watches which, moreover, give precisely the right time at these moments. This result of Walsh has been generalized by various authors. As an example, we state the following elegant formulation due to Deutsch [9].

Theorem 3. *Let Y be a dense linear subspace of the linear topological space X and let $L_1, \dots, L_n$ be continuous linear functionals on X. Then for each $x \in X$ and each neighbourhood U of x, there is a $y \in Y$ such that $y \in U$ and $L_i(y) = L_i(x), i = 1, \dots, n$.*

If, in the above theorem, we replace the space X by the closure of the space Y, then we obtain an "individual element" version of this theorem. Namely, suppose $x \in X$ lies in the closure of the subspace Y. Let U be any neighbourhood of x and let $L_1, \dots, L_n$ be continuous linear functionals on X. Then, there is a $y \in Y$ such that $y \in U$ and $L_i(y) = L_i(x), i = 1, \dots, n$. This version bears a greater resemblance to Walsh's original formulation and dispenses with the hypothesis of denseness.

The qualitative theory is being slowly extended to other operators than the Cauchy-Riemann operator and the Laplacian and to other norms than the uniform norm. Given a differential operator T and a particular norm, the problem is to ascertain whether a given function on a compact set K can be approximated in the given norm by solutions of the equation $T_n = 0$. We shall not discuss, here, such results. The interested reader may consult, for example, [18] and [22] and the bibliographies therein, for further references.

After his retirement from Harvard, Walsh held the prestigious position of distinguished professor at the University of Maryland, a position to which was appended a research associate. It has been a great privilege for me to be the last such research associate to J. L. Walsh and to triauthor, together with Alice Roth, his last paper with him. Alice Roth, whom I affectionately called "Alice in Switzerland", had, many years earlier, earned a prominent place in the history of rational approximation by being the first to construct an example of a compact nowhere dense subset of the complex plane on which not every continuous function can be approximated by rational functions. This famous example is known as a Swiss cheese (due to its holeyness rather than the nationality of the cheesemaker).

As Walsh's young protégé, I often had lunch at the faculty club with Professor Walsh and some of his venerable colleagues in various fields. The conversations of these sages were of great interest to me. I particularly remember the stories about a very very young undergraduate student whom several of the professors (from non mathematical disciplines) seated about the luncheon table considered as one of the best students they had encountered: Charles Fefferman. Of course, the mathematicians had even higher praise for "their" prodigy, but that is well known. During my year as Walsh's associate, I was also positively influenced by Bruce Reinhart, Reinhold Remmert, Edoardo Vesentini and Larry Zalcman, all of whom were at the University of Maryland. That was also the year that Walsh introduced me to his (then recent) student Ed Saff who would honor the memory of Walsh by his distinguished career and by editing the present opus.

While I was at Maryland, the chrysostomic Larry Zalcman gave a lecture on the history of approximation. This lecture, as I recall it, was in fact nothing more (and nothing less) than an eloquent depiction of the beauty and importance of Walsh's Jordan-arc and Jordan-domain theorems. Walsh, of course, was present, and Larry began with "I feel like the preacher who, upon beginning this sermon, looked up and noticed that God was sitting in the church." Speaking of the Almighty, I thank Him-Her for my good fortune of having been associated with a mathematical giant. But in no small measure, my thanks also go directly to J. L. Walsh.

References

[1] Antoine, L. (1921): *Sur l'homéomorphie de deux figures et de leurs voisinages*, J. Math. Pure Appl. (8) **4**, 221–325.

[2] Andrus, G. F., Nishimura, T. (1980): *Stochastic approximation of random functions*, Rend. Mat. (6) **13**, 593–615.

[3] Bagby, T., Cornea, A., Gauthier, P. M. (unpublished). Presented at A. M. S. summer meeting, Orono, 1991.

[4] Bagby, T., Gauthier, P. M.: *Uniform approximation by global harmonic functions*, Approximation by Solutions of Partial Differential Equations, Quadrature Formulae, and Related Topics. Ed.: M. Goldstein, W. Haussmann. Kluwer, 15–26.

[5] Brown, L., Schreiber, B. M. (1989): *Approximation and extension of random functions*, Monatsh. Math. **107**, 111–123.

[6] Constantinescu, C., Cornea, A. (1972): *Potential Theory on Harmonic Spaces*, Springer.

[7] Denjoy, A. (1909): *Sur les ensembles parfaits discontinus*, C. R. Acad. Sci. Paris **149**, 1048–1050.

[8] Denjoy, A. (1910): *Continu et discontinu*, C. R. Acad. Sci. Paris **151**, 138–140.

[9] Deutsch, F. (1966): *Simultaneous interpolation and approximation in topological linear spaces*, SIAM J. Appl Math. **14**, 1180–1190.

[10] Gamelin, T. W. (1969): *Uniform Algebras*, Prentice-Hall.

[11] Hartogs, F., Rosenthal, A. (1931): *Über Folgen analytischer Funktion*, Math. Ann. **104**, 606–610.

[12] Keldysh, M. A. (1941): *On the solubility and stability of the Dirichlet problem*, Engl. transl.: Amer. Math. Soc. Transl. (2) **51** (1966), 1–73.

[13] Khenkin, G. M. (1985): *The method of integral representations in complex analysis*, Several Complex Variables I. Ed.: A. G. Vitushkin. Engl. transl.: Springer (1990), 19-116.

[14] Landkof, N. S. (1966): *Foundations of Modern Potential Theory*, Engl. transl.: Springer (1972).

[15] Lavrentiev, M. A. (1936): *Sur Les Fonctions d'une Variable Complexe Représentables par des Séries de Polynomes*, Hermann.

[16] Marden, M. (1975): *Joseph L. Walsh in memoriam*, Bull. Amer. Math. Soc. **81**, No. 1, 45–65.

[17] Mergelyan, S. N. (1952): *Uniform approximations to functions of a complex variable*, Engl. transl.: Translations Amer. Math. Soc. **3** (1962), 294–391.

[18] Paramonov, P., Verdera, J. (1994): *Approximation by solutions of elliptic equations on relatively closed subsets of* R^n, Math. Scand. **74**, 249–259.
[19] Range, R. M. (1986): *Holomorphic Functions and Integral Representations in Several Complex Variables*, Springer.
[20] Rudin, W. (1956): *Subalgebras of spaces of continuous functions*, Proc. Amer. Math. Soc. **7**, 825–830.
[21] Runge, K. (1885): *Zur Theorie der eindeutigen analytischen Funktionen*, Acta Math. **6**, 228–244.
[22] Tarkhanov, N. N. (1993): *Approximation on compact sets by solutions of systems with surjective symbols*, Uspekhi Mat. Nauk. Engl. transl.: Russ. Math. Surv. **48**, 103–145.
[23] Vitushkin, A. G. (1966): *Conditions on a set which are necessary and sufficient in order that any continuous function, analytic at its interior points, admit uniform approximation by rational fractions*, Engl. transl.: Soviet Math. Dokl. **7** (1966), 1622–1625.
[24] Walsh, J. L. (1926): *Über die Entwicklung einer analytischen Funktion nach Polynomen*, Math. Ann. **96**, 430–436.
[25] Walsh, J. L. (1926): *Über die Entwicklung einer Funktion einer komplexen Veränderlichen nach Polynomen*, Math. Ann. **96**, 437–450.
[26] Walsh, J. L. (1928): *On the expansion of analytic functions in series of polynomials and in series of other analytic functions*, Trans. Amer. Math. Soc. **30**, 307–332.
[27] Walsh, J. L. (1928): *Über die Entwicklung einer harmonischen Funktion nach harmonischen Polynomen*, J. Reine Angew. Math. **159**, 197–209.
[28] Walsh, J. L. (1929): *The approximation of harmonic functions by harmonic polynomials and by harmonic rational functions*, Bull. Amer. Math. Soc. **35**, 499–544.
[29] Walsh, J. L. (1935): *Interpolation and Approximation by Rational Functions in the Complex Domain*, Amer. Math. Soc., 5th ed. 1969.
[30] Weierstrass, K. (1885): *Über die analytische Darstellbarkeit sogenannter willkürlicher Functionen einer reellen Veränderlichen*, Sitz.ber. Akad. Wiss. Berlin, 633–639, 789–805.
[31] Wermer, J. (1955): *Polynomial approximation on an arc in* C^3, Ann. Math. **62**, 269–270.

Département de mathématiques et de statistiques,
Centre de rechèrches mathématiques,
Université de Montréal
Montréal
Canada

Chapter 4

Conformal Mapping

ON THE SHAPE OF LEVEL CURVES OF GREEN'S FUNCTION*

By J. L. WALSH, Harvard University

The level curves of Green's function—that is to say, the images of concentric circles in a smooth conformal map—have been widely studied, particularly with reference to convexity [**1**]†, star-shapedness [**2**, p. 82], and centers of curvature [**3**, **4**]. It is the object of the present note to obtain some new results on the behavior of a level curve in the large, notably its approximation to the shape of a circle.

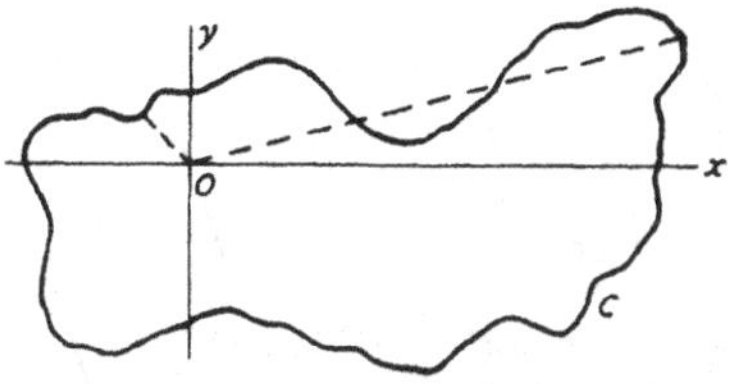

FIG. 1

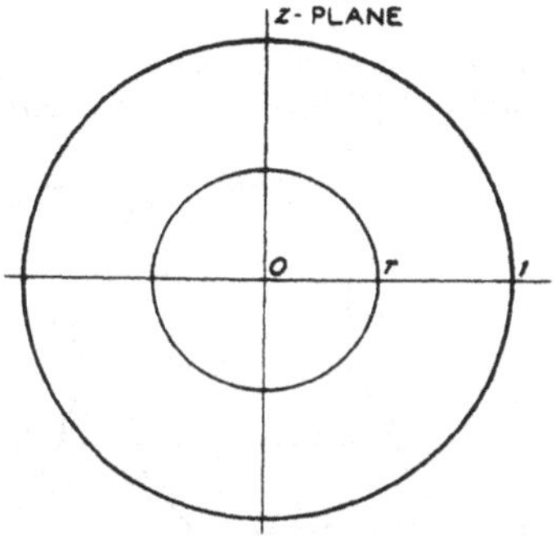

FIG. 2

1. *The circularity of a curve.* Let C be a closed curve with or without multiple points (Figure 1), or more generally an arbitrary closed limited point set which separates the point O from the point at infinity. *By the circularity* $K(C)$ *of* C *(with respect to* O*) we understand the ratio*

$$K(C) = \frac{\text{Distance from } O \text{ to nearest point of } C}{\text{Distance from } O \text{ to farthest point of } C}. \qquad (1)$$

The quantity $K(C)$ obviously depends on O, is always positive, and is never greater than unity. The quantity $K(C)$ is a rough measure of the closeness with which C approximates to the shape of a circle with center O. Obviously the circularity $K(C)$ equals unity when and only when C is a circle whose center is O.

For convenience we formulate here a well known result [**2**, pp. 2, 103; or **5**, p. 39] to which we shall make frequent reference:

SCHWARZ'S LEMMA. *Let* $\Phi(z)$ *be analytic and bounded:* $|\Phi(z)| \leqq M$ *for* $|z| < 1$, *with* $\Phi(0) = 0$. *Then we have*

$$|\Phi(z)| \leqq rM \quad \textit{for} \quad |z| \leqq r < 1, r > 0; \qquad (2)$$

the strong inequality is valid in (2) *unless* $\Phi(z)/z$ *is a constant of modulus* M.

* Presented to the American Mathematical Society, December, 1936.

† Bold faced numbers in square brackets refer to references listed at the end of the paper.

2. *Monotonic character of* $K(C_r)$. Our first result on the shape of level curves is

THEOREM 1. *Let the function* $f(z)$ *be analytic and bounded for* $|z|<1$, *with* $f(0)=0$, $f'(0)\neq 0$, *and* $f(z)$ *bounded from zero except in the neighborhood of the origin. Let* $C_r(0<r\leqq 1)$ *denote the boundary in the w-plane of the image of the region* $|z|<r$ *under the map* $w=f(z)$. *Then* $K(C_r)$, *the circularity of* C_r *with respect to* $w=0$, *increases monotonically as* r *decreases*:

$$K(C_{r_1}) \geqq K(C_{r_2}) \quad if \quad r_1 < r_2, \tag{3}$$

and $K(C_r)$ *approaches unity as* r *approaches zero. The strong inequality holds in* (3) *unless* $f(z)/z$ *is a constant—that is to say, unless* $K(C_r)$ *is identically unity.*

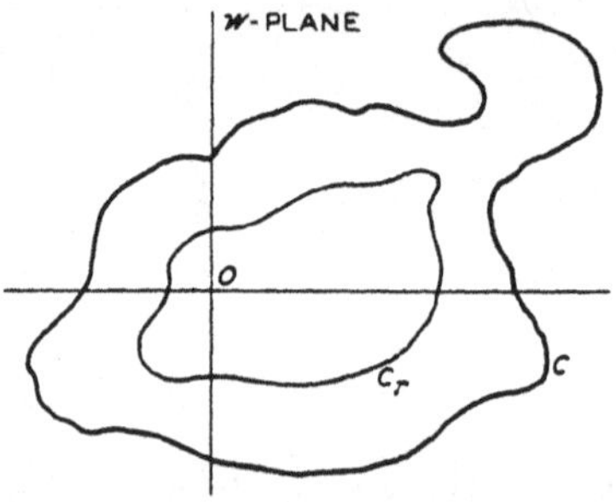

FIG. 3

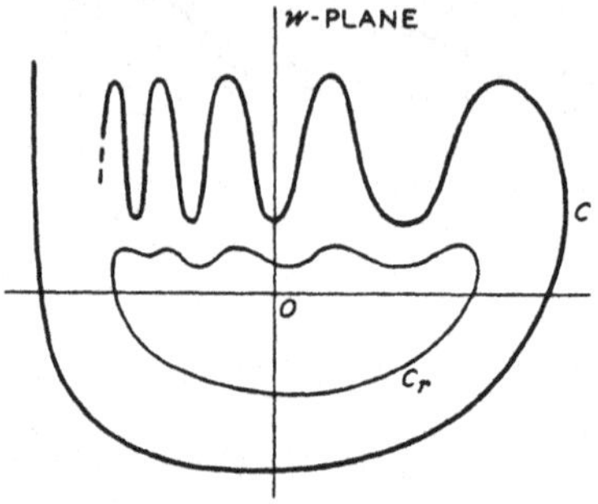

FIG. 4

Figure 2 represents the situation in the z-plane, with Figures 3 and 4 as suggestions of the numerous possibilities in the w-plane.

Here and in the sequel we denote by M_r and m_r the greatest and least values of $|f(z)|$ on the circle $|z|=r<1$; by M and m we denote the superior and inferior limits of $|f(z)|$ as z approaches the circumference $|z|=1$. By hypothesis m is greater than zero. Thus we have

$$K(C_r) = \frac{m_r}{M_r}, \qquad K(C_1) = \frac{m}{M}; \tag{4}$$

it is to be noticed that the neighborhood of $z=0$ is mapped by the transformation $w=f(z)$ onto the neighborhood of $w=0$, so the boundary C_r must separate $w=0$ from the point at infinity. By Schwarz's Lemma we have

$$\frac{M_r}{r} \leqq M. \tag{5}$$

The function $z/f(z)$, when suitably defined at the origin, is analytic and different from zero for $|z|<1$. When $|z|$ approaches unity, the function $|z/f(z)|$ can approach no limit greater than $1/m$, so by the principle of maximum modulus for an analytic function [**6**, p. 146; or **7**, p. 136; or **8**, p. 4], we have for $|z|<1$

$$\left|\frac{z}{f(z)}\right| \leqq \frac{1}{m},$$

$$\frac{r}{m_r} \leqq \frac{1}{m}. \tag{6}$$

By equations (4) and inequalities (5) and (6) we now have

$$\frac{m_r}{M_r} \geqq \frac{m}{M}, \quad \text{or} \quad K(C_r) \geqq K(C_1). \tag{7}$$

We notice that C_{r_1}, the image of the circle $|z|=r_1<r_2$ under the transformation $w=f(z)$, is the image of the circle $|z'|=r_1/r_2<1$ under the transformation $w=f(r_2z')$, which transforms $|z'|<1$ into a region bounded by C_{r_2}. Inequality (3) now follows from the second of inequalities (7). If $f(z)/z$ is not a constant, the strong inequality holds in (5), hence also in (7) and (3).

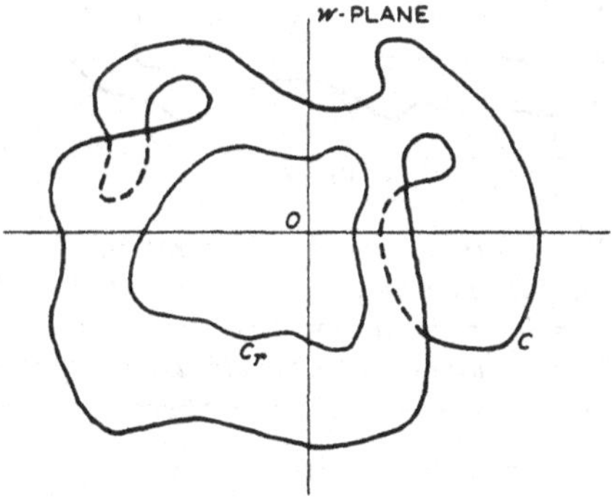

FIG. 5

For all values of z, $|z|<1$, we can write

$$\begin{aligned} f(z) &= a_1z + a_2z^2 + a_3z^3 + \cdots, \qquad a_1 \neq 0, \\ &= a_1z + z^2\chi(z), \end{aligned}$$

where $\chi(z)$ is analytic. Since $f(z)$ is bounded for $|z|<1$, so also is $\chi(z)$. If we suppose $|\chi(z)| \leqq h$, we have for $r<1$

$$M_r \leqq |a_1| r + hr^2, \qquad m_r \geqq |a_1| r - hr^2.$$

Thus we have

$$K(C_r) = \frac{m_r}{M_r} \geqq \frac{|a_1| r - hr^2}{|a_1| r + hr^2} = \frac{|a_1| - hr}{|a_1| + hr},$$

which approaches unity as r approaches zero. Then $K(C_r)$ also approaches unity as r approaches zero, so Theorem 1 is established.

In Theorem 1 the image in the w-plane of the region $|z|<1$ may overlap itself as in Figure 5. That is to say, the function $f(z)$ of Theorem 1 may take on the same value w in several distinct points z, $|z|<1$; the function $f(z)$ is not necessarily *univalent* (*schlicht*) for $|z|<1$.

3. *Direct study of green's function.* Theorem 1 can be proved by methods of conformal mapping, as above, or by methods of pure potential theory, as we proceed to indicate. We now direct our attention to the plane of $w=x+iy$ itself, without necessarily referring to the mapping onto the z-plane.

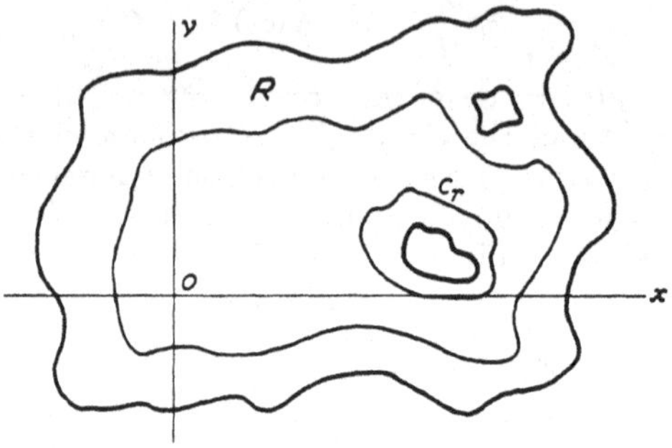

FIG. 6

THEOREM 2. *Let R* (Figure 6) *be a bounded region of the* (x, y)*-plane whose Green's function $G(x, y)$ with pole in the interior point O exists. Let C_r* $(0<r\leqq 1)$ *denote the locus $G(x, y)=\log r$ in the closed region R, so that C_r for $r<1$ consists of a finite number of Jordan curves. Then the circularity of C_r increases monotonically as r decreases, and approaches unity as r approaches zero.*

By Green's function $G(x, y)$ for R with pole at O: $(0, 0)$, we mean the unique function which is harmonic interior to R except at O, which approaches zero whenever (x, y) in R approaches a point of the boundary of R, and which in the neighborhood of O can be expressed

$$G(x, y) = G_0(x, y) + \log \rho, \qquad \rho = (x^2 + y^2)^{1/2}, \tag{8}$$

where $G_0(x, y)$ is harmonic throughout the neighborhood of O. The locus C_r obviously separates O from the boundary of R, hence separates O from the point at infinity.

If the region R of the plane for $w=x+iy$ is simply connected, and is mapped conformally in a one-to-one manner onto the region $|z|<1$ of the z-plane by the function $w=f(z)$ with $f(0)=0$, then [**9**, p. 717; or **2**, p. 57; or **10**, p. 365] this transformation can also be written $z=e^{G(x,y)+iH(x,y)}$, where the function $H(x, y)$ is conjugate harmonic to $G(x, y)$ interior to R. That is to say, we have under this transformation $|z|=e^{G(x,y)}$, so the locus C_r: $G(x, y)=\log r$ defined in Theorem 2 is precisely the locus C_r (the transform of $|z|=r$) defined in Theorem 1.

Under the hypothesis of Theorem 2, equation (8) defines a function $G_0(x, y)$ harmonic throughout the interior of R, continuous in the corresponding closed region. Let D_r denote the greatest distance and d_r the least distance from O to a point of C_r. By the definition of $C_r: G(x, y) = \log r$ we have

$$\begin{aligned} \max\,[G_0(x, y), \text{ on } C_r] &= \log r - \log d_r, \\ \min\,[G_0(x, y), \text{ on } C_r] &= \log r - \log D_r. \end{aligned} \tag{9}$$

The function $G_0(x, y)$ is harmonic in the region bounded by C_{r_2} $(r_2 \leqq 1)$ and containing O, continuous in the corresponding closed region; that region contains in its interior the locus C_{r_1}, $r_1 < r_2$, so we have

$$\begin{aligned} \max\,[G_0(x, y), \text{ on } C_{r_1}] &\leqq \max\,[G_0(x, y), \text{ on } C_{r_2}], \\ \min\,[G_0(x, y), \text{ on } C_{r_1}] &\geqq \min\,[G_0(x, y), \text{ on } C_{r_2}]. \end{aligned} \tag{10}$$

These inequalities, by virtue of equations (9), can be written

$$\begin{aligned} \log r_1 - \log d_{r_1} &\leqq \log r_2 - \log d_{r_2}, \\ \log r_1 - \log D_{r_1} &\geqq \log r_2 - \log D_{r_2}, \end{aligned}$$

from which we derive

$$\begin{aligned} \log d_{r_1} - \log D_{r_1} &\geqq \log d_{r_2} - \log D_{r_2}, \\ \log K(C_{r_1}) &\geqq \log K(C_{r_2}), \end{aligned}$$

$$K(C_{r_1}) \geqq K(C_{r_2}), \tag{11}$$

as we were to prove. The strong inequality holds in (10) and therefore in (11) unless $G_0(x, y)$ is identically constant, that is to say, unless R is the interior of a circle.

For r sufficiently small, the locus C_r is a small Jordan curve containing O in its interior, and which approaches O monotonically when r approaches zero. If we set $G_0(0, 0) = g$, equations (9) can be written respectively

$$\begin{aligned} \log d_r &= \log r - (g + \epsilon_1), \\ \log D_r &= \log r - (g - \epsilon_2), \end{aligned}$$

where ϵ_1 and ϵ_2 depend on r and approach zero with r. Then we have

$$\log K(C_r) = \log \frac{d_r}{D_r} = -\epsilon_1 - \epsilon_2,$$

which approaches zero with r, so $K(C_r)$ approaches unity as r approaches zero. Theorem 2 is established.

The proof and conclusion of Theorem 2 apply even if the region R overlaps itself, as in Figure 5 or Figure 7, provided each point of the neighborhood of O is covered but once. That is to say, the region R need not be smooth (schlicht), if each point of the neighborhood of O is covered only once. Under these circumstances, the distances ρ, d_r, D_r from O to a point of R are as before merely the rectilinear distances, measured as if R were entirely in one plane.

The proof and conclusion of Theorem 2 remain valid even if the region R overlaps itself a finite number of times throughout the neighborhood of O in such a way that several sheets come together at O (that is, if the Riemann surface on which R lies has an algebraic branch point at O), provided *all* the sheets of the surface covering O are connected at O, as in Figure 8. But the proof of Theorem 2 breaks down if the neighborhood of O is covered more than once by sheets that are not connected at O, for in that case the function $G_0(x, y)$ defined by (8) is no longer harmonic in R except at a single point of R; the reader may notice that the conclusion also fails in this case.

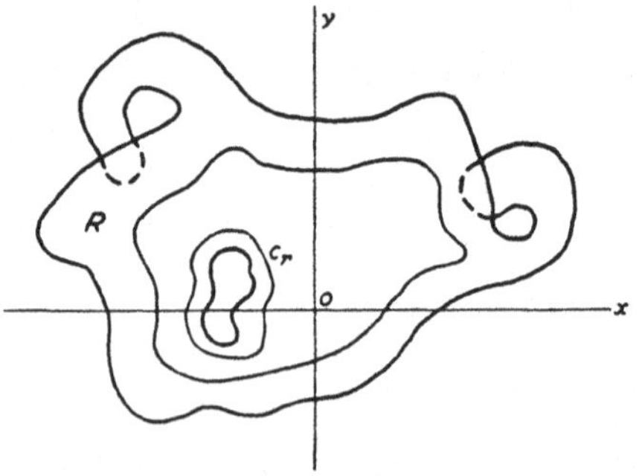

FIG. 7

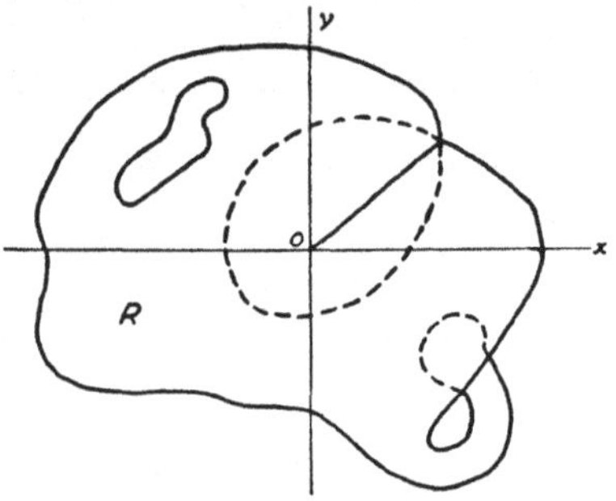

FIG. 8

Theorem 2 as formulated has the advantage over Theorem 1 of applying in the case of an arbitrary multiply connected region, provided merely that $G(x, y)$ exists. But Theorem 1 can also be extended to apply more generally to the mapping of a multiply connected region; let us assume $f(z)$ not necessarily single-valued but bounded and analytic except perhaps for algebraic branch points for $|z| < 1$, with $f(0) = 0$, $f'(0) \neq 0$, $f(z)$ bounded from zero except in the neighborhood of the origin, and let us assume $|f(z)|$ single valued for $|z| < 1$. The proof and conclusion hold without change.

Under the hypothesis of Theorem 1 fairly precise bounds can be obtained for the dependence of $K(C_r)$ on r. But here as in Theorem 1 itself the methods are again those of analytic functions. As a preliminary to the obtaining of these bounds, we need to make a detailed study of a certain conformal map.

4. *Mapping of an annulus on a circle.* Let A denote the annulus in the W-plane: $0 < m < |W| < M$, $m < 1 < M$; we shall map A conformally onto the interior of the unit circle in the Z-plane so that $W = 1$ corresponds to $Z = 0$. For the map to be possible and one-to-one, A must be interpreted as a simply-connected region; we therefore consider A as an infinite strip of width $M - m$

which winds in both senses from the line segment $m<W<M$ infinitely many times around the outside of the circle $|W|=m$.

The function

$$W_1 = \log \frac{W}{M}$$

maps the region A as just interpreted onto the strip*

$$\log \frac{m}{M} < \Re(W_1) < 0, \tag{12}$$

and the point $W=1$ corresponds to $W_1 = -\log M$.

The function

$$W_2 = \exp\left[\pi i\left(1 - \frac{W_1}{\log m - \log M}\right)\right] = \exp\left[\pi i \frac{\log m - \log W}{\log m - \log M}\right]$$

maps the strip (12) onto the upper half plane $\Im(W_2)>0$. The point $W_1 = -\log M$ corresponds to the point

$$W_2 = \exp\left[\pi i \frac{\log m}{\log m - \log M}\right] = e^{i\theta} = \alpha, \qquad 0 \leqq \theta \leqq \pi. \tag{13}$$

The line segment $\Im(W)=0$, $m<W<M$, corresponds to the segment $\Im(W_1)=0$, $\log(m/M)<W_1<0$, which corresponds to the semicircle S satisfying the conditions $|W_2|=1$, $0<-i\log W_2<\pi$.

The function

$$Z = \frac{W_2 - \alpha}{W_2 - \bar{\alpha}}$$

transforms the upper half plane $\Im(W_2)>0$ onto the interior of the circle $|Z|<1$ in such a way that $W_2=\alpha$ corresponds to $Z=0$. The semicircle S corresponds to the line segment L joining the two points

$$\beta = \frac{1-\alpha}{1-\bar{\alpha}} = -\alpha \quad \text{and} \quad -\beta.$$

The point $Z=\beta r$ $(0<r<1)$ on L corresponds to the point on S

$$W_2 = \frac{\bar{\alpha}Z - \alpha}{Z - 1} = \frac{1}{\alpha}\frac{r+\alpha}{r+\bar{\alpha}}. \tag{14}$$

* The notations $\Re[\Gamma]$ and $\Im[\Gamma]$ indicate respectively the real part of Γ and the quotient by i of the pure imaginary part of Γ.

If ω denotes the angle (argument, amplitude) of the quantity $r+\alpha$, we compute at once by (13)

$$\omega = \tan^{-1}\frac{\sin\theta}{r+\cos\theta},$$

whence equation (14) can be replaced by

$$\begin{aligned} W_2 &= \exp\left\{i\left[2\tan^{-1}\frac{\sin\theta}{r+\cos\theta}-\theta\right]\right\} \\ &= \exp\left\{2i\tan^{-1}\frac{(1-r)\sin\theta}{(1+r)(1+\cos\theta)}\right\} = e^{i\delta}, \end{aligned} \tag{15}$$

where we set

$$\delta = 2\tan^{-1}\frac{(1-r)\sin\theta}{(1+r)(1+\cos\theta)}, \qquad 0<\delta<\pi. \tag{16}$$

In a similar way, we find that the point $Z=-\beta r$ on L corresponds to the point on S

$$W_2 = e^{i\Delta}, \tag{17}$$

with the notation

$$\Delta = 2\tan^{-1}\frac{(1+r)\sin\theta}{(1-r)(1+\cos\theta)}, \qquad 0<\Delta<\pi. \tag{18}$$

The circle $|Z|=r<1$ cuts L orthogonally in the points βr and $-\beta r$, hence corresponds to a circle in the W_2-plane which cuts S orthogonally in the two points $e^{i\delta}$ and $e^{i\Delta}$. This circle in the W_2-plane therefore lies in the (closed) sector whose vertex is the origin and whose bounding rays make respective angles of δ and Δ with the positive direction of the axis of reals. This sector $\delta \leqq \mathfrak{I}(\log W_2) \leqq \Delta$ corresponds to the vertical strip

$$\left(1-\frac{\delta}{\pi}\right)\log\frac{m}{M} \leqq \mathfrak{R}(W_1) \leqq \left(1-\frac{\Delta}{\pi}\right)\log\frac{m}{M},$$

which corresponds to the annulus

$$m\left(\frac{M}{m}\right)^{\delta/\pi} \leqq |W| \leqq m\left(\frac{M}{m}\right)^{\Delta/\pi}. \tag{19}$$

Of course the entire closed region $|Z|\leqq r$ corresponds to a closed region in the W-plane which lies in the closed region (19).

5. *Application to the modulus of an analytic function*. The properties of the conformal map just analyzed are to be used in the proof of

THEOREM 3. *Let the function $F(z)$ be analytic for $|z|<1$, with $F(0)=1$. If we have*

$$0 < m \leqq |F(z)| \leqq M \quad \textit{for} \quad |z| < 1,$$

then we have also ($r>0$)

$$m\left(\frac{M}{m}\right)^{\delta/\pi} \leqq |F(z)| \leqq m\left(\frac{M}{m}\right)^{\Delta/\pi} \textit{for} \quad |z| \leqq r < 1, \tag{20}$$

where δ and Δ are given by (13), (16), (18).

Let the functions $W(Z)$ and $Z(W)$ map the Z-plane onto the W-plane in the manner already described (§4). Then the function $Z[F(z)]$ is analytic and of modulus less than unity for $|z|<1$, with $Z[F(0)]=0$. By Schwarz's Lemma, stated in §1, we have

$$|Z[F(z)]| \leqq r \quad \text{for} \quad |z| \leqq r < 1. \tag{21}$$

But every point of the region $|Z| \leqq r$ corresponds to a point of the region (19), so inequality (21) implies (19) for the values

$$W = W\{Z[F(z)]\} \equiv F(z),$$

and (20) is established.

The bounds that occur in (20) are actually taken on by an admissible function, so inequality (20) cannot be improved without restricting the class of admissible functions.

6. *Application to conformal mapping*. Theorem 3 is to be used in the proof of

THEOREM 4. *Let the function*

$$w = f(z) \equiv z + a_2 z^2 + a_3 z^3 + \cdots \tag{22}$$

be analytic for $|z|<1$, be different from zero for $0<|z|<1$, and map $|z|<1$ onto a region R of the w-plane whose boundary is C. Let $M<\infty$ and $m>0$ be respectively the greatest and least distances from O to a point of C.

Let the function (22) *map $|z|<r<1$ onto a region R_r whose boundary is C_r, and let M_r and m_r be respectively the greatest and least distances from O to a point of C_r. Then we have*

$$mr\left(\frac{M}{m}\right)^{\delta/\pi} \leqq m_r \leqq M_r \leqq mr\left(\frac{M}{m}\right)^{\Delta/\pi}, \tag{23}$$

where δ and Δ are given by (13), (16), (18).

The function $f(z)$ is not assumed univalent (schlicht) in the region $|z|<1$, but points of the neighborhood of the origin $w=0$ cannot be covered more than once in the map.

The function $F(z) \equiv f(z)/z$, when suitably defined for $z=0$, satisfies the hypothesis of Theorem 3. Inequalities (23) follow from (20).

Theorem 4 was established by a similar but somewhat different method by Koebe [11]. The present method is likewise similar to one that had been previously employed by Carathéodory [12]. The details are provided here for the purpose of deriving the inequalities (25) and (26) below.

7. *Inequalities for* $K(C_r)$. Under the hypothesis of Theorem 4 we find from (23) one of our main results:

$$K(C_r) = \frac{m_r}{M_r} \geqq \left(\frac{m}{M}\right)^{(\Delta-\delta)/\pi} \geqq \frac{m}{M}; \tag{24}$$

this first inequality is more precise than (7), and can be used to give a complete proof of Theorem 1. The first inequality in (24) can also be written

$$K(C_r) \geqq \left(\frac{m}{M}\right)^{\eta/\pi}, \qquad \eta = 2\tan^{-1}\frac{2r\sin\theta}{1-r^2}. \tag{25}$$

The strong inequalities are valid in (20) unless $Z[F(z)] \equiv z$ or $F(z) \equiv W(z)$; the strong inequalities are valid in (23), (24), and (25) unless $f(z) \equiv zW(z)$.

Inequality (25) is exact, in the sense that it cannot be improved for the totality of functions of the kind admitted in Theorem 4. But a simpler inequality is also of interest. We note the relations

$$\eta = \Delta - \delta = 2\tan^{-1}\frac{2r\sin\theta}{1-r^2} \leqq 2\tan^{-1}\frac{2r}{1-r^2} = 4\tan^{-1} r.$$

Consequently we may write from (25) our fundamental inequality

$$K(C_r) = \frac{m_r}{M_r} \geqq \left(\frac{m}{M}\right)^{(4/\pi)\tan^{-1}r} = [K(C_1)]^{(4/\pi)\tan^{-1}r}, \tag{26}$$

an inequality of relatively simple form, which does not involve explicitly $f'(0)$, and involves m and M only through their ratio $K(C_1)$. The quantities $K(C_r)$ are all invariant under a transformation of the form $w'=kw$, where $k\neq 0$ is constant. The situation of Theorem 1 [$f'(0)\neq 0$] can be transformed by such a transformation into the situation of Theorem 4 [$f'(0)=1$], and therefore *inequality* (26) *is valid under the conditions of Theorem* 1. By (13), the relation $\sin\theta=1$ is satisfied whenever we have $m=1/M$, so (26) is exact in the sense that it cannot be improved for the class of functions which satisfy the hypothesis of Theorem 1.

Under the hypothesis of Theorem 4, it follows from a general inequality [11, p. 73] on the modulus of an analytic function that we have

$$\frac{M_r}{r} \leqq M^{2r/(1+r)}, \qquad \frac{r}{m_r} \leqq \frac{1}{m^{2r/(1+r)}}.$$

Thus we may write also

$$K(C_r) = \frac{m_r}{M_r} \geqq \left(\frac{m}{M}\right)^{2r/(1+r)} = [K(C_1)]^{2r/(1+r)},$$

an inequality only slightly less favorable than (26).

Let now an arbitrary function $w=f(z)$ analytic for $|z|<1$ with $f(0)=0$, $f'(0)=1$, map $|z|<1$ smoothly (schlicht) onto a limited or unlimited region of the w-plane; we have the well known inequalities [2, p. 78] for $|z| \leqq r<1$:

$$\frac{r}{(1+r)^2} \leqq |f(z)| \leqq \frac{r}{(1-r)^2},$$

so we may write

$$\frac{r}{(1+r)^2} \leqq m_r, \qquad M_r \leqq \frac{r}{(1-r)^2}, \qquad K(C_r) \geqq \left(\frac{1-r}{1+r}\right)^2,$$

an inequality which has the advantage of not involving $K(C_1)$. Indeed, under the present circumstances the point set C_1 need not separate the point $w=0$ from the point at infinity, so $K(C_1)$ need not be defined. This last inequality for $K(C_r)$ obviously remains valid if the assumption $f'(0)=1$ is omitted.

8. *Mapping of infinite regions.* If the region $|z|<1$ is mapped conformally onto an *infinite* region of the w-plane so that $z=0$ corresponds to $w=\infty$, results can be obtained similar to those already given. Indeed, *the circularity $K(C)$ of a point set C in the z-plane with respect to $z=0$ is invariant under the transformation $z'=1/z$.* For let C' denote the transformed set. The distance from O to the nearest point of C is the reciprocal of the distance from O to the farthest point of C', and the distance from O to the farthest point of C is the reciprocal of the distance from O to the nearest point of C'. Thus from (1) we have

$$K(C) = K(C').$$

For simplicity we formulate the result analogous to Theorem 1 and to (26) rather than (25):

THEOREM 5. *Let the function*

$$f(z) \equiv \frac{a}{z} + a_0 + a_1 z + a_2 z^2 + \cdots, \qquad a \neq 0,$$

be analytic for $0<|z|<1$ and bounded except in the neighborhood of the origin. Let C_r $(0<r\leqq 1)$ denote the boundary in the w-plane of the image of the region $|z|<r$ under the map $w=f(z)$. Suppose a neighborhood of the point $w=\gamma$ is left uncovered by the image of $|z|<1$. Then $K(C_r)$, the circularity of C_r with respect to $w=\gamma$, increases monotonically as r decreases:

$$K(C_{r_1}) \geqq K(C_{r_2}) \quad \text{if} \quad r_1 < r_2, \tag{3}$$

and $K(C_r)$ approaches unity as r approaches zero. The strong inequality holds in (3) *unless $z[f(z)-\gamma]$ is a constant. Inequality* (26) *is valid.*

We emphasize the fact that Theorem 5 involves circularity with respect to an *arbitrary* point $w=\gamma$, provided a neighborhood of γ is left uncovered by the image of $|z|<1$.

References

1. Eduard Study, Konforme Abbildung einfach zusammenhängender Bereiche, Leipzig, 1913.
2. Ludwig Bieberbach, Lehrbuch der Funktionentheorie, vol. II, Leipzig, 1931.
3. Ernst Peschl, Mathematische Annalen, vol. 106, 1932, pp. 574–594.
4. J. L. Walsh, this Monthly, vol. 42, 1935, pp. 1–17.
5. Constantin Carathéodory, Conformal Representation, Cambridge Tracts, No. 28, 1932.
6. Ludwig Bieberbach, Lehrbuch der Funktionentheorie, vol. I, Leipzig, 1930.
7. Pólya-Szegö, Aufgaben und Lehrsätze, vol. I, Berlin, 1925.
8. J. L. Walsh, Interpolation and Approximation, New York, 1935.
9. W. F. Osgood, Funktionentheorie, vol. I, Leipzig, 1928.
10. O. D. Kellogg, Potential Theory, Berlin, 1929.
11. Paul Koebe, Mathematische Zeitschrift, vol. 6, 1920, pp. 52–84.
12. Constantin Carathéodory, Mathematische Annalen, vol. 72, 1912, pp. 107–144.

NOTE ON THE CURVATURE OF ORTHOGONAL TRAJECTORIES OF LEVEL CURVES OF GREEN'S FUNCTIONS

J. L. WALSH

The writer has recently established the following theorem:*

THEOREM I. *Let R be a simply connected region of the extended (x, y)-plane whose boundary B contains at least two points. Let $G(x, y)$ be Green's function for R with pole in the point O. Let $\{T\}$ denote the set of orthogonal trajectories to the level curves $G(x, y) = \log r$, $(0 < r < 1)$. The totality of circles each osculating a curve T at O is precisely the set of circles through O and through another fixed point D depending on O and R. There exists no circle that separates both O and D from B.*

The curves T are of course the images of the radii under the conformal mapping of a circle onto R, so that the center of the circle corresponds to O.

In Theorem I, the term "circle" is used in the extended sense, to include straight lines.

In the situation of Theorem I, we shall call D the *conjugate* of O with respect to R. This terminology seems justified, because in the case that B is a line, the point D is the reflection of O in B; and in the case that B is a proper circle, the point D is the inverse of O in B.

It is the object of the present note to establish the following theorem:

THEOREM II. *Under the hypotheses of Theorem* I *the point D may be chosen arbitrarily exterior to R; if D is so chosen there exists at least one point O interior to R whose conjugate with respect to R is D.*

In the proof of Theorem II it is sufficient, by the use of a suitably chosen linear transformation, to treat the special case that D is chosen at infinity.†

THEOREM III. *Let R be an arbitrary limited simply connected region R. There exists interior to R at least one point O whose conjugate D lies at infinity. That is to say, there exists at least one point O of R such that the orthogonal trajectories to the level curves of Green's function with pole at O have at O zero curvature.*

* Proceedings of the National Academy of Sciences, vol. 23 (1937), pp. 166–169.

† Choice of O at infinity yields the theorem: *Let R be a simply connected region, and let D be a point exterior to R. There exists a linear transformation of the plane which carries R into an infinite region for which the image of D is the conformal center of gravity.*

The proof of Theorem III is to be set forth with the aid of several lemmas. If R is a limited simply connected region of the z-plane, the *inner radius* of R with respect to an interior point $O: z=a$ is defined as the unique positive number $r(a)$ such that R can be mapped conformally onto the region $|w|<r(a)$ by a function of the form

$$z - a = w + c_2w^2 + c_3w^3 + \cdots, \qquad |w| < r(a). \tag{1}$$

LEMMA I. *If R is a limited simply connected region of the z-plane, if $O: z=a$ lies interior to R, and if $\delta(a)$ denotes the distance from O to the boundary of R, then we have*

$$4\delta(a) \geqq r(a). \tag{2}$$

Let the function (1) map R onto $|w|<r(a)$. Then the function

$$\zeta = \frac{1}{r(a)}\left[r(a)\omega + c_2[r(a)]^2\omega^2 + c_3[r(a)]^3\omega^3 + \cdots\right], \qquad |\omega| < 1,$$

maps the region R_1 of the ζ-plane, obtained from R by the transformation $\zeta=(z-a)/r(a)$, onto the region $|\omega|<1$. The distance from the point $\zeta=0$ to the boundary of R_1 is $\delta(a)/r(a)$, so inequality (2) follows at once from the Verzerrungssatz.* An immediate conclusion is the following lemma:

LEMMA II. *Under the hypothesis of Lemma* I *the function $r(a)$ approaches zero whenever the point $O: z=a$ approaches the boundary of R.*†

* See for instance Pólya-Szegö, *Aufgaben und Lehrsätze aus der Analysis*, vol. 2, Berlin, 1925, p. 27, exercise 152. Lemma II is given there explicitly (p. 19, exercise 112) for the case that R is bounded by an analytic curve.

† Let the function $z=f(w)$ map conformally $|w|<1$ onto R, with $a=f(b)$; then the function $z=\phi(w)=f[(w+b)/(1+\bar{b}w)]$ maps $|w|<1$ onto R with $a=\phi(0)$, and we have $r(a)=|\phi'(0)|=|f'(b)|(1-|b|^2)$, so Lemma I is precisely the inequality

$$4\delta(a) \geqq |f'(b)|(1-|b|^2);$$

Lemma II asserts precisely the conclusion

$$\lim |f'(b)|(1-|b|^2) = 0, \tag{A}$$

as a approaches the boundary of R (or, what is the same thing, as $|b|\to 1$).

Still another method of proof of Lemma II is the following. Instead of keeping the region R fixed and allowing a sequence of interior points $z=a$ to approach the boundary of R, we may study $r(a)$ by keeping the point $z=a$ fixed and allowing the region R to vary by a sequence of translations in such a way that the boundary of the variable R approaches $z=a$. Again by the use of $\phi(w)$, Carathéodory's theory of the conformal mapping of variable regions then yields Lemma II.

These two distinct methods in connection with the study of $\phi(w)$, namely (i) use of various inequalities such as (2), and (ii) Carathéodory's theory of variable regions, can be employed not merely for the proof of (A) but can also be used to study higher derivatives of $f(w)$, as will be shown on another occasion.

The following lemma will also be useful:

LEMMA III. *If $r(a)$ has a relative maximum at the point O: $|z| = a$, then for the corresponding mapping function* (1) *we have $c_2 = 0$, and the conjugate of O lies at infinity.*

Under the present hypothesis the relation $a_2 = 0$ is not difficult to establish,* where the inverse of (1) is given by

$$w = (z - a) + a_2(z - a)^2 + a_3(z - a)^3 + \cdots .$$

By the usual formulas for the inversion of the power series (1), we then have $c_2 = 0$, from which it follows† that the conjugate of O lies at infinity.

We are now in a position to establish Theorem III. The function $r(a)$ is positive and continuous‡ at each point O: $z = a$ of R. Whenever the point O in R approaches the boundary of R, the function $r(a)$ approaches zero, by Lemma II. The function $r(a)$, when suitably defined on the boundary, is then continuous in the corresponding closed region and possesses an absolute maximum in that region. That maximum occurs interior to R and is necessarily a relative maximum of $r(a)$. Theorem III follows from Lemma III.

We now give an example to show that *when D is given the point O of which D is the conjugate need not be unique.* For convenience we choose the situation of Theorem III. Let R be the region of the z-plane formed by the interiors of the two circles

$$|z \pm (.99)^{1/2}| = 1;$$

the two points $z = \pm i/10$ lie on the boundary of R. In the notation already introduced, we have $r(0) \leqq .4$, by Lemma I. But R contains in its interior the region $|z - (.99)^{1/2}| < 1$, so we have§ $r(.99)^{1/2} \geqq 1$. Thus the point $z = 0$ does not furnish an absolute maximum for the function $r(a)$; such an absolute maximum exists interior to R; such an absolute maximum cannot be unique, for if $z = \alpha$ furnishes an absolute maximum, so also does $z = -\alpha \neq \alpha$; and every such absolute maximum has its conjugate at infinity.

In the example just given, the point $z = 0$ has its conjugate at infinity even though it does not give an absolute maximum to the function $r(a)$.

---

* Pólya-Szegö, op. cit., p. 19, exercise 113.

† Walsh, loc. cit.

‡ The continuity follows for instance from the formula for $r(a)$ in terms of a fixed $r(b)$; Pólya-Szegö, op. cit., p. 19, exercise 110.

§ Pólya-Szegö, op. cit., p. 21, exercise 121.

A slight modification of the example given shows that if N is chosen arbitrarily, there exists a limited region* having at least N distinct points O whose conjugates D lie at infinity.†

Theorem III becomes false if in the hypothesis the region R is not assumed limited, for the reader may verify that no point O of the region R has its conjugate at infinity if R is the entire plane slit along the positive half of the axis of reals from the point $z=0$ to infinity.

HARVARD UNIVERSITY

* For the *unlimited* region R: $|y| \leqq b > 0$ of the (x, y)-plane, every point $(x, 0)$ has as conjugate the point at infinity.

† The referee points out that for any region the set of points O whose conjugates D lie at infinity is identical with the set of critical points of the function $r(a)$.

ON THE CIRCLES OF CURVATURE OF THE IMAGES OF CIRCLES UNDER A CONFORMAL MAP*

J. L. WALSH, Harvard University

Under a smooth conformal map $w=f(z)$ of the region $|z|<1$ of the z-plane onto a region R of the w-plane, the circles $|z|=r<1$ correspond to level curves of Green's function for R with pole in the point $w=f(0)$; the lines through $z=0$ correspond to lines of flow, or orthogonal trajectories of these level curves. The geometric properties of both sets of curves in the w-plane have been widely studied,† including the location of their centers of curvature. But so far as the writer is aware, there has been no systematic study of the circles of curvature *with especial reference to intersection of such circles with the boundary*. It is the object of the present paper to initiate such a systematic study, by considering the circles of curvature of the images of an arbitrary circle under both the transformation $w=f(z)$ and its inverse, especially intersections with boundaries in the respective planes. The present paper does not, however, claim to be complete; in fact possible classes of regions R suggest themselves in such variety that it is doubtful whether any treatment of the subject could be regarded as entirely exhaustive.

1. Curvature of images. As a first step in our discussion we establish

THEOREM 1. *Let the function $f(z)$ with the expansion for $|z|$ sufficiently small*

$$w = f(z) \equiv z + a_2z^2 + a_3z^3 + \cdots \tag{1}$$

be meromorphic for $|z|<1$ and map the region $|z|<1$ onto a smooth region R of the extended w-plane. Let B denote the boundary of R. Let C_ρ be a circle of radius ρ in the z-plane which passes through the origin, and let A_ρ be its image in the w-plane, or to be more exact, the image in the w-plane of the portion of C_ρ which lies in $|z|<1$. Let T_ρ denote the circle of curvature of A_ρ at the origin in the w-plane. Then T_ρ depends on a_2 but not on a_k for $k>2$.

In Theorem 1 and below we use the term *circle* in the extended sense, to include straight line.

Theorem 1 seems reasonable, by virtue of the fact that curvature depends on second derivatives of rectangular coördinates but not on higher derivatives.

---

* Presented to the American Mathematical Society, February 1939.

† See for instance the following:

Scheffers, Jahresbericht. d. deut. Math-Vereinigung, vol. 31, 1922, pp. 170–174;

Neubauer, same journal, vol. 32, 1923, p. 310;

Lilienthal, same journal, vol. 33, 1925, pp. 127–139;

Ringleb, same journal, vol. 45, 1935, pp. 57–60;

Tzitzéica, Paris Comptes Rendus, vol. 195, 1932, pp. 476–478;

Ginzel, Deutsche Mathematik, vol. 2, 1937, pp. 401–416.

Thus Theorems 1 and 2 of the present paper are not to be regarded as novel; compare especially Ginzel, *loc. cit.*, and mention there of results due to Böhmer. Introduction of the comparison transformation (3) seems to be due to Böhmer.

Nevertheless, certain expressions involving second derivatives may fail to be defined for such a function as $f(z)$, and may require the use of higher derivatives in their evaluation, so we proceed to give a formal proof of Theorem 1 based on computation of T_ρ.

Let C_ρ be the circle

$$|z - \lambda\rho| = \rho, \qquad |\lambda| = 1, \tag{2}$$

so that C_ρ can be represented as

$$z = \lambda\rho + \rho e^{i\theta}, \qquad \theta \text{ real}.$$

Then we have

$$dz = i\rho e^{i\theta} d\theta.$$

The transformation is given by $w = f(z)$, whence

$$dw = f'(z)dz = i\rho f'(z)e^{i\theta}d\theta,$$

$$\arg(dw) = \arg[f'(z)] + \theta + \frac{\pi}{2}, \qquad |dw| = \rho|f'(z)|\,d\theta.$$

If κ denotes the curvature of T_ρ we have

$$\kappa = \frac{d[\arg(dw)]}{|dw|}.$$

We proceed to compute

$$d[\arg(dw)] = d[\arg f'(z)] + d\theta = \Im\{d \log [f'(z)]\} + d\theta$$
$$= \Re\left[1 + \rho e^{i\theta}\frac{f''(z)}{f'(z)}\right]d\theta.$$

The normal to C_ρ at $z=0$ has the direction $\arg \lambda$, and the normal to T_ρ at $w=0$ has this same direction; at $z=0$ we have $\lambda = -e^{i\theta}$, whence for T_ρ at $w=0$ the center of curvature is

$$\frac{\lambda}{\kappa} = \frac{\rho\lambda}{\Re[1 - 2\rho\lambda a_2]},$$

an expression whose independence of a_k for $k>2$ establishes Theorem 1. Of course the case of a line C_ρ or T_ρ is not properly included in the formulas given, but can be treated by an obvious limiting process. Henceforth the value $\rho = \infty$ is not excluded.

2. Circles of curvature under transformation.

THEOREM 2. *Under the hypothesis of Theorem 1 the set of all circles T_ρ for a given ρ is the set of circles through $w=0$ tangent to the circle $|(1+a_2w)/w| = 1/2\rho$ of the coaxal family determined by $w=0$ and $w=-1/a_2$ as null circles.*

By virtue of Theorem 1, the circles T_ρ are the same whether we set $w=f(z)$ or

$$(3) \qquad w = \frac{z}{1 - a_2 z} = z + a_2 z^2 + a_2^2 z^3 + \cdots .$$

For the transformation (3), the circles A_ρ and T_ρ are identical. Under the transformation

$$(4) \qquad \zeta = \frac{1}{w} = \frac{1 - a_2 z}{z} \equiv \frac{1}{z} - a_2,$$

the points $z=0$ and $z=\infty$ correspond to $\zeta=\infty$ and $\zeta=-a_2$, respectively. The circle $|z|=2\rho$ (to which all C_ρ are tangent) corresponds to the circle $|\zeta+a_2|=1/2\rho$, to which all the transforms of the C_ρ are tangent. Thus all the T_ρ under the transformation (1) or (3) pass through $w=0$ and are tangent to the circle

$$\left| \frac{1}{w} + a_2 \right| = \frac{1}{2\rho},$$

and the proof is complete.

The case $a_2=0$ is especially interesting, for in that case the center and radius of T_ρ in the w-plane are the same as the center and radius of C_ρ in the z-plane. Whenever the region R is bounded, there exists* at least one point interior to it for which the mapping function (either for R or for a region found from R by shrinking or stretching) is of form (1) with $a_2=0$.

3. **Intersection of circles of curvature with boundary.** We shall now continue to study the circles S_ρ in the ζ-plane, which are simultaneously the images in the ζ-plane of the circles C_ρ in the z-plane under the transformation (4), and also the image of the circles T_ρ in the w-plane under the transformation $\zeta=1/w$, where w is given by (1); the circles S_ρ are precisely the straight lines (*i.e.*, circles through the point at infinity) tangent to the circle

$$K_\rho: |\zeta + a_2| = \frac{1}{2\rho}\cdot$$

Under the transformation $\zeta=1/w$ the region R of the w-plane corresponds to a region R_1 of the ζ-plane whose boundary B_1 is bounded. The region R_1 is the image of $|z|<1$ under the univalent transformation

$$\zeta = \frac{1}{w} = \frac{1}{f(z)} = \frac{1}{z} - a_2 + b_1 z + b_2 z^2 + \cdots .$$

It is consequently well known† that if the origin $\zeta=0$ is not interior to R_1 we have

* Walsh, Bulletin American Mathematical Society, vol. 44, 1938, pp. 520–523.

† The results are due mainly to Koebe, Bieberbach, Faber, Pick, and Löwner. See, for instance, Faber, Münchner Berichte, 1920, pp. 49–64. Or see Pólya and Szegö, Aufgaben und Lehrsätze, Berlin, 1925, vol. 2, p. 25.

$$|a_2| \leqq 2, \tag{5}$$

and that in any case B_1 lies in the closed region

$$|\zeta + a_2| \leqq 2; \tag{6}$$

the equality signs can be omitted in (5) and (6) unless B_1 is a line segment, necessarily of length 4. The point set B_1 has points exterior to the circle $|\zeta+a_2| = 1$ unless B_1 is identical with that circle.

The point $\zeta = -a_2$ is called the *conformal center of gravity of* R_1, and is the actual center of gravity in the ζ-plane of each curve corresponding to a circle $|z| = r < 1$, weighted according to the weight function $|dz/d\zeta|$.

If R contains the point $w = \infty$ in its interior, or if $w = \infty$ lies on B, or if B is limited, then in the respective cases the point $\zeta = 0$ is interior to R_1, or is on B_1, or does not lie on B_1 but is separated by B_1 from the point $\zeta = \infty$.

We are now in a position to prove

THEOREM 3. *Under the hypothesis of Theorem* 1 *there exist two numbers* ρ_1 *and* ρ_2 *depending on* R, *with* $\frac{1}{4} \leqq \rho_1 \leqq \rho_2 \leqq \infty$ *such that for* $\rho_2 \leqq \rho \leqq \infty$ *each* T_ρ *cuts* B; *for* $\rho_1 \leqq \rho < \rho_2$ *a given* T_ρ *need not cut* B, *but for each* ρ *some* T_ρ *cuts* B; *for* $\rho < \rho_1$ *no* T_ρ *cuts* B. *If* B *is a circle* (*of the extended plane*) *we have* $\rho_1 = \rho_2 = \frac{1}{2}$; *in every other case* $\rho_1 < \rho_2$ *and* $\rho_1 < \frac{1}{2}$.* *If* B *is an arc of a circle through* 0, *we have* $\rho_1 = \frac{1}{4}$, $\rho_2 = \infty$; *in every other case we have* $\frac{1}{4} < \rho_1$, $\rho_2 < \infty$.

If B_1 is not a line segment, the convex hull of B_1 [*i.e.*, the smallest convex set (necessarily closed) containing B_1] has the point $\zeta = -a_2$ as an *interior* point.† This is a consequence of the interpretation of $\zeta = -a_2$ as center of gravity of the weighted transforms of the curves $|z| = r < 1$ in the ζ-plane. Let

$$|\zeta + a_2| = \frac{1}{2\rho_2} > 0$$

be the largest circle with $\zeta = -a_2$ as center contained in the convex hull of B_1. Then any line tangent to the circle

$$|\zeta + a_2| = \frac{1}{2\rho}, \qquad \rho \geqq \rho_2, \tag{7}$$

* It might be conjectured that we always have $\rho_1 \leqq \frac{1}{2} \leqq \rho_2$, but *this conjecture is false*, as is illustrated by the function $w = z/(1+z^8)^{1/4}$. The corresponding point set B_1 of the ζ-plane consists of eight equally spaced line segments each of length $2^{1/4}$ radiating from $\zeta = 0$. Thus we have

$$\frac{1}{2\rho_2} = 2^{1/4} \cos 22\tfrac{1}{2}° = 1.0987,$$

whence $\rho_2 < \frac{1}{2}$.

† This fact is intuitively obvious. The writer expects to publish a formal proof in another connection, in a paper to appear in the Bulletin of the American Mathematical Society.

cuts B_1, so for $\rho_2 \leqq \rho \leqq \infty$ each T_ρ cuts B. But for every $\rho < \rho_2$ there exists a line tangent to (7) which does not cut B_1; so there exists a T_ρ which does not cut B. Let the largest circle with $\zeta = -a_2$ as center passing through a point of B_1 be

$$|\zeta + a_2| = \frac{1}{2\rho_1} > 0;$$

we obviously have $\rho_1 \leqq \rho_2$. Any circle

$$|\zeta + a_2| = \frac{1}{2\rho}, \qquad \rho_1 \leqq \rho < \rho_2, \tag{8}$$

cuts B_1, but fails to lie completely in the convex hull of B_1. Thus some lines tangent to (8) cut B_1, and some lines tangent to (8) fail to cut B_1. There exists a T_ρ which cuts B and there exists a T_ρ which fails to cut B. For $\rho < \rho_1$ the circle

$$|\zeta + a_2| = \frac{1}{2\rho}$$

contains B_1 in its interior, so no line tangent to this circle can cut B_1; no circle T_ρ can cut B. The fact that B_1 lies in the region (6) shows that $\frac{1}{4} \leqq \rho_1$; unless B_1 is a line segment, that is to say, unless B is the arc of a circle through 0, we have $\frac{1}{4} < \rho_1$.

If $\rho_1 = \rho_2$, each point of B_1 lies on or within the circle

$$|\zeta + a_2| = \frac{1}{2\rho_1},$$

and every line tangent to that circle cuts B_1. Then every point of that circle is a point of B_1, and R_1 consists of the exterior of the circle. Consequently B_1 is a circle, necessarily of radius unity, by the form of the mapping function $\zeta = 1/f(z)$; and we have $\rho_1 = \rho_2 = \frac{1}{2}$.

The fact that not every point of B_1 can lie interior to the circle $|\zeta + a_2| = 1$ and that if B_1 is not that circle points of B_1 lie exterior to the circle, shows that we have $\rho_1 < \frac{1}{2}$ unless B_1 is that circle, in which case $\rho_1 = \frac{1}{2}$. Hence if B is a circle we have $\rho_1 = \frac{1}{2}$, otherwise $\rho_1 < \frac{1}{2}$.

We mention explicitly a result of some intrinsic interest:

COROLLARY. *Under the conditions of Theorem 3, no circle T_ρ with $\rho < \frac{1}{4}$ can cut B.*

The fact that every T_∞ cuts B has been established previously.*

Although Theorem 3 has been established for a region R which corresponds to a mapping function of form (1), it is clear that the requirements $f(0) = 0$, $f'(0) = 1$ may be ignored for the purposes of that theorem; the point of the extended w-plane which corresponds to $z = 0$, and which may be $w = \infty$, is denoted by 0.

---

* Walsh, this MONTHLY, vol. 42, 1935, pp. 1–17; Proceedings National Academy of Sciences, vol. 23, 1937, pp. 166–169.

4. Numerical relations involving circles T_ρ. Let us now proceed to determine which circles in the w-plane having the origin as center can be cut by a given T_ρ. We establish

THEOREM 4. *Under the conditions of Theorem 1, every T_ρ cuts the circle $|w|=r$ if*

$$r \leqq \frac{2\rho}{1+2|a_2|\rho}; \tag{9}$$

no T_ρ cuts the circle $|w|=r$ if

$$r > \frac{2\rho}{1-2|a_2|\rho}, \qquad 1-2|a_2|\rho > 0. \tag{10}$$

In particular, if $w=\infty$ is not an interior point of R, it is sufficient for (9) *if*

$$r \leqq \frac{2\rho}{1+4\rho}, \tag{11}$$

and it is sufficient for (10) *if*

$$r > \frac{2\rho}{1-4\rho}, \qquad 1-4\rho > 0. \tag{12}$$

If $w=\infty$ is not an exterior point of R, we denote by M ($M=\infty$ is not excluded) the greatest distance from $w=0$ to a point of B. It is sufficient for (9) *if*

$$r \leqq \frac{2M\rho}{M+2(1+2M)\rho}, \tag{13}$$

and it is sufficient for (10) *if*

$$r > \frac{2M\rho}{M-2(1+2M)\rho}, \qquad M-2(1+2M)\rho > 0. \tag{14}$$

If R is not the extended plane cut along the arc of a circle through $w=0$, the first sign $>$ in (12) *and* (14) *may be replaced by $\geqq$.*

Under the conditions of Theorem 4, with (9) fulfilled, the circle in the ζ-plane

$$|\zeta + a_2| = \frac{1}{2\rho} \tag{15}$$

is internally tangent to or is completely interior to the circle

$$|\zeta| = \frac{1}{r} \geqq |a_2| + \frac{1}{2\rho}. \tag{16}$$

Consequently every line tangent to the circle (15) must cut the circle (16).

That is to say, every T_ρ must cut $|w| = r$ if (9) is fulfilled.

The circle

$$(17) \qquad |\zeta| = \frac{1}{r} < \frac{1}{2\rho} - |a_2|, \qquad \frac{1}{2\rho} - |a_2| > 0,$$

lies interior to the circle (15), so no line tangent to (15) can cut the circle (17). That is to say, no T_ρ can cut $|w| = r$ if (10) is fulfilled.

We remark also that if the sign $<$ is replaced by $=$ in (17), the corresponding circle can have no point exterior to the circle (15), so if the first sign $>$ in (10) is replaced by the sign $=$, no T_ρ can cut the circle $|w| = r$ in more than one point. This remark applies similarly in connection with (12) and (14).

If $w = \infty$ is not an interior point of R, the point $\zeta = 0$ is not an interior point of R_1, and it follows from (5), that (11) and (12) are sufficient for (9) and (10), respectively.

If $w = \infty$ is not an exterior point of R, the point $\zeta = 0$ is not an exterior point of R_1; by virtue of the fact that B_1 lies in the circle (6) it follows that $|a_2|$ is not greater than 2 plus the distance from 0 to B_1, namely $1/M$. Consequently (13) and (14) are sufficient for (9) and (10), respectively.

If $w = \infty$ is not an interior point of R, the point $\zeta = 0$ is not an interior point of R_1, and if B_1 is a line segment it must be the segment of a line through 0; if B_1 is not such a line segment (that is to say, if R is not the extended plane cut along the arc of a circle through $w = 0$), the equality sign may be omitted in (5) and hence may be inserted in the first of inequalities (12). If $w = \infty$ is not an exterior point of R, the point $\zeta = 0$ is not an exterior point of R_1. If $\zeta = 0$ is a point of B_1 and if B_1 is not a line segment we have $|a_2| < 2$. If $\zeta = 0$ is not a point of B_1 and if B_1 is not a line segment through $\zeta = 0$, either B_1 satisfies $|\zeta + a_2| < 2$, or the shortest distance $1/M$ from $\zeta = 0$ to B_1 does not lie along the line joining $\zeta = 0$ and $\zeta = -a_2$; in either case we have $|a_2| < 2 + 1/M$, and the equality sign may be inserted in the first inequality in (14).

5. Applications concerning T_ρ. It is a consequence of (10) that whenever we have

$$\rho < \frac{1}{2|a_2|}$$

the circle T_ρ cannot cut $|w| = r$ provided r is sufficiently large. Hence T_ρ cannot be a straight line, and A_ρ cannot have a point of inflection at the point $w = 0$. In particular, if we have $|a_2| \leq 1$, any circle C_ρ which lies interior to $|z| = 1$ corresponds to a curve A_ρ which fails to have a point of inflection at $w = 0$. If the region R is convex, the function which maps $|z| < 1$ onto R with $z = 0$ corresponding to an arbitrary point w_0 of R is of the form

$$w - w_0 = \mu[z + a_2 z^2 + a_3 z^3 + \cdots], \qquad \mu \neq 0,$$

with $|a_2| \leq 1$. Consequently the image of every circle interior to $|z| = 1$ under a specific map of $|z|<1$ onto R can have no point of inflection in R, a theorem due to Study* for the case of circles $|z|=r<1$ and to Carathéodory† in the general case.

We formulate explicitly the

COROLLARY. *Under the conditions of Theorem 4, the curve A_ρ has no point of inflection at $w=0$ if we have*

$$\rho < \frac{1}{2|a_2|};$$

if $w=\infty$ is not an interior point of R it is sufficient if we have

$$\rho < \tfrac{1}{4};$$

if $w=\infty$ is not an exterior point of R, and if M denotes the greatest distance from $w=0$ to a point of B, it is sufficient if we have

$$\rho < \frac{M}{2(1+2M)}.$$

We have thus far used, in addition to the obvious geometric properties of the situation in the w-plane, only inequalities (5) and (6). It is clear that various other results related to (5) and (6) that already occur in the literature and others that may be established in the future can be used to sharpen our results. For instance,‡ under the hypothesis of Theorem 4 let R lie in the region $|w|<M$. Under the present circumstances we have the inequality§

$$|a_2| \leqq 2\left(1-\frac{1}{M}\right) \quad \text{or} \quad M \geqq \frac{2}{2-|a_2|}.$$

Consequently (9) and (10) are fulfilled if we have, respectively,

$$r \leqq \frac{2M\rho}{M+4(M-1)\rho}, \tag{18}$$
$$r > \frac{2M\rho}{M-4(M-1)\rho}, \qquad M-4(M-1)\rho > 0.$$

In particular, no circle T_ρ cuts $|w|=M$ provided we have either of the two inequalities

* Konforme Abbildung Einfach-Zusammenhängender Bereiche, Leipzig, 1913, p. 110.

† Mathematische Annalen, vol. 79, 1919, p. 402.

‡ Other instances are: (i) determination of the sharp lower bound for ρ_2 in connection with Theorem 3; (ii) determination of limits for ρ_1 and ρ_2 when R lies interior to a given circle $|w|=M$; (iii) determination of limits for ρ_1 and ρ_2 when B has no point interior to a given circle $|w|=M$; (iv) improvement of (6) and hence of Theorem 3 for convex R.

§ See Pick, Wiener Berichte, vol. 126, 1917, pp. 247–263.

Inspection of the situation in the ζ-plane yields only the weaker inequality $|a_2|\leqq 2-1/M$.

$$\rho < \frac{M}{4M - 2}$$

(a condition which implies the second of inequalities (18) for $r = M$), or

$$\rho < \frac{1}{2 + |a_2|}$$

(a condition which implies (10) directly); it is of course sufficient if $\rho < \frac{1}{4}$.

6. Properties of U_ν. We have studied the images in the w-plane of circles through the origin in the z-plane. We now proceed to consider the inverse problem. The results already established enable one to deduce results on the circle of curvature U_ν in the z-plane of the image in the z-plane of a circle of radius ν through the origin in the w-plane.

The inverse of (1) can be written

$$z = F(w) \equiv w - a_2 w^2 + c_3 w^3 + \cdots \tag{19}$$

which is precisely of form (1) with the sign of the second coefficient of the last member reversed. For convenience in reference we formulate the analog of Theorem 1, which is essentially merely Theorem 1 itself:

THEOREM 5. *Under the hypothesis of Theorem* 1 *let D_ν be a circle of radius ν in the w-plane which passes through the origin, and let B_ν be its image in the z-plane (or to be more exact, let B_ν be the image in the z-plane of that portion of D_ν which lies in R). Let U_ν denote the circle of curvature of B_ν at the origin in the z-plane. Then U_ν depends on a_2 but not on a_k for $k > 2$.*

We proceed next to the discussion of Theorem 4. The entire reasoning used in the proof of Theorem 4 is still valid if a_2 is replaced by $-a_2$; such substitution leaves unchanged the formal statement, so we have.

THEOREM 6. *Under the conditions of Theorem* 4 *every U_ν cuts the circle $|z| = s$ if*

$$s \leqq \frac{2\nu}{1 + 2|a_2|\nu}; \tag{20}$$

no U_ν cuts the circle $|z| = s$ if

$$s > \frac{2\nu}{1 - 2|a_2|\nu}, \qquad 1 - 2|a_2|\nu > 0. \tag{21}$$

In particular, if $w = \infty$ is not an interior point of R, it is sufficient for (20) *if*

$$s \leqq \frac{2\nu}{1 + 4\nu}, \tag{22}$$

and it is sufficient for (21) *if*

$$s > \frac{2\nu}{1 - 4\nu}, \qquad 1 - 4\nu > 0. \tag{23}$$

If $w=\infty$ is not an exterior point of R, we denote by M ($M=\infty$ is not excluded) the greatest distance from $w=0$ to a point of B. It is sufficient for (20) *if*

$$s \leqq \frac{2M\nu}{M + 2(1+2M)\nu},$$

and it is sufficient for (21) *if*

$$s > \frac{2M\nu}{M - 2(1+2M)\nu}, \qquad M - 2(1+2M)\nu > 0.$$

Some special situations under Theorem 6 deserve comment. If R is convex, we have $|a_2| \leqq 1$, and it follows from (20) that every U_∞ cuts the circle $|z|=1$. The geometric properties are independent of the condition $f'(0)=1$, so it follows that *whenever R is convex, each circle of curvature in the z-plane to the curve in the z-plane which is the image of a line segment interior to R in the w-plane,—every such circle of curvature cuts the circle $|z|=1$*, a remark due to Carathéodory, *loc. cit.** Whether or not R is convex, if $w=\infty$ is not an interior point of R we have $|a_2| \leqq 2$, so it follows from (20) that every U_∞ cuts the circle $|z|=\frac{1}{2}$. On the other hand if $w=\infty$ is not an interior point of R it follows from (23) that no U_ν cuts $|z|=1$ if we have $\nu < \frac{1}{8}$.

The test analogous to (18), and similarly proved, is: *Let R lie in the region $|w| < M$; then sufficient conditions for* (20) *and* (21) *respectively are*:

$$s \leqq \frac{2M\nu}{M + 4(M-1)\nu},$$

$$s > \frac{2M\nu}{M - 4(M-1)\nu}, \qquad M - 4(M-1)\nu > 0.$$

7. Applications concerning U_ν. A corollary of Theorem 6 is analogous to the corollary to Theorem 4:

COROLLARY. *Under the conditions of Theorem 4 no U_ν is orthogonal to $|z|=1$ provided we have*

$$\nu < \frac{1}{2|a_2|}. \tag{24}$$

In particular if the diameter of D_ν is less than the distance from $w=0$ to B, this condition (24) *is fulfilled.*

If $w=\infty$ is not an interior point of R it is sufficient if we have $\nu < \frac{1}{4}$.

---

* Even if R is not convex, every $|z|=r<r_1<1$ for sufficiently small r_1 is transformed into a convex curve; see Theorem 8 below. Hence at every point of $|z| \leqq r_1$, a circle of curvature to the image in the z-plane of a line segment in R must cut $|z|=r_1$; this remark also is due to Carathéodory.

Whether R is convex or not, it is a consequence of Theorem 6 that every U_∞ cuts $|z|=1$ provided $|a_2| \leqq 1$.

Of course U_ν is orthogonal to $|z|=1$ when and only when U_ν cuts every $|z|=s$. It follows from (21) that (24) is sufficient to ensure that U_ν shall not cut every $|z|=s$.

Let m denote the distance in the w-plane from 0 to B. In the ζ-plane as used in the proof of Theorem 4, the point set B_1 then lies in the closed region $|\zeta|\leq 1/m$, and for every $\epsilon>0$ lies interior to $|\zeta|=\epsilon+1/m$. The transform in the ζ-plane of some circumference $|z|=r<1$ lies also interior to $|\zeta|=\epsilon+1/m$; hence the center of gravity $\zeta=-a_2$ of that weighted transform also lies interior to $|\zeta|=\epsilon+1/m$. Consequently we have*

$$|a_2| \leq \frac{1}{m}, \tag{25}$$

and the condition

$$2\nu < m$$

implies (24).

The last part of the corollary follows from (5).

We formulate another consequence of Theorem 6 as follows:

THEOREM 7. (i) *Under the conditions of Theorem 4, suppose*

$$2\nu = \frac{M}{\lambda}, \qquad \lambda > 0,$$

where M is defined as in Theorem 4 and where $w=\infty$ is not an interior point of R. Then every U_ν cuts the circle

$$|z| = \frac{M}{\lambda + 2M - 2}; \tag{26}$$

no U_ν cuts the circle $|z|=s$ if

$$s > \frac{M}{\lambda - 2M + 2}, \qquad \lambda - 2M + 2 > 0. \tag{27}$$

(ii) *Under the conditions of Theorem 4, suppose*

$$2\nu = \frac{m}{\lambda}, \qquad \lambda > 0,$$

where m is the distance in the w-plane from $w=0$ to B. Then every U_ν cuts the circle

$$|z| = \frac{m}{1+\lambda}; \tag{28}$$

no U_ν cuts the circle $|z|=s$ if

* Walsh, Proceedings National Academy of Sciences, vol. 23, 1937, pp. 166–169.

$$s > \frac{m}{\lambda - 1}, \qquad \lambda > 1. \tag{29}$$

In connection with (18) we have used the inequality

$$|a_2| \leq 2\left(1 - \frac{1}{M}\right).$$

Under the hypothesis of (i), the second member of (20) is not less than the second member of (26); the second member of the first inequality in (27) is not less than the second member of (21); then (i) is established.

Under the hypothesis of (ii), it follows from (25) that the second member of (20) is not less than the second member of (28); the second member of (29) is not less than the second member of (21); hence (ii) is established.

Some numerical cases of Theorem 7 are of interest. Thus if $2\nu \geq M/2$, that is to say if the diameter of D_ν is at least half the greatest distance from $w=0$ to a point of B, and if $w=\infty$ is not an interior point of R, then every U_ν cuts $|z| = \frac{1}{2}$. If $2\nu \geq m$, that is to say if the diameter of D_ν is at least the distance from $w=0$ to B, whether or not $w=\infty$ is an interior point of R, then every U_ν cuts $|z| = m/2$; if $w=\infty$ is not an interior point of R, and if D_ν cuts B (or more generally if the diameter of D_ν is no less than the distance from 0 to B) we have $2\nu \geq m \geq \frac{1}{4}$, so every U_ν cuts $|z| = \frac{1}{8}$; a more favorable result is obtained by taking $2\nu \geq \frac{1}{4}$, for then it follows from (22) that every U_ν cuts $|z| = \frac{1}{6}$. Also a part of the corollary to Theorem 6 follows from (ii).

8. Circles not passing through origin. Thus far we have obtained results on circles of curvature at the origin $w=0$ or $z=0$ of the images of circles. These results can be applied in the study of circles of curvature of such images at points other than the origin, by the aid of a transformation of the region $|z|<1$ into itself. But of course the number a_2 that figures so prominently in our results must be adjusted to accord with the new configuration. In certain cases the number a_2 enters only implicitly by means of a standard inequality, in which case the new results may be of relatively simple form. Let us proceed to give several illustrations of this method.

LEMMA. *Let the circle $|z-r_1| = r_1 < \frac{1}{2}$ be the transform of the circle $|z| = r < 1$ under the transformation*

$$z_1 = \frac{z + r}{1 + rz} \tag{30}$$

which leaves the region $|z| < 1$ invariant. Then we have

$$r_1 = \frac{r}{1 + r^2}, \tag{31}$$

$$r = \frac{1 - \sqrt{1 - 4r_1^2}}{2r_1}. \tag{32}$$

The transformation (30) carries the points $z = -r$, 0, r into the respective points $z_1 = 0$, r, $2r_1$, where r_1 is given by (31); hence (30) carries the circle $|z| = r$ into the circle $|z - r_1| = r_1$. Equation (32) is merely the solution of (31) for r, with the restriction $r < 1$.

The value $r_1 = \frac{1}{4}$ corresponds by means of (32) to the value $r = 2 - \sqrt{3} = 0.26+$. Thus we have from Theorem 3 and its corollary:

THEOREM 8. *Under a smooth conformal map of the region $|z| < 1$ onto a region R of the extended plane with boundary B, no circle of curvature of the image of a circle $|z| = r < 2 - \sqrt{3}$ can cut B.*

Unless B is the arc of a circle through the image of $z = 0$, the requirement $r < 2 - \sqrt{3}$ can be replaced by $r \leqq 2 - \sqrt{3}$.

If B is the arc of a circle through the image of $z = 0$, one and only one circle of curvature of the image of the circle $|z| = 2 - \sqrt{3}$ cuts B, and that in but a single point of B.

A consequence of the first part of Theorem 8 is the well known fact that if R does not contain the point at infinity the image of every circle $|z| = r < 2 - \sqrt{3}$ is convex; for no circle of curvature can cut B, hence no circle of curvature can pass through the point at infinity, hence the image of $|z| = r$ can have no point of inflection.

Parts of Theorem 7 are independent of the conditions $f(0) = 0$, $f'(0) = 1$. We shall prove:

THEOREM 9. *Let the region $|z| < 1$ be mapped smoothly onto the region R with boundary B of the w-plane by the function $w = f(z)$, analytic for $|z| < 1$. If w_0 is a point of R, with $w_0 = f(z_0)$, $|z_0| < 1$, if D_r is a circle through w_0 whose diameter is at least half the greatest distance from w_0 to a point of B, if D_r is transformed into a curve B_r of the z-plane, and if U_r is the circle of curvature of B_r at z_0, then under a transformation of the form*

$$z_1 = \frac{z - \alpha}{1 - \bar{\alpha} z}$$

which carries U_r into a circle whose center is the origin $z_1 = 0$, U_r corresponds to a circle of radius at least $2 - \sqrt{3}$.

It follows from Theorem 7 that the radius of U_r is at least $\frac{1}{4}$, if we have $f(0) = w_0$, and the value $r_1 = \frac{1}{4}$ in the lemma corresponds to $r = 2 - \sqrt{3}$.

THEOREM 10. *Let the region $|z| < 1$ be mapped smoothly onto the region R with boundary B of the w-plane, by the function $w = f(z)$, analytic for $|z| < 1$. If w_0 is a point of R, with $w_0 = f(z_0)$, $|z_0| < 1$, if D_r is a circle through w_0 which cuts B (or more generally whose diameter is at least as great as the distance from w_0 to B),* *if the transform of D_r is the curve B_r of the z-plane, and if U_r is the circle of curvature of B_r at z_0, then under a transformation of the form*

$$z_1 = \frac{z - \alpha}{1 - \bar{\alpha} z}$$

which carries U_r into a circle whose center is the origin $z_1 = 0$, U_r corresponds to a circle of radius at least $6 - \sqrt{35}$.

It follows from Theorem 7, as we have already remarked, that the radius of U_r is at least 1/12, if we have $f(0) = w_0$; and the value $r_1 = 1/12$ in the lemma corresponds to $r = 6 - \sqrt{35}$.

9. Invariant formulation. Throughout the present paper we have studied the circles of curvature of the transforms of circles. It is of course true, as the reader may prove, that if two curves have the same circle of curvature at $z = 0$, then their images under the transformation (1) have the same circle of curvature at $w = 0$. This remark can be applied in connection with virtually all of the preceding results.

Thanks to this remark, such results as Theorems 8, 9, 10 are conveniently expressed in general form in terms of non-euclidean geometry, for this formulation is independent of the center of the circle in the z-plane. Thus Theorems 8, 9, 10 are the respective parts of

THEOREM 11. *Let R be an arbitrary simply-connected region of the w-plane whose boundary B has more than one point, and let R be provided with the usual non-euclidean metric by a conformal map onto the region $|z| < 1$ of the z-plane.*

If the non-euclidean curvature of a curve at a point of R is numerically greater than 4, the euclidean circle of curvature at that point cannot cut B.*

If $w = \infty$ is not interior to R, and if the euclidean circle of curvature of a curve at a point w_0 of R has as diameter at least half the greatest distance from w_0 to a point of B, then the non-euclidean curvature of the curve at w_0 is at most 4.

If $w = \infty$ is not interior to R, and if the euclidean circle of curvature of a curve at a point w_0 of R has a diameter at least as great as the distance from w_0 to B, then the non-euclidean curvature of the curve at w_0 is at most 12.

Various other results of the present paper are conveniently expressed in terms of non-euclidean geometry; for instance, it follows from Theorem 3 that under the conditions of Theorem 11 if the non-euclidean curvature of a curve at a point of R is zero, then the euclidean circle of curvature at that point must cut B. Further similar formulations should present no difficulty to the reader.

---

* Defined as in Carathéodory, Conformal Representation (Cambridge Tract No. 28), §44.

NOTE ON THE CURVATURE OF ORTHOGONAL TRAJECTORIES OF LEVEL CURVES OF GREEN'S FUNCTION. III

J. L. WALSH

If R is a simply connected region of the extended (x, y)-plane with boundary B, and if Green's function $G(x, y)$ exists for R with pole in the finite point O, we denote by $\{T\}$ the set of orthogonal trajectories to the level curves $G(x, y) = \log r$, $0 < r < 1$, in R. The totality of circles each osculating at O one of the set of curves T passing through O consists precisely of the set of circles through O and through another fixed point D, depending on O and R. The point D is called the *conjugate of O with respect to R*. The term "circle" is here and below used in the extended sense, to include straight line, unless otherwise noted.

In a series of papers[1] the writer has recently studied some of the properties of the point D, notably (in M and I) that every circle through O and D cuts B; and (in II) that every point exterior to R is the conjugate with respect to R of a suitably chosen point O interior to R. It is the object of the present note to establish still further properties of the conjugate, namely the following theorems:

THEOREM 1. *Let R be a simply connected region of the w-plane with at least two boundary points. Let C be a circle intersecting the boundary of R in the finite point α. Let C be the boundary of a circular region R' (a half-plane, interior of a circle, or exterior of a circle, boundary points not included) whose points lie in R, and let T be a triangle contained in R', with the vertex α. Let the sequence of points $w_1, w_2, \cdots$ lie in T and approach α. Then the conjugate of w_n with respect to R also approaches α as n becomes infinite.*

THEOREM 2. *Let R be a simply connected region of the w-plane with at least two boundary points, and let w_0 be a boundary point of R. Then there exists a sequence of points $w_1, w_2, \cdots$ in R approaching w_0 such that the conjugate of w_n with respect to R approaches w_0.*

THEOREM 3. *There exists a limited Jordan region R of the w-plane, a boundary point w_0 of R, and a sequence $w_1, w_2, \cdots$ of points of R approaching w_0 such that the conjugate of w_n with respect to R becomes infinite with n.*

---

[1] American Mathematical Monthly, vol. 42 (1935), pp. 1–17; Proceedings of the National Academy of Sciences, vol. 23 (1937), pp. 166–169; this Bulletin, vol. 44 (1938), pp. 520–523. We shall refer to these papers as M, I, II respectively.

Let $w=f(z)$ map $|z|<1$ onto R. Then the conjugate of w with respect to R is the point

$$w' = f(z) - \frac{2[f'(z)]^2(1-z\bar{z})}{f''(z)(1-z\bar{z}) - 2\bar{z}f'(z)};$$

this is the function whose continuity (as a function of w) we are studying.

1. **Proof of Lemma** 1. As a step in the proof of Theorem 1 we establish a preliminary result.

LEMMA 1. *Let the finite point O lie interior to the simply connected region R whose Green's function with respect to O exists, and let each point of the boundary B of R lie on or to the right of the vertical line Oy. Let at least one point A of B lie to the right of Oy. Then the conjugate D of O with respect to R is finite and lies to the right*[2] *of Oy.*

In the proof it is convenient to transform O to infinity by a linear transformation of the complex variable, so the boundary B of R can now be supposed finite. We preserve the original notation. The conjugate of O is a finite point D, the conformal center of gravity[3] of R. There exists a line L with (for suitable orientation of the plane) each point of the boundary B of R lying on or to the right of L, and with at least one point A of B lying to the right of L. For definiteness let A be the point (or one of the points) of B farthest from L.

If no boundary point of R lies on L, a suitably chosen level curve of Green's function $G(x, y)$ for R with pole in O lies to the right of L. The center of gravity of this level curve with positive mass distribution defined by the equation

$$d\sigma = \frac{\partial G}{\partial n} ds$$

is the conjugate of O with respect to R, and lies to the right of L. The conclusion of Lemma 1 follows in this case. Henceforth we suppose at least one point of B to lie on L.

An arbitrary cut in the region R (assumed to lie in the w-plane) corresponds under the conformal map $\zeta=\zeta(w)$ of R onto $|\zeta|>1$ to a cut in the region[4] $|\zeta|>1$. Let a circle K be drawn with center A and radius one-half the distance from A to L. A certain simply connected

[2] It follows (see II) that O cannot be a maximum or even a critical point of the function $r(\alpha)$.

[3] See M and I.

[4] See for instance Carathéodory, *Conformal Representation* (Cambridge Tract, no. 28), p. 83.

region S consisting of points of R and containing some points of R lying to the right of A, lies interior to K and is bounded entirely by points of B and of K. That is to say, we define S as the set of points P each of which can be joined to a fixed point H in K and R to the right of A, by a Jordan arc PH not cutting B or K. The boundary of S contains at least one arc Q of K whose end-points are points of B but which otherwise lies in R; all such arcs Q (by hypothesis belonging to the boundary of S) are finite or denumerably infinite in number. Under the conformal map each arc Q corresponds to a Jordan arc Q_1 in the region $|\zeta| > 1$ whose end-points are distinct points of $|\zeta| = 1$, whether or not the two end-points of Q are distinct.[5] Under the conformal map the region S corresponds to a region S_1 bounded in part by arcs Q_1. We now prove that *the boundary of S_1 must contain at least one arc of the circle $|\zeta| = 1$.*

Two distinct arcs Q_1 may have a terminal point T_1 in common, but if so the corresponding arcs Q have a terminal point T in common, and a suitably chosen neighborhood of T contains no boundary points of S interior to K. For there exists a sequence of Jordan arcs J_1 in S_1 with their end-points on the respective arcs Q_1; and the end-points of the J_1 approach T_1. Let a point M_1 of S_1, whose transform is the point M of S, be separated by each of these Jordan arcs plus the boundary of S_1 from the neighborhood of T_1 in S_1. The corresponding sequence of Jordan arcs J in S have their end-points on the respective arcs Q, and each arc J together with the boundary of S separates M from the neighborhood of at least one boundary point of S. If each of the Jordan arcs J together with the boundary of S separates a neighborhood in S of more than one boundary point from M, let E denote the totality of such boundary points. No point of E not an end-point of an arc Q can be a limit point of a sequence of boundary points of S not lying on E; for two nonintersecting arcs J together with the two subarcs of arcs Q intercepted between their terminal points bound a Jordan region in S which contains then no point of the boundary of S in its interior; of course S is simply connected, and no boundary point of S lies exterior to K. If E contains more than one point, it contains either a point interior to K or a point on K not an end-point of arcs Q (hence lying on an arc of K belonging to B), and in any case contains a point N with the property that the mapping function $\zeta = \zeta(w)$ for R is continuous in some neighborhood of N in S and constant (equal to the value of ζ in the point T_1) at every boundary point of S in the neighborhood, which is impossible.[6] It follows that E consists of a

[5] Carathéodory, op. cit., p. 85.
[6] Carathéodory, op. cit., p. 82.

single point T; a suitably chosen neighborhood of T contains no boundary point of S interior to K.

The method of proof just given shows also that a point T_1 of $|\zeta|=1$, not an end-point of an arc of $|\zeta|=1$ which is part of the boundary of S_1 but which is a limit point of an infinity of Jordan arcs Q_1, corresponds in the sense just considered to a single point T on B and K; a suitably chosen neighborhood of T contains no boundary point of S interior to K. The point A is not an interior point of S, so it follows that there exists some point of K which is a limit point of boundary points of S interior to K, whence not all points of B on K can be points T as considered above, and the boundary of S_1 must contain at least one arc of the circle $|\zeta|=1$, say of length $\sigma>0$, as we desired to prove.

Denote by d the distance from A to L. Choose r, $(0<r<1)$, so near unity that an arc of the Jordan curve B_r: $G(x, y)=\log r$ interior to K and S corresponds to an arc of $|\zeta|=1/r$ in S_1 of length greater than $\sigma/2r$, which is possible by inspection of the situation in the ζ-plane; and also choose r so near to unity that no point of the curve B_r lies at a greater distance from B than some positive number

$$d' < \sigma d/(8\pi - 2\sigma).$$

Then no point of the curve B_r lies at a distance greater than d' to the left of L. The weight of that part of B_r interior to K is at least $\sigma/2$. The center of gravity of the weighted locus B_r lies to the right of L at a distance not less than

$$\left\{\frac{\sigma}{2}\frac{d}{2}-\left(2\pi-\frac{\sigma}{2}\right)d'\right\}\Big/2\pi,$$

which is positive. That is to say, the center of gravity of the weighted locus B_r (that is, the conformal center of gravity of R, or the conjugate of O with respect to R) lies to the right of L; so Lemma 1 follows.

2. **Proof of Theorem** 1. For convenience suppose the plane oriented so that α lies at the right-hand extremity of a horizontal diameter of C. Let R_n be the region of the w-plane obtained from R by stretching without rotation so that final and initial lengths are in the ratio $1:r(w_n)$, with the new transform of w_n corresponding to the origin. Here (as in II) we use $r(w_n)$ to denote the *inner radius* of R with respect to the point w_n. From each sequence R_n can be extracted a subsequence of regions converging to a kernel in the sense of Carathéodory. For the corresponding functions $w=f_n(z)$ which map $|z|<1$ onto R_n are univalent with $f_n(0)=0$ and $|f_n'(0)|=1$, hence

form a normal family in $|z|<1$. Let us suppose a subsequence of the R_n to converge to the kernel R_0, with the corresponding mapping functions $w=f_n(z)$ converging continuously in $|z|<1$ to the mapping function $w=f_0(z)$ for R_0, with $f_0(0)=0$, $|f_0'(0)|=1$. The kernel R_0 is obviously not a single point. Then we have for the subsequence

$$\lim_{n\to\infty} -2\frac{[f_n'(0)]^2}{f_n''(0)} = -2\frac{[f_0'(0)]^2}{f_0''(0)}, \tag{1}$$

or in other words (compare I) the conjugate of the origin with respect to R_n approaches the conjugate of the origin with respect to R_0.

We shall henceforth assume that the sequence R_n itself converges to R_0 and that (1) is valid. This assumption involves no loss of generality, for if the conclusion of the theorem is false it is false for a sequence of indices such that R_n converges to a kernel and such that (1) is valid for that sequence.

We shall now assume that the angle (argument, amplitude) $\angle(\alpha-w_n)$ approaches a limit γ as n becomes infinite; this assumption likewise involves no loss of generality, for if the theorem is false there exists a subsequence of the w_n for which this angle approaches a limit and whose conjugates do not approach α. Of course γ is not the angle $\pi/2$ or $-\pi/2$.

Denote by t_n the boundary point of R nearest to w_n; then t_n lies on or exterior to C:

$$|w_n - t_n| \geqq a - |w_n|, \tag{2}$$

where C is the circle $|w|=a$; here we assume also that C is a proper circle in whose interior T lies. This assumption involves no loss of generality. The inequality

$$|w_n - t_n| \leqq r(w_n) \tag{3}$$

is well known.[7] Moreover for suitably chosen $\delta>0$ independent of n we have

$$\delta|w_n - \alpha| \leqq a - |w_n| \tag{4}$$

by virtue of the fact that w_n lies in the triangle T. We obviously have, by (2), (3), and (4),

$$|w_n - \alpha|/r(w_n) \leqq 1/\delta.$$

There is no boundary point of R_n or of R_0 closer to O than the dis-

[7] See for instance Pólya-Szegö, *Aufgaben und Lehrsätze aus der Analysis*, vol. 2, Berlin, 1925, p. 21, exercise 121.

tance 1/4, by the distortion theorem (Verzerrungssatz). But by the inequality just established there is a boundary point of R_n (namely the transform of the point α) on the line segment from O in the direction $\angle(\alpha - w_n)$ of length $1/\delta$, so there is a boundary point of the kernel R_0 on the line segment from O in the direction γ of length $1/\delta$. Consequently R_0 has at least one boundary point which lies *to the right* of the vertical line through O.

It follows from Lemma 1 that the conjugate of O with respect to R_0 is not infinite, follows from (1) that the conjugate of O with respect to R_n approaches a finite limit, and follows from the relation $r(w_n) \to 0$ (proved in II) that the distance from w_n to its conjugate with respect to R (which is $r(w_n)$ multiplied by the distance from O to its conjugate with respect to R_n) approaches zero. Theorem 1 is established.

It is clear from the proof of Theorem 1 that the essential part of the proof is that every kernel R_0 of a subsequence of the regions R_n should have the property that the conjugate of the origin with respect to R_0 shall be finite; whenever the sequence R_n has this property, the condition $w_n \to \alpha$ implies that the conjugate of w_n with respect to R also approaches α.

3. **Proof of Theorem** 2. Thanks to Theorem 1, the proof of Theorem 2 is extremely simple. For definiteness suppose w_0 finite (the contrary case can be reduced to this by a linear transformation). Let ω_k be a point of R whose distance from w_0 is less than $1/2^k$. Let γ_k be the circle whose center is ω_k and whose radius is the distance from ω_k to the boundary of R; this distance is necessarily less than $1/2^k$. The interior of this circle lies in R, but at least one point α of the circumference is a boundary point of R. A triangle T satisfying the requirements of Theorem 1 can be constructed. By Theorem 1 there exists a point w_k interior to γ_k whose conjugate with respect to R lies in the circle $|w - w_0| = 1/2^{k-1}$, and we have $|w_k - w_0| \leq 1/2^{k-1}$. The sequence w_k satisfies the requirements of Theorem 2.

4. **Proof of Theorem** 3. The region R whose existence is asserted in Theorem 3 is now to be constructed by the following method.[8] We consider the sequence of circles C_n in the $w(=x+iy)$-plane, each tangent to its predecessor:

$$C_0:\quad x^2 + y^2 = 1, \qquad C_1:\quad (x - 3/2)^2 + y^2 = 1/4,$$
$$C_n:\quad (x - 3(2^n - 1)/2^n)^2 + y^2 = 1/2^{2n}.$$

[8] This method is quite similar to one employed for a somewhat different purpose in a forthcoming paper by Seidel and Walsh, of which an abstract was published in Proceedings of the National Academy of Sciences, vol. 24 (1938), pp. 337–340.

The interiors of these circles are to be joined by canals in the neighborhoods of the points

$$1,\ 2,\ 5/2,\ \cdots,\ \alpha_n: 3 - 1/2^{n-1},\ \cdots$$

so as to form a Jordan region; the banks of the canals are short segments of lines parallel to the axes of reals, a short distance above and below that axis near the points α_n of tangency of successive circles; the arcs of the original circles intercepted between those lines are to be suppressed, and the banks of the canals are to be terminated by successive circumferences. We proceed to indicate in more detail the construction of the canals.

Let the canal in the neighborhood of the point α_n be bounded by the lines $y = \pm\delta_n$, with $\delta_n > 0$. Let R be the region formed by the interior of the circles C_n together with the canals, and let the point $w = 0$ correspond to the point $\zeta = 0$ when R is mapped onto $|\zeta| < 1$ with directions at the origins unaltered. When δ_1 approaches zero, the *kernel* of the variable region R in the sense of Carathéodory (op. cit., p. 75) is precisely the interior of C_0, *independently of the values of* $\delta_2, \delta_3, \cdots$. The function $w = f_0(\zeta)$ which maps $|\zeta| < 1$ onto R with $f_0(0) = 0$, $f_0'(0) > 0$, approaches the function $w = \phi_0(\zeta) \equiv \zeta$ which maps $|\zeta| < 1$ onto R with $\phi_0(0) = 0$, $\phi_0'(0) > 0$; convergence is uniform in every $|\zeta| \leqq r < 1$. The conjugate of the origin with respect to R is $-2[f_0'(0)]^2/f_0''(0)$, which approaches the conjugate of the origin with respect to the interior of C_0, namely the point at infinity. Consequently it is possible to choose δ_1 so small independently of $\delta_2, \delta_3, \cdots$ that the conjugate of the point $w_0 = 0$ with respect to R lies exterior to the circle $|w| = 1$.

Introduce the notation for the center of C_n: $w_n = 3(2^n - 1)/2^n$. When δ_1 and δ_2 approach zero, the kernel of the variable region R is the interior of C_1 independently of $\delta_3, \delta_4, \cdots$, if the region $|\zeta| < 1$ is mapped onto R by the function $w = f_1(\zeta)$ with $f_1(0) = w_1$, $f_1'(0) > 0$. Then the function $w = f_1(\zeta)$ approaches the function $w = \phi_1(\zeta) \equiv w_1 + \zeta/2$ which maps $|\zeta| < 1$ onto the interior of C_2 with $\phi_1(0) = w_1$, $\phi_1'(0) > 0$; convergence is uniform in every $|\zeta| \leqq r < 1$. The conjugate of w_1 with respect to R is $-2[f_1'(0)]^2/f_1''(0)$, which approaches the conjugate of w_1 with respect to the interior of C_1, namely the point at infinity. Consequently it is possible to choose δ_1 and δ_2 so small independently of $\delta_3, \delta_4, \cdots$ that the conjugate of the point w_1 with respect to R lies exterior to the circle $|w| = 2$. The number δ_1 has been subjected to a similar restriction in connection with the conjugate of the point w_0, and is to be subjected to no further condition.

We continue now in the way thus commenced. The numbers δ_n and

δ_{n+1} are to be chosen independently of $\delta_{n+2}, \delta_{n+3}, \cdots$ in such manner that the conjugate of the point w_n with respect to R lies exterior to the circle $|w| = 2^{n+1}$; each δ_n (for $n > 1$) is subjected then to two conditions, and the numbers δ_n can be determined in succession. The resulting region R is a Jordan region. The sequence w_n approaches the boundary point $w = 3$ of R, and the conjugate of w_n with respect to R becomes infinite with n. Theorem 3 is established.

Harvard University

ON THE SHAPE OF THE LEVEL LOCI OF HARMONIC MEASURE*

By

J. L. WALSH** AND W. J. SCHNEIDER***

in College Park, Md., U.S.A. *in Syracuse, N.Y., U.S.A.*

In the study of the shape of level loci of Green's function of a region in the w-plane, a measure of the global nearness of the shape of a Jordan curve Γ with origin 0 interior to Γ to the shape of a circle with center 0, may be called [1] the *circularity* of Γ:

$$\kappa(\Gamma) = \frac{\min|w| \text{ on } \Gamma}{\max|w| \text{ on } \Gamma} \leqq 1. \tag{1}$$

We have $\kappa(\Gamma) = 1$ when and only when Γ is a circle with center 0.

The object of the present note is to study this and other measures of the shape of level loci of harmonic measure, for an annular region, a region of higher connectivity, an infinite strip, an infinite sector, and a wedge. These results are relatively easy to prove, but elegant and striking in application.

§1. Circularity for simply and doubly connected domains. We have (loc. cit.)

Theorem 1.1. *Let D be a bounded Jordan region of the w-plane containing the origin 0, whose boundary is Γ_1. Let Γ_λ denote the image in the w-plane of the circle $C_\lambda: |z| = \lambda \leqq 1$ when the interior of $C_1: |z| = 1$ is mapped onto the interior of Γ_1 so that $z = 0$ corresponds to $w = 0$. Then $\kappa(\Gamma_\lambda)$ varies monotonically with λ, and approaches unity when $\lambda \to 0$.*

* Abstract published in Amer. Math. Soc. Notices, vol. 16 (1969), p. 569.

** Research sponsored by the Air Force Office of Scientific Research, Office of Aerospace Research, United States Air Force, under AFOSR Grant No. 1130-66.

*** This research was partially supported by the National Science Foundation, Grant G.P. 8787.

Let the mapping function be $z = \phi(w)$, $\phi(0) = 0$, and let m_λ and M_λ denote respectively $\min|w|$ and $\max|w|$ for w on C_λ. On the curve Γ_1 and hence throughout D we have

$$\left|\frac{\phi(w)}{w}\right| \leqq \frac{1}{m_1}, \qquad \left|\frac{w}{\phi(w)}\right| \leqq M_1, \tag{1.1}$$

and on Γ_λ

$$\left|\frac{\phi(w)}{w}\right| \leqq \frac{\lambda}{m_\lambda} \leqq \frac{1}{m_1}, \qquad \left|\frac{w}{\phi(w)}\right| \leqq \frac{M_\lambda}{\lambda} \leqq M_1, \tag{1.2}$$

$$1 \leqq \frac{M_\lambda}{m_\lambda} = \frac{1}{\kappa(\Gamma_\lambda)} \leqq \frac{M_1}{m_1} = \frac{1}{\kappa(\Gamma_1)}. \tag{1.3}$$

This shows $\kappa(\Gamma_\lambda) \geqq \kappa(\Gamma_1)$. There follows the monotonic increasing property of $\kappa(\Gamma_\lambda)$ as λ decreases; for we may replace Γ_1 by Γ_{λ_1} and Γ_λ by Γ_λ, $0 < \lambda < \lambda_1 < 1$. Equality throughout (1.3) occurs when and only when Γ_1 is a circle with center 0.

If we set (as we may) $\phi(w) = aw + a_1 w^2 + \cdots$, $a \neq 0$, we have $\phi(w)/w = a + a_1 w \cdots$, so m_λ and M_λ approach a as λ approaches zero, whence $\kappa(\Gamma_\lambda) \to 1$ as λ approaches zero.

Theorem 1.1 admits an extension to a region of higher connectivity:

Theorem 1.2. *Let D be a bounded annular region of the w-plane whose boundary consists of two Jordan curves Γ_1 and Γ_ρ, with Γ_ρ interior to Γ_1 and containing the origin $w = 0$. Suppose D is mapped one-to-one and conformally onto the annulus $0 < \rho < |z| < 1$ by $z = \phi(w)$ (the number ρ is uniquely determined) and let m_λ and M_λ denote respectively* $\min|w|$ *and* $\max|w|$ *for w on the image Γ_λ of $|z| = \lambda$, $\rho \leqq \lambda \leqq 1$. Then we have for $\rho < \lambda < 1$*

$$\kappa(\Gamma_\lambda) \geqq \min[\kappa(\Gamma_\rho), \kappa(\Gamma_1)], \tag{1.4}$$

and the logarithm of $\kappa(\Gamma_\lambda)$ is a concave function of $\log \lambda$.

We apply the Hadamard Three-Circle Theorem [5, p. 126] in the annulus $\rho < |z| < 1$, in the z-plane, $z = \phi(w)$. We have then $(\rho < \lambda < 1)$

$$\log M_\lambda \leqq \frac{\log\lambda}{\log\rho}\log M_\rho + \frac{\log\rho - \log\lambda}{\log\rho}\log M_1,$$

$$\log m_\lambda \geqq \frac{\log\lambda}{\log\rho}\log m_\rho + \frac{\log\rho - \log\lambda}{\log\rho}\log m_1,$$

$$\log\kappa(\Gamma_\lambda) = \log m_\lambda - \log M_\lambda$$

$$\geqq \log\left(\frac{m_\rho}{M_\rho}\right)^\mu + \log\left(\frac{m_1}{M_1}\right)^{1-\mu}, \quad \mu = \frac{\log\lambda}{\log\rho},$$

$$\kappa(\Gamma_\lambda) \geqq \left(\frac{m_\rho}{M_\rho}\right)^\mu \cdot \left(\frac{m_1}{M_1}\right)^{1-\mu}. \tag{1.5}$$

The conclusion of Theorem 1.2 follows at once from (1.5).

An interesting special case of Theorem 1.2 is

Theorem 1.3. *Under the conditions of Theorem 1.2, suppose Γ_ρ is the circle $|w| = \rho$, namely $\kappa(\Gamma_\rho) = 1$. Then for arbitrary λ, $\rho \leqq \lambda \leqq 1$, we have $\kappa(\Gamma_\rho) = 1 \geqq \kappa(\Gamma_\lambda) \geqq \kappa(\Gamma_1)$. The relation $\kappa(\Gamma_\lambda) = \kappa(\Gamma_1)$ occurs when and only when Γ_1 is a circle with center 0, necessarily the circle $|w| = 1$.*

Of course the region D can be mapped onto an annulus if and only if the ratio of the radii of the two bounding circles of the annulus is $1:\rho$.

Another theorem is somewhat similar to Theorem 1.2, as we now indicate.

Theorem 1.4. *Let D be a bounded annular region of the w-plane whose boundary consists of two Jordan curves Γ_1 and Γ_ρ, with Γ_ρ interior to Γ_1 and containing the origin. Suppose D is mapped conformally onto the annulus $0 < \rho < |z| < 1$ by $z = \phi(w)$, and let m_λ and M_λ denote respectively $\min|w|$ and $\max|w|$ for w on the image Γ_λ of $|z| = \lambda$, $\rho \leqq \lambda \leqq 1$. Suppose the function $|\phi(w)/w|$ takes its maximum and minimum in the closure of D on Γ_1. Then for arbitrary λ, $\rho \leqq \lambda \leqq 1$, we have $\kappa(\Gamma_\lambda) \geqq \kappa(\Gamma_1)$.*

We write here

$$\max_{\Gamma_1}\left|\frac{\phi(w)}{w}\right| = \frac{1}{m_1}, \qquad \min_{\Gamma_1}\left|\frac{\phi(w)}{w}\right| = \frac{1}{M_1},$$

$$\max_{\Gamma_\rho}\left|\frac{\phi(w)}{w}\right| = \frac{\rho}{m_\rho}, \qquad \min_{\Gamma_\rho}\left|\frac{\phi(w)}{w}\right| = \frac{\rho}{M_\rho}.$$

The hypothesis of Theorem 1.4 then yields

$$\frac{1}{m_1} \geqq \frac{\rho}{m_\rho}, \qquad \frac{1}{M_1} \leqq \frac{\rho}{M_\rho},$$

whence

$$\frac{m_1}{M_1} \leqq \frac{m_\rho}{M_\rho}, \qquad \kappa(\Gamma_1) \leqq \kappa(\Gamma_\rho),$$

and the conclusion follows from Theorem 1.2.

Theorem 1.4 can also be proved directly by the method of proof of Theorem 1.1.

In connection with Theorems 1.2–1.4, it is of interest to note that generically $\kappa(\Gamma)$ is unchanged by an inversion with center 0.

Theorem 1.1 deals with the shape of the level loci of Green's function for D with pole in 0. Theorems 1.2–1.4 deal with the shape of the level loci of the harmonic measure $\omega(z;\Gamma_1,D)$, or of $1-\omega(z;\Gamma_1,D) = \omega(z;\Gamma_\rho,D)$. In Theorems 1.2–1.4 the origin $w = 0$ may be chosen arbitrarily interior to Γ_ρ.

§2. Circularity for regions of higher connectivity.

Theorems 1.2 and 1.3 extend to the case of a multiply connected region D of the w-plane bounded by a finite number (> 2) of mutually disjoint Jordan curves that are divided into two disjoint sets Γ_1 and Γ_ρ, $0 < \rho < 1$. Let $h(w)$ be a function harmonic in D, continuous in the closure of D, equal to zero and $\log \rho$ (< 0) on Γ_1 and Γ_ρ respectively, and let $k(w)$ be the (possibly multiple-valued) conjugate to $h(w)$ in D. Then $z = \phi(w) = \exp[h(w) + ik(w)]$, maps D conformally but not simply onto the annulus $\rho < |z| < 1$, and $|z| = \exp[h(w)]$ is single valued in D. Let Γ_λ, $\rho < \lambda < 1$, denote generically the locus $|\phi(w)| = \lambda$ in D, the image of $|z| = \lambda$, and suppose the function $|\phi(w)/w|$ assumes on Γ_1 both its maximum and minimum values in the closure of D. Then for arbitrary λ, $\rho \leqq \lambda \leqq 1$, we have $\kappa(\Gamma_\lambda) \geqq \kappa(\Gamma_1)$. The proof is as before, and the reasoning can be reapplied (if the hypothesis is satisfied) to Γ_λ and Γ_{λ_1}, $\rho \leqq \lambda \leqq \lambda_1 \leqq 1$. Compare here [1] and Green's functions. The function $\kappa(\Gamma_\lambda)$ is concave in $\log \lambda$.

We now show by means of an example that in an annular region *the circularity $\kappa(\Gamma_\lambda)$ need not always vary monotonically with λ.* Let the annulus $\frac{1}{2} < |z| < 1$ be mapped onto an annular region D of the w-plane bounded by an ellipse $\Gamma_{\frac{1}{2}}$ and the unit circle Γ_1, where the ellipse has center $w = 0$ and is not a circle. Invert D in the unit circle in the w-plane, so that D is transformed into some region D'. The annular region $D + \Gamma_1 + D'$ is the image of the annulus $\frac{1}{2} < |z| < 2$ under the previous map (extended) of $\frac{1}{2} < |z| < 1$. We have $\kappa(\Gamma_{\frac{1}{2}}) = \kappa(\Gamma_2) < 1$, $\kappa(\Gamma_1) = 1$. Indeed the circularity $\kappa(\Gamma_\lambda)$, $\frac{1}{2} \leqq |z| \leqq 2$ considered as a function of λ has a maximum for $\lambda = 1$.

Remark. *There exists an annular region D in the w-plane bounded by analytic Jordan curves Γ_1 and Γ_ρ such that the level loci Γ_λ of the harmonic measure of Γ_ρ with respect to D have a maximum circularity $\kappa(\Gamma_\lambda)$ in D for which that maximum is not unity.* We choose Γ_1 and Γ_ρ as disjoint similar ellipses having the same center 0 and same orientation of axes, yet so that no circle with center 0 separates them. It follows from the concavity of $\kappa(\Gamma_\lambda)$ as a function of $\log\lambda$ that for all λ, $\rho < \lambda < 1$, we have $\kappa(\Gamma_\lambda) \geqq \kappa(\Gamma_1) = \kappa(\Gamma_\rho)$, so for some λ, $\rho < \lambda < 1$, $\kappa(\Gamma_\lambda)$ is a maximum but $\kappa(\Gamma_\lambda) < 1$.

§3. **Ellipticity.** Both Theorems 1.2 and 1.3 can be generalized by a conformal map so as to apply to an arbitrary doubly connected region D in the w-plane, and to the shape of the corresponding level loci of the harmonic measure of one of the bounding continua with respect to the region, compared with the shape of that bounding continuum itself as standard. If D is the extended plane minus two disjoint continua Γ_1 and Γ_2, then we map the complement of Γ_2 onto the exterior D' of the circle $C: |z| = 1$ in the z-plane. We then compare by the methods already developed, the shape of the curves in the z-plane (transforms of level loci in the w-plane) with the shape of the level loci (i.e., circles) of Green's function for D' whose pole lies at infinity. This comparison of shapes can be interpreted in the w-plane. Rather than formulate a general result (which is left to the reader) we formulate explicitly an interesting special case.

Denote by S the segment $-1 \leqq u \leqq 1$ of the $w(= u + iv)$-plane, and let

Γ be an arbitrary Jordan curve containing S in its interior. We compare the shape of Γ with that of the family of ellipses Γ_ρ whose common foci are $+1$ and -1, and define the *ellipticity* of Γ (nearness of the shape of Γ to that of the ellipses E_ρ) as

$$E(\Gamma) = \frac{\begin{array}{c}\text{sum of semi-axes of largest ellipse } E_\rho\\ \text{having no point exterior to } \Gamma\end{array}}{\begin{array}{c}\text{sum of semi-axis smallest ellipse } E_\rho\\ \text{containing } \Gamma \text{ in the closure of its interior}\end{array}}. \tag{3.1}$$

This quantity is less than unity unless Γ is an ellipse of the family E_ρ. The family E_ρ is the set of ellipses $u = \frac{1}{2}(\rho + \rho^{-1})\cos\theta$, $v = \frac{1}{2}(\rho - \rho^{-1})\sin\theta$, $\rho > 1$, and the transformation

$$w = \tfrac{1}{2}(z + z^{-1}) \tag{3.2}$$

maps the w-plane onto $|z| \geqq 1$, where $z = \rho e^{i\theta}$. This transformation carries the ellipse E_ρ just mentioned (whose foci are $+1$ and -1 and whose semi-axes are $\frac{1}{2}(\rho + \rho^{-1})$ and $\frac{1}{2}(\rho - \rho^{-1})$) into the circumference $|z| = \rho$; the sum of the semi-axes is ρ.

Another form of (3.1) is ($z = z(w)$, $|z| > 1$)

$$E(\Gamma) = \frac{\min |z| \text{ for } w \text{ on } \Gamma}{\max |z| \text{ for } w \text{ on } \Gamma} = \kappa(\text{image of } \Gamma \text{ in } z\text{-plane}),$$

a form that emphasizes the analogy (indeed the identity under the transformation (3.2)) between ellipticity in the w-plane and circularity with respect to $z = 0$ in the z-plane.

If we define $E(\Gamma)$ generically by (3.1), we have by Theorem 1.3

Theorem 3.1. *Let the annular region D of the w-plane be bounded by $S: -1 \leqq w \leqq 1$ and by a Jordan curve Γ containing S in its interior. Then each level locus L of the harmonic measure of S with respect to D is a Jordan curve whose ellipticity $E(L)$ is greater than or equal to that of Γ:*

$$E(L) \geqq E(\Gamma); \tag{3.3}$$

$E(L)$ varies monotonically with L, and approaches unity as L approaches S. The equality sign holds in (3.3) for $L \neq \Gamma$ when and only when Γ is an ellipse of the family E_ρ.

The limiting case of Theorem 3.1 as Γ becomes infinite is itself a limiting case of the result of inverting the configuration of Theorem 1.1.

§4. **Another measure of circularity.** In considering the shape of a Jordan curve and the nearness of the shape to that of a circle, one may emphasize global dimensions as in Theorems 1.1, 1.2, and 1.3, or one may emphasize local infinitesimal properties. This contrast occurs for instance in comparing the shape of an ellipse which is smooth but whose eccentricity is large, with the shape of a gear wheel with small and sharp but numerous teeth. The latter shape can be described in terms of the angle ψ now to be mentioned, first studied in this connection by H. Grunsky [4], by P. Davis and H. Pollak [3], and later by Walsh [2].

If Γ is a smooth Jordan curve in the w-plane containing the origin in its interior, the angle ψ is defined as the angle at w measured from the radius vector extended through w, to the directed tangent to Γ at w in the counterclockwise sense on Γ, and ψ is to vary continuously with w. In particular if Γ_r, $0 < r < 1$, is the image in the w-plane of the circle $|z| = r$ in the z-plane, when the interior of Γ is mapped onto $|z| < 1$ so that $w = 0$ and $z = 0$ correspond to each other, then ψ is harmonic in w and z, and we have

$$[\min \psi, w \text{ on } \Gamma_1] \leqq [\min \psi, w \text{ on } \Gamma_r] \leqq [\max \psi, w \text{ on } \Gamma_r] \leqq [\max \psi, w \text{ on } \Gamma_1].$$

If Γ_1 is star-shaped with respect to 0, so also is Γ_r. Thus $[\min \psi, w \text{ on } \Gamma_r]$ and $[\max \psi, w \text{ on } \Gamma_r]$ give [2] a measure of the nearness of the shape of Γ_r to that of a circle, a measure that varies monotonically and approaches $\pi/2$ as r decreases and approaches zero. We now extend this property to level loci of harmonic measure.

Theorem 4.1. *Let D be a doubly connected region of the w-plane whose boundary consists of two smooth Jordan curves Γ_1 and Γ_ρ, with Γ_ρ interior to Γ_1 and $w = 0$ interior to Γ_ρ. Let D be mapped by $w = f(z)$, $z = \phi(w)$*

onto the annulus $\rho < |z| < 1$ *so that the curve* $|z| = r, \rho \leqq r \leqq 1$, *is carried into a curve* Γ_r. *Then we have*

$$(4.1) \qquad \min[\psi, w \text{ on } \Gamma_1 + \Gamma_\rho] \leqq \min[\psi, w \text{ on } \Gamma_r]$$

$$\leqq \max[\psi, w \text{ on } \Gamma_r] \leqq \max[\psi, w \text{ on } \Gamma_1 + \Gamma_\rho].$$

These inequalities are strong unless both Γ_1 *and* Γ_ρ *are circles with center* 0.

It is especially remarkable here that the position of $w = 0$ interior to Γ_ρ is completely arbitrary. Of course Γ_r is a level locus of the harmonic measure of Γ_1 with respect to D.

We have now $z = re^{i\theta}$ on $|z| = r$, $dz = iz d\theta$, $dw = f'(z)dz = izf'(z)d\theta$, and for w on Γ_r

$$\psi = \arg dw - \arg w = \arg \frac{izf'(z)d\theta}{f(z)} = \frac{\pi}{2} + \operatorname{Im}\left(\log \frac{zf'(z)}{f(z)}\right).$$

The function $f(z)$ is single valued and analytic in $\rho < |z| < 1$, and $f'(z)$ is continuous in the closure of D. Also, since $z \neq 0$, $f(z) \neq 0$, $\log(zf'(z)/f(z))$ is locally harmonic in D. When z traces the circle $|z| = r$, $\arg z$ increases by 2π, $\arg f(z)$ increases by 2π, and $f'(z) \neq 0$, so $\arg(zf'(z)/f(z))$ does not change. Thus $\log(zf'(z)/f(z))$ is single valued and continuous for $\rho \leqq r \leqq 1$, harmonic for $\rho < r < 1$. The maximum of ψ occurs on the boundary of D, as does $\min \psi$, so (4.1) follows. Equality holds if and only if $\psi \equiv \pi/2$.

On each curve Γ_r we have $\min \psi \leqq \pi/2 \leqq \max \psi$, so if Γ_1 or Γ_ρ is a circle whose center is 0 (but not both) we may omit that curve in the inequality (4.1); in that case, both $\min \psi$ and $\max \psi$ vary monotonically on Γ_r as r varies. As measures of ψ in the large, one can use $\min \psi$, $\max \psi$, or indeed the difference $\Delta = \max \psi - \min \psi$.

§5. **Infinite strips.** In this section we shall be concerned with infinite strip domains in the w-plane of the form $D = \{w = u + iv \mid -\infty < u < +\infty, 0 \leqq g(u) < v < f(u)\}$ where $f(u)$ and $g(u)$ are given continuous functions. We shall denote by $\omega(w)$ the harmonic measure of $\{w \mid -\infty < u < +\infty, v = f(u)\}$ with respect to the domain D. For the level curves $\omega(w) = \lambda$,

$0 \leqq \lambda \leqq 1$, a natural analog of the concept of *circularity* is the *flatness* of the λ curve which we define as

$$F_\lambda = \frac{\inf\limits_{\omega(w)=\lambda} \operatorname{Im} w}{\sup\limits_{\omega(w)=\lambda} \operatorname{Im} w} = \frac{I_\lambda}{S_\lambda} \quad (S_\lambda \neq 0).$$

We note that flatness depends on the position of the axis $v = 0$.

The following theorem concerning the flatness of the λ-curves of an infinite strip domain is analogous to our Theorem 1.2 concerning the circularity of the λ-curves in an annular domain.

Theorem 5.1. *Let D be an infinite strip domain bounded by the continuous curves $v = f(u)$ and $v = g(u)$ $(<f(u))$. Further suppose that S_f $(=\sup\limits_{-\infty<u<\infty} f(u))$ and S_g $(=\sup\limits_{-\infty<u<\infty} g(u))$ are finite and that I_g $(=\inf\limits_{-\infty<u<\infty} g(u))$ is positive. Under these conditions if $F_f \leqq F_g$ then $F_\lambda \geqq F_f$.*

Let $w = c(z)$ be a conformal map from the domain

$$\tilde{D} = \{z = x + iy \mid -\infty < x < +\infty,\ 0 < y < 1\}$$

onto the domain D which takes the upper boundary of $\tilde{D}$ onto the upper boundary of D and the lower boundary of $\tilde{D}$ onto the lower boundary of D. Let $h(z)$ be the harmonic function which solves the Dirichlet problem on $\tilde{D}$ with the constant boundary values S_f on $y = 1$ and S_g on $y = 0$. Since on $y = 1$ we have $\operatorname{Im} c(z) \leqq S_f$ and on $y = 0$ we have $\operatorname{Im} c(z) \leqq S_g$ we have by the maximum principle that $\operatorname{Im} c(z) \leqq h(z)$ throughout $\tilde{D}$. This inequality combined with the facts i) $c(z)$ takes the line $y = \lambda$ in the z-plane onto the curve $\omega = \lambda$ in the w-plane and ii) on the line $y = \lambda$ we have $h(z) = \lambda S_f + (1-\lambda)S_g$, leads us to the inequality

$$S_\lambda \leqq \lambda S_f + (1-\lambda)S_g. \tag{5.1}$$

By a similar minorization argument we are led to

$$I_\lambda \geqq \lambda I_f + (1-\lambda)I_g. \tag{5.2}$$

Therefore we have

$$F_\lambda = \frac{I_\lambda}{S_\lambda} \geqq \frac{\lambda I_f + (1-\lambda)I_g}{\lambda S_f + (1-\lambda)S_g}$$

$$= \frac{\lambda\left(\frac{I_f}{I_g}\right) + (1-\lambda)}{\lambda\left(\frac{S_f}{I_g}\right) + (1-\lambda)\frac{S_g}{I_g}} \geqq \frac{\left(\lambda\frac{I_f}{I_g}\right) + (1-\lambda)}{\left(\lambda\frac{I_f}{I_g}\right)\frac{S_f}{I_f} + (1-\lambda)\frac{S_f}{I_f}} = F_f.$$

We note that the following theorem is implicit in the proof of Theorem 5.1:

Theorem 5.2. *Let $h(z)$ be a function positive and harmonic on $\tilde{D} = \{z = x + iy \mid -\infty < x < \infty,\ 0 < y < 1\}$ and continuous on the closure of D. Let $\tilde{S}_\lambda = \sup_{-\infty<x<\infty} h(x + i\lambda)$, $\tilde{I}_\lambda = \inf_{-\infty<x<\infty} h(x + i\lambda)$ (with $I_\lambda > 0$) and and $\Delta_\lambda = I_\lambda/S_\lambda$.*

Under these conditions if $\Delta_1 \leqq \Delta_0$ then $\Delta_\lambda \geqq \Delta_1$.

We shall have cause to mention later this important ratio Δ_λ, and shall refer to it as the *min-max ratio.*

Theorem 5.1 is the analog of Theorem 1.2. We proceed to prove the analog of Theorem 1.4; here $w = c(z)$ maps D onto $\rho < y < 1$ in the z-plane with $w = \pm\infty$ corresponding to $z = \pm\infty$:

Theorem 5.3. *Let D satisfy the conditions of Theorem 5.1, and suppose $v(w)/y(z)$, where $w = c(z)$, takes on Γ_1 its supremum and infimum in the closure $\bar{D}$ of D. Then we have $F_\lambda \geqq F_1$, $0 < \lambda \leqq 1$.*

By hypothesis we have $\sup_{\Gamma_1}[v(w) - S_1 y(z)] \leqq 0$, hence throughout $\bar{D}$

$$v(w) - S_1 y(z) \leqq 0, \quad w = c(z); \tag{5.3}$$

hence there follows for w on Γ_ρ

$$S_\rho - S_1\rho \leqq 0. \tag{5.4}$$

In particular, (5.3) holds also for w on Γ_λ, $\rho \leqq \lambda \leqq 1$,

$$S_\lambda - S_1\lambda \leqq 0. \tag{5.5}$$

Likewise by hypothesis we have $\inf_{\Gamma_1}[v(w) - I_1 y(z)] \geqq 0$, hence throughout $\tilde{D}$

$$v(w) - I_1 y(z) \geqq 0, \quad w = c(z); \tag{5.6}$$

$$I_\lambda - I_1 \lambda \geqq 0. \tag{5.7}$$

From (5.5) and (5.7) we now deduce

$$F_\lambda = \frac{I_\lambda}{S_\lambda} \geqq \frac{I_1}{S_1} = F_f, \qquad \rho \leqq \lambda \leqq 1,$$

provided F_λ is defined.

Corollary 5.1. *If we have $g(v) \equiv 0$ in Theorem 5.3, and $\tilde{D}$ is $0 < y < 1$, then $F_\lambda \geqq F_1$ without any hypothesis on $v(w)/y(z)$.*

Under the new hypothesis, (5.4) is obvious for $\rho = 0$, and (5.3) holds on Γ_1, so (5.5) follows. Also (5.6) is valid on Γ_1, and clearly valid on Γ_0, hence valid throughout $\tilde{D}$ and on Γ_λ. Thus (5.6) holds for $0 < \lambda \leqq 1$, namely (5.7), and the Corollary follows.

If $v(w) \not\equiv S_1$ on Γ_1 or $v(w) \not\equiv I_1$ on Γ_1, then the strong inequality holds in (5.5), (5.7), and in $F_\lambda \geqq F_1$.

Two remarks are in order.

First, there are genuine difficulties in extending Theorem 5.1 to the case where S_g (and hence I_g) is equal to zero. In this case since $g(v) \equiv 0$, one might be tempted to define F_g to be equal to one since F_g would equal one for $g(v)$ identically equal to any other non-negative constant. Also as we will see later (Corollary 5.3) $\lim_{\lambda \to 1} F_\lambda$ always exists. The problem is that $\lim_{\lambda \to 1} F_\lambda$ need not equal one, as the following strip domain D shows. Let D be bounded by the curve $v \equiv 0$ and the curve $v = f(u)$, where $1 \leqq f(u) \leqq 2$ and $f(u)$ non-constant. Let $\omega(z)$ be the harmonic measure of $\{w \mid -\infty < u < +\infty, v = f(u)\}$ relative to the domain D. By ω_v we shall mean $\partial\omega/\partial v$. Since $f(u)$ is non-constant, the function ω_v (which is well defined by the reflection principle) is non-constant on the u-axis. This follows from the fact that if $\omega_v \equiv k$ on the u-axis then $\omega \equiv kv$ (otherwise the critical points of $\omega - kv$ would be isolated). Let u_1 and u_2 be such that $\omega_v(u_1) > \omega_v(u_2)$. From the local behavior of ω this implies for all λ less than some fixed λ_0 we have $S_\lambda \geqq [3\omega_v(u_1) + \omega_v(u_2)] \cdot \lambda/4$ and $I_\lambda \leqq [3\omega_v(u_2) + \omega_v(u_1)] \cdot \lambda/4$. This implies that F_λ is bounded away from one.

Second, if we modify the hypothesis so that D is now bounded by more general disjoint Jordan arcs C_1 and C_2 parametrized in the form $C_i(i = 1,2) = \{w = u + iv \mid u = \alpha_i(t),\ v = \beta_i(t),\ -1 < t < 1,\ \beta_i(t) \geqq 0,\ \lim_{t\to\pm 1}\alpha_i(t) = \pm\infty\}$ the proof goes through essentially without change.

The following three corollaries are all almost immediate consequences of Theorem 5.1.

Corollary 5.2. *Under the hypothesis of Theorem* 5.1, *if* $F_g = 1$ *then* F_λ *is a monotone decreasing function of* λ *which approaches unity as* λ *approaches zero.*

If $F_g = 1$ then from Theorem 5.1 it follows that

$$F_f \leqq F_\lambda \leqq F_g . \tag{5.8}$$

However the proof of that theorem applies if either of the boundary curves (upper or lower) is an interior λ-curve; hence from (5.8) we have:

$$F_{\lambda_1} \leqq F_\lambda \leqq F_{\lambda_2} \quad \text{if} \quad 0 < \lambda_1 < \lambda < \lambda_2 < 1 .$$

To state Corollary 5.3 we need a definition, namely that a continuous function $\alpha(t)$ is *bi-monotone increasing* in $(0,1)$ if there exists a number $c(0 < c < 1)$ such that $\alpha(t)$ is monotone increasing on $(0,c]$ and $\alpha(t)$ is monotone decreasing $[c,1)$.

Corollary 5.3. *Under the hypotheses of Theorem* 5.1, F_λ *is in* $(0,1)$ *either*: i) *monotone or* ii) *bi-monotone increasing.*

We first note that the corollary follows immediately if we can show that it holds in any arbitrary closed interval $[L,R]$ contained in $(0,1)$.

We know by the continuity of F_λ that it has an absolute maximum on $[L,R]$.

Case I: The absolute maximum occurs at the right-hand end point R. In this case F_λ must be monotone increasing. Assume not, then there exist points λ_1 and λ_2 with $\lambda_1 < \lambda_2 < R$ and $F_{\lambda_2} < F_{\lambda_1}$. By applying Theorem 5.1 to the interval $[\lambda_1, R]$ we obtain a contradiction.

Case II: The absolute maximum occurs at the left-hand end-point L. In this case F_λ must be monotone decreasing. The proof is essentially the same as in Case I.

Case III: The absolute maximum occurs at a point λ_0 with $L < \lambda_0 < R$. In this case F_λ is monotone increasing in $[L, \lambda_0]$ and monotone decreasing in $[\lambda_0, L]$. The proof is essentially the same as in Case I.

Finally as an immediate consequence of (5.1) and (5.2) we obtain

Corollary 5.4. *Under the hypotheses of Theorem* 5.1 *the curve* $\omega(w) = \lambda$ *lies entirely within the closed strip*

$$\{w = u + iv \mid -\infty < u < +\infty,\ \lambda I_f + (1-\lambda)I_g \leqq v \leqq \lambda S_f + (1-\lambda)S_g\}.$$

One may also obtain in an infinite strip domain theorems analogous to our results in §4. A sample theorem of this sort is the following

Theorem 5.4. *Let D be an infinite strip domain bounded by the continuously differentiable disjoint curves $v = 0$ and $v = f(u)\,(> 0)$ and let $\omega(w)$ be the harmonic measure of $\{w \mid -\infty < u < +\infty,\ v = f(u)\}$ with respect to D. For each λ, $0 < \lambda < 1$, let $T(\lambda)$ equal the supremum of the values of the angles the tangents to the curve $\omega(w) = \lambda$ make with the line $v = 0$. Under these conditions $T(\lambda)$ is a monotone increasing function of λ.*

The proof follows immediately by applying the maximum principle to the harmonic function $\arg c'(z)$ where $c(z)$ is the conformal map considered in Theorem 5.1.

We conclude this section with two examples.

The condition that a function be "monotone or bi-monotone increasing" is quite reminiscent of the condition that a function be convex downward. This is, in fact, a necessary condition for a function to be convex downward. In §7 we shall show that the circularity is both "monotone or bi-monotone increasing" and convex downward as a function of λ. The following example shows this is not the case with flatness:

Example 5.1. *There exists an infinite strip domain for which F_λ is not convex with respect to λ.*

Let D be the domain consisting of the union of the two sets

$$\{w = u + iv \mid -\infty < u < \infty,\ 1 < v < 49\}$$

and

$$\{w = u + iv \mid -N < u < N,\ 49 \leqq v < 50\}.$$

By making N large we can make $S_{\frac{1}{2}}$ as close as we like to $51/2$. Independent of our choice of N, $I_{\frac{1}{2}} = 25$. Therefore we have $F_0 = 1$, $F_{\frac{1}{2}}$ very close to $50/51$ and $F_1 = 49/50$. However for F_λ to be convex downward $F_{\frac{1}{2}}$ would have to be greater than or equal to $99/100$, but $50/51 < 99/100$.

If one has an infinite strip domain whose lower boundary is the line $v = 0$ and whose upper boundary is not necessarily a single valued function of u one might wonder whether the λ-curves, other than $\lambda = 0$, represent single-valued functions of u. The following example shows that in some quite general cases very good estimates can be made.

Example 5.2. *Let D be an infinite strip domain whose lower boundary is the line $v = 0$ and whose upper boundary has a continuously turning tangent which is parallel to the u-axis at one point and which never turns more than π radians (plus or minus) from zero. Under these conditions all the λ-curves are single-valued functions of u for $\lambda \leqq 1/2$.*

Let $c(z)$ be the conformal map considered in the proof of Theorem 5.1. Now all we need to do is note that we can strictly majorize [minorize] the function $\arg c'(z)$ by the harmonic function $h(x,y) = \pi y[h(x,y) = -\pi y]$. Hence for $\lambda = 1/2$ we have $-\pi/2 < \arg c'(x + i\lambda) < \pi/2$, which implies for these values of λ that the image of $y = \lambda$ under $c(z)$ is single-valued.

§6. Wedge domains. In this section we shall consider domains in the first quadrant of the W-plane ($W = U + iV$) which are bounded by two Jordan arcs J and K joining the origin to the point at infinity with $J \subset \{W \mid \alpha_1 \leqq \arg W \leqq \alpha_2\}$ and $K \subset \{W \mid \alpha_3 \leqq \arg W \leqq \alpha_4\}$ where $0 \leqq \alpha_1 < \alpha_2 < \alpha_3 < \alpha_4 \leqq \pi/2$. Such domains will be called *wedge domains.*

By means of the conformal map $w = \log W$ we may interpret a number of the results in the w-plane of §5 as theorems for wedge domains.

From Corollary 5.4 we obtain the following

Theorem 6.1. *Let D be a wedge domain with notation as introduced above. Also let $\omega(W)$ be the harmonic measure of K with respect to the domain D. Under these conditions the curve $\{W \mid \omega(W) = \lambda\}$ must lie in the sector*

$$\{W \mid (1-\lambda)\alpha_1 + \lambda\alpha_3 \leqq \arg W \leqq (1-\lambda)\alpha_2 + \lambda\alpha_4\}.$$

A theorem closely related to Theorem 6.1 is the following theorem which extends the Carathéodory Theorem on the behavior of a conformal map at a corner [7, p. 104]:

Theorem 6.2. *Let $T = c(W)$ map the open upper half W-plane conformally onto a Jordan domain D_0 in the T-plane which has the origin as a boundary point. Also suppose that $c(0) = 0$ under the extension of $c(W)$ to a continuous map on the closed upper half plane. In addition let there exist some number $U_0(>0)$ such that the extension of $c(W)$ takes the interval $\{W = U + iV \mid 0 \leqq U \leqq U_0,\ V = 0\}$ into $\{T \mid \alpha_1 \leqq \arg T \leqq \alpha_2\}$ and the interval $\{W = U + iV \mid -U_0 \leqq U \leqq 0,\ V = 0\}$ into $\{T \mid \alpha_3 \leqq \arg T \leqq \alpha_4\}$. Under these conditions the image of $\arg W = \lambda/2$ is eventually (in some neighborhood of $T = 0$) contained in the sector*

$$\{T \mid (1-\lambda/2\pi)\alpha_1 + (\lambda/2\pi)\alpha_2 \leqq \arg T \leqq (1-\lambda/2\pi)\alpha_3 + (\lambda/2\pi)\alpha_4\}.$$

The argument follows the proof of Theorem 6.1; we note that the local behavior of $c(U+iV)$ in the neighborhood of the origin depends only on the local behavior of the image of $[-U_0, U_0]$ under $T = c(U)$. (This follows from the fact that our method reduces the problem to one concerning harmonic functions and their local behavior).

For λ-level curves in a wedge the natural analog of circularity is *angular flatness* which we define as

$$A_\lambda = \frac{\inf\limits_{\omega(W)=\lambda} \arg W}{\sup\limits_{\omega(W)=\lambda} \arg W} = \frac{I_\lambda}{S_\lambda} \quad (S_\lambda \neq 0).$$

Angular flatness can be easily studied by the methods of §5.

§7. Another form of min-max deviation. In previous sections we have considered min-max deviations related to the geometry of the level curves of harmonic measures. In this section we shall consider a min-max deviation for certain functions on the level curves of harmonic measures of two particularly simple domains. More general cases will be considered in the next section.

If $h(Z)$ is a bounded harmonic function in the strip $\{Z = X + iY \mid -\infty < X < \infty,\ C_1 < Y < C_2\}$ we define the strip *max-min difference* to be

$$\sigma_\gamma = \sup_{-\infty < X < \infty} h(X + i\gamma) - \inf_{-\infty < X < \infty} h(X + i\gamma).$$

If $k(Z)$ is a bounded harmonic function in the annulus $\{Z \mid R_1 < |Z| < R_2\}$ we define the *max-min difference* to be

$$\mu_\rho = \max_{|Z| = e^\rho} k(Z) - \min_{|Z| = e^\rho} k(Z).$$

The functions σ_γ and μ_ρ are convex functions of γ and ρ respectively. This follows, in each case, by applying the two-constant theorem [5, p. 126] to each term in the difference.

As an application of the fact that μ_ρ is convex we prove the following

Theorem 7.1. *Let A be an annular domain in the S-plane ($S = P + iQ$) with inner boundary A_1 and outer boundary A_2. Let the origin in the S-plane be in the interior of the curve A_1. Denote by $S = f(Z)$ the conformal map from $\{Z \mid 1 < |Z| < R_0\}$ onto A and by $\kappa(r)$ the circularity of the image of the circle $|Z| = r$ under the map. Let generically $\tau(\rho) = 1/\kappa(e^\rho)$ ($0 < \rho < \log R_0$). Under these conditions $\tau(\rho)$ is a convex function of ρ.*

Let $k(Z) = \log|f(Z)|$. Since $f(Z) \neq 0$, $k(Z)$ is harmonic and hence $\max_{|Z| = e^\rho} \log|f(Z)| - \min_{|Z| = e^\rho} \log|f(Z)|$ is a convex function of ρ. Elementary properties of the logarithm lead us to

$$\max_{|Z|=e^\rho} \log|f(Z)| - \min_{|Z|=e^\rho} \log|f(Z)|$$

$$= \log[\max_{|Z|=e^\rho} |f(Z)|] - \log[\min_{|Z|=e^\rho} |f(Z)|]$$

$$= \log\left[\frac{\max_{|Z|=e^\rho} |f(Z)|}{\min_{|Z|=e^\rho} |f(Z)|}\right] = \log(1/\kappa(e^\rho)) = \log\tau(\rho).$$

Therefore $\tau(\rho) = e^{c(\rho)}$, where $c(\rho)$ is a convex function of ρ. Since the exponential function is monotone and convex, it follows that $\tau(\rho)$ is convex.

Let ω be the harmonic measure of A_1 with respect to A. Theorem 7.1 has an immediate corollary concerning the circularity of the level curves $\omega = \lambda$ as a function of λ:

Corollary 7.1. *Let A and ω be defined as above. Let $\tilde{\kappa}(\lambda)$ be the circularity of the level curve $\omega(S) = \lambda$. Under these conditions the function $\tilde{\kappa}_{-1}(\lambda) = 1/\tilde{\kappa}(\lambda)$ is a convex function of λ.*

We know the r-circle in the Z-plane corresponds to the level curve $\omega(S) = \log r/\log R_0$. Therefore $\tilde{\kappa}(\lambda) = \kappa(e^{\lambda \log R_0})$ and $\tilde{\kappa}_{-1}(\lambda) = \tau(\lambda \log R_0)$. Since convex functions remain such under a linear change of variable, it follows that $\tilde{\kappa}(\lambda)$ is a convex function of λ.

§8. Further generalizations. In this last section we shall consider some generalizations of our previous work, first in the direction of theorems on more general plane domains and second in the direction of theorems in higher dimensions.

We will now outline how one can generalize Theorem 5.2 to more general plane domains. It will then be clear, in principle, how to extend much of our other work in this direction.

Theorem 8.1. *Let G be a domain in the z-plane which is bounded by finitely many mutually disjoint Jordan curves and let A consist of finitely many subarcs of ∂G. Let $\omega(z)$ be the harmonic measure of A with respect to the domain G and let $h(z)$ be a function positive and harmonic on G and continuous on the closure of G. In addition let*

$$\tilde{S}_\lambda = \sup_{z \in \{\omega = \lambda\}} h(z), \quad \tilde{I}_\lambda = \inf_{z \in \{\omega = \lambda\}} h(z)$$

and min-max ratio $\Delta_\lambda = \tilde{I}_\lambda / \tilde{S}_\lambda$. *Under these conditions if* $\Delta_1 \leqq \Delta_0$ *then* $\Delta_\lambda \geqq \Delta_1$.

To start the proof of the theorem the following topological lemma is needed:

Lemma 8.1. *Under the hypotheses of Theorem* 8.1 *the set* $\{z \mid \alpha < \omega(z) < \beta\}$ *consists of the union of finitely many mutually disjoint domains, each of which is bounded by the union of finitely many Jordan arcs or curves on which* $\omega(z) = \alpha$ *or* $\omega(z) = \beta$ [*except possibly for end points*].

For domains bounded by analytic arcs the lemma follows from the fact that the closure of G is compact and from applications of the theory of the local behavior of the level curves of harmonic measures [6, §4.1]. To prove the lemma for domains bounded by arbitrary Jordan curves one uses the fact that such domains are always conformally equivalent to ones bounded by *analytic* Jordan curves.

The proof now follows by applying, together with the two-constant theorem [5, p. 125], the methods of the proof of Theorem 5.1 to each of the domains in $\{z \mid \alpha < \omega(z) < \beta\}$.

As an indication of how a number of our theorems can be generalized to 3-dimensions we conclude by proving a 3-dimensional analog of Theorem 1.1.

Theorem 8.2. *Let* S *be a bounded,* 3-*dimensional, simply connected domain containing the origin. Further suppose there exists a Green's function* $G(p)$ *for* S *relative to the origin. Let* $\kappa(\lambda)$ *be the circularity relative to the origin of the* λ-*level surface of* $G(p)$. *Then* $\kappa(\lambda)$ *is a monotone decreasing function of* $\lambda(-\infty < \lambda \leqq 0)$.

By a Green's function for S relative to the origin we mean a function with the following properties: i) $G(p) + 1/\|p\|$ is harmonic throughout S (where $\|p\|$ denotes the distance from p to the origin), ii) $\lim_{p \to \partial S} G(p) = 0$. (In 3-dimensions some authors consider the Green's function to be the negative of our Green's function.)

Let M_λ be the maximum distance and m_λ be the minimum distance from a surface $S_\lambda : G(p) = \lambda$ to the origin. The circularity of S_λ is defined as $\kappa(\lambda) = m_\lambda / M_\lambda$. We now consider the two functions

$$(8.1)\qquad \begin{cases} H(p) = G(p) + [1/\|p\| - 1/M_0], \\ K(p) = G(p) + [1/\|p\| - 1/m_0]. \end{cases}$$

Both $H(p)$ and $K(p)$ are harmonic in S, hence by the maximum and minimum principles we have

$$(8.2)\qquad H(p) \geqq 0, \quad K(p) \leqq 0, \quad p \in S.$$

The equations (8.1) and (8.2) lead us to the following inequalities for p on the λ-level surface:

$$\lambda + 1/\|p\| - 1/M_0 \geqq 0,$$

$$\lambda + 1/\|p\| - 1/m_0 \leqq 0.$$

This implies in particular that

$$(8.3)\qquad \begin{cases} \lambda + 1/M_\lambda - 1/M_0 \geqq 0, \\ \lambda + 1/m_\lambda - 1/m_0 \leqq 0. \end{cases}$$

Rewriting (8.3) we obtain

$$1/M_\lambda \geqq 1/M_0 - \lambda,$$

$$1/m_\lambda \leqq 1/m_0 - \lambda,$$

This implies that

$$\kappa(\lambda) = \frac{m_\lambda}{M_\lambda} \geqq \frac{m_0(1 - M_0\lambda)}{M_0(1 - m_0\lambda)}.$$

Since λ is negative we have $\kappa(\lambda) \geqq m_1/M_1$ (i.e. $\kappa(\lambda) \geqq \kappa(0)$). The same argument applied to any λ_1-surface instead of ∂S leads to the inequality $\kappa(\lambda_2) \geqq \kappa(\lambda_1)$ if $\lambda_2 \leqq \lambda_1$. This completes the proof.

References

1. J. L. Walsh, On the shape of level curves of Green's function, *Amer. Math. Monthly*, **44** (1937), 202–213.

2. J. L. Walsh, Note on the shape of level curves of Green's function, *Amer. Math. Monthly*, **60** (1953), 671–674.

3. P. Davis and H. Pollak, On the zeros of total sets of polynomials, *Trans. Amer. Math. Soc.*, **72** (1952), 82–103.

4. H. Grunsky, Zwei Bemerkungen zur konformen Abbildung, *Jahresber. d.d. Math. Vereinigung*, **43** (1933), 140–143.

5. L. Bieberbach, Lehrbuch der Funktionentheorie II, (Leipzig, 1931).

6. J. L. Walsh, Interpolation and Approximation, *Amer. Math. Soc. Coll.*, **20** (1935), Providence, R.I.

7. C. Carathéodory, Theory of Functions of a Complex Variable II, Chelsea, New York City, 1954.

University of Maryland
College Park, Md., U.S.A.
and
Syracuse University
Syracuse, N. Y., U.S.A.

Zur Methode der variablen Gebiete bei der Randverzerrung

Von J. L. Walsh und D. Gaier in Cambridge, Mass.

1. Einleitung

Gegeben sei ein einfach zusammenhängendes Gebiet $\mathfrak{G}$ der w-Ebene, das durch eine in $\Re(z) < 1$ reguläre Funktion $w = f(z)$ konform auf die Halbebene $\mathfrak{H}: \Re(z) < 1$ abgebildet werde; der Punkt $w = 1$ sei erreichbarer Randpunkt von $\mathfrak{G}$, und das $w = 1$ enthaltende Primende entspreche dem Punkt $z = 1$.

Um das Verhalten von $f(z)$ in der Nähe von $z = 1$ zu studieren, kann man folgendermaßen vorgehen. Man unterwerfe $\mathfrak{G}$ und $\mathfrak{H}$ je einer gewissen Folge von Ähnlichkeitstransformationen $W = W_n(w)$, $Z = Z_n(z)$, wodurch die Gebiete $\mathfrak{G}$ und $\mathfrak{H}$ von ihren Randpunkten $w = 1$ und $z = 1$ aus gestreckt werden. $\mathfrak{H}$ wird dabei in sich selbst übergeführt, während $\mathfrak{G}$ in ein Gebiet $\mathfrak{G}_n$ übergeht, dessen Abbildungsfunktion $f_n(Z)$ auf $\Re(Z) < 1$ sich leicht durch $f(z)$ ausdrücken läßt. Wenn nun die variablen Gebiete $\mathfrak{G}_n$ gegen ihren Kern $\mathfrak{K}$ konvergieren, so konvergiert die Funktionenfolge $f_n(Z)$ in $\Re(Z) < 1$ gegen eine Grenzfunktion $F(Z)$ (vorausgesetzt, daß die $f_n(Z)$ geeignet normiert sind), und dies wiederum läßt auf das Verhalten von $f(z_n)$ für $z_n \to 1$ in $\Re(z) < 1$ schließen. Was hier über die Untersuchung von $f(z)$ gesagt wurde, gilt ebenso für andere Ausdrücke wie z. B. $\arg(f(z) - 1)$; auf diesen Fall werden wir in § 3 zurückkommen.

Diese Methode, das Verhalten der Abbildungsfunktion $f(z)$ am Rande durch Betrachtung einer Funktionenfolge zu untersuchen, wurde zuerst von Montel entwickelt ([6]), der so unter anderem Sätze über die Zuordnung von Primenden bei konformer Abbildung beweisen konnte. Weitere Anwendungen der Methode stammen von Ferrand ([2], S. 79—82, und [3]), die damit stark vereinfachte Beweise des Ostrowskischen Hauptsatzes über die Winkeltreue ([8], S. 447) sowie des ersten Faltensatzes ([8], S. 458) geben konnte, deren ursprüngliche Beweise auf zum Teil tiefliegenden Sätzen über das harmonische Maß beruhten. Der erste Faltensatz ist bekanntlich insbesondere für die Untersuchung des Ausdrucks $\frac{f(z)-1}{z-1}$ bei beliebiger Annäherung von z an $z = 1$ bedeutsam. In den letzten Jahren wurde die Methode verwandt von Walsh ([10]) (Winkeltreue am Rande, Satz von Visser) sowie von Lelong-Ferrand ([5]) und Walsh und Rosenfeld ([11]) (konforme Abbildung von Streifen).

*) Diese Arbeit wurde zum Teil mit Unterstützung des US Office of Naval Research durchgeführt.

Wir wollen hier weitere Anwendungen der Methode der variablen Gebiete geben. Nach der Einführung einiger benützter Begriffe und Tatsachen in § 2 untersuchen wir in § 3 die Frage, wie sich $\arg(f(z)-1)$ bei Annäherung von z an $z=1$ „im Winkelraum" verhält, wenn der Rand von $\mathfrak{G}$ bei $w=1$ zwischen zwei Paaren von Grenzstützen oszilliert. Diese Frage wurde von OSTROWSKI ([7], S. 173) und WARSCHAWSKI ([12], S. 674) mit potentialtheoretischen Mitteln behandelt. Die erhaltenen Schwankungsgrenzen für $\arg(f(z)-1)$ sind scharf, was neu zu sein scheint; die hierbei verwandte Methode (Methode der variablen Gebiete, gekoppelt mit der Benützung des harmonischen Maßes) wäre ebenfalls bei Vorliegen von Grenzstützen anwendbar. — Ist $\mathfrak{G}$ ein JORDANgebiet, dessen Rand R an der Stelle $w=1$ eine Tangente aufweist, so gilt nach dem Randverzerrungssatz von VISSER

$$f'(z):\frac{f(z)-1}{z-1}\to 1 \tag{1.1}$$

für $z\to 1$ im Winkelraum (VISSER [9], S. 37). In § 4 beschäftigen wir uns allgemeiner mit der Frage, notwendige und hinreichende Bedingungen über die Randstruktur von $\mathfrak{G}$ anzugeben, damit (1.1) erfüllt ist. Es wird sich ergeben, daß die Winkeltreue in $z=1$ der durch $w=f(z)$ vermittelten Abbildung (was sich ja durch die Randstruktur von $\mathfrak{G}$ ausdrücken läßt) für das Bestehen von (1.1) hinreicht, jedoch nicht notwendig ist. Der Schluß von (1.1) auf die Winkeltreue ist vielmehr genau dann richtig, wenn irgend eine Kurve C_z in $\mathfrak{H}$, längs der $\arg(z-1)\to\gamma$ strebt $\left(\frac{\pi}{2}<\gamma<\frac{3\pi}{2}\right)$, durch $w=f(z)$ auf eine Kurve C_w in $\mathfrak{G}$ abgebildet wird, längs der $\arg(w-1)$ einen Grenzwert hat. Eine einfache Folgerung aus (1.1) über die Abbildung von Kurven, die mit einer L-Tangente versehen im Winkelraum in $z=1$ einmünden, beschließt die Arbeit.

2. Vorbemerkungen

Sei $\{\mathfrak{G}_n\}$ eine Folge einfach zusammenhängender, endlicher Gebiete der w-Ebene, die $w=0$ enthalten und für die der kleinste Abstand des Punktes $w=0$ vom Rand von $\mathfrak{G}_n$ beschränkt ist. Gibt es keine Kreisscheibe um $w=0$, die in allen $\mathfrak{G}_n$ liegt, so heißt $w=0$ der Kern $\mathfrak{K}$ der Gebietsfolge; andernfalls wird das größte, $w=0$ enthaltende Gebiet $\mathfrak{K}$ mit der Eigenschaft, daß jeder kompakte Teil von $\mathfrak{K}$ in $\mathfrak{G}_n$ liegt für $n>n_0$, als Kern der Gebietsfolge bezeichnet. $\{\mathfrak{G}_n\}$ konvergiert gegen $\mathfrak{K}$, wenn jede Teilfolge von $\{\mathfrak{G}_n\}$ den Kern $\mathfrak{K}$ hat.

Sei $f_n(z)$ die normierte (d. h. $f_n(0)=0$, $f_n'(0)>0$) Abbildungsfunktion von $\Re(z)<1$ auf $\mathfrak{G}_n$ und $\{\mathfrak{G}_n\}$ konvergiere gegen den Kern $\mathfrak{K}\neq 0$, dessen normierte Abbildungsfunktion $F(z)$ sei. Dann besagt der Satz von CARATHÉODORY über die Abbildung variabler Gebiete ([1], S. 76): Es gilt $f_n(z)\to F(z)$, gleichmäßig in jedem abgeschlossenen Teil von $\Re(z)<1$. Gilt umgekehrt für unsere, diesmal nur durch $f_n(0)=0$ normierten Abbildungsfunktionen $f_n(z)\to F(z)$, gleichmäßig in jedem abgeschlossenen Teil von $\Re(z)<1$, so konvergieren die $\mathfrak{G}_n$ gegen ihren Kern, das

Bild von $\Re(z) < 1$ unter der Abbildung $w = F(z)$. Konvergiert $\{\mathfrak{G}_n\}$ gegen $w = 0$, so gilt $f_n(z) \to 0$, wiederum gleichmäßig in jedem abgeschlossenen Teil von $\Re(z) < 1$.

Das Gebiet $\mathfrak{G}$ hat im Randpunkt $w = 1$ das Richtungsintervall (α, β), wenn es zu jedem Winkel γ des offenen Intervalls (α, β) ein von $w = 1$ aus ins Innere von $\mathfrak{G}$ mündendes Stück des Halbstrahls $\arg(w-1) = \gamma$ gibt, und wenn (α, β) das größte Intervall dieser Art ist.

Die zwei von $w = 1$ ausgehenden Zweige des Randes R von $\mathfrak{G}$ seien mit C_1 und C_2 bezeichnet. Dann hat R im Randpunkt $w = 1$ von $\mathfrak{G}$ die Paare (α_1, β_1) und (α_2, β_2) von Grenzstützen, wenn zu jedem $\varepsilon > 0$ ein $r > 0$ existiert derart, daß für alle in $|w - 1| < r$ gelegenen Punkte von C_1 bzw. C_2 gilt: $\alpha_1 - \varepsilon < \arg(w-1) < \beta_1 + \varepsilon$ bzw. $\alpha_2 - \varepsilon < \arg(w-1) < \beta_2 + \varepsilon$. Ist $\alpha_1 = \beta_1$, $\alpha_2 = \beta_2$, und $\alpha_2 - \alpha_1 = \pi$, so sagen wir, R habe im Punkt $w = 1$ eine Tangente (genauer W-Tangente, vgl. [14], S. 46).

Eine in $z = 1$ [bzw. $z = \infty$] einmündende JORDANkurve $y = y(x)$ hat dort eine L-Tangente mit der Steigung γ, wenn $(y(x_1) - y(x_2))/(x_1 - x_2)$ gegen den Grenzwert γ strebt, gleichmäßig für $x_1, x_2 \to 1$ [bzw. $x_1, x_2 \to +\infty$].

Eine Punktfolge $\{w_n\}$ mit $w_n \to 1$ heißt dicht an der Stelle $w = 1$, wenn $|w_{n-1} - 1| / |w_n - 1|$ gegen 1 strebt, und mit der Richtung δ versehen, wenn $\arg(w_n - 1) \to \delta$ gilt.

3. Vorliegen von Grenzstützen

Das einfach zusammenhängende, endliche Gebiet $\mathfrak{G}$ habe im Randpunkt $w = 1$ das Richtungsintervall $\left(\frac{\pi}{2} + \alpha_1, \frac{3\pi}{2} - \alpha_2\right)$ und die Grenzstützen $\left(\frac{\pi}{2}, \frac{\pi}{2} + \alpha_1\right)$ und $\left(\frac{3\pi}{2} - \alpha_2, \frac{3\pi}{2}\right)$ mit $\alpha_1 \geqq 0$, $\alpha_2 \geqq 0$, $\alpha_1 + \alpha_2 < \pi$. Durch $w = f(z)$ mit $f(1) = 1$ werde eine konforme Abbildung von $\mathfrak{H}$: $\Re(z) < 1$ auf $\mathfrak{G}$ vermittelt. Wir untersuchen das Verhalten von $\arg(f(z) - 1)$ für $z \to 1$ im Winkelraum, wozu es genügt, die Abbildung der Halbstrahlen h: $\arg(z-1) = \beta$ $\left(\frac{\pi}{2} < \beta < \frac{3\pi}{2}\right)$ zu kennen.

Satz 1. *Es gilt*

$$(3.1) \qquad \underline{L} = \beta - \alpha_2 \frac{2\beta - \pi}{2\pi} \leqq \varliminf_{\substack{z \to 1 \\ h}} \arg(f(z) - 1) \leqq \beta + \alpha_1 \frac{3\pi - 2\beta}{2\pi} = \overline{L},$$

und die Größen $\underline{L}$ *und* $\overline{L}$ *sind bestmöglich* [1]).

[1]) Daraus folgt für die Umkehrfunktion $z = f^{-1}(w)$

$$\underline{L}' = \max\left(\frac{\pi}{2}, \frac{\pi}{2}\,\frac{2\beta' - 3\alpha_1}{\pi - \alpha_1}\right) \leqq \overline{\lim} \arg(z_n - 1) \leqq \min\left(\frac{3\pi}{2}, \frac{\pi}{2}\,\frac{2\beta' - \alpha_2}{\pi - \alpha_2}\right) = \overline{L}',$$

wenn $w_n = f(z_n)$ mit $\arg(w_n - 1) \to \beta'$ $\left(\frac{\pi}{2} \leqq \beta' \leqq \frac{3\pi}{2}\right)$ in $\mathfrak{G}$ gegen $w = 1$ strebt und wenn dabei auch $z_n = f^{-1}(w_n)$ gegen $z = 1$ strebt. Dabei sind die Größen $\underline{L}'$ und $\overline{L}'$ bestmöglich.

Zum Beweis von (3.1) behandeln wir zuerst den Fall, daß $\mathfrak{G}$ zur u-Achse symmetrisch ist ($w = u + iv$), sodann den Fall, daß ein Randzweig von $\mathfrak{G}$ eine Halbgerade ist, und schließlich den allgemeinen Fall. Zunächst beweisen wir einen Hilfssatz.

Hilfssatz. *$\mathfrak{G}$ sei zur u-Achse symmetrisch, und R liege vollständig zwischen den Halbgeraden:* $\left(\arg(w-1) = \frac{\pi}{2}, \arg(w-1) = \frac{\pi}{2} + \alpha\right)$ *und* $\left(\arg(w-1) = \frac{3\pi}{2} - \alpha, \arg(w-1) = \frac{3\pi}{2}\right)$ *mit* $0 \leqq \alpha < \frac{\pi}{2}$. *Wird $\mathfrak{G}$ durch $w = f(z)$ konform auf $\mathfrak{H}$ abgebildet* $[f(0) = 0,\ f(1) = 1]$, *so gilt für alle z auf* $\arg(z-1) = \beta$

$$\beta \leqq \arg(f(z)-1) \leqq \beta + \frac{2\alpha}{\pi}(\pi - \beta) \qquad (\pi/2 < \beta \leqq \pi). \tag{3.2}$$

Beweis: Die linke Seite von (3.2) folgt aus der Tatsache, daß der Sektor $\beta < \arg(z-1) < 2\pi - \beta$ bei unserer Abbildung in ein zur u-Achse symmetrisches Teilgebiet des Sektors $\beta < \arg(w-1) < 2\pi - \beta$ übergeht; dies ist eine einfache Folgerung aus dem SCHWARZschen Lemma, wie in [1], S. 53 ausgeführt ist. Zum Beweis der rechten Seite setzen wir $(1-w_1) = (1-w)^{\pi/(\pi - 2\alpha)}$, wodurch der in $\mathfrak{G}$ liegende Sektor $\frac{\pi}{2} + \alpha < \arg(w-1) < \frac{3\pi}{2} - \alpha$ in $\Re(w_1) < 1$ übergeht. Die Funktion $z = f^{-1}(w) = \Phi(w_1)$ bildet daher wieder den Sektor $\beta < \arg(w_1 - 1) < 2\pi - \beta$ in ein zur x-Achse symmetrisches Teilgebiet von $\beta < \arg(z-1) < 2\pi - \beta$ ab, so daß das Bild von $\arg(z-1) = \beta$ jedenfalls in $0 < \arg(w_1 - 1) \leqq \beta$ liegt, d. h. es gilt $\arg(w-1) \leqq \beta + \frac{2\alpha}{\pi}(\pi - \beta)$.

a) Beweis von Satz 1 im symmetrischen Fall. Wir betrachten eine beliebige Punktfolge $\{z_n\}$ auf $\arg(z-1) = \beta\ \left(\frac{\pi}{2} < \beta \leqq \pi\right)$ mit $z_n = x_n + iy_n \to 1$; ferner sei $\lim \arg(f(z_n) - 1) = \delta$ angenommen. Sodann führen wir die linearen Transformationen

$$\text{(a)}\ Z = Z_n(z) = \frac{z - x_n}{1 - x_n} \quad \text{und} \quad \text{(b)}\ W = W_n(w) = \frac{w - f(x_n)}{1 - f(x_n)} \tag{3.3}$$

aus und erhalten in

$$W = \frac{f(Z(1 - x_n) + x_n) - f(x_n)}{1 - f(x_n)} = f_n(Z) \tag{3.4}$$

eine Folge von Abbildungsfunktionen mit $f_n(0) = 0$, $f_n'(0) > 0$. Durch (3.3a) geht $\mathfrak{H}$ in $\Re(Z) < 1$ über, während $\mathfrak{G}$ durch (3.3b) in ein Gebiet $\mathfrak{G}_n$ transformiert wird, das man durch Streckung von $\mathfrak{G}$ vom Punkte $w = 1$ aus erhält. Die Folge $\{f_n(Z)\}$ ist normal ([1], S. 73), und es sei $\{f_{n_k}(Z)\}$ eine in $\Re(Z) < 1$ gegen $F(Z)$ konvergente Teilfolge. $\{\mathfrak{G}_{n_k}\}$ konvergiert dann gegen einen Kern $\mathfrak{K} \neq 0$, der zur U-Achse symmetrisch liegt ($W = U + iV$), und dessen Rand auf Grund unserer Annahmen vollständig zwischen den im Hilfssatz genannten Halbgeraden liegt ($\alpha_1 = \alpha_2 = \alpha$). Wir haben also für die Grenzfunktion $F(Z)$

$$\beta \leqq \arg(F(Z) - 1) \leqq \beta + \frac{2\alpha}{\pi}(\pi - \beta),$$

das heißt

$$\beta \leqq \lim_{n_k} \arg(f_{n_k}(Z) - 1) \leqq \beta + \frac{2\alpha}{\pi}(\pi - \beta);$$

nsbesondere gilt dies für den Punkt $Z = \frac{i\, y_{n_k}}{1 - x_{n_k}}$. Das liefert

$$\beta \leqq \lim_{n_k} \arg(f(z_{n_k}) - 1) \leqq \beta + \frac{2\alpha}{\pi} (\pi - \beta),$$

also $\beta \leqq \delta \leqq \beta + \frac{2\alpha}{\pi} (\pi - \beta)$. Wir haben somit im symmetrischen Fall

$$\beta \leqq \overline{\lim_{\substack{z \to 1 \\ h}}} \arg(f(z) - 1) \leqq \beta + \frac{2\alpha}{\pi} (\pi - \beta) \qquad \left(\frac{\pi}{2} < \beta \leqq \pi\right); \tag{3.5}$$

dies ist sogar schärfer als der entsprechende Sonderfall von (3.1), was verständlich ist, weil die bloße Forderung $\alpha_1 = \alpha_2 = \alpha$ schwächer ist als die (3.5) zugrunde liegende Symmetrie.

b) Zweiter Fall. Wir nehmen nun an, einer der beiden von $w = 1$ ausgehenden Randzweige bestehe aus der Halbgeraden $u = 1$, $v < 0$, während der zweite Zweig die Grenzstützen $\left(\frac{\pi}{2}, \frac{\pi}{2} + \alpha_1\right)$ habe, und wir wollen die Abbildung von $\arg(z-1) = \beta \left(\frac{\pi}{2} < \beta < \frac{3\pi}{2}\right)$ in der Umgebung von $z = 1$ verfolgen. Dabei können wir ohne Einschränkung annehmen, daß sich die Halbgeraden $u = 1$, $v < 0$ und $x = 1$, $y < 0$ entsprechen. Wir schlitzen die w- und z-Ebenen längs dieser Halbgeraden auf und wenden je eine Quadratwurzeloperation an ($w_1 = u_1 + i v_1$, $z_1 = x_1 + i y_1$), wonach der Quadrant $x_1 < 1$, $y_1 > 0$ auf ein in $v_1 > 0$ gelegenes Gebiet abgebildet wird, das als eine Begrenzungslinie die Gerade $u_1 < 1$, $v_1 = 0$ hat. Spiegelung an den Geraden $y_1 = 0$ und $v_1 = 0$ führt auf den schon diskutierten Fall a). Durch Rücktransformation erhält man auf einfache Weise

$$\beta \leqq \overline{\lim} \arg(f(z) - 1) \leqq \beta + \frac{\alpha}{2\pi} (3\pi - 2\beta). \tag{3.6}$$

c) Allgemeiner Fall. Wir können annehmen, $\mathfrak{G}$ sei nicht beschränkt; andernfalls werfen wir durch eine lineare Abbildung einen Randpunkt $\neq 1$ von $\mathfrak{G}$ nach ∞, wobei $w = 1$ und ein durch $w = 1$ gehendes Linienelement festbleibe. Das neue Gebiet hat in der Umgebung von $w = 1$ dieselben Eigenschaften wie $\mathfrak{G}$ und ist nicht beschränkt. Für ein beliebiges $\varepsilon > 0$ $\left(\varepsilon < \frac{\pi}{2}\right)$ ziehen wir nun die Halbgerade g: $\arg(w-1) = \frac{\pi}{2} - \varepsilon$. Sie trifft R entweder gar nicht (im Endlichen) oder in einem ersten, von $w = 1$ verschiedenen Punkt P. Im letzteren Fall wende man wie oben eine lineare Abbildung an, die P nach ∞ wirft, so daß wir annehmen können, g treffe R nicht. Nun durchlaufe man den Teil von R mit den Grenzstützen $\left(\frac{3\pi}{2} - \alpha_2, \frac{3\pi}{2}\right)$ von $w = 1$ aus bis zu einem ersten unendlichfernen Punkt; dieser Teil von R sei σ und g und σ begrenzen das Gebiet $\mathfrak{G}_1 \supset \mathfrak{G}$. $\mathfrak{G}_1$ werde auf $\Re(w_1) < 1$ abgebildet, wobei $w = 1$ in $w_1 = 1$ und $\mathfrak{G}$ in einen Teil von $\Re(w_1) < 1$ übergeht. Schließlich schlitzen wir noch die w-Ebene längs g auf und bilden den durch g und $u = 1$, $v < 0$ begrenzten Sektor auf $\Re(w_2) < 1$ ab; das zweite Paar von Grenzstützen ist dann $\left(\frac{3\pi}{2} - \alpha_2', \frac{3\pi}{2}\right)$.

Nun verfolgen wir die Abbildung von $\arg(z-1) = \beta$ in der w-Ebene unter Zwischenschaltung der w_1- und w_2-Ebenen. $z \to w_1$: Diese Abbildung ist von der im Fall 2 beschriebenen Form und wir haben daher mit (3.6) $\underline{\lim} \arg(w_1 - 1) \geqq \beta$. $w_1 \to w_2$: $\Re(w_1) < 1$ wird auf das der Transformation $w \to w_2$ unterworfene Gebiet $\mathfrak{G}_1$ abgebildet, so daß wieder die Überlegungen von Fall 2 angewandt werden können; diesmal sind nur die beiden Randzweige zu vertauschen. Das Verhalten des Bildes von $\arg(w_1-1)=\beta$ in der w_2-Ebene wird daher durch $\underline{\lim}\arg(w_2-1)\geqq\beta+$ $+\frac{\alpha_2'}{2\pi}(\pi-2\beta)=\beta_2$ gegeben. In der w-Ebene bedeutet dies $\underline{\lim}\arg(w-1)\geqq\beta_2+\varepsilon\left(\frac{\beta}{\pi}-\frac{3}{2}\right)$. Führt man diese Konstruktion aus für $\varepsilon\to 0$, so geht $\alpha_2'\to\alpha_2$ und man erhält

$$\underline{\lim}_{\substack{z\to 1\\ h}}\arg(f(z)-1)\geqq\beta+\frac{\alpha_2}{2\pi}(\pi-2\beta).$$

Die rechte Seite von (3.1) erhält man, indem man statt g die Halbgerade $\arg(w-1)=\frac{3\pi}{2}+\varepsilon$ zieht und entsprechend verfährt. Damit ist (3.1) bewiesen.

Um zu sehen, daß die in (3.1) angegebenen Zahlen $\underline{L}$ und $\overline{L}$ bestmöglich sind, wählen wir $\mathfrak{G}$ folgendermaßen. K_1 und K_2 seien die in $v>0$ und $v<0$ gelegenen Hälften von $|w|=1$, K_3 und K_4 zwei durch ± 1 gehende Kreisbogen, die unter dem Winkel α_1 bzw. α_2 gegen K_1 bzw. K_2 von $\Re(w)<1$ aus in $w=1$ einmünden; das von K_1 und K_4 begrenzte Gebiet heiße $\mathfrak{B}$. Von K_1 seien bezüglich $w=1$ kreisförmige Einschnitte E bis K_3 gezogen, während von K_4 aus dünne Kanäle K bis K_2 führen, die zu $\mathfrak{B}$ hinzugefügt das Gebiet $\mathfrak{G}$ ergeben sollen. Die E und K sollen sich dabei in $w=1$ häufen, aber so dünn verteilt sein, daß man nach Ausführung der Abbildung $W=\frac{w-u_n}{1-u_n w}$ $(0<u_n\to 1-0)$ aus $\mathfrak{G}$ ein $\mathfrak{G}_n$ erhält, für das $\mathfrak{G}_n\cap\{|W+1|>\delta_n\}\cap\{|W-1|>\delta_n\}=\mathfrak{G}_n'$ keine E und K mehr enthält, wobei noch $\delta_n\to 0$ gehen soll. Schließlich sei noch D_n der Durchmesser von $\mathfrak{G}_n$ und $D_n\leqq D$. Nun betrachten wir die in $\mathfrak{B}\cap\mathfrak{G}_n$ harmonische Funktion ($\omega=$ harmonische Maße)

$$U(W)=\omega(W,\mathfrak{G}_n,R_n^+)-\omega(W,\mathfrak{B},K_1)+u_1(W,\delta_n)+u_2(W,\delta_n),$$

wobei R_n^+ der oberhalb K_3 liegende Teil des Randes von $\mathfrak{G}_n$ ist und

$$u_1(W,\delta_n)=\frac{\log D-\log|W-1|}{\log D-\log\delta_n},\quad u_2(W,\delta_n)=\frac{\log D-\log|W+1|}{\log D-\log\delta_n}$$

gesetzt ist. Nun stellt man leicht fest, daß $\lim U(W)\geqq 0$ ist, wenn W gegen den Rand von $\mathfrak{G}_n'$ strebt. Daher gilt insbesondere für jedes feste W auf $\Re(W)=0$ in $\mathfrak{B}$

$$\omega(W,\mathfrak{G}_n,R_n^+)-\omega(W,\mathfrak{B},K_1)\geqq -u_1(W,\delta_n)-u_2(W,\delta_n),$$

und für $n\to\infty$ strebt die rechte Seite gegen Null. Benützt man eine analoge Abschätzung nach oben, so kommt für die genannten W-Werte

$$\omega(W,\mathfrak{G}_n,R_n^+)\to\omega(W,\mathfrak{B},K_1)\qquad(n\to\infty);$$

also gilt auf Grund der Invarianz des harmonischen Maßes bei konformer Abbildung

$$\omega(w_n,\mathfrak{G},R^+)\to\omega(W,\mathfrak{B},K_1)\qquad(n\to\infty),\tag{3.7}$$

wenn R^+ der oberhalb K_3 liegende Teil des Randes von $\mathfrak{G}$ ist. W und w_n liegen auf demselben, durch $+1$ und -1 führenden Kreisbogen, der unter dem Winkel γ gegen K_1 in $w=1$ einmünden möge. Wird daher $\Re(z)<1$ durch $w=f(z)$ auf $\mathfrak{G}$ abgebildet und $\Re(z)<1$ durch $W=g(z)$ auf $\mathfrak{B}$ (wobei R^+ und K_1 in $\arg(z-1)=\frac{\pi}{2}$ übergehen), so bedeutet (3.7)

$$\lim\arg(f^{-1}(w_n)-1)=\arg(g^{-1}(W)-1)=\frac{\pi}{2}+\frac{\pi\gamma}{\pi-\alpha_2}.$$

Nun nehme man $z_n=f^{-1}(w_n)$ und setze in (3.1) $\beta=\frac{\pi}{2}+\frac{\pi\gamma}{\pi-\alpha_2}$ ein, was

$$\underline{\lim}\arg(w_n-1)\geqq\beta-\alpha_2\cdot\frac{2\beta-\pi}{2\pi}=\frac{\pi}{2}+\gamma$$

ergibt, und gemäß unserer Wahl von w_n haben wir hierin tatsächlich das Gleichheitszeichen. Analog schließt man, daß $\overline{L}$ bestmöglich ist, womit Satz 1 bewiesen ist.

4. Der Satz von Visser

Gegeben sei wieder ein einfach zusammenhängendes endliches Gebiet $\mathfrak{G}$ der w-Ebene, für das $w=1$ erreichbarer Randpunkt sei; $w=f(z)$ mit $f(1)=1$ vermittle eine konforme Abbildung von $\mathfrak{H}$: $\Re(z)<1$ auf $\mathfrak{G}$. Unter welchen Annahmen über den Rand von $\mathfrak{G}$ gilt der Randverzerrungssatz von Visser

$$f'(z):\frac{f(z)-1}{z-1}\to 1 \qquad (z\to 1 \text{ im Winkelraum})? \tag{4.1}$$

Satz 2. a. *Besitzt* $\mathfrak{G}$ *in* $w=1$ *ein Richtungsintervall* $(\alpha, \alpha+\pi)$ *und hat der Rand* R *von* $\mathfrak{G}$ *zwei an der Stelle* $w=1$ *dichte Punktfolgen* $\{w_r\}$ *und* $\{w_r'\}$ *mit den Richtungen* α *bzw.* $\alpha+\pi$, *so gilt* (4.1).

b. *Die in* a *gemachten Annahmen über* $\mathfrak{G}$ *und* R *folgen aus* (4.1) *genau dann, wenn irgend eine* Jordan*kurve* C_z *in* $\mathfrak{H}$, *längs der* $\arg(z-1)\to\gamma$ *strebt* $\left(\frac{\pi}{2}<\gamma<\frac{3\pi}{2}\right)$, *durch* $w=f(z)$ *auf eine* Jordan*kurve* C_w *in* $\mathfrak{G}$ *abgebildet wird, längs der* $\arg(w-1)$ *einen Grenzwert hat.*

Benützt man die Tatsache, daß die genannten Eigenschaften von $\mathfrak{G}$ und R äquivalent sind mit der Winkeltreue unserer Abbildung $w=f(z)$ in $z=1$ ([8], S. 447), so erhält man die in der Einleitung formulierte Fassung von Satz 2.

Beweis. **a.** Ohne Einschränkung der Allgemeinheit können wir $\alpha=\frac{\pi}{2}$ annehmen; andernfalls ersetzen wir $f(z)$ durch $\overline{f}(z)=1+e^{-i(\alpha-\pi/2)}(f(z)-1)$. Für irgend eine Folge $\{x_n\}$ positiver Zahlen mit $x_n\to 1-0$ bilden wir die in $\Re(Z)<1$ regulären Funktionen

$$f_n(Z)=\frac{f(Z(1-x_n)+x_n)-f(x_n)}{1-f(x_n)}. \tag{4.2}$$

Durch $W=f_n(Z)$ wird $\Re(Z)<1$ auf ein Gebiet $\mathfrak{G}_n$ abgebildet, das aus $\mathfrak{G}$ durch Anwendung der linearen Abbildung $W-1=(w-1)/(1-f(x_n))$ entsteht. Wegen $f_n(0)=0$ und $f_n(1)=1$ ist $\{f_n(Z)\}$ in $\Re(Z)<1$ normal ([1], S. 73), und es sei $\{f_{n_k}(Z)\}\to F(Z)$ in $\Re(Z)<1$.

Wir zeigen nun: $\{\mathfrak{G}_{n_k}\}\to\mathfrak{K}$: $\Re(W)<1$. Zunächst gilt wegen der schon erwähnten Winkeltreue der Abbildung in $z=1$ neben $1-f(x_n)\to 0$ auch $\arg(1-f(x_n))\to 0$. Unsere Voraussetzung über das Richtungsintervall von $\mathfrak{G}$ in $w=1$ hat also jedenfalls $\mathfrak{K}\supseteq\{\Re(W)<1\}$ zur Folge. Weiter betrachten wir die den beiden Randpunktfolgen $\{w_r\}$ und $\{w_r'\}$ entsprechenden Folgen $\{W_r^{(n)}\}$ und $\{W_r'^{(n)}\}$ von $\mathfrak{G}_n$, die mit den Richtungen $\frac{\pi}{2}+\varepsilon_n$ bzw. $\frac{3\pi}{2}+\varepsilon_n'$ versehen sind $(\varepsilon_n, \varepsilon_n'\to 0,\ n\to\infty)$. Für beliebiges, aber festes $R>0$ gilt in $|W-1|<R$

$$|W_r^{(n)}-W_{r+1}^{(n)}|=\bigl|\,|W_r^{(n)}-1|-|W_{r+1}^{(n)}-1|\,\bigr|+o(1) \tag{4.3}$$

für $n\to\infty$, gleichmäßig für alle r, sobald $W_r^{(n)}$ und $W_{r+1}^{(n)}$ in $|W-1|<R$ liegen. Ferner ist

$$|W_r^{(n)}-1|=M_n\,|w_r-1| \quad\text{mit}\quad M_n=|1-f(x_n)|^{-1}.$$

6*

Wir haben also für (4.3)

$$M_n \, \big|\, |w_r - 1| - |w_{r+1} - 1| \,\big| + o(1) \leqq R \left| 1 - \frac{w_{r+1} - 1}{w_r - 1} \right| + o(1) = o(1)$$

für $n \to \infty$, da mit $n \to \infty$ auch $r \to \infty$ strebt, wenn wir nur die $W_r^{(n)}$ in $|W - 1| < R$ betrachten. Dies hat zur Folge, daß jeder Punkt von $U = 1$, $V \geqq 0$ Häufungspunkt von Randpunkten der Gebietsfolge $\{\mathfrak{G}_n\}$ ist. Da $\{x_n\}$ beliebig war, gilt dasselbe für $\{\mathfrak{G}_{n_k}\}$, d. h. kein Punkt von $U = 1$, $V \geqq 0$ kann zu $\mathfrak{K}$ gehören, oder, wenn man dieselbe Überlegung für $\{w_r'\}$ anstellt, $\Re(W) < 1$ ist der Kern der Folge $\{\mathfrak{G}_{n_k}\}$.

Damit ist $F(Z)$ eine Abbildungsfunktion von $\Re(Z) < 1$ auf $\Re(W) < 1$ mit $F(0) = 0$, und man sieht weiter auf Grund von $\arg(f(x) - 1) \to \pi\,(x \to 1 - 0)$ sofort, daß z. B. der Punkt $Z = \frac{1}{2}$ ein reelles Bild haben muß, d. h. es ist $F(Z) = Z$. Da $F(Z)$ nicht von $\{x_{n_k}\}$ abhängt, gilt somit

$$f_n(Z) \to Z \qquad (\Re(Z) < 1,\ n \to \infty), \tag{4.4}$$

gleichmäßig in jedem kompakten Teil von $\Re(Z) < 1$. Differentiation liefert für diese Z

$$f_n'(Z) \to 1 \qquad (\Re(Z) < 1,\ n \to \infty). \tag{4.5}$$

Jetzt betrachten wir eine beliebige Punktfolge $\{z_n\}$ in $\Re(z) < 1$ mit $z_n = x_n + iy_n \to 1$ im Winkelraum und setzen in (4.2) $x_n = \Re(z_n)$, $Z = Z_n = iy_n/(1 - x_n)$ ein. Aus (4.4) und (4.5) folgt dann

$$\frac{f(z_n) - f(x_n)}{1 - f(x_n)} : \frac{iy_n}{1 - x_n} = A_n \to 1 \quad \text{und} \quad \frac{f'(z_n)(1 - x_n)}{1 - f(x_n)} = B_n \to 1,$$

also

$$f'(z_n) : \frac{f(z_n) - 1}{z_n - 1} = \frac{B_n}{A_n + (1 - A_n)\left(\frac{1 - x_n}{1 - z_n}\right)} \to 1,$$

da $(1 - x_n)/(1 - z_n)$ beschränkt ist. Damit ist der erste Teil von Satz 2 bewiesen.

b. Umgekehrt sei nun (4.1) erfüllt, insbesondere für die Punktfolge $z_n = x_n + Z_0(1 - x_n)$ mit $x_n \to 1 - 0$ und festem Z_0 auf $\Re(Z) = 0$, so daß also $\{z_n\}$ im Winkelraum liegt. Die Familie $\{f_n(Z)\}$ ist wieder normal in $\Re(Z) < 1$, und es sei $\{f_{n_k}(Z)\} \to F(Z)$, gleichmäßig in jedem kompakten Teil von $\Re(Z) < 1$; Differentiation liefert $\{f_{n_k}'(Z)\} \to F'(Z)$ in $\Re(Z) < 1$. Setzt man $Z = Z_0$ ein, so heißt das $(n_k = m)$

$$\frac{f(z_m) - f(x_m)}{1 - f(x_m)} = C_m \to F(Z_0) \quad \text{und} \quad \frac{f'(z_m)(1 - x_m)}{1 - f(x_m)} = D_m \to F'(Z_0),$$

und wenn (4.1) gilt, muß demnach

$$\frac{D_m}{1 - C_m} \cdot \frac{1 - z_m}{1 - x_m} = \frac{D_m}{1 - C_m}(1 - Z_0) \to 1$$

streben, also muß für $\Re(Z) = 0$ und daher für $\Re(Z) < 1$

$$F'(Z)(1 - Z) = 1 - F(Z)$$

gelten. Berücksichtigt man $F(0) = 0$ und integriert, so folgt $F(Z) = Z$, es gilt also (4.4).

Wir nehmen nun an, es gebe eine JORDANkurve C_z von der in **b** beschriebenen Art, und wir wählen für unsere weiterhin beliebige Folge $\{x_n\}$ eine Punktfolge $\{z_n\}$ mit $z_n = x_n + iy_n$ auf C_z. Setzt man diese x_n und $Z = Z_n = iy_n/(1-x_n)$ in (4.4) ein, so ergibt sich

$$\arg(f(Z_n)-1) - \arg(Z_n-1) \to 0,$$

also

$$\arg(f(z_n)-1) - \arg(f(x_n)-1) \to \gamma - \pi \qquad (n\to\infty). \tag{4.6}$$

Hierin hat $\arg(f(z_n)-1)$ einen Grenzwert, woraus dasselbe für $\arg(f(x_n)-1)$ folgt, und da $\{x_n\}$ beliebig war, haben wir für eine gewisse Konstante δ

$$\arg(f(x)-1) \to \delta \qquad (x\to 1-0). \tag{4.7}$$

Daraus läßt sich nun die Winkeltreue unserer Abbildung in $z=1$ erschließen. Sei nämlich $z_n = x_n - iy_n \to 1\,(n\to\infty)$ und $\arg(z_n-1)\to\beta\left(\frac{\pi}{2}<\beta<\frac{3\pi}{2}\right)$, so liefern analoge Überlegungen wie oben, zusammen mit (4.7), daß $\arg(f(z_n)-1)$ gegen $\beta+(\delta-\pi)$ strebt, d. h. es liegt Winkeltreue unserer Abbildung in $z=1$ vor. Damit ist Satz 2 bewiesen.

Wir kommen nun zu der in der Einleitung gemachten Behauptung, daß *die Gültigkeit der* VISSER*schen Relation* (4.1) *allein noch nicht die Winkeltreue der Abbildung* $w=f(z)$ *in* $z=1$ *nach sich zieht.*

Um dies zu beweisen, bilden wir die z- und w-Ebenen (längs $x>1$ und $u>1$ aufgeschlitzt) durch $z_1 = -\log(1-z)$ und $w_1 = -\log(1-w)$ ab. Aus unserer Abbildung $w=f(z)$ wird daher eine Abbildung des Streifens $|\mathfrak{J}(z_1)| < \frac{\pi}{2}$ auf ein Gebiet $\mathfrak{G}_1$ der w_1-Ebene vermöge

$$w_1 = -\log(1-f(1-e^{-z_1})) = f_1(z_1).$$

Bildet man $f_1'(z_1)$, so erhält man gerade den in (4.1) auftretenden Ausdruck:

$$f_1'(z_1) = \frac{f'(z)\,(1-z)}{1-f(z)}.$$

Die Winkeltreue von $w=f(z)$ in $z=1$ bedeutet andererseits für $f_1(z_1)$, daß z. B. jede Gerade $\mathfrak{J}(z_1)=c\left(-\frac{\pi}{2}<c<+\frac{\pi}{2}\right)$ in eine JORDANkurve C_{w_1} transformiert wird, längs der $\mathfrak{J}(w_1)$ einem Grenzwert zustrebt. Wählt man aber für $f_1(z_1)$ die Abbildung des von den beiden Kurven $v_1 = \frac{\pi}{4}\cos\sqrt{|u_1|} \pm \frac{\pi}{2}$ $(w_1 = u_1 + iv_1)$ begrenzten L-Streifens $\mathfrak{G}_1$ auf $|\mathfrak{J}(z_1)| < \frac{\pi}{2}$, so strebt einerseits $f_1'(z_1)\to 1$ für $\mathfrak{R}(z_1)\to+\infty$, gleichmäßig in $|\mathfrak{J}(z_1)| \leqq \frac{\pi}{2}-\delta$ für jedes $\delta>0$ ([13], S. 281); andererseits wird aber $\mathfrak{J}(z_1)=0$ auf eine JORDANkurve abgebildet, die sich für größer werdendes z_1 der Kurve $v_1 = \frac{\pi}{4}\cos\sqrt{|u_1|}$ unbegrenzt nähert ([13], S. 281), also besitzt die Ordinate dieser Kurve keinen Grenzwert. Damit ist unsere Behauptung bewiesen.

Aus Satz 2 folgt noch leicht der folgende Zusatz.

Zusatz. *Ist unsere Abbildung* $w=f(z)$ *in* $z=1$ *winkeltreu, so wird jede* JORDAN*kurve* C_z, *die im Winkelraum verläuft und in* $z=1$ *mit einer L-Tangente einmündet, in eine* JORDAN*kurve* C_w *mit denselben Eigenschaften abgebildet.*

Zum Beweis betrachten wir z. B. eine Kurve C_z, die in $z = 1$ die Steigung 0 hat, und längs deren Bild C_w also $\arg(w-1)$ gegen einen Grenzwert strebt, den wir (eventuell nach Rotation von $\mathfrak{G}$ um $w = 1$) als Null annehmen dürfen. Nun sieht man leicht, daß C_z in $z = 1$ genau dann eine L-Tangente mit der Steigung 0 hat, wenn die Steigung jeder der in einem Punkt $z \in C_z$ gezogenen vier Haupttangenten gegen Null strebt mit $z \to 1$. Diese Eigenschaft ist aber invariant bei unserer Abbildung vermöge $w = f(z)$, da wegen (4.1) $\arg f'(z) \to 0$ strebt für $z \to 1$ längs C_z. Also hat auch C_w eine L-Tangente mit der Steigung 0 in $z = 1$.

Literaturverzeichnis

[1] C. Carathéodory, Conformal representation. Cambridge 1932.

[2] J. Ferrand, Étude de la représentation conforme au voisinage de la frontière. Ann. sci. École norm. sup. (3) **59**, 43—106 (1942).

[3] J. Ferrand, Nouvelle démonstration d'un théorème de M. Ostrowski. C. R. Acad. Sci., Paris **220**, 550—551 (1945).

[4] C. Gattegno et A. Ostrowski, Représentation conforme à la frontière. Mémorial des Sciences Mathématiques. Bd. 109—110. Paris 1949.

[5] J. Lelong-Ferrand, Sur la représentation conforme des bandes. J. d'analyse math. **2**, 51—71 (1952).

[6] P. Montel, Sur la représentation conforme. J. Math. pur. appl. (7) **3**, 1—54 (1917).

[7] A. Ostrowski, Über den Habitus der konformen Abbildung am Rande des Abbildungsbereiches. Acta math. **64**, 81—184 (1935).

[8] A. Ostrowski, Zur Randverzerrung bei konformer Abbildung. Prace mat.-fiz. **44**, 371—471 (1936).

[9] C. Visser, Über beschränkte analytische Funktionen und die Randverhältnisse bei konformen Abbildungen. Math. Ann. **107**, 28—39 (1932).

[10] J. L. Walsh, On distortion at the boundary of a conformal map. Proc. nat. Acad. Sci. USA **36**, 152—156 (1950).

[11] J. L. Walsh and L. Rosenfeld, On the boundary behavior of a conformal map. Erscheint in den Trans. Amer. math. Soc.

[12] S. Warschawski, On the preservation of angles at a boundary point in conformal mapping. Bull. Amer. math. Soc. **42**, 674—680 (1936).

[13] S. Warschawski, On conformal mapping of infinite strips. Trans. Amer. math. Soc. **51**, 280—335 (1942).

[14] J. Wolff, Démonstration d'un théorème sur la conservation des angles dans la représentation conforme d'un domaine au voisinage d'un point frontière. Proc. Akad. Amsterdam **38**, 46—50 (1935).

Eingegangen am 16. 3. 1954

ON THE BOUNDARY BEHAVIOR OF A CONFORMAL MAP

BY

J. L. WALSH AND L. ROSENFELD

The object of this paper is to indicate the immediate usefulness of Carathéodory's theory of the conformal mapping of variable regions in the study of boundary behavior of a fixed but arbitrary conformal map. We study especially the mapping of an infinite strip and its behavior at infinity.

Studies of an infinite strip by other methods have previously been made, especially by Ahlfors, Ferrand, Ostrowski, and Warschawski; for a résumé see Gattegno and Ostrowski [6]. The present results are more directly geometric than these previous ones in both method and conclusion; broadly speaking, they are in some respects more and in other respects less general than the previous ones.

The theory of conformal mapping of variable regions introduced by Carathéodory in 1912 was employed by Montel in 1917 to study the properties of prime ends under conformal mapping. That theory has more recently been used in the study of boundary behavior of conformal maps by Ferrand [7; 8], emphasized by Walsh [2], and used for the study of strips by Madame Lelong-Ferrand [3]. The essential difference between the latter and the present paper is that here we consistently use both translation and stretching of the original region to obtain a sequence of variable regions possessing a kernel, whereas Madame Lelong-Ferrand uses primarily translation. The present results are thus more general, both where the width of the strip (the variable $2\phi(u)$, in the notation of §2 below) has an infinite number of limit values, and more especially where that limit is zero or infinite.

We introduce a new condition (property B, below) on the boundary of an infinite strip, which is useful in the study of conformal mapping of the strip. The intrinsic properties of this condition are studied in §1, and applications to conformal mapping in successively more general situations are considered in §§2–5. The extension of property B and its implications for finite boundary points are studied in §6, and the relation of property B to various other conditions on the boundary of an infinite strip is investigated in §7.

Property B is exhibited as a sufficient (but not necessary) condition for the fundamental asymptotic relations, such as (2.1), a less restrictive condition than that for an L-strip.

1. **Property B.** *If $\phi(u)$ is a real function of the real variable u defined for*

Presented to the Society, December 27, 1951 and April 25, 1953; received by the editors October 28, 1953.

$u_1 \leqq u < +\infty$ $(-\infty \leqq u_1 < +\infty)$, *and if we have uniformly in every interval I*: $|U| \leqq U_0$ *the equation*

$$\lim_{u \to +\infty} \frac{\phi[U\phi(u) + u]}{\phi(u)} = 1, \tag{1.1}$$

then $\phi(u)$ *is said to have property* B *(at infinity)*. We do not require that $\phi(u)$ be single-valued, but do require that (1.1) hold for any choice of values of $\phi(u)$. It will appear from our use of (1.1) that the same value $\phi(u)$ should be used in the denominator and inside the square bracket in the numerator. But an arbitrary value of $\phi[U\phi(u)+u]$ may be used, even in the case $U=0$. In particular, we deduce

$$\lim_{u \to +\infty} \frac{\max\ \phi(u)}{\min\ \phi(u)} = 1,$$

where the maximum and minimum are taken over $\phi(u)$ for each fixed u.

When we add (or subtract) two multiple-valued functions of this kind, we add (or subtract) all functional values for each value of the independent variable.

Property B is the fundamental requirement that we impose on the boundary of an infinite strip. We now develop for reference some easily proved intrinsic consequences of the definition, without reference to conformal mapping. We retain the notation used in the definition.

THEOREM 1.1. *If* $\phi(u)$ *has property* B *at infinity, and if* $\alpha(u, t)$ *defined for* $u > u_1$, $t_0 \leqq t \leqq t_1$ $(-\infty \leqq t_0 < t_1 \leqq \infty)$, *has the property*

$$\left| \frac{\alpha(u, t) - u}{\phi(u)} \right| \leq M, \qquad u \geqq u_1,\ t_0 \leqq t \leqq t_1,$$

then we have

$$\lim_{u \to +\infty} \frac{\phi[\alpha(u, t)]}{\phi(u)} = 1$$

uniformly for all t, $t_0 \leqq t \leqq t_1$.

We set

$$\alpha(u, t) \equiv B(u, t)\phi(u) + u, \tag{1.2}$$

whence $|B(u, t)| \leqq M$ for $u \geqq u_1$, $t_0 \leqq t \leqq t_1$. Thus we have

$$\frac{\phi[\alpha(u, t)]}{\phi(u)} \equiv \frac{\phi[B(u, t)\phi(u) + u]}{\phi(u)}, \tag{1.3}$$

and the theorem follows from (1.1).

The proof of Theorem 1.2 is similar to the proof just given and is omitted:

THEOREM 1.2. *If $\phi(u)$ has property* B *at infinity, and if $u_n \to +\infty$, $u'_n \to +\infty$ with*

$$\limsup_{n\to\infty} \left| \frac{u_n - u'_n}{\phi(u_n)} \right| < \infty,$$

then we have

$$\lim_{n\to\infty} \frac{\phi(u_n)}{\phi(u'_n)} = 1.$$

It can be verified immediately that the function $\phi(u) \equiv u^{1/2}$ has property B at infinity. Nevertheless, that property limits the order of $\phi(u)$ as $u \to +\infty$:

THEOREM 1.3. *If $\phi(u)$ has property* B *at infinity, then*

$$\lim_{u\to\infty} \phi(u)/u = 0.$$

If this conclusion is false, there exists a sequence $u_n \to \infty$ such that

$$\lim_{n\to\infty} \frac{u_n}{\phi(u_n)} = \lambda, \tag{1.4}$$

with $|\lambda| < \infty$. Of course we have $\phi(u_n) \to +\infty$, whence by (1.4) and (1.1)

$$\lim_{n\to\infty} \frac{\phi[((a - u_n)/\phi(u_n))\phi(u_n) + u_n]}{\phi(u_n)} = 1, \tag{1.5}$$

where the constant a is arbitrary. The numerator in the first member of (1.5) reduces to $\phi(a)$, so $\phi(u)$ is identically constant (not zero, by (1.1)), in contradiction to (1.4).

THEOREM 1.4. *If $\phi(u)$ has property* B *at infinity and if $\psi(u)$ is a real function of u defined for $u \geqq u_1$, which satisfies*

$$\lim_{u\to+\infty} \psi(u)/\phi(u) = 1, \tag{1.6}$$

then $\psi(u)$ has property B *at infinity.*

It follows from (1.6) and from Theorem 1.3 that $\psi(u)/u \to 0$ as $u \to \infty$. Then for U in I: $|U| \leqq U_0$ we have uniformly

$$U\psi(u) + u = u\left(U\frac{\psi(u)}{u} + 1\right) \to +\infty.$$

We write

$$\frac{\psi[U\psi(u) + u]}{\psi(u)} = \frac{\psi[U\psi(u) + u]}{\phi[U\psi(u) + u]} \frac{\phi[U(\psi(u)/\phi(u))\phi(u) + u]}{\phi(u)} \frac{\phi(u)}{\psi(u)}; \tag{1.7}$$

the first factor in the second member approaches unity as $u\to+\infty$ by (1.6) because the common argument of ψ and ϕ becomes infinite. Thus each factor in the second member of (1.7) approaches unity uniformly for U in I, and the theorem is established.

It is clear from (1.1) that if $\phi(u)$ has property B at infinity and if the constant $\lambda(\neq 0)$ is arbitrary, then $\lambda\phi(u)$ has property B at infinity. From this remark and Theorem 1.4 we have

THEOREM 1.5. *Theorem* 1.4 *remains valid if* (1.6) *is replaced by*

$$\lim_{u\to+\infty} \psi(u)/\phi(u) = \lambda \neq 0. \tag{1.8}$$

Theorem 1.5 cannot be strengthened merely by replacing (1.8) by

$$0 < a \leqq \left|\frac{\phi(u)}{\psi(u)}\right| \leqq b < \infty, \qquad u \geqq u_1,$$

as the reader may show from the counter-example $\phi(u)\equiv 1$, $\psi(u)\equiv 1+2^{-1}\sin u$.

Property B persists under addition, if an auxiliary condition is provided:

THEOREM 1.6. *If $\phi_1(u)$ and $\phi_2(u)$ have property* B *at infinity, and if we have for $u\geqq u_1$*

$$0 < a \leqq \phi_1(u)/\phi_2(u) \leqq b < \infty, \tag{1.9}$$

then $\phi(u)\equiv\phi_1(u)+\phi_2(u)$ has property B *at infinity.*

Theorem 1.6 follows from the identity

$$\frac{\phi[U\phi(u)+u]}{\phi(u)} - 1 \equiv \frac{[\alpha_1(u)-1]\phi_1(u)/\phi_2(u) + \alpha_2(u) - 1}{1+\phi_1(u)/\phi_2(u)},$$

$$\alpha_1(u) \equiv \frac{\phi_1[(U+U\phi_2(u)/\phi_1(u))\phi_1(u)+u]}{\phi_1(u)},$$

$$\alpha_2(u) \equiv \frac{\phi_2[(U+U\phi_1(u)/\phi_2(u))\phi_2(u)+u]}{\phi_2(u)},$$

where $\alpha_1(u)\to 1$ and $\alpha_2(u)\to 1$ uniformly for U in I.

Since we have not required singlevaluedness of a function in order that it have property B, it may occur in Theorem 1.6 that the locus $v=\phi_1(u)+\phi_2(u)$ separates the plane even when neither of the loci $v=\phi_1(u)$ and $v=\phi_2(u)$ separates the plane. For instance, suppose in the interval I_0: $|u|<2$ the function $\phi_1(u)$ consists of $v=1$ together with the segment of the line $v=1+u$, $0\leqq u\leqq 1$, and $\phi_2(u)$ in I_0 consists of $v=1$ together with the segment of the line $v=2-u$, $0\leqq u\leqq 1$. Then $\phi_1(u)+\phi_2(u)$ consists of $v=2$, the two line segments $v=2+u$ and $v=3-u$, $0\leqq u\leqq 1$, and also the segment $v=3$, $0\leqq u\leqq 1$. This ambiguity leads to no difficulty, by virtue of our convention that such relations as (1.1) and (1.9) hold for all choices of functional values.

The locus $v=\phi(u)$, $-\infty \leqq u_0 \leqq u < +\infty$, is said to have *a horizontal L-tangent (at infinity)* if we have ($u_1\to\infty$, $u_2\to\infty$)

$$\phi(u_2) - \phi(u_1) \over u_2 - u_1 \to 0. \tag{1.10}$$

It follows that $\phi(u)/u \to 0$ as $u\to\infty$, for if u_n is any sequence ($u_n\to\infty$) for which $\phi(u_n)/u_n$ approaches a limit finite or infinite we have by (1.10)

$$0 = \lim_{u\to\infty}\left[\lim_{u_n\to\infty} \frac{\phi(u_n)/u_n - \phi(u)/u_n}{1 - u/u_n}\right] = \lim_{u_n\to\infty} \frac{\phi(u_n)}{u_n}.$$

If $v=\phi(u)$ has a horizontal L-tangent at infinity, then $\phi(u)$ has property B *at infinity*, for with $0<|U|<U_0$ and $u_2=U\phi(u_1)+u_1$ we have by (1.10)

$$U\frac{\phi[U\phi(u_1)+u_1] - \phi(u_1)}{U\phi(u_1)} \to 0$$

uniformly with respect to U since $u_1\to\infty$ implies $u_2\to\infty$ uniformly; of course $U=0$ in (1.1) is trivial.

2. **Symmetric infinite strips.** We turn now to our main purpose, study of the conformal mapping of infinite strips. Symmetric strips illustrate the method in its simplest form, and are important as a preliminary topic. We postpone detailed references to the literature until the more general results are established.

THEOREM 2.1. *Let S be a simply connected region of the $w(=u+iv)$-plane which contains the segment $u_1\leqq u$ of the axis of reals and is symmetric in the axis of reals. Let the boundary of S in the half-plane $u_1\leqq u$ consist of the two loci $v=\phi(u)$, $v=-\phi(u)$, where $\phi(u)\geqq 0$ and has property* B *at infinity. Let $w=f(z)\equiv u(z)+iv(z)$ map the infinite strip Σ: $|y|<\pi/2$ of the $z(=x+iy)$-plane onto S in such a way that $f(z)$ is real when z is real and $\lim_{x\to\infty} u(x+iy)=+\infty$. Then if x_n is any sequence of real numbers with $x_n\to+\infty$, we have*

$$\lim_{n\to\infty} \frac{f(z+x_n) - f(x_n)}{\phi[u(x_n)]} = \frac{2z}{\pi} \tag{2.1}$$

for z in Σ, uniformly on any closed bounded subset of Σ.

We suppose here and in similar cases below that the region S is actually a strip, in the sense that each point $w=u+iv$ of S lies interior to a vertical finite line segment bounded by some point of each of the loci $v=\pm\phi(u)$. The purpose of this hypothesis is to exclude such a region as the w-plane slit along the two infinite segments $u\geqq 0$, $v=\pm 1$; even such a slit region, however, can be treated by our methods with little or no modification.

The function $f(z)$ is uniquely determined except for an arbitrary horizontal translation of Σ, namely $z'=z-\alpha$ where α is real. We introduce the notation

$$(2.2)\qquad f_n(z) \equiv u_n(z) + iv_n(z) \equiv \frac{f(z+x_n)-f(x_n)}{\phi[u(x_n)]}, \qquad w_n = f(x_n).$$

Since $w=f(z+x_n)$ maps Σ onto S, the transformation $w'=f(z+x_n)-f(x_n)$ maps Σ onto S_n', namely S horizontally translated so that w_n in S becomes $w'=0$ in S_n'. The denominator in (2.2) has the property of changing the size but not the orientation of S_n' (leaving the origin fixed) so that the boundary of the new image S_n of Σ under the transformation $W=U+iV=f_n(z)$ passes through the point $W=i$. *We prove that the kernel of the regions S_n in the sense of Carathéodory consists of the strip* $|V|<1$.

If we set $w_n=f(x_n)=u(x_n)+iv(x_n)$, the strip S_n' in the plane of $w'=u'+iv'$ is bounded by the curves whose ordinates are $v'=\pm\phi[u'+u(x_n)]$, respectively. We now set $W=w'/\phi[u(x_n)]$, $V=v'/\phi[u(x_n)]$, $U=u'/\phi[u(x_n)]$, so S_n is bounded by the curves whose ordinates are

$$V = \pm\,\phi\,[U\phi[u(x_n)] + u(x_n)]/\phi[u(x_n)].$$

Since by (1.1) these ordinates approach the respective values $V=\pm 1$ uniformly on every interval $|U|\leqq U_0$, it follows that any region $|V|<1-\delta$, $|U|<U_0$, $\delta>0$, lies in all S_n for n sufficiently large; but no region containing both $W=0$ and a point exterior to $|V|<1$ lies in all S_n for n sufficiently large. Consequently the regions S_n converge to the kernel $|V|<1$. Equation (2.1) now follows by Carathéodory's results, for we have $f_n(0)=0$, $f_n'(0)>0$.

Theorem 2.1 has various applications:

THEOREM 2.2. *With the hypothesis of Theorem 2.1 let $z_n=x_n+iy_n$ and $z_n'=x_n'+iy_n'$ be two sequences of points in Σ such that $|y_n|<a<\pi/2$, $|y_n'|<a<\pi/2$, $|x_n-x_n'|<b<\infty$. Then we have*

$$(2.3)\qquad \lim_{n\to\infty}\frac{\phi[u(z_n')]}{\phi[u(z_n)]} = 1,$$

$$(2.4)\qquad \lim_{n\to\infty}\left[\frac{f(z_n)-f(z_n')}{\phi[u(z_n)]} - \frac{2}{\pi}(z_n - z_n')\right] = 0.$$

Substitution $z=iy_n$ in an equation equivalent to (2.1) yields

$$(2.5)\qquad \lim_{n\to\infty}\left[\frac{f(z_n)-f(x_n)}{\phi[u(x_n)]} - \frac{2i}{\pi}y_n\right] = 0,$$

and the corresponding equation for z_n' is

$$(2.6)\qquad \lim_{n\to\infty}\left[\frac{f(z_n')-f(x_n')}{\phi[u(x_n')]} - \frac{2i}{\pi}y_n'\right] = 0;$$

substitution $z=x_n'-x_n$ in an equation equivalent to (2.1) yields

$$\lim_{n\to\infty}\left[\frac{u(x_n')-u(x_n)}{\phi[u(x_n)]}-\frac{2}{\pi}(x_n'-x_n)\right]=0. \tag{2.7}$$

The real part of the first member of (2.5) is

$$\lim_{n\to\infty}\frac{u(z_n)-u(x_n)}{\phi[u(x_n)]}=0. \tag{2.8}$$

Theorem 1.2 in conjunction with (2.8) and (2.7) respectively yields

$$\lim_{n\to\infty}\frac{\phi[u(z_n)]}{\phi[u(x_n)]}=\lim_{n\to\infty}\frac{\phi[u(x_n)]}{\phi[u(x_n')]}=1,$$

and we similarly have

$$\lim_{n\to\infty}\frac{\phi[u(z_n')]}{\phi[u(x_n')]}=1.$$

Equation (2.3) is now immediate, and (2.4) is a consequence of (2.5), (2.6), (2.7).

A consequence of (2.1) or (2.4) is especially significant geometrically; in (2.5) we suppose $y_n\to y_0$, $|y_0|<\pi/2$, whence

$$\lim_{n\to\infty}\frac{v(z_n)}{\phi[u(x_n)]}=\frac{2y_0}{\pi}; \tag{2.9}$$

the ratio of the ordinates y_n and $\pi/2$ approaches the same limit as the ratio of the corresponding ordinates in the w-plane. By way of a converse, we prove

THEOREM 2.3. *With the hypothesis of Theorem 2.1 let γ be a curve in S defined for $u>u_1$ by the equation $v=\psi(u)$, with $\lim_{u\to\infty}\psi(u)/\phi(u)=\lambda$. If Γ is the image in the z-plane of γ, we have on Γ*

$$\lim_{x\to\infty} y=\frac{\pi\lambda}{2}\cdot \tag{2.10}$$

We suppose first $|\lambda|<1$, and choose $\delta(>0)$ so that $|\lambda\pm\delta|<1$. Denote by $\Gamma_{\lambda+\delta}$ and $\Gamma_{\lambda-\delta}$ the lines $y=\pi(\lambda+\delta)/2$ and $y=\pi(\lambda-\delta)/2$, respectively, and by $\gamma_{\lambda+\delta}$ and $\gamma_{\lambda-\delta}$ their images in the w-plane. By (2.9) we have for z on $\Gamma_{\lambda\pm\delta}$

$$\lim_{z\to\infty}\frac{v(z)}{\phi[u(x)]}=\lambda\pm\delta \tag{2.11}$$

respectively. For u sufficiently large it follows from (2.11) that γ lies between $\gamma_{\lambda+\delta}$ and $\gamma_{\lambda-\delta}$, so it follows from the topological properties of the conformal map that for u (or x) sufficiently large, Γ lies between $\Gamma_{\lambda+\delta}$ and $\Gamma_{\lambda-\delta}$, which is equivalent to (2.10).

3. Nonsymmetric strips. Theorem 2.1 is to be regarded as preliminary; we shall prove several analogues under varying hypotheses which apply to strips not necessarily symmetric.

THEOREM 3.1. *Let S be an infinite simply connected strip in the $w(=u+iv)$-plane which contains the segment $u>u_1$, $v=0$, and whose boundaries can be represented in the form $v=\phi_+(u)$ and $v=\phi_-(u)$, with $\phi_+(u)>0$, $\phi_-(u)<0$, $\lim_{u\to\infty}\phi_-(u)/\phi_+(u)=\lambda$, where $\phi_+(u)$ and $\phi_-(u)$ are not necessarily single valued. Suppose $\theta(u)\equiv\phi_+(u)-\phi_-(u)$ has property* B *at infinity. Let $w=f(z)=u(z)+iv(z)$ map the infinite strip Σ_z: $|y|<\pi/2$ onto S so that $\lim_{z\to\infty}u(z)=+\infty$. Then if x_n is any sequence of real points with $x_n\to+\infty$, we have*

$$\lim_{n\to\infty}\frac{f(z+x_n)-f(x_n)}{\theta[u(x_n)]}=\frac{z}{\pi} \tag{3.1}$$

throughout Σ_z, uniformly on any closed bounded subset of Σ_z.

The case $\lambda=-\infty$ is treated by interchanging the roles of $\phi_+(u)$ and $\phi_-(u)$, whence we have $\lambda=0$. The case $\lambda=0$ is to be treated separately by special methods later, so for the present we assume $0>\lambda>-\infty$.

From the conditions on $\phi_+(u)$ and $\phi_-(u)$ we have

$$\lim_{u\to+\infty}\frac{\phi_+(u)-\phi_-(u)}{\phi_+(u)}=1-\lambda,$$

and from Theorem 1.5 it follows that $\phi_+(u)$ has property B; similarly it follows that $\phi_-(u)$ has property B.

For each u sufficiently large we have $\min\theta(u)>\max\phi_+(u)$. Indeed, we note

$$\frac{\min\theta(u)}{\max\phi_+(u)}=\frac{\min\phi_+(u)}{\max\phi_+(u)}-\frac{\max\phi_-(u)}{\max\phi_+(u)};$$

the first term of the second member approaches unity and the second term approaches $-\lambda$ as u becomes infinite. Similarly we have for u sufficiently large $\min\theta(u)>-\min\phi_-(u)$.

The asymptotic behavior of $f(z)$ depends only on the behavior of the boundary of S in the neighborhood of $z=\infty$, so it is no essential loss of generality to *assume*, as we do, *the relation* $\min\theta(u)>\max\phi_+(u)$ *for all u for which $\theta(u)$ is defined.*

Denote by S^* the infinite strip in the w-plane which contains S and whose boundaries are contained in the loci $v=\pm\theta(u)$. The locus $v=\pm\theta(u)$, where all allowable values of $\phi_+(u)$ and $\phi_-(u)$ are admitted in the equation $\theta(u)\equiv\phi_+(u)-\phi_-(u)$, may separate the plane into even an infinite number of regions. One such region contains S and is denoted by S^*. Let Σ_ζ in the $\zeta(=\xi+i\eta)$-plane be the strip $|\eta|<\pi/2$, and let $w=g(\zeta)$ map Σ_ζ onto S^* so that $\xi=+\infty$ corresponds to $u=+\infty$. Denote by γ_+ and γ_- the loci $\eta=\eta_+(\xi)$

and $\eta=\eta_-(\xi)$, namely the respective images in the ζ-plane of the loci $v=\phi_+(u)$ and $v=\phi_-(u)$. By Theorem 2.3 we have

$$\lim_{\xi\to\infty}\eta_+(\xi)=\frac{1}{1-\lambda}\frac{\pi}{2},\qquad \lim_{\xi\to\infty}\eta_-(\xi)=\frac{\lambda}{1-\lambda}\frac{\pi}{2}. \tag{3.2}$$

Let Ω_ζ denote the image of S in the ζ-plane under the map $w=g(\zeta)$, and let finally

$$z=-\log(1-s),\qquad s=\sigma+i\tau, \tag{3.3}$$

where z is real when s is real, and

$$\zeta=-\log(1-t), \tag{3.4}$$

where ζ is real when t is real. The transformation (3.3) maps Σ_z onto the half-plane $\sigma<1$, and it follows from (3.2) that (3.4) maps Ω_ζ onto a region in the t-plane whose boundary has an angle $\pi/2$ at $t=1$ and which contains some interval $t_1<t<1$, $t_1<1$.

Let x_n be an arbitrary sequence of real points in the z-plane with $x_n\to+\infty$, and set $s_n=s(x_n)$, $w_n=f(x_n)$, $\zeta_n=\xi_n+i\eta_n=\zeta(w_n)$, $t_n=t(\zeta_n)$. It is well known (Lindelöf) that in the transformation from the t-plane to the s-plane angles are transformed proportionally, so we have from (3.4) and (3.2)

$$\lim_{n\to\infty}\eta_n=-\lim_{n\to\infty}\arg(1-t_n)=\frac{1+\lambda}{1-\lambda}\frac{\pi}{4}. \tag{3.5}$$

From (3.5) and Theorem 2.3, we have

$$\lim_{n\to\infty}\frac{v(\zeta_n)}{\theta[u(\zeta_n)]}=\frac{1}{2}\frac{1+\lambda}{1-\lambda},$$

which with a slight change of notation is written

$$\lim_{n\to\infty}\frac{v(x_n)}{\theta[u(x_n)]}=\frac{1}{2}\frac{1+\lambda}{1-\lambda}. \tag{3.6}$$

We proceed further using the method of proof of Theorem 2.1, by setting here

$$W=U+iV=f_n(z)\equiv\frac{f(z+x_n)-f(x_n)}{\theta[u(x_n)]}; \tag{3.7}$$

the denominator here is not the same as in (2.2); moreover in (2.2) the number $f(x_n)$ is real but not necessarily so in (3.7); in (2.2) we have $f_n'(0)>0$ but not necessarily in (3.7). We write

$$\frac{dw}{dz}=\frac{dw}{d\zeta}\cdot\frac{d\zeta}{dt}\cdot\frac{dt}{ds}\cdot\frac{ds}{dz}=g'(\zeta)(1-t)^{-1}t'(s)e^{-z},$$

$$\arg f'(x_n) = \arg g'(\zeta_n) - \arg(1 - t_n) + \arg t'(s_n). \tag{3.8}$$

Theorem 2.1 applies to the map $w = g(\zeta)$, so from differentiation of (2.1) we have $\arg g'(\zeta_n) \to 0$. From this fact and Visser's formula (see for instance [2, equation (8)]) we deduce by (3.8) that $\arg f'(x_n) \to 0$. Then by (3.6) the image of S under the sequence of transformations (3.7) consists of a sequence of regions converging to a kernel, precisely the infinite strip $|V| < 1/2$. Equation (3.1) follows by the Carathéodory theory.

We treat now the case $\lambda = 0$, hitherto excluded. From the assumption $\phi_-(u)/\phi_+(u) \to 0$ it follows that for each u sufficiently large we have $\min \phi_+(u) > \max [-\phi_-(u)]$; since the asymptotic behavior of $f(z)$ as $x \to \infty$ depends only on the behavior of the boundary of S in the neighborhood of infinity, it is no loss of generality for us to assume, as we do, $\min \phi_+(u) > \max [-\phi_-(u)]$ for all u for which these functions are defined. As $u \to \infty$ we have $\phi_+(u)/\theta(u) \to 1$, so in order to prove (3.1) it is sufficient to prove

$$\lim_{n\to\infty} \frac{f(z + x_n) - f(x_n)}{\phi_+[u(x_n)]} = \frac{z}{\pi};$$

the function $\phi_+(u)$ has property B, by Theorem 1.4; the method already used now applies without essential change, where $\phi_+(u)$ takes the role previously taken by $\theta(u)$, and this completes the proof of Theorem 3.1.

A strip bounded by two loci $v = \phi_+(u)$ and $v = \phi_-(u) < \phi_+(u)$ is called an L-strip if each of these loci has an L-tangent at infinity.

When S is an L-strip, the real part of (3.1) is given by Warschawski [4, (18.7)]. Indeed, for this case (3.1) can be readily proved by his methods. If $z_1 = x_1 + iy_1$, $z_2 = x_2 + iy_2$, $x_2 > x_1$, we have

$$f(z_2) - f(z_1) = \int_{z_1}^{z_2} f'(z)dz,$$

where the path of integration is a line segment. By [4, Theorem II] if $|z - z_1| \leq M$,

$$f'(z)/f'(z_1) \to 1 \quad \text{as} \quad x_1 \to +\infty,$$

uniformly for $|y_1|$, $|y| \leq \beta < \pi/2$. But by [4, Theorem X(ii)]

$$f'(z_1) = (1/\pi)\theta[u(x_1)][1 + o(1)] \quad \text{as} \quad x_1 \to +\infty,$$

uniformly for $|y_1| \leq \beta$. Combination of these relations yields

$$\frac{f(z_2) - f(z_1)}{\theta[u(x_1)]} = \frac{z_2 - z_1}{\pi} + o(1),$$

uniformly for $|z_1 - z_2| \leq M$, $|y_1| \leq \beta$. Equation (3.1) follows with $z_1 = x_n$ $z_2 = x + x_n$.

Theorem 3.1 has applications to the further study of the behavior of the mapping function.

THEOREM 3.2. *With the hypothesis of Theorem* 3.1 *let* $z_n = x_n + iy_n$ *be a sequence of points of* Σ_z *such that* $x_n \to +\infty$, $\limsup |y_n| < \pi/2$. *Then we have*

$$\lim_{n\to\infty} \frac{f'(z_n)}{\theta[u(z_n)]} = \frac{1}{\pi}, \tag{1}$$

$$\lim_{n\to\infty} \frac{f^{(\nu)}(z_n)}{\theta[u(z_n)]} = 0, \qquad \nu > 1, \tag{2}$$

$$\lim_{n\to\infty} \arg f'(z_n) = 0. \tag{3}$$

Equations (1) and (3) were proved by Warschawski [4, pp. 315, 288] with the stronger hypothesis that S is an L-strip, but without our restrictions $\phi_-(u)/\phi_+(u) \to \lambda$ and that a segment $u_1 < u < +\infty$, $v = 0$ lie in S. He proves (3) uniformly in $|y| < \pi/2$ without the restriction $\limsup |y_n| < \pi/2$.

We differentiate the two members of (3.1) and then set $z = iy_n$; this set of points is contained in a closed bounded subset of S, so we have

$$\lim_{n\to\infty} \frac{f'(z_n)}{\theta[u(x_n)]} = \frac{1}{\pi}. \tag{3.9}$$

Equation (3.1) yields

$$\frac{u(z_n) - u(x_n)}{\theta[u(x_n)]} \to 0,$$

from which Theorem 1.2 and (3.9) yield (1). Equation (3) follows at once from (1), and (2) follows from (3.1) by repeated differentiation.

We leave to the reader a proof similar to that of Theorem 3.2 which establishes

THEOREM 3.3. *With the hypothesis of Theorem* 3.1, *let* $z_n = x_n + iy_n$, $z_n' = x_n' + iy_n'$ *be two sequences of points of* Σ_z *such that*

$$x_n \to +\infty$$

with $\limsup_{n\to\infty} |x_n - x_n'| < +\infty$, $\limsup_{n\to\infty} |y_n| < \pi/2$, $\limsup_{n\to\infty} |y_n'| < \pi/2$. *Then we have*

$$\lim_{n\to\infty} \left[\frac{u(z_n) - u(z_n')}{\theta[u(z_n)]} - \frac{x_n - x_n'}{\pi} \right] = 0; \tag{1}$$

If $\lim_{n\to\infty} y_n = \beta$, $|\beta| < \pi/2$, *then*

$$\lim_{n\to\infty} \frac{v(z_n)}{\theta[u(z_n)]} = \frac{1}{2} \frac{1+\lambda}{1-\lambda} + \frac{\beta}{\pi}. \tag{2}$$

Part (1) is proved by Warschawski [4, p. 324] and part (2) also by Warschawski [4, p. 315]; he supposes that S is an L-strip and proves (2) uniformly for $|\beta| < \pi/2$, without some of our restrictions.

4. **Some extensions.** We proceed to prove an extension of Theorem 3.1:

THEOREM 4.1. *Theorem* 3.1 *remains valid if we replace* $\lim \phi_-(u)/\phi_+(u) = \lambda$ *by the condition*

$$0 < a \leqq |\phi_-(u)/\phi_+(u)| \leqq b < +\infty$$

and require that both $\phi_+(u)$ *and* $\phi_-(u)$ *have property* B.

It follows here by Theorem 1.6 that $\theta(u) \equiv \phi_+(u) - \phi_-(u)$ has property B· For each u sufficiently large, we have $\min \theta(u) > \max \phi_+(u)$. Indeed, in the equation

$$\frac{\min \theta(u)}{\max \phi_+(u)} = \frac{\min \phi_+(u)}{\max \phi_+(u)} - \frac{\min [-\phi_-(u)]}{\max \phi_+(u)},$$

the first term of the second member approaches unity as u becomes infinite and the second term is not less than $a(>0)$ for all u. As in the proof of Theorem 3.1 we assume $\min \theta(u) > \max \phi_+(u)$, $\min \theta(u) > -\min \phi_-(u)$ for all u, which involves no essential loss of generality.

We proceed to prove Theorem 4.1 in several steps, by successive specializations of the region S.

LEMMA 1. *Theorem* 4.1 *is true if* $\phi_+(u) \equiv \pi/2$.

The function $\theta(u) \equiv \pi/2 - \phi_-(u)$ has property B, by Theorem 1.6. Then we can apply Theorem 2.1 to the function $w' = u' + iv' = g(z')$, $z' = x' + iy'$, which maps the strip $|y'| < \pi$ onto the symmetric strip of the w'-plane whose boundary consists of the loci $v' = \pm\theta(u')$. Lemma 1 follows by combining the transformations $w' = g(z')$, $w' = w - i\pi/2$, $z' = z - i\pi/2$.

LEMMA 2. *With the hypothesis of Theorem* 2.1 *let* γ_σ *be a curve in* S *defined for* $u_1 < u < +\infty$ *by the equation* $v = \sigma(u)$, *and suppose* $\sigma(u)$ *to have property* B. *Suppose* $\liminf_{u\to+\infty} |\sigma(u)/\theta(u)| > 0$ *and* $\limsup_{u\to\infty} |\sigma(u)/\theta(u)| < 1/2$, *with* $\theta(u) \equiv 2\phi(u)$. *Let* Γ_σ *be the image in the z-plane of* γ_σ *under the mapping* $w = f(z)$, *represented by* $y = y_\sigma(x)$. *Then* $y_\sigma(x)$ *has property* B *at* $x = +\infty$.

It follows from Theorem 2.3 that the ordinates of Γ_σ are in absolute value bounded from zero and $\pi/2$, hence follows from (2.1) that we have

$$\lim_{n\to\infty} \left[\frac{\sigma(u)}{\theta(u)} - \frac{y_\sigma}{\pi}\right] = 0.$$

Our hypothesis on $\liminf |\sigma(u)/\theta(u)|$ now yields

(4.1) $$\lim_{x\to\infty} \frac{y_\sigma(x)}{\pi} \Big/ \frac{\sigma(u)}{\theta(u)} = 1.$$

Let I_0 be the interval $|X| < X_0$, where X_0 is arbitrary. Set $x' = Xy_\sigma(x) + x$, $u' = u[x' + iy_\sigma(x')]$. By the method of proof of (4.1) we have

(4.2) $$\lim_{x\to\infty} \frac{y_\sigma(x')}{\pi} \Big/ \frac{\sigma(u')}{\theta(u')} = 1$$

uniformly for X in I_0. By virtue of (4.1) and (4.2) it is now sufficient for the proof of the lemma to show

(4.3) $$\lim_{x\to\infty} \frac{\sigma(u)}{\sigma(u')} = \lim_{x\to\infty} \frac{\theta(u)}{\theta(u')} = 1,$$

uniformly for X in I_0.

By Theorem 2.2 we have

(4.4) $$\lim_{n\to\infty} \left[\frac{u' - u}{\theta(u)} - \frac{Xy_\sigma(x)}{\pi}\right] = 0.$$

Since X is in I_0 and $|y_\sigma(x)| < \pi/2$, we conclude by Theorem 1.1

$$\lim_{x\to\infty} \frac{\theta(u)}{\theta(u')} = 1,$$

the second part of (4.3). Equation (4.4) suggests the form

$$u' = M(x) \frac{\theta(u)}{\sigma(u)} \sigma(u) + u;$$

since $M(x)$ is uniformly bounded and $\lim\inf |\sigma(u)/\theta(u)| > 0$, and since $\sigma(u)$ has property B at $+\infty$ the first part of (4.3) follows. Lemma 2 is established.

LEMMA 3. *With the hypothesis of Lemma* 1 *let* x_n *be any sequence of real points with* $\lim_{n\to\infty} x_n = +\infty$. *Then with the notation* $\psi(u) \equiv [\phi_-(u) - \pi/2]/2$ *we have*

$$\lim_{n\to\infty} \frac{v(x_n) - \psi[u(x_n)]}{\theta[u(x_n)]} = 0.$$

This lemma follows in a manner analogous to the derivation of (3.6).

LEMMA 4. *If* $\phi_+(u)$ *and* $\phi_-(u)$ *satisfy the hypothesis of Theorem* 4.1 *and also the inequalities* $0 < a_1 \leqq \phi_+(u) \leqq a_2 < \pi/2$, $a_1 \leqq -\phi_-(u) \leqq a_2$ *for* $u_1 < u < +\infty$, *and if* $\lim_{u\to+\infty} \theta(u) = \pi/2$, *then the conclusion of Theorem* 4.1 *is valid.*

Let S' be the strip of the w-plane bounded by the two loci $v = \phi_-(u)$, $v = \pi/2$, $u_1 < u < +\infty$, and let $w = g(\zeta)$, $\zeta = \xi + i\eta$ map the strip S_ζ: $|\eta| < \pi/2$

onto S' so that $w=+\infty$ corresponds to $\zeta=+\infty$. Let $\eta=\eta_+(\xi)$ denote the image of the curve $v=\phi_+(u)$ under the inverse of the mapping $w=g(\zeta)$; from Lemmas 1 and 3 with the notation $\psi_0(u)\equiv\pi/2-\phi_-[u(\xi)]$, we have

$$\lim_{\xi\to+\infty}\left[\frac{\phi_+[u(\xi)]}{\psi_0(u)}-\frac{1}{2}\,\frac{\pi/2+\phi_-[u(\xi)]}{\psi_0(u)}-\frac{\eta_+(\xi)}{\pi}\right]=0,$$
(4.5)
$$\lim_{\xi\to+\infty}\left[\frac{\phi_+[u(\xi)]}{\psi_0(u)}-\frac{2\eta_+(\xi)}{\pi}\right]=0.$$

From the inequality $\phi_+[u(\xi)]\geqq a_1>0$ it follows by the method of proof of Lemma 2 that $\eta_+(\xi)$ has property B.

Let $\zeta=h(z)$ map S_z onto the strip of the ζ-plane bounded by $\eta_+(\xi)$ and $\eta=-\pi/2$, by the composition $g(\zeta)\equiv g[h(z)]\equiv f(z)$. For any sequence of real points x_n, $x_n\to+\infty$, we have from Lemma 1

$$\lim_{n\to\infty}\frac{h(z+x_n)-h(x_n)}{\pi/2+\eta_+[\xi(x_n)]}=\frac{z}{\pi} \tag{4.6}$$

throughout S_z, uniformly on any closed bounded subset of S_z. Likewise for any sequence of real points ξ_n, $\xi_n\to\infty$, we have

$$\lim_{n\to\infty}\frac{g(\zeta+\xi_n)-g(\xi_n)}{\psi_0[u(\xi_n)]}=\frac{\zeta}{\pi} \tag{4.7}$$

throughout S_ζ, uniformly on any closed bounded subset of S_ζ.

With the notation $\zeta_n=h(x_n)=\xi_n+i\eta_n$, $\zeta_n'=h(z+x_n)=\xi_n'+i\eta_n'$, we have from (4.6) applied to $\zeta=h(z)$ the relation $\lim\sup_{n\to\infty}|\xi_n'-\xi_n|<\infty$, whence from (4.7)

$$\lim_{n\to\infty}\left[\frac{g(\xi_n')-g(\xi_n)}{\psi_0[u(\xi_n)]}-\frac{\xi_n'-\xi_n}{\pi}\right]=0, \tag{4.8}$$

$$\lim_{u\to\infty}\left[\frac{g(\zeta_n)-g(\xi_n)}{\psi_0[u(\xi_n)]}-\frac{i\eta_n}{\pi}\right]=0, \tag{4.9}$$

$$\lim_{n\to\infty}\left[\frac{g(\zeta_n')-g(\xi_n')}{\psi_0[u(\xi_n')]}-\frac{i\eta_n'}{\pi}\right]=0. \tag{4.10}$$

Theorem 1.2 enables us to replace $\psi_0[u(\xi_n')]$ in (4.10) by $\psi_0[u(\xi_n)]$, so by (4.8), (4.9), (4.10) we obtain

$$\lim_{n\to\infty}\left[\frac{g(\zeta_n')-g(\zeta_n)}{\psi_0[u(\xi_n)]}-\frac{\zeta_n'-\zeta_n}{\pi}\right]=0. \tag{4.11}$$

Again by Theorem 1.2 we can here replace $\psi_0[u(\xi_n)]$ by $\psi_0[u(\zeta_n)]$. Thus from

(4.11), (4.5), and (4.6) with $\theta(u)\to\pi/2$ we have

$$\lim_{n\to\infty}[f(z+x_n)-f(x_n)]=z/2$$

throughout S_z, and the uniformity on any closed bounded subset of S_z follows readily. This completes the proof of Lemma 4.

We are now in a position to complete the proof of Theorem 4.1. Let S' be the symmetric strip in the w-plane whose boundary is the loci $v=\pm\theta(u)$, $u_1<u<+\infty$. It follows from Theorem 1.6 that $\theta(u)$ has property B. We use the notation $w=g(\zeta)$, $\zeta=\xi+i\eta$ for the function which maps S_ζ: $|\eta|<\pi/2$ onto S' such that $\xi=+\infty$ corresponds to $u=+\infty$. If ξ_n is any sequence of real points with $\xi_n\to+\infty$ as $n\to\infty$, we have from Theorem 2.1

$$\lim_{n\to\infty}\left[\frac{g(\zeta+\xi_n)-g(\xi_n)}{\theta[u(\xi_n)]}\right]=\frac{2\zeta}{\pi} \tag{4.12}$$

throughout S_ζ, uniformly on any closed bounded subset of S_ζ.

By Lemma 2 the images of $v=\phi_-(u)$ and $v=\phi_+(u)$ under the inverse of the mapping $w=g(\zeta)$ are curves $\eta=\eta_-(\xi)$ and $\eta=\eta_+(\xi)$ which have property B. Moreover both of the latter curves satisfy the additional hypothesis of Lemma 4. For each ξ we set $\zeta_+=\xi+i\eta_+(\xi)$, $\zeta_-=\xi+i\eta_-(\xi)$, whence

$$\begin{aligned}&\lim_{\xi\to+\infty}\left[\frac{\phi_+[u(\zeta_+)]}{\theta[u(\xi)]}-\frac{2\eta_+(\xi)}{\pi}\right]=0,\\&\lim_{\xi\to+\infty}\left[\frac{\phi_-[u(\zeta_-)]}{\theta[u(\xi)]}-\frac{2\eta_-(\xi)}{\pi}\right]=0.\end{aligned} \tag{4.13}$$

From (4.12) it is readily shown that

$$\lim_{\xi\to+\infty}\left[\frac{u(\zeta_+)-u(\xi)}{\theta[(u(\xi)]}\right]=0, \tag{4.14}$$

whence by Theorem 1.1

$$\lim_{\xi\to+\infty}\left[\frac{\phi_+[u(\zeta_+)]-\phi_-[u(\zeta_-)]}{\theta[u(\xi)]}\right]=1. \tag{4.15}$$

Equations (4.15) and (4.13) yield

$$\lim_{\xi\to+\infty}[\eta_+(\xi)-\eta_-(\xi)]=\pi/2. \tag{4.16}$$

We now define the mapping $\zeta=h(z)$ by the equations $g(\zeta)\equiv g[h(z)]\equiv f(z)$. If x_n is any sequence of real points of S_z with $x_n\to+\infty$ as $n\to\infty$, we have by (4.16) and Lemma 4

$$\lim_{n\to\infty}[h(z+x_n)-h(x_n)]=z/2 \tag{4.17}$$

throughout S_z, uniformly on any closed bounded subset of S_z. By combining (4.12) and (4.17) as in the proof of Lemma 4 we have (3.1) throughout S_z, uniformly on any closed bounded subset of S_z, which establishes Theorem 4.1.

It is apparent that Theorems 3.2 and 3.3, proved from Theorem 3.1, have analogues similarly proved from Theorem 4.1:

THEOREM 4.2. *With the hypothesis of Theorem* 4.1 *let* $z_n = x_n + iy_n$ *be a sequence of points of* S_z *with* $x_n \to +\infty$, $\limsup_{n\to\infty} |y_n| < \pi/2$. *Then we have*

$$(1) \qquad \lim_{n\to\infty} \frac{f'(z_n)}{\theta[u(z_n)]} = \frac{1}{\pi},$$

$$(2) \qquad \lim_{n\to\infty} \frac{f^{(\nu)}(z_n)}{\theta[u(z_n)]} = 0, \qquad \nu = 2, 3, \cdots,$$

$$(3) \qquad \lim_{n\to\infty} \arg f'(z_n) = 0.$$

THEOREM 4.3. *With the hypothesis of Theorem* 4.1 *let* $z_n = x_n + iy_n$, $z_n' = x_n' + iy_n'$ *be two sequences of points of* S_z *with* $x_n \to +\infty$, $x_n' \to +\infty$, $\limsup |x_n - x_n'| < \infty$, $\limsup |y_n| < \pi/2$, $\limsup |y_n'| < \pi/2$. *Then*

$$(1) \qquad \lim_{n\to\infty} \left[\frac{u(z_n') - u(z_n)}{\theta[u(z_n)]} - \frac{x_n' - x_n}{\pi} \right] = 0.$$

If $\lim_{n\to\infty} y_n = \beta$, $|\beta| < \pi/2$, *then*

$$(2) \qquad \lim_{n\to\infty} \frac{v(z_n) - v(x_n)}{\theta[u(z_n)]} = \frac{\beta}{\pi}.$$

Parts (1) and (3) of Theorem 4.2 were obtained by Warschawski [4, pp. 315, 288] for the case that S is an L-strip but without our restriction on $|\phi_-(u)/\phi_+(u)|$. Warschawski does not assume $\limsup |y_n| < \pi/2$ nor that S contains the segment $u > u_1$, $v = 0$. Likewise part (1) of Theorem 4.3 was obtained by Warschawski [4, p. 324] when S is an L-strip without our restriction on $\limsup |y_n|$ and $\limsup |y_n'|$. Part (2) of Theorem 4.3 was obtained by Warschawski for L-strips [4, (17.10)].

From Theorems 4.2 and 4.3 we shall obtain also

THEOREM 4.4. *With the hypothesis of Theorem* 4.3 *we have*

$$(1) \qquad \lim_{n\to\infty} \theta[u(z_n)]/\theta[u(z_n')] = 1,$$

$$(2) \qquad \lim_{n\to\infty} f'(z_n)/f'(z_n') = 1.$$

Part (1) is a consequence of part (1) of Theorem 4.3 and Theorem 1.2; part (2) is a consequence of part (1) and part (1) of Theorem 4.2. Warschawski [4, p. 288] proved part (2) for an L-strip.

5. **A further extension.** Thus far we have envisaged only strips S containing an infinite segment $u_1 < u < +\infty$ of the axis of reals. We now remove that restriction.

THEOREM 5.1. *Let S be a strip in the plane of $w = u + iv$ whose boundary consists of the two loci $v = \phi_+(u)$, $v = \phi_-(u)$, $-\infty < u_1 < u < +\infty$, where $\phi_+(u)$ is positive and has property* B. *Let $w = f(z)$, $z = x + iy$, map the strip S_z: $|y| < \pi/2$ onto S such that $u(z) \to +\infty$ when $x \to +\infty$. If x_n is any sequence of real points with $x_n \to +\infty$, and if* $\lim_{u\to+\infty} \phi_-(u)/\phi_+(u) = \lambda$, $0 \leqq \lambda < 1$, *then*

$$\lim_{n\to\infty} \frac{f(z + x_n) - f(x_n)}{\theta[u(x_n)]} = \frac{z}{\pi} \tag{5.1}$$

for z in S_z, uniformly on any closed bounded subset of S_z, where $\theta(u) \equiv \phi_+(u) - \phi_-(u)$.

We shall not give a detailed proof, for the technique is essentially the same as that used in the proof of Theorem 3.1. Analogous supplementary conditions are imposed, such as $\phi_+(u) \geqq |\phi_-(u)|$ for $u_1 < u < +\infty$. The proof is based on the ability to replace S by a symmetric region containing S to which we can apply the results of the previous sections.

Let $w = g(\zeta)$, $\zeta = \xi + i\eta$, map the strip S_ζ: $|\eta| < \pi/2$ onto the strip S' whose boundary consists of the two loci $v = \pm\phi_+(u)$, $u_1 < u < +\infty$. Theorem 2.1 applies to the mapping function $g(\zeta)$. By Theorem 2.3 and Theorem 1.6′the image of S under the inverse of the mapping $w = g(\zeta)$ is a region in the ζ-plane, which is transformed by the vertical translation $\zeta' = \zeta - (1 + \lambda)\pi i/4$ into a region satisfying the hypothesis of Theorem 3.1. The composite mapping can then be treated by the methods already used, and the theorem follows.

Applications of Theorem 5.1 analogous to those of Theorem 3.1 are

THEOREM 5.2. *With the hypothesis of Theorem* 5.1 *let $z_n = x_n + iy_n$ be a sequence of points of S_z such that $x_n \to +\infty$ and* $\limsup_{n\to\infty} |y_n| < \pi/2$. *Then we have*

$$\lim_{n\to\infty} \frac{f'(z_n)}{\theta[u(z_n)]} = \frac{1}{\pi}, \tag{1}$$

$$\lim_{n\to\infty} \frac{f^{(\nu)}(z_n)}{\theta[u(z_n)]} = 0, \qquad \nu = 2, 3, \cdots, \tag{2}$$

$$\lim_{n\to\infty} \arg f'(z_n) = 0. \tag{3}$$

THEOREM 5.3. *With the hypothesis of Theorem* 5.1, *let $z_n = x_n + iy_n$ and $z_n' = x_n' + iy_n'$ be two sequences of points of S_z with $x_n \to +\infty$, $x_n' \to +\infty$,* $\limsup |x_n - x_n'| < +\infty$, $\limsup_{n\to\infty} |y_n| < \pi/2$, $\limsup |y_n'| < \pi/2$. *Then we have*

(1) $$\lim_{n\to\infty}\left[\frac{u(z_n)-u(z_n')}{\theta[u(z_n)]}-\frac{x_n-x_n'}{\pi}\right]=0;$$

if $\lim_{n\to\infty} y_n=\beta$, $|\beta|<\pi/2$, *then*

(2) $$\lim_{n\to\infty}\frac{v(z_n)}{\theta[u(z_n)]}=\frac{1}{2}\frac{1+\lambda}{1-\lambda}+\frac{\beta}{\pi}\cdot$$

The relation of Theorem 3.1 to Theorem 4.1 corresponds to the relation of Theorem 5.1 to

THEOREM 5.4. *The conclusion of Theorem* 5.1 *is valid if* $\phi_-(u)$ *has property* B *and the condition* $\lim_{u\to\infty}\phi_-(u)/\phi_+(u)=\lambda$ *is replaced by*

$$0<a_1\leqq|\phi_-(u)/\phi_+(u)|\leqq a_2<1.$$

Analogues of Theorems 5.2 and 5.3 are

THEOREM 5.5. *With the hypothesis of Theorem* 5.4, *let* $z_n=x_n+iy_n$ *be a sequence of points of* S_z *such that* $x_n\to\infty$, $\limsup|y_n|<\pi/2$. *Then the conclusion of Theorem* 4.2 *is valid.*

THEOREM 5.6. *If* S *satisfies the conditions of Theorem* 5.4 *rather than those of Theorem* 4.1, *the conclusions of Theorems* 4.3 *and* 4.4 *are valid.*

Our previous references to Warschawski apply also to the present section.

6. **Zero angles at a finite point.** We consider in this section the analogues of our previous results, but where the significant boundary points are finite rather than at infinity. We choose rather to apply our previous results than the methods used to prove them, but this choice is merely a matter of convenience.

A function $\phi(u)$ defined in the interval $u_1<u\leqq 1$ $(-\infty\leqq u_1<1)$ is said *to have property* B *at* $u=1$ *if* $\phi(u)\to 0$ *as* $u\to 1^-$ *and if uniformly in any interval* I: $|U|\leqq U_0$ *we have*

(6.1) $$\lim_{u\to 1^-}\frac{\phi[U\phi(u)+u]}{\phi(u)}=1.$$

We do not require that $\phi(u)$ be single valued, but do require that (6.1) hold for any choice of values of $\phi(u)$.

It is obvious that only a slight modification of the proofs of §1 yields results on functions having property B at $u=1$ analogous to those of §1. For instance we have $\phi(u)/(1-u)\to 0$ as $u\to 1$. It is perhaps more important that property B is invariant under inversion:

THEOREM 6.1. *Suppose the function* $y=\phi(x)$ *has property* B *at* $x=1$. *Set*

(6.2) $$x=1-u/(u^2+v^2),\ y=v/(u^2+v^2).$$

Then the function $v=h(u)$ *defined by* $y=\phi(x)$ *and* (6.2) *has property* B *at*

$u=+\infty$. *Conversely, if* $v=\Phi(u)$ *has property* B *at* $u=+\infty$, *then the function* $y=H(x)$ *defined by* $v=\Phi(u)$ *and* (6.2) *has property* B *at* $x=1$.

We omit the proof of Theorem 6.1, which is relatively straightforward and depends on the results of §1 and their analogues.

An immediate consequence of Theorems 6.1 and 3.1 is

THEOREM 6.2. *Let* $w=f(z)$ *with* $w=u+iv$, $z=x+iy$ *map the half plane* R_z: $x<1$ *onto the simply connected region* R *of the w-plane which contains the interval* $u_1<u<1$ *but not the point at infinity, with* $f(1)=1$. *Suppose that the boundary of* R *consists of two loci* $v=\phi_+(u)$, $v=\phi_-(u)$ *for* $u_2<u<1$, *that* $\phi_+(u)>0$ *and* $\phi_-(u)<0$ *for* $u_2<u<1$, *and that* $\phi_+(u)$ *has property* B *at* $u=1$. *Suppose also* $\lim_{u\to 1^-}\phi_-(u)/\phi_+(u)=\lambda$, $-\infty<\lambda\leqq 0$. *Then if* $0<x_n<1$ *is any sequence of real points with* $\lim_{n\to\infty} x_n=1$, *we have*

$$\lim_{n\to\infty}\frac{f[(1-x_n)z+x_n]-f(x_n)}{\theta[u(x_n)]}=-\frac{1}{\pi}\log(1-z) \tag{6.3}$$

throughout R_z, *uniformly on any closed bounded subset of* R_z, *where* $\theta(u)\equiv\phi_+(u)-\phi_-(u)$.

Both for Theorem 6.2 and Theorem 6.6 below we assume $\min\theta(u)>\{\max\phi_+(u)-\min\phi_-(u)\}$ for each u for which $\theta(u)$ is defined. This assumption involves no essential loss of generality, as is indicated for property B at $u=+\infty$ instead of $u=1^-$ in connection with Theorems 3.1 and 4.1.

An immediate consequence of Theorem 6.2 is

THEOREM 6.3. *With the hypothesis of Theorem* 6.2 *let* z_n *and* z_n' *be two sequences of points of* R_z *such that* $z_n\to 1$ *and* $z_n'\to 1$ *"in angle," namely so that* $\limsup_{n\to\infty}|\arg(1-z_n)|<\pi/2$, $\limsup_{n\to\infty}|\arg(1-z_n')|<\pi/2$, *and also so that we have* $0<c_1\leqq|(1-z_n)/(1-z_n')|\leqq c_2<\infty$. *Then we have*

$$\lim_{n\to\infty}\frac{1-f(z_n)}{1-f(z_n')}=1, \tag{1}$$

$$\lim_{n\to\infty}\frac{f'(z_n)(1-z_n)}{f'(z_n')(1-z_n')}=1. \tag{2}$$

Since $z_n=x_n+iy_n\to 1$ in angle, the set of points $iy_n/(1-x_n)$ is contained in a closed bounded subset of R_z. By (6.3) we obtain

$$\lim_{n\to\infty}\left[\frac{f(z_n)-f(z_n)}{\theta[u(x_n)]}+\frac{1}{\pi}\log\frac{1-z_n}{1-x_n}\right]=0;$$

the analogue of Theorem 1.3 is $\phi(u)/(1-u)\to 0$, from which we have

$$\lim_{n\to\infty}\frac{1-f(z_n)}{1-f(x_n)}=1.$$

In a similar manner we have

$$\lim_{n\to\infty} \frac{1 - f(z_n')}{1 - f(x_n')} = 1.$$

From the last part of the hypothesis it follows that the sequence

$$(x_n' - x_n)/(1 - x_n)$$

lies in a closed bounded subset of R_z; thus we write

$$\lim_{n\to\infty} \left[\frac{f(x_n') - f(x_n)}{\theta[u(x_n)]} + \frac{1}{\pi} \log \frac{1 - x_n'}{1 - x_n}\right] = 0,$$

$$\lim_{n\to\infty} \frac{1 - f(x_n)}{1 - f(x_n')} = 1.$$

Part (1) now follows immediately. To obtain part (2) we differentiate (6.3) and obtain

$$\lim_{n\to\infty} \frac{(1 - x_n)(1 - z)f'[(1 - x_n)z + x_n]}{\theta[u(x_n)]} = \frac{1}{\pi}, \tag{6.4}$$

uniformly on any closed bounded subset of R_z. By the method used in part (1) and from the relation $\theta[u(z_n)]/\theta[u(z_n')]\to 1$ we deduce part (2) from (6.4).

Ostrowski [5, p. 176] proved part (1) of Theorem 6.3 with the weaker hypothesis that the boundary of R forms a zero angle at the accessible point $w=1$. He proved part (2) without the "in angle" restriction provided the boundary curves have an L-tangent at $w=1$, and under those conditions showed that the "in angle" restriction can be removed from part (1).

Similar methods can be used to establish

THEOREM 6.4. *With the hypothesis of Theorem 6.2 let z_n be a sequence of points of R_z such that $z_n\to 1$ in angle. Then we have*

(1) $$\lim_{n\to\infty} [\arg f'(z_n) + \arg (1 - z_n)] = 0,$$

(2) $$\liminf_{n\to\infty} \frac{\log |f'(z_n)|}{\log |1 - z_n|} \geq -1,$$

(3) $$\lim_{n\to\infty} \frac{f^{(\nu)}(z_n)(1 - z_n^{\nu})}{\theta[u(z_n)]} = \frac{(\nu - 1)!}{\pi}, \qquad \nu = 1, 2, \cdots,$$

(4) $$\lim_{n\to\infty} \frac{f^{(\nu)}(z_n)(1 - z_n)^{\nu}}{1 - f(z_n)} = 0, \qquad \nu = 1, 2, \cdots,$$

(5) $$\lim_{n\to\infty} \frac{f^{(\nu+1)}(z_n)(1 - z_n)^{\nu}}{f'(z_n)} = \nu!, \qquad \nu = 0, 1, \cdots.$$

Ostrowski proved parts (1), (2) [5, p. 184] and (5) [5, p. 130] for the case that the boundary of R has an L-tangent at $w=1$, without the "in angle" assumption for part (1) and with equality for part (2). He also proved (4) [5, p. 130] under the same assumptions as he used for part (1) of Theorem 6.3.

Theorem 6.5 is another immediate consequence of Theorem 6.2; the conclusion is an analogue for the zero angle case of the Lindelöf theorem on proportionality of angles. Theorem 6.5 is stronger in conclusion but more restrictive in hypothesis than the corresponding result due to Warschawski [4, p. 327].

THEOREM 6.5. *With the hypothesis of Theorem* 6.3 *and the added conditions* $\arg(1-z_n)\to\alpha$, $|\alpha|<\pi/2$, $\arg(1-z_n')\to\beta$, $|\beta|<\pi/2$, $\beta\neq\pi(1+\lambda)/2(1-\lambda)$, *we have*

$$\lim_{n\to\infty}\frac{\arg[1-f(z_n)]}{\arg[1-f(z_n')]}=\frac{2(1-\lambda)\alpha-\pi(1+\lambda)}{2(1-\lambda)\beta-\pi(1+\lambda)}.$$

We can modify Theorem 6.2 by applying Theorem 4.1 instead of Theorem 3.1:

THEOREM 6.6. *Theorem* 6.2 *remains true if the condition*

$$\lim_{u\to 1^-}\frac{\phi_-(u)}{\phi_+(u)}=\lambda$$

is replaced by the inequalities $0<a\leqq|\phi_-(u)/\phi_+(u)|\leqq b<\infty$ *and the condition that* $\phi_-(u)$ *and* $\phi_+(u)$ *have property* B *at* $u=1$.

Theorems 6.3 and 6.4 likewise can be modified by replacing Theorem 6.2 by Theorem 6.6 in the hypothesis of each.

Theorem 5.1 has an analogue, which we proceed to formulate.

We say that a simply connected region R of the w-plane is *admissible* if either of the following sets of conditions is satisfied:

(1) the boundary of R consists of the two loci $v=\phi_+(u)$, $v=\phi_-(u)$, $-\infty\leqq u_1<u<1$.

(2) the function $\phi_+(u)$ is positive $u_1<u<1$.

(3) $\phi_+(u)$ has property B at $u=1$.

(4) $\lim_{u\to 1^-}\phi_-(u)/\phi_+(u)=\lambda$ $(0\leqq\lambda<1)$.

(3′) $\phi_-(u)$ and $\phi_+(u)$ have property B at $u=1$.

(4′) $\liminf_{u\to 1^-}\phi_-(u)/\phi_+(u)>0$, $\limsup_{u\to 1^-}\phi_-(u)/\phi_+(u)<1$.

Use of 6.1 in conjunction with Theorems 5.1 and 5.4 yields

THEOREM 6.7. *Let* $w=f(z)$ *map the half plane* R_z: $x<1$ *onto the admissible region* R *such that* $z\to 1$ *implies* $f(z)\to 1$. *Then if* x_n *is any sequence of real points with* $0<x_n<1$, $x_n\to 1$, *we have*

$$\lim_{n\to\infty} \frac{f[(1-x_n)z+x_n]-f(x_n)}{\theta[u(x_n)]} = -\frac{1}{\pi}\log(1-z)$$

throughout R_z, *uniformly on any closed bounded subset of* R_z, *where* $\theta(u)\equiv\phi_+(u)-\phi_-(u)$.

THEOREM 6.8. *With the hypothesis of Theorem* 6.7 *let* z_n, z_n' *be two sequences of points of* R_z *with* $z_n\to 1$ *and* $z_n'\to 1$ *in angle, and* $0<c_1\leqq|(1-z_n)/(1-z_n')|\leqq c_2<\infty$. *Then we have*

$$\lim_{n\to\infty} \frac{1-f(z_n)}{1-f(z_n')} = 1, \tag{1}$$

$$\lim_{n\to\infty} \frac{f'(z_n)(1-z_n)}{f'(z_n')(1-z_n')} = 1. \tag{2}$$

THEOREM 6.9. *With the hypothesis of Theorem* 6.7 *let* z_n *be a sequence of points of* R_z *such that* $z_n\to 1$ *in angle. Then we have* (1), (2), (3), (4), (5) *of Theorem* 6.4.

7. **Conditions for convergence to kernel.** We have based the previous discussion (§§1–5) on condition (1.1) rather than on the condition

$$\lim_{u_n\to\infty} \frac{\phi[U\phi(u_n)+u_n]}{\phi(u_n)} = 1 \tag{7.1}$$

uniformly for U in every interval I; this is of course merely a matter of choice, corresponding to the limit of the continuous variable u rather than the limit of a sequence u_n. Condition (7.1) might be termed *property* B *with respect to a particular sequence* u_n; it is obviously the more discriminating condition, and can be made fundamental in our treatment instead of (1.1). Results thus obtained are wholly analogous to those already formulated in detail.

As an illustration here (compare [1]), let the strip S consist of the circles $|w-n|<1/4$ ($n=0, 1, 2, \cdots$) joined by horizontal canals of respective widths 1/8, 1/16, $\cdots$, constant for each canal, the entire strip S being symmetric in the u-axis. If the points u_n in (7.1) are chosen as the midpoints of the canals, the kernel of the regions S_n (notation of §2) consists of the strip $|V|<1$; here (2.1) is valid, where $u_n=f(x_n)$. If the points u_n in (7.1) are chosen as the points $n=0, 1, 2, \cdots$, $\phi(u_n)=1/4$, the kernel of the regions S_n consists of the circle $|W|<1$; here the second member of (2.1) is to be replaced by the function which maps Σ onto $|W|<1$ with $z=0$ corresponding to $W=0$ and with directions at those points invariant.

Neither (1.1) nor (7.1) implies the existence of an L-tangent; indeed, given a smooth curve $v=\phi(u)>0$ with an L-tangent we may modify $\phi(u)$ by placing a countable infinity of small vertical cuts extending downward from the curve at the abscissas $u=1, 2, \cdots$, or at the points u_n if the latter are given. The length of each cut divided by the original ordinate at the same abscissa shall approach zero as the (variable) abscissa becomes infinite. Then the new ordinate $\phi_1(u)$ has property B; the new ordinate is in part multiple valued, and the locus $v=\phi_1(u)$ does not possess an L-tangent. The existence of an L-tangent is frequently required by Ostrowski and Warschawski.

In the special case that $\phi(u)$ or $\phi(u_n)$ is bounded and bounded from zero, (1.1) and (7.1) can be written in the respective forms

$$\phi[U+u]-\phi(u)\to 0, \tag{7.2}$$

$$\phi[U+u_n]-\phi(u_n)\to 0, \tag{7.3}$$

uniformly for U in every interval I. It is thus clear by the mean value theorem that if $\phi'(u)$ exists for every u and approaches zero as $u\to\infty$, these conditions (7.2) and (7.3) are satisfied. The condition that $\phi'(u)$ exists and approaches zero of course implies that S has an horizontal L-tangent at infinity. Under these circumstances the regions S_n (we assume for simplicity S to be symmetric in the u-axis) converge to a kernel which is an infinite strip bounded by two lines parallel to and equidistant from the U-axis; equation (2.1) can be written in the form

$$\lim_{n\to\infty}[f(z+x_n)-f(x_n)]/\phi[u(x_n)]=\lambda z,$$

where the constant λ is suitably chosen.

However, compare Madame Lelong-Ferrand [3, §13], *it is not sufficient for this conclusion that* (7.3) *hold nonuniformly with respect to* U *in* I, as we indicate by a counter-example.

We denote by $\psi_k(u)$ the continuous even function which takes the values $0,\ 1-1/k,\ 0,\ 0$, in the respective points $u=\pm 1$, $u=\pm(1+1/2k)$, $u=\pm(1+1/k)$, $\pm\infty$, and which is linear in the successive intervals of $-\infty<u<+\infty$ bounded by these points. We note the relation $\psi_k(u)\to 0$ as $k\to\infty$ for every u. We now define

$$\phi(u)\equiv 1-\sum_{k=1}^{\infty}\psi_k(u-2^{k+1});$$

of course this infinite series contains for each u at most one nonzero term. The graph of the function $\phi(u)$ is the profile of a saw of infinite length whose teeth become sharper and approach unity in height as the distance from the origin becomes infinite. We set here $u_n=2^n$, whence

$$\phi(u_n+U)-\phi(u_n)\equiv\sum_{k=1}^{\infty}\psi_k(2^n-2^{k+1})-\sum_{k=1}^{\infty}\psi_k(U+2^n-2^{k+1}),$$

so that (7.1) and (7.3) are satisfied for every U but not uniformly for U in every interval I. We have $\phi(u_n)=1$, and if the region S is bounded by the loci

$v=\pm\phi(u)$, the regions S_n (notation of §2) converge to a kernel which consists of the square $|U|<1$, $|V|<1$. Equation (2.1) is not valid but becomes valid if the second member of (2.1) is replaced by the function which maps Σ onto this kernel so that the origins correspond to each other and directions at the origins are invariant.

Condition (7.1) *holding uniformly is sufficient but not necessary that the regions S_n converge to the kernel* $|V|<1$, and hence that (2.1) be valid; this is shown by the counter-example of a region S bounded by the loci $v=\pm\phi(u)$,

$$\phi(u) \equiv 1 + \sum_{k=1}^{\infty} \psi_k(u - 2^{k+1}), \tag{7.4}$$

where ψ_k is defined above; we have $(U_0>1)$

$$\lim_{n\to\infty} \{\max [\phi(U+u_n), \ |U| \leqq U_0]\} = 2.$$

Nevertheless a necessary condition that regions S_n found as in §2 from a region S bounded by $v=\pm\phi(u)$ converge to the kernel $|V|<1$ is for every $U_0(>0)$

$$\lim_{u_n\to\infty} \frac{\min \{\phi[U\phi(u_n)+u_n], \ |U| \leqq U_0\}}{\phi(u_n)} = 1;$$

for n sufficiently large all the S_n must contain the region $|V|<1-\delta$, $|U|<U_0$ where $\delta(>0)$ is arbitrary; for n sufficiently large all the S_n can contain no region $|V|<1+\delta$, $|U|<U_0$, $\delta>0$. This necessary condition combined with the condition

$$\lim_{u_n\to\infty} \frac{\phi[U\phi(u_n)+u_n]}{\phi(u_n)} = 1$$

for a set of values U everywhere dense on $-\infty<U<+\infty$ is sufficient that the regions S_n found from a region S bounded by $v=\pm\phi(u)$ converge to the kernel $|V|<1$.

A necessary and sufficient condition for (2.1), if S is a region bounded by $v=\pm\phi(u)$, is that the regions S_n converge to the kernel $|V|<1$. Of course, condition (7.3) without uniformity is not necessary for the S_n to converge to the kernel $|V|<1$ in the case $\phi(u_n)\to 1$, as is shown by the counter-example (7.4).

Condition (7.1) is readily expressed in terms of a mean value of $\phi'(u)$, provided $\phi'(u)$ exists and $\phi(u)$ can be expressed as an indefinite integral of $\phi'(u)$. We have

$$\frac{\phi[U\phi(u_n)+u_n]-\phi(u_n)}{\phi(u_n)} = \frac{1}{\phi(u_n)}\int_{u_n}^{u_n+U\phi(u_n)} \phi'(u)\,du,$$

so the condition

$$\lim_{n\to\infty} \frac{1}{\phi(u_n)} \int_{u_n}^{u_n+U\phi(u_n)} \phi'(u)du = 0 \tag{7.5}$$

uniformly for U in every I is sufficient for (7.1). Of course (7.5) is satisfied if the mean value relation

$$\lim_{n\to\infty} \frac{1}{U\phi(u_n)} \int_{u_n}^{u_n+U\phi(u_n)} \phi'(u)du = 0$$

is satisfied uniformly for U in every I.

Madame Lelong-Ferrand points out that if $\int^\infty [\phi'(u)]^2 du$ exists, we may write

$$[\phi(u_n + l) - \phi(u_n)]^2 < l \int_{u_n}^{u_n+l} [\phi'(u)]^2 du;$$

this last member approaches zero uniformly for bounded l as $u_n \to \infty$, so (7.3) is satisfied uniformly for U in I. If in addition lim inf $\phi(u_n) > 0$, condition (7.1) is satisfied uniformly for U in I.

BIBLIOGRAPHY

1. C. Carathéodory, *Conformal representation*, Cambridge University Press, 1932.

2. J. L. Walsh, *On distortion at the boundary of a conformal map*, Proc. Nat. Acad. Sci. U.S.A. vol. 36 (1950) pp. 152–156.

3. J. Lelong-Ferrand, *Sur la représentation conforme des bandes*, Journal d'Analyse Mathématique vol. 2 (1952) pp. 51–71.

4. S. E. Warschawski, *On conformal mapping of infinite strips*, Trans. Amer. Math. Soc. vol. 51 (1942) pp. 280–335.

5. A. Ostrowski, *Über den Habitus der konformen Abbildung am Rande des Abbildungsbereiches*, Acta Math. vol. 64 (1935) pp. 81–184.

6. C. Gattegno and A. Ostrowski, *Représentation conforme à la frontière: domaines généraux; domaines particuliers*, Mémorial des Sci. Math., nos. 109 and 110, Paris, 1949.

7. J. Ferrand, *Étude de la représentation conforme au voisinage de la frontière*, Ann. École Norm. (3) vol. 59 (1942) pp. 43–106.

8. ———, *Nouvelle demonstration d'un théorème de M. Ostrowski*, C.R. Acad. Sci. Paris vol. 220 (1945) pp. 550–551.

HARVARD UNIVERSITY,
CAMBRIDGE, MASS.
MELPAR, INC., CAMBRIDGE, MASS.

ON THE CONFORMAL MAPPING OF MULTIPLY CONNECTED REGIONS(1)

BY

J. L. WALSH

It is the object of this paper to set forth a proof of Theorem 1 below, to the effect that an arbitrary plane region D bounded by a finite number of mutually disjoint Jordan curves can be mapped one to one and conformally onto a region Δ bounded by two level loci of a function which is the product of linear factors with exponents (positive and negative) not necessarily rational; the boundary curves of D can be divided arbitrarily into two classes, which correspond respectively to the two level loci bounding Δ. We establish also a limiting case (Theorem 3) of Theorem 1, in which such an arbitrary region D can be mapped one to one and conformally onto a region Δ bounded by a single level locus of a function of the kind already mentioned, where arbitrary points interior to D can be made to correspond to the zeros of that function. Theorem 1 includes the classical theorem on the mapping of a region D bounded by two disjoint Jordan curves onto a circular annulus, and Theorem 3 includes the Riemann mapping theorem on the mapping of a Jordan region. In both Theorem 1 and Theorem 3 the maps are essentially unique (Theorems 2 and 4).

Both Theorems 1 and 3 present new canonical maps for multiply connected regions, maps that are especially useful in the study of level loci of various functions, namely Green's functions, linear combinations of Green's functions, and harmonic measures—and in the study of the orthogonal trajectories of these loci. It is desirable to develop methods for the effective determination of these maps. The maps are useful certainly in the study of approximation [Walsh, 1955], although the applications have not yet been carried as far as possible, and the maps are presumably useful also in other parts of analysis.

De la Vallée Poussin [1930] and later Julia [1934] have studied the conformal mapping of an arbitrary multiply connected region onto a region bounded by the whole or (more commonly) parts of one or several lemniscates, namely level loci of a polynomial. Julia uses also the level loci of certain rational functions, especially those with fundamental circles, not considered here. De la Vallée Poussin states [1931] our Theorem 1 for the case $\nu=1$, namely the case that the boundary curves of D are divided into two classes

Received by the editors September 12, 1955.

(1) This research was done under the auspices of the U. S. Air Force, Office of Scientific Research of the Air Research and Development Command.

one of which contains but a single curve; however, his proof refers for details back to his preceding papers [1930, 1930a] which are not without flaw [de la Vallée Poussin, 1931a], so even for his case a new treatment would seem to be appropriate.

The present paper is the elaboration of two notes [Walsh, 1954] which contain the outlines of the proofs of Theorems 1 and 3, but without indication of Theorems 2 and 4.

I

THEOREM 1. *Let D be a region of the extended z-plane whose boundary consists of mutually disjoint Jordan curves $B_1, B_2, \cdots, B_\mu; C_1, C_2, \cdots, C_\nu$, $\mu\nu \neq 0$. There exists a conformal map of D onto a region Δ of the extended Z-plane, one to one and continuous in the closures of the two regions, where Δ is defined by*

$$(1)\quad 1 < |T(Z)| < e^{1/\tau}, \qquad T(Z) \equiv \frac{A(Z-a_1)^{M_1}(Z-a_2)^{M_2}\cdots(Z-a_\mu)^{M_\mu}}{(Z-b_1)^{N_1}(Z-b_2)^{N_2}\cdots(Z-b_\nu)^{N_\nu}},$$

$$\sum M_j = \sum N_j = 1, \ \tau > 0.$$

The exponents M_j and N_j are positive but need not be rational. The locus $|T(Z)| = 1$ consists of μ mutually disjoint Jordan curves, respective images of the B_j, which separate Δ from the a_j; the locus $|T(Z)| = e^{1/\tau}$ consists of ν mutually disjoint Jordan curves, respective images of the C_j, which separate Δ from the b_j.

We have written $T(Z)$ in the form where all the a_j and b_j are finite, but it is evident that an arbitrary linear transformation of the Z-plane can be made, so that in particular an a_j or a b_j may be infinite. In such a case, the form of $T(Z)$ is to be modified by simply omitting the factor corresponding to the point at infinity. In the sequel we use the form of $T(Z)$ and of similar functions as given in (1), with the tacit understanding that a zero or infinity at the infinite point is not excluded, even though the form in (1) then requires modification.

We postpone treatment of the case $\mu = \nu = 1$, for which the conclusion is classical, and suppose $\mu \geqq 2$, which involves if necessary interchanging the roles of the B_j and the C_j. We take also the Jordan curves B_j and C_j analytic, which is possible by a preliminary conformal transformation, and we suppose that D lies interior to C_1.

The function $T(Z)$ is transcendental, if the exponents are irrational. As a first part of the proof, we now show that D can be approximated by a region bounded by two level loci of a *rational* function.

There exists a unique function $u(z)$ harmonic in D, continuous in the closure $\overline{D}$ of D, equal to zero and unity on the B_j and C_j respectively. Then $u(z)$ is harmonic also slightly beyond the curves B_j and C_j, let us say in the closure of a finite region D' which contains $\overline{D}$ and whose boundary consists of $\mu+\nu$ analytic Jordan curves, respectively $B_1', B_2', \cdots, B_\mu', C_1', C_2', \cdots, C_\nu'$

near the B_j and C_j. We can choose this boundary as $B' = \sum B_j' : u(z) = -\delta_1$ (… and $C' = \sum C_j' : u(z) = -\delta = 1+\delta_1$. The conjugate $v(z)$ of $u(z)$ is not si… valued in D'; nevertheless Green's formula valid for z in D' can be writ… [Walsh, 1935, §§8.7, 9.12; 1950, §§7.1, 8.1]

$$
(2)\qquad
\begin{aligned}
u(z) &\equiv \int_0^{\tau} \log|z-t|\,d\sigma_t - \int_{\tau}^{2\tau} \log|z-t|\,d\sigma_t - \delta,\\
d\sigma &= |dv|/2\pi, \qquad \tau = \int_{B'} d\sigma = \int_{C'} d\sigma > 0.
\end{aligned}
$$

The integrals in (2) are to be taken over B' and C' respectively. The numb… τ, a kind of modulus of D, and a conformal invariant, does not depend on δ. We can suppose that the derivative of the function $u(z)+iv(z)$ does n… vanish in the closure of the point sets $-\delta_1 < u(z) < 0$, $1 < u(z) < -\delta$ in D'. Then on the boundary of D' we have $\partial u/\partial s = 0$, $\partial v/\partial s \neq 0$; thus $v(z)$ is mono-tonic on B' and on C'. On the boundary of D', if η denotes interior norm… we have $(\partial u/\partial \eta)ds = -(\partial v/\partial s)ds = -dv$; on B' we have $\partial u/\partial\eta > 0$, and on C' we have $\partial u/\partial \eta < 0$.

We express the integrals in (2) as the limits of their Riemann sums; let the α_k and β_k, depending on n, divide B' and C' respectively in n equal parts—equal with respect to the parameter σ, not necessarily equal with respect to arc length. For z on any compact in D' we have uniformly

$$
u_n(z) \equiv \frac{\tau}{n}\sum_1^n \log|z-\alpha_k| - \frac{\tau}{n}\sum_1^n \log|z-\beta_k| - \delta \to u(z);
$$

the uniformity of convergence is a consequence of the uniform continuity of the functions involved, which in turn is a consequence of the boundedness of those harmonic functions.

It follows that D *is approached by* $D_n: 0 < u_n(z) < 1$, *a region bounded by* μ *Jordan curves forming one level locus and by* ν *Jordan curves forming another level locus of a rational function*(2), approached in the sense that if $\mu+\nu$ arbitrary mutually disjoint annular regions (neighborhoods of the B_j and C_j) are given containing the B_j and C_j respectively, then for n sufficiently large the $\mu+\nu$ bounding curves of D_n lie in, and separate the boundary curves of these annular regions. For instance let us choose as a neighborhood of B_1 an annular region α in D' bounded by one Jordan curve of each of the loc…

(2) We have here a general result, that arbitrary finite sets of Jordan curves B_j, C_j bounding a region D can be approximated by respective level loci of a rational function whose zeros and poles lie exterior to D; the analyticity of the curves B_j and C_j is not essential. The limiting case where $\mu=\nu=1$ and C_1 reduces to the point at infinity was established by Hilbert in 1897; the rational function is a polynomial. The present method is a generalization of that of Hilbert involving rational functions more general than polynomials, and has been used a number of times by the present writer (loc. cit.).

$u(z)=\delta_2$ and $u(z)=-\delta_2$, $0<\delta_2<\delta_1<1/2$. For n sufficiently large we have in the closure of α the inequality $|u(z)-u_n(z)|<\delta_2/2$, hence on these respective curves $\delta_2/2<u_n(z)<3\delta_2/2$, $-3\delta_2/2<u_n(z)<-\delta_2/2$. The locus $u_n(z)=0$ obviously separates these two curves, that locus cuts neither curve, and (by the principle of maximum for harmonic functions) that locus consists in α of a single analytic Jordan curve. Application of a similar argument simultaneously for all curves B_j, C_j, shows that for n sufficiently large the locus $u_n(z)=0$ consists of μ curves respectively near the B_j, and the locus $u_n(z)=1$ consists of ν curves respectively near the C_j; these loci can contain no other points; for instance, in the interior of the curve $u_n(z)=0$ which lies near B_1 there lie certain points α_k but no points β_k, so we have $u_n(z)<0$ throughout that interior, again by the principle of maximum for harmonic functions. Each compact K in D lies in all the D_n for n sufficiently large (n depending on K), and each compact K exterior to D lies exterior to all the D_n for n sufficiently large (n depending on K).

The region D_n is defined by the inequalities

$$(3)\qquad 1<|R_n(z)|<e^{n/\tau},\qquad R_n(z)\equiv e^{-n\delta/\tau}\prod_1^n\frac{z-\alpha_k}{z-\beta_k}\equiv e^{n(u_n+iv_n)/\tau},$$

where $v_n(z)$ is conjugate to $u_n(z)$ in D'; thus D_n is bounded by two level loci of a *rational* function. As a second step in the proof, we proceed to show that D_n can be mapped onto a plane region bounded by two level loci of a simpler rational function likewise of degree n, having precisely μ distinct zeros and ν distinct poles.

The equation $w=R_n(z)$ defines the conformal map of the extended z-plane onto a Riemann surface σ_0 of n sheets over the extended w-plane, and the image of D_n is the totality of points w of σ_0 satisfying $1<|w|<e^{n/\tau}$, a connected set bounded wholly by points of the circles $|w|=1$ and $|w|=e^{n/\tau}$.

LEMMA 1. *If the function $f(z)$ is analytic except for precisely p poles and if $f(z)\neq 0$ in the closure of a bounded region E bounded by a finite number of Jordan curves, and if we have $|f(z)|=m$ on the entire boundary of E, then $f(z)$ takes every value w_0, $|w_0|>m$, precisely p times in E. The image of E under the transformation $w=f(z)$ is the portion $|w|>m$ of the extended w-plane covered precisely p times.*

When z traces the boundary of E, arg $[f(z)-w_0]$ for $|w_0|>m$ has zero net increase, but when z traces small circles about the respective poles of $f(z)$, on each of which we have $|f(z)|>|w_0|$, then arg $[f(z)-w_0]$ increases in totality $2\pi p$, so $f(z)$ takes on the value w_0 in E precisely p times. By the principle of maximum modulus, $f(z)$ takes on no value of modulus less than or equal to m in E, so the lemma is established. It is a corollary that *if $f(z)$ is analytic in such a region E and has there precisely p zeros, and if we have $|f(z)|=m$ on the entire boundary of E, then $f(z)$ takes every value w_0, $|w_0|<m$,*

precisely p times in E. The image of E under the transformation $w=f(z)$ is the portion $|w|<m$ of the w-plane covered precisely p times. For the proof, we merely apply the lemma to the function $1/f(z)$.

If E in Lemma 1 is bounded by a finite number of analytic Jordan curves, and if $f(z)$ is still analytic on the boundary, we can apply Lemma 1 to a region E_1 containing the closure $\bar{E}$ of E, E_1 being bounded by a finite number of Jordan curves on which we have $|f(z)|=m_1$. We must have $m_1<m$, for the boundary of E lies interior to E_1. The entire locus $|f(z)|=m$ in E_1 is the boundary of E, which therefore under the transformation $w=f(z)$ covers the circle $|w|=m$ precisely p times.

For n sufficiently large, the complement of D_n (with respect to the extended z-plane) consists of μ regions containing respectively, let us say, $m_1, m_2, \cdots, m_\mu$ points α_j, and ν regions containing respectively $n_1, n_2, \cdots, n_\nu$ points β_j; the m_k and n_k depend on n. It is easily proved that we have ($\sum m_j=\sum n_j=n$).

$$(4)\qquad \frac{m_j}{n}\to\frac{1}{\tau}\int_{B'_j}d\sigma=M_j,\qquad \frac{n_j}{n}\to\frac{1}{\tau}\int_{C'_j}d\sigma=N_j,\qquad \sum M_j=\sum N_j=1;$$

equations (4) define the M_j and N_j and the integrals may be taken over the B_j and C_j instead of the B'_j and C'_j.

The $\mu+\nu$ regions of the z-plane complementary to D_n are simply connected and are mapped by $w=R_n(z)$ onto μ simply connected subregions of σ_0 covering the disc $|w|<1$ precisely m_j times and ν subregions of σ_0 covering the region $|w|>e^{n/\tau}$ precisely n_j times, by Lemma 1. Since $u(z)+iv(z)$ has no critical points on or near the boundary of D, neither does $R_n(z)$, and the boundaries of these $\mu+\nu$ subregions of σ_0 consist of circles traced monotonically m_j and n_j times. Indeed, on these boundaries arg $w=nv_n/\tau$ varies monotonically.

Following a method used by Julia [1934, p. 82], we now form a new Riemann surface σ_1 by replacing the jth one of the μ subregions of σ_0 by a subregion containing m_j sheets with a single branch point $w=0$ of the Riemann surface for the inverse of $w=z^{m_j}$ covering the region $|w|<1$, and replacing the jth one of the ν subregions of σ_0 by a subregion containing n_j sheets with a single branch point $w=\infty$ of the Riemann surface for the inverse of $w=z^{n_j}$ covering $|w|>e^{n/\tau}$. This replacement can be made continuously along the boundaries $|w|=1$ and $|w|=e^{n/\tau}$ so that σ_1 is smooth in each of its n sheets above these boundaries, except of course for the branch lines. Since σ_0 is the image of the extended plane, and since each of the $\mu+\nu$ subregions of σ_0 is simply connected and replaced continuously by a simply connected region, also σ_1 is topologically the image of the extended plane, and (Schwarz) can be mapped conformally and one to one onto the extended Z-plane. Since σ_1 covers each point of the extended w-plane precisely n times, this mapping function is rational of degree n, necessarily of the form

$$w = S_n(Z) \equiv \frac{A_n(Z - a_1')^{m_1}(Z - a_2')^{m_2} \cdots (Z - a_\mu')^{m_\mu}}{(Z - b_1')^{n_1}(Z - b_2')^{n_2} \cdots (Z - b_\nu')^{n_\nu}}, \qquad \sum m_j = \sum n_j = n;$$

the a_j and b_j (depending on n) are distinct, exterior to the image $\Delta_n : 1 < |S_n(Z)| < e^{n/\tau}$ of D_n in the Z plane.

The great advantage of replacing σ_0 by σ_1 is this relatively simple form of the function $S_n(Z)$ which defines the region Δ_n. We have now mapped the region D_n approximating D onto Δ_n. It remains to allow n to become infinite, and by studying the conformal maps of these variable regions D_n and Δ_n complete the proof of Theorem 1.

By a suitable linear transformation of the Z-plane we may choose $a_1' = 0$, $a_2' = 1$, $b_1' = \infty$ independent of n. As $n \to \infty$ there exists a partial sequence of n such that all the numbers $A_n^{1/n}$, a_j', b_j' approach limits A, a_j, b_j; only this partial sequence of the n is to be considered henceforth; by virtue of (4) the inequalities which define Δ_n approach the form (1), defining a region Δ. We proceed to show that D can be mapped onto Δ.

The transformation $R_n(z) = S_n(Z)$ of D_n onto Δ_n can be written $Z = Z_n(z)$. The functions $Z_n(z)$ admit in D_n the exceptional values 0, 1, ∞, hence even if not defined throughout D form a normal family in every closed subregion of D and also in D. Henceforth we consider only a subsequence of values n such that the $Z_n(z)$ uniformly approach a limit $Z_0(z)$ on every compact in D. This function $Z_0(z)$ cannot be identically constant; for instance if we have $Z_0(z) \equiv g \neq 0$, we choose a Jordan curve Γ in D near B_1 and surrounding B_1. The image of Γ under $Z = Z_n(z)$ surrounds $Z = 0$, whence

$$[\arg Z_n(z)]_\Gamma = 2\pi,$$

which contradicts $Z_n(z) \to g$ uniformly on Γ. Similarly (by use of an auxiliary linear transformation) $Z_n(z) \to \infty$ in D is seen to be impossible. Then the function $Z = Z_0(z)$ is univalent in D, and maps D onto some region Δ_0 of the Z-plane. The region Δ_0 like D is bounded by $\mu + \nu$ mutually disjoint continua, the respective images of the B_j and C_j. If Γ_j is an analytic Jordan curve in D near B_j and surrounding B_j, we have

$$(5) \qquad \frac{1}{2\pi i}\int_{\Gamma_j} \frac{Z_n'(z)dz}{Z_n(z) - a_k'} = \begin{cases} 0, & j \neq k, \\ 1, & j = k, \end{cases}$$

and this integral approaches the value

$$\frac{1}{2\pi i}\int_{\Gamma_j} \frac{Z_0'(z)dz}{Z_0(z) - a_k} = \begin{cases} 0, & j \neq k, \\ 1, & j = k. \end{cases}$$

It is to be noted that $Z_0(z)$ does not assume the value a_k interior to D, as we shall shortly prove (Lemma 2), so the denominator of the integrand in (5) is bounded from zero. Consequently the image of Γ_j under $Z = Z_0(z)$ contains a_j in its interior but contains none of the points a_k with $k \neq j$. Similar reasoning regarding the points b_k shows that the points a_j and b_k are all distinct, that none lies in Δ_0, and that a Jordan curve in D separating B_j or C_j from all the other curves B_k and C_k has for its image in Δ_0 a Jordan curve which separates the image of B_j or C_j from the images of all the other curves B_k and C_k.

In our further detailed study of the mapping of variable regions, we make use of the ideas of Carathéodory [1932, §§120–123]; compare also Bieberbach [1931, p. 13].

LEMMA 2. *Let the functions $f_n(z)$ analytic and univalent in a region D^0 converge uniformly on every compact in D^0 to a function $f_0(z)$ which is not identically constant and therefore univalent. Let $w=f_0(z)$ map D^0 onto a region Δ_0 of the w-plane, and let Δ_0' be a closed subregion of Δ_0. Then for n sufficiently large, the image of D^0 by $w=f_n(z)$ covers Δ_0'.*

Hurwitz's theorem assures us that if w_0 is any point in Δ_0, then for n sufficiently large every $f_n(z)$ takes the value w_0 in D^0; Lemma 2 refers to *uniformity* with respect to all w_0 in Δ_0'.

Let Γ be a Jordan curve or a sum of a finite number of Jordan curves in D^0 whose image under the map $w=f_0(z)$ lies in $\Delta_0-\Delta_0'$ and separates Δ_0' from the boundary of Δ_0. For z on Γ and for all n sufficiently large we have uniformly for all w_0 in Δ_0'.

$$\left|\frac{f_n(z)-f_0(z)}{f_0(z)-w_0}\right|<1,$$

so by Rouché's theorem $f_n(z)-w_0$ has the same number of zeros in the region bounded by Γ as does $f_0(z)-w_0$, namely one. Lemma 2 is established.

The functions $Z_n(z)$ are not necessarily defined throughout the whole region D of Theorem 1, but are defined (for n sufficiently large) on every closed subregion. The function $Z_0(z)$ cannot assume the value a_k interior to D, for if it did it would assume interior to D all values in a fixed neighborhood of a_k, hence by Lemma 2 also $Z_n(z)$ would assume for n sufficiently large in a closed subregion of D all values in a neighborhood of a_k, in particular $Z_n(z)$ would assume there the value a_k', contrary to our definition of Δ_n.

All the regions Δ_n for n sufficiently large contain the image of an arbitrary closed subregion of D under the transformation $Z=Z_0(z)$. The distinct points a_j' and b_j' used to define Δ_n approach distinct limits a_j and b_j, so the regions Δ_n approach their kernel, the region Δ defined by (1); of course the function $|S_n(Z)|^{1/n}$ for fixed Z is a continuous function of the variables $A_n^{1/n}$, a_j', b_j', m_j, n_j. The term *kernel* more precisely consists of the neighborhood of a specific point Z_1 plus fixed closed regions containing that neighborhood which lie in all the Δ_n for n sufficiently large. In the present case, any point Z_1 of Δ can be chosen, for two arbitrary points Z_1 and Z_2 of Δ lie in some closed region which is contained in all the Δ_n for n sufficiently large. The function $Z=Z_0(z)$ maps D onto some subregion Δ_0 of Δ.

We turn now to the consideration of the inverses $z=z_n(Z)$ of the functions $Z=Z_n(z)$. The $z_n(Z)$ are univalent in the respective regions Δ_n, hence for sufficiently large n are univalent in an arbitrary closed subregion of Δ. The functional values $z=z_n(Z)$ lie in D_n, hence are bounded, so in Δ the sequence forms a normal family; henceforth we consider only a subsequence which converges to some function $\zeta(Z)$ in Δ, uniformly in any closed subregion of Δ. The function $\zeta(Z)$ is either identically constant or univalent in Δ. If z_0 is any fixed point of D, we write

$$(6) \qquad Z_n(z_0) = Z_n \to Z_0(z_0), \qquad z_n(Z_n) = z_0,$$

and the points Z_n lie in Δ_0, for sufficiently large index, because $Z_0(z_0)$ lies in Δ_0. Since the Z_n lie in Δ_0 for n sufficiently large, and since they approach $Z_0(z_0)$ in Δ_0, we have, by (6), $z_n(Z_n)=z_n[Z_n(z_0)]=z_0\to\zeta[Z_0(z_0)]$. That is to say, $\zeta(Z)$ is the inverse function of $Z_0(z)$ for z in D and Z in Δ_0. By a similar argument it follows that $Z_0(z)$ is the inverse function of $\zeta(Z)$, for Z in Δ and for z in a suitable subregion of D, namely the image of Δ under the transformation $z=\zeta(Z)$; compare Lemma 2.

The transformation $Z=Z_0(z)$ maps D onto a subregion Δ_0 of Δ, but every value Z_1 in Δ is taken on by $Z_0(z)$ at some point $z_1=\zeta(Z_1)$ in D. Since $Z_0(z)$ is univalent in D, it follows that $Z=Z_0(z)$ maps D conformally and one to one onto Δ.

The well known studies of the boundary behavior of the conformal mapping of Jordan regions, due to Carathéodory and Montel, establishing continuity and one-to-oneness in the closed regions, apply with no essential change in the present conditions, and show that the conformal map of D onto Δ can be extended so as to be one to one and continuous in the closed regions involved. Theorem 1 is established.

The proof of Theorem 1 as given does not include the classical case $\mu=\nu=1$, but the construction of the Riemann surface σ_1 is valid, and the transformation $R_n(z)=S_n(Z)$ maps D_n onto Δ_n, which may be taken as $1<|S_n(Z)|<e^{n/\tau}$,

$$S_n(Z) \equiv \frac{A_n(Z-a_1')^n}{(Z-b_1')^n},$$

which with the choice $a_1'=0$, $b_1'=\infty$ becomes $S_n(Z)\equiv A_nZ^n$. Then Δ_n can be expressed as

$$1<|A_nZ^n|<e^{n/\tau};$$

two such regions for different values of A_n can be mapped onto each other by a dilatation with the origin fixed, so finally we choose Δ_n as $1<|Z|<e^{1/\tau}$. This region is independent of n, and the remainder of the proof of Theorem 1, continued study of the transformation $R_n(z)=S_n(Z)\equiv Z^n$, can be carried through even more simply than in the general case.

Indeed, the case $\mu=\nu=1$ is included in a larger category which, as we now show, can be treated still more simply. Here is included for instance also the case $\mu=2$, $\nu=1$, $M_1=M_2$, for which it is sufficient that B_1 and B_2 be mutually symmetric in a line L in which C_1 is symmetric.

If the numbers M_j and N_j of Theorem 1 defined by (4) are all rational, the region Δ is bounded by two level loci of a *rational* function. For if the integer N is suitably chosen, the numbers NM_j and NN_j are all integers, $[T(Z)]^N$ is a rational function of Z of degree N, and (1) can be written

$$1 < |[T(Z)]^N| < e^{N/\tau}. \tag{1'}$$

For this case Theorem 1 can be established without use of the two infinite sequences of regions D_n and Δ_n. Since the numbers

$$NM_j = \frac{N}{\tau}\int_{B_j} d\sigma, \qquad NN_j = \frac{N}{\tau}\int_{C_j} d\sigma,$$

are integers, the increase in $Nv(z)/2\pi\tau$ as z traces any one of the curves B_j, C_j is integral. Then the function $w=e^{N(u+iv)/\tau}$ is single valued in D and maps D conformally onto part of a Riemann surface σ_0; the image D_0 of D is the annulus $1<|w|<e^{N/\tau}$ covered precisely N times. We use here an extension of Lemma 1, which concerns an arbitrary function $f(z)$ analytic in a closed multiply connected region E bounded by a finite number of mutually disjoint Jordan curves, and we suppose $|f(z)|=m_0$ on several components P_1 of the boundary of E, $|f(z)|=m_1(>m_0)$ on the remaining part P_2 of the boundary, $f(z)\neq 0$ in E; it follows that each value w, $m_0<|w|<m_1$, is assumed by $f(z)$ in E the same number of times, namely the quotient by 2π of the increase in $\arg f(z)$ over P_1. We return to the region D; as z traces $\sum C_j$, $\arg e^{N(u+iv)/\tau}$ is increased by precisely $2\pi N$. Moreover, as z traces each of the curves B_j or C_j, $\arg e^{N(u+iv)/\tau}$ changes monotonically and is increased by precisely $-2\pi NM_j$ or $2\pi NN_j$, so w traces each of the $\mu+\nu$ boundary curves $|w|=1$ or $|w|=e^{N/\tau}$ monotonically $-NM_j$ or NN_j times. The Riemann surface σ_0 contains the image not merely of $\overline{D}$ but even of D', but is not known to cover the entire extended w-plane. Nevertheless continuous adjunction to D_0 of $\mu+\nu$ closed simply connected subregions $|w|\leq 1$ and $|w|\geq e^{N/\tau}$ of the Riemann surfaces for the inverses of $w=z^{NM_j}$ and $w=z^{NN_j}$ constructs a new Riemann surface σ_1 over the entire extended w-plane which is the topological image of the extended z-plane, namely D plus the closures of $\mu+\nu$ Jordan regions exterior to D whose boundaries are respectively the B_j and C_j. Schwarz's theorem now asserts that σ_1 can be mapped one to one and conformally onto the extended Z-plane. The mapping function is of the form $w=[T(Z)]^N$, and the one to one conformal image of D is the region (1').

The proof just given is especially simple in the case $\mu=\nu=1$, for then $N=1$ and $T(Z)$ is of degree one; the extended w-plane and extended Z-plane may be taken as identical. The proof is also simple in the case $\mu=2$, $\nu=1$, $M_1=M_2$, for then $N=2$ and we may choose for instance $[T(Z)]^2\equiv A_0(Z^2-1)$.

II

It is evident that the region Δ of Theorem 1 is not uniquely determined, for one can still transform the Z-plane by an arbitrary linear transformation. Indeed, any possible region Δ, an image of D, may be found from a single such region Δ by a linear transformation:

THEOREM 2. *Let D be a region of the z-plane defined by the inequality*

$$1 < |R(z)| < e^{1/\tau}, \qquad R(z) \equiv \frac{A(z-a_1)^{m_1}\cdots(z-a_\mu)^{m_\mu}}{(z-b_1)^{n_1}\cdots(z-b_\nu)^{n_\nu}},$$

$$\sum m_j = \sum n_j = 1,$$

and whose boundary consists of mutually disjoint Jordan curves $B_1, B_2, \cdots, B_\mu$, $C_1, C_2, \cdots, C_\nu$, where B_j separates a_j from D and C_j separates b_j from D. Let Δ be a region of the Z-plane defined by

$$1 < |R_1(Z)| < e^{1/\tau'}, \qquad R_1(Z) \equiv \frac{A'(Z-a_1')^{m_1'}\cdots(Z-a_\mu')^{m_\mu'}}{(Z-b_1')^{n_1'}\cdots(Z-b_\nu')^{n_\nu'}},$$

$\sum m_j' = \sum n_j' = 1$, *and whose boundary consists of mutually disjoint Jordan curves $B_1', B_2', \cdots, B_\mu', C_1', C_2', \cdots, C_\nu'$, where B_j' separates a_j' from Δ and C_j' separates b_j' from Δ. If there exists a one to one conformal transformation of D onto Δ so that each B_j corresponds to B_j' and each C_j corresponds to C_j', then the transformation is a linear transformation of the complex variable z, defined throughout the extended planes z and Z. We have $\tau=\tau'$, $m_j=m_j'$, $n_j=n_j'$. If* arg A' *is suitably chosen, we have $R(z)\equiv R_1(Z)$.*

If D is an arbitrary region whose boundary consists of $\mu+\nu$ Jordan curves, $\mu+\nu>2$, those curves can be separated into two classes containing respectively μ and ν curves in a wide variety of ways, and so also with Δ; compare Theorem 1. Theorem 2 is not concerned with all possible maps of D onto Δ, but with merely a certain map where we suppose that two classes of curves bounding D correspond respectively to two classes of curves bounding Δ.

If there exists a one-to-one conformal transformation of D onto Δ, that transformation can be extended so as to be one-to-one and continuous in the closed regions, so it is possible to consider the correspondence of boundary curves. Indeed, the Jordan curves bounding both D and Δ are analytic, so the transformation is even one to one and analytic in larger regions D' and Δ' containing respectively the closures $\overline{D}$ and $\overline{\Delta}$ of D and Δ.

For the present we suppose that D and Δ contain respectively the points $z=\infty$ and $Z=\infty$. Let $U(z)$ be the unique function harmonic in D, continuous in $\overline{D}$, equal to zero and unity on $B=\sum B_j$ and $C=\sum C_j$ respectively. Similarly let $U'(Z)$ be the unique function harmonic in Δ, continuous in $\overline{\Delta}$, equal to zero and unity on $B'=\sum B_j'$ and $C'=\sum C_j'$ respectively. Let $V(z)$ and $V'(Z)$ be conjugate to $U(z)$ and $U'(Z)$. If the given transformation is $Z=Z(z)$, we have $U'[Z(z)] \equiv U(z)$, $V'[Z(z)] \equiv V(z)$, if the determinations of the multiform functions $V(z)$ and $V'(Z)$ are suitably chosen. The function $\log |R(z)|$ is harmonic in D and takes the values zero and $1/\tau$ on B and C, so for z in D we have $|R(z)| \equiv e^{U(z)/\tau}$, $R(z) \equiv e^{[U(z)+iV(z)]/\tau}$, and similarly for Z in Δ we have $|R_1(Z)| \equiv e^{U'(Z)/\tau'}$, $R_1(Z) \equiv e^{[U'(Z)+iV'(Z)]/\tau'}$, again for suitable determination of $V(z)$ and $V'(Z)$.

The numbers a_j, b_j, a_j', b_j' are all finite; for the positive directions on the boundary curves with respect to D and Δ we have

$$[\arg R(z)]_{B_j} = -2\pi m_j, \qquad [\arg R(z)]_{C_j} = 2\pi n_j,$$
$$[\arg R_1(Z)]_{B_j'} = -2\pi m_j', \qquad [\arg R_1(Z)]_{C_j'} = 2\pi n_j'.$$

On the other hand we have

$$[\arg R(z)]_{B_j} = [V(z)/\tau]_{B_j}, \qquad [\arg R(z)]_{C_j} = [V(z)/\tau]_{C_j},$$

with similar equations for functions of Z, whence

$$[V(z)/\tau]_B = -2\pi \sum m_j = -2\pi,$$
$$[V'(Z)/\tau']_{B'} = -2\pi \sum m_j' = -2\pi,$$
$$\tau = \tau', m_j = m_j', n_j = n_j'.$$

If now D and Δ are bounded, say interior to C_1 and C_1' respectively, the functions $R(z)$ and $R_1[Z(z)]$ are analytic in $\bar{D}$, different from zéro there, with

$$\left| \frac{R(z)}{R_1[Z(z)]} \right| = 1$$

on the boundary $B+C$ of D. Then this quotient is identically constant in D, and by changing $\arg A'$ if necessary we may write

$$R(z) \equiv R_1[Z(z)]; \tag{7}$$

this identity is valid not merely in $\bar{D}$ but also throughout D', and indeed wherever the functions involved may be continued analytically from D'. We now keep A and A' fixed.

The functions $U(z)$ and $U'(Z)$ have each precisely [Walsh, 1950, p. 274] $\mu+\nu-2$ critical points in D and Δ; this is the total number of critical points of $R(z)$ and $R_1(Z)$, so the latter functions have all their critical points in D and Δ respectively.

Let D_1 be the region whose boundary is B_1 which belongs to the complement of $\bar{D}$, and Δ_1 the region whose boundary is B_1' which belongs to the complement of $\bar{\Delta}$. When z moves from D across B_1 into D_1, the point $Z(z)$ moves from Δ across B_1' into Δ_1; in D_1 we have $|R(z)| < 1$ and in Δ_1 we have $|R_1(Z)| < 1$, so by (7) the point z cannot leave D_1 if Z remains in Δ_1, nor can Z leave Δ_1 if z remains in D_1.

The correspondence between z and Z defined by (7) involves the equation $|R(z)| = |R_1(Z)|$, and the level curves in D and Δ of these two functions $R(z)$ and $R_1(Z)$ correspond to each other. These level curves pass one through each point of D_1 and Δ_1 (except of course $z=a_1$ and $Z=a_1'$), have no multiple points in D_1 and Δ_1 (because $R(z)$ and $R_1(Z)$ have no critical points there), and by the principle of maximum modulus each level curve is an analytic Jordan curve containing a_1 or a_1' in its interior. When z or Z traces such a level curve in the positive sense, arg $[R(z)]$ or arg $[R_1(Z)]$ increases by $2\pi m_1 = 2\pi m_1'$. The orthogonal trajectories to these level curves are the loci arg $[R(z)]=$const and arg $[R_1(Z)]=$const, which can be considered as Jordan arcs in D_1 and Δ_1 from a_1 and a_1' to B_1 and B_1'; any two such loci have no point of intersection in D_1 and Δ_1 except a_1 or a_1' unless they coincide throughout. One such locus extends from a_1 to each point of B_1 and one from a_1' to each point of B_1', and cuts each level curve in D_1 and Δ_1 once and only once.

Equation (7) *sets up a one-to-one analytic correspondence between the points of* D_1 *and those of* Δ_1. For fixed θ, let α and α' be respective loci arg $R(z)=\theta$ in D_1 and arg $R_1(Z)=\theta$ in Δ_1; on the subarc of α in D' and its image under the conformal map we have points z and Z corresponding to each other by (7). All along α and α' we now set up a correspondence by requiring $|R(z)| = |R_1(Z)|$; this correspondence is one to one in D_1 and Δ_1. We extend this correspondence continuously from α and α' in the same sense along each level locus in D_1 and the corresponding level locus in Δ_1, allowing arg $R(z)$ and arg $R_1(Z)$ to vary continuously and setting up the correspondence by means of the equation arg $R(z)=$arg $R_1(Z)$; the total change of both arg $R(z)$ and arg $R_1(Z)$ along each level curve is precisely $2\pi m_1$. We thus have defined a one-to-one correspondence between the points z of D_1-a_1 and the points Z of Δ_1-a_1' which coincides in $D_1\cdot D'$ and $\Delta_1\cdot\Delta'$ with the analytic extension of the given conformal map. This correspondence in D_1-a_1 and Δ_1-a_1' is defined locally by equation (7), and is analytic, for if we set $R(z)=w=R_1(Z)$, then w is an analytic function of z and Z is an analytic function of w, since $R_1'(Z)\neq 0$ in Δ_1; hence Z is an analytic function of z. This function is continuous even at $z=a_1$ if we define $Z(a_1)$ as a_1', hence (Riemann) is analytic also for $z=a_1$.

The reasoning just given applies to each of the $\mu+\nu$ regions into which the z-plane is separated by $\overline{D}$, and to the analogous $\mu+\nu$ regions of the Z-plane, and shows that equation (7) defines a one to one conformal map not merely of D onto Δ but of the extended z-plane onto the extended Z-plane. Such a map is necessarily defined by a linear transformation, so Theorem 2 is established.

If one chooses $a_1=a_1'=0$, $b_1=b_1'=\infty$, this linear transformation must be of the form $Z=\lambda z$, $\lambda\neq 0$, so D and Δ are similar figures—this condition of similarity is both necessary and sufficient that D and Δ be conformally representable on each other, with B_1 and C_1 corresponding respectively to B_1' and C_1'.

Theorem 2 naturally applies to the conformal transformations of a region D into itself, where $\sum B_j$ is invariant. If we again suppose D in the canonical form of Theorem 2 with $a_1 = a_1' = 0$, $b_1 = b_1' = \infty$ (the conditions $a_1 = a_1'$, $b_1 = b_1'$ are equivalent to requiring that B_1 and C_1 shall be invariant), a transformation which carries D into itself must be of the form $Z = \lambda z$ with $|\lambda| = 1$; otherwise a point z_0 of D could be carried into the points $\lambda z_0, \lambda^2 z_0, \lambda^3 z_0, \cdots$, which approach 0 or ∞, which is impossible. Then a necessary and sufficient condition that D admit a conformal transformation into itself (with $a_1 = a_1' = 0$, $b_1 = b_1' = \infty$) other than the identity is that D admit a rotation about 0 into itself. A necessary condition is that the sets of numbers $m_2, m_3, \cdots, m_\mu$ and $n_2, n_3, \cdots, n_\nu$ each fall into ρ identical groups, where ρ is an integer greater than unity; the rotation about 0 will then be through the angle $2\pi/\rho$; but this condition is not sufficient.

It is clear that in the case $\mu = 1$, $\nu = 2$ with $n_1 = n_2$ (or $\mu = 2$, $\nu = 1$ with $m_1 = m_2$) the region D *always admits* a conformal transformation into itself, for one can fix a_1 while interchanging b_1 and b_2 (or fix b_1 while interchanging a_1 and a_2) by a linear transformation.

III

We proceed to show the validity of the limiting case of Theorem 1 where in the hypothesis the numbers M_j are kept fixed but the curves B_j are allowed to shrink to points:

THEOREM 3. *Let D be a region of the z-plane whose boundary consists of mutually disjoint Jordan curves $C_1, C_2, \cdots, C_\nu$, let $\alpha_1, \alpha_2, \cdots, \alpha_\mu$ be arbitrary distinct points of D, and let $M_1, M_2, \cdots, M_\mu$ be arbitrary positive numbers with $\sum M_j = 1$. Then there exists a conformal map of D onto a region Δ of the Z-plane, one-to-one and continuous in the closed regions, where Δ is defined by*

$$|T(Z)| < 1, \qquad T(Z) \equiv \frac{A(Z-a_1)^{M_1}(Z-a_2)^{M_2}\cdots(Z-a_\mu)^{M_\mu}}{(Z-b_1)^{N_1}(Z-b_2)^{N_2}\cdots(Z-b_\nu)^{N_\nu}}, \tag{8}$$
$$N_j > 0, \quad \sum N_j = 1.$$

The a_j are the respective images of the α_j; the locus $|T(Z)| = 1$ consists of ν analytic Jordan curves, which are respective images of the C_j, and which separate Δ from the b_j.

Theorem 3 is of particular interest in the case $\mu = 1$, for if then we choose $a_1 = \infty$, Δ is defined by

$$|(Z-b_1)^{N_1}(Z-b_2)^{N_2}\cdots(Z-b_\nu)^{N_\nu}| > |A|.$$

The exponents N_j are not necessarily rational, but if they are rational, with

$\mu=1$ and $a_1=\infty$, the boundary of Δ is a lemniscate. Theorem 3 includes the Riemann mapping theorem for a region bounded by a single Jordan curve.

The proof of Theorem 3 is so similar to that of Theorem 1 that we omit most of the details. We omit for the present the classical case $\mu=\nu=1$ and thanks to a preliminary transformation we suppose the Jordan curves C_j to be analytic, with D interior to C_1.

If $g_j(z)$ is Green's function for D with pole in α_j, we set

$$u(z) \equiv M_1 g_1(z) + M_2 g_2(z) + \cdots + M_\mu g_\mu(z),$$

a function harmonic in D except in the points α_j, and which can be extended harmonically across each C_j so as to be harnonic in a closed region D' which contains the closure $\overline{D}$ of D and whose boundary $C': u(z)=-\delta(<0)$ consists of ν analytic Jordan curves near the respective C_j. If $v(z)$ denotes the conjugate of $u(z)$, we assume the derivative of $u(z)+iv(z)$ not to vanish in the closure of the annular regions $D'-\overline{D}$. The formula of Green, valid for z in D', can be written [Walsh, 1935, p. 215]

$$(9) \qquad u(z) = \int_0^1 \log|z-t|\,d\sigma - \sum_1^\mu M_j \log|z-\alpha_j| - \delta,$$

$$d\sigma = dv/2\pi, \qquad \int_{C'} d\sigma = 1.$$

The integral in (9) is to be taken over C'.

Let the β_k (depending on n) divide C' into n parts equal with respect to the parameter σ, and let integers m_j (depending on n) be chosen so that $\sum m_j = n$, $m_j/n \to M_j$. On any closed set in the region $D'-\sum\alpha_j$ we have uniformly as $n\to\infty$

$$u_n(z) \equiv \frac{1}{n}\sum_1^n \log|z-\beta_k| - \frac{1}{n}\sum_1^\mu m_j \log|z-\alpha_j| - \delta \to u(z).$$

As $n\to\infty$, the region D is approximated by the region $D_n: u_n(z)>0$, a region bounded by ν Jordan curves near (for n sufficiently large as close as desired to) the C_j. Thus D_n is defined by

$$|R_n(z)| > 1, \qquad R_n(z) \equiv \frac{e^{-n\delta}(z-\beta_1)(z-\beta_2)\cdots(z-\beta_n)}{(z-\alpha_1)^{m_1}(z-\alpha_2)^{m_2}\cdots(z-\alpha_\mu)^{m_\mu}}$$
$$\equiv e^{n(u_n+iv_n)},$$

where $v_n(z)$ is conjugate to $u_n(z)$ in D'. Each compact K of D lies in all the D_n for n sufficiently large, and each compact K exterior to D lies exterior to all the D_n sufficiently large.

The equation $w=R_n(z)$ defines the conformal map of the extended z-plane onto an n-sheeted Riemann surface σ_0 over the extended w-plane, and the image $|w|>1$ of D_n is connected, having μ branch points at infinity of respective orders m_j-1, and having for boundary ν circumferences $|w|=1$ of respective multiplicities n_j, namely the numbers of the β_k on the respective components of C', with $n_j/n \to \int_{C_j} d\sigma = N_j$ (definition of N_j). The ν regions of the z-plane complementary to D_n have as images ν simply connected regions of σ_0, covering $|w|<1$ respectively n_j times; the boundaries of these regions consist of the circle $|w|=1$ traced monotonically n_j times.

We construct a new Riemann surface σ_1 over the w-plane by replacing continuously each of these ν regions of σ_0 by the portion $|w|<1$ of the n_j-sheeted Riemann surface for the inverse of $w=z^{n_j}$, having then but a single branch point, namely $w=0$. Thus σ_1 like σ_0 is the topological image of the extended plane and covers each point of the w-plane precisely n times, hence can be mapped conformally onto the extended Z-plane, by the transformation $w=S_n(Z)$,

$$S_n(Z) \equiv \frac{A_n(Z-b_1')^{n_1}(Z-b_2')^{n_2}\cdots(Z-b_\nu')^{n_\nu}}{(Z-a_1')^{m_1}(Z-a_2')^{m_2}\cdots(Z-a_\mu')^{m_\mu}},$$

$\sum m_j = \sum n_j = n$; the a_j' and b_j' (depending on n) are all distinct, and the b_j' are exterior to the image Δ_n: $|S_n(Z)|>1$ of D_n, while a_j' lies interior to Δ_n as the image of α_j. By a linear transformation of the Z-plane let us choose $a_1'=0$, $b_1'=1$, $b_2'=\infty$ independent of n (if $\nu=1$ we choose $a_2'=\infty$ instead of $b_2'=\infty$). We write the transformation $R_n(z)=S_n(Z)$ as $Z=Z_n(z)$.

If we denote by D_0 the region $D_n-\alpha_1$ (if $\nu=1$ we choose $D_n-\alpha_1-\alpha_2$), the functions $Z_n(z)$ admit the exceptional values 0, 1, ∞ in D_0, hence form a normal family in $D-\alpha_1-\alpha_2$. We have also

$$[\arg Z_n(z)]_\Gamma = 2\pi,$$

where Γ is a curve in D surrounding α_1 but not surrounding any other α_j, so no limit function of the sequence $Z_n(z)$ can be identically constant in D_0; every such limit function must be univalent in $D-\alpha_1$ and indeed in D.

As $n\to\infty$, the inequality $|S_n(Z)|>1$ defining Δ_n approaches the form (8), if we choose a subsequence of the n such that all the numbers $A_n^{1/n}$, a_j', b_j' approach limits. The remainder of the proof of Theorem 3 now follows closely that of Theorem 1, and is left to the reader.

The proof of Theorem 3 as indicated requires but minor modifications to apply to the case $\mu=\nu=1$, for then $S_n(Z)$ is of the form

$$\frac{A_n(Z-b_1')^n}{(Z-a_1')^n}.$$

The choice $a_1'=0$, $b_1'=\infty$ yields a family of regions Δ_n depending on A_n which are all trivially conformally equivalent, so we may choose $A_n=1$ and complete the proof using a region Δ_n which does not depend on n. However, in this proof of the Riemann mapping theorem for a region bounded by a Jordan curve, we have used the fact that such a curve may be taken as analytic, and have also used Schwarz's theorem on the conformal map of an arbitrary Riemann surface topologically equivalent to the sphere.

In Theorem 3 (compare Theorem 1) the numbers M_j and N_j may all be rational, in which case the region Δ is bounded by a level locus of the *rational* function $[T(Z)]^N$, where the integer N is so chosen that the numbers NM_j and NN_j are all integers. It follows from this choice of N that the function $w=e^{N(u+iv)}$ is single valued in D and maps D onto the subregion $|w|>1$ of an N-sheeted Riemann surface σ_0 over a portion of the extended w-plane. The Riemann surface σ_0 can be replaced by a simpler N-sheeted Riemann surface σ_1 over the extended w-plane without modifying the image of D, and the proof of Theorem 3 (like that of Theorem 1) can be completed without use of the infinite sequences of regions D_n and Δ_n.

IV

The map of Theorem 3, like that of Theorem 1, is uniquely determined except for a possible linear transformation:

THEOREM 4. *Let D be a region of the z-plane defined by the inequality*

$$|R(z)| < 1, \qquad R(z) \equiv \frac{A(z-a_1)^{m_1}(z-a_2)^{m_2}\cdots(z-a_\mu)^{m_\mu}}{(z-b_1)^{n_1}(z-b_2)^{n_2}\cdots(z-b_\nu)^{n_\nu}}, \qquad m_j > 0,$$

$$\sum m_j = \sum n_j = 1,$$

and whose boundary consists of mutually disjoint Jordan curves $C_1, C_2, \cdots, C_\nu$, which separate the b_j respectively from D. Let Δ be a region of the Z-plane defined by the inequality

$$|R_1(Z)| < 1, \qquad R_1(Z) \equiv \frac{A'(Z-a_1')^{m_1}(Z-a_2')^{m_2}\cdots(Z-a_\mu')^{m_\mu}}{(Z-b_1')^{n_1'}(Z-b_2')^{n_2'}\cdots(Z-b_\nu')^{n_\nu'}},$$

$$\sum n_j' = 1,$$

and whose boundary consists of mutually disjoint Jordan curves $C_1', C_2', \cdots, C_\nu'$ which separate the b_j' respectively from Δ. If there exists a one to one conformal transformation of D onto Δ so that the a_j correspond respectively to the a_j', then this correspondence can be continued beyond D and Δ as a linear transformation of the extended z-plane onto the extended Z-plane. We have $n_j=n_j'$, and if arg A' *is suitably chosen,* $R(z)\equiv R_1(Z)$.

An element of contrast between Theorem 1 and Theorem 3 is that in the former the numbers m_j are determined initially by the properties of D, whereas in the latter theorem the (positive) m_j are entirely arbitrary, subject merely to the restriction $\sum m_j=1$. Theorem 4 is false if the respective exponents in the numerator of $R_1(Z)$ are not supposed equal to those in the numerator of $R(z)$, as we shall prove. We choose an arbitrary region D_0 bounded by ν mutually disjoint Jordan curves, and points $\alpha_1, \alpha_2, \cdots, \alpha_\mu$ in D_0; then we can choose two different sets of exponents $m_1, m_2, \cdots, m_\mu$ and $m_1', m_2', \cdots, m_\mu'$ with $\sum m_j = \sum m_j' = 1$, and there exist by Theorem 3 two distinct maps of D_0 onto the regions D and Δ of Theorem 4 but with the exponents m_j' instead of the m_j in the definition of $R_1(Z)$. That is to say, this region D can be mapped one to one and conformally onto Δ, with coincidence of the two images of each α_j, but clearly the map cannot be extended so as to be a linear transformation of the extended z-plane onto the extended Z-plane.

If $g_j(z)$ denotes Green's function for D with pole in a_j, in D we have $\log |R(z)| \equiv \sum m_j g_j(z)$, and if $g_j'(z)$ denotes Green's function for Δ with pole in a_j', in Δ we have $\log |R_1(Z)| \equiv \sum m_j g_j'(Z)$. However, Green's function is invariant under a one-to-one conformal transformation, so if the given transformation of Theorem 4 is $Z = Z(z)$, throughout D we have $g_j'[Z(z)] \equiv g_j(z)$, whence $|R_1[Z(z)]| \equiv |R(z)|$. The quotient $R(z)/R_1[Z(z)]$ is analytic throughout D and does not vanish there, so by a change in the argument of A' if necessary we may write throughout D

$$R(z) \equiv R_1[Z(z)]. \tag{10}$$

If ϵ (>0) is sufficiently small, the loci $|R(z)| = \epsilon$ and $|R_1(Z)| = \epsilon$ lie in D and Δ respectively and consist each of μ small curves about the points a_j and the points a_j'. It follows from (10) that the given conformal map $Z = Z(z)$ transforms the region $1 < |1/R(z)| < 1/\epsilon$ onto the region $1 < |1/R_1(Z)| < 1/\epsilon$, so Theorem 4 follows from Theorem 2.

We add a few further remarks relative to the conformal transformations of a multiply connected region D into itself, where $\sum B_j$ is invariant, a topic to which Theorem 2 applies, and we naturally use the canonical form of Theorem 2 for D. A linear transformation which carries D into itself can be neither parabolic, hyperbolic, nor loxodromic, for none of these transformations is periodic; each such transformation carries an analytic Jordan curve B_j or C_j which is not a circle through the fixed points (or for a parabolic transformation through the fixed point) into an infinity of distinct images, hence cannot transform D into itself.

We consider then an elliptic transformation T which transforms D into D, where now $\mu+\nu>2$, and we choose the fixed points of T as zero and infinity. If ρ (>1) is the smallest integer for which T^ρ is the identity, an integer which must exist by virtue of $\mu+\nu>2$, each point a_j or b_j other than 0 and ∞ belongs to a set of ρ points a_j or b_j, a set which is transformed into itself by T (a rotation about 0), and for which m_j or n_j does not depend on j. Each curve B_j or C_j which is not a Jordan curve with ρ-fold rotational symmetry about 0 containing 0 in its interior thus belongs to a set of ρ curves B_j or C_j, a set invariant under T. No curve B_j or C_j can be a circle with center 0, for if it were the corresponding a_j or b_j would be 0 or ∞, and Schwarz reflection of the harmonic function $\log |R(z)|$ in this circle with center 0 shows that $\mu+\nu\leqq 2$. If 0 is not a point a_j or b_j, it is a zero of the derivative $R'(z)$, for in the field of force [Walsh, 1950, p. 89] which determines the zeros of $R'(z)$ the point 0 is a point of zero force for each set of ρ points a_j and b_j. One sees by interchanging the roles of 0 and ∞ that ∞ also is either a point a_j or b_j or a zero of $R'(z)$.

Each zero of $R'(z)$ lies in D, for D contains $\mu+\nu-2$ such zeros [Walsh, 1950, p. 274], and $R'(z)$ has precisely $\mu+\nu-2$ zeros. The conditions on Γ that we have established as necessary are easily seen to be sufficient, so in conclusion we have: *If $\mu+\nu>2$, the region D in canonical form admits a conformal transformation into itself with $\sum B_j$ invariant if and only if D and $\sum B_j$ are invariant with respect to an elliptic transformation T whose ρth power is the identity; each fixed point of T is either an a_j, a b_j, or a zero of $R'(z)$, the latter being necessarily in D.*

When D is given, there exist at most $\mu+\nu-2$ distinct zeros of $R'(z)$. The number ρ is a divisor of μ, $\mu-1$, or $\mu-2$ according as 0, 1, or 2 fixed points of T are points a_j, and ρ is likewise a divisor of ν, $\nu-1$, or $\nu-2$ according as 0, 1, or 2 fixed points of T are points b_j. Then for any given D, only relatively few transformations T are possible, especially in view of the equality necessary among the m_j and n_j. It should not be difficult to examine all the possible transformations to identify those which carry D into itself.

Of course this study of the transformations of D into itself is essentially the study of the automorphisms of the function $|R(z)|$; each such automorphism corresponds to an elliptic transformation T, essentially involving a multiple symmetry with each fixed point of T either an a_j, a b_j, or a critical point of $R(z)$.

In the study of interpolation or approximation to a given analytic function $f(z)$, it is often necessary to cover a given region by the level curves of a harmonic function, a function which depends on the particular procedure of interpolation or approximation. Both the degree of approximation to $f(z)$ and the regions of convergence of the approximating functions depend primarily on the regions of analyticity of $f(z)$ with respect to this system of curves. Theorem 3 is especially useful for this purpose of interpolation and approximation because if a given function $f(Z)$ is analytic in Δ, the a_j can be chosen as points of interpolation for a series of rational functions whose poles lie in the b_j. Theorem 1 is likewise useful, because a function $f(Z)$ analytic in Δ is the sum of two component functions analytic in the respective regions $|T(Z)|>1$ and $|T(Z)|<e^{1/\tau}$; interpolation series of rational functions for these two components using first the b_j as points of interpolation and the a_j as poles of the rational functions, then interchanging these roles of the a_j and the b_j,—such a series is analogous to and even a generalization of Laurent's series, and gives a useful representation of the given $f(Z)$ [compare Walsh, 1955].

BIBLIOGRAPHY

1931. L. Bieberbach, *Lehrbuch der Funktionentheorie*, Leipzig.

1932. C. Carathéodory, *Conformal representation*, Cambridge Tracts in Mathematics and Mathematical Physics, No. 28.

1934. G. Julia, *Sur la représentation conforme des aires multiplement connexes*, Paris.

1930. C. J. de la Vallée Poussin, *Sur la représentation des aires multiplement connexes*, Ann. École Norm. (3) vol. 47, pp. 267–309.

1930a. ———, *Sur la représentation conforme des aires planes multiplement connexes*, C.R. Acad. Sci. Paris, vol. 191, pp. 1414–1418.

1931. ———, *Sur la représentation conforme des aires planes multiplement connexes*, C.R. Acad. Sci. Paris, vol. 192, pp. 128–131.

1931a. ———, *Sur la représentation conforme des aires planes multiplement connexes*, Bull. de l'Académie royale de Belgique (5), vol. 17, pp. 10–27.

1935. J. L. Walsh, *Interpolation and approximation*, Amer. Math. Soc. Colloquium Publications, vol. 20.

1950. ———, *Location of critical points*, Amer. Math. Soc. Colloquium Publications, vol. 34.

1954. ———, *Sur la représentation conforme des aires multiplement connexes*, C.R. Acad. Sci. Paris, vol. 239, pp. 1572–1574, 1756–1758.

1955. ———, *Sur l'approximation par fonctions rationnelles et par fonctions holomorphes bornées*, Annali di Matematica Pura ed Applicata (4), vol. 39, pp. 267–277.

HARVARD UNIVERSITY,
CAMBRIDGE, MASS.

Commentary by Dieter Gaier

Throughout his mathematical career, J. L. Walsh was always interested in geometric aspects of complex analysis. We begin our report by pointing out two of his papers which mark the beginning and the end of his engagement in questions of conformal mapping.

In his paper [26-b*] Walsh proved his fundamental theorem concerning polynomial approximation of functions f holomorphic in a Jordan domain D and continuous in $\bar{D}$. Walsh was on leave of absence from Harvard to do research in Munich where he met the constellation Faber/Carathéodory. The latter suggested to him to use a theorem of Courant (1914) on the conformal mapping of variable regions to obtain the above mentioned result, and indeed Walsh was successful. Here for the first time one of the more elaborate theorems in conformal mapping was applied to prove a theorem in complex approximation.

In one of his last papers [73-c] Walsh pays tribute to Osgood's role in proving the Riemann mapping theorem. In 1900, Osgood gave the first complete and rigorous proof of that result, and now, in 1973, Walsh traces Osgood's steps carefully using Green's function, exhaustion of the domain, and Harnack's principle for sequences of harmonic functions.

In the following we have classified Walsh's papers related to conformal mapping into three topics:

- Geometry of level curves and related topics;
- Conformal mapping near the boundary;
- Conformal mapping of multiply connected domains.

Our aim is not to give a detailed description of his papers but rather to point out the later developments following from or connected with Walsh's work in conformal mapping.

Topic I: Geometry of level curves and related topics

About ten papers of Walsh deal with the geometry of level curves of Green's function and of lemniscates. If D is a possibly multiply connected domain and G its Green's function with respect to a point $z_0 \in D$, level curves are sets of points $z \in D$ where $G(z; z_0)$ is a fixed positive constant. Lemniscates are point sets where $|(z - z_1)(z - z_2) \cdots (z - z_n)|$ is constant.

One is interested in matters like circularity, center of curvature, orthogonal trajectories, multiple points, convexity, points of inflexion.

Relevant papers [35-a], [37-a], [37-d*], [38-a*], [39-d*], [40-a*], [53-d], [53-e], [61-a], [70-a*].

There is an enormous body of literature concerning level curves. In the following we pick out three types of problems that are closely related to Walsh's work. Further works can be found in the comprehensive bibliography of schlicht functions by Bernardi [2], Sections T7 and T35.

I.1. Domains convex in one direction

A simply connected domain $D \subset \mathbf{C}$ is called convex in the direction of the imaginary axis (CIA) if its intersection with any vertical line is either connected or empty. If f maps $\mathbf{D} = \{z : |z| < 1\}$ conformally onto D, normalized by $f(0) = 0$ and $f'(0) = 1$, then f is called a function in CIA. For $0 < r < 1$ we let $C_r = \{f(z) : |z| = r\}$ and $D_r = \text{int } C_r$.

Quite often properties (like convexity) of D are inherited by D_r, and it came as a surprise when Hengartner and Schober discovered in 1973 that $D \in$ CIA need not imply $D_r \in$ CIA for all $r \in (0, 1)$. Goodman and Saff (1979) studied a concrete example and showed that for every $r \in (\sqrt{2} - 1, 1)$ there is a domain $D \in$ CIA such that $D_r \notin$ CIA, and they conjectured that $D_r \in$ CIA whenever $r \leq \sqrt{2} - 1$ (Goodman-Saff conjecture). This was finally proved by Prokhorov [28] using an integral representation for functions in CIA, and a little later by Ruscheweyh and Salinas [32].

Styer and Wright [36] and Brown [4] showed that there exist ρ_1, ρ_2 with $\sqrt{2} - 1 < \rho_1 < \rho_2 < 1$ and domains $D \in$ CIA such that $D_{\rho_1} \notin$ CIA but $D_{\rho_2} \in$ CIA; membership of D_r to CIA may therefore change when r increases from $\sqrt{2} - 1$ to 1.

I.2. Length and area problems

Here we assume $f \in S$, i.e. f is schlicht in $\mathbf{D}$ and normalized by $f(0) = 0$, $f'(0) = 1$. For $0 < r < 1$ we let $C_r = \{f(z) : |z| = r\}$ and $D_r = \text{int } C_r$, further $L(r) =$ length of C_r and $A(r) =$ area of D_r. If $f(\mathbf{D})$ is starshaped with respect to $w = 0$, $f \in S_s$, Keogh (1959) obtained the best possible estimate $L(r) = O(\log(1/(1-r)))$ $(r \to 1-)$, provided f is bounded in $\mathbf{D}$. Thomas [37] replaced the latter condition by the requirement that $f(\mathbf{D})$ has finite area. And if f is close-to-convex in $\mathbf{D}$, the image of $|z| = r$ under f has maximal length if f is the Koebe function $k(z) = z/(1-z)^2$:

$$L_f(r) \le L_k(r) \cong \frac{4}{(1-r)^2};$$

see Clunie and Duren [5]. Whether this is true for all $f \in S$ seems to be still an open problem. In the paper [19] by Krzyż and Radziszewski the relation between $L(r)$ and $A(r)$ is studied; $L^2(r) - 4\pi A(r)$ increases with r or vanishes identically.

I.3. On the geometry of lemniscates

In a series of papers D. B. Shaffer, who was Walsh's Ph.D. student in 1962, studies the geometry of lemniscates

$$L_\mu : |(z - z_1)(z - z_2)\cdots(z - z_n)| = \mu,$$

in particular of L_1, and more generally of the curves where

$$R(z) = \frac{\prod_{j=1}^{n}(z - \alpha_j)}{\prod_{j=1}^{m}(z - \beta_j)}$$

has constant absolute value. For example, if all α_j are in a disc D_1 and all β_j are in a disc D_2 with $D_1 \cap D_2 = \emptyset$, precise upper and lower bounds for the curvature of a lemniscate through a point P exterior to $D_1 \cup D_2$ can be given [33]. Further she asks: When is L_1 convex? Let $z_0 = \frac{1}{n}\sum_{j=1}^{n} z_j$, and assume that $|z_j - z_0| \le a$ for all $j = 1, \ldots, n$. Then L_1 is convex if $4(1-a^2)^{3/2}(1-2a^2) \ge a$. If only $|z_j| \le a$, the lemniscate L_μ will be convex if it lies in $|z| \ge a\sqrt{2}$; see [34]. This follows from certain estimates for the curvature of L_μ at a point z which can even be improved if the z_j show certain symmetries: $\sum_{j=1}^{n} z_j^l = 0$ for $l = 1, 2, \ldots, p$.

In [35] Shaffer studies equipotential surfaces in $\mathbf{R}^n$, in the sense of Kahane, and their geometric properties. The sets $\{\alpha_j\}$ and $\{\beta_j\}$ are now replaced by disjoint sets E_1, E_2 that are compact and convex, and on an equipotential surface V_λ we have $V(x) = \lambda > 0$. One result is as follows. Let E_1 be contained in a ball B with center 0. Then any surface V_λ lying outside B is starshaped with respect to 0.

Topic II: Conformal mapping near the boundary

For many questions in complex approximation theory the behaviour of a conformal map near the boundary is important. Walsh has contributed to this topic with three papers.

Relevant papers [50-a], [55-a*], [56-b*].

In these papers, Walsh and his coauthors demonstrate that the study of a conformal map f near a boundary point can, in some cases, be reduced to the study of a sequence $\{f_n\}$ of mapping functions generated from f by making simple transformations of the argument. This sequence $\{f_n\}$ is then discussed by means of the Carathéodory kernel theorem.

This idea goes back to Montel (1917); shortly before Walsh it was used also by Lelong-Ferrand (1952) in conjunction with the study of strip mappings.

In the sequel we report about some problems in this area that are related to Walsh's work.

II.1. Conformal mapping of strip domains

In the study of a conformal map f near a boundary point z_0, it is often convenient to assume $z_0 = \infty$. We therefore assume that f maps the parallel strip

$$S = \left\{z = x + iy : -\infty < x < +\infty, |y| < \frac{\pi}{2}\right\}$$

conformally onto a domain $D = f(S)$ such that

$$\mathrm{Re} f(x) \to +\infty \quad \text{as} \quad x \to +\infty;$$

the image domain thus extends to ∞. Walsh's papers deal with the general problem: Find geometric properties of D which permit to derive properties of f as $z \to \infty$ in S. In particular, attention has focussed on the problem of the angular derivative: Find conditions on D such that

$$\lim[z - f(z)] = C \tag{1.1}$$

exists with $-\infty < C < +\infty$, where the limit is for $x \to +\infty$, $|y| < (\pi/2) - \delta$ and for each δ with $0 < \delta < (\pi/2)$. C is then called the angular derivative of f at $+\infty$.

In the last 20 years, several authors have made important contributions to this old problem: Eke [5a,b], Rodin and Warschawski [30], [31], Jenkins and Oikawa [18], and Oikawa [24]. In particular, necessary and sufficient conditions that (1.1) is true were given in [31] and [18]. To formulate these, assume that D contains the positive real axis; for $u > 0$ let θ_u be the crosscut in D which intersects the real axis and which lies on $\{w : \mathrm{Re}\, w = u\}$. If $0 < u' < u''$, then $D \backslash (\theta_u \cup \theta_{u''})$ consists of three components; that component which contains each θ_u with $u' < u < u''$ will be denoted by $Q(u', u'')$. Finally, the extremal length of all arcs in $Q(u', u'')$ which join $\theta_{u'}$ to $\theta_{u''}$ is denoted by $\lambda(u', u'')$.

In this setting, Rodin and Warschawski [31] and Jenkins and Oikawa [18] have reduced our problem to an extremal length problem. The mapping f has an angular derivative at ∞ if and only if:

For each δ with $0 < \delta < \frac{\pi}{2}$, D contains a half-strip

$$\left\{w = u + iv : u > u_0, \quad |v| < \frac{\pi}{2} - \delta\right\};$$

$$\lambda(u', u'') = u'' - u' + o(1), \text{ where } o(1) \to 0 \text{ as } u'' > u' \to \infty.$$

This result can be combined with Euclidean estimates for extremal length to yield angular derivative criteria of a geometric nature.

This situation becomes particularly simple if $D \subset S$ or $S \subset D$ is assumed. Here necessary and sufficient conditions in terms of subdivisions $\{u_n\}$ of D or of certain coverings can be derived which improve earlier results of Ferrand (1944). In [24], Oikawa gives a nice necessary and sufficient condition for (1.1) in case that $S \subset D$. We introduce

$$\delta^+(u) = \operatorname{dist}\left(u + i\frac{\pi}{2}, \partial D\right) \quad \text{and} \quad \delta^-(u) = \operatorname{dist}\left(u - i\frac{\pi}{2}, \partial D\right).$$

Then f has an angular derivative at $+\infty$ if and only if

$$\int^{\infty} \delta^+(u)\,du < \infty \quad \text{and} \quad \int^{\infty} \delta^-(u)\,du < \infty.$$

In [29], Rodin and Warshawski considered mappings f of S onto L-strips. These are strip domains of the form

$$D = \{w = u + iv : -\infty < u < +\infty, \varphi_-(u) < v < \varphi_+(u)\},$$

where φ_+, φ_- are continuous on $\mathbf{R}$ such that

$$\frac{\varphi_+(u_2) - \varphi_+(u_1)}{u_2 - u_1} \to 0 \quad \text{and} \quad \frac{\varphi_-(u_2) - \varphi_-(u_1)}{u_2 - u_1} \to 0$$

as $u_2 > u_1 \to +\infty$. Of course we assume that $f(x) \to +\infty$ as $x \to +\infty$. Under these assumptions $\arg f'(z) \to 0$ as $\operatorname{Re} z \to +\infty$, uniformly in S, and

$$\lim \frac{|f'(z)|}{\theta(u(z))} = \frac{1}{\pi} \quad \text{as} \quad \operatorname{Re} z \to +\infty$$

uniformly for $|y| \le \frac{\pi}{2} - \delta$ $(\delta > 0)$; here $\theta(u) = \varphi_+(u) - \varphi_-(u)$. For the inverse function $g = f^{-1}$ we have the asymptotic formula

$$q(w) \sim \pi \int^{u} \frac{dt}{\theta(t)} \quad \text{as} \quad \operatorname{Re} w \to +\infty,$$

uniformly in D. Results of this kind are contained in Walsh and Rosenfeld [56-b*] but the new proofs using extremal length methods are simpler than the orginial ones.

II.2. Hölder continuity of the mapping function

Here we assume that f maps the unit disc $\mathbf{D}$ onto a Jordan domain D, and we let $g = f^{-1}$ be its inverse function. If $\partial D = C$ is Dini-smooth, i.e. if the tangent angle $\nu(s)$ satisfies a Dini condition, then f' and $1/f'$ have continuous extensions to $\bar{\mathbf{D}}$, and so f and g satisfy Lipschitz condtions in $\bar{\mathbf{D}}$ and $\bar{D}$, respectively:

$$|f(z_1) - f(z_2)| \leq M|z_1 - z_2| \quad \text{and} \quad |g(w_1) - g(w_2)| \leq N|w_1 - w_2|.$$

Several authors have asked for weaker assumptions on C which ensure the Hölder continuity of f or that of g, that is

$$|f(z_1) - f(z_2)| \leq M|z_1 - z_2|^{\alpha} \quad (z_1, z_2, \in \bar{\mathbf{D}}) \tag{2.1}$$

or

$$|g(w_1) - g(w_2)| \leq N|w_1 - w_2|^{\beta} \quad (w_1, w_2, \in \bar{D}). \tag{2.2}$$

Foremost here is a result of Lesley [22]. We say that C satisfies an interior α-wedge condition if there exist $r > 0$ and $\alpha \in (0, 1)$ such that, for each $w_0 \in C$, a closed circular sector of radius r and opening $\alpha\pi$ lies in $\bar{D}$, with vertex at w_0. The exterior α-wedge condition is defined in a similar way. Lesley shows: If C satisfies an interior α-wedge condition, then f is Hölder continuous in $\bar{D}$ with exponent α, and if C satisfies an exterior α-wedge condition, then g is Hölder continuous in $\bar{D}$ with exponent $1/(2-\alpha)$. In both cases, the exponents are best possible. In the proof $\mathbf{D}$ and D are mapped onto strips, and Ahlfors' distortion theorem can be applied.

Another geometric condition on C is the arc length – chord length condition $\Delta s \leq c|w_1 - w_2|$ which ensures the Hölder continuity of g in $\bar{D}$ with a certain $\beta = \beta(c)$; see Warschawski [38] and Gaier [9].

Explicit estimates of the form (2.2) are given in Gaier [9]. Assume, for example, that $0 \in D$, $1 \in \partial D$, so that we can normalize g by $g(0) = 0$ and $g(1) = 1$. If now the line segment $(w, 1)$ connecting w to 1 is contained in D, then $|g(w) - 1| < 4|w - 1|^{\frac{1}{2}}$, and 4 cannot be replaced by a constant < 2.

If C is a quasicircle, i.e. the image of the unit circle under a quasiconformal mapping of the plane, then f and g satisfy Hölder conditions. Quantitative results are due to Näkki and Palka [23]. On the other hand, it may happen that f, g and their respective exterior maps satisfy Hölder conditions without C being a quasicircle; see Becker and Pommerenke [1].

FitzGerald and Lesley [7] investigate the integrability of f' and of $1/f'$ using geometric assumptions about C, and Pommerenke [27] studies the questions when $\log f' \in H^1$.

Topic III: Conformal mapping of multiply connected domains

Given a domain D of connectivity n, a classical problem in geometric function theory requires that D should be mapped conformally onto some standard domain.

Many such standard domains are known; see, for example, the classical results in Golusin [10] and Jenkins [17]. Walsh contributed to this area by introducing a new canonical map of D onto a "generalized lemniscate domain".

Relevant papers [54-g], [54-h], [56-d*], [59-k].

III.1. Walsh's new canonical map

We assume that D is a domain in the extended plane $\hat{\mathbf{C}}$, that $\infty \in D$, and denote its boundary components by $C_1, C_2, \ldots, C_n$ with $n \geq 2$. We fix $m \in \mathbf{N}$ with $1 \leq m < n$. Then there exists a conformal map $w = f(z)$ of D onto a domain Δ, normalized at ∞ by

$$(*) \qquad f(z) = z + \frac{a_1}{z} + \frac{a_2}{z^2} + \cdots,$$

such that the points of Δ are described by

$$1 < |T(w)| < e, \quad \text{where} \quad T(w) = A \prod_{k=1}^{n} (w - w_k)^{\alpha_k}.$$

Here A and the α_k are certain real constants with $\sum_{k=1}^{n} \alpha_k = 0$, $1 < A < e$, and the $\alpha_1, \cdots, \alpha_m$ are positive, the $\alpha_{m+1}, \cdots, \alpha_n$ are negative. Each w_k is a certain point in the interior of $f(C_k)$. All images $f(C_k)$ are analytic Jordan curves, and we have

$$|T(w)| = 1 \quad \text{on} \quad f(C_1), \ldots, f(C_m)$$

while

$$|T(w)| = e \quad \text{on} \quad f(C_{m+1}), \ldots, f(C_n).$$

With the normalization $(*)$, the mapping f is unique.

Walsh's result [56-d*] contains the earlier results by de La Vallée Poussin and Julia (1934) in which the canonical domain is bounded by level lines of a certain polynomial P of degree $n - 1$, i.e. the boundary is given by $\{w : |P(w)| = \text{constant}\}$. Therefore Walsh's domains can be called generalized lemniscate domains. Limiting cases, in which some of the boundary components of D may shrink to points, or some of the C_j may instersect (Jordan configuration), were also considered by Walsh and Landau [59-k].

III.2. New approaches to Walsh's theorem

Soon after the publication of Walsh's results, Grunsky [11] and Jenkins [16] gave new proofs for Walsh's mapping theorem. Grunsky introduces linear combinations of harmonic measures and Green's function of D, completes D suitably to an elliptic Riemann surface and uses uniformization to map it onto the Riemann sphere. Jenkins relies on previous results on quadratic differentials by himself and Spencer to prove existence and uniqueness of the mapping.

Landau's approach [21] relies on the observation that the canonical domain Δ has the property that the harmonic measure of $\bigcup_{j>m} f(C_j)$ with respect to Δ can be continued harmonically into the complex plane except for n points w_k each lying in the interior of $f(C_k)$. The mapping f itself can be obtained by composition of n mappings of simply connected domains: $f = H_1 \circ H_2 \circ \cdots \circ H_n$, and the H_j can be constructed by an iterative process. See also the exposition in Gaier [8], p. 238. Landau further characterizes the canonical map by extremal properties.

In Pirl's paper [25], a general mapping theorem is derived, using Koebe's classical continuity method, and thus a new proof for Walsh's theorems is given.

In Kühnau's Bemerkungen [20] several mapping theorems are derived from the solution of an extremal problem due to Golusin (1947) in which the functional

$$\operatorname{Re} \sum_{i,k=1}^{n} \gamma_{ik} \log \frac{w_i - w_k}{z_i - z_k}, \qquad w_k = f(z_k),$$

has to be maximized in the class $\Sigma(D)$ of all functions f univalent in D and normalized by $(*)$; the γ_{ik} and the $z_k \in D$ are given. First a limiting case of Golusin's theorem is proved from which Walsh's results follow if $\gamma_{ik} = \gamma_i \cdot \gamma_k$ is chosen.

In the same spirit, Harrington [12], [13] solves an extremal problem and shows that its solution gives a slight generalization of Walsh's mapping theorem. This extremal problem is further studied in the class Σ_K of quasiconformal mappings of D, and naturally variational methods of Schiffer and Schober are applied.

III.3. General canonical domains

Given a domain D of connectivity $n \geq 2$ the question is here: What type of canonical domain can we prescribe? One well known result is due to Courant, Manel and Schiffmann (1940) but it was overlooked that already Grötzsch (1935) had a rather general result, using Koebe's continuity method. In both cases, the image curves had to satisfy certain restrictions.

After 1956, the subject was taken up again by several authors. Pirl [25], using again Koebe's method, gives a very general result for canonical domains for which there exists a uniqueness theorem.

Brandt [3] uses a different approach to obtain the following. Given n simply connected domains $D_1, D_2, \ldots, D_n$ in $\hat{\mathbb{C}}$ each containing ∞, there exist n linear mappings $l_j(w) = a_j w + b_j$ and a conformal map f of D, normalized as in (*), such that $l_j(D_j)$ will be the full exterior of the j-th component of the boundary of $f(D)$. In his proof he considers $\Sigma(D)$ as a topological space and shows that every continuous map T from $\Sigma(D)$ into $\Sigma(D)$ possesses a fixed point, i.e. an $f \in \Sigma(D)$ with $Tf = f$; here an extension of Schauder's fixed point theorem is used. Futhermore, factorization is used: Every $f \in \Sigma(D)$ can be written as $f = h_1 \circ h_2 \circ \cdots \circ h_n$, where the h_j are normalized as in $(*)$.

The same general result is proved by Harrington [14] using a homotopy method, complex potentials and Walsh's mapping theorem. This author asserts that for a large class of canonical maps, his method is constructive.

We finally mention the following works: Jenkins [15] and Pirl [26].

References

[1] Becker, J. and Ch. Pommerenke: Hölder continuity of conformal mappings and non-quasiconformal Jordan curves. *Comment. Math. Helv.* **57** (1982), 221–225. MR **84a** # 30014.

[2] Bernardi, S. D.: *Bibliography of schlicht functions.* Mariner Publ. Co., Inc., Tampa, 1982. MR **84k** # 00014.

[3] Brandt, M.: Ein Abbildungssatz für endlich-vielfach zusammenhängende Gebiete. *Bull. Soc. Sci. Lett. Lódz* **30** (1980), no. 3. MR **83g** # 30009.

[4] Brown, J. E.: Level sets for functions convex in one direction. *Proc. Amer. Math. Soc.* **100** (1987), 442–446. MR **88h** # 30011.

[5] Clunie, J. and P. L. Duren: Addendum: An arclength problem for close-to-convex functions. *J. London Math. Soc.* **41** (1966), 181–182. MR **32** # 7725.

[5a] Eke, B. G.: On the differentiability of conformal maps at the boundary. *Nagoya Math. J.* **41** (1971), 43–53. MR **43** # 2201.

[5b] Eke, B. G.: Comparison domains for the problem of the angular derivative. *Comment. Math. Helv.* **46** (1971), 98–112. MR **45** # 2150.

[6] FitzGerald, C. H. and F. D. Lesley: Boundary regularity of domains satisfying a wedge condition. *Complex Variables Theory Appl.* **5** (1986), 141–154. MR **88e** # 30020

[7] FitzGerald, C. H. and F. D. Lesley: Integrability of the derivative of the Riemann mapping function for wedge domains. *J. Analyse Math.* **49** (1987), 271–292. MR **89c** # 30018.

[8] Gaier, D.: *Konstruktive Methoden der konformen Abbildung.* Springer, Berlin, 1964. MR **33** # 7507.

[9] Gaier, D.: Estimates of conformal mappings near the boundary. *Indiana Univ. Math. J.* **21** (1971/72), 581–595. MR **45** # 2151.

[10] Golusin, G. M.: *Geometrische Funktionentheorie.* Deutscher Verlag der Wissenschaften, Berlin, 1957. MR **19**, 735.

[11] Grunsky, H.: Über konforme Abbildungen, die gewisse Gebietsfunktionen in elementare Funktionen transformieren. I and II. *Math. Z.* **67** (1957), 129–132 and 223–228. MR **19**, 538.

[12] Harrington, A. N.: Some extremal problems in conformal and quasiconformal mapping. *Mich. Math. J.* **27** (1980), 95–116. MR **81c** # 30041.

[13] Harrington, A. N.: Extremal conformal mappings of many-component sets onto sets bounded by generalized lemnicscates. *Indiana Univ. Math. J.* **30** (1981), 703–712. MR **82k** # 30026.

[14] Harrington, A. N.: Conformal mappings onto domains with arbitrarily specified boundary shapes. *J. Analyse Math.* **41** (1982), 39–53. MR **84a** # 30015.

[15] Jenkins, J. A.: Some new canonical mappings for multiply-connected domains. *Ann. of Math.* (2) **65** (1957), 179–196. MR **18**, 568.
[16] Jenkins, J. A.: On a canonical conformal mapping of J. L. Walsh. *Trans. Amer. Math. Soc.* **88** (1958), 207–213. MR **19**, 538.
[17] Jenkins, J. A.: Univalent functions and conformal mapping. Springer, Berlin, 1958. MR **20** # 3288.
[18] Jenkins, J. A. and K. Oikawa: Conformality and semi-conformality at the boundary. *J. Reine Angew. Math.* **291** (1977), 92–117. MR **55** # 12924.
[19] Krzyż, J. and K. Radziszewski: Isoperimetrical defect and conformal mapping. *Ann. Univ. Mariae Curie-Sklodowska. Sect. A.* **10** (1956), 49–56. MR **20** # 1764.
[20] Kühnau, R.: Bemerkungen zu einigen neueren Normalabbildungen in der Theorie der konformen Abbildung. *Math. Z.* **97** (1967), 21–28. MR **35** # 355.
[21] Landau, H. J.: On canonical conformal maps of multiply connected domains. *Trans. Amer. Math. Soc.* **99** (1961), 1–20. MR **22** # 12212.
[22] Lesley, F. D.: Conformal mappings of domains satisfying a wedge condition. *Proc. Amer. Math. Soc.* **93** (1985), 483–488. MR **86a** # 30010.
[23] Näkki, R. and B. Palka: Quasiconformal circles and Lipschitz classes. *Comment. Math. Helv.* **55** (1980), 485–498. MR **82a** # 30029.
[24] Oikawa, K.: Remarks on conformality at the boundary. *Tôhoku Math. J.* (2) **35** (1983), 313–319. MR **85b** # 30045.
[25] Pirl, U.: Zum Normalformenproblem für endlich-vielfach zusammenhängende schlichte Gebiete. Wiss. Z. Martin-Luther-Univ. Halle-Wittenberg. *Math. Nat. Reihe* **6** (1956/57), 799–802. MR **22** # 98.
[26] Pirl, U.: Normalformen für endlich-vielfach zusammenhängende in einen Kreisring eingelagerte Gebiete. *Math. Nachr.* **76** (1977), 181–194. MR **58** # 17061.
[27] Pommerenke, Ch.: One-sided smoothness conditions and conformal mapping. *J. London Math. Soc.* (2) **26** (1982), 77–88. MR **83j** # 30006.
[28] Prokhorov, D. V.: Level curves of functions convex in the direction of an axis. *Mat. Zametki* **44** (1988), 767–769. MR **89m** # 30030.
[29] Rodin, B. and S. E. Warschawski: On conformal mapping of L-strips. *J. London Math. Soc.* (2) **11** (1975), 301–307. MR **53** # 3281.
[30] Rodin, B. and S. E. Warschawski: Extremal length and the boundary behavior of conformal mappings. *Ann. Acad. Sci. Fenn. Ser. A I Math.* **2** (1976), 467–500. MR **57** # 6394.
[31] Rodin, B. and S. E. Warschawski: Extremal length and univalent functions. I. The angular derivative. *Math. Z.* **153** (1977), 1–17, MR **58** # 28461.
[32] Ruscheweyh, St. and L. C. Salinas: On the preservation of direction-convexity and the Goodman-Saff conjecture. *Ann. Acad. Sci. Fenn. Ser. A I Math.* **14** (1989), 63–73. MR **90h** # 30041.
[33] Shaffer, D. B.: Distortion theorems for the level curves of rational functions and harmonic functions. *J. Math. Mech.* **19** (1969/70), 41–48. MR **41** # 444.
[34] Shaffer, D. B.: The curvature of level curves. *Trans. Amer. Math. Soc.* **158** (1971), 143–150. MR **43** # 3428.

[35] Shaffer, D. B.: Lemniscates and equipotentials. *J. Approx. Theory* **6** (1972), 431–438. MR **50** # 2524.
[36] Styer, D. and D. J. Wright: Level curves for regions convex in one direction. *Complex Variables Theory Appl.* **9** (1988), 373–379. MR **89f** # 30013.
[37] Thomas, D. K.: On starlike and close-to-convex univalent functions. *J. London Math. Soc.* **42** (1967), 427–435. MR **35** # 6812.
[38] Warschawski, S. E.: On Hölder continuity at the boundary in conformal maps. *J. Math. Mech.* **18** (1968/69), 423–427. MR **38** # 1242.

Chapter 5

Polynomial Approximation

ON POLYNOMIAL INTERPOLATION TO ANALYTIC FUNCTIONS WITH SINGULARITIES*

BY J. L. WALSH

Méray has given† the following illustration to show that polynomials formed from a given function by interpolation do not necessarily converge to that function. Interpolate to the function $f(z) = 1/z$ by means of the polynomials $p_n(z)$ of respective degrees‡ $n = 1, 2, 3, \cdots$, required to coincide with $f(z)$ in the $(n+1)$th roots of unity; this condition defines the polynomials $p_n(z)$ uniquely. Moreover, we have

$$(1) \qquad p_n(z) \equiv z^n,$$

because the equation $p_n(z) = 1/z$ is satisfied provided z is one of the $(n+1)$th roots of unity. Even though the sequence $p_n(z)$ is defined by interpolation from the function $f(z) = 1/z$, the polynomials $p_n(z)$ do not approach the function $f(z)$ for $|z| < 1$, as n becomes infinite, but approach the limit zero. It is naturally not surprising that these polynomials should fail to approach $f(z)$ for $|z| < 1$, since $f(z)$ has a singularity in that region; this sequence of polynomials fails to approach the limit $f(z)$ even in a neighborhood of the curve $|z| = 1$ on which interpolation takes place.

In this connection, it is worth while to recall Runge's result§ that if $f(z)$ is analytic for $|z| \leqq 1$, then the sequence of interpolating polynomials $p_n(z)$ of respective degrees n which coincide with $f(z)$ in the $(n+1)$th roots of unity converges to the limit $f(z)$ for $|z| \leqq 1$.

---

* Presented to the Society, March 25, 1932.

† Annales de l'École Normale Supérieure, (3), vol. 1 (1884), pp. 165–176. This illustration is also presented by Montel (after Méray), in his *Séries de Polynomes*, 1910, p. 51.

‡ A polynomial of the form $a_0z^n + a_1z^{n-1} + \cdots + a_n$ is said to be of degree n.

§ *Theorie und Praxis der Reihen*, 1904, p. 137. This method of Runge's has been more systematically developed by Fejér, Göttinger Nachrichten, 1918, pp. 319–331, and by L. Kalmár, *Mathematikai és Physikai Lapok*, 1926, pp. 120–149, but only for interpolation in points on the boundary of a region of a function known to be analytic in that region.

Méray's illustration is of such simplicity, directness, and beauty, and apparently has stood alone for so long, that it seems worth while to furnish it with some companionship. The purpose of the present note is to provide such companionship by the proof of the following theorem.

THEOREM 1. *Let $f(z)$ be an arbitrary function continuous for $|z|=1$. Let the polynomials $p_n(z)$ of respective degrees n, $(n=1, 2, \cdots)$, be defined by the requirement of coinciding with $f(z)$ in the $(n+1)$th roots of unity. Then the sequence $p_n(z)$ approaches the limit*

$$(2) \qquad f_1(z) = \frac{1}{2\pi i}\int_{|z|=1} \frac{f(t)dt}{t-z}$$

for $|z|<1$, uniformly for $|z| \leqq r<1$.

Lagrange's interpolation formula for the polynomial $p_n(z)$ of degree n which takes on the values $K_1, K_2, \cdots, K_{n+1}$ at the $n+1$ distinct points $z_1, z_2, \cdots, z_{n+1}$ is

$$p_n(z) = \sum_{k=1}^{n+1} \frac{K_k}{p'(z_k)} \frac{p(z)}{z-z_k},$$

where $p(z)=(z-z_1)(z-z_2)\cdots(z-z_{n+1})$; the polynomial $p_n(z)$ is uniquely determined by these requirements. Under the circumstances of the theorem, we set $z_k=\omega^k$, where $\omega=e^{2\pi i/(n+1)}$, $p(z)=z^{n+1}-1$. It follows that we have

$$(3) \qquad p_n(z) = \sum_{k=1}^{n+1} f(\omega^k)\frac{\omega^k(z^{n+1}-1)}{(n+1)(z-\omega^k)}.$$

With the exception of the term z^{n+1} in the numerator, which approaches zero, equation (3) suggests computation of the integral which appears in (2) by division of the circle C defined by $|z|=1$, at the points ω^k. We have

$$(4) \qquad f_1(z) = \lim_{n\to\infty} \frac{1}{2\pi i} \sum_{k=1}^{n+1} \frac{f(\omega^k)(\omega^{k+1}-\omega^k)}{\omega^k - z},$$

$$(5) \quad \lim_{n\to\infty} [f_1(z) - p_n(z)]$$

$$= \lim_{n\to\infty}\left[\frac{1}{2\pi i} + \frac{z^{n+1}-1}{(n+1)(\omega-1)}\right] \sum_{k=1}^{n+1} \frac{\omega^k(\omega-1)f(\omega^k)}{\omega^k - z}.$$

By (4), the summation on the right-hand side approaches the limit $2\pi i f_1(z)$, which is continuous for $|z|<1$, and the limit is approached uniformly for $|z|\leqq r<1$.* The quantity $(n+1)(\omega-1)$ approaches as its limit $2\pi i$, for we have

$$\omega = \cos\frac{2\pi}{n+1} + i\sin\frac{2\pi}{n+1},$$

$$\frac{(n+1)(\omega-1)}{2\pi i} = \frac{\cos\frac{2\pi}{n+1} - 1}{\frac{2\pi i}{n+1}} + \frac{\sin\frac{2\pi}{n+1}}{\frac{2\pi}{n+1}},$$

which approaches the limit unity. The square bracket in the right-hand member of (5) thus approaches zero for $|z|<1$, uniformly for $|z|\leqq r<1$, and the factor of this bracket is bounded uniformly in z and n, $|z|\leqq r<1$, so the proof of our theorem is complete.

The theorem and proof are obviously valid if the given function $f(z)$ is not supposed continuous on C: $|z|=1$, but merely integrable in the sense of Riemann.

The sequence $p_n(z)$ clearly converges uniformly in a region containing the curve C in its interior when and only when the given function $f(z)$ or its analytic extension is analytic on and within C. The first part of this result was proved by Féjer (loc. cit.), or compare Walsh.† Reciprocally, if the sequence $p_n(z)$ converges uniformly for $|z|\leqq\rho>1$, we shall prove $f(z)$ analytic on and within C. The limit of the sequence $p_n(z)$ is analytic for $|z|<\rho$. The obvious equation $p_{m(n+1)}(\omega^k)=f(\omega^k)$, $(m=1, 2, \cdots)$, where $\omega=e^{2\pi i/(n+1)}$, implies the convergence of the sequence $p_n(z)$ to the function $f(z)$ at each point ω^k, hence at a set of points everywhere dense on C. The continuity of $f(z)$ for $|z|=1$ and the analytic character of the limit of the sequence $p_n(z)$ for $|z|<\rho$ then implies the identity on C of those two functions, hence the fact that $f(z)$ or its analytic extension is analytic for $|z|<\rho$. If the sequence $p_n(z)$ converges uniformly for $|z|\leqq 1$, the function $f(z)$ is analytic for $|z|<1$, and continuous for $|z|\leqq 1$.

* Compare Runge, *Acta Mathematica*, vol. 6 (1885), pp. 229–244; Montel, loc. cit., p. 57; or Osgood, *Funktionentheorie*, 1928, pp. 579–581.

† Transactions of this Society, vol. 34 (1932), pp. 22–74, §11.

The following result is a not uninteresting complement to the main theorem already proved.

THEOREM 2. *The function* $f_1(z)$ *which is the limit for* $|z|<1$ *of the sequence* $p_n(z)$ *of the interpolating polynomials for* $f(z)$ *is also the limit for* $|z|<1$, *uniformly for* $|z|\leqq r<1$, *of the sequence of polynomials* $P_n(z)$ *of respective degrees* $n=1, 2, \cdots$, *of best approximation to* $f(z)$ *on* C *in the sense of least squares.*

The polynomial $P_n(z)$ is that polynomial of degree n for which

$$\int_C | f(z) - P_n(z) |^2 | dz |$$

is least. Such a polynomial is known to exist and be unique; it is defined* by

$$(6)\qquad \begin{aligned} P_n(z) &= c_0 + c_1 z + \cdots + c_n z^n, \\ c_n &= \frac{1}{2\pi}\int_C f(z)\bar{z}^n | dz | = \frac{1}{2\pi i}\int_C f(z)\frac{dz}{z^{n+1}}. \end{aligned}$$

We have the relation

$$f_1(z) = \frac{1}{2\pi i}\int_C \frac{f(t)dt}{t-z} = \frac{1}{2\pi i}\int_C f(t)dt\left[\frac{1}{t}+\frac{z}{t^2}+\frac{z^2}{t^3}+\cdots\right].$$

The infinite series converges uniformly in z and t for t on C, $|z|\leqq r<1$, and hence may be integrated term by term. Thus we have from (6) $f_1(z)=c_0+c_1z+c_2z^2+\cdots$, a series which is uniformly convergent for $|z|\leqq r<1$; this includes the relation to be proved:

$$f_1(z) = \lim_{n\to\infty} P_n(z), \qquad |z| < 1.$$

The polynomial $P_n(z)$ is the sum of the first $n+1$ terms of the Maclaurin development of $f_1(z)$.

Several particular cases are worth mentioning. If there exists a function analytic for $|z|<1$, continuous for $|z|\leqq 1$, which coincides with the given function $f(z)$ on C, then $f_1(z)$ naturally coincides with this function. If there exists a function $F(z)$ analytic for $|z|>1$ (including the point $z=\infty$), continuous for $|z|\geqq 1$, which coincides with the given function $f(z)$ on C, then each $c_n(n>0)$ vanishes, and we have

* See for instance Kowalewski, *Determinantentheorie*, 1909, §137.

$$\lim_{n\to\infty} p_n(z) = \lim_{n\to\infty} P_n(z) = c_0 = F(\infty), \ |z| < 1;$$

here is included the case $f(z)=1/z^k$, k a positive integer ($k=1$ is Méray's example), and we have $f_1(z)\equiv 0$. More generally, if on C the function $f(z)$ can be expressed as $f(z)=f_1(z)+f_2(z)$, where $f_1(z)$ is analytic for $|z|<1$, continuous for $|z|\leqq 1$, and where $f_2(z)$ is analytic for $|z|>1$ including the point at infinity, is continuous for $|z|\geqq 1$, and vanishes at infinity, then we have

$$\lim_{n\to\infty} p_n(z) = \lim_{n\to\infty} P_n(z) = f_1(z), \ |z| < 1,$$

uniformly for $|z|\leqq r<1$. If this function $f_1(z)$ is analytic for $|z|<\rho>1$, then we have

$$\lim_{n\to\infty} P_n(z) = f_1(z), \ |z| < \rho, \text{ uniformly for } |z| \leqq \rho' < \rho,$$

no matter what $f_2(z)$ may be, but we have

$$\lim_{n\to\infty} p_n(z) = f_1(z), \ |z| < \rho, \text{ uniformly for } |z| \leqq \rho' < \rho,$$

if and only if $f_2(z)$ vanishes identically.

Another illustration of the close connection between interpolation in the $(n+1)$th roots of unity and approximation on C in the sense of least squares has recently been indicated by the present writer (loc. cit.).

If the function $f(z)$ is analytic for $|z|<T>1$, then for the polynomials $p_n(z)$ and $P_n(z)$ already defined we have

$$\lim_{n\to\infty} [p_n(z) - P_n(z)] = 0, \ |z| < T^2, \textit{ uniformly for } |z| \leqq R < T^2$$

even though $f(z)$ has singularities for $T\leqq|z|<T^2$.

It is not to be supposed (compare Kalmár, loc. cit.) that interpolation in points *arbitrarily* chosen on C is always equivalent in the sense illustrated to approximation on C in the sense of least squares, even when the points of interpolation $z_n^{(k)}$ for the interpolating polynomial $p_n(z)$ of degree n are such that the limit of the maximum distance between successive points $z_n^{(k)}$ on C for a given n approaches zero with $1/n$. We give an example to illustrate this fact, where the $n+1$ points of interpolation for the polynomial $p_n(z)$ are the roots of

$$\left(\frac{1-\alpha z}{\alpha - z}\right)^{n+1} = 1, \ \alpha > 1. \tag{7}$$

These points of interpolation are thus the transforms in the z-plane of the $(n+1)$th roots of unity in the w-plane, under the transformation

$$w = \frac{1-\alpha z}{\alpha - z}, \tag{8}$$

which leaves C invariant and transforms the interior and exterior of C respectively into the interior and exterior of C. It is simpler to study the situation in the w-plane. We interpolate to the function $f(z)=1/(w+T)$, $0<T<1$, in the points w, $w^{n+1}=1$, by rational functions $p_n(z)=F_n(w)$ of respective degrees (in z or w) n whose poles coincide in the points $w=\alpha$. The reader can easily verify the formula

$$\begin{aligned} p_n(z) &= F_n(w) \\ &= \frac{(T+\alpha)^n(w^{n+1}-1) + [(-1)^n + T^{n+1}](w-\alpha)^n}{[(-1)^n + T^{n+1}](w+T)(w-\alpha)^n}; \end{aligned} \tag{9}$$

the expression in terms of z is found from (8). It appears from (9) that one may write

$$p_n(z) = \frac{1}{w+T} + \frac{(T+\alpha)^n(w^{n+1}-1)}{[(-1)^n + T^{n+1}](w+T)(w-\alpha)^n},$$

so that, for $|w|<1$, we have actual convergence when and only when the condition

$$T + \alpha < |w - \alpha| \tag{10}$$

is satisfied. Condition (10) is equivalent to the condition that w should lie exterior to a certain circle whose center is α and which cuts $C:|w|=1$. Thus the sequence $p_n(z)$ converges in only a part of the unit circle $|z|\leqq 1$, and in that part converges to the original function $1/(w+T)$.

It would be of interest to extend the main theorem of this note to the study of curves other than the unit circle; compare the references already given to Fejér and Kalmár.

HARVARD UNIVERSITY

NOTE ON THE RELATION BETWEEN CONTINUITY AND DEGREE OF POLYNOMIAL APPROXIMATION IN THE COMPLEX DOMAIN*

BY J. L. WALSH AND W. E. SEWELL

1. *Introduction.* It is the purpose of the present note to establish the following theorems:

THEOREM I. *Let C be an analytic Jordan curve in the z-plane and let $f(z)$ be defined in $\overline{C}$, the closed limited point set bounded by C. For each n, $n=1, 2, \cdots$, let a polynomial $P_n(z)$ of degree n in z exist such that*

$$(1)\qquad |f(z) - P_n(z)| \leqq \frac{M}{n^{p+\alpha}}, \qquad z \text{ in } \overline{C}, \qquad 0 < \alpha \leqq 1,$$

where M is a constant independent of n and z, and p is a non-negative integer. Then $f(z)$ is analytic in C and continuous in $\overline{C}$; the pth derivative $f^{(p)}(z)$ exists on C in the one-dimensional sense and satisfies the condition

$$(2)\qquad |f^{(p)}(z_1) - f^{(p)}(z_2)| \leqq L|z_1 - z_2|^\alpha |\log|z_1 - z_2||^\beta, \qquad z_1, z_2 \text{ on } C,$$

where $\beta=0$ if $\alpha<1$, and $\beta=1$ if $\alpha=1$, and where L is a constant independent of z_1 and z_2.

THEOREM II. *Let E, with boundary C, be a closed limited point set in the z-plane whose complement K is connected, and is regular in the sense that there exists a function $w=\phi(z)$ which maps K conformally but not necessarily uniformly onto $|w|>1$ so that the points at infinity in the two planes correspond to each other. Let the locus C_R: $|\phi(z)|=R>1$, consist of a finite number of mutually exterior analytic Jordan curves. Let $f(z)$ be defined in E, and for each n, $n=1, 2, \cdots$, let a polynomial $P_n(z)$ of degree n in z exist such that*

$$(3)\qquad |f(z) - P_n(z)| \leqq \frac{M}{n^{p+\alpha+1}R^n}, \qquad z \text{ in } E, \qquad 0 < \alpha \leqq 1,$$

* Presented to the Society, March 27, 1937.

where M is a constant independent of n and z, and p is a non-negative integer. Then $f(z)$ when suitably defined exterior to E is analytic in C_R and continuous in $\overline{C}_R$; the pth derivative $f^{(p)}(z)$ exists on C_R in the one dimensional sense and satisfies the condition

$$(4) \qquad |f^{(p)}(z_1) - f^{(p)}(z_2)| \leqq L|z_1 - z_2|^\alpha |\log|z_1 - z_2||^\beta, \qquad z_1, z_2 \text{ on } C_R,$$

where $\beta = 0$ if $\alpha < 1$, and $\beta = 1$ if $\alpha = 1$, and L is a constant independent of z_1 and z_2.

In these theorems the case $p=0$ is not excluded:

$$f^{(0)}(z) \equiv f(z).$$

Sewell* has already proved the slightly less general results that under the hypotheses of Theorems I and II the function $f^{(p)}(z)$ satisfies a Lipschitz condition† of every order $\alpha' < \alpha$ on C and C_R, respectively.

J. Curtiss has shown‡ that if the boundary C of a closed limited point set E consists of a finite number of mutually exterior analytic Jordan curves, if $f(z)$ is analytic in the interior points of E and continuous in E, and if $f^{(p)}(z)$ exists in the one-dimensional sense on C and satisfies on C a Lipschitz condition of order α, $0 < \alpha \leqq 1$, then there exist polynomials $P_n(z)$ of respective degrees n such that (1) is valid for z in E with $\beta = 0$ even for $\alpha = 1$. Thus *Theorem* I *is for* $0 < \alpha < 1$ *an exact converse of Curtiss's result.* We show by an example that for $\alpha = 1$ an exact converse is impossible.

To be sure, Curtiss did not state his result in the above form, *but assumed $f^{(p)}(z)$ continuous in E.* Nevertheless the lighter assumption is sufficient for his purposes, because we prove below (Theorem III) that this lighter assumption implies also the heavier assumption.

In the notation of Theorem II suppose $|f(z) - P_n(z)| \leqq \epsilon_n$, z on E, $n = 1, 2, \cdots$, where ϵ_n approaches zero as n becomes infinite. The study of the relation between ϵ_n and the continuity

---

* Transactions of this Society, vol. 41 (1937), pp. 84–123.

† The function $f(z)$ is said to satisfy a Lipschitz condition of order α, $0 < \alpha \leqq 1$, on the set E if for z_1 and z_2 on E, we have $|f(z_1) - f(z_2)| \leqq L|z_1 - z_2|^\alpha$, where L is a constant independent of z_1 and z_2.

‡ This Bulletin, vol. 42 (1936), pp. 873–878.

properties of $f(z)$ on E is called Problem α, and the study of the relation between ϵ_n and the continuity properties of $f(z)$ on C_R is called* Problem β. Thus in Theorem I we have a result on Problem α and in Theorem II a result on Problem β.

The method of proof of Theorems I and II is an application of conformal mapping to de la Vallée Poussin's† results on trigonometric approximation.

2. *Proof of Theorem* I. The analyticity of $f(z)$ interior to C and the continuity of $f(z)$ in $\bar{C}$ follow directly from inequality (1). We consider first C to be the unit circle: $|z|=1$. Let $f(e^{i\theta})\equiv u(\theta)+iv(\theta)$; then inequality (1) on C implies

$$(5)\qquad | u(\theta) - p_n(\theta) | \leqq \frac{M}{n^{p+\alpha}}, \qquad | v(\theta) - q_n(\theta) | \leqq \frac{M}{n^{p+\alpha}},$$

where $P_n(e^{i\theta})\equiv p_n(\theta)+iq_n(\theta)$. But $p_n(\theta)$ and $q_n(\theta)$ are trigonometric sums of order n and hence (de la Vallée Poussin, loc. cit., pp. 57 and 61–62) we have

$$(6)\qquad \begin{aligned} | u^{(p)}(\theta_1) - u^{(p)}(\theta_2) | &\leqq L_1 | \theta_1 - \theta_2 |^{\alpha} | \log | \theta_1 - \theta_2 | |^{\beta}, \\ | v^{(p)}(\theta_1) - v^{(p)}(\theta_2) | &\leqq L_2 | \theta_1 - \theta_2 |^{\alpha} | \log | \theta_1 - \theta_2 | |^{\beta}, \end{aligned}$$

where $\beta=0$ if $\alpha<1$ and $\beta=1$ if $\alpha=1$. Here $u^{(p)}(\theta)$ denotes the pth derivative of $u(\theta)$ with respect to θ; $u^{(0)}(\theta)\equiv u(\theta)$. We have

$$f'(e^{i\theta}) = \frac{df(e^{i\theta})}{de^{i\theta}} = \frac{d}{d\theta}(u(\theta) + iv(\theta))\cdot\frac{d\theta}{de^{i\theta}} = (u'(\theta) + iv'(\theta))\frac{1}{ie^{i\theta}},$$

$$f''(e^{i\theta}) = \frac{df'(e^{i\theta})}{de^{i\theta}} = -\frac{1}{ie^{2i\theta}}\left[i(u''(\theta) + iv''(\theta)) + (u'(\theta) + iv'(\theta))\right],$$

and similarly for higher derivatives.

Now $u^{(k)}(\theta)$, $k=0, 1, 2, \cdots, p-1$, satisfies a Lipschitz condition of order 1, as does the function $e^{-ki\theta}$. Thus we have

$$| f^{(p)}(e^{i\theta_1}) - f^{(p)}(e^{i\theta_2}) | \leqq L_3 | \theta_1 - \theta_2 |^{\alpha} | \log | \theta_1 - \theta_2 | |^{\beta},$$

which through the properties of $e^{i\theta}$ implies inequality (2) for z on C; the proof of Theorem I is complete for C the unit circle.

* A more general formulation of these problems is given by Sewell, loc. cit., along with an extensive bibliography.

† Ch. J. de la Vallée Poussin, *Leçons sur l'Approximation des Fonctions d'une Variable Réele*, Paris, 1919; see especially Chap. IV.

Now let C be an arbitrary analytic Jordan curve, and let the function $w=\Phi(z)$, $z=\Psi(w)$, map $\overline{C}$ conformally onto $|w|\leqq 1$; the function $\Psi(w)$ is analytic and univalent in some closed region $|w|\leqq R_0>1$. Inequality (1) implies

$$(7) \qquad |f[\Psi(w)] - P_n[\Psi(w)]| \leqq \frac{M}{n^{p+\alpha}}, \qquad |w| \leqq 1.$$

The polynomials $P_n(z)$ are uniformly bounded on C for all n, say $|P_n(z)|\leqq M_1$; then by a well known theorem* we have (notation of Theorem II) $|P_n(z)|\leqq M_1R^n$ for z on C_R. Let Γ_{R_1} be the curve $|\Phi(z)|=R_1<R_0$, $R_1>1$, in the z-plane. Choose $R>1$ so that Γ_{R_1} lies interior to C_R. Then $|P_n[\Psi(w)]|\leqq M_1R^n$ for $|w|\leqq R_1$. We have

$$P_n[\Psi(w)] - Q_m(w) = \frac{1}{2\pi i}\int_{|t|=R_1} \frac{P_n[\Psi(t)]w^{m+1}dt}{(t-w)t^{m+1}},$$
$$|w| < R_1 < R_0,$$

where $Q_m(w)$ is the sum of the first $m+1$ terms of the Taylor development of $P_n[\Psi(w)]$ about the origin. This yields

$$|P_n[\Psi(w)] - Q_m(w)| \leqq \frac{M_1R^n r^{m+1}}{(R_1 - r)R_1^m}, \qquad |w| \leqq r < R_1.$$

Let us choose r, $1<r<R_1$, and choose (this method is similar to that used by Curtiss, loc. cit.) $m=qn$ where q is a positive integer such that $R(r^q/R_1^q)=r_1<1$; then we have

$$(8) \quad |P_n[\Psi(w)] - Q_{qn}(w)| \leqq \frac{M_1 r}{R_1 - r} r_1^n, \qquad |w| \leqq r > 1.$$

Inequalities (7) and (8) yield

$$|f[\Psi(w)] - Q_{qn}(w)| \leqq \frac{M}{n^{p+\alpha}} + \frac{M_1 r}{R_1 - r} r_1^n, \qquad |w| \leqq 1;$$

but we have $r_1<1$, and $(M_1r)/(R_1-r)$ is a constant independent of n, so for suitably chosen M_2 we have

$$|f[\Psi(w)] - Q_{qn}(w)| \leqq \frac{M_2}{n^{p+\alpha}} = \frac{M_2 q^{p+\alpha}}{(qn)^{p+\alpha}}, \qquad |w| \leqq 1.$$

* See, for example, J. L. Walsh, *Interpolation and Approximation*, Colloquium Publications of this Society, vol. 20 (1935), pp. 77–78.

Now choose $L_N(w)\equiv 0$, $N=1, 2, \cdots, q-1$, $L_{qN'+h}(w)\equiv Q_{qN'}(w)$, $h=0, 1, 2, \cdots, q-1$, $N'=1, 2, \cdots$. Thus we have

$$|f[\Psi(w)] - L_N(w)| \leqq \frac{M_3}{N^{p+\alpha}}, \quad |w| \leqq 1, N = 1, 2, \cdots,$$

since when M_3 is suitably chosen we have $M_2q^{p+\alpha}/N_1^{p+\alpha} \leqq M_3/(N_1+h)^{p+\alpha}$, $0\leqq h\leqq q-1$. Hence inequality (2) is satisfied by $d^pf[\Psi(w)]/dw^p$ in the one-dimensional sense on $|w|=1$. Since $\Psi(w)$ is analytic on $|w|=1$ it follows by the method used above* that inequality (2) is valid for z on C; the proof of Theorem I is complete.

For the function $f(w)=\sum_{k=1}^{\infty}w^k/k(k-1)$ we have $(|w|\leqq 1)$

$$\left|f(w) - \sum_{k=1}^{n}\frac{w^k}{k(k-1)}\right| \leqq \sum_{k=n+1}^{\infty}\frac{1}{k(k-1)}$$
$$= \sum_{k=n+1}^{\infty}\left(\frac{1}{k-1} - \frac{1}{k}\right) = \frac{1}{n}, \qquad n = 1, 2, \cdots;$$

but $f(w)$ does not satisfy a Lipschitz condition of order 1 on $|w|=1$ since the derivative $f'(w)\equiv -\log(1-w)$ becomes infinite as w approaches 1. Thus for $\alpha=1$ the hypothesis of Theorem I does not imply a Lipschitz condition of order α; an exact converse of Curtiss's theorem for $\alpha=1$ is impossible.

3. *Proof of Theorem* II. Let us set

$$(9)\quad f(z) = P_1(z) + [P_2(z) - P_1(z)] + [P_3(z) - P_2(z)] + \cdots.$$

Inequality (3) implies

$$|P_{n+1}(z) - P_n(z)| \leqq \frac{2M}{n^{p+\alpha+1}R^n}, \qquad z \text{ in } E,$$

$$|P_{n+1}(z) - P_n(z)| \leqq \frac{2MR}{n^{p+\alpha+1}}, \qquad z \text{ in } \overline{C}_R.$$

Thus we have from equation (9) even exterior to E

$$|f(z) - P_n(z)| \leqq \frac{M'}{n^{p+\alpha}}, \qquad z \text{ in } \overline{C}_R.$$

---

* Or W. E. Sewell, this Bulletin, vol. 41 (1935), pp. 111–117; especially p. 117.

Theorem II now follows from Theorem I.

If we set for $|w| \leqq R > 1$

$$f(w) = \sum_{k=1}^{\infty} \frac{w^k}{R^k k(k-1)},$$

we have for $|w| \leqq 1$

$$\left| f(w) - \sum_{k=1}^{n} \frac{w^k}{R^k k(k-1)} \right| \leqq \sum_{k=n+1}^{\infty} \frac{1}{R^k k(k-1)}$$

$$< \frac{1}{n(n+1)} \sum_{k=n+1}^{\infty} \frac{1}{R^k} < \frac{1}{R-1} \frac{1}{n^2 R^n}.$$

Yet $f(w)$ does not satisfy a Lipschitz condition of order 1 on $|w| = R$ since, as above, the derivative $f'(w)$ becomes infinite as w approaches R. As in §2 we consequently see that for $\alpha = 1$ inequality (3) does not imply a Lipschitz condition of order α.

4. *Continuity of the Derivative in* $\overline{C}$. To establish further properties of the functions $f(z)$ of Theorems I and II, and to relate these theorems to the results of Curtiss, we prove a third theorem:

THEOREM III. *Let C be an analytic Jordan curve, let the function $f(z)$ be analytic interior to C and continuous in the corresponding closed region $\overline{C}$, and let the pth derivative $f^{(p)}(z)$ exist and be continuous on C in the one-dimensional sense. Then $f^{(p)}(z)$ defined interior to C in the usual two-dimensional sense and defined on C in the one-dimensional sense is continuous throughout $\overline{C}$.*

It is sufficient to prove the theorem for the case $p = 1$, for the proof extends automatically by induction.

Let C be the unit circle; the more general case can be transformed by conformal mapping to this special case. The original hypothesis fulfilled for an analytic curve C implies the corresponding hypothesis for the unit circle, and the conclusion proved for the case of the unit circle implies the more general conclusion stated in the theorem.

The function $f(z)$ then satisfies the conditions

$$\int_C f(z) z^k dz = 0, \qquad k = 0, 1, 2, \cdots, \tag{10}$$

by Cauchy's integral theorem. If on C we set $z=e^{i\theta}$, $dz=ie^{i\theta}d\theta$, the Fourier development of $f(z)$ on C by (10) can be written

$$(11)\qquad \begin{aligned} f(z) &\sim \sum_{n=0}^{\infty} a_n e^{in\theta} = \sum_{n=0}^{\infty} a_n z^n, \\ a_n &= \frac{1}{2\pi}\int_C f(z)e^{-in\theta}d\theta = \frac{1}{2\pi i}\int_C \frac{f(z)}{z^{n+1}}\,dz. \end{aligned}$$

By partial integration we may write, for $k=0, 1, -1, 2, -2, \cdots$,

$$(12)\qquad \int_C f'(z)z^k dz = [z^k f(z)]_C - k\int_C f(z)z^{k-1}dz.$$

The first term on the right vanishes because $f(z)$ is single-valued and continuous on C. The Fourier development on C of $f'(z)$ is therefore by (10) and (12)

$$(13)\qquad f'(z) \sim \sum_{n=1}^{\infty} na_n e^{i(n-1)\theta} = \sum_{n=1}^{\infty} na_n z^{n-1},$$

where the a_n are given by (11).

In (11) and (13) the series represent thus far only formal developments on C. But by the continuity of $f(z)$ in $\overline{C}$, the series $\sum_{n=0}^{\infty} a_n z^n$ is precisely the Taylor development of $f(z)$ valid throughout the interior of C. By differentiation it follows that $\sum_{n=1}^{\infty} na_n z^{n-1}$ is the Taylor development of $f'(z)$, likewise valid throughout the interior of C; it follows by inspection that this series is formally identical with the Fourier development (13) on C of the derivative $f'(z)$. By the continuity of $f'(z)$ on C, the second form of the development (13) is valid uniformly on C when summed by the method of arithmetic means; the corresponding sequence converges uniformly in $\overline{C}$ to the function $f'(z)$ on C and to the function $f'(z)$ interior to C. Consequently the function $f'(z)$ is continuous throughout $\overline{C}$, and the proof is complete.

Various refinements of Theorem III exist and are to be published by the present writers on another occasion.

Harvard University
Georgia School of Technology

NOTE ON DEGREE OF TRIGONOMETRIC AND POLYNOMIAL APPROXIMATION TO AN ANALYTIC FUNCTION§

J. L. WALSH AND W. E. SEWELL

1. **Introduction.** Well known results|| relate the continuity properties of a real function $f(x)$ to the degree of approximation to $f(x)$ by trigonometric sums and by polynomials in x. In more recent years further results¶ have related the continuity properties of a complex function $f(z)$ to the degree of approximation to $f(z)$ by polynomials in the complex variable z. The object of the present note is to obtain some new results lying on the border line of these two general fields of research.

To be more explicit, if $f(z)$ is analytic in the annulus $\rho > |z| > 1/\rho < 1$, the degree of convergence on $|z| = 1$ of the Laurent development of

† O. Veblen, *A conformal wave equation*, Proceedings of the National Academy of Sciences, vol. 21 (1935), p. 484.

‡ P. A. M. Dirac, *Wave equations in conformal space*, Annals of Mathematics, (2), vol. 37 (1936), p. 429.

§ Presented to the Society, September 6, 1938.

|| Due especially to S. Bernstein, Jackson, Lebesgue, Montel, and de la Vallée Poussin.

¶ Due especially to J. Curtiss, Sewell, and Walsh.

$f(z)$ is intimately connected with the continuity properties of $f(z)$ on the two circles $|z|=\rho$ and $|z|=1/\rho$. This fact is already known (Bernstein, de la Vallée Poussin), if $f(z)$ has poles or certain other singularities on those two circles, and is established in the present note if $f(z)$ or one of its derivatives satisfies a Lipschitz condition on $|z|=\rho$ and $|z|=1/\rho$.

Once the latter connection is established, standard methods involving suitable conformal transformations enable us to study the relation between polynomial approximation to an analytic function $f(z)$ on the segment $-1\leqq z\leqq 1$ and the continuity properties of $f(z)$ on the largest ellipse whose foci are $+1$ and -1 within which $f(z)$ is analytic. We also make application to the relation between trigonometric approximation on $y=0$ (with $z=x+iy$) to an analytic function $f(z)$ with period 2π and the continuity properties of the function on the lines $y=\pm b$ bounding a region within which $f(z)$ is analytic.

2. **Approximation on the unit circle.** We prove the following theorem:

THEOREM 1. *Let $f(\theta)$ be periodic with period 2π, and suppose the numbers a_{nk} and b_{nk} (not necessarily real) are given so that*

$$s_n(\theta) = \frac{a_{n0}}{2} + \sum_{k=1}^{n} (a_{nk}\cos k\theta + b_{nk}\sin k\theta),$$

with the relation, for $n=1, 2, \cdots$ and for all θ,

$$|f(\theta) - s_n(\theta)| \leqq M/n^{p+\alpha+1}\rho^n, \qquad 0<\alpha\leqq 1,\ \rho>1, \tag{1}$$

where p is a non-negative integer and M is a constant. Then the function

$$F(z) \equiv \lim_{n\to\infty}\left[c_{n0} + \sum_{k=1}^{n}(c_{n,-k}z^{-k} + c_{nk}z^k)\right], \tag{2}$$

$$2c_{nk} = a_{nk} - ib_{nk}, \qquad 2c_{n,-k} = a_{nk} + ib_{nk},$$

*coincides with $f(\theta)$ on the circle $|z|=1$, with $z=\cos\theta+i\sin\theta$, and $F(z)$ is analytic in the annulus $\rho>|z|>1/\rho$ and continuous in the corresponding closed region. For z_1 and z_2 on $|z|=\rho$ or $1/\rho$ we have**

$$|F^{(p)}(z_1) - F^{(p)}(z_2)| \leqq L\cdot|z_1-z_2|^{\alpha}\cdot\big|\log|z_1-z_2|\big|^{\beta}, \tag{3}$$

where $\beta=0$ if $\alpha<1$, and $\beta=1$ if $\alpha=1$, and where L is a constant independent of z_1 and z_2.

* The notation $F^{(p)}(z)$ indicates the pth derivative of $F(z)$, if $p>0$, and the function $F(z)$ itself for $p=0$. Here and below, such derivatives on the boundaries of regions of analyticity are considered in the one-dimensional sense.

Inequality (1) written for successive indices implies

$$| s_{n+1}(\theta) - s_n(\theta) | \leqq 2M/n^{p+\alpha+1}\rho^n$$

for all θ, an inequality which we write in the form

$$| p_{n+1}(z) - p_n(z) | \leqq 2M/n^{p+\alpha+1}\rho^n, \qquad \text{for } |z| = 1,$$

where $p_n(z)$ is the expression in square brackets in (2), a polynomial in z and $1/z$ of degree n, equal to $s_n(\theta)$ on $|z| = 1$. An easily proved lemma* then yields

$$| p_{n+1}(z) - p_n(z) | \leqq 2M\rho/n^{p+\alpha+1}, \qquad \text{for } \rho \geqq |z| \geqq 1/\rho.$$

In particular, on the two circles $|z| = \rho$ and $|z| = 1/\rho$ we may write

$$| F(z) - p_n(z) | \leqq 2M\rho\left[\frac{1}{n^{p+\alpha+1}} + \frac{1}{(n+1)^{p+\alpha+1}} + \cdots\right] < M_1/n^{p+\alpha}.$$

The function $p_n(z)$ (not necessarily real) is, on $|z| = \rho$ and on $|z| = 1/\rho$, a trigonometric polynomial in θ of order n with $z = \rho e^{i\theta}$ or $z = \rho^{-1}e^{i\theta}$, so that by the results of de la Vallée Poussin† the function $F(z)$ satisfies on $|z| = \rho$ and $|z| = 1/\rho$ a condition *with respect to* θ of form (3); hence (3) itself is fulfilled.‡

It is a corollary to our proof of Theorem 1 that the sequence $s_n(\theta)$, if defined off the circumference $|z| = 1$ as the function $p_n(z)$, converges uniformly to $F(z)$ in the closed annulus $\rho \geqq |z| \geqq 1/\rho$, with an error not greater than $M_1/n^{p+\alpha}$.

In the direction of the converse of Theorem 1 we prove the following theorem:

THEOREM 2. *Let the function $F(z)$ be analytic in the annular region bounded by the circles $|z| = \rho > 1$ and $|z| = 1/\rho$ and continuous in the corresponding closed region, and let $F^{(p)}(z)$, p a non-negative integer, satisfy a Lipschitz condition§ of order α on $|z| = \rho$ and $|z| = 1/\rho$. Let us set*

$$F(z) \equiv \sum_{k=-\infty}^{\infty} c_k z^k, \qquad \rho > |z| > 1/\rho; \tag{4}$$

* Walsh, *Interpolation and Approximation by Rational Functions in the Complex Domain*, American Mathematical Society Colloquium Publications, vol. 20, New York, 1935, p. 259.

† *Leçons sur l'Approximation*, Paris, 1919, chap. 4. The results in question are equally valid for real and for complex valued functions.

‡ See for instance Walsh and Sewell, this Bulletin, vol. 43 (1937), pp. 557–563.

§ That is to say, let (3) be satisfied with $\beta = 0$.

then with the notation $a_k = c_k + c_{-k}$, $b_k = i(c_k - c_{-k})$, *we have on* $|z| = 1$, *with* $z = e^{i\theta}$, *the relation*

$$(5)\qquad \left| F(e^{i\theta}) - \left[\frac{a_0}{2} + \sum_{k=1}^{n} (a_k \cos k\theta + b_k \sin k\theta)\right]\right| \leqq M/n^{p+\alpha}\rho^n,$$

where M is a constant.

In the usual proof of the validity of the Laurent development (4), it is established that we may write for $\rho > |z| > 1/\rho$ the equation $F(z) \equiv F_1(z) + F_2(z)$, where

$$F_1(z) = \sum_{k=0}^{\infty} c_k z^k, \quad |z| < \rho; \qquad F_2(z) = \sum_{k=-1}^{-\infty} c_k z^k, \quad |z| > 1/\rho.$$

Under the present conditions on $F(z)$, the function $F_2(z)$ is analytic on $|z| = \rho$, so that the function $F_1^{(p)}(z) \equiv F^{(p)}(z) - F_2^{(p)}(z)$ is continuous on $|z| = \rho$ and satisfies on $|z| = \rho$ a Lipschitz condition of order α. Similarly the function $F_2^{(p)}(z)$ is continuous on $|z| = 1/\rho$ and satisfies on $|z| = 1/\rho$ a Lipschitz condition of order α.

By the continuity of $F_1(z)$ on $|z| = \rho$ and of $F_2(z)$ on $|z| = 1/\rho$, we may write

$$c_k = \frac{1}{2\pi i}\int_{|t|=\rho} \frac{F_1(t)dt}{t^{k+1}}, \quad k \geqq 0; \qquad c_k = \frac{1}{2\pi i}\int_{|t|=1/\rho} \frac{F_2(t)dt}{t^{k+1}}, \quad k < 0;$$

whence for $|z| < \rho$, by the uniform convergence of the series involved, we have

$$(6)\qquad F_1(z) - \sum_{k=0}^{n} c_k z^k = \frac{1}{2\pi i}\int_{|t|=\rho} F_1(t)\left(\sum_{k=n+1}^{\infty} \frac{z^k}{t^{k+1}}\right) dt.$$

If $P_n(t)$ is an arbitrary polynomial in t of degree n, we have

$$\int_{|t|=\rho} \frac{P_n(t)dt}{t^{k+1}} = 0, \qquad k > n,$$

so that, for $|z| < \rho$, (6) may be written

$$(7)\qquad F_1(z) - \sum_{k=0}^{n} c_k z^k = \frac{1}{2\pi i}\int_{|t|=\rho} [F_1(t) - P_n(t)]\left(\sum_{k=n+1}^{\infty} \frac{z^k}{t^{k+1}}\right) dt.$$

A consequence of the Lipschitz condition on $F_1^{(p)}(z)$ on $|z| = \rho$ is that there exist* polynomials $P_n(t)$ of respective degrees n with

* J. Curtiss, this Bulletin, vol. 42 (1936), pp. 873–878. The proof is based on methods due to Bernstein, Jackson, and de la Vallée Poussin.

$$|F_1(t) - P_n(t)| \leqq M'/n^{p+\alpha}, \qquad |t| = \rho,$$

where M' is independent of n and of t. Thus equation (7) yields

$$\left| F_1(z) - \sum_{k=0}^{n} c_k z^k \right| \leqq M''/n^{p+\alpha}\rho^n, \qquad |z| = 1,$$

where M'' is independent of n and of z. A similar and similarly proved relation for $F_2(z)$ yields the inequality (5) and the theorem.

A remark due to Sewell* concerning degree of convergence of Taylor developments applies to the degree of convergence of the sequence $\sum_{k=0}^{n} c_k z^k$ to $F_1(z)$ on the circle $|z| = \rho$, and also to the degree of convergence of the sequence $\sum_{k=-1}^{-n} c_k z^k$ to $F_2(z)$ on the circle $|z| = 1/\rho$, so that we obtain

$$\left| F(z) - \sum_{k=-n}^{n} c_k z^k \right| \leqq (M''' \log n)/n^{p+\alpha}, \qquad \text{for } \rho \geqq |z| \geqq 1/\rho,$$

where M''' is independent of n and z.

It is a matter of indifference whether in Theorem 1 we prove and in Theorem 2 assume that $F^{(p)}(z)$ is continuous *merely on the circles* $|z| = \rho$ *and* $|z| = 1/\rho$ or that $F^{(p)}(z)$ is continuous *in the closed region* $\rho \geqq |z| \geqq 1/\rho$, for the one condition implies the other.† A similar remark applies to Theorems 3–6.

3. **Approximation on the segment** $-1 \leqq z \leqq 1$. The analog of Theorem 1 is the following theorem:

THEOREM 3. *Let $f(z)$ be defined on the segment $-1 \leqq z \leqq 1$, and let, for $n = 1, 2, \cdots$, a polynomial $P_n(z)$ in z of degree n exist such that on the segment $-1 \leqq z \leqq 1$*

$$|f(z) - P_n(z)| \leqq M/n^{p+\alpha+1}\rho^n, \qquad \rho > 1, \ 0 < \alpha \leqq 1,$$

where p is a non-negative integer. Then the function $f(z)$, if suitably defined, is analytic throughout the interior of the ellipse γ whose foci are -1 and $+1$ and whose semi-sum of axes is ρ; moreover $f(z)$ is continuous in the closed interior of γ and on γ satisfies the condition

$$(8) \qquad |f^{(p)}(z_1) - f^{(p)}(z_2)| \leqq L \cdot |z_1 - z_2|^{\alpha} \cdot |\log|z_1 - z_2||^{\beta},$$

where $\beta = 0$ if $\alpha < 1$ and $\beta = 1$ if $\alpha = 1$, and where L is a constant independent of z_1 and z_2.

---

* This Bulletin, vol. 48 (1935), pp. 111–117, Theorem 4.

† This follows by consideration of the functions $F_1(z)$ and $F_2(z)$ from the results of Walsh and Sewell, loc. cit.

We map the z-plane onto the w-plane by the transformation $z=(w+w^{-1})/2$. Under this transformation the image in the w-plane of the segment $-1\leqq z\leqq 1$ counted twice is the unit circle $|w|=1$, the image in the w-plane of γ in the z-plane counted twice consists of the two circles $|w|=\rho$ and $|w|=1/\rho$, and the image in the w-plane of the interior of γ in the z-plane counted twice is the annular region $\rho>|w|>1/\rho$. The polynomial $P_n(z)$ corresponds to a polynomial in w and $1/w$ of degree n; that is to say, considered as a function of w it is precisely of the form of the function $p_n(z)$ introduced in the proof of Theorem 1, with $c_{nk}=c_{n,-k}$. It follows from Theorem 1 that the transform of $f(z)$ (considered as a function of w) defined in the annulus as the limit of the sequence $P_n(z)$ (considered as a function of w) satisfies a condition of form (3) with respect to w on the circles $|w|=\rho$ and $|w|=1/\rho$ and is symmetric in w and $1/w$. The function $f(z)$, the transform in the z-plane of the limit in the w-plane of the sequence $P_n(z)$, is single-valued interior to γ, is obviously analytic interior to γ except perhaps for $-1\leqq z\leqq 1$, and is analytic on that segment because continuous there in the two-dimensional sense. By the analyticity of the transformation $z=(w+w^{-1})/2$ on $|w|=\rho$ and $|w|=1/\rho$, inequality (8) follows, and the proof is complete.

As a corollary to this proof we remark that the sequence $P_n(z)$ itself converges uniformly to $f(z)$ on and within γ, with an error not greater than $M_1/n^{p+\alpha}$.

In the direction of the converse of Theorem 3, immediate application of Theorem 2 yields the following theorem:

THEOREM 4. *Let γ denote the ellipse whose foci are -1 and $+1$ and whose semi-sum of axes is ρ, let the function $f(z)$ be analytic interior to γ and continuous in the corresponding closed region, and let $f^{(p)}(z)$, p a non-negative integer, satisfy a Lipschitz condition of order α, $(0<\alpha\leqq 1)$, on γ. Then for $n=1, 2, \cdots$ there exists a polynomial $P_n(z)$ of degree n in z such that $|f(z)-P_n(z)|\leqq M/n^{p+\alpha}\rho^n$, $(-1\leqq z\leqq 1)$, where M is a constant independent of n and z.*

Under the transformation $z=(w+w^{-1})/2$, the function $f(z)$ corresponds to a function of w which is analytic in the neighborhood of $|w|=1$ except perhaps on that circumference, continuous in the two-dimensional sense at every point of $|w|=1$, and hence analytic on $|w|=1$ and throughout the annulus $\rho>|w|>1/\rho$. The function $f[(w+w^{-1})/2]$ is symmetric in w and $1/w$, so that the corresponding Laurent polynomials used in the proof of Theorem 2 are in this case symmetric in w and $1/w$ and hence are polynomials in z; Theorem 4 follows from Theorem 2.

It is not without interest to notice that the successive approximating Laurent polynomials that here present themselves in the w-plane are of the form $\sum_{k=0}^{n} c_k(w^k+w^{-k})$, where c_k is independent of n, so that the approximating polynomials $P_n(z)$ of Theorem 4 are of the form

$$\sum_{k=0}^{n} c_k T_k(z),$$

where c_k is independent of n, and where $T_k(z) \equiv w^k+w^{-k}$ is a polynomial of the set studied by Tschebycheff: $T_0(z)=2$, $T_1(z) \equiv 2z$, $T_2(z) \equiv 4z^2-2$, $T_3(z) \equiv 8z^3-6z$, $\cdots$ which are mutually orthogonal on the interval $-1 \leqq z \leqq 1$ with respect to the norm function $(1-z^2)^{-1/2}$ and are also orthogonal* on γ and on every ellipse confocal with γ with respect to the norm function $|1-z^2|^{-1/2}$.

As in the proof of Theorem 2, we have, for z on and within γ, the relation

$$\left| f(z) - \sum_{k=0}^{n} c_k T_k(z) \right| \leqq (M' \log n)/n^{p+\alpha},$$

where M' is independent of n and z. The expansion $\sum_{k=0}^{\infty} c_k T_k(z)$ of $f(z)$ on γ converges uniformly to $f(z)$ on γ and hence is an expansion of the usual form in terms of orthogonal polynomials:

$$c_k \int_\gamma | T_k(z) |^2 \cdot | 1 - z^2 |^{-1/2} \cdot | dz | = \int_\gamma f(z)\overline{T}_k(z) | 1 - z^2 |^{-1/2} \cdot | dz |.$$

Results analogous to Theorems 3 and 4 for the case that $f(z)$ is uniformly bounded interior to γ or is analytic interior to γ with poles or certain other singularities on γ are due to Bernstein† and to de la Vallée Poussin (op. cit., chaps. 8 and 9).

4. **Approximation to a periodic function on the axis of reals.** A conformal transformation different from that used in Theorems 3 and 4 will now give further results from Theorems 1 and 2.

THEOREM 5. *Let the function $f(z)$ be periodic with period 2π, and let there exist trigonometric polynomials $t_n(z)$ of respective degrees n such that we have for all real $z=x+iy$*

$$| f(z) - t_n(z) | \leqq M/n^{p+\alpha+1}\rho^n, \qquad 0 < \alpha \leqq 1, \ \rho > 1,$$

where p is a non-negative integer. Then the function $f(z)$ can be analytically extended so that it is analytic throughout the band $|y| < \log \rho$, is

* Walsh, this Bulletin, vol. 40 (1934), pp. 84–88.

† *Leçons sur les Propriétés Extrémales*, Paris, 1926, chap. 3.

continuous in the corresponding closed region, and on the lines $y = \pm \log \rho$ *satisfies condition* (8).

The transformation $w = e^{iz}$ carries the line $y=0$ into the unit circle $|w| = 1$, the band $|y| < \log \rho$ into the annulus $\rho > |w| > 1/\rho$, the function $f(z)$ into a function analytic and single-valued in that annulus, and the trigonometric polynomial $t_n(z)$ on $y=0$ into the trigonometric polynomial $t_n(\theta)$ on $|w| = 1$ with $w = e^{i\theta}$. The hypothesis of Theorem 1 is satisfied, and an inequality of form (3) on the circles $|w| = \rho$ and $|w| = 1/\rho$ leads to the conclusion of Theorem 5.

The given sequence $t_n(z)$ can be expressed on $y=0$ as a sequence of polynomials in $w = e^{iz}$ and $1/w$, with $z = x = \theta$:

$$\sin kz = -\frac{i}{2}(e^{ikz} - e^{-ikz}) = -\frac{i}{2}(w^k - w^{-k}),$$

$$\cos kz = \tfrac{1}{2}(e^{ikz} + e^{-ikz}) = \tfrac{1}{2}(w^k + w^{-k}).$$

These equations are then valid even if y is different from zero.

It follows from the proof of Theorem 1 that the sequence $t_n(z)$, expressed in trigonometric form, converges in the closed region $|y| \leqq \log \rho$, with an error not greater than $M_1/n^{p+\alpha}$.

THEOREM 6. *Let the function $f(z)$ be periodic with period 2π, let $f(z)$ be analytic in the band $|y| < \log \rho$, where $z = x + iy$, and continuous in the corresponding closed region, and let $f^{(p)}(z)$, p a non-negative integer, satisfy a Lipschitz condition of order α, $(0 < \alpha \leqq 1)$, on the lines $y = \pm \log \rho$. Then there exist trigonometric polynomials $t_n(z)$ of respective degrees n such that we have for all real z*

$$|f(z) - t_n(z)| \leqq M/n^{p+\alpha}\rho^n,$$

where M is a constant independent of n and z.

The detailed proof of Theorem 6 is readily supplied by the reader by use of the same transformation $w = e^{iz}$ and the method of proof of Theorem 4. It is of interest to note that the function $t_n(z)$ appears also for complex values of z as the sum of the first $n+1$ terms of a series of the form

$$\frac{1}{2}a_0 + \sum_{k=1}^{\infty}(a_k \cos kz + b_k \sin kz).$$

As in Theorem 2, we have

$$|f(z) - t_n(z)| \leqq (M' \log n)/n^{p+\alpha}, \qquad \text{for } |y| \leqq \log \rho,$$

where M' is independent of n and z.

From the orthogonality of the functions w^k on $|w|=\rho$ it follows that the transformed functions e^{ikz}, $k=\cdots,-1,0,1,2,\cdots$, form an orthogonal set on the line segment $y=-\log\rho$, $(0\leqq x\leqq 2\pi)$, with respect to the norm function $|dw/dz|=|de^{iz}/dz|=e^{-y}$. This norm function is a nonvanishing constant on the segment and may therefore be omitted. Thus

$$\int_{-i\log\rho}^{2\pi-i\log\rho} e^{ikz}\cdot\overline{e^{ilz}}\cdot dx = 0, \qquad k\neq l,$$

where k and l are integers, positive, negative, or zero. Of course this orthogonality condition holds on any interval $x_0\leqq x\leqq x_0+2\pi$, $y=y_0$, the respective limits of integration being x_0+iy_0 and $x_0+2\pi+iy_0$.

The set e^{ikz} is closed (with respect to the class of continuous functions) on the interval $x_0\leqq x\leqq x_0+2\pi$, $y=y_0$, as is seen by transformation to the w-plane. On every interval $x_0\leqq x\leqq x_0+2\pi$, $y=y_0$, $|y_0|\leqq\log\rho$, the sequence

$$t_n(z) = \frac{1}{2}a_0 + \sum_{k=1}^{n}(a_k\cos kz + b_k\sin kz) = \sum_{k=-n}^{n} c_k e^{ikz},$$

$$2c_k = a_k - ib_k, \qquad 2c_{-k} = a_k + ib_k,$$

is the sum of the first $2n+1$ terms of the uniformly convergent formal development on that interval of the function $f(z)$ in terms of the orthogonal functions e^{ikz}, so that the coefficients are given by formulas of the usual type:

$$c_k = \frac{e^{2ky_0}}{2\pi}\int_{x_0+iy_0}^{x_0+2\pi+iy_0}\overline{e^{ikz}}\cdot f(z)\cdot dz,$$

where on $y=y_0$ we have $\overline{e^{ikz}}=e^{-ik\bar{z}}=e^{-ky_0}(\cos kx - i\sin kx)$.

Analogs of Theorems 5 and 6 have been established by de la Vallée Poussin (op. cit., chaps. 8 and 9) for the case that $f(z)$ is bounded for $|y|<\log\rho$ or has poles or other singularities on the lines $y=\pm\log\rho$.

There is an obvious discrepancy of unity in the exponents of n that appear in Theorems 1 and 2, in Theorems 3 and 4, and in Theorems 5 and 6. This discrepancy is inherent in the nature of the problem, as is shown by examples that the writers will publish on another occasion.

Harvard University and
Georgia School of Technology

ON THE DERIVATIVE OF A POLYNOMIAL AND CHEBYSHEV APPROXIMATION[1]

T. S. MOTZKIN AND J. L. WALSH

Introduction. The location of the zeros of the derivative of a polynomial has been much studied, as has the location of the zeros of the Chebyshev polynomial. In §1 of the present note we set forth in a direct and elementary manner the equivalence of these two problems in a suitably specialized situation. This conclusion is mentioned (for integral λ_i) with an indication of the proof by Fekete and von Neumann [4],[2] and the conclusion is related to a much deeper investigation due to Fekete [3]. We obtain (§9) some new results on zeros of approximating polynomials and (§§2, 3, 8) on the argument of the deviation. In §10 we consider approximation by an arbitrary linear family. Throughout the paper we study primarily approximation on a set of n points by a polynomial of degree $n-2$. In §§4–7 weight functions with infinities and Chebyshev rational functions are introduced.

1. **Determination of special Chebyshev polynomials with weight function.** If E is a closed bounded point set of the z-plane on which the weight function $\mu(z)$ is positive and continuous, *the Chebyshev polynomial* $T_m(z)$ *of degree* m *for* E *with weight function* $\mu(z)$ is defined as that polynomial of the form

$$z^m + A_1 z^{m-1} + \cdots + A_m \tag{1}$$

for which the *norm*

$$\max\,[\mu(z)\,|\,T_m(z)\,|,\ z \text{ on } E]$$

is least. It can be shown that $T_m(z)$ exists and is unique. In the special case about to be considered the existence and uniqueness of $T_m(z)$ will be established.

THEOREM. *Let the point set* E *consist of the distinct points* $z_1, z_2, \cdots, z_n$ $(n>1)$ *and let us set* $\omega(z) \equiv (z-z_1)(z-z_2)\cdots(z-z_n)$. *Then the unique Chebyshev polynomial* $T_{n-1}(z)$ *of degree* $n-1$ *for* E *with the weight function* $\mu(z_i) \equiv \mu_i = 1/\lambda_i|\omega'(z_i)|$, *where* $\lambda_i > 0$, $\sum_1^n \lambda_i = 1$, *is the*

Presented to the Society, September 7, 1951; received by the editors January 25, 1952.

[1] The preparation of this paper was sponsored (in part) by the Office of Scientific Research, USAF.

[2] Figures in brackets indicate the literature references at the end of this paper.

polynomial

$$T_{n-1}(z) \equiv \omega(z) \sum_1^n \frac{\lambda_i}{z - z_i}. \tag{2}$$

Note that we consider here essentially the most general positive weight function on E, for multiplication of the weight function by a positive constant does not alter the Chebyshev polynomial.

We observe that $T_{n-1}(z)$ as defined by (2) is of form (1), and that from the definition of $\omega'(z_i)$ as a limit we have $T_{n-1}(z_i) = \lambda_i \omega'(z_i)$. The norm of $T_{n-1}(z)$ on E is thus

$$\max \left[\mu_i \,|\, T_{n-1}(z_i) \,|\, \right] = 1.$$

The Lagrange interpolation formula represents (as may be verified directly) an arbitrary polynomial of degree[3] $n-1$ in terms of its values in the n points z_i, and in the present case becomes

$$T_{n-1}(z) \equiv \sum_1^n \frac{T_{n-1}(z_i)}{\omega'(z_i)} \frac{\omega(z)}{z - z_i} \equiv \omega(z) \sum_1^n \frac{\lambda_i}{z - z_i},$$

which is (2).

An arbitrary polynomial of the form $P(z) \equiv z^{n-1} + B_1 z^{n-2} + \cdots + B_{n-1}$ is expressed by Lagrange's formula

$$P(z) \equiv \sum_1^n \frac{P(z_i)}{\omega'(z_i)} \frac{\omega(z)}{z - z_i};$$

inspection of the terms of highest degree in z yields

$$1 = \sum_1^n \frac{P(z_i)}{\omega'(z_i)}. \tag{3}$$

The norm of $P(z)$ on E is

$$\max \left[\mu_i \,|\, P(z_i) \,|\, \right] = \max \left[\,|\, P(z_i)/\lambda_i \omega'(z_i) \,|\, \right]. \tag{4}$$

We contrast (3) with the equation $1 = \sum_1^n \lambda_i$, where we have $\lambda_i > 0$. Either for some value of i we have $|P(z_i)/\omega'(z_i)| > \lambda_i$, in which case the norm (4) of $P(z)$ is greater than unity, the norm of $T_{n-1}(z)$, or we have $|P(z_i)/\omega'(z_i)| = \lambda_i$ for every i, in which case all the quotients $P(z_i)/\omega'(z_i)$ are real and positive and we have $P(z_i)/\omega'(z_i) = \lambda_i$ for every i, whence $P(z) \equiv T_{n-1}(z)$. That is to say, we have shown that $T_{n-1}(z)$ as defined by (2) is the unique polynomial of form (1)

[3] As is customary in the study of approximation, we define a polynomial of degree $n-1$ as an arbitrary function of the form $A_0 z^{n-1} + A_1 z^{n-2} + \cdots + A_{n-1}$.

of least norm on E; the theorem is established.

We note that $T_{n-1}(z)$ is $\omega(z)$ multiplied by the logarithmic derivative of

$$\prod_1^n (z - z_i)^{\lambda_i}, \qquad \sum_1^n \lambda_i = 1, \tag{5}$$

where of course the λ_i need not be commensurable. Thus *the study of the zeros of the derivative of* (5) *is identical with the study of the zeros of* $T_{n-1}(z)$.

2. Properties of the arguments of the Chebyshev polynomial. The equation $T_{n-1}(z_i) = \lambda_i \omega'(z_i)$ shows that *for each* z_i *the arguments of the* $T_{n-1}(z_i)$ *are fixed independently of the choice of the* λ_k:

$$\arg [T_{n-1}(z_i)] = \arg [\omega'(z_i)].$$

By a new application of Lagrange's formula we have a representation of the polynomial unity:

$$1 \equiv \sum_1^n \frac{1}{\omega'(z_i)} \frac{\omega(z)}{z - z_i};$$

inspection of the terms of highest degree gives ($n > 1$)

$$0 = \sum_1^n \frac{1}{\omega'(z_i)},$$

which we write in the form

$$\sum_1^n \frac{\lambda_i}{\lambda_i \omega'(z_i)} = 0, \tag{6}$$

and by taking conjugates write this also in the form

$$\sum_1^n \lambda_i \omega'(z_i) \frac{1}{\lambda_i \, | \omega'(z_i) |^2} = 0. \tag{7}$$

Equation (7) states that if the points $\lambda_i \omega'(z_i)$ are interpreted in the w-plane, then the origin is the center of gravity of those points considered with suitably chosen positive weights, namely $1/\lambda_i |\omega'(z_i)|^2$. Consequently *the origin lies in the smallest convex polygon containing* (i.e., *lies in the convex hull of*) *the points* $\lambda_i \omega'(z_i) = T_{n-1}(z_i)$, *and lies interior to this polygon except when the polygon degenerates to a line segment*; in the latter case the origin is an interior point of the segment. This conclusion is familiar in the case that all the points z_i are real and each λ_i equal to $1/n$; here it follows from (2) that we have $T_{n-1}(z) \equiv \omega'(z)/n$; the values of $\lambda_i\omega'(z_i) = \omega'(z_i)/n$ are alternately positive and negative if $z_1 < z_2 < \cdots < z_n$; the origin lies interior to the smallest line segment containing the points $\lambda_i\omega'(z_i)$.

Equation (6) obviously shows that the origin lies in the convex hull of the points $1/\lambda_i\omega'(z_i)$.

3. **Examples.** It is of interest to add some examples where μ_i, λ_i, and $|\omega'(z_i)|$ are all independent of i; thus $\lambda_i = 1/n$, and $T_{n-1}(z) \equiv \omega'(z)/n$. In each example E is defined as the set of zeros of $\omega(z)$.

EXAMPLE 1. $\omega(z) \equiv z^n - 1$, $T_{n-1}(z) \equiv \omega'(z)/n \equiv z^{n-1}$.

EXAMPLE 2. $\omega(z) \equiv (z^n - \alpha)(z^n - \bar{\alpha})$, $\alpha \neq 0$, $T_{2n-1}(z) \equiv \omega'(z)/2n$.

EXAMPLE 3. $\omega(z) \equiv z(z^n - \alpha)(z^n - \bar{\alpha})$, $|\alpha| = 1$, $|\alpha - \bar{\alpha}| = 1/n$, $T_{2n}(z) \equiv \omega'(z)/(2n+1)$.

4. **Weight functions with infinities.** Let now E no longer necessarily consist of n points. It is entirely permissible not to require that the given weight function $\mu(z)$ be continuous on the given set E; for instance $\mu(z)$ may be of the form

$$\mu(z) \equiv \frac{\mu_1(z)}{|(z - \alpha_1) \cdots (z - \alpha_k)|}, \tag{8}$$

where $\mu_1(z)$ is positive and continuous on E and the points $\alpha_1, \alpha_2, \cdots, \alpha_k$ not necessarily distinct belong to E. The norm of a polynomial $T_m(z)$ is as before max $[\mu(z)|T_m(z)|$, z on $E]$; even if α_i is an isolated point of E, the value $\mu(\alpha_i)|T_m(\alpha_i)|$ is to be interpreted as

$$\lim_{z \to \alpha_i} \mu_1(\alpha_i) \, | T_m(z) | / | (z - \alpha_1) \cdots (z - \alpha_k) |$$

provided the limit (finite or infinite) exists. For the Chebyshev polynomial $T_m(z)$ properly to exist, the norm of $T_m(z)$ is to be finite, we must have $m \geqq k$, and all the points α_i must be zeros of $T_m(z)$. If we write

$$T_m(z) \equiv (z - \alpha_1)(z - \alpha_2) \cdots (z - \alpha_k) S_m(z),$$

the norm of $T_m(z)$ is max $[\mu_1(z)|S_m(z)|$, z on $E]$, where $S_m(z) \equiv z^{m-k} + A_1 z^{m-k-1} + \cdots + A_{m-k}$ is of degree $m-k$; the problem of studying the Chebyshev polynomial $T_m(z)$ of degree $m(\geqq k)$ with weight function $\mu(z)$ is identical with the problem of studying the Chebyshev polynomial $S_m(z)$ of degree $m-k$ with weight function $\mu_1(z)$.

5. **Chebyshev rational functions.** The use of a weight function of form (8) is related to Chebyshev rational functions with prescribed poles. For simplicity let E be a closed and bounded point set of the

z-plane on which the weight function $\mu(z)$ is positive and continuous, and let the prescribed points $\alpha_1, \alpha_2, \cdots, \alpha_k$ be disjoint from E. The *Chebyshev rational function* $R_m(z)$ is the function of form

$$R_m(z) \equiv \frac{z^m + B_1 z^{m-1} + \cdots + B_m}{(z-\alpha_1)(z-\alpha_2)\cdots(z-\alpha_k)} \tag{9}$$

of least norm. Here we have precisely the problem of the Chebyshev polynomial $T_m(z)$ on E with the norm function

$$\frac{\mu(z)}{|(z-\alpha_1)(z-\alpha_2)\cdots(z-\alpha_k)|}.$$

It is of course imperative either to require that the numerator of $R_m(z)$ should be of the form indicated in (9) or in some other way to exclude as inadmissible the trivial function $R_m(z) \equiv 0$.

6. **Fejér's theorem.** We return to the general Chebyshev polynomial T_m given by (1) for a set E containing at least m points. Beyond the case where all points of E are collinear, the first geometric result on the zeros of $T_m(z)$ is due to Fejér [2]: *All zeros of $T_m(z)$ lie in the convex hull of E.* The proof is not difficult. Let the point z_0 lie exterior to K, the convex hull of E, and let β be the point of K nearest z_0. Then we have in each point z of E

$$|z - (z_0 + \beta)/2| < |z - z_0|. \tag{10}$$

For any polynomial $Q(z) \equiv z^{m-1} + \cdots$, we therefore have at each point z of E with $Q(z) \neq 0$

$$|[z - (z_0 + \beta)/2]Q(z)| < |(z - z_0)Q(z)|.$$

Thus $(z - z_0)Q(z)$ cannot be a Chebyshev polynomial for E with positive weight function.

As Fekete and von Neumann (loc. cit.) point out, by considering (5) there follows the theorem of Lucas, that *the convex hull of the zeros of a polynomial contains the zeros of the derived polynomial.*

7. **Poles of extremal rational functions.** The theorem of Fejér implies that the finite zeros of the C.r.f. $R_m(z)$ of form (9) lie in the convex hull of E. If we consider even a more general extremal problem for rational functions, where now the α_i are allowed to vary in suitable regions, it is clear that the C.r.f. for this more general problem is also the C.r.f. for the former problem where the α_i are suitably determined and considered fixed. It is still true that the finite zeros of $R_m(z)$ lie in the convex hull K of E; moreover the reasoning used in

connection with (10) now applies in reverse: if α_i exterior to K is a pole of the C.r.f. and if L is the half line through α_i from the nearest point of K, then α_i must be on L as far away from K as possible.

8. **Approximation to an arbitrary function.** The term *Chebyshev polynomial* is also used in a sense different from that defined in §1. If E is a closed bounded point set, if $\mu(z)$ is a weight function positive and continuous on E, and if $f(z)$ is continuous on E, then the *Chebyshev polynomial of best approximation* $t_m(z)$ is the polynomial of degree m such that the norm

$$\max \left[\mu(z) \mid f(z) - t_m(z) \mid, \ z \text{ on } E\right] \tag{11}$$

is least. It can be shown that $t_m(z)$ exists and is unique; we shall prove this result in the special case which concerns us below.

Suppose that E consists of precisely n distinct points. Let $f_1(z) \equiv B_0 z^{n-1} + B_1 z^{n-2} + \cdots + B_{n-1}$ be the unique polynomial of degree $n-1$ which coincides with $f(z)$ on E. In the case $m = n-2$, the norm (11) becomes

$$\max \left[\mu(z) \mid f_1(z) - t_{n-2}(z) \mid, \ z \text{ on } E\right]. \tag{12}$$

If $B_0 = 0$, the polynomial $f_1(z)$ is itself a polynomial of degree $n-2$, admissible in the consideration of all approximating polynomials, whence $t_{n-2}(z) \equiv f_1(z)$. Henceforth we suppose $B_0 = \sum_1^n f(z_i)/\omega'(z_i) \neq 0$, so that (12) can be written

$$\begin{aligned} \max \big[\mid B_0 \mid \mu(z) \mid z^{n-1} + B_1 z^{n-2}/B_0 + \cdots \\ + B_{n-1}/B_0 - t_{n-2}(z)/B_0 \mid, \ z \text{ on } E\big]. \end{aligned} \tag{13}$$

The problem of minimizing (13) is precisely the problem of determining the Chebyshev polynomial $T_{n-1}(z)$ for E with weight function $|B_0|\mu(z)$, a problem that we have already discussed. In particular it follows that $t_{n-2}(z)$ exists and is unique. If we define the numbers

$$\lambda_i = \frac{1}{\mu(z_i) \mid \omega'(z_i) \mid} \Big/ \sum_1^n \frac{1}{\mu(z_i) \mid \omega'(z_i) \mid}, \tag{14}$$

we have by (2)

$$\frac{f_1(z) - t_{n-2}(z)}{B_0} \equiv T_{n-1}(z) \equiv \omega(z) \sum_1^n \frac{\lambda_i}{z - z_i},$$

from which $t_{n-2}(z)$ can be determined; the second factor in this last member is the logarithmic derivative of (5).

The norm (12) can now be computed by setting $T_{n-1}(z_i) = \lambda_i \omega'(z_i)$:

$$\mu(z_i)\,|\,f(z_i) - t_{n-2}(z_i)\,| = |\,B_0\,|\,\mu(z_i)\,|\,T_{n-1}(z_i)\,|$$
$$= |\,B_0\,| \Big/ \sum_1^n \frac{1}{\mu(z_i)\,|\,\omega'(z_i)\,|},$$

which is obviously independent of i. Again, as in §2, *the value of* $\arg\{[f(z)-t_{n-2}(z)]/B_0\}=\arg[T_{n-1}(z)]=\arg[\omega'(z)]$ *in each point of* E *is independent of the particular choice of* $\mu(z)$ *and of* $f(z)$. *The origin lies in the convex hull of the points* $[f(z_i)-t_{n-2}(z_i)]$, *and lies interior to this polygon except when the polygon degenerates to a line segment.*

In the special examples given in §3 the value of $\arg\{[f(z)-t_{n-2}(z)]/B_0\}=\arg[\omega'(z)]$ can be written down at once.

All zeros of the polynomial $f_1(z)-t_{n-2}(z)$ *lie in the convex hull of* E (*if* $B_0\neq 0$).

Specific formulas for the norm (12) have been obtained by de la Vallée Poussin [1]. He also proves that a necessary and sufficient condition that given numbers δ_i, all different from zero, be deviations of the approximating polynomial $t_{n-2}(z)$ from some function $f(z)$ in the n points z_i is that for no polynomial $p_{n-2}(z)$ of degree $n-2$ all the quotients $p_{n-2}(z_i)/\delta_i$ have positive real parts. Thus we may now state: *a necessary and sufficient condition that for given* $\delta_i(\neq 0)$ *no polynomial* $p_{n-2}(z)$ *of degree* $n-2$ *exists such that all the quotients* $p_{n-2}(z_i)/\delta_i$ *have positive real parts is* $\arg\delta_i=\arg\omega'(z_i)$. Hence if this condition is satisfied, the origin O lies in the convex hulls of the points $p_{n-2}(z_i)/\delta_i$, and of those points $\delta_i/p_{n-2}(z_i)$ which are finite. This includes the result of §2.

This characterization of the arguments of δ_i is an extension to the complex domain of the alternation of signs of the δ_i in the real domain; compare §2. However, as de la Vallée Poussin indicates, there is an essential difference in that approximation on an arbitrary closed bounded set E in the real domain by polynomials of degree $n-2$ is equivalent to approximation on a subset of n points of E, whereas such approximation in the complex domain is equivalent to approximation on a subset of E containing m points, $n\leqq m\leqq 2n-1$. In the present paper we consider only the case $m=n$; for other m the arguments of the δ_i are no longer uniquely determined.

9. **Location of zeros of approximating polynomials.** We proceed to devote some attention to the zeros of the polynomials that occur in §8, with $B_0\neq 0$, using the notation of §8.

The function $f(z)$ considered on the set E: $(z_1, z_2, \cdots, z_n)$ is entirely arbitrary, so the polynomial $f_1(z)$ is an arbitrary polynomial of degree $n-1$ except for the restriction $B_0\neq 0$, and nothing can be deduced concerning its zeros without further hypothesis. Likewise the polynomial $t_{n-2}(z)$ given by

$$t_{n-2}(z) \equiv f_1(z) - B_0 T_{n-1}(z) \tag{15}$$

can be considered an arbitrary polynomial of degree $n-2$, provided $f(z)$ is arbitrary. However, under suitable conditions we can derive certain conclusions regarding these polynomials:

Let the zeros of the polynomial $f_1(z)$ *lie in the closed interior (respectively exterior)* C_1 *of the circle* $|z-\alpha| = r_1$, *and let the zeros of* $T_{n-1}(z)$ *lie in the closed interior* C_2 *of the circle* $|z-\beta| = r_2$ *(for which it is sufficient that* E *lie in* C_2*), where* C_1 *and* C_2 *are disjoint. Then all zeros of* $t_{n-2}(z)$ *in* (15) *lie in the* $n-2$ *closed regions*

$$\left| z - \frac{\alpha - \epsilon\beta}{1 - \epsilon} \right| \leqq \frac{r_1 + r_2}{|1 - \epsilon|}, \tag{16}$$

$$\left| z - \frac{\alpha - \epsilon\beta}{1 - \epsilon} \right| \geqq \frac{r_1 - r_2}{|1 - \epsilon|}, \tag{17}$$

respectively, where ϵ *takes all the values except unity of the* $(n-1)$*st roots of unity. If the circles* (16) *are mutually exterior, they contain each one zero of* $t_{n-2}(z)$.

Inequalities (16) follow [6] from equation (15), since the coefficient of z^{n-1} in $f_1(z)$ is B_0, and inequalities (17) can be similarly proved.

In the special case in which (15) is used to define $T_{n-1}(z)$ itself, we set $f_1(z) \equiv z^{n-1}$, $B_0 = 1$, $\alpha = 0$, $r_1 = 0$.

It is essential to assume that C_1 and C_2 are disjoint; otherwise we may have, e.g., $t_{n-2}(z) \equiv 0$.

All the circles (16) (resp. (17)) lie [7] in the closed exterior of the hyperbola (ellipse) C whose foci are α and β, and whose transverse (major) axis is $r_1 + r_2$ (resp. $r_1 - r_2$); the centers of the circles are equidistant from α and β, and the circles are doubly tangent to C.[4]

10. Approximation by functions of a linear family. We turn now to best approximation on the set E: $(z_1, z_2, \cdots, z_n)$, $n > 1$, to the function $f(z)$ by functions of the form $\sum_1^{n-1} c_j f_j(z)$, where the $f_j(z)$ are arbitrary functions defined on E and the complex constants c_j are to be determined so that

$$\max \left[\left| f(z) - \sum_1^{n-1} c_j f_j(z) \right|, z \text{ on } E \right] \tag{18}$$

---

[4] If C is an ellipse, the double tangency may occur merely in the formal algebraic sense.

is least. We suppose the functions $f_j(z)$ to be linearly independent on E. Since only the values of the $f_j(z)$ on E concern us, we can suppose these functions and $f(z)$ to be polynomials of degree $n-1$. Then the family of polynomials $f(z) - \sum_1^{n-1} c_j f_j(z)$ is the totality of polynomials of degree $n-1$ with precisely one linear relation between their coefficients. If this relation is homogeneous the polynomial zero will belong to the family and $f(z)$ can be exactly represented on E; this case is henceforth excluded. In any case not excluded, the relation can be written as $\Gamma[\phi(z)]=1$, where $\phi(z)$ is an arbitrary polynomial of degree $n-1$ and Γ is a linear homogeneous function of the coefficients of $\phi(z)$.

With the continued notation $\omega(z)=\prod(z-z_i)$, we define (complex) weights

$$w_i = \frac{1}{\Gamma[\omega_i(z)]}, \qquad \omega_i(z) \equiv \omega(z)/(z-z_i), \tag{19}$$

supposing no denominator to vanish; we discuss later the exceptional case. Consider the polynomials $\phi(z)$ of degree $n-1$ such that

$$\phi(z_i) = w_i(z_i^{n-1} + d_1 z_i^{n-2} + \cdots + d_{n-1}), \tag{20}$$

for arbitrary coefficients $d_1, d_2, \cdots, d_{n-1}$. These polynomials are again defined by a single linear relation among their coefficients since the rows

$$w_1 z_1^k, \ w_2 z_2^k, \cdots, w_n z_n^k \qquad (k = 0, 1, \cdots, n-2)$$

are linearly independent; the latter fact is a consequence of the identical vanishing of every polynomial of degree $n-2$ which vanishes on E.

This linear relation is precisely $\Gamma(\phi)=1$. For we have by Lagrange's formula

$$\phi(z) \equiv \sum_1^n \frac{\phi(z_i)}{\omega'(z_i)}\omega_i(z),$$

$$\Gamma[\phi(z)] = \sum_1^n \frac{\phi(z_i)}{\omega'(z_i)}\Gamma[\omega_i(z)] = \sum_1^n \frac{\phi(z_i)}{\omega'(z_i) w_i} = 1; \tag{21}$$

the last equation is again a consequence of Lagrange's formula

$$z^{n-1} + d_1 z^{n-2} + \cdots + d_{n-1} \equiv \sum_1^n \frac{z_i^{n-1} + d_1 z_i^{n-2} + \cdots + d_{n-1}}{\omega'(z_i)}\omega_i(z),$$

by inspection of the terms of degree $n-1$ in z.

The original problem of minimizing (18) is thus reduced to the problem of minimizing max $[|\phi(z)|, z \text{ on } E]$, with $\phi(z)$ defined as in (20). The minimizing polynomial $T_{n-1}(z)$ is (by §1) unique and is given by (2) with

$$\lambda_i = \frac{1}{|w_i||\omega'(z_i)|\sum 1/|w_i||\omega'(z_i)|},$$

namely

$$T_{n-1}(z) \equiv \sum_1^n \frac{\omega_i(z)}{|w_i||\omega_i(z_i)|\sum 1/|w_i||\omega'(z_i)|}.$$

Moreover we have the equations for every i

$$\arg \frac{\phi(z_i)}{w_i} = \arg [\omega'(z_i)].$$

Thus we have the values at z_i of the minimizing polynomials $T_{n-1}(z)$ and $\phi_0(z)$

$$T_{n-1}(z_i) = \frac{1}{|w_i||\omega'(z_i)|} \frac{\omega'(z_i)}{\sum 1/|w_i||\omega'(z_i)|},$$

$$\phi_0(z_i) = w_i T_{n-1}(z_i).$$

THEOREM. *The minimizing polynomial $\phi_0(z)$ is unique and given by*

$$\phi_0(z) \equiv \sum_1^n \frac{w_i\omega_i(z)}{|w_i\omega_i(z_i)|} \Big/ \sum 1/|w_i\omega'(z_i)|.$$

The requirement that (19) shall have a meaning, $i=1, 2, \cdots, n$, is essential, as we illustrate by an example. Choose $n=2$, $z_1=0$, $z_2=1$, hence $\omega_1(z)\equiv z-1$. Choose also $\Gamma(a_0z+a_1)\equiv a_0+a_1$, whence $\Gamma[\omega_1(z)]=0$. The functions $\phi(z)$ with $\Gamma[\phi(z)]=1$ are the polynomials $(1-a_1)z+a_1$. Their values at z_1 and z_2 are a_1 and 1 respectively. Here there is no unique polynomial $\phi(z)$ with least maximum modulus. The least possible value for the maximum modulus is unity, and is attained for all $\phi(z)$ with $|a_1|\leqq 1$. Note that for all $\phi(z)$ with $|a_1|<1$, the maximum of the modulus is taken on but once. Thus families of this kind behave very differently from the general type of linear families.

A formula for the most general linear functional $\Gamma[\phi(z)]$ of the coefficients of an arbitrary polynomial $\phi(z)$ of degree $n-1$ is readily found as in (21):

$$\phi(z) \equiv \sum_1^n \frac{\phi(z_i)}{\omega'(z_i)} \omega_i(z), \qquad \Gamma[\phi(z)] \equiv \sum_1^n \frac{\Gamma[\omega_i(z)]}{\omega_i(z_i)} \phi(z_i),$$

which is the desired relation.

The special case treated in §§1 and 2 is that of a linear family $\phi(z)$ defined by the relation

$$\sum_1^n \frac{\phi(z_i)}{w_i\omega_i(z_i)} = 1,$$

where the w_i are positive with sum unity. The Chebyshev polynomial $T_{n-1}(z)$ is not necessarily a member of this family, but related to it by the formula

$$T_{n-1}(z_i) = \phi_0(z_i)/w_i,$$

where $\phi_0(z)$ is the minimizing polynomial.

We turn to a special case of approximation by a linear family of functions. Let E_0: $(\alpha_1, \alpha_2, \cdots, \alpha_m)$ be a set of points not necessarily all distinct, and let the functional values $\beta_1, \beta_2, \cdots, \beta_m$ be assigned, in the sense that at a point α_j of multiplicity m_j the corresponding functional values assigned shall be interpreted as the values of $\phi(\alpha_j), \phi'(\alpha_j), \cdots, \phi^{(m_j-1)}(\alpha_j)$. Let E, disjoint from E_0, be the point set $z_1, z_2, \cdots, z_n$ of distinct points, let $\mu(z)$ be an arbitrary positive weight function defined on E, and let $f(z)$ be an arbitrary function defined on E. *We consider best approximation on E to $f(z)$ by polynomials $p(z)$ of degree $m+n-2$ which take the prescribed values in the points of E_0.*

Let the polynomial $P(z)$ of degree $m-1$ take the values β_j in the points α_j; it is well known that $P(z)$ exists and is unique. We set $\omega_0(z) \equiv (z-\alpha_1)(z-\alpha_2) \cdots (z-\alpha_m)$. Then any polynomial $p(z)$ with the required properties can be written as

$$p(z) \equiv P(z) + \omega_0(z)q(z),$$

where $q(z)$ is a suitable polynomial of degree $n-2$. Moreover, if $q(z)$ is an arbitrary polynomial of degree $n-2$, this equation defines a polynomial $p(z)$ of the class prescribed.

The problem of best approximation is to minimize

$$\max [\mu(z) \mid f(z) - p(z) \mid, z \text{ on } E]$$
$$= \max \left[\mu(z) \mid \omega_0(z) \mid \left| \frac{f(z) - P(z)}{\omega_0(z)} - q(z) \right|, z \text{ on } E \right];$$

thus our problem has been reduced to the problem of approximating on E to the function $[f(z)-P(z)]/\omega_0(z)$ by an arbitrary polynomial $q(z)$ of degree $n-2$, with the positive weight function $\mu(z)|\omega_0(z)|$. All results of §§1–9 can then be applied, even without our special discussion of general linear families of functions.

As a concluding remark we stress the possibility of extensions of our results to approximation by very general nonlinear families. A natural tool for the investigation of nonlinear families of functions is the linear family obtained by differentiating, in the neighborhood of an approximating function, with respect to the parameters of the family. It should turn out, as directly established [5] for certain families of functions of one real variable, that the behavior of the minimizing function on the given set is essentially the same as in the linear case.

References

1. Ch.-J. de la Vallée Poussin, *Sur les polynomes d'approximation à une variable complexe*, Bulletin de l'Académie Royale de Belgique (classe des Sciences) vol. 3 (1911) pp. 199–211.

2. L. Fejér, *Über die Lage der Nullstellen von Polynomen, die aus Minimumforderungen gewisser Art entspringen*, Math. Ann. vol. 85 (1922) pp. 41–48.

3. M. Fekete, *On the structure of extremal polynomials*, Proc. Nat. Acad. Sci. U.S.A. vol. 37 (1951) pp. 95–103.

4. M. Fekete and J. L. von Neumann, *Über die Lage der Nullstellen gewisser Minimumpolynome*, Jber. Deutschen Math. Verein. vol. 31 (1922) pp. 125–138.

5. Th. Motzkin, *Approximation by curves of a unisolvent family*, Bull. Amer. Math. Soc. vol. 55 (1949) pp. 789–793.

6. J. L. Walsh, *On the location of the roots of certain types of polynomials*, Trans. Amer. Math. Soc. vol. 24 (1922) pp. 163–180; Theorem IVa.

7. ———, *A certain two-dimensional locus*, Amer. Math. Monthly vol. 29 (1922) pp. 112–114.

National Bureau of Standards, Los Angeles,
University of California, Los Angeles, and
Harvard University

PACIFIC JOURNAL OF MATHEMATICS
Vol. 44, No. 1, 1973

EQUILIBRIUM OF INVERSE-DISTANCE FORCES IN THREE-DIMENSIONS

T. S. MOTZKIN AND J. L. WALSH

This paper is about the field of force in three dimensions due to particles that repel according to the inverse-distance law, the analog of the field in the complex plane used by C. F. Gauss in studying the zeros of the derivative of a polynomial. In particular it deals with the analog in three dimensions of Jensen's theorem and related theorems in the plane, shows that the positions of equilibrium in the field of force when the total mass is zero are invariant under spherical transformation, and gives a geometric construction to find the force due to an axially symmetric distribution of matter.

To be more explicit, it is classical that C. F. Gauss indicated that in the complex plane the zeros of the derivative $p'(z)$ of a polynomial $p(z)$ (not multiple zeros of $p(z)$) are precisely the positions of equilibrium in the field of force due to particles at the zeros of $p(z)$, where each particle z_0 has as mass the multiplicity of z_0 as a zero of $p(z)$, and where each particle repels with a force equal to its mass times the inverse distance. Gauss's theorem has been widely used in the study of the location of the zeros of $p'(z)$ by F. Lucas, Bôcher, Jensen, and later writers [see for instance Marden, 2; Walsh, 6]. Also in three-dimensions, G. Sz. Nagy [4] has introduced the concept of *distance polynomial*, namely as

$$F(x, y, z) \equiv C \prod_{1}^{n} [(x - x_k)^2 + (y - y_k)^2 + (z - z_k)^2] , \quad C > 0 ,$$

whose *derivative* is defined

$$F'(x, y, z) \equiv \sum_{1}^{n} (F_x^2 + F_y^2 + F_z^2)/4F ,$$

when the n points P_k: (x_k, y_k, z_k) are given. Sz. Nagy then sets up the Gaussian field of force, namely a set of (possibly multiple) particles at the respective zeros of $F(x, y, z)$, where each particle repels with a force equal to its mass times the inverse distances; each zero of $F'(x, y, z)$ not a multiple zero of $F(x, y, z)$ is then a position of equilibrium in the corresponding field of force.

This field of force may now be regarded as an appropriate object of independent study, and is so regarded in the present note. New results are obtained beyond those already in the literature. Applications to distance polynomials are immediate, and are left to the reader.

F. Lucas's theorem asserts that *the smallest convex set of the plane containing the zeros of a nontrivial polynomial $p(z)$ contains also the zeros of $p'(z)$.* The analog clearly holds in higher dimension, as Sz. Nagy indicates, and the results that follow are complementary to the analog of Lucas's theorem.

1. An extension of Jensen's theorem. We recall the statement of Jensen's theorem:

THEOREM 1 (*Jensen's theorem*). *Let $p(z)$ be a nonconstant real polynomial, let all circles (Jensen circles) be constructed, each having as diameter the line segment joining a conjugate pair of zeros of $p(z)$. Then each nonreal zero of the derivative $p'(z)$ lies in the closed interior of a Jensen circle.*

The first published proof of Jensen's theorem was given by Walsh [6], who also laid the foundation for the deeper study of the zeros of the derivative of a real polynomial. Under the hypothesis of Jensen's theorem, the force in the Gaussian field due to two equal conjugate imaginary particles has [Walsh, 7, §1.4.1] a nonzero component toward the axis of reals, parallel to that axis, or away from that axis at every non-real point respectively interior to the Jensen circle, on the Jensen circle, or exterior to the Jensen circle. Thus at each non-real point P exterior to all Jensen circles the force due to each conjugate pair of particles has a nonzero component away from the axis of reals, and at each such point P the force due to each particle at a real zero has also such a nonzero component, so the total force at P is not zero. Thus P is neither a point of equilibrium in the field of force nor a multiple zero of $p'(z)$, hence is not a zero of $p'(z)$, and Jensen's theorem is established.

This same proof gives at once:

THEOREM 2. *In three dimensions let there be a finite number of positive Gaussian particles situated symmetrically with respect to the (x, y)-plane, and let all (Jensen) spheres be constructed whose diameters are line segments joing pairs of points symmetric in the (x, y)-plane. Then each position of equilibrium lies in the closed interior of some Jensen sphere.*

Theorem 2 is given by Sz. Nagy [4].

Another result in three dimensions is analogous to a result [Walsh, 7, §2.2.2, Theorem 3] in the plane:

THEOREM 3. *In three dimensions, let a positive distribution of total mass n consist of two particles* $(0, 0, \pm 1)$ *each of mass k, and particles of total mass* $n - 2k$ *in the* (x, y)*-plane. Then each equilibrium point lies either in the* (x, y)*-plane or in the closed sphere whose center is the origin and radius* $[(n - 2k)/n]^{1/2}$.

In the proof we shall need

LEMMA 1. *If the force at a point P due to m variable unit particles in a plane* π *not containing P has the direction of a line* λ *through P then the force is not greater than it would be if all m particles were concentrated at the intersection of* π *and* λ.

Denote by π' the inverse of π in the unit sphere S whose center is P. The force at P due to a particle at Q is in magnitude, direction, and sense the vector $Q'P$, where Q' is the inverse of Q in S. The force at P due to the m given particles is equivalent to the force at P due to an m-fold particle on λ situated at the inverse Q_0 of the center of gravity Q_0' of the points Q', inverses of the given particles Q on π. Since the points Q lie on π, their inverses Q' lie in π', and Q_0' lies on or within π' and on λ. The greatest force $Q_0'P$ is exerted if Q_0' lies at the intersection (other than P) of π' and λ, which implies that all m particles are concentrated at the intersection of π and λ.

The present Lemma 1 combined with [Walsh, 7, Lemma 2, § 2.2.2] and the method of proof of [Walsh, 7, § 2.2.2, Theorem 3] now establishes the present Theorem 3.

2. Second extension of Jensen's theorem. We turn now to an analog of [Walsh, 7, § 2.5, Theorem 1]:

THEOREM 4. *Let a positive distribution of mass consist of a pair* $(0, 0, \pm 1)$ *of equal particles, plus a distribution of mass in the* (x, y)*-plane in the disk* $x^2 + y^2 \leqq r$. *Then all positions of equilibrium not in the* (x, y)*-plane lie in the spindle-shaped region whose boundary is formed by the rotation about* $0z$ *of the shorter arc bounded by* $(0, 0, \pm 1)$ *of the circle through those two points tangent at those points to the lines joining those points to* $(r, 0, 0)$.

It will be noticed [Walsh, 7, § 2.2.1, Theorem 1] that the circular arc (except for $z = 0$ or ± 1) is precisely the locus of points of equilibrium in the (x, y)-plane due to the two given particles and to a single particle of suitably variable mass in the point $(x_0, 0, 0)$ as $x_0 > 0$ varies. If P: (x_0, y_0, z), $z_0 \neq 0$, is a point exterior to the spindle, then the line of force at P due to the two given particles lies in the plane $y_0x - x_0y = 0$ and passes through the point $(x_1, y_1, 0)$, where $x_1^2 + y_1^2 > r^2$ (loc. cit.). On the other hand, the line of resultant force at P due to the given particles in the (x, y)-plane cuts that plane in a point $(x_2, y_2, 0)$ where $x_2^2 + y_2^2 \leqq r^2$, for that fact is true for the line of force at P due to each given particle in the (x, y)-plane. Consequently the total force at P is not zero, and Theorem 4 is proved.

By way of contrast to Theorem 4 and to [Walsh, 7, §2.5, Theorem 1] we now indicate that the replacement of the region $x^2 + y^2 \leqq r^2$ in the (x, y)-plane of Theorem 4 by the region $x^2 + y^2 \geqq r^2$ in the (x, y)-plane gives no diminution in the interior of the Jensen sphere (Theorem 2) as the locus of positions of equilibrium. Let P be a point not in the (x, y)-plane interior to the Jensen sphere corresponding to the two equal particles $(0, 0, \pm 1)$; we shall show that P is a position of equilibrium in the field of force due to the those particles and a suitable distribution in the (x, y)-plane satisfying $x^2 + y^2 > r^2$. Choose for definiteness P in the (x, z)-plane, $z > 0$. The force at P due to the two given particles is represented by a certain vector λ in that plane with a nonzero negative z component. The inverse in the sphere of radius unity whose center is P, of the entire (x, y)-plane is a spherical shell S of which P is the point furthest from that plane. The inverse of the region $x^2 + y^2 \geqq r^2$ is a spherical cap of S containing P and bounded by a circle Γ which is the inverse of the circle $x^2 + y^2 = r^2$ in the (x, y)-plane. The spherical cap and the plane of Γ bound a closed region R which is a convex subregion of the closed interior of S. Any set of particles in the region $x^2 + y^2 \geqq r^2$ of the (x, y)-plane have inverses which lie on the spherical cap, and the corresponding force which they exert at P is represented by a single nonzero vector of corresponding multiplicity whose initial point Q in R is the center of gravity of those inverses and whose terminal point is P. Then Q is a point in the convex closed region R, and conversely any point Q of R may be chosen as the initial point of the vector QP by suitably choosing the position and (integral) masses of the distribution in the region $x^2 + y^2 \geqq r^2$ of the (x, y)-plane. In particular the distribution may be so chosen that the corresponding force exerted at P is the negative of λ, so P is a point of equilibrium, as we have asserted.

Theorems 3 and 4 are in a sense mutually complementary; each limits the possible point sets on which the equilibrium points are situated, the one based on magnitude of the masses on and off the (x, y)-plane, the other based wholly on the position of those masses; they both apply in any specific case when the masses at $(0, 0, \pm 1)$ (supposed to be the only masses exterior to the (x, y)-plane) and those in the (x, y)-plane are given.

Of course the analog of Lucas's theorem is available too, for the purpose of further limiting the possible point sets where the equilibrium points are located; thus if a closed dihedral angle less than π contains all masses, it contains also all equilibrium points.

With the hypothesis of Theorem 3, a point P: (x_1, y_1, z_1) not $(0, 0, \pm 1)$ but on the surface of the sphere mentioned, is a point of equilibrium when and only when all the $(n - 2k)$ units of mass in the (x, y)-plane are concentrated at the point of the (x, y)-plane where the tangent line at P to the circle through P and $(0, 0, \pm 1)$ cuts the (x, y)-plane; compare [Walsh, 7, §2.2.1].

Likewise, with the hypothesis of Theorem 4, a point P of the surface of the spindle (other than $(0, 0, \pm 1)$ is an equilibrium point when and only when the mass in the (x, y)-plane is concentrated in a single point in the (x, y)-plane on the circle $x^2 + y^2 = r^2$, in the plane through Oz and P, and that mass has a single suitable value (not necessarily integral) depending on the masses at $(0, 0, \pm 1)$.

3. **Transformation of force field.** The study of the location of critical points of a polynomial may be interpreted as a chapter of circle geometry, and the study of the location of equilibrium points in three dimensions as a chapter of sphere geometry. Circle and sphere transformations may be a convenient tool in the respective theories, as we now indicate. The following theorem is essentially due to Bôcher (1904) for two dimensions; we use the the phraseology for three dimensions.

THEOREM 5. *In any finite number of dimensions, the direction of the force (including sense) in a field due to a finite number of positive and negative particles of total mass zero which repel according to the inverse distance law is invariant under spherical tranformation.*

If such a field of force is given, and the total mass is not zero, we may adjoin suitable particles at infinity so that the total mass becomes zero. These new particles then persist under a spherical transformation that carries the point at infinity into a finite point.

Let P be a point in the field of force due to particles $Q_1, Q_2, \cdots, Q_k$ of respective positive masses $m_1, m_2, \cdots, m_k$. The force at P due to the particle Q_1 is represented in magnitude, direction, and sense by the m_1-fold vector $Q_1'P$, where Q_1' is the inverse of Q_1 in the unit sphere whose center is P. The force at P due to all the particles Q_i is represented by the sum of the vectors $Q_i'P$, of respective multiplicities m_i, where Q_i' is the inverse of Q_i in that same sphere. The total force at P due to the particles Q_i is thus represented by a vector $Q_0'P$ of multiplicity $m_1 + m_2 + \cdots + m_k = M$ where Q_0' is the center of gravity of the weighted points Q_i', so the total force at P due to the particles Q_i is equivalent to the force at P due to a particle of mass M situated at Q_0, the inverse of Q_0' in the same unit sphere.

We introduce the concept of *center of gravity with respect to* P of the weighted points Q_i. This shall be the ordinary center of gravity of the Q_i when P lies at infinity, and otherwise is defined to be invariant under any spherical transformation, in particular under a transformation that carries P to infinity and carries the Q_i into finite points. It will be noted that the point Q_0 previouly defined is precisely the center of gravity with respect to P of the given points Q_i.

If now the field of force of Theorem 5 is due to k positive particles at points Q_i of respective masses m_1, $m_2, \cdots, m_k$, and to j negative particles R_i of respective masses $n_1, n_2, \cdots, n_j$ $(n_i < 0)$, where $m_1 + m_2 + \cdots + m_k = M = -n_1 - n_2 - \cdots - n_j$, the total force at P is the force due to an M-fold particle situated at the center of gravity Q_0 with respect to P of the Q_i, plus the force exerted at P by a $(-M)$-fold particle situated at the center of gravity R_0 with respect to P of the R_i. The total force at P is thus the force at P due to the positive particle of mass M at Q_0 plus the force at P due to the negative particle of mass $(-M)$ at R_0, which is known [Walsh, 7, §4.1.2] to have the direction and sense at P of the circular arc Q_0PR_0, on which P separates Q_0 and R_0. Infinity is defined to be a position of equilibrium if it can be carried into such a finite position under spherical transformation of the given masses. The direction of the total force at P (which is zero if and only if Q_0 and R_0 coincide) is thus expressed in a form invariant under spherical transformation, and Theorem 5 is established.

An equivalent statement of the theorem is that the lines of force (including sense) are invariant under spherical transformation. Thus the surfaces orthogonal to the lines of force are also invariant. We have too the

COROLLARY. *Under the conditions of Theorem* 5, *positions of equilibrium are invariant under spherical transformation of space.*

Positions of equilibrium are points different from the given particles where the force is zero, hence where direction of the force is not defined.

It is obvious that the theorem extends to distributions of matter not in discrete particles, but such that the field of force can be approximated by a field due to such particles. Proofs of Theorem 5 for stereographic projections of the plane are given not merely by Bôcher but also by Marden [2, §11] and Walsh [7, §4.1.2].

4. Axial symmetry. As an application of Theorem 5, we prove

THEOREM 6. *With force proportional to inverse distance in three dimensions, let C be a uniform circumference of center* 0, *radius ρ, and total mass unity in a plane π, and let the axis of C be α. Then the total force at a point P of π interior to C is zero, that at a point P of π exterior to C is $1/\overline{OP}$, and that at a point not in π nor α has a nonzero component sensed away from α.*

At a point P in π, the logarithmic potential due to mass C is a constant or $\log \overline{OP}$ plus a constant according as P is interior or exterior to C, for the potential is harmonic and behaves like $\log \overline{OP}$ when $\overline{OP}$ is sufficiently large; thus the force at P is as asserted. If now P lies on one side of π, let O_1 be the point of α farthest from π on the sphere through P and C and separated by π from P. Let the line O_1P cut π at some point P_1 interior to C. The total force, at P is due to C of mass unity and to a particle at infinity of mass minus unity. Invert the space in the sphere S of center O_1 passing through C. Then O_1, C, P, S are transformed into ∞, C, P_1, and π respectively. By Theorem 5, the force at P_1 due to the transformed mass is proportional to the force (zero) at P_1 due to C, plus the force at P_1 due to the (negative) particle at O_1, in direction and sense P_1O_1. Then the original force (reversed in sense by inversion) exerted at P has direction and sense O_1P, hence has a nonzero component at P directed away from α, so Theorem 6 is established.

The proof just given determines the equivalent particle, preserving the force exerted by C at P, hence the components of that force too are immediately determined by construction.

In the plane π_1 through α and the same point P, the line PO_1 bisects the angle at P between the two chords PC_1 and PC_2 from P through C in π_1, where C_1 and C_2 are distinct and lie in π_1 on C. Thus the line of force at P has the direction of a hyperbola in π_1 with foci C_1 and C_2. We state the

COROLLARY 1. *If a distribution of positive mass M has axial symmetry in a line α and lies in a plane π orthogonal to α, then there is no position of equilibrium except in π; at any point not on α nor π the force has a nonzero component away from α. There is no position of equilibrium in π except interior simultaneously to all the circumferences in M having α as axis.*

COROLLARY 2. *If a distribution of positive mass M has axial symmetry in a line α and lies in more than one plane orthogonal to*

α, let the planes be in order $\pi_1, \pi_2, \cdots, \pi_\nu$. Then there is no position of equilibrium on π_1 or π_ν. Except on α there is no position of equilibrium, for the total force has a nonzero component directed away from α.

It is obvious that if a finite number of circular wire distributions of uniform positive masses lie in a plane, and if the corresponding circular discs have a region D in common, then each point of D is a position of equilibrium.

Under the conditions of Theorem 6, each line of force is a hyperbola whose axes are α and a line in π through α, with force on C, which may degenerate to α or a line in π through O. The orthogonal surfaces to the lines of force are oblate ellipsoids with α as axis of revolution; a degenerate case is the disk of C.

There are naturally numerous other geometric results which correspond to known and even to new theorems in the plane, compare [Walsh, 7, Ch. 3]. As simple examples corresponding to new theorems in the plane we mention:

With the inverse distance law, let pairs of equal positive particles mutually inverse in the (x, y)-plane lie on or interior to the sphere, $x^2 + y^2 + z^2 = 1$, and perhaps other positive particles lie in the (x, y)-plane. Then all position of equilibrium lie in the closed interior of the ellipsoid $x^2 + y^2 + 2z^2 = 2$.

With the same law, let pairs of equal positive particles mutually inverse in the (x, y)-plane lie on or interior to the circumference $x^2 + z^2 = 1$, $y = 0$, and perhaps other positive particles lie in the (x, y)-plane. Then all positions of equilibrium lie in the closed interior of the ellipsoid $x^2 + 2y^2 + 2z^2 = 2$.

In these respective cases, all closed Jensen spheres for the given particles lie in the closed ellipsoids defined. Indeed, the new theorems just proved determine the actual *loci* of the position of equilibrium under the several hypotheses on the given particles, now no longer required to be of integral masses.

These ellipsoids are the analogs of the Jensen ellipses, first introduced by J. L. W. V. Jensen. For proofs of the theorems compare [Walsh, 7, §3.8]. The former of these theorems is valid for several given spheres containing particles, and several ellipsoids similar to $x^2 + y^2 + 2z^2 = 2$ are constructed; other positive particles are permitted in the (x, y)-plane. Likewise the latter of these theorems is valid for several given circumferences whose centers lie in the (x, y)-plane, and ellipsoids are constructed similar to $x^2 + 2y^2 + 2z^2 = 2$. Here of course the word *similar* is used in the technical sense.

We remark that results in the plane concerning critical points of rational functions [Walsh, 7, Ch. 4 and 5] have as analogs in higher spaces results involving both positive and negative distributions of matter.

5. Extrema of potential. Inverse-distance forces in any number d of dimensions correspond to the logarithmic potential $\log r$. This potential for a positive mass distribution and exterior to it, is superharmonic for $d = 1$, harmonic for $d = 2$, and subharmonic for $d > 2$. Hence its critical points (positions of equilibrium in the corresponding force field) cannot be strong minima for $d \leqq 2$ nor strong maxima for $d \geqq 2$, but may be weak minima or weak maxima for $d = 2$. For $d = 1$ two positive mass-points produce a maximum on the segment joining them; for two mass-points of unequal sign not of total mass zero, an extremum occurs outside the segment. For a homogeneous circular wire of positive mass in three-space each point of the closed disk bounded by the wire gives a weak global minimum of the potential.

To obtain a strong minimum of the potential for $d = 3$ consider the vertices $(\pm 1, 0, 0)$, $(0, \pm 1, 0)$, $(0, 0, \pm 1)$ of a regular octahedron as point unit masses. The potential at (x, y, z) is

$$H(x, y, z) \equiv \frac{1}{2} \log \{[(x-1)^2 + y^2 + z^2] \cdot [(x+1)^2 + y^2 + z^2] \cdots\}$$

$$\equiv \frac{1}{2} \log [(x^2 + y^2 + z^2 + 1) - 4x^2] + \cdots ,$$

so the first partial derivative of the potential is

$$H_x \equiv \frac{2x(x^2 + y^2 + z^2 + 1) - 4x}{(x^2 + y^2 + z^2 + 1)^2 - 4x^2} + \frac{2x(x^2 + y^2 + z^2 + 1)}{(x^2 + y^2 + z^2 + 1)^2 - 4y^2}$$

$$+ \frac{2x(x^2 + y^2 + z^2 + 1)}{(x^2 + y^2 + z^2 + 1)^2 - 4z^2} .$$

Hence $H_{xx}(0, 0, 0) = (2 - 4) + 2 + 2 = 2$, $H_{xy}(0, 0, 0) = 0$, $H_{xz}(0, 0, 0) = 0$. Thus H has a strong minimum at $(0, 0, 0)$.

Since $H(x, 0, 0)$, $0 \leqq x \leqq 1$, has a local minimum at $x = 0$ and is $-\infty$ at $x = 1$, it must have a maximum at some intermediate point. It is easy to verify (by the formula for H_x) that such a maximum occurs for $x^2 = 1/3$ and only there. Obviously this is a point of equilibrium of the force-field, yet not an extremum of the potential, for the latter is subharmonic.

For a mass distribution which is spherically symmetric with respect to a point 0 and lies exterior to some sphere with center 0, the point 0 must be an extremum of the potential. Hence for $d = 2$, the potential, being harmonic, is constant in a neighborhood of 0. In the case of a positive mass distribution, 0 is necessarily a local maximum of the (superharmonic) potential if $d = 1$, and a local minimum if $d \geqq 3$.

To study the force-field generated by a homogeneous distribution of total mass one on the surface of the unit sphere, consider a point $(0, 0, a)$, $0 < a \neq 1$. The vertical component of the force exerted at this point is

$$\frac{1}{4\pi}\int_S \frac{a - z}{x^2 + y^2 + (a - z)^2} dS = \frac{1}{4\pi}\int_{-\pi}^{\pi}\int_{-\frac{\pi}{2}}^{\frac{\pi}{2}} \frac{a - \sin\phi}{a^2 + 1 - 1 - 2a\sin\phi} \cos\phi \, d\phi \, d\psi$$

$$= \frac{1}{2}\int_{-1}^{1} \frac{a - u}{a^2 + 1 - 2au} du$$

$$= \frac{1}{2a} - \frac{a^2 - 1}{8a^2} \log\left(\frac{a^2 + 1}{2a}\right)\Big|_{-1}^{1}$$

$$= \frac{1}{2a} - a^2 - 1 \log \frac{|1 - a|}{1 + a} .$$

For $a < 1$, this force is

$$\frac{a}{1\cdot 3} + \frac{a^3}{3\cdot 5} + \frac{a^5}{5\cdot 7} + \cdots .$$

Thus when a increases from zero to one, this force increases from zero to one-half. The potential has a single minimum, namely at the center of the sphere. Both force and potential are continuous throughout space; this is in contrast to potential and force due to a circular wire, where the force is not continuous.

References

1. M. Marden, *A generalization of a theorem of Bôcher*, SIAM Journal of Numerical Analysis, **3** (1966), 269-275.
2. ———, *The geometry of the zeros of a polynomial in a complex variable*, Math. Surveys, vol. III. Amer. Math. Soc., 1949.
3. G. v. Sz. Nagy, *Zur Theorie der algebraischen Gleichungen*, Jber. d. deutschen Math.-Verein., **31** (1922), 238-251.
4. ———, *Über die Lage der Nullstellen eines Abstandspolynoms und seiner Derivierten*, Bull. Amer. Math. Soc., **55** (1949), 329-342.
5. A. Schurrer, *On the location of the zeros of the derivative of rational functions of distance polynomials*, Trans. Amer. Math. Soc., **89** (1958), 100-112.
6. J. L. Walsh, *On the location of the roots of the derivative of a polynomial*, Annals of Math., **22** (1920), 128-144.
7. ———, *The location of critical points of analytic and harmonic functions*, Amer. Math. Soc. Coll. Pub., **34** (1950).

Received August 10, 1971. Sponsorship: Research supported (in part) by the U. S. Air Force Office of Scientific Research, under Contract AFOSR 1690-69.

UNIVERSITY OF CALIFORNIA, LOS ANGELES
AND
UNIVERSITY OF MARYLAND

NOTE ON THE ORTHOGONALITY OF TCHEBYCHEFF POLYNOMIALS ON CONFOCAL ELLIPSES*

BY J. L. WALSH

In the study of polynomial expansions of analytic functions in the complex plane, two different definitions of orthogonality are current:

$$(1)\qquad \int_C n(z)p_j(z)p_k(z)\,dz = 0, \text{ or } \int_C p_j(z)q_k(z)dz = 0, \quad (j \neq k),$$

$$(2)\qquad \int_C n(z)p_j(z)\overline{p_k(z)}\,|\,dz\,| = 0, \qquad (j \neq k).$$

Definition (1) in one form or the other (the second form of (1) may be called biorthogonality) is of frequent use, for instance in connection with the Legendre polynomials,† and has the great advantage that if the functions involved are analytic, the contour of integration C may be deformed without altering the orthogonality property. Definition (2) is of importance—indeed inevitable—when one wishes to study approximation on C in the sense of least squares, and it is entirely with definition (2) that we shall be concerned in the present note. More explicitly, condition (2) may be described as orthogonality with respect to the norm function $n(z)$, ordinarily chosen as continuous and positive or at least non-negative on C.

An illustration of (1), where C is the unit circle $|z| = 1$, is the set of functions $1, z, z^2, \cdots, n(z) \equiv 1$. An illustration of (2), where C is the unit circle, is the set of functions $\cdots, z^{-2}, z^{-1}, 1, z, z^2, \cdots, n(z) \equiv 1$:

$$\int_C z^j \bar{z}^k\,|\,dz\,| = \int_C \frac{z^j}{z^k}\frac{dz}{iz} = 0, \qquad (j \neq k).$$

The connection of orthogonality in the sense of (2) with ap-

* Presented to the Society, October 28, 1933.

† The reader may refer to Heine, *Kugelfunktionen*, 1878; Darboux, Journal de Mathématiques, (3), vol. 4 (1878), pp. 5–56, and pp. 377–416; Geronimus, Transactions of this Society, vol. 33 (1931), pp. 322–328.

proximation in the sense of least squares has long been known.* The application to polynomial expansions on curves in the complex plane is due to Szegö† in the case $n(z)\equiv 1$ and to Walsh‡ in the more general case. The most important theorem which concerns us here is the following, which is due to Szegö ($n(z)\equiv 1$, C analytic), Smirnoff ($n(z)\equiv 1$, C rectifiable and subject to auxiliary condition), and Walsh, loc. cit., ($n(z)$ positive and continuous, C rectifiable).

Let C be a rectifiable Jordan curve and let the function $w=\phi(z)$ map the exterior of C onto the exterior of $|w|=1$ so that the points at infinity correspond to each other. Let the curve $|\phi(z)|=R>1$ be generically denoted by C_R.

If the function $f(z)$ is analytic interior to C_ρ but has a singularity on C_ρ, and if

$$(3)\qquad \begin{gathered} f(z)\sim \sum a_k p_k(z),\\ a_k\int_C n(z)p_k(z)\overline{p_k(z)}\,|dz| = \int_C n(z)f(z)\overline{p_k(z)}\,|dz|, \end{gathered}$$

is the formal expansion of $f(z)$ in terms of the polynomials $p_k(z)$ of respective degrees k orthogonal on C with respect to the function $n(z)$ positive and continuous on C, then series (3) converges to $f(z)$ interior to C_ρ, uniformly on any closed point set interior to C_ρ, and converges uniformly in no region containing in its interior a point of C_ρ.

There is no series other than (3) of the form $\sum b_k p_k(z)$ which converges to $f(z)$ uniformly on C.

This theorem is valid also in the limiting case that C is a line segment or other rectifiable Jordan arc. The case that C is a line segment has long been studied and by numerous writers; for instance the weight function $n(z)\equiv 1$ leads to expansions in Legendre polynomials treated for complex values of the argument by C. Neumann in 1862. When C is a line segment, the curves C_R are ellipses whose foci are the ends of the segment.

The theorem naturally raises the question as to whether (3) can be both the formal expansion of $f(z)$ on C and the formal expansion of $f(z)$ on some $C_{\rho'}$, $\rho'<\rho$, or in other words whether

* See for instance Kowalewski, *Determinantentheorie*, 1909, p. 335.

† Mathematische Zeitschrift, vol. 9 (1921), pp. 218–270.

‡ Transactions of this Society, vol. 32 (1930), pp. 794–816.

the same set of polynomials $p_k(z)$ can result from orthogonalization (in the sense corresponding to (2)) of the set $1, z, z^2, \cdots$, on two different curves. An obvious illustration is the case that C is the unit circle. The polynomials $1, z, z^2, \cdots$ are orthogonal with weight function unity on *every* circle whose center is the origin, and the formal expansion (3) of a function $f(z)$ analytic at the origin is the same on every such circle containing on or within it no singularity of $f(z)$.

It is the object of the present note to show that *the Tchebycheff polynomials found by orthogonalization of* $1, z, z^2, \cdots$ *on the line segment* $C: -1 \leqq z \leqq +1$ *with respect to the norm function* $(1-z^2)^{-1/2}$ *are also orthogonal with respect to suitable norm functions on all the corresponding curves* C_R, *which are ellipses with the common foci* $(-1, +1)$.

We shall find it more convenient to transform our problem to the w-plane. The exterior of $C: -1 \leqq z \leqq +1$ is transformed onto the exterior of $\gamma: |w| = 1$ by the transformation

$$z = \frac{1}{2}\left(w + \frac{1}{w}\right), \tag{4}$$

so that the points at infinity correspond to each other. Let the polynomials $p_0(z) \equiv 1, p_1(z), p_2(z), \cdots$ result from orthogonalizing on γ the linearly independent set $1, z, z^2, \cdots$ with norm function unity. We shall prove that the polynomials $p_k(z)$ form an orthogonal set on $\Gamma: |w| = R > 1$ with norm function unity. For the sake of reference we write the equations

$$\begin{aligned} z &= \frac{1}{2}\left(w + \frac{1}{w}\right), \quad z^2 = \frac{1}{4}\left(w^2 + 2 + \frac{1}{w^2}\right), \\ z^3 &= \frac{1}{8}\left(w^3 + 3w + \frac{3}{w} + \frac{1}{w^3}\right), \cdots, \end{aligned} \tag{5}$$

whence, on Γ,

$$\begin{aligned} \bar{z} &= \frac{1}{2}\left(\frac{R^2}{w} + \frac{w}{R^2}\right), \quad \bar{z}^2 = \frac{1}{4}\left(\frac{R^4}{w^2} + 2 + \frac{w^2}{R^4}\right), \\ \bar{z}^3 &= \frac{1}{8}\left(\frac{R^6}{w^3} + \frac{3R^2}{w} + \frac{3w}{R^2} + \frac{w^3}{R^6}\right), \cdots. \end{aligned} \tag{6}$$

We need merely prove that the polynomial $p_n(z)$ is orthogonal to each of the polynomials $p_0(z)$, $p_1(z)$, $\cdots$, $p_{n-1}(z)$, and it is sufficient to show that $p_n(z)$ is orthogonal to each of the functions $1, z, \cdots, z^{n-1}$. We shall prove this by induction.

We have by hypothesis

$$\int_\gamma p_n(z)\,|dw| = 0, \tag{7}$$

and we show first that we have also

$$\int_\Gamma p_n(z)\,|dw| = 0. \tag{8}$$

Direct computation gives us $|dw| = R\,dw/(iw)$ for w on Γ. When this substitution is made in (7) and (8), it is seen that the integral in (8) is except for the factor R precisely the integral in (7); the function $p_n(z)$ has no singularities in the w-plane except at the origin and at infinity. Hence equation (8) follows at once.

Let us now suppose

$$\int_\Gamma p_n(z)\,|dw| = 0, \quad \int_\Gamma p_n(z)\bar{z}\,|dw| = 0, \cdots, \tag{9}$$

$$\int_\Gamma p_n(z)\bar{z}^{k-1}\,|dw| = 0, \qquad (k-1 < n-1).$$

We are to prove that

$$\int_\Gamma p_n(z)\bar{z}^k\,|dw| = 0. \tag{10}$$

From inspection of equations (6) and by virtue of equations (9), it follows that the integral in (10) can be written

$$\frac{1}{2^k}\int_\Gamma p_n(z)\left(\frac{R^{2k}}{w^k} + \frac{w^k}{R^{2k}}\right)\frac{R\,dw}{iw}. \tag{11}$$

If we consider the various terms in $p_n(z)$ as expressed by means of (5), and omit such of those terms as obviously make no contribution to the integral (11), we see that (11) can be written

$$A_{nk}\int_\Gamma \left(w^k + \frac{1}{w^k}\right)\left(\frac{R^{2k}}{w^k} + \frac{w^k}{R^{2k}}\right)\frac{R\,dw}{iw} \tag{12}$$

$$= 2\pi A_{nk}R\left[R^{2k} + \frac{1}{R^{2k}}\right],$$

where A_{nk} is a suitable constant independent of R, easy to compute in terms of the coefficients of the powers of z in $p_n(z)$. By hypothesis the polynomials $p_n(z)$ are orthogonal on the circle γ. Hence the integrals corresponding to (10), (11), and (12) vanish for $R=1$. Thus we have $A_{nk}=0$, so (10) is established and the orthogonality on Γ of the set $p_k(z)$ with respect to the norm function unity is completely proved.

It remains to study the norm function in the z-plane, and to identify our present polynomials with the polynomials of Tchebycheff. From (4) we have for $|w|=R$,

$$dz = \frac{1}{2}\left(1 - \frac{1}{w^2}\right) dw, \qquad \left| \frac{dw}{dz} \right| = \frac{2R}{\left| w - \dfrac{1}{w} \right|} = \frac{R}{|(1 - z^2)^{1/2}|} .$$

The circle $|w|=1$ corresponds to the segment $-1 \leqq z \leqq +1$ (counted twice, or in the study of orthogonality, only once if we prefer, for the norm function is single-valued on the segment). We clearly have

$$\int_{-1}^{1} n(z) p_n(z) \overline{p_k(z)} \, | dz | = \int_{\gamma} n(z) p_n(z) \overline{p_k(z)} \, | dw | \left| \frac{dz}{dw} \right| ,$$

so the corresponding norm function on the segment $-1 \leqq z \leqq +1$ is $1/|(1-z^2)^{1/2}|$. The norm function on an arbitrary ellipse whose foci are $+1$ and -1 is $R/|(1-z^2)^{1/2}|$, where the ellipse is represented by $|z-1|+|z+1|=R+1/R$, $R>1$. The norm function on any curve can be modified by any non-vanishing constant factor, so if we prefer we can still express the norm function on any ellipse C_R as $|1-z^2|^{-1/2}$.

In the present note we have studied orthogonality on a particular set of confocal ellipses; a simple transformation yields the analogous results for orthogonality of corresponding polynomials on any set of confocal ellipses.

The writer is not aware of any case other than the present one, where orthogonalization in the sense of (2) of the set $1, z, z^2, \cdots$, on a curve C_1 with respect to a norm function $n_1(z)$, is equivalent to orthogonalization of that set on another curve C_2 with respect to a norm function $n_2(z)$, except where C_1 and C_2 are concentric circles.

HARVARD UNIVERSITY

NOTE ON THE COEFFICIENTS OF OVERCONVERGENT POWER SERIES

J. L. WALSH

M. B. Porter gave the first known example of an overconvergent power series, that is to say, of a power series in the complex variable with finite radius of convergence such that a suitable sequence of partial sums converges uniformly in a region containing in its interior both points inside and points outside the circle of convergence. Bourion has recently published[1] a general exposition of the theory of overconvergence to which the reader is referred for further historical and technical details.

Ostrowski established the surprising result that a power series $\sum_{n=0}^{\infty} a_n z^n$ of which the partial sums $s_{m_k} \equiv \sum_{n=0}^{m_k} a_n z^n$ exhibit overconvergence, can be expressed as the sum of a power series $\sum_0^{\infty} a_n' z^n$ with a larger radius of convergence and a power series of the form

$$\sum_0^{\infty} a_n'' z^n, \qquad a_n'' = 0, \quad \text{whenever } m_k < n \leqq n_k \tag{1}$$

where n_k and λ are suitably chosen, with $m_k < \lambda n_k$, $0 < \lambda < 1$. Here we have $a_n = a_n' + a_n''$, $a_n' \cdot a_n'' = 0$; the partial sums $s_{m_k}''(z) = \sum_{n=0}^{m_k} a_n'' z^n$ of (1) also exhibit overconvergence.

It is the object of the present note to employ methods already known in the literature to make Ostrowski's result slightly more precise, especially to indicate that in series (1) the gaps cannot be uniquely defined with abrupt initial and terminal elements impossible of alteration by Ostrowski's process of writing the series as the sum of a series with a larger radius of convergence and a series with larger gaps which exhibits overconvergence. The moduli of the coefficients a_n'' must taper off gradually before the gap (m_k, n_k), and must increase gradually after the end of the gap; this remark is to be understood first in the sense that there is an upper limit to the moduli of the coefficients near the ends of a gap, a limit which increases as one moves away from the gap.

Presented to the Society April 27, 1940, under the title *Note on overconvergent power series*; received by the editors January 30, 1941, and, in revised form July 26, 1941.

[1] *L'Ultraconvergence dans les Séries de Taylor*, Actualités Scientifiques et Industrielles, no. 472, Paris, 1937.

THEOREM 1. *Let the series*

$$(2)\qquad \sum_{n=0}^{\infty} c_n z^n$$

whose radius of convergence is unity:

$$(3)\qquad \limsup_{n\to\infty} |c_n|^{1/n} = 1,$$

have the gaps $(m_1, n_1), (m_2, n_2), \cdots$ *in the sense that* $c_n=0$ *whenever* $m_k<n\leqq n_k$, *and let the sequence of partial sums* $s_{m_k}(z)\equiv\sum_{n=0}^{m_k}c_nz^n$ *exhibit overconvergence. If* $R_0>1$ *is arbitrary, there exists* σ *depending on* R_0 *with* $0<\sigma<1$ *such that*

$$(4)\qquad \limsup_{\mu_k\to\infty}\left[|c_{\mu_k}|,\ \mu_k\leqq m_k\right]^{1/\mu_k}\leqq R_0^{\sigma(\limsup m_k/\mu_k)-1};$$

if $r_0<1$ *is arbitrary, there exists* τ *depending on* r_0 *with* $\tau>1$ *such that*

$$(5)\qquad \limsup_{\nu_k\to\infty}\left[|c_{\nu_k}|,\ \nu_k>n_k\right]^{1/\nu_k}\leqq r_0^{\tau(\liminf n_k/\nu_k)-1}.$$

The only novelty in Theorem 1 is its emphasis on (4) and (5) for the series (2) which overconverges and which possesses gaps, rather than for a series which overconverges and into which gaps may be introduced by Ostrowski's process; compare Bourion loc. cit., chap. 1, §2.

With the general notation

$$f(z)\equiv\sum_0^{\infty}c_nz^n,\qquad s_m(z)\equiv\sum_0^{m}c_nz^n,\qquad r_m(z)\equiv f(z)-s_m(z),$$

Cauchy's inequality yields

$$(6)\quad [\max|s_n(z)|,\ \text{for } |z|=R_0>1]\geqq|c_0|,\ |c_1|R_0,\cdots,\ |c_n|R_0^n;$$

$$(7)\quad [\max|r_n(z)|,\ \text{for } |z|=r_0<1]\geqq|c_{n+1}|r_0^{n+1},\ |c_{n+2}|r_0^{n+2},\cdots.$$

Under the hypothesis of Theorem 1 we have for suitable σ and τ (these inequalities follow from the fact of overconvergence by the use of a suitable harmonic majorant)

$$(8)\qquad \limsup_{m_k\to\infty}[\max|s_{m_k}(z)|,\ \text{for } |z|=R_0]^{1/m_k}=R_0^{\sigma},$$

$$(9)\qquad \limsup_{n_k\to\infty}[\max|r_{n_k}(z)|,\ \text{for } |z|=r_0]^{1/n_k}=r_0^{\tau}.$$

By virtue of (6) and (8) we have

$$\limsup_{m_k\to\infty} \left[\, |c_{\mu_k}| \cdot R_0^{\mu_k},\ \mu_k \leqq m_k \right]^{1/m_k} \leqq R_0^{\sigma}, \tag{10}$$

which implies (4); by virtue of (7) and (9) we have

$$\limsup_{n_k\to\infty} \left[\, |c_{\nu_k}| \cdot r_0^{\nu_k},\ \nu_k > n_k \right]^{1/n_k} \leqq r_0^{\tau}, \tag{11}$$

which implies (5). Theorem 1 is established. The first member of (4) is less than unity so long as we have lim sup $m_k/\mu_k < 1/\sigma$, and the first member of (5) is less than unity so long as we have lim inf $n_k/\nu_k > 1/\tau$.

A further description of the tapering-off of the moduli of the coefficients can be elaborated as follows. Under the conditions of Theorem 1, there exists a sequence c_{p_k} with $\lim_{p_k\to\infty} |c_{p_k}|^{1/p_k} = 1$; if necessary we change the notation of m_k, n_k, p_k so that we have also $m_1 < n_1 < p_1 < m_2 < n_2 < p_2 < \cdots$. It is now more convenient to employ (10) rather than (4); by setting $\mu_k = p_{k-1}$ we find

$$\liminf m_k/p_{k-1} \geqq 1/\sigma;$$

consequently the numbers $m_k - p_{k-1}$ cannot be small relative to p_{k-1}. In a similar manner we find from (11) with $\nu_k = p_k$

$$\limsup n_k/p_k \leqq 1/\tau;$$

consequently the numbers $p_k - n_k$ cannot be small relative to p_k. It will be noticed that with our present notation the moduli of the coefficients do taper off from the $|c_{p_k}|$, at least immediately before and after the gaps, because the second members of (4) and (5) are less than unity for $\mu_k = m_k$ and for $\nu_k = n_k + 1$. But we have not shown, nor is it true, that the moduli of the coefficients necessarily taper off monotonically.

As an application of Theorem 1 we prove (compare Bourion, chap. 2, §4) the following theorem:

THEOREM 2. *Let the series* (2) *have the radius of convergence unity, so that* (3) *is satisfied. Let unity be an isolated limit point of the set* $\{|c_n|^{1/n}\}$. *Let one of the following conditions be satisfied*:

(a) *the series has gaps of relative lengths bounded from zero, in the sense that* $c_n = 0$ *whenever* $m_k < n \leqq n_k$, *with* $m_k < \lambda n_k$, $\lambda < 1$;

(b) *for some* $R_0 > 1$ *and* σ, $0 < \sigma < 1$, *and for some sequence* m_k, *equation* (8) *is valid*;

(c) *for some* $r_0 < 1$ *and* $\tau > 1$ *and for some sequence* n_k, *equation* (9) *is valid.*

Then the unit circle is a natural boundary for the series (2).

If the unit circle is not a natural boundary for the series (2), the function $f(z)$ represented is analytic along some arc A of the unit circle, and Ostrowski has shown that the conditions (a), (b), (c) imply overconvergence of the respective sequences $s_{m_k}(z)$, $s_{m_k}(z)$, $s_{n_k}(z)$ across the arc A.

Let us suppose that no limit point of the sequence $\{|c_n|^{1/n}\}$ other than unity lies in some interval $(1, 1-\eta)$, $\eta>0$; we set

$$c_n' = c_n, \qquad \text{if } |c_n|^{1/n} > 1 - \tfrac{1}{2}\eta,$$
$$c_n' = 0, \qquad \text{if } |c_n|^{1/n} \leqq 1 - \tfrac{1}{2}\eta,$$
$$c_n'' = c_n - c_n',$$

$$f_1(z) \equiv \sum_0^\infty c_n' z^n, \qquad f_2(z) \equiv \sum_0^\infty c_n'' z^n, \qquad f(z) \equiv f_1(z) + f_2(z).$$

The series defining $f_2(z)$ has a radius of convergence greater than unity; any overconvergent sequence for $f(z)$ is an overconvergent sequence for $f_1(z)$; it follows from Theorem 1 that $f_1(z)$ has no overconvergent sequence. Theorem 2 is established.

A necessary condition that a series (2) with (3) satisfied exhibit overconvergence is therefore that unity be a non-isolated limit point of the set $\{|c_n|^{1/n}\}$.

It is instructive in considering Theorem 1 to compare such an example as $\sum[z(z+1)]^{3^n}$, suggested by Bourion as a special case of Porter's original formulas; the function represented has the lemniscate $|z(z+1)|=1$ as a natural boundary; the Taylor development about the origin is convergent in the circle $|z|<\frac{1}{2}5^{1/2}-\frac{1}{2}=0.6$; the coefficients exhibit the characteristics described in Theorem 1.

HARVARD UNIVERSITY

Reprinted from the Proceedings of the NATIONAL ACADEMY OF SCIENCES,
Vol. 37, No. 12, pp. 821–826. December, 1951

NOTE ON APPROXIMATION BY BOUNDED ANALYTIC FUNCTIONS

BY J. L. WALSH

DEPARTMENT OF MATHEMATICS, HARVARD UNIVERSITY

Communicated October 10, 1951

The primary object of this note is to establish a result on the topic indicated in the title:

THEOREM 1. *Let C be an analytic Jordan curve, which together with its interior lies in a region D. Let the sequence of functions* $f_n(z)$ *be analytic in D, satisfy the inequality* $(R > 1)$

$$|f_n(z)| \leqq A_1 R^n, \ z \text{ in } D, \tag{1}$$

and likewise for some function $f(z)$ *satisfy the inequality*

$$|f(z) - f_n(z)| \leqq A_2/n^{p+\alpha}, \ z \text{ on } C, \tag{2}$$

where p is a non-negative integer and we have $0 < \alpha \leqq 1$*; here and below, all numbers* A_j *represent positive constants, independent of n and z. Then the pth derivative* $f^{(p)}(z)$ *satisfies on C a Lipschitz condition of order* α *if we have* $0 < \alpha < 1$*, and is of class* Λ^* *with respect to arc length on C if we have* $\alpha = 1$.

We assume that D is a Jordan region bounded by a curve C_ρ, namely the image of the circle γ_ρ: $|w| = \rho(>1)$ when the exterior of C is mapped conformally by $z = \psi(w)$ onto the exterior of γ: $|w| = 1$, so that the points at infinity correspond to each other; this assumption obviously does not diminish the generality of the proof. Let $P_0(z)$, $P_1(z)$, $P_2(z)$, ..., be the Faber polynomials for C; any function $F(z)$ analytic and bounded interior to C_ρ can be expanded interior to C_ρ:

$$F(z) \equiv \sum_{k=0}^{\infty} a_k P_k(z), \quad a_k = \frac{1}{2\pi i}\int_{\gamma_\rho} \frac{F[\psi(w)]\,dw}{w^{k+1}}, \tag{3}$$

where in the integral the Fatou boundary values are used on γ_ρ. The polynomials $P_k(z)$ are uniformly bounded on C.

Thus if we write

$$f_n(z) \equiv a_{n0}P_0(z) + a_{n1}P_1(z) + a_{n2}P_2(z) + \ldots, \tag{4}$$

we have by (1) and (3)

$$|a_{nk}| \leqq A_1 R^n/\rho^k,$$

and if $S_{n,N}(z)$ denotes the sum of the first $N + 1$ terms of the second member of (4), we have by the boundedness of the $P_n(z)$ on C,

$$|f_n(z) - S_{n,N}(z)| \leqq \sum_{k=N+1}^{\infty} A_3 R^n/\rho^k = A_4 R^n/\rho^N, \ z \text{ on } C. \tag{5}$$

We now choose the positive integer λ so that $\rho^\lambda > R$, whence

$$|f_n(z) - S_{n,\lambda n}(z)| \leqq A_4(R/\rho^\lambda)^n, \; z \text{ on } C, \tag{6}$$

with $R/\rho^\lambda < 1$. From (2) and (6) we may write

$$|f(z) - S_{n,\lambda n}(z)| \leqq A_5/n^{p+\alpha}, \; z \text{ on } C. \tag{7}$$

The polynomials $S_{n,\lambda n}$ of respective degrees λn are not defined for every degree, but we may set

$$p_k(z) \equiv S_{n,\lambda n}(z), \; \lambda n \leqq k < \lambda(n+1).$$

Then the polynomials $p_k(z)$ are defined for all positive integral values of k, we have from (7)

$$|f(z) - p_k(z)| \leqq A_6/k^{p+\alpha}, \; z \text{ on } C, \tag{8}$$

and the conclusion follows[1, 2] from (8).

In determining the polynomials $S_{n,\lambda n}(z)$ of degrees λn in (6) we have used for convenience the Faber polynomial expansion of the given functions $f_n(z)$. Various other polynomial expansions may also be used, not necessarily of the form $\Sigma c_k P_k(z)$ in terms of given polynomials $P_k(z)$. It is sufficient if there can be determined polynomials $S_{n,\lambda n}(z)$ of respective degrees λn approximating on C to an arbitrary function $f_n(z)$ analytic and in modulus not greater than M_n interior to C_ρ, with measure of approximation indicated by the inequality

$$|f_n(z) - S_{n,\lambda n}(z)| \leqq A_7 M_n/\rho^{\lambda n}, \; z \text{ on } C.$$

Numerous kinds of polynomials, defined for instance by interpolation, satisfy this condition;[3] it is essential that C possess suitable continuity properties, but C need not be analytic; Theorem 1 admits a corresponding extension.

To Theorem 1 we add an easily established converse:

THEOREM 2. *Let C be an analytic Jordan curve, which together with its interior lies in a bounded region D. Let $f(z)$ be analytic interior to C, continuous in the corresponding closed region, and let $f^{(p)}(z)$ satisfy on C a Lipschitz condition of order $\alpha(0 < \alpha < 1)$ or be of class Λ^* with respect to arc-length on C $(\alpha = 1)$. Then there exist polynomials $f_n(z)$ of respective degrees n so that (1) and (2) are satisfied.*

The existence of polynomials $f_n(z)$ satisfying (2) is known.[1, 2] For these polynomials we have by (2) and by the boundedness of $f(z)$ on C

$$|f_n(z)| \leqq A_8, \; z \text{ on } C. \tag{9}$$

We choose $\rho(>1)$ so large that, in the notation already introduced, D lies interior to C_ρ. Then (9) implies by the generalized Bernstein lemma[4]

$$|f_n(z)| \leqq A_8\rho^n, \ z \text{ interior to } C_\rho,$$

so (1) follows with $R = \rho$.

The discussion just given also includes the

COROLLARY. *If the polynomials $f_n(z)$ of respective degrees n are given so that (2) is valid, and if D is bounded, then (1) is also valid.*

If C, D, $f(z)$ satisfy the conditions of Theorem 1, and if a positive number M is given, it follows from Montel's theory of normal families *that among the functions analytic and in modulus not greater than M in D, there exists at least one function $F_M(z)$ of best approximation to $f(z)$ on C*, namely, such that

$$\mu_M = \max. [|f(z) - F_M(z)|, z \text{ on } C] \qquad (10)$$

is least. Theorems 1 and 2 are of significance in the study of such functions:

THEOREM 3. *Let D be a bounded region containing in its interior the closed interior of the analytic Jordan curve C, and let the function $f(z)$ be analytic interior to C, continuous in the corresponding closed region. For each positive M, let $F_M(z)$ be the (or a) function analytic and of modulus not greater than M in D, of best approximation to $f(z)$ on C in the sense that μ_M given by (10) is least. Then a necessary and sufficient condition that $f^{(p)}(z)$ satisfy a Lipschitz condition of order $\alpha(0 < \alpha < 1)$ on C or be of class Λ^* $(\alpha = 1)$ for arc-length on C is that*

$$\log M \cdot \mu_M^{1/(p + \alpha)} \qquad (11)$$

be bounded as $M \to \infty$.

If (11) is bounded, we choose the sequence $M = e^n$, whence for the functions $F_M(z)$ and a suitably chosen A_9

$$\mu_M^{1/(p + \alpha)} \leqq A_9/n, \ \mu_M \leqq A_9^{p + \alpha}/n^{p + \alpha},$$

and the conclusion follows from Theorem 1.

Conversely, let $f(z)$ be given with the requisite properties on C. We compare the measure of approximation μ_M with the corresponding measure of approximation for the polynomial $f_n(z)$ of Theorem 2, for specific values of M not less than the modulus of $f_n(z)$ in D. We obviously have $\mu_M \leqq A_2/n^{p + \alpha}$, provided n is defined as a function of M by the inequality $A_1 R^n \leqq M < A_1 R^{n + 1}$, whence

$$\mu_M^{1/(p + \alpha)} \leqq A_2^{1/(p + \alpha)}/n, \ \log M < \log A_1 + (n + 1) \log R,$$

from which it follows that (11) is bounded. Theorem 3 is established.

An alternate form of the condition on (11) is to require that

$$\log \mu_M + (p + \alpha) \log \log M$$

be bounded as M becomes infinite. If (for $0 < \alpha < 1$) the function $f^{(p)}(z)$

satisfies a Lipschitz condition of order α but of no greater order, or (for $\alpha = 1$) if $f^{(p)}(z)$ is of class Λ^* but $f^{(p+1)}(z)$ does not satisfy a Lipschitz condition of any order, *then* $p + \alpha$ *is the greatest number* γ *for which*

$$\log \mu_M + \gamma \log \log M \qquad (12)$$

is bounded as $M \to \infty$, where μ_M is the measure of approximation for the function $F_M(z)$ of best approximation. Indeed, $p + \alpha$ is the greatest number γ for which (12) is bounded as M becomes infinite, not passing through all values but passing through an arbitrary sequence of values of the form $M = A_0R_1^n$, $n = 1, 2, 3, \ldots$.

Alternative (and simpler) proofs of Theorem 1, and hence of Theorem 3 can be given by mapping the *interior* of C onto the interior of $\gamma: |w| = 1$, and making use of the special properties of the Taylor development in the closed interior of γ. Of course for the unit circle the Faber development is identical with the Taylor development, so this alternate proof of Theorem 1 is not greatly different in form from the proof already given. This alternate proof has no immediate analog, however, if we replace the Jordan curve C by a Jordan arc.

Theorems 1, 2 and 3 extend to the case where C is an analytic Jordan arc.

THEOREM 4. *Let C be an analytic Jordan arc, let D be a region containing C, and let the functions $f_n(z)$ be analytic in D and satisfy both (1) and (2). Then $f^{(p)}(z)$ satisfies a Lipschitz condition of order α if $0 < \alpha < 1$ and is of class Λ^* for arc-length on C if $\alpha = 1$, on any closed subarc of C containing no endpoint of C.*

Map C conformally by $z = \varphi(w)$ onto the line segment $S: -1 \leqq w \leqq 1$; conditions (1) and (2) in suitably modified form persist. We approximate in the w-plane on S to the functions $f_n[\varphi(w)]$ by polynomials in w which interpolate to $f_n[\varphi(w)]$ in the zeros of the Tchebycheff polynomials for S. The second (outlined) proof of (6) making use of polynomials defined by interpolation is then valid.[3] Introduce now the classical transformation $w = \cos\theta$ which maps S onto the axis of reals $-\infty < \theta < \infty$ and transforms a polynomial in w of degree n into a trigonometric polynomial in θ of order n. From a theorem due to S. Bernstein and de la Vallée Poussin in the case $0 < \alpha < 1$ and from a theorem due to Zygmund in the case $\alpha = 1$, it now follows that $f[\varphi(\cos\theta)]$ has the requisite properties as a function of θ. The class Λ^* is invariant under suitable transformation,[2] so Theorem 4 follows.

If C is of length $2c$, and arc-length s is suitably measured on C algebraically from the mid-point, it readily follows by a slight extension of the reasoning just given, that the function $f(z) \equiv F(s/c) \equiv F(\cos\omega)$ has on C a pth derivative with respect to ω which satisfies a Lipschitz condition of order $\alpha(0 < \alpha < 1)$ or is of class $\Lambda^*(\alpha = 1)$ with respect to ω. The reciprocal of this reasoning is also valid, including the proof of (1) and (2).

THEOREM 5. *Let C be an analytic Jordan arc which lies in a bounded region D, and let the function f(z) be given on C. For each positive M, let* $F_M(z)$ *be the (or a) function analytic and of modulus not greater than M in D, whose measure of approximation to f(z) on C as given by* (10) *is least. Let C be of length 2c and let arc-length s on C be measured algebraically from the mid-point; we set* $f(z) \equiv F(s/c) \equiv F(\cos \omega)$ *on C. Then a necessary and sufficient condition that* $d^pF(\cos \omega)/d\omega^p$ *satisfy a Lipschitz condition of order* $\alpha(0 < \alpha < 1)$ *or be of class* $\Lambda^*(\alpha = 1)$ *with respect to* ω *is that* (11) *be bounded as M becomes infinite.*

A further result is similar to Theorems 4 and 5:

THEOREM 6. *Let f(x) be a real or complex-valued function of the real variable x, periodic with period* 2π. *Let* $d(>0)$ *be arbitrary but fixed. For each M* (>0), *among the functions analytic and in modulus not greater than M in the region D:* $|y| < d$, *periodic with period* 2π, *let* $F_M(z)$ *denote the (or a) function of best approximation to f(x) on C:* $-\infty < x < \infty$ *in the sense that*

$$\mu_M = lub[|f(x) - F_M(x)|, -\infty < x < \infty]$$

is least. Then a necessary and sufficient condition that $f^{(p)}(x)$ *satisfy a Lipschitz condition of order* α $(0 < \alpha < 1)$ *or be of class* Λ^* $(\alpha = 1)$ *on C is that (11) be bounded as* $M \to \infty$.

The sufficiency of this condition follows from Theorem 4 or Theorem 5. To prove the necessity of the condition, we employ as comparison sequence as in the proof of Theorem 3, the known sequence of trigonometric polynomials $f_n(z)$ of respective orders n which satisfy (2). For this sequence inequality (1) is readily established by setting $w = e^{iz}$, which maps C onto the circumference $|w| = 1$: a trigonometric polynomial in z of order n is a polynomial in w and $1/w$ of degree n, and it is convenient to apply a lemma[5] concerning the latter type of polynomial. Theorem 6 remains valid if D is no longer the region $|y| < d$ but a subregion of the latter containing C, and also remains valid if the $F_M(z)$ are not required to be periodic.

If we modify the hypothesis of Theorem 1 by merely requiring that C be a continuum (not a single point) interior to D, and by replacing the second member of (2) by A_2/R_2^n, $R_2 > 1$, it follows that the sequence $f_n(z)$ converges uniformly and its limit $f(z)$ is analytic in some subregion of D containing C [compare Walsh, *Trans. Am. Math. Soc.*, **47**, 293–304 (1940)]. A similar remark applies to Theorems 4 and 6.

The results of the present note concern Problem α, namely the study of the connection between measure of approximation to $f(z)$ on C and continuity properties of $f(z)$ on C. A further study involves Problem β, the connection between measure of approximation to $f(z)$ on C and continuity properties of $f(z)$ on a related point set such as C_ρ; results on the latter problem will be forthcoming.

[1] Sewell, W. E., "Degree of Approximation by Polynomials in the Complex Domain," *Ann. Math. Studies*, 9 (Princeton, 1942).

[2] Walsh, J. L., and Elliott, H. M., *Trans. Am. Math. Soc.*, **68**, 183–203 (1950). The class Λ^* was first introduced into the theory of approximation by Zygmund.

[3] Walsh, J. L., and Sewell, W. E., *Duke Math. Jnl.*, **6**, 658–705 (1940). See especially Theorems 3.2 and 4.7, and §5.

[4] Walsh, J. L., *Interpolation and Approximation by Rational Functions in the Complex Domain*, New York, 1935, p. 77.

[5] Page 259 of reference 4.

APPROXIMATION BY BOUNDED ANALYTIC FUNCTIONS: UNIFORM CONVERGENCE AS IMPLIED BY MEAN CONVERGENCE[1]

BY

J. L. WALSH

In three recent notes [1], [2], [3] I have discussed uniform convergence by polynomials (in the complex variable) to a given function as a consequence of convergence in the mean of those polynomials to the given function, and also convergence in the mean of one order as a consequence of convergence in the mean of a lower order. The present note contains analogs of those results, but now for approximation by bounded analytic functions. As a first illustration of the new results, we have

THEOREM 1. *Let Γ be an analytic Jordan curve contained in the simply-connected region D of the z-plane, and suppose we have for some function $f(z)$ continuous on Γ and functions $f_n(z)$ analytic in D*

$$\int_\Gamma |f(z)-f_n(z)|^q\,|dz| \leqq A/n^{q\alpha}, \qquad q>0, \tag{1}$$

$$|f_n(z)| \leqq AR^n, \qquad z \text{ in } D. \tag{2}$$

Then for $\alpha+1/p-1/q>0$ and $0<q<p\leqq\infty$ we have for the pth power norm on Γ

$$\|f(z)-f_n(z)\|_p \leqq A/n^{\alpha+(1/p)-(1/q)}. \tag{3}$$

Here and below the constants A are independent of n and z, and may change from one inequality to another.

For $p=\infty$ we consider the first member of (3) as the Tchebycheff (uniform) norm of $[f(z)-f_n(z)]$ on Γ, with a similar interpretation in later formulas. As is usual in the study of convergence by bounded analytic functions, we note (see for instance [4, §2.2]) that there exist for each n and N polynomials $P_{n,N}(z)$ of respective degrees N such that we have

$$|f_n(z)-p_{n,N}(z)| \leqq AR^n/R_1^N, \qquad z \text{ on } \Gamma, \quad R_1>1. \tag{4}$$

If we choose the integer λ so large that $R_1^\lambda > R$, there follow

$$|f_n(z)-p_{n,\lambda n}(z)| \leqq A(R/R_1^\lambda)^n, \qquad z \text{ on } \Gamma, \tag{5}$$

$$\int_\Gamma |f_n(z)-p_{n,\lambda n}(z)|^q\,|dz| \leqq A/n^{q\alpha}. \tag{6}$$

Presented to the Society, August 23, 1966; received by the editors September 9, 1966.

[1] Research sponsored (in part) by the Air Force Office of Scientific Research.

Standard algebraic inequalities depending on q yield by (1) and (6)

$$\int_\Gamma |f(z)-p_{n,\lambda n}(z)|^q\,|dz| \leqq A_0/n^{q\alpha}. \tag{7}$$

The polynomials $p_{n,\lambda n}(z)$ are defined only for the degrees $\lambda n=\lambda, 2\lambda, 3\lambda, \ldots$, but to obtain polynomials $P_m(z)$ for all degrees we may set $P_m(z)=p_{n,\lambda n}(z)$ for $\lambda n \leqq m < \lambda(n+1)$, whence for $m=1, 2, 3, \ldots$

$$\int_\Gamma |f(z)-P_m(z)|^q\,|dz| \leqq \frac{A_0}{n^{q\alpha}} \leqq \frac{A_1}{[\lambda(n+1)]^{q\alpha}} \leqq \frac{A_1}{m^{q\alpha}},$$

provided $A_1 \geqq A_0\lambda^{q\alpha}(n+1)^{q\alpha}/n^{q\alpha}$ for all n. Consequently $f(z)$ has various known properties on Γ. Thus by [3, Theorem 11] we have (since $\alpha+1/p>1/q$)

$$\|f(z)-p_{n,\lambda n}(z)\|_p \leqq A/n^{\alpha+(1/p)-(1/q)}. \tag{8}$$

Inequality (8) together with (5) now yields (3).

The reader may notice the validity of

COROLLARY 1. *In Theorem* 1 *the second member of* (1) *may be replaced by* $A\varepsilon_n^q$, *where* ε_n (>0) *is monotonic nonincreasing as n increases, is such that* $r^n=o(\varepsilon_n)$ *for every* r (<1), *with the property* $\varepsilon_n=O(\varepsilon_{\lambda n})$ *whenever integral* $\lambda>1$, *and where the expression* $(2^{m-1}\leqq n<2^m)$,

$$\frac{2^{mr}\varepsilon_n+2^{(m+1)r}\varepsilon_{2^m}+2^{(m+2)r}\varepsilon_{2^{m+1}}+\cdots}{n^r\varepsilon_n}, \qquad p \geqq 1, \tag{9}$$

$$\frac{(2^m)^{pr}\varepsilon_n^p+(2^{m+1})^{pr}\varepsilon_{2^m}^p+(2^{m+2})^{pr}\varepsilon_{2^{m+1}}^p+\cdots}{n^{pr}\varepsilon_n^p}, \qquad p<1, \tag{10}$$

where $r=1/q-1/p$, *has a meaning and is bounded as* $n\to\infty$; *the second member of* (3) *is to be replaced by* $An^{1/q-1/p}\varepsilon_n$, *assumed to approach zero.*

In the proof, the second members of (6), (7), and (8) are to be replaced by $A\varepsilon_n^q$, $A_0\varepsilon_n^q$, $An^{1/q-1/p}\varepsilon_n$ respectively.

Both Theorem 1 and Corollary 1 can be modified in hypothesis and conclusion so that the first member of (1) is a double integral taken over the interior of Γ, as we now indicate.

First we state a result [3, Theorem 14] on degree of convergence by polynomials:

THEOREM 2. *Let E be the closed interior of an analytic Jordan curve, and let a function* $f(z)$ *continuous on E and polynomials* $p_n(z)$ *of respective degrees n be given such that we have for the qth power norm on E*

$$\|f(z)-p_n(z)\|_q' \leqq \varepsilon_n, \qquad q>0 \tag{11}$$

and where ε_n *has the first three properties of Corollary 1. Let us suppose the expression* (9) *or* (10) *with r replaced by* $s=2/q-2/p$ *exists and is bounded as* $n\to\infty$, *where* $2^{m-1}\leqq n<2^m$. *Then we have for* $0<q<p\leqq\infty$,

$$\|f(z)-p_n(z)\|_p' \leqq An^{2/q-2/p}\varepsilon_n, \tag{12}$$

where the second member is supposed to approach zero. In particular we may choose $\varepsilon_n = n^{-\alpha}$, $\alpha > 2/q - 2/p$.

Second, we indicate the analog of Theorem 2 for approximation by bounded analytic functions, which is thus an extension of Theorem 2, in the spirit of Theorem 1 and its Corollary as an extension of [1, Theorem 2].

THEOREM 3. *Let E be the closed interior of an analytic Jordan curve contained in the simply-connected region D, and suppose some function* $f(z)$ *analytic interior to E, continuous on E, and functions* $f_n(z)$ *analytic throughout D satisfy*

$$\iint_E |f(z)-f_n(z)|^q \, dS \leqq A/n^{q\alpha}, \qquad q > 0, \tag{13}$$

$$|f_n(z)| \leqq AR^n, \qquad z \text{ in } D. \tag{14}$$

Then for $\alpha > 2/q - 2/p$ *and* $0 < q < p \leqq \infty$ *we have*

$$\|f(z)-f_n(z)\|_p' \leqq A/n^{\alpha+2/p-2/q}. \tag{15}$$

Theorem 3 follows by the methods of proof of [1, Theorem 4] and the present Theorem 2. Like Theorem 1, Theorem 3 can be generalized in a suitable corollary:

COROLLARY 1. *In Theorem 3 the second member of* (13) *may be replaced by* $A\varepsilon_n^q$ *where* ε_n (>0) *is arbitrary monotonic nonincreasing, and is such that* $r^n = o(\varepsilon_n)$ *for every* r (<1), *with the property* $\varepsilon_n = O(\varepsilon_{\lambda n})$ *whenever integral* $\lambda > 1$, *and where the expression* (9) *or* (10) *with r replaced by s has a meaning and is bounded as* $n \to \infty$, *with* $2^{m-1} \leqq n < 2^m$. *The second member of* (15) *is to be replaced by* $An^{2/q-2/p}\varepsilon_n$, *and is assumed to approach zero.*

The preceding results, primarily relating to approximation by bounded analytic functions, have an analog for approximation on a curve rather than in a region:

THEOREM 4. *Let* Γ *be an analytic Jordan curve contained in a region D not necessarily simply-connected, and suppose we have for some function* $f(z)$ *continuous on* Γ *and functions* $f_n(z)$ *analytic in D*

$$\int_\Gamma |f(z)-f_n(z)|^q \, |dz| \leqq A\varepsilon_n^q, \qquad q > 0, \tag{16}$$

$$|f_n(z)| \leqq AR^n, \qquad z \text{ in } D, \tag{17}$$

where ε_n (>0) *is monotonic nonincreasing, is such that* $r^n = o(\varepsilon_n)$ *for every* r (<1), *and with the property* $\varepsilon_n = O(\varepsilon_{\lambda n})$ *whenever integral* $\lambda > 1$, *and where the expression* (9) *or* (10) *has a meaning and is bounded as* $n \to \infty$, *with* $2^{m-1} \leqq n < 2^m$. *Then if* $n^{1/q-1/p}\varepsilon_n \to 0$ *and* $0 < q < p \leqq \infty$, *we have for the pth power norm on* Γ

$$\|f(z)-f_n(z)\|_p \leqq An^{1/q-1/p}\varepsilon_n. \tag{18}$$

In particular we may choose $\varepsilon_n = n^{-\alpha}$, $\alpha > 1/q - 1/p$.

In the proof of Theorem 4, we assume the origin to lie interior to Γ, approximate the $f_n(z)$ on Γ by polynomials in z and $1/z$, and use the method of [3]. Details are left to the reader.

Theorem 4 applies to approximation on the unit circumference Γ to a real or complex function $f(z)$ by real or complex bounded analytic functions $f_n(z)$, or with the substitution $z=e^{i\theta}$, approximation on the real line $-\infty<\theta<\infty$ to a function with period 2π by bounded analytic functions with period 2π in a strip containing the line. In particular if $f_n(z)$ is a polynomial in z and $1/z$ of degree n satisfying (16), then (17) follows if D is an annulus containing Γ in its interior with boundary components having 0 as center, and $f_n(e^{i\theta})$ is a trigonometric polynomial of order n. Compare here [2, Theorems 6–9].

Theorem 4 suggests approximation by bounded analytic functions in a multiply connected region, as measured by a line integral over the boundary:

THEOREM 5. *Let E be a closed bounded region whose boundary Γ consists of a finite number of mutually disjoint analytic Jordan curves, and which lies in a region D. Suppose for some function $f(z)$ analytic interior to E and continuous on E and for functions $f_n(z)$ analytic in D we have* (16) *and* (17), *where ε_n satisfies the conditions of Theorem* 4. *Then if $n^{1/q-1/p}\varepsilon_n\to 0$ and $0<q<p\leqq\infty$ we have* (18) *for the pth power norm on Γ. In particular we may choose $\varepsilon_n=n^{-\alpha}$, $\alpha>1/q-1/p$.*

To prove Theorem 5, we merely apply Theorem 4 to each component of Γ and of $f(z)$.

Our primary topic in the foregoing theorems is degree of uniform convergence of the $f_n(z)$ to $f(z)$, so it is natural to assume those functions continuous in the closed regions considered. Some comments on uniform convergence in subregions as a consequence of mean convergence on the boundary or over a region are made in [5, §5.8].

We proceed to study the analog of Theorem 5, using as norm a double integral, whose proof is more involved than that of Theorem 5:

THEOREM 6. *Let E be a closed bounded region whose boundary Γ consists of a finite number of mutually disjoint analytic Jordan curves, and which lies in a region D. Suppose for some function $f(z)$ analytic interior to E, continuous on E, and for functions $f_n(z)$ analytic in D we have*

$$\iint_E |f(z)-f_n(z)|^q\, dS \leqq A\varepsilon_n^q, \qquad q>0, \tag{19}$$

and (14), *where ε_n satisfies the conditions of Corollary* 1 *to Theorem* 3. *Then if $n^{2/q-2/p}\varepsilon_n\to 0$ and $0<q<p\leqq\infty$ we have*

$$\|f(z)-f_n(z)\|_p' \leqq An^{2/q-2/p}\varepsilon_n, \tag{20}$$

where we assume the second member approaches zero. In particular we may choose $\varepsilon_n=n^{-\alpha}$, $\alpha>2/q-2/p$.

Let the components of Γ be $\Gamma_1, \Gamma_2, \ldots, \Gamma_\nu$ where Γ_1 bounds a closed finite region E_1 containing E, and Γ_j $(j>1)$ bounds a closed infinite region E_j containing E. Let Γ_j' be a variable analytic Jordan curve interior to E $(j=1, 2, \ldots, \nu)$ which together with Γ_j bounds a closed annular region G_j, where the G_j are mutually disjoint. Since the curve Γ_j' lies in E, there follows from (19) by [5, §5.3, Lemma 2]

$$|f(z)-f_n(z)| \leqq A\varepsilon_n, \quad z \text{ on } \Gamma_j', \tag{21}$$

where A varies with Γ_j'.

If z is an arbitrary point interior to E, the Γ_j' can be chosen so that z lies exterior to the G_j, and indeed z lies interior to the region bounded by all ν of the Γ_j'. For this point z, the value of $f(z)$ is represented by the Cauchy integral of $f(z)$ over $\sum \Gamma_j'$, so we may write $f(z) = \sum f^{(j)}(z)$ for z interior to E, and similarly $f_n(z) \equiv \sum f_n^{(j)}(z)$ for z interior to E, where the ν components $f^{(j)}(z)$ and $f_n^{(j)}(z)$ of $f(z)$ and $f_n(z)$ are represented by the Cauchy integrals of $f(z)$ and $f_n(z)$ over the respective Γ_j' but are independent of the Γ_j' having the required properties. By inequality (21) we have for z on any closed subset of E disjoint from G_j

$$|f^{(j)}(z)-f_n^{(j)}(z)| \leqq A\varepsilon_n \qquad (j = 1, 2, \ldots, \nu). \tag{22}$$

The functions $f^{(j)}(z)$ and $f_n^{(j)}(z)$ are defined throughout the interior of E_j and inequality (22) is valid also for z on $E_j - G_j$ minus a neighborhood of Γ_j'.

It is natural to attempt to use (22) to obtain an inequality on the functions $f^{(j)}(z)-f_n^{(j)}(z)$ on each E_k, but this procedure is complicated by the fact that $\nu-1$ of these regions are infinite and the surface integral norm cannot be used directly.

We may choose points $\alpha_1=\infty, \alpha_2, \ldots, \alpha_\nu$ fixed in the respective regions $D_1, D_2, \ldots, D_\nu$ exterior to E bounded by $\Gamma_1, \Gamma_2, \ldots, \Gamma_\nu$, and choose in each D_j and in D an analytic Jordan curve Γ_j'' separating α_j from E but so that the region D_0 bounded by $\sum \Gamma_j''$ contains no point not in D. The components of $f_n(z)$ already defined can be represented by Cauchy integrals of $f_n(z)$ over the curves Γ_j'', and we have by (14)

$$|f_n^{(j)}(z)| \leqq AR^n, \qquad z \text{ in } D_j^0, \tag{23}$$

where D_j^0 is a suitable closed region containing E_j in its interior and separated by Γ_j'' from α_j.

We fasten our attention now on $E_1, f^{(1)}(z)$, and $f_n^{(1)}(z)$. Inequality (22) yields

$$|f^{(j)}(z)-f_n^{(j)}(z)| \leqq A\varepsilon_n, \qquad z \text{ on } G_1, \quad j > 1,$$

$$\iint_{G_1} \sum_{j>1} |f^{(j)}(z)-f_n^{(j)}(z)|^q \, dS \leqq A\varepsilon_n^q,$$

and by (19) with the integral over G_1 there follows

$$\iint_{G_1} |f^{(1)}(z)-f_n^{(1)}(z)|^q \, dS \leqq A\varepsilon_n^q. \tag{24}$$

The point set G_1 is to some extent variable, so we deduce also by (22) and by the finiteness of the area of E_1,

$$\iint_{E_1-G_1} |f^{(1)}(z)-f_n^{(1)}(z)|^q \, dS \leqq A\varepsilon_n^q, \tag{25}$$

where the new E_1-G_1 contains in its interior the partial boundary Γ_1' of the G_1 in (24). Then by (24) and (25) we have

$$\iint_{E_1} |f^{(1)}(z)-f_n^{(1)}(z)|^q \, dS \leqq A\varepsilon_n^q. \tag{26}$$

By (23) and (26) we are in a position to apply Corollary 1 to Theorem 3, which establishes

$$\iint_{E} |f^{(1)}(z)-f_n^{(1)}(z)|^p \, dS \leqq An^{2p/q-2}\varepsilon_n^p; \tag{27}$$

the integral may be taken over E_1 or E. This proof does not apply directly to (27) with 1 replaced by j (>1) because the area of E_j is then infinite.

However, for $j>1$ we make a linear transformation $w=\phi(z)$ that carries α_j to infinity, which then transforms E_j into a finite region of the w-plane. By the method of proof of (24) we establish

$$\iint_{G_j} |f^{(j)}(z)-f_n^{(j)}(z)|^q \, dS \leqq A\varepsilon_n^q, \qquad dS = dS_z.$$

With the transformation $w=\phi(z)$, $z=\psi(w)$, we may set $dS_w=|\phi'(z)|^2 \, dS_z$, where $|\phi'(z)|$ is bounded and bounded from zero except near $z=\alpha_j$ and $z=\infty$ and their images, whence for the integral over the image of G_j

$$\iint |f^{(j)}[\psi(w)]-f_n^{(j)}[\psi(w)]|^q \, dS_w \leqq A\varepsilon_n^q. \tag{28}$$

By (22) we may write (28) for the integral over the image of a new E_j-G_j containing the partial boundary Γ_j' of the previously used G_j (by the boundedness of the area of the image of E_j). There follows for the integral over the image of E_j this same inequality (28).

By virtue of (23) interpreted in the w-plane, we can now apply Corollary 1 to Theorem 3, which proves for the integral over the image of E_j or E

$$\iint |f^{(j)}[\psi(w)]-f_n^{(j)}[\psi(w)]|^p \, dS_w \leqq An^{2p/q-2}\varepsilon_n^p.$$

We use this integral over the image of E, on which $\psi'(w)$ is bounded and bounded from zero, so there follows ($j>1$)

$$\iint_{E} |f^{(j)}(z)-f_n^{(j)}(z)|^p \, dS_z \leqq An^{2p/q-2}\varepsilon_n^p,$$

and (27) yields (20), which completes the proof of Theorem 6.

We add now some general comments on the theorems already proved. If the $f_n(z)$ of Theorem 1 are polynomials of respective degrees n satisfying (1), inequality (2) is a consequence of (1). For inequality (1) implies the boundedness ($n \to \infty$) of

$$\int_\Gamma |f_n(z)|^q \, |dz|, \tag{29}$$

and (2) follows where D is an arbitrary finite region bounded by a level locus Γ_R, by [5, §5.2, Lemma]. Here we denote generically by Γ_ρ ($\rho > 1$) the locus $|\phi(z)| = \rho$ in the complement K of E, where $w = \phi(z)$ maps K onto $|w| > 1$, $\phi(\infty) = \infty$. A more general remark can be made:

REMARK. *Let E be a closed limited point set whose complement is simply connected and whose boundary Γ has positive linear measure. If the rational functions $f_n(z)$ of respective degrees n satisfy* (1), *and if the poles of the $f_n(z)$ have no limit point on E, then for a suitably chosen region D containing E, inequality* (2) *is satisfied.*

An inequality

$$\int_\Gamma |f_n(z)|^q \, |dz| \leqq L^q, \qquad q > 0, \tag{30}$$

follows by the method of treatment of (29). If the $f_n(z)$ have no poles on or interior to Γ_B, $B > 1$, then [5, §9.8, Lemma III] we have for z on and within Γ_Z

$$|f_n(z)| \leqq AL[(BZ-1)/(B-Z)]^n, \qquad 1 < Z < B, \tag{31}$$

so we may choose D as the closed interior of Γ_Z by identifying (31) with (2).

The Remark just established deserves a number of additional comments.

1°. It is immaterial whether the hypothesis of the Remark is chosen as (1) or as the replacement of (1) as in Corollary 1 to Theorem 1. In either case we obtain (30) at once.

2°. Let the hypothesis (1) of the Remark be replaced by the inequality

$$|f(z) - f_n(z)| \leqq A\varepsilon_n, \qquad z \text{ on } \Gamma. \tag{32}$$

The uniform boundedness of the rational functions $f_n(z)$ follows on Γ, and an appropriate lemma [5, §9.7, Lemma I] yields (31) for z on or within Γ_Z if all poles of the $f_n(z)$ lie exterior to Γ_B, $1 < Z < B$. Thus D can be chosen as the interior of Γ_Z. This comment is of interest in connection with approximation also in the real domain, as in [6].

3°. The hypothesis of the Remark may be replaced by an inequality for the double integral:

$$\iint_E |f(z) - f_n(z)|^q \, dS \leqq A\varepsilon_n^q, \tag{33}$$

say under the hypothesis of Theorem 2, where the rational functions $f_n(z)$ of respective degrees n have no limit point of poles on E. We obtain the boundedness of the integrals

$$\iint_E |f_n(z)|^q \, dS,$$

hence [5, §5.3, Lemma II] there follows on an arbitrary closed region E' interior to E the uniform boundedness of the $f_n(z)$. Let the poles of the $f_n(z)$ have no limit point on or exterior to E_ρ, $\rho > 1$. Then for E' sufficiently large in E, the locus $(E')_\rho$ can be chosen as near E_ρ as desired (but interior to E_ρ), so in particular we can choose E' so that $(E')_\rho$ contains in its interior some E_B, $B > 1$, which contains E in its interior. If $|f_n(z)| \leqq L$ for z on E', we have for z on $(E')_Z$ (chosen to contain E and be contained in E_B)

$$|f_n(z)| \leqq AL[(\rho Z - 1)/(\rho - Z)]^n, \qquad 1 < Z < \rho,$$

by [5, §9.7, Lemma I]. The region D can be chosen as the interior of $(E')_Z$.

4°. The Remark can be extended so as to apply even if the complement of E is not simply connected, provided the boundary of E consists of a finite number of mutually disjoint analytic Jordan curves. We assume that $f_n(z)$ is a sequence of rational functions of respective degrees n whose poles have no limit point on E; it follows for instance that inequality (16) implies (17). Compare here Theorem 4 and [2, Theorems 6, 7, and 8].

5°. The reasoning involved in the Remark may apply even if the approximating functions $f_n(z)$ are no longer rational functions, provided each $f_n(z)$ is meromorphic with not more than n poles in each of one or more suitable regions. For instance we might consider approximation on a Jordan curve E containing in its interior a closed simply connected region E_0, where the functions $f_n(z)$ are respectively meromorphic with no more than n poles in the complement E_1 of E_0, continuous and bounded on the boundary of E_0.

Throughout this paper we have assumed for simplicity that the Jordan curves involved are analytic. That assumption can be somewhat weakened, as by the use of curves of type B in [1], and of type B' in [2].

References

1. J. L. Walsh, *Approximation by polynomials: uniform convergence as implied by mean convergence*, Proc. Nat. Acad. Sci. U.S.A. **55** (1966), 20–25.

2. ———, *Approximation by polynomials: uniform convergence as implied by mean convergence*. II, Proc. Nat. Acad. Sci. U.S.A. **55** (1966), 1405–1407.

3. ———, *Approximation by polynomials: uniform convergence as implied by mean convergence*. III, Proc. Nat. Acad. Sci. U.S.A. **56** (1966), 1406–1408.

4. ———, *Approximation by bounded analytic functions*, Mémor. Sci. Math., Fasc. 144, Gauthier-Villars, Paris, 1960.

5. ———, *Interpolation and approximation*, Colloq. Publ., Vol. 20, Amer. Math. Soc., Providence, R. I., 1935.

6. ———, *Degree of approximation by rational functions and polynomials*, Michigan Math. J. (to appear).

University of Maryland,
College Park, Maryland

HISTORY OF THE RIEMANN MAPPING THEOREM

J. L. WALSH, University of Maryland

The Riemann mapping theorem, that *an arbitrary simply connected region of the plane can be mapped one-to-one and conformally onto a circle*, first appeared in the Inaugural dissertation of Riemann (1826–1866) in 1851. The theorem is important, for by it a result proved for the circle can often be transformed from the circle to a more general region. The proof is difficult, as involving both behavior of a function in the small (conformal mapping) and behavior in the large (one-to-one mapping). Riemann's proof was open to criticism and in the following decades numerous mathematicians sought for a proof, e.g., H. A. Schwarz (1843–1921), A. Harnack (1851–1888), H. Poincaré (1854–1912), etc., until the first rigorous proof was given in 1900 by W. F. Osgood. The proof of Osgood represented, in my opinion, the "coming of age" of mathematics in America. Until then, numerous American mathematicians had gone to Europe for their doctorates, or for other advanced study, as indeed did Osgood. But the mathematical productivity in this country in quality lagged behind that of Europe, and no American before 1900 had reached the heights that Osgood then reached.

William Fogg Osgood (1864–1943) was born in Boston in 1864, graduated from Harvard College in 1886, stayed in Cambridge for a year of graduate work, and then went to Göttingen with a Harvard fellowship for further study, especially with Felix Klein (1849–1925). According to gossip, Osgood became so enamored of a Göttingen lady that his work suffered and Klein sent him to Erlangen for his doctorate. In any case, he was accorded the degree from Erlangen in 1890 for a thesis on Abelian integrals, and one or two days later he married the girl in Göttingen, and one or two days still later they sailed for the United States of America. His

---

Professor Walsh received his Harvard Ph. D. under Maxime Bôcher and George David Birkhoff. He continued at Harvard as Instructor through Perkins, Professor of Mathematics and became Professor Emeritus in 1966; since then he has been at the Univ. of Maryland. He has spent leaves of absence at the Sorbonne, the Univ. of Munich, the Institute for Advanced Study, and has spent several sabbatical leaves in Paris and Jerusalem.

He is a Fellow of the American Academy of Arts and Sciences and a Member of the National Academy of Sciences. Both the SIAM Journal on Numerical Analysis and the Journal of Approximation Theory have dedicated volumes to Joseph Walsh. His main research is on zeros, extremal problems, and approximations by polynomials and orthogonal functions. He is widely known for his invention of the Walsh functions.

His publications include *Interpolation and Approximation* (Amer. Math. Soc. Coll. Series, 1935, Russian Tranlation — 1961), *Location of Critical Points of Analytic and Harmonic Functions* (Amer. Math. Soc. Coll. Series, 1950), *Approximation by Polynomials* (Paris, 1935), *Approximation by Bounded Analytic Functions* (Paris 1960). *The Theory of Splines and their Application* (with J. H. Ahlberg and E. N. Nilson, Academic Press, New York, 1967), *A Bibliography on Orthogonal Polynomials* (with J. S. Shohat and Einar Hille, National Research Council, Bulletin, Washington, D. C. 1940), and *A Rigorous Treatment of Maximum-Minimum Problems in the Calculus* (Heath, 1962). *Editor.*

early mathematical work was also of high quality. During the 1890's he was Lebesgue's forerunner in the study of sequences of functions of a real variable. Osgood taught at Harvard from 1890 until his retirement in 1933.

Osgood seems not to have received the recognition for his work that he deserves. For instance, C. Carathéodory and G. Julia each wrote a book on conformal mapping without mention of the name of Osgood.

We proceed now with the proof of Riemann's theorem!

By a **simply connected region** Riemann understood a region bounded by a simple closed curve, and before him special mappings by simple functions were well known. We assume the given region to be bounded, which may require an elementary preliminary transformation. Let us examine Riemann's proof (based on Dirichlet's Principle) and postpone discussion of its validity.

Mapping of a region T onto a circle is equivalent to the existence of **Green's function for** T, namely a function $G(z)$ such that

(1) $G(z)$ is harmonic in T except at the origin 0, assumed interior to T;

(2) in the neighborhood of 0 the function takes the form $G(z) \equiv G_1(z) + \log r$, where $r = |z|$ and $G_1(z)$ is harmonic throughout T;

(3) $G(z)$ is continuous and equal to zero at every point of the boundary C of T.

These three conditions determine $G(z)$ uniquely. Green's function for a region T is invariant under one-to-one conformal mapping of T.

If the function $w = \phi(z)$ maps T (Figure 1) onto $|w| < 1$ so that $\phi(0) = 0$, then we clearly have

$$\phi(z) = e^{G(z)+iH(z)}$$

where $H(z)$ is conjugate to $G(z)$ in T, for each of the conditions (1), (2), (3), is satisfied by $G(z)$ as thus defined. Conversely, if $G(z)$ is Green's function for T with pole in 0, then every point of T is transformed by $w = \phi(z)$ into a point $|w| < 1$. Each locus L_r: $|\phi(z)| = r$, $0 < r < 1$ in T bounds two subregions of T, where $G(z) > \log r$ and $G(z) < \log r$ respectively; the locus L_r has no multiple points and

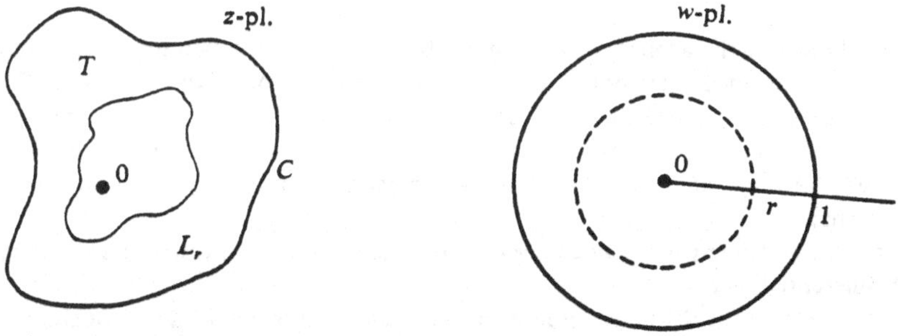

Fig. 1

separates 0 and C. On L_r we have $\partial G/\partial n \neq 0$, where n is the interior normal for the latter subregion, whence

$$\int_{L_r} \frac{\partial G}{\partial n} ds = \arg w \,|_{L_r} = \int_{L_r} \frac{\partial H}{\partial s} ds = \int_{L_r} \frac{\partial \log r}{\partial n} ds = 2\pi,$$

so the transformation $w = \phi(z)$ defines a one-to-one map of T onto $|w| < 1$.

If T is given, the determination of $G(z)$ requires the solution of the Dirichlet problem for T with the prescribed boundary values $\log r$ on C, a problem that Riemann treated by means of Dirichlet's principle. The physical evidence for the existence of $G(z)$ is great, for in the steady two-dimensional flow of heat, the temperature is a harmonic function provided T is a uniform body whose continuous boundary temperatures on C are prescribed.

The Dirichlet integral defined for a function $u(x, y)$ given in a region T is defined as

$$D(u) = \iint_T \left[\left(\frac{\partial u}{\partial x}\right)^2 + \left(\frac{\partial u}{\partial y}\right)^2\right] dx\, dy \quad (\geqq 0).$$

We compare this integral with the corresponding integral where $u(x, y)$ is replaced by $u(x, y) + \varepsilon \cdot v(x, y)$, where $v(x, y)$ vanishes on the boundary C of T. Thus we have, *to study the function* $u(x, y)$ *with given boundary values minimizing* $D(u)$,

$$D(u + \varepsilon v) = \iint_T \left[\left(\frac{\partial u}{\partial x}\right)^2 + \left(\frac{\partial u}{\partial y}\right)^2\right] dx\, dy + 2\varepsilon \iint_T \left(\frac{\partial u}{\partial x}\frac{\partial v}{\partial x} + \frac{\partial u}{\partial y}\frac{\partial v}{\partial y}\right) dx\, dy$$
$$+ \varepsilon^2 \iint_T \left[\left(\frac{\partial v}{\partial x}\right)^2 + \left(\frac{\partial v}{\partial y}\right)^2\right] dx\, dy.$$

Considered as a function of ε, this second term on the right must be zero, namely,

$$\iint_T \frac{\partial}{\partial x}\left[\left(v \frac{\partial u}{\partial x}\right) + \frac{\partial}{\partial y}\left(v \frac{\partial u}{\partial y}\right)\right] dx\, dy - \iint_T v \nabla^2 u \, dx\, dy = 0$$

for all choices of the arbitrary function v. The former of these two integrals reduces to two contour integrals over C with v (= 0 on C) as a factor of the integrand. Thus for the function u minimizing $D(u)$, $\nabla^2 u = 0$ throughout T, and u is harmonic in T. "The function solving the boundary value problem is the function minimizing $D(u)$."

This "proof," although accepted by Riemann, is obviously open to various objections:

(1) The treatment has a meaning only if C has certain properties of smoothness and differentiability.

(2) The fact that $D(u)$ has a non-negative greatest lower bound does not show the existence of a *minimum* (Weierstrass).

(3) The fact that $D(u) < \infty$ for some $u(x, y)$ satisfying the given boundary values needs to be shown (Prym 1871, Hadamard 1906).

It is convenient to assume that T is bounded; if not, we may use the transformation $w = \sqrt{(z-\alpha)/(z-\beta)}$, where α and β are two distinct boundary points of T. Then T in the z-plane corresponds to two regions T_1 and T_2 on the w-sphere, one-to-one conformal images of T, which have no common point. If two such regions do not exist, a point w_1 in T_1 can be joined to a point w_1 by a path in T_1 separating $w = 0$ and $w = \infty$, so there is a closed curve in T separating α and β, and T is not simply connected. Inversion of a point of T_2 to infinity now maps T_1 onto a bounded region.

We mention here several results that we shall need for discussion of Osgood's proof.

(1) **Axel Harnack's Theorem (1887).** If a function u_n is harmonic in a region T for all sufficiently large values of n, and if u_n increases at all points of T when n increases; if furthermore at a single point of T u_n approaches a (finite) limit when n becomes infinite; then u_n converges at all points of T, to a function harmonic throughout T. (It is reported that when Harnack first told Felix Klein of this theorem, the latter refused to believe its validity.)

(2) **H. A. Schwarz.** Green's function exists for a simply connected region T bounded by a finite number of analytic arcs. (Schwarz used the alternating method, due to C. Neumann.)

(3) **Lemma.** If the bounded region T contains the closure of the region T_1, and if O lies in T_1, then the respective Green's functions g and g_1 with poles in O for T and T_1 satisfy the inequality $g > g_1 > 0$ in T_1. For the difference $g - g_1$ is harmonic in T_1, and $g_1 = 0$, $g > g_1$, on the boundary of T_1, whence $g - g_1 > 0$, $g - g_1 \not\equiv 0$, throughout T_1.

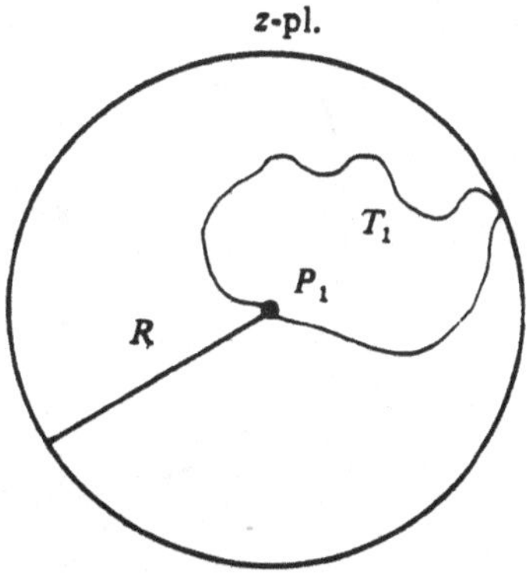

FIG. 2

(4) Given a region T_1, it can be exhausted by a monotonic sequence of subregions, composed for instance of adjacent squares whose sides are parallel to the coordinate axes.

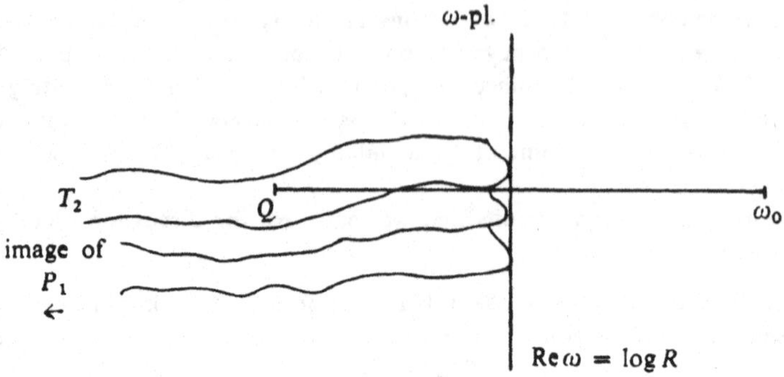

FIG. 3

Given, now, (Figure 2) a bounded simply connected region T_1 of the z-plane, we show that T_1 can be transformed into a region T of the w-plane in such a manner that a given boundary point P_1 of T_1 corresponds to a point P of a circle Γ which contains T. Let T_1 be considered to lie on the Riemann surface for $\omega = \log z$ with P_1 at $z = 0$. The image T_2 in the ω-plane of T_1 consists (Figure 3) of an infinite number of images of T_1, each the translation of another such region by the vector

z	ω	w
P_1	∞	P
T_1	T_2	T
	ω_0	∞
	Q	O_w

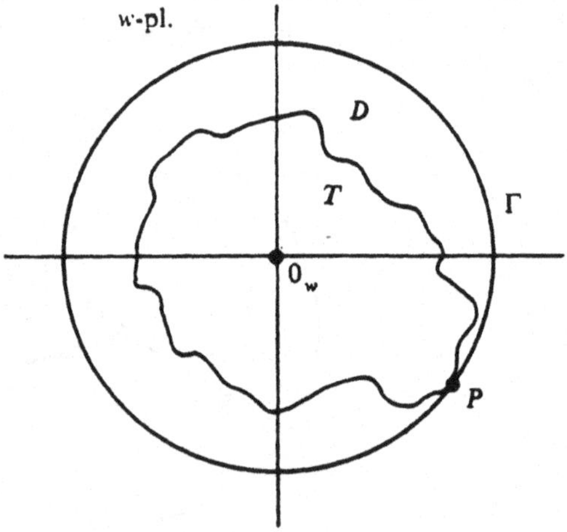

FIG. 4

$\omega = \pm 2\pi i$. In each such region the point at infinity $\omega = \infty$ corresponds to P_1, for all boundary points of T_1 in the region $|z| < \varepsilon$ correspond to points ω with $\operatorname{Re}\omega < \log \varepsilon$. Let $|z| < R$ be the smallest circular disk with center P_1 containing T_1; then all points of T_2 lie in the half-plane $\operatorname{Re}\omega < \log R$. A linear transformation carrying to infinity in the w-plane a finite point ω_0, $\operatorname{Re}\omega_0 > \log R$; carries T_2 into a region T of the w-plane (Figure 4) which lies in a circular disk D (image of $\operatorname{Re}\omega < \log R$) whose boundary passes through the image P of P_1.

It may be noted too that an arbitrary point Q of T_2 with $\operatorname{Im} Q = \operatorname{Im}\omega_0$ can be chosen so that Q is simultaneously carried into O_w in the w-plane. The point O_w in T is then the center of D.

Let D_n be a monotonic sequence of subregions of T containing O_w, and each with a Green's function g_n with pole in O_w, with $D_{n+1} \supset D_n$, exhausting T. Let g_0 be Green's function for D with pole in O_w; then $g_n < g_0$ in D_n. Let $g = \lim_{n\to\infty} g_n$, defined (Harnack) and harmonic throughout T except in 0. Then g is Green's function for T; we have $0 < g_n < g_0$, $0 < g \leqq g_0$. Suppose $P_k \in T$, $P_k \to P$. Since

$$\lim_{P_k \to P} g_0(P_k) = 0 \text{ for } P_k \in D,$$

we also have

$$\lim_{P_k \to P} g(P_k) = 0, \; g(P) = 0,$$

and this shows the existence of Green's function for T and thus completes Osgood's proof of Riemann's theorem.

We have not mentioned the work of Hilbert (1862–1943), who gave a treatment of Riemann's theorem in weakened form by new methods of the Calculus of Variations, commencing about 1900. This general problem in the Calculus of Variations was presented as Problem 20 in his famous Paris lecture of 1900. He suggested in particular thesis topics on the subject for several American doctoral students in Göttingen: C. A. Noble, E. R. Hedrick, and Max Mason. However, Hilbert's method required certain smoothness properties of the boundary and of the limit function, and was thus less general than the idea of the original Dirichlet principle and less general than Osgood's proof. A new method of proof, based on function theoretic rather than potential theoretic properties, was developed by F. Riesz and L. Fejér, published in 1923 by T. Radò. Montel's theory of normal families was used, and a lemma due to Koebe. This is the standard modern proof.

Research supported in part by U. S. Air Force of Scientific Research, Grant AF 69–1690.

References

1. B. Riemann, Grundlagen für eine allgemeine Theorie der Funktionen einer veränderlichen complexen Grösse. Inauguraldissertation, Göttingen, 1851.

2. A. Harnack, Logarithmisches Potential, Leipzig, 1887, p. 167.

3. H. A. Schwarz, Zur Integration der partiellen Differentialgleichung $\Delta u = 0$, Ges. Math. Abhandlungen vol. 11. pp. 144–171. See also Osgood, Funktionentheorie, Fifth Ed. Leipzig 1928, Ch. 14 § 4.

4. Henri Poincaré, Sur un théorème de la théorie générale des fonctions, Bull. Soc. Math. France, 11 (1883) 112–125.

5. ———, Sur l'uniformisation des fonctions analytiques, Acta Mathematica, 31 (1907) 1–63.

6. W. F. Osgood, On the existence of Green's function for the most general simply connected plane region, Trans. Amer. Math. Soc., 1 (1900) 310–314.

7. ———, Funktionentheorie, Fifth Ed. Leipzig 1928, Ch. 14 § 5.

8. D. Hilbert, Über das Dirichletsche Prinzip, Jber. Deutsch. Math-Verein., 8 (1900) 184–188.

9. ———, Über das Dirichletsche Prinzip, Math. Annalen, 59 (1904) 161–186.

Commentary by T.J. Rivlin

Comments on [32-c*]

In this short paper in 1932 Walsh repeats more clearly (p. 293, equivalently page 387 of this selected works) a result that appeared in Transactions of the American Mathematics Society, vol. 34, 1932, pp. 22–74, Section 11. A formal presentation of this interesting result appears in 1935 in the First Edition of his "Interpolation and Approximation by Rational Functions in the Complex Domain", A.M.S. Colloquium Publications, vol. XX, page 153. Attention seems not to have been attracted to this report, but in 1980 in a paper entitled "Interpolation in the roots of unity: An extension of a theorem of J. L. Walsh" appeared in Resultate der Mathematik (3), 155-191, authored by A. S. Cavaretta Jr., A. Sharma and R. S. Varga. They present "a well-known and beautiful result of J. L. Walsh [which is Theorem 1 on p. 153]" and then proceed to extend Walsh's Theorem 1 into many new directions in the next 33 pages.

In a paper of 1986 entitled "Some Recent Results on Walsh Theory of Equiconvergence", by A. Sharma, 23 references are new "recent literature which has sprung around this classic theorem of Walsh". (Among the authors were T. J. Rivlin and E. B. Saff.) The interest in Theorem 1 continued apace, but by the mid nineties it had almost disappeared. The result was probably about 200 published papers!

Comments on [37-g*] and [38-d*]

Professor Walsh formally retired from Harvard University in 1966. A retirement party was arranged in March of 1967 at the famous Luchow's Restaurant on 14th

Street in New York City. Since I was living in New York at the time, I was asked to help with the arrangement of the dinner. Many other Ph.D. students of Walsh were present, as were some of his former colleagues at Harvard. At the party Mrs. Walsh sat on one side of her husband, and his favorite, Colonel Walter E. Sewell (Ph.D., 1936; Colonel in the U.S. Army in World War II) on the other side. Another of his favorite students, whom he had not seen for many years, Zehman I. Mosesson (Ph.D., 1937; Distinguished Actuary and Vice-President of Prudential Insurance) sat across the table from him. Walsh and Sewell wrote quite a few papers together. I have chosen two of their early papers about polynomial approximation.

It might be of interest to the readers of this volume that both of these papers (among many others) refer to the famous Belgian mathematician as "de la Vallée Poussin", as did Walsh and many other mathematicians. However, the spelling of the famous Belgian's name is incorrect (except in France). In the death notice of the Belgian mathematician, by Paul Montel, (Cr. Acad. Sci. Paris, Vol. 254, 1962, pp. 2473-2476) the name is correctly given as "Charles de La Vallée Poussin". Montel remarks that the name "Poussin" came from Charles' great grandfather Lavallée, a painter of the early 18th century related to Nicolas Poussin, who decided to conflate their names.

Works related to [37-g*] and [38-d*] are described below by V. Maymeskul.

Results like Theorems I and II of [37-g*] are now called *inverse* theorems of approximation theory. Along with *direct* theorems, estimating the rate of possible polynomial approximation to a function depending on its smoothness, they give so-called *constructive characterizations* of classes of analytic functions (provided, of course, their assertions are mutually inverse). Theorem I and Curtiss's result mentioned in the paper give such a description for domains with analytic boundary, which is similar to that for the unit circle (or the trigonometric case). The problem becomes more delicate if the boundary of a domain (or, generally, of a bounded continuum $\mathcal{M}$ with connected complement) is *not* analytic. In this case, inverse theorems as well as direct ones should involve information about the geometrical structure of $\mathcal{M}$. It turns out that the geometry of sets is nicely described by the quantity $\rho_u(z)$, $u > 1$, which denotes the distance from a point $z \in \partial\mathcal{M}$ to the uth level line of the Riemann mapping function $\Phi : \overline{\mathbb{C}}\setminus\mathcal{M} \to \{|w| > 1\}$ normalized at infinity. Assuming *uniform* lower estimates on $\rho_u(z)$, inverse theorems were obtained by Sewell in [18] for Lip α classes in Jordan domains bounded by quasi-smooth curves (i.e., curves whose subarc length and chord length are of the same order). For domains with piecewise smooth boundaries, some inverse results (still based on uniform approximation and sharp to within ϵ) were obtained by Mergelyan in [15], [16]. The first inverse results using (implicitly) the *local* behavior $\rho_{1+1/n}(z)$ were obtained for $\mathcal{M} = [-1, 1]$ by Dzjadyk in [10] (for Lip α classes) and by Timan in [24] (for arbitrary moduli of continuity). The function $\rho_{1+1/n}(z)$ was introduced explicitly by Dzjadyk in [9]. For continua with connected complement whose boundaries consist of a finite number of curves with continuous curvature, inverse theorems were obtained in [9] for Lip α classes and in [8] for classes defined by general moduli of continuity. These results were generalized by Lebedev and Tamrazov in [13], [14], where inverse theorems were proved for arbitrary compact sets whose complements are regular with respect to the Dirichlet problem.

The above mentioned results deal mainly with continuity properties of functions or their derivatives on $\partial\mathcal{M}$. To derive conclusions about the smoothness of a function on whole set $\mathcal{M}$ ([18], [23]) one needs the so-called contour-solid statements about moduli of continuity of analytic functions that are discussed below in connection with Theorem III. Note also that in all the results discussed above conclusions about *global* moduli of continuity are based on pointwise estimates of polynomial approximation. There are other approaches such as estimates of local moduli of smoothness of f on $\partial\mathcal{M}$ or those of the composition $f \circ \Phi^{-1}$ on $\{|w| = 1\}$ based on a rate of *uniform* polynomial approximation on $\mathcal{M}$. See, for instance, [1], [4], [11].

Theorem III of [37-g*] is one of the first results studying a relation between continuity properties of analytic functions and their derivatives on the boundary $\partial\mathcal{M}$ and on $\mathcal{M}$ itself. In the case when $\mathcal{M}$ is a closed Jordan domain the equality Lip α $(\partial\mathcal{M})$ = Lip α $(\mathcal{M})$, $0 < \alpha < 1$, was shown by Walsh and Sewell [25] and by Warschawski [26]. For classes of functions defined via the first modulus of continuity, the most general results in this direction were obtained by Tamrazov in [20], [21]. In particular, it was shown that the inequality $\omega_1(f;t)_{\mathcal{M}} \leq c \cdot \omega_1(f;t)_{\partial\mathcal{M}}$ holds for any bounded continuum $\mathcal{M}$ whose complement is connected and regular with respect to the Dirichlet problem, provided the first modulus $\omega_1(f;t)_{\mathcal{M}}$ satisfies a very natural regularity condition. Similar results were proved by Shevchuk in [19] for classes of functions defined via the second modulus of continuity.

For a more detailed review and recent results on direct and inverse theorems, see [5], [7], [22].

Similar problems have also been considered for some other classes of functions (for example, [2], [3], [6], [12], [17]).

References

[1] Andrievskii, V.V. *A description of classes of functions with a given rate of decrease of their best uniform polynomial approximations*, Ukrainian Math. J., **36** (1984), 447–450.

[2] Andrievskii, V.V. *A constructive characterization of harmonic functions in domains with quasiconformal boundary*, Math. USSR-Izv., **34** (1990), 441–454.

[3] Andrievskii, V.V. *Body-contour properties of harmonic functions*, Soviet Math. (Iz. VUZ), **34** (1990), 14–23.

[4] Andrievskii, V.V. *Uniform polynomial approximation of analytic functions on a quasidisk*, J. Approx. Theory, **73** (1993), 136–148.

[5] Andrievskii, V.V., Belyi V.I., and Dzjadyk, V.K. Conformal Invariants in Constructive Theory of Functions of Complex Variable. World Federation Publ., Atlanta, GA, 1995.

[6] Andrievskii, V.V., Belyi, V.I., and Maimeskul, V.V. *Approximation of solutions of the equation* $\bar{\partial}^j f = 0,\ j \geq 1$, *in domains with quasiconformal boundary*, Math. USSR Sb., **68** (1991), 303–323.

[7] Dzjadyk, V.K. Introduction to the Theory of Uniform Approximation of Functions by Polynomials. "Nauka", Moscow, 1977. (Russian)

[8] Dzjadyk, V.K. *Inverse theorems of approximation theory in complex domains*, Ukrain. Mat. Zh., **15** (1963), 365–375. (Russian)

[9] Dzjadyk, V.K. *On a problem of S.M. Nikol'skii in complex domain*, Izv. Akad. Nauk SSSR, ser. Matem., **23** (1959), 697–736. (Russian)

[10] Dzjadyk, V.K. *On constructive characteristic of functions, satisfying* Lip α, $0 < \alpha < 1$, condition on a finite interval on real axis, Izv. Akad. Nauk SSSR, ser. Matem., **20** (1956), 623–642. (Russian)

[11] Dzjadyk, V.K., Alibekov, G.A. *Uniform approximation of functions of a complex variable on closed sets with corners*, Math. USSR-Sb., **4** (1968), 463–517.

[12] Gagua, M.B. *On a theorem of Hardy and Littlewood*, Uspehi Matem. Nauk (N.S.), **8** (1953), 121–125. (Russian)

[13] Lebedev, N.A., Tamrazov, P.M. *Inverse theorems of approximation on closed sets of the complex plane*, Soviet Math. Doklady, **9** (1968), 498–501.

[14] Lebedev, N.A., Tamrazov, P.M. *Inverse approximation theorems on regular compact subsets of the complex plane*, Math. USSR-Izv., **4** (1970), 1355–1405.

[15] Mergelyan, S.N. *Certain questions of the constructive theory of functions*, Proc. Steklov Inst. Math., **37** (1951), 1–91. (Russian)

[16] Mergelyan, S.N. *Uniform approximations to functions of a complex variable*, Amer. Math. Soc. Translations, ser. 1, **3** (1962), 294–391.

[17] Rubel, L.A., Shields, A.L., and Taylor, B.A. *Mergelyan sets and the modulus of continuity of analytic functions*, J. Approx. Theory, **15** (1975), 23–40.

[18] Sewell, W.E. Degree of Approximation by Polynomials in the Complex Domain. Princeton Univ. Press, Princeton, 1942.

[19] Shevchuk, I.A. *Solid and contour properties of analytic functions in terms of the second modulus of continuity*, In: "Problems in Approximation Function Theory and its Applications", Inst. Math. of Ukrain. Acad. of Sci., Kiev (1976), 202–204. (Russian)

[20] Tamrazov, P.M. *Boundary and solid properties of holomorphic functions in a complex domain*, Soviet Math. Doklady, **13** (1972), 725–732.

[21] Tamrazov, P.M. *Contour and solid structure properties of holomorphic functions*, Russ. Math. Surveys, **28** (1973), 141–173.

[22] Tamrazov, P.M. Smoothness and Polynomial Approximations. "Naukova Dumka", Kiev, 1975. (Russian).

[23] Tamrazov, P.M. *Solid inverse theorems of polynomial approximation for regular compacta in the complex plane*, Soviet Math. Doklady, **12** (1971), 855–858.
[24] Timan, A.F. *Converse theorems in the constructive theory of functions given on a finite segment on the real axis*, Dokl. Akad. Nauk SSSR (N.S.), **116** (1957), 762–765. (Russian)
[25] Walsh, J.L., Sewell, W.E. *Sufficient conditions for various degrees of approximation by polynomials*, Duke Math. J., **6** (1940), 658–705.
[26] Warschawski, S. *Über das Verhalten der Ableitung der Abbildungsfunktion bei konformer Abbildung*, Math. Zeitschr., **38** (1934), 669–683.

Comments on [53-c*] and [73-a*]

Theodore S. Motzkin (1908-1970) collaborated with Walsh in 15 published papers from 1953 to 1973 - more than any of Walsh's other collaborators. The list of these papers is: 53-c*, 55-b, 56-e, 57-a, 57-b, 59-b, 59-f, 59-i, 60-e, 61-b, 62-d, 63-a, 66-a, 68-c, 73-a*. The first and last of these papers are presented in this book.

Motzkin received his Doctorate at Basel, Switzerland in 1934 as a student of Alexander Ostrowski. He published 139 papers; 5 in German, 7 in French, 11 in Hebrew and the rest in English. His thesis (in English translation), "Contributions to the Theory of Linear Inequalities", became a basis for linear programming in later years. The "Selected Papers of Theodore S. Motzkin", edited with an introduction by David Cantor, Basil Gordon and Bruce Rothschild, appeared in a volume of Contemporary Mathematicians, Birkhäuser, Boston, Mass., 1983, xxvi + 530 pp.

Walsh's collaboration with Motzkin commenced at the Institute for Numerical Analysis (INA) at the University of California at Los Angeles. Motzkin had joined the senior staff for the summer of 1950, and Walsh appeared on the scene in the summer of 1951. By 1953 there appeared the first collaborative publication of this pair. Motzkin became a member of the Department of Mathematics at UCLA in 1954.

In the course of the next 20 years, 15 papers were published by them jointly. Walsh lived to see their last collaborative publication [73-a*] in 1973 (3 years after Motzkin's death).

Comments on [34-c*]

This extremely clear and simple presentation does not require a long remark. The basic result is clearly presented as follows: "It is the object of the present note to show that the Tchebycheff polynomials found by orthogonalization of $1, z, z^2, \dots$ on the line segment $C : -1 \leq z \leq +1$ with respect to the norm function $(1-z^2)^{-1/2}$ are also orthogonal with respect to suitable norm functions on all the corresponding curves C_R, which are ellipses with the common foci $-1, 1$."

Comments on [42-a*]

Professor Walsh became acquainted with the book Georges Bourion, "L'Ultra-convergence dan les séries de Taylor", Actualités scientifiques et industrielle, no. 472, Hermann & Cie., Paris, 1937, a bit before 1940. The first known example of an overconvergent power series, as Walsh remarks, is due to M. B. Porter in 1907. Porter's paper, "On the polynomial convergents of a power series", appeared in Annals of Mathematics, 2nd series, vol. 8, 1907, pp. 189-192.

Porter's work seems to have passed unnoticed and the general theory of over-convergence of power series is due mainly to Alexander Ostrowski, whose first published paper in this area was "Uber eine Eigenschaft gewisser Potenzreihen mit unendlich vielen verschwindenden Koeffizienten", Sitzungsberichte der Preussischen Akademie der Wissenschaften, vol. 34, 1921, pp. 557-565. Ostrowski's paper put overconvergence on the map, and was followed by five papers on this topic in 1922, 1923, 1926 and 1930. The Walsh paper presented here is a small twist on Ostrowski's work. (My Ph.D. thesis under Walsh's tutelage, a further minor elaboration of the Porter, Ostrowski, Walsh material was my first published paper in 1955.)

Comments on [51-c*]

As Walsh says, the object of this note is to consider some results about approximation by bounded analytic functions. Theorem 1 and its easily established converse, Theorem 2, and its corollary, are clearly presented. Further similar themes are continued in Theorems 3 through 6. At the conclusion of this note, Walsh remarks that a further study involving "Problem α", and that a further study involving "Problem β", will be forthcoming. Two further publications involving "Problem β" appear in the next year: (1) "Degree of Approximation to Functions on Jordan Curve", Trans. American Math. Soc., Vol. 73, pp. 447-458, November, 1952; and (2) "Degree of Approximation on a Jordan Curve", Proceedings of the National Academy of Sciences, Vol. 38, pp. 1058-1066, December, 1952, (together with H. Margaret Elliot, one of Walsh's Ph.D. students).

Comments on [73-c*]

Among miscellaneous literary remains, a biographical sketch of William Fogg Osgood was found in Walsh's papers in the Harvard University Archives. This biographical sketch was printed in "A Century of Mathematics in America", Part II, published by the American Mathematical Society, P.O. Box 6248, Providence, Rhode Island, 02940.

The "History of the Riemann Mapping Theorem" is further evidence of Walsh's high opinion of Osgood, who had not been recognized abroad. As Walsh remarks, "the first rigorous proof (of the Riemann mapping theorem) was given by W. F. Osgood in 1900. The proof of Osgood represented, in my opinion, the "coming of age" of mathematics in America". Having given his opinion of Osgood, Walsh then turns to various aspects of the Riemann Mapping Theorem.

Chapter 6

Rational Approximation

ON THE OVERCONVERGENCE OF CERTAIN SEQUENCES OF RATIONAL FUNCTIONS OF BEST APPROXIMATION.

By

J. L. WALSH

of Cambridge, Mass.

1. **Introduction.** For many purposes, arbitrary rational functions are more useful in approximating to given analytic functions of a complex variable than are polynomials. For instance it is shown by Runge in his classical paper on approximation by polynomials[1] that a function $f(z)$ analytic in a closed region of the z-plane bounded by a finite number of non-intersecting Jordan curves can be uniformly approximated in that region as closely as desired by a rational function of z.[2] Such approximation by a *polynomial* may not be possible. It is the purpose of the present paper to show that in the study of two other phases of approximation it may also be more advantageous to use general rational functions than polynomials, namely 1) degree of approximation, that is, asymptotic properties of the measure of approximation of the sequence of functions of best approximation, and 2) overconvergence, the phenomenon that a sequence of functions approximating a given function in a given region frequently converges to that given function (or its analytic extension) not merely in the given region but also in a larger region containing the given region in its interior. The term overconvergence has recently been used by Ostrowski in a somewhat different connection.

A rational function of the form

[1] Acta mathematica vol. 6 (1885), pp. 229—244.

[2] For more detailed results, compare Walsh, Mathematische Annalen vol. 96 (1926), pp. 437—450 and Transactions of the American Mathematical Society vol. 31 (1929), pp. 477—502.

$$\frac{a_0 z^n + a_1 z^{n-1} + \cdots + a_n}{b_0 z^n + b_1 z^{n-1} + \cdots + b_n}$$

where the denominator does not vanish identically, is said to be of degree n.

We shall deal with the entire plane of the complex variable z, closed by the adjunction of a single point at infinity. The *derivative*, or more explicitly the *first derivative* of an arbitrary point set E is the set E' composed of the limit points of E. The second derivative of E is the first derivative E'' of E', and in general the k-th derivative $E^{(k)}$ of E is similarly defined as the first derivative of the $(k-1)$-st derivative of E. The principal result of the present paper is

Theorem I. *Suppose $f(z)$ is an analytic function of z whose singularities form a set E one of whose derivatives $E^{(k)}$ is empty. Suppose C is a closed point set with no point in common with E. Then a sequence of rational functions $r_n(z)$ of respective degrees n of best approximation to $f(z)$ on C such that the poles of $r_n(z)$ lie in E, converges to the function $f(z)$ over the entire plane except on the set E. The convergence is uniform on any closed point set containing no point of E, and on any such point set the convergence is better than that of any geometric series.*

The term *best approximation* deserves some explanation. There are various measures of approximation of the function $r_n(z)$ to the given function $f(z)$ defined in a region C, for instance $\max [|f(z) - r_n(z)|, z \text{ on } C]$, $\int |f(z) - r_n(z)|^2 |dz|$ taken over the boundary of C, or $\iint_C |f(z) - r_n(z)|^2 dS$ taken over the area of C. Let us consider a particular measure of approximation and a particular value of n, and call *admissible* any rational function of degree n whose poles lie in E. Then a rational function $r_n(z)$ of degree n *of best approximation* to $f(z)$ on C such that the poles of $r_n(z)$ lie in E is that admissible function $r_n(z)$ or one of those admissible functions whose measure of approximation to $f(z)$ on C is less than the measure of approximation to $f(z)$ on C of any other admissible function. It is not obvious but can be shown without great difficulty that such a function of best approximation always exists, for the various measures of approximation that we shall use, although it need not be unique.[1]

[1] See Walsh, Transactions of the American Mathematical Society vol. 33 (1931).

The existence af a function of best approximation depends essentially on the closure of the set E.

If E contains but a finite number of points, there are for a given n but a finite number of possible distributions of the orders of the poles of $r_n(z)$ among the points of E. For each such

In Theorem I the set C may be 1) any closed point set not a single point whose complement is simply connected, approximation being measured in the sense of Tchebycheff; 2) any closed set not a single point whose complement is simply connected, approximation being measured by integration over the circle $\gamma : |w| = 1$ *when the complement of C is mapped onto the exterior of* γ; *3) any limited closed set C whose boundary is a rectifiable Jordan arc or curve, or more generally any limited set C whose boundary C′ is of positive linear measure and whose complement is simply connected, approximation being measured by a line integral over C′; 4) any simply connected region, approximation being measured by integration on the circle* $\gamma : |w| = 1$ *when C is mapped onto the interior of* γ; *5) any region or point set with at least one interior point and having positive area, approximation being measured by a double integral over C.*[1]

By *approximation in the sense of Tchebycheff* we understand that the measure of approximation of $r_n(z)$ to a given function $f(z)$ on a point set C is

$$\max [|f(z) - r_n(z)|, \; z \text{ on } C].$$

In this measure of approximation it is a slight generalization to insert a weight or norm function $n(z)$ positive and continuous on C and to use as the measure of approximation

$$\max [n(z)|f(z) - r_n(z)|, \; z \text{ on } C].$$

This introduction of a norm function presents no difficulty, and for the sake of simplicity we do not make the introduction for the Tchebycheff measure of approximation. We do introduce a norm function, however, for the integral measures of approximation 2)—5).

The measures of approximation 1)—5) have recently been used by the present writer in the study of approximation to given analytic functions by polynomials,[2] and results analogous to Theorem I have been established. It is to be noticed that in Theorem I the case that $f(z)$ is an entire function leads

distribution there is loc. cit.) but a single rational function of best approximation, and hence independently of this distribution there are but a finite number of functions $r_n(z)$ of best approximation.

[1] The reader may notice from the discussion which follows that in all of these cases the reasoning we give is valid or can be modified so as to be valid even if the complement of C is finitely multiply connected, provided that C contains no isolated point.

[2] Transactions of the American Mathematical Society, vol. 32 (1930), pp. 794—816, and vol. 33 (1931), pp. 370—388.

precisely to the approximation of $f(z)$ by polynomials, which case has been treated with others in the papers just mentioned.

The results just referred to are perhaps worth stating in detail so that they can be compared with Theorem I. Special cases of these results are naturally due to various other writers; we shall have occasion later to mention the special case due to S. Bernstein. Let C be an arbitrary limited closed point set of the z-plane and denote by D the set of all points each of which can be joined to the point at infinity by a broken line which does not meet C. We suppose D to be simply connected. Let the function $w = \varphi(z)$ map D onto the exterior of $|w| = 1$ so that the points at infinity correspond to each other and denote by C_R the curve $|\varphi(z)| = R > 1$ in the z-plane, namely the image of the circle $|w| = R$. If the function $f(z)$ is analytic interior to C_R but has a singularity on C_R, the sequence of polynomials $\pi_n(z)$ of best approximation to $f(z)$ on C (measured in any one of the ways 1)—5) provided that in 5) the point set C is a closed region), converges to $f(z)$ for z interior to C_R, uniformly for z on any closed point set interior to C_R, and converges uniformly in no region containing in its interior a point of C_R. If $R_1 < R$ and if the measure of approximation μ_n involves the p-th power of $|f(z) - \pi_n(z)|$, $p > 0$, then the inequality

$$\mu_n \leq \frac{M}{R_1^{np}}, \qquad n = 1, 2, \ldots,$$

is valid, where M depends on R_1 but not on n, but this inequality is valid for no choice of $R_1 > R$.

A somewhat trivial but nevertheless illuminating illustration of the difference between polynomials and more general rational functions when used for the approximation of a given function, occurs for a function $f(z)$, approximated in the sense of least squares on the unit circle C: $|z| = 1$ and having a single singularity in the plane, namely at the point $z = a$ whose modulus is greater than unity. The sequence of polynomials of best approximation to $f(z)$ on C in the sense of least squares is the sequence of partial sums of the Taylor development of $f(z)$ at the origin. This sequence $\{\pi_n(z)\}$ converges in such a way that we have

$$\int_C |f(z) - \pi_n(z)|^2 |dz| \leq \frac{M}{R^{2n}}, \qquad n = 1, 2, \ldots,$$

where R is an arbitrary number less than $|a|$, but this inequality holds for no choice of M when R is greater than $|a|$. The sequence $\{\pi_n(z)\}$ converges for $|z| < |a|$ and diverges for $|z| > |a|$.

On the other hand, if we study the best approximation to $f(z)$ on C in the sense of least squares by rational functions $r_n(z)$ of respective degrees n whose poles lie in the point $z = a$, the inequality

$$\int_C |f(z) - r_n(z)|^2 |dz| \leqq \frac{M}{R^{2n}}, \qquad n = 1, 2, \ldots,$$

is satisfied for *an arbitrary* R, provided that a suitable M (depending naturally on R) is chosen. This fact is easily proved for itself by help of the transformation $w = (1 - \bar{a} z)/(z - a)$ and indeed follows from Theorem I, as does the fact that the corresponding sequence $\{r_n(z)\}$ converges to the sum $f(z)$ at every point of the plane other than $z = a$. Thus the degree of approximation is not so great for approximation to $f(z)$ on C by polynomials as for approximation by rational functions with poles in $z = a$, and in the latter case the region of convergence is also greater.

Theorem I is true in the trivial case that C is the entire plane, for in this case $f(z)$ must be a constant and all the approximating rational functions $r_n(z)$ are this same constant. Approximation on C can be measured by either of the methods 1) or 5). Henceforth this trivial case is excluded.

2. **Degree of Approximation.** A preliminary theorem which we shall apply is

Theorem II. *Suppose $f(z)$ is an analytic function of z whose singularities form a set E one of whose derivatives $E^{(k)}$ is empty. Suppose C is a closed point set with no point in common with E. Then there exists a sequence of rational functions $r_n(z)$ of respective degrees n whose poles lie on E such that for an arbitrary R we have*

$$|f(z) - r_n(z)| \leqq \frac{M}{R^n}, \qquad z \text{ on } C,$$

where M depends on R but not on n.

We prove Theorem II first for the case that E' is empty, so that E consists of a finite number of points $A_1, A_2, \ldots, A_\nu$.

The function $f(z)$ can be expressed as the sum of ν functions, each analytic on the entire extended plane except in a point A_k. In fact, let us assume that

A_1 is the point at infinity; this is no restriction of generality. Then Cauchy's integral

$$f(z) = \sum_{k=1}^{\nu} \frac{1}{2\pi i} \int_{\gamma_k} \frac{f(t)\,dt}{t-z} \tag{2.1}$$

gives this expression directly, if γ_1 is a circle containing $A_2, A_3, \ldots, A_\nu$, and $\gamma_k\,(k > 1)$ is a circle about the point A_k but containing no other point A_j. Equation (2.1) is valid if z lies in the region bounded by these ν circles, integration being taken in the positive sense with respect to this region. The integrals in (2.1) are all independent of the particular circles γ_k chosen, provided merely that the circle γ_1 is sufficiently large and the other circles are sufficiently small; each integral defines a function analytic over the entire extended plane except at a point A_k. Let us introduce the notation

$$f_k(z) = \frac{1}{2\pi i} \int_{\gamma_k} \frac{f(t)\,dt}{t-z}, \tag{2.2}$$

it being understood that the circle γ_k is so chosen as to separate z and A_k, but not to separate z and any other point A_j. The function $f_k(z)$ is thus defined and analytic at every point of the extended plane except at A_k.

The function $f_k(z)$ can be uniformly approximated in the sense of Tchebycheff on the point set C by a sequence of rational functions $r_n^{(k)}(z)$ of respective degrees n whose poles lie in A_k and such that we have

$$|f_k(z) - r_n^{(k)}(z)| \leq \frac{M_k}{R_k^n}, \quad z \text{ on } C, \tag{2.3}$$

where $R_k > 1$ is arbitrary and M_k depends on R_k. In fact, if we transform A_k into the point at infinity by a linear transformation of the complex variable, the successive convergents of degree n of the Taylor development of the transformed $f_k(z)$ about the new origin yield by transformation back to the original situation a suitable set of functions $r_n^{(k)}(z)$. The rational function

$$r_{\nu n}(z) = \sum_{k=1}^{\nu} r_n^{(k)}(z)$$

may be considered of degree νn, so we may write by addition of inequalities (2.3)

$$|f(z) - r_{\nu n}(z)| \leq \frac{M'}{R_1^n}, \quad z \text{ on } C,$$

where all the numbers R_k are chosen the same and M' is the sum of the M_k. This inequality does not yet hold for rational functions of *all* degrees, but we may write

$$(2.4) \qquad |f(z) - r_m(z)| \leq \frac{M}{R^m}, \quad z \text{ on } C,$$

where we set $R = R_1^{1/\nu}$, where we set

$$r_m(z) = r_{\nu n}(z),$$

νn being the smallest multiple of ν not less than m, and where we have $M = M' R_1$. Inequality (2.4) thus holds for all m, where $r_m(z)$ is a rational function of degree m and where $R > 1$ is arbitrary.

This completes the proof of Theorem II in the case that E' is empty. Let us treat next the case that E'' is empty, so that E' consists of a finite number of points $A_1, A_2, \ldots, A_\nu$; we assume that A_1 is the point at infinity. Let $R > 1$ be given. Let γ_1 be a large circle containing $A_2, A_3, \ldots, A_\nu$ in its interior and let $\gamma_2, \gamma_3, \ldots, \gamma_\nu$ be small circles about the points $A_2, A_3, \ldots, A_\nu$ respectively. Let δ denote half the maximum diameter of C. Then the radius of γ_1 is to be chosen larger than $\delta R^{\nu+1}$. The radius of γ_2 is to be chosen so small that when A_2 is transformed to infinity by a linear transformation of the complex variable the radius of the corresponding circle (transform of γ_2) is larger than the product of $R^{\nu+1}$ by half the maximum diameter of the transform of C. The radii of $\gamma_3, \gamma_4, \ldots, \gamma_\nu$ are to be chosen correspondingly. None of these circles γ_i shall pass through a singularity of $f(z)$.

Cauchy's integral

$$f(z) = \sum_{k=1}^{\nu+1} \frac{1}{2\pi i} \int_{\gamma_k} \frac{f(t)\,dt}{t-z}, \quad z \text{ on } C,$$

where $\gamma_{\nu+1}$ is an arbitrary curve or curves separating C from none of the circles $\gamma_1, \ldots, \gamma_\nu$ but separating C from all the singular points of $f(z)$ interior to γ_1 and exterior to $\gamma_2, \gamma_3, \ldots, \gamma_\nu$, expresses $f(z)$ as the sum of $\nu + 1$ functions which are analytic respectively interior to γ_1, exterior to $\gamma_2, \gamma_3, \ldots, \gamma_\nu$, and exterior to $\gamma_{\nu+1}$. The function

$$\frac{1}{2\pi i} \int_{\gamma_{\nu+1}} \frac{f(t)\,dt}{t-z}$$

is independent of the particular curve or curves $\gamma_{\nu+1}$ chosen, and the only singularities of this function are the points of E not exterior to γ_1 or interior to $\gamma_2, \gamma_3, \ldots, \gamma_\nu$.

There exist rational functions $r_n^{(i)}(z)$ of respective degrees n such that we have

$$\left| \frac{1}{2\pi i} \int_{\gamma_i} \frac{f(t)\, dt}{t-z} - r_n^{(i)}(z) \right| \leqq \frac{M^{(i)}}{R^{n(\nu+1)}}, \quad z \text{ on } C, \quad i = 1, 2, \ldots, \nu; \tag{2.5}$$

in fact the rational function $r_n^{(i)}(z)$ may be chosen so as to have all its poles in the point A_i, and the function $r_n^{(i)}(z)$ may be chosen as the sum of the first $n+1$ terms of the Taylor development of the function approximated, about a suitable point, when A_i is transformed to infinity by a suitable linear transformation of the complex variable. It follows from the particular choice of the circles γ_i that the inequality (2.5) will be satisfied by these particular rational functions. The function

$$\frac{1}{2\pi i} \int_{\gamma_{\nu+1}} \frac{f(t)\, dt}{t-z}$$

has as its only singularities in the plane the points of E not exterior to γ_1 or interior to $\gamma_2, \gamma_3, \ldots, \gamma_\nu$, and these singularities of this function are finite in number. Then by the part of Theorem II already established (i. e. E' empty), there exist rational functions $r_n^{(\nu+1)}(z)$ of respective degrees n such that we have

$$\left| \frac{1}{2\pi i} \int_{\gamma_{\nu+1}} \frac{f(t)\, dt}{t-z} - r_n^{(\nu+1)}(z) \right| \leqq \frac{M^{(\nu+1)}}{R^{n(\nu+1)}}, \quad z \text{ on } C.$$

If we set

$$r_{(\nu+1)n}(z) = \sum_{k=1}^{\nu+1} r_n^{(k)}(z),$$

we have a rational function of degree $(\nu+1)n$ with the property

$$|f(z) - r_{(\nu+1)n}(z)| \leqq \frac{M_1}{R^{(\nu+1)n}}, \quad z \text{ on } C.$$

We now make the definition

$$r_m(z) = r_{(\nu+1)n}(z),$$

where $(\nu+1)n$ is the smallest multiple of ν not less than m, so by setting $M = M_1 R^{\nu+1}$ we have the inequality

$$|f(z) - r_m(z)| \leqq \frac{M}{R^m}, \quad z \text{ on } C,$$

which holds for all values of m, $R > 1$, and Theorem II is established in the case that E'' is empty.

A formal proof of Theorem II in the general case that $E^{(k)}$ is empty follows directly the proof just given, by the use of mathematical induction, and the details are left to the reader.

In the present paper we are primarily concerned with the rational functions of degree n of *best* approximation. By Theorem II there exists *some* sequence of rational functions $r_n(z)$ of respective degrees n whose poles lie on the set E such that (2.4) is satisfied. It follows directly that the same inequality must be valid for the sequence of rational functions $r_n(z)$ of *best* approximation in the sense of Tchebycheff whose poles lie on E.

3. **A Theorem on Overconvergence.** Another preliminary theorem which we shall have occasion to use is

Theorem III. *If the sequence of rational functions $r_n(z)$ of respective degrees n converges in a region C' (containing no limit point of poles of the $r_n(z)$) in such a way that we have for every R*

$$(3.1) \qquad |f(z) - r_n(z)| \leqq \frac{M}{R^n}, \quad z \text{ in } C',$$

where M depends on R but not on n, then the sequence $\{r_n(z)\}$ converges and $f(z)$ is analytic at every point of the extended plane except the limit points of poles of the functions $r_n(z)$ and except points separated from C' by such limit points. Convergence is uniform on any closed region C'' containing no such limit point, and for z on C'' an inequality of form (3.1) holds for an arbitrary R provided that M (depending on R) is suitably chosen.

In the proof of Theorem III we need to apply a lemma of which a special case was first used by S. Bernstein. The proof of the present lemma is inspired directly by the proof of Bernstein's special case given by Marcel Riesz in a letter to Mittag-Leffler.[1] The entire discussion of the present paper is analogous to Bernstein's discussion in which his lemma was proved. His chief result in this connection is that if a function $f(z)$ is analytic on and within the ellipse with foci 1 and -1 and semi-axes a and b, then there exist polynomials $p_n(z)$ of

[1] Acta mathematica vol. 40 (1916), pp. 337—347.

respective degrees n such that we have

$$|f(z) - p_n(z)| \leqq \frac{M}{\varrho^n}, \qquad -1 \leqq z \leqq +1, \quad \varrho = a + b. \tag{3.2}$$

Reciprocally, if there exist polynomials $p_n(z)$ of respective degrees n such that (3.2) is satisfied for $-1 \leqq z \leqq +1$, for a certain value of ϱ, then the function $f(z)$ is analytic interior to the ellipse described. Our Theorem II is the analogue of the first part of Bernstein's theorem, and our Theorem III is the analogue of the second part of that theorem.

The following lemma has already been established elsewhere,[1] although in a slightly less general form, but the proof is simple and typical of other proofs to be given, and so will be repeated.

Lemma I. *Let Γ be an arbitrary closed limited point set of the z-plane whose complement is simply connected, and denote by $w = \Phi(z)$, $z = \Psi(w)$, a function which maps the complement of Γ onto the exterior of the unit circle γ in the w-plane so that the two points at infinity correspond to each other. Let Γ_R denote the curve $|\Phi(z)| = R > 1$ in the z-plane, the transform of the circle $|w| = R$. If $P(z)$ is a rational function of degree n whose poles lie exterior to Γ_ϱ, $\varrho > 1$, and if we have*

$$|P(z)| \leqq L, \quad z \text{ on } \Gamma,$$

then we have likewise

$$|P(z)| \leqq L\left(\frac{\varrho R_1 - 1}{\varrho - R_1}\right)^n, \quad z \text{ on } \Gamma_{R_1}, \quad R_1 < \varrho.$$

In the statement of Lemma I we have, as a matter of convenience, required that Γ should be limited and that in the conformal mapping the point at infinity in the z-plane should be transformed into the point at infinity in the w-plane. The result can naturally be phrased in terms of an arbitrary closed point set Γ, where in the conformal mapping an arbitrary point of the complement of Γ is transformed into the point at infinity in the w-plane.

The function $P[\Psi(w)]$ has at most n poles for $|w| \geqq 1$ and these all lie exterior to $|w| = \varrho$. For convenience in exposition we suppose that there are precisely n poles $\alpha_1, \alpha_2, \ldots, \alpha_n$, not necessarily all distinct, and that none of them lies at infinity. If there are less than n poles, or if infinity is also a pole,

---

[1] Walsh, Transactions of the American Mathematical Society, vol 30 (1928), pp. 838—847; p. 842.

there are only obvious modifications to be made in the discussion. The function

$$\pi(w) = P[\Psi(w)] \frac{(w-\alpha_1)(w-\alpha_2)\cdots(w-\alpha_n)}{(1-\bar{\alpha}_1 w)(1-\bar{\alpha}_2 w)\cdots(1-\bar{\alpha}_n w)} \tag{3.3}$$

is analytic for $|w| > 1$. When $w\,(|w| > 1)$ approaches γ, z approches C, and all limiting values of $|P[\Psi(w)]|$ are less than or equal to L; the function $(w-\alpha_i)/(1-\bar{\alpha}_i w)$ is continuous and has the modulus unity on γ, from which it follows that the limiting values of $|\pi(w)|$ for w approaching $\gamma\,(|w| > 1)$ are not greater than L. Then we have

$$|\pi(w)| \leqq L \tag{3.4}$$

for $|w| > 1$, since the function $|\pi(w)|$ can have no maximum for $|w| > 1$.

The transformation $\zeta = (w-\alpha_i)/(1-\bar{\alpha}_i w)$ transforms $|w| = R_1$ into the circle $|(\zeta+\alpha_i)/(1+\bar{\alpha}_i\zeta)| = R_1$, so we have

$$|\zeta| \geqq \frac{|\alpha_i| - R_1}{R_1|\alpha_i| - 1} \geqq \frac{\varrho - R_1}{R_1\varrho - 1} \qquad \text{for } |w| = R_1 < \varrho.$$

Thus we find from (3.3) and (3.4),

$$|P[\Psi(w)]| \leqq L\prod_{i=1}^{n}\left|\frac{1-\bar{\alpha}_i w}{w-\alpha_i}\right| \leqq L\left(\frac{R_1\varrho - 1}{\varrho - R_1}\right)^n$$

for $|w| = R_1 < \varrho$, and Lemma I is established.

Lemma I in the form in which we have considered it, is not expressed so as to be invariant under all linear transformations of the complex variable. That it to say, a suitable linear transformation yields a new result. One way in which we shall apply Lemma I is in proving the following remark:

If the sequence of rational functions $P_n(z)$ of respective degrees n satisfy the inequality

$$|P_n(z)| \leqq \frac{M_1}{R_1^n}, \qquad z \text{ in a circular region } K,$$

for every value of R_1, where M_1 depends on R_1, and if the circular region K' contains K but contains on or within it no limit point of poles of the functions $P_n(z)$, then the inequality

$$|P_n(z)| \leq \frac{M_2}{R_2^n}, \quad z \text{ in } K'.$$

is satisfied for every value of R_2, where M_2 depends on R_2.

In the sense here considered, a circular region is the closed interior or exterior of a circle, or a closed half-plane. The proof of the remark follows directly from the lemma, by transforming the given circular regions K and K' into two regions bounded by concentric circles of respective radii 1 and $\varrho_1 > 1$. This transformation is naturally to be a linear transformation of the complex variable, and to prove the remark we need merely set

$$\frac{1}{R_2} = \frac{1}{R_1} \frac{\varrho \varrho_1 - 1}{\varrho - \varrho_1}, \quad M_1 = M_2,$$

where the circle concentric with K and K' of radius ϱ contains K' but contains no limit point of poles of the functions $P_n(z)$.

Theorem III follows directly from Lemma I and from the remark just made. From the inequalities

$$|f(z) - r_{n-1}(z)| \leq \frac{M}{R^{n-1}}, \quad z \text{ in } C',$$

$$|f(z) - r_n(z)| \leq \frac{M}{R^n}, \quad z \text{ in } C',$$

we derive

$$|r_n(z) - r_{n-1}(z)| \leq \frac{N}{R^n}, \quad z \text{ in } C',$$

where $N = M(1 + R)$. The function $r_n(z) - r_{n-1}(z)$ is rational of degree $2n - 1$, so if the point set C' is limited, and this situation can be reached by a linear transformation, we obtain from the lemma

$$|r_n(z) - r_{n-1}(z)| \leq \frac{N}{R^n} \left(\frac{\varrho R_1 - 1}{\varrho - R_1}\right)^{2n-1}, \quad z \text{ on } C'_{R_1}, \quad R_1 < \varrho,$$

for n sufficiently large, where C'_ϱ contains on or within it no limit point of poles of the functions $r_n(z)$, and it is to be remembered that R is arbitrary. It follows that the sequence $\{r_n(z)\}$ converges interior to any curve C'_{R_1} which contains on or within it no limit point of poles of that sequence, and that in any such curve C'_{R_1} the inequality

$$|r_n(z) - r_{n-1}(z)| \leq \frac{M}{R^n} \tag{3.5}$$

holds for an arbitrary choice of R.

By a method entirely analogous to that of analytic extension[1] it can now be shown from the remark following Lemma I that this same inequality holds in the region C'' of Theorem III. Inequality (3.5), holding in some region C'_{R_1}, holds also in any circular region containing no limit point of poles of the $r_n(z)$ but having a subregion in common with C'_{R_1}. The process of extending step by step the domain of known validity of (3.5) can be stopped only by limit points of poles of the $r_n(z)$, and any point set such as the C'' prescribed in Theorem III can be included in this domain by a finite number of steps. The uniform convergence on C'' of the sequence $\{r_n(z)\}$ follows directly from (3.5), and the identity of the limit function with $f(z)$ (or its analytic extension) follows from (3.1) for z in C' and hence for z on C''. There is no difficulty in deriving (3.1) for z on C'' from (3.5) for z on C'', so the proof of Theorem III is complete.

We have now a proof of Theorem I in the case 1), that approximation is measured in the sense of Tchebycheff. Inequality (3.1) holds for the sequence of rational functions of best approximation whose poles lie in E, as we have already indicated, and the conclusion of Theorem I follows from Theorem III.

4. **Approximation measured after Conformal Mapping.** We now take up the measure 2) of approximation to $f(z)$ on C, that C is an arbitrary closed set not a single point whose complement is simply connected, and approximation is measured in the sense of weighted p-th powers ($p > 0$) by integration on the circle $\gamma: |w| = 1$ when the complement of C is mapped onto the exterior of γ. This measure of approximation naturally depends on the particular point O' of the complement of C chosen to correspond to the point at infinity in the w-plane, but the problem of best approximation for a particular choice of O' with a particular choice of the norm function $n(w)$ is equivalent to the problem of best approximation for an arbitrary choice of O' with a suitable norm function $n(w)$. In the present paper we suppose O', once determined, to be fixed. A similar

[1] Ostrowski has indicated the close analogy between analytic extension and the study of regions of convergence of certain series. See for instance Abhandlungen aus dem Mathematischen Seminar der Hamburgischen Universität, Vol. 1 (1922), pp. 327—350.

remark applies to the norm function and to the particular map chosen for our other measures of approximation involving conformal mapping.

We shall need the following lemma:

Lemma II. *Let C be an arbitrary limited closed point set of the z-plane, not a single point, whose complement is simply connected, and denote by $w = \Phi(z)$, $z = \Psi(w)$ a function which maps the complement of C onto the exterior of the unit circle γ in the w-plane so that the points at infinity correspond to each other. Let C_R denote the curve $|\Phi(z)| = R > 1$ in the z-plane. If $P(z)$ is a rational function of degree n whose poles lie exterior to C_ϱ, $\varrho > 1$, and if we have*

$$(4.1) \qquad \int_\gamma |P(z)|^p |dw| \leqq L^p, \quad p > 0,$$

then we have likewise

$$|P(z)| \leqq LL' \left(\frac{\varrho R_1 - 1}{\varrho - R_1}\right)^n, \quad z \text{ on } C_{R_1}, \quad R_1 < \varrho,$$

where L_1 depends on R_1 but not on $P(z)$.

Properly speaking, the function $P[\Psi(w)]$ is not defined on γ, and therefore the use of the integral (4.1) requires some explanation. The function $\Psi(w)/w$ is analytic and uniformly limited for $|w| > 1$, and therefore by Fatou's theorem[1] this function and hence the function $\Psi(w)$ approaches a limit almost everywhere on γ when w remains exterior to γ and approaches γ along a radius. When w approaches γ, the function $z = \Psi(w)$ approaches a boundary point of C and hence $P[\Psi(w)]$ approaches a limit. It is these values of $P[\Psi(w)]$, which therefore exist almost everywhere on γ, that are intended to be used in the integral in (4.1). A similar fact holds for the other measures of approximation that we shall use which depend on conformal mapping.

The proof of Lemma II is quite similar to the proof of Lemma I. The function $P[\Psi(w)]$ has at most n poles for $|w| \geqq 1$, and these all lie exterior to $|w| = \varrho$. For convenience in exposition we shall suppose that there are precisely n poles $\alpha_1, \alpha_2, \ldots, \alpha_n$, not necessarily all distinct, and that none lies at infinity. If there are less than n poles, or if infinity is also a pole, there are only obvious modifications to be made in the discussion. In the latter case, for instance, we consider in the right-hand member of (4.2) the function found by taking the

[1] Acta mathematica, vol. 30 (1906), pp. 335—400.

limit as one or more of the α_i become infinite. Similarly let $\beta_1, \beta_2, \ldots, \beta_n$ denote the zeros of $P[\Psi(w)]$ exterior to γ. The function

$$(4.2)\quad \pi(w) = P[\Psi(w)]\frac{(w-\alpha_1)(w-\alpha_2)\cdots(w-\alpha_n)(1-\bar{\beta}_1 w)(1-\bar{\beta}_2 w)\cdots(1-\bar{\beta}_n w)}{(1-\bar{\alpha}_1 w)(1-\bar{\alpha}_2 w)\cdots(1-\bar{\alpha}_n w)(w-\beta_1)(w-\beta_2)\cdots(w-\beta_n)}$$

is analytic and different from zero for $|w| > 1$, and on $\gamma: |w| = 1$ we have for the values taken on by normal approach to γ,

$$|\pi(w)| = |P[\Psi(w)]|.$$

The hypothesis of Lemma II is therefore

$$\int_\gamma |\pi(w)|^p |dw| \leqq L^p, \qquad p > 0.$$

We transform now by the substitution $w = 1/w'$; the function $\pi(1/w')$ is analytic and different from zero for $|w'| < 1$, and so also is the function $[\pi(1/w')]^p$, if we consider a suitable determination of the possibly multiple valued function. Cauchy's formula

$$[\pi(1/w')]^p = \frac{1}{2\pi i}\int_\gamma \frac{[\pi(1/w'')]^p\, dw''}{w'' - w'}$$

yields the inequality

$$|\pi(1/w')|^p \leqq \frac{1}{2\pi}\int_\gamma |\pi(1/w'')|^p \frac{|dw''|}{1-r}, \qquad \text{for } |w'| \leqq r < 1,$$

or

$$|\pi(1/w')|^p \leqq \frac{L^p}{2\pi(1-r)}, \qquad |w'| \leqq r < 1,$$

which is the same as

$$|\pi(w)|^p \leqq \frac{L^p}{2\pi(1-r)}, \qquad |w| \geqq \frac{1}{r} > 1.$$

The function $(1-\bar{\beta}_i w)/(w-\beta_i)$ has a modulus greater than unity for $|w| < 1$, so this last inequality implies

$$\left| P[\Psi(w)]\frac{(w-\alpha_1)(w-\alpha_2)\cdots(w-\alpha_n)}{(1-\bar{\alpha}_1 w)(1-\bar{\alpha}_2 w)\cdots(1-\bar{\alpha}_n w)}\right| \leqq \frac{L}{[2\pi(1-r)]^{\frac{1}{p}}}, \qquad |w| \geqq \frac{1}{r} > 1.$$

It is readily shown that

$$\left|\frac{1 - \bar{\alpha}_i w}{w - \alpha_i}\right| \leq \frac{R_1 \varrho - 1}{\varrho - R_1}, \qquad \text{for } |w| = R_1 < \varrho,$$

and from this inequality Lemma II follows immediately.

Lemma II is in reality more general than Lemma I, in the sense that it yields an easy proof of Lemma I, but we shall find it nevertheless convenient to have Lemma I for reference.

Let us now prove Theorem I in case approximation is measured by the method 2). By Theorem II there exists a sequence of rational functions $r_n(z)$ of respective degrees n with their poles in the set E such that we have for an arbitrary R

$$|f(z) - r_n(z)| \leq \frac{M}{R^n}, \quad z \text{ on } C,$$

where M depends on R but not on n. The present measure of approximation of $r_n(z)$ to $f(z)$ is

$$\int_\gamma n(w)\,|f(z) - r_n(z)|^p\,|dw|, \qquad p > 0,$$

where $n(w)$ is continuous and positive on γ. An inequality of the form

$$\int_\gamma n(w)\,|f(z) - r_n(z)|^p\,|dw| \leq \frac{M'}{R^{np}} \tag{4.3}$$

is satisfied for the particular rational functions $r_n(z)$ just mentioned, and so this same inequality holds for the sequence of rational functions $r_n(z)$ of best approximation. If we have $0 < n' < n(w)$ for w on γ, inequality (4.3) implies

$$\int_\gamma |f(z) - r_n(z)|^p\,|dw| \leq \frac{M'}{n' R^{np}}. \tag{4.4}$$

We are now in a position to use inequality (4.4) for two successive values of n and to apply the general inequalities

$$\begin{aligned} \int |\chi_1 + \chi_2|^p\,dx &\leq 2^{p-1}\int |\chi_1|^p\,dx + 2^{p-1}\int |\chi_2|^p\,dx, \qquad p > 1, \\ \int |\chi_1 + \chi_2|^p\,dx &\leq \int |\chi_1|^p\,dx + \int |\chi_2|^p\,dx, \qquad 0 < p \leq 1. \end{aligned} \tag{4.5}$$

There results the inequality

$$\int_\gamma |r_n(z) - r_{n-1}(z)|^p \, |dw| \leqq \frac{M_1}{R^{np}},$$

where R is arbitrary and M_1 depends on R. Our conclusion follows now from Lemma II by the method used in connection with Lemma I.

It will be noted that the function $F(z)$ to which the sequence $r_n(z)$ converges must coincide with $f(z)$ on γ, for the inequality

$$\int_\gamma |F(z) - r_n(z)|^p \, |dw| \leqq \frac{M}{R^{np}},$$

which is a consequence of (3. 5), yields by (4. 4) and (4. 5)

$$\int_\gamma |F(z) - f(z)|^p \, |dw| \leqq \frac{N}{R^{np}}.$$

Hence the integral on the left is zero and the functions $F(z)$ and $f(z)$ coincide almost everywhere on γ. Thus $F(z)$ and $f(z)$ coincide at an infinity of points of C and are identical.

5. **Approximation measured by a Line Integral.** Let us now turn to method 3) as a measure of approximation, namely that C is an arbitrary closed limited point set whose boundary is a rectifiable Jordan arc or curve or other point set of positive linear measure, and whose complement (i. e. of C) is simply connected; approximation is measured in the sense of weighted p-th powers ($p > 0$) by a line integral over C. In particular C may be a region bounded by a rectifiable Jordan curve — in this case the proof of Theorem I is especially simple — or may be composed of even a suitable infinity of such regions, together with Jordan arcs abutting on and exterior to them. This measure of approximation (for $p = 2$) has been used by Szegö for approximation of given functions by polynomials[1] in case C is either a Jordan curve or arc.

We shall need the following lemma:

Lemma III. *Let C be an arbitrary closed limited point set whose boundary C' has positive linear measure, whose complement is simply connected, and denote by $w = \Phi(z)$, $z = \Psi(w)$ a function which maps the complement of C onto the exterior*

[1] See particularly Mathematische Zeitschrift, vol. 9 (1921), pp. 218—270.

of the unit circle γ in the w-plane so that the points at infinity correspond to each other. Let C_R denote the curve $|\Phi(z)| = R > 1$ in the z-plane. If $P(z)$ is a rational function of degree n whose poles lie exterior to C_ϱ, $\varrho > 1$, and if we have

$$\int_{C'} |P(z)|^p |dz| \leqq L^p, \qquad p > 0, \tag{5.1}$$

then we have likewise

$$|P(z)| \leqq LL' \left(\frac{\varrho R_1 - 1}{\varrho - R_1}\right)^n, \qquad z \text{ on } C_{R_1}, \quad R_1 < \varrho,$$

where L' depends on R_1 but not on $P(z)$.

The boundary C' is composed of a connected set consisting of a finite or infinite number of Jordan curves and arcs, and we shall need later to consider the plane cut along C'. For the truth of Lemma III and of Theorem I in case approximation is measured by 3), it is immaterial whether in such an integral as (5.1) [or (5.3)] we consider the cut plane or uncut plane; in the cut plane Jordan arcs belonging to C' not parts of Jordan curves belonging to C' are naturally to be counted twice in the integral. However, we shall later use Cauchy's integral formula for the region D complementary to C. If an integral is extended over a curve K in D and if K varies monotonically so that every point of D is exterior to some position of K, then K approaches as a limiting position the point set C', *where the plane is cut along* C' — that is, where each arc of C' not part of the boundary of a region belonging to C is counted doubly. As a matter of convenience, then, we shall suppose that in considering integrals over C', each arc of C' not part of the boundary of a region belonging to C is counted doubly. The weight function $n(z)$ used below may, if desired, be considered to have two distinct values at points of such an arc C', corresponding to the double valence.

The function $\Phi(z)$ is continuous in the z-plane cut along the point set C'. Let the poles and zeros of $P(z)$ on the complement of C be respectively $\alpha_1, \alpha_2, \ldots, \alpha_n$ and $\beta_1, \beta_2, \ldots, \beta_n$. We may have less than n zeros or poles or both, but that requires only a slight and obvious modification in the reasoning now to be used. The function

$$\pi(z) = P(z) \frac{\Phi(z) - \Phi(\alpha_1)}{1 - \bar{\Phi}(\alpha_1)\Phi(z)} \frac{\Phi(z) - \Phi(\alpha_2)}{1 - \bar{\Phi}(\alpha_2)\Phi(z)} \cdots \frac{\Phi(z) - \Phi(\alpha_n)}{1 - \bar{\Phi}(\alpha_n)\Phi(z)}$$
$$\cdot \frac{1 - \bar{\Phi}(\beta_1)\Phi(z)}{\Phi(z) - \Phi(\beta_1)} \frac{1 - \bar{\Phi}(\beta_2)\Phi(z)}{\Phi(z) - \Phi(\beta_2)} \cdots \frac{1 - \bar{\Phi}(\beta_n)\Phi(z)}{\Phi(z) - \Phi(\beta_n)}$$

is analytic and different from zero on the complement D of C, and so therefore is $[\pi(z)]^p$. On C', the two functions $\pi(z)$ and $P(z)$ have the same modulus.

The function $[\pi(z)]^p/\Phi(z)$ is analytic at infinity and vanishes there, so we have

$$\frac{[\pi(z)]^p}{\Phi(z)} = \frac{1}{2\pi i}\int_{C'} \frac{[\pi(t)]^p}{\Phi(t)} \frac{dt}{t-z}, \quad z \text{ in } D,$$

(5.2)

$$\left|\frac{[\pi(z)]^p}{\Phi(z)}\right| \leqq \frac{L^p}{2\pi\delta}, \quad z \text{ on } C'_{R_1},$$

where δ is suitably chosen. The integral over C' is the ordinary integral, in the positive direction with respect to D.

The function $[1 - \bar{\Phi}(\beta_i)\Phi(z)]/[\Phi(z) - \Phi(\beta_i)]$ has a modulus greater than unity for z on C_{R_1}, and for z on C_{R_1} the function $[\Phi(z) - \Phi(\alpha_i)]/[1 - \bar{\Phi}(\alpha_i)\Phi(z)]$ has a modulus not less than $(\varrho - R_1)/(\varrho R_1 - 1)$, so Lemma III follows at once.

The method of application of Lemma III to the proof of Theorem I is quite similar to that of Lemma II. By Theorem II there exists some sequence of rational functions $r_n(z)$ of respective degrees n with their poles in the set E such that we have for an arbitrary R

$$|f(z) - r_n(z)| \leqq \frac{M}{R^n}, \quad z \text{ on } C.$$

Our present measure of approximation is

$$\int_{C'} n(z)\,|f(z) - r_n(z)|^p\,|dz|, \qquad p > 0,$$

where $n(z)$ is continuous and positive on C'. An inequality of the form

(5.3)
$$\int_{C'} n(z)\,|f(z) - r_n(z)|^p\,|dz| \leqq \frac{M'}{R^{np}}$$

is satisfied for this particular set of rational functions $r_n(z)$ and so the same inequality holds for the sequence of rational functions of best approximation. If we have $0 < n' < n(z)$ for z on C', inequality (5.3) implies

$$\int_{C'} |f(z) - r_n(z)|^p\,|dz| \leqq \frac{M'}{n' R^{np}},$$

which, used for two succesive values of n, implies by the use of inequality (4. 5)

$$\int_{C'} |r_n(z) - r_{n-1}(z)|^p |dz| \leq \frac{M_1}{R^{np}}.$$

This inequality is of the precise form for application of Lemma III, and by the methods already used yields Theorem I for the measure of approximation which we have been considering.

6. **Approximation in a Region; Conformal Mapping.** Method 4) of measuring approximation is next to be studied, namely that C is an arbitrary simply connected region and approximation is measured in the sense of weighted p-th powers ($p > 0$) by integration over the circle $\gamma : |w| = 1$ when the interior of C is mapped conformally onto the interior of γ. This method (without the use of a weight·function and for $p > 1$) has recently been used by Julia[1] in the study of approximation of harmonic functions by harmonic polynomials.

If we are dealing with either of the measures of approximation 4) or 5), lemmas precisely analogous to those already established may be used, but it is just as convenient to proceed in a somewhat different way. Let us prove[2]

Lemma IV. *If each of the functions $P_n(z)$ is analytic and bounded interior to the simply connected region C and if we have*

$$(6.\ 1) \qquad \int_{\gamma} |P_n(z)|^p |dw| \leq L^p, \qquad p > 0,$$

where the interior of C is mapped onto the interior of $\gamma : |w| = 1$, then we have

$$|P_n(z)| \leq L' L,$$

for z on an arbitrary closed point set C' interior to C, where L' depends on C' but not on $P_n(z)$.

In the integral in (6. 1) the value of $|P_n(z)|$ on γ is naturally to be taken in the sense of normal approach to γ; these boundary values are known to exist.

Let the zeros of $P_n(z)$, if any, interior to C be $\alpha_1, \alpha_2, \ldots$ We assume $P_n(z)$ not identically zero, for the lemma is obviously true so far as concerns such functions. Consider the function

[1] Acta Litterarum ac Scientiarum (Szeged), vol. 4 (1929), pp. 217—226.

[2] Compare Walsh, Transactions of the American Mathematical Society, vol. 33 (1931), pp. 370—388.

$$F_n(z) = \frac{P_n(z)\, \Pi\, |\varphi(\alpha_i)|}{\Pi \dfrac{\varphi(z) - \varphi(\alpha_i)}{\varphi(z) - \dfrac{1}{\bar{\varphi}(\alpha_i)}}}, \tag{6. 2}$$

where $w = \varphi(z)$, $z = \psi(w)$ is a function which maps the interior of C conformally onto the interior of γ. There may be an infinity of points α_i but if so the infinite products here converge, by Blaschke's theorem. We assume $\varphi(\alpha_i) \neq 0$, which involves no loss of generality, for the following reasoning concerning $P_n(z)$ may be applied to the quotient by $[\varphi(z)]^k$ of a given $P_n(z)$, where k is the order of the zero of the given $P_n(z)$ at the point $z = \psi(0)$. The function $F_n(z)$ is analytic and different from zero interior to C, and has the same modulus as $P_n(z)$ on C or on γ. The function $[F_n(z)]^p$ is likewise analytic and uniformly bounded interior to C and γ, if we consider an arbitrary determination of the p-th power at an arbitrary point interior to C or γ and is analytic extension, so we have Cauchy's integral

$$\{F_n[\psi(w)]\}^p = \frac{1}{2\pi i} \int_\gamma \{F_n[\psi(t)]\}^p \frac{dt}{t - w};$$

Cauchy's integral is naturally valid here, for the boundary values of $\psi(w)$ and hence of $F_n(z)$ for normal approach to γ exist almost everywhere.

It follows now that we have

$$|F_n(z)|^p \leqq \frac{1}{2\pi(1-r)} \int_\gamma |F_n(z)|^p |dw| \leqq \frac{L^p}{2\pi(1-r)}, \qquad |\varphi(z)| \leqq r < 1.$$

Each function

$$\frac{|\varphi(\alpha_i)|}{\dfrac{\varphi(z) - \varphi(\alpha_i)}{\varphi(z) - \dfrac{1}{\bar{\varphi}(\alpha_i)}}}$$

is of absolute value greater than unity for z interior to C, so we have from (6 2)

$$|P_n(z)|^p \leqq |F_n(z)|^p \leqq \frac{L^p}{2\pi(1-r)}, \qquad |\varphi(z)| \leqq r < 1,$$

and the proof of the lemma is complete.

The application of Lemma IV is immediate. By Theorem III there exist rational functions $r_n(z)$ of respective degrees n with poles on the set E such that we have

$$|f(z) - r_n(z)| \leqq \frac{M}{R^n}, \quad z \text{ on } C,$$

so the inequality

$$\int_\gamma n(w)\,|f(z) - r_n(z)|^p\,|dw| \leqq \frac{M'}{R^{np}},$$

where the weight function $n(w)$ is positive and continuous on γ, is satisfied for this particular set of rational functions and hence for the rational functions of best approximation. This leads in turn to inequalities of the form

$$\int_\gamma |f(z) - r_n(z)|^p\,|dw| \leqq \frac{M'}{n'\,R^{np}}.$$

This last inequality yields by Lemma IV

$$|f(z) - r_n(z)| \leqq \frac{M_2}{R^n}, \quad z \text{ on } C',$$

where C' is an arbitrary closed point set interior to C and where R is arbitrary, whence Theorem I follows by Theorem III for the measure of approximation that we are here considering.

7. **Approximation in a Region; Surface Integrals.** Method 5) of measuring approximation involves the use of a double integral,

$$\iint_C n(z)\,|f(z) - r_n(z)|^p\,dS, \qquad p > 0,$$

and this method has been used by Carleman[1] in considering the approximation to an analytic function by polynomials. We shall find it convenient to prove

Lemma V. *If each of the functions $P_n(z)$ is analytic interior to an arbitrary region C, and if we have*

(7.1) $$\iint_C |P_n(z)|^p\,dS \leqq L^p \qquad p > 0,$$

[1] Arkiv för Matematik, Astronomi och Fysik, vol. 17 (1922—23).

then we have

$$|P_n(z)| \leqq L' L \tag{7.2}$$

for z on an arbitrary closed point set C'' interior to C, where L' depends on C'' but not on $P_n(z)$.

The integral

$$\frac{1}{2\pi}\int_0^{2\pi} |P_n(z_0 + re^{i\theta})|^p \, d\theta, \qquad p > 0,$$

is well known to be a non-decreasing function of r, in an arbitrary circle K which together with its interior lies interior to C. Here (r, θ) are polar coordinates with pole at the point z_0. The limit of this integral as r approaches zero is obviously $|P_n(z_0)|^p$, from which follows the inequality

$$|P_n(z_0)|^p \leqq \frac{1}{2\pi}\int_0^{2\pi} |P_n(z_0 + re^{i\theta})|^p \, d\theta.$$

We multiply both members of this inequality by $r\,dr$ and integrate from zero to k, the radius of K. The resulting inequality is

$$\frac{k^2}{2}|P_n(z_0)|^p \leqq \frac{1}{2\pi}\iint_K |P_n(z)|^p \, dS,$$

so we may write by virtue of (7. 1)

$$|P_n(z_0)|^p \leqq \frac{1}{\pi k^2}\iint_K |P_n(z)|^p \, dS \leqq \frac{1}{\pi k^2}\iint_C |P_n(z)|^p \, dS \leqq \frac{L^p}{\pi k^2}.$$

This inequality holds for every point z_0 interior to C provided merely that the distance from z_0 to the boundary of C is not less than k. The inequality therefore holds for proper choice of k for z_0 on an arbitrary closed point set C' interior to C and is equivalent to (7. 2) for z on C', so the lemma is completely established.

The application of Lemma V in the proof of Theorem I does not differ materially from the application of Lemma IV and is left to the reader. Theorem I is now completely proved.

55—31104. *Acta mathematica.* 57. Imprimé le 4 septembre 1931.

8. **Further Remarks.** There are three problems, distinct from those already treated, which are intimately connected with the discussion given. We mention merely the statement of the problems and leave the details to the reader. In each of the three cases some new results can be found directly from our previous work, while other new results lie but little deeper.

1. The given function $f(z)$ may be meromorphic instead of analytic on C and the approximating rational functions $r_n(z)$ of respective degrees n are permitted to have poles in all the singularities E of $f(z)$, in particular in the poles of $f(z)$ belonging to C. Under certain conditions it is still true that the sequence of rational functions of best approximation whose poles lie in the singularities of $f(z)$, converges to the limit $f(z)$ on the entire plane except at the singularities of $f(z)$.[1]

2. The given function $f(z)$ may be analytic or meromorphic on C and the given rational functions $r_n(z)$ may be required or not to satisfy auxiliary conditions interior to C, those conditions being the prescription of the values of $r_n(z)$ with perhaps some of its derivatives at various points interior to C; indeed, the functions $r_n(z)$ may be allowed to be meromorphic interior to C, and have their principal parts prescribed at various points interior to C. These auxiliary conditions need have no relation to the given function $f(z)$. If the auxiliary conditions do not depend on n, if the limit function $F(z)$ (which is uniquely determined by $f(z)$ and the auxiliary conditions) of the sequence $r_n(z)$ has all of its singularities in a point set E one of whose derivatives is empty, and if the poles of the approximating functions $r_n(z)$ are merely restricted to lie on E, then under suitable simple restrictions on C, the sequence of rational functions $r_n(z)$ of best approximation to $f(z)$ on the boundary of C in the sense of Tchebycheff and satisfying the auxiliary conditions, converges to the function $F(z)$ on the entire plane except at the singularities of $F(z)$.[2] If the prescribed auxiliary conditions involve merely the coincidence of the values of $r_n(z)$ and the given function $f(z)$ at certain points interior to C, then under suitable conditions we have the conclusion of Theorem I satisfied: the sequence $r_n(z)$ approaches the function $f(z)$ at every point of the plane not on E, uniformly on any closed

[1] Compare Walsh, Transactions of the American Mathematical Society, vol. 30 (1928), pp. 838—847.

[2] Compare Walsh, Transactions of the American Mathematical Society, vol. 32 (1930), pp. 335—390.

point set containing no point of E, where approximation is measured by any of the methods 1)—5).

3. The results of the present paper have application to the study of approximation of harmonic functions by harmonic rational functions. If a suitably restricted harmonic function $u(x, y)$ is given, the function

$$f(z) = u(x, y) + i\,v(x, y), \tag{8.1}$$

where $v(x, y)$ is a function conjugate to $u(x, y)$, satisfies the hypothesis of Theorem I. Approximation to $f(z)$ by rational functions $r_n(z) = r'_n(x, y) + i\,r''_n(x, y)$ implies approximation to $u(x, y)$ by the harmonic rational functions $r'_n(x, y)$. Even if the given function $u(x, y)$ is not so simple that an equation of form (8.1) is valid, where $f(z)$ satisfies the hypothesis of Theorem I, it may be possible to approximate $u(x, y)$ by harmonic rational functions plus harmonic functions involving the logarithms of distances. Such methods of approximation have already been used to some extent by the present writer.[1]

[1] Bulletin of the American Mathematical Society, vol. 35 (1929), pp. 499—544.

On Approximation to an Analytic Function by Rational Functions of Best Approximation[1].

By

J. L. Walsh of Cambridge (Mass.)

1. Introduction. Let C be an arbitrary closed Jordan region in the plane of the complex variable z, and let the function $f(z)$ be analytic interior to C, continuous in the corresponding closed region. For each pair of values of m and n, there exists (see § 6, below) a rational function of the form

$$(1)\quad r_{m,n}(z) = \frac{a_0 z^m + a_1 z^{m-1} + \dots + a_m}{b_0 z^n + b_1 z^{n-1} + \dots + b_n}, \quad b_0 z^n + b_1 z^{n-1} + \dots + b_n \not\equiv 0,$$

which approximates best in the sense of Tchebycheff to $f(z)$ in C. That is to say, there exists a rational function $r_{mn}(z)$ for which the expression

$$\max |f(z) - r_{mn}(z)|, \quad z \text{ in } C,$$

takes on its least value. We denote by $R_{mn}(z)$ this rational function of best approximation; the function $R_{mn}(z)$ need not be unique, but the notation is used to represent any such function. Thus there results a table of double entry not necessarily uniquely defined

$$(2)\quad \begin{array}{llll} R_{00}(z), & R_{01}(z), & R_{02}(z), & \dots, \\ R_{10}(z), & R_{11}(z), & R_{12}(z), & \dots, \\ \dots & \dots & \dots & \dots \end{array}$$

and it is of interest to study the convergence of various sequences formed from the table (2). In the present note it is our purpose to prove the most elementary results regarding such convergence.

Table (2) is entirely analogous to the table of rational fractions of form (1) constructed for a given analytic function $\varphi(z)$ by Padé, with the requirement that the fraction $\varrho_{m,n}(z)$ of the table should be the fraction of form (1) which has contact of highest order with $\varphi(z)$ at the origin. The study of the convergence of various sequences of the fractions $\varrho_{mn}(z)$ is of much interest but quite difficult. Padé considers various continued fractions which can be formed from sequences of

[1]) Presented to the Amer. Math. Soc. Oct. 29, 1932.

functions chosen from the $\varrho_{mn}(z)$. Similarly, continued fractions can be made from various sequences of the functions $R_{mn}(z)$.

In our present table (2) it is to be noted that any function $R_{pq}(z)$ is also a function $r_{p+k,\,q+l}(z)$, $k > 0$, $l > 0$. Thus if we introduce the notation

$$\mu_{m,n} = \max |f(z) - R_{m,n}(z)|, \quad z \text{ in } C,$$

we have the doubly infinite array corresponding to (2):

$$(3) \qquad \begin{array}{llll} \mu_{00}, & \mu_{01}, & \mu_{02}, & \cdots, \\ \mu_{10}, & \mu_{11}, & \mu_{12}, & \cdots, \\ \cdot & \cdot & \cdot & \cdot \end{array}$$

in which the relation

$$(4) \qquad \mu_{p+k,\,q+l} \leqq \mu_{p,q}$$

holds whenever k and l are positive.

2. Special Sequences. The sequence $R_{00}(z)$, $R_{10}(z)$, $R_{20}(z)$, ... is precisely the sequence of polynomials of best approximation to $f(z)$ on C in the sense of Tchebycheff. Under the hypothesis made on $f(z)$ it is known[2]) that we have

$$\lim_{m\to\infty} R_{m0}(z) = f(z)$$

uniformly for z in C. This fact can be expressed as

Lemma I. *We have*

$$\lim_{m\to\infty} \mu_{m0} = 0.$$

The sequence $R_{00}(z)$, $R_{01}(z)$, $R_{02}(z)$, ... consists of zeros or the reciprocals of polynomials. A function $R_{0n}(z)$ which does not vanish identically cannot vanish interior to C. Hence, if the function $f(z)$ does not vanish identically but has one or more zeros interior to C, the equation

$$(5) \qquad \lim_{n\to\infty} R_{0n}(z) = f(z), \text{ uniformly for } z \text{ in } C,$$

is impossible. For by a well known theorem due to Hurwitz, condition (5), where $f(z)$ does not vanish identically, implies that $R_{0n}(z)$ actually takes on the value zero interior to C for n sufficiently large. However, if the function $f(z)$ does not vanish on or within C, we can readily establish equation (5):

---

[2]) Walsh, Math. Annalen 96 (1926), S. 430—436.

Lemma II. *If $f(z)$ is different from zero in the closed region C, then we have*

$$\lim_{n\to\infty} \mu_{0n} = 0.$$

The proof is easy; the function $F(z) = 1/f(z)$ is analytic interior to C and continuous in the corresponding closed region. Hence (by Lemma I) there exist polynomials $P_n(z) = c_{0n} z^n + c_{1n} z^{n-1} + \dots + c_{nn}$ such that we have

$$\lim_{n\to\infty} [\max|f(z) - P_n(z)|, \ z \text{ in } C] = 0.$$

Consequently we may set $r_{0n}(z) = 1/P_n(z)$, at least for n sufficiently large, whence follows

$$\lim_{n\to\infty} r_{0n}(z) = 1/F(z) = f(z), \quad \text{uniformly for } z \text{ in } C.$$

Since this equation holds for the functions $r_{0n}(z)$, the equation holds also if the functions $r_{0n}(z)$ are replaced by the $R_{0n}(z)$, and this is merely another way of stating the conclusion of Lemma II.

Lemma II can be slightly generalized, to admit the case that $f(z)$ vanishes on the boundary of C:

Lemma III. *If $f(z)$ does not vanish interior to C, then we have*

$$\lim_{n\to\infty} \mu_{0n} = 0. \tag{6}$$

Let the interior of C be mapped onto the interior of the unit circle γ of the w-plane by a transformation $w = \varphi(z)$, $z = \psi(w)$. The function $\Psi(w) = f[\psi(w)]$ is continuous on and within C; it may vanish on the circumference γ but does not vanish interior to γ. The function $\Psi(rw)$, where r is positive and slightly less than unity, is analytic and different from zero on and within γ and approximates uniformly the function $\Psi(w)$ in the closed interior of γ. More explicitly, if an arbitrary ϵ be given, then r exists, $0 < r < 1$, such that we have $|\Psi(w) - \Psi(rw)| < \epsilon$ for w on and within γ.

For approximation by rational functions to the function $\Psi[r\varphi(z)]$, which is analytic interior to C, continuous and different from zero in the corresponding closed region, we have by Lemma II,

$$\lim_{n\to\infty} \mu'_{0n} = 0,$$

where μ'_{0n} refers to the approximation not of $f(z)$ but of $\Psi[r\varphi(z)]$. But for suitable choice of r we have

$$|f(z) - \Psi[r\varphi(z)]| < \epsilon, \quad \text{for } z \text{ in } C.$$

Thus we have

$$\overline{\lim_{n\to\infty}} \mu_{0n} \leqq \epsilon,$$

which implies (6).

We are now in a position to prove

Theorem I. *If the function $f(z)$ is analytic and different from zero interior to the Jordan region C and is continuous in the corresponding closed region, then for any infinite sequence of distinct functions formed from* (2) *we have*

$$\lim R_{mn}(z) = f(z) \tag{7}$$

uniformly for z in the closed region C.

If Theorem I is not true, there exists some sequence of numbers μ_{mn} from the array (3) which does not approach zero. There exists some subsequence of this sequence which likewise does not approach zero and which is such that either the first subscript becomes infinite or the second subscript becomes infinite. This contradicts (4) and Lemma I or Lemma III. The proof of Theorem I is complete.

It may be noticed too that equation (7) is valid uniformly for z in the closed region C in the sense of the double limit as m and n become infinite. The proof follows immediately out of (4).

3. Functions Which Vanish in the Region of Approximation. We shall now establish a generalization of Theorem I, to include the case that $f(z)$ vanishes interior to C:

Theorem II. *If the function $f(z)$ is analytic interior to the Jordan region C, continuous in the corresponding closed region, and has precisely ν zeros interior to C, then for any infinite sequence of functions formed from the array*

$$\begin{array}{llll} R_{\nu 0}(z), & R_{\nu 1}(z), & R_{\nu 2}(z), & \ldots, \\ R_{\nu+1,0}(z), & R_{\nu+1,1}(z), & R_{\nu+1,2}(z), & \ldots, \\ \cdots & \cdots & \cdots & \cdots \end{array} \tag{8}$$

we have (7) *valid uniformly for z in the closed region C.*

In Theorem II each root of $f(z)$ is supposed to be counted according to its multiplicity. Let us denote these roots by $\alpha_1, \alpha_2, \ldots, \alpha_\nu$ respectively. In the array (2) formed for approximation to the function[3])

$$F(z) = \frac{f(z)}{(z-\alpha_1)(z-\alpha_2)\ldots(z-\alpha_\nu)}$$

we know by Theorem I that any sequence, say $\sigma_{m,n}(z)$, approaches $F(z)$ uniformly on C. Then we have

$$\lim\,[(z-\alpha_1)(z-\alpha_2)\ldots(z-\alpha_\nu)\,\sigma_{mn}(z)] = f(z) \tag{9}$$

[3]) To be sure, this function has artificial singularities at $z = \alpha_1, \alpha_2, \ldots, \alpha_\nu$, but we consider the function defined so that these artificial singularities are removed. Similarly below.

uniformly for z in C. The rational function in the left-hand member of (9) is precisely of the form of the rational function of the array (8), in the $(m+1)$st row and $(n+1)$st column. The obvious inequality

$$\max [|f(z) - (z - \alpha_1)(z - \alpha_2) \ldots (z - \alpha_\nu)\sigma_{mn}(z)|, z \text{ on } C]$$
$$\geqq \max [|f(z) - R_{m+\nu, n}(z)|, z \text{ on } C],$$

now yields Theorem II.

Another way of expressing Theorem II is the following:

If the function $f(z)$ is analytic interior to the Jordan region C, continuous in the corresponding closed region, and has precisely ν zeros interior to C, then for any infinite sequence of functions formed from (2) *such that the first subscript has no limit value less than ν, we have* (7) *valid uniformly for z in the closed region C.*

It is clear from Hurwitz's theorem that if $f(z)$ satisfies the hypothesis of Theorem II, no sequence of functions formed from (2) can imply (7) uniformly in the closed region C, unless the first subscript of the functions of the sequence has no limit value less than ν.

4. Functions with Infinitely Many Zeros. This new formulation of Theorem II suggests another theorem; here the function $f(z)$ may have infinitely many zeros interior to C:

Theorem III. *If the function $f(z)$ is analytic interior to the Jordan region C and continuous in the corresponding closed region, then for any infinite sequence of functions formed from the array* (2) *such that the first subscript becomes infinite, equation* (7) *is valid uniformly for z in the closed region C.*

Let us assume the theorem false; we shall reach a contradiction. There exists some infinite subsequence μ'_{mn} of the μ_{mn} such that the first subscript becomes infinite and with the property

$$(10) \qquad \overline{\lim}\, \mu'_{mn} = \mu > 0.$$

We introduce, as in the proof of Lemma III, the functions $\Psi(w) = f[\varphi(w)]$ and $\Psi(rw) = f[\varphi(rw)]$. The number r is to be chosen so that we have $|\Psi(w) - \Psi(rw)| < \mu/2$ for w on and within γ. The function $\Psi(rw)$ has at most a finite number ν of zeros interior to γ.

The function $\Psi[r\varphi(z)]$ has at most ν zeros interior to C, so we have by Theorem II for any subsequence $\sigma_{mn}(z)$ of the array (8) corresponding to this function $\Psi[r\varphi(z)]$,

$$\lim [\max |\Psi[r\varphi(z)] - \sigma_{mn}(z)|, z \text{ on } C] = 0.$$

The relation $|f(z) - \Psi[r\varphi(z)]| < \mu/2$, z on C, then yields at once

$$\overline{\lim}\, \mu'_{mn} \leqq \mu/2,$$

in contradiction to (10). This contradiction completes the proof of Theorem III.

It is clear that if the function $f(z)$ of Theorem III actually has infinitely many zeros interior to C, then equation (7) can be valid uniformly for z in the closed region C for a particular infinite sequence of functions formed from (2) only if the first subscript becomes infinite.

Theorems I, II and III can be combined into a single statement: *Let $f(z)$ be analytic interior to the Jordan region C, and continuous in the corresponding closed region. Then any infinite sequence of the $R_{mn}(z)$, for which the first subscript has no lower limit less than the number of zeros of $f(z)$ interior to C, converges to $f(z)$ uniformly in the closed region C.*

The formation of the table (2) does not depend essentially on the fact that C is a region or that $f(z)$ is analytic. By means of Theorem I can be proved without great difficulty for the functions $R_{mn}(z)$ in the obvious extended sense,

Theorem IV. *If the function $f(z)$ is continuous on the Jordan arc C, then for any infinite sequence of functions formed from* (2) *we have*

$$\lim R_{mn}(z) = f(z)$$

uniformly for z on C.

We leave the proof to the reader.

5. Degree of Convergence. If the function $f(z)$ considered in § 1 is analytic in the *closed* region C, it is of interest to study the degree of convergence of various sequences [that is, the asymptotic character of the μ_{mn} of (3)] and the regions of convergence of the sequences. This we do in some detail for the case $m = 0$.

Let the exterior of C be mapped conformally onto the exterior of the unit circle of the w-plane so that the points at infinity correspond to each other. Let the transform in the z-plane of the circle $|w| = R > 1$ be denoted by C_R. Then if $f(z)$ is analytic on and within C_R, the inequality

$$\lim_{m \to \infty} R_{m0}(z) = f(z)$$

is valid for z on and within C_R, and the inequality

$$\mu_{m0} \leqq M/R^m \tag{11}$$

is valid for suitably chosen M [4]).

---

[4]) Faber, Crelle's Journal 150 (1920), S. 79—106; S. 105. Walsh, Trans. Amer. Math. Soc. 32 (1930), S. 335—390: § 6.

Indeed, it can be proved[5]) that the inequality

$$|f(z) - r_{m0}(z)| \leqq M/R^m, \quad z \text{ on } C,$$

always implies

(12) $$\lim_{m\to\infty} r_{m0}(z) = f(z)$$

for z within C_R, uniformly for z on any closed set interior to C_R.

The general extension of the results just mentioned to more general sequences $R_{mn}(z)$ and μ_{mn} seems to be difficult, but we can at least make a beginning by proving

Theorem V. *Let C be an arbitrary closed Jordan region and $f(z)$ a function analytic in C and meromorphic and different from zero on and within C_R. Then there exist polynomials $p_n(z)$ of respective degrees n such that we have*

(13) $$|f(z) - 1/p_n(z)| \leqq M/R^n, \quad R > 1,$$

for z on C.

If polynomials $p_n(z)$ of respective degrees n exist such that (13) *is valid for z on C, and if $f(z)$ does not vanish identically on C, then the function $f(z)$ is analytic in C and meromorphic and different from zero interior to C_R. The sequence $1/p_n(z)$ approaches $f(z)$ for z interior to C_R, uniformly for z on any closed point set interior to C_R on which $f(z)$ has no poles.*

Under the hypothesis made in the first part of this theorem, the function $1/f(z)$ is analytic on and within C, so by (11) there exist polynomials $p_n(z)$ such that we have

(14) $$|p_n(z) - 1/f(z)| \leqq M_1/R^n, \quad z \text{ on } C.$$

Thus we have

$$|1/p_n(z) - f(z)| = \left|\frac{f(z)}{p_n(z)}\left[p_n(z) - \frac{1}{f(z)}\right]\right|.$$

For n sufficiently large the quotient $1/p_n(z)$ approaches $f(z)$ uniformly and hence is uniformly limited for z on C. If some of the early polynomials $p_n(z)$ vanish on C, they may be replaced by polynomials of the same degree which do not vanish on C, with possible modification of the constant M_1 in (14). The totality of these new polynomials is bounded from zero on C. Thus (14) implies (13).

Reciprocally, let the polynomials $p_n(z)$ be given so that (13) is valid. Let us suppose first that $f(z)$ does not vanish on C. We can write

(15) $$\left|p_n(z) - \frac{1}{f(z)}\right| = \left|\frac{p_n(z)}{f(z)}\left[\frac{1}{p_n(z)} - f(z)\right]\right|.$$

---

[5]) Walsh, Münchner Berichte 1926, S. 223—229.

The function $f(z)$ does not vanish on C, and hence the polynomials $p_n(z)$ are uniformly limited on C. Thus we have

$$\left| p_n(z) - \frac{1}{f(z)} \right| \leq \frac{M}{R^n}, \quad z \text{ on } C.$$

It follows by (12) that we have

$$\lim_{n \to \infty} p_n(z) = \frac{1}{f(z)} \tag{16}$$

for z interior to C_R, uniformly on any closed point set interior to C_R. The function $f(z)$ can have no poles on C, by (13) for $n = 0$. Hence the conclusion follows as stated, under the assumption that $f(z)$ does not vanish on C.

The function $f(z)$ cannot vanish *interior* to C, by Hurwitz's theorem, for the function $1/p_n(z)$ cannot vanish and hence its limit function interior to C either vanishes identically or does not vanish at all. The case that $f(z)$ conceivably vanishes on the boundary of C remains to be considered. Denote by C' an arbitrary Jordan curve interior to C. Then on C' inequality (13) is valid and $f(z)$ does not vanish, so we can apply the reasoning used for (15). It follows that (16) is valid for z interior to C'_R, uniformly on any closed point set interior to C'_R; the notation C'_R naturally indicates the transform of $|w| = R > 1$ when the exterior of C' is mapped onto the exterior of $|w| = 1$ so that the points at infinity in the w- and z-planes correspond to each other. But if C' interior to C is allowed to approach the boundary of C uniformly, the curve C'_R also approaches C_R uniformly, and this gives a complete proof of Theorem V.

In Theorem V we have insisted on inequality (13) for *all* values of n, in particular for $n = 0$, which implies the analyticity of $f(z)$ on C. This requirement is in accord with our discussion of § 1. If inequality (13) is required merely for n sufficiently large, it is then possible to study approximation to functions which are meromorphic (instead of analytic) on C, with results analogous to Theorem V.

The reader may notice that Theorem V is true even if C is not required to be a Jordan region, but is allowed to be a Jordan arc or even a much more general point set.

It is rather remarkable that there seems to be no analogue of the second part of Theorem V in the case that $f(z)$ vanishes identically. Indeed, if we set

$$p_n(z) = n!\, n^{3n} \left(z - 1 - \frac{1}{n}\right)^n,$$

then we have

$$\left| \frac{1}{p_n(z)} \right| \leq \frac{1}{n!\, n^n}, \qquad \text{for } |z| \leq 1.$$

so that $1/p_n(z)$ approaches zero with fair rapidity, yet the equation

$$\lim_{n \to \infty} \frac{1}{p_n(z)} = 0$$

is uniformly valid in no region $|z| \leq r > 1$. This example can clearly be modified so as to give an *arbitrary* preassigned degree of approximation of $p_n(z)$ to the function zero, for $|z| \leq 1$, yet so that the equation

$$\lim_{n \to \infty} \frac{1}{p_n(z)} = 0$$

is valid uniformly in no region $|z| \leq r > 1$.

Theorem V has immediate and obvious application to the study of the numbers μ_{mn} and the functions $R_{mn}(z)$. The first and second parts of Theorem V yield respectively the first and second parts of

Theorem VI. *If C is an arbitrary closed Jordan region and $f(z)$ is a function analytic in C and meromorphic and different from zero on and within C_R, then we have*

$$\mu_{0n} \leq \frac{M}{R^n}, \tag{17}$$

where M is suitably chosen, and

$$\lim_{n \to \infty} R_{0n}(z) = f(z),$$

uniformly for z on any closed point set on which $f(z)$ is analytic on and within C_R.

Thus, if $f(z)$ is analytic on C and meromorphic and different from zero interior to C_ϱ but has a zero or a non-polar singularity on C_ϱ, then (17) holds for an arbitrary R less than ϱ but for no R greater than ϱ, and the equation

$$\lim_{n \to \infty} R_{0n}(z) = f(z)$$

holds uniformly on any closed point set on which $f(z)$ is analytic interior to C_ϱ; this equation can be shown to hold uniformly in no region containing in its interior a point of C_ϱ [6]).

The second part of Theorem VI is the precise analogue of the corresponding result for the Padé table. If $\varphi(z)$ is analytic and different from zero at the origin, then the fraction $\varrho_{0n}(z)$ of Padé is the reciprocal of the first $n+1$ terms of the Maclaurin development of $1/\varphi(z)$. The relation

$$\lim_{n \to \infty} \frac{1}{\varrho_{0n}(z)} = \frac{1}{\varphi(z)}$$

[6]) Compare Walsh, Trans. Amer. Math. Soc. 33 (1931), S. 370—388; § 9.

is valid uniformly on and within any circle whose center is the origin and which has on or within it no singularity of $1/\varphi(z)$. Then the relation

$$\lim_{n\to\infty} \varrho_{0n}(z) = \varphi(z)$$

is valid uniformly on and within any circle whose center is the origin and which has on or within it no singularity and no zero of $\varphi(z)$, and indeed is valid uniformly on any closed point set which contains no pole of $\varphi(z)$ and which lies within the largest circle whose center is the origin and whose interior contains no zero and no non-polar singularity of $\varphi(z)$.

6. The rational function $R_{mn}(z)$ exists but need not be unique. We shall not give a proof of the following theorem, but content ourselves with the remark that the proof can be given easily by methods used elsewhere [7]) by the writer in similar situations.

If the function $f(z)$ is defined and continuous on a point set C which is dense in itself, and if for given values of m and n there exists at least one rational function $r_{mn}(z)$ such that $|f(z) - r_{mn}(z)|$ is limited on C, then there exists at least one rational function $R_{mn}(z)$ — namely a function $r_{mn}(z)$ such that

$$\overline{\text{bound}}\,[|f(z) - r_{mn}(z)|,\ z \text{ on } C]$$

is least.

For the particular case $n = 0$, Tonelli has shown the uniqueness of the function $R_{mn}(z)$, provided the point set C contains at least $m+1$ points and is closed and limited. Even those restrictions on C do not ensure the uniqueness of $R_{mn}(z)$ for general values of n, and this situation is not improved if we restrict C to be a closed Jordan region in which $f(z)$ is analytic. We illustrate the lack of uniqueness of $R_{mn}(z)$ by an example.

Denote by γ the unit circle $|z| = 1$, by a the point on γ in the second quadrant whose abscissa is $-.9$, by b the point on γ in the third quadrant whose abscissa is $-.9$. Denote by C_1 the circular arc terminated by a and b and passing through the point $(.9, 0)$. Denote by C_2 the circular arc terminated by a and b and passing through the point $(10/9, 0)$. The region C shall be the Jordan region bounded by these two circular arcs, and the function $f(z)$ shall be $z + 1/z$. It will be noted that this entire configuration is unchanged if we make the substitution $w = 1/z$.

---

[7]) Trans. Amer. Math. Soc. 33 (1931), S. 668—689, especially §§ 1—4.

For the particular situation just described, we do not have $R_{11}(z)$ identically equal to a constant. Indeed, a constant $R_{11}(z)$ would involve a maximum error

$$|f(z) - R_{11}(z)|$$

of at least 1.8, for we have $f(a) = -1.8$, $f(10/9) = 181/90 > 1.8$. Yet the function $r_{11}(z) = 1/z$ involves a maximum error

$$|f(z) - r_{11}(z)| = |z|$$

of no more than 10/9. No matter what the function $R_{11}(z)$ may be, then, the function $R_{11}(1/z)$, is also a function $R_{11}(z)$, by the properties of the original configuration. The two functions $R_{11}(z)$ and $R_{11}(1/z)$ are distinct, for any two functions $r_{11}(z)$ and $r_{11}(1/z)$ are distinct unless constant. That is to say, there can be no unique function $R_{11}(z)$.

The example just given disposes of the case that C is a Jordan region; if a region bounded by an *analytic* Jordan curve is desired, for which the function $R_{11}(z)$ is not unique, it is readily obtained by a slight modification of the example just given. It is perhaps interesting, however, to indicate in detail that the example can be modified to show that if $f(z)$ is analytic on an analytic Jordan arc C, then the function $R_{mn}(z)$ is not necessarily unique.

The Jordan arc C shall be the arc of γ teminated by a and b and containing the point $z = 1$, and again we choose $f(z) = z + 1/z$. The configuration is unchanged by the substitution $w = 1/z$. The function $R_{11}(z)$ cannot be constant, for a constant $R_{11}(z)$ would involve a maximum error $|f(z) - R_{11}(z)|$ of at least 1.8, since $f(a) = -1.8$, $f(1) = 2 > 1.8$. Yet the function $r_{11}(z) = 1/z$ involves a maximum error

$$|f(z) - r_{11}(z)| = |z|$$

of no more than unity. No matter what the function $R_{11}(z)$ may be, then, the function $R_{11}(1/z)$ is also a function $R_{11}(z)$ and distinct from the original one, so the function $R_{11}(z)$ cannot be unique.

We have thus given examples to show that the function $R_{mn}(z)$ need not be unique. That function may of course be unique for all values of m and n in particular non-trivial cases, such as

$$f(z) \equiv z, \qquad C\colon\ |z| \leq 1.$$

7. Further Problems. The present paper clearly makes only a beginning in the study of the main problems, the convergence of sequences formed from (2) and the asymptotic behavior of table (3). There are many other problems suggested by the discussion given, some of which we shall now mention.

12*

1. The study in more detail [in addition to (4), (11), (17)] of the asymptotic character of table (3). This asymptotic character depends at least to some extent on the location of the singularities of the function $f(z)$ and on the zeros (both interior and exterior to C) of $f(z)$. The methods of § 5 apply to the study of $\mu_{\nu n}$ and $R_{\nu n}(z)$, where $f(z)$ is analytic on and within C and has precisely ν zeros interior to C.

2. The study of the overconvergence of various sequences from (2), when $f(z)$ is known to be analytic in the closed region C. Such a study would presumably involve both problem 1. and the determination of the asymptotic location of the poles of the functions $R_{mn}(z)$. At any rate, certain results do exist [8]) which are based on knowledge of the asymptotic character of the maximum error and the asymptotic location of the poles of the rational functions, and which yield information on overconvergence. Indeed, knowledge of the asymptotic behavior of the poles may be used alone [9]) to establish results on overconvergence.

We are here using the term *overconvergence* in the sense connected with (12), namely that a certain dégree of convergence on a point set may necessarily imply convergence on a larger point set.

It is noteworthy that merely *a rough study of table* (3) *without the knowledge of the asymptotic behavior of the poles of the fonctions* $R_{mn}(z)$, *cannot be used to yield general results on overconvergence.* This we illustrate by a specific example.

Let $f(z)$ be an arbitrary entire function not a polynomial, and C an arbitrary closed Jordan region. Then polynomials $p_n(z)$ of respective degrees n exist such that we have

$$|f(z)-p_n(z)| \leqq \mu_n, \qquad \text{for } z \text{ on } C,$$

where $\mu_n \leqq M/R^n$, R greater than unity is arbitrary, M depends on R. Let the points $\alpha_1, \alpha_2, \ldots$ be exterior to C, distinct from the roots of the $p_n(z)$, and everywhere dense in the plane exterior to C. There exist rational functions

$$r_{n+1,1}(z) = \frac{(z-\alpha_n')\,p_n(z)}{z-\alpha_n}, \qquad \alpha_n' \neq \alpha_n,$$

such that we have

$$|p_n(z)-r_{n+1,1}(z)| \leqq \mu_n, \quad z \text{ on } C.$$

Hence we have

$$(18) \qquad |f(z)-r_{n+1,1}(z)| \leqq 2\,\mu_n, \quad z \text{ on } C.$$

---

[8]) Walsh, Trans. Amer. Math. Soc. 30 (1928), S. 838—847; Amer. Journ. Math. 54 (1932), S. 559—570; Acta Math. 57 (1931), S. 411—435, especially Theorem III.

[9]) Walsh, Trans. Amer. Math. Soc. 34 (1932), S. 22—74, especially Theorem IV.

The effective poles of the rational functions $r_{n+1,1}(z)$ are everywhere dense in the plane exterior to C.

That is to say, inequality (18) is valid, and expresses fairly rapid convergence of the sequence $r_{n1}(z)$ to $f(z)$ on C — more rapid convergence than the convergence of any non-trivial geometric series — but the sequence $r_{n1}(z)$ converges to $f(z)$ uniformly in no region exterior to C.

3. As we have already pointed out, if the function $f(z)$ has ν zeros interior to C, no infinite sequence of rational functions chosen from the set

$$\begin{array}{llll} R_{00}(z), & R_{01}(z), & R_{02}(z), & \ldots, \\ R_{10}(z), & R_{11}(z), & R_{12}(z), & \ldots, \\ \multicolumn{4}{c}{\ldots\ldots\ldots\ldots\ldots\ldots\ldots\ldots} \\ R_{\nu-1,0}(z), & R_{\nu-1,1}(z), & R_{\nu-1,2}(z), & \ldots, \end{array}$$

can approach $f(z)$ uniformly in C. *What can be said of the convergence of such sequences of rational functions and of the corresponding values of μ_{pq}? In general, what can be said of approximation in a region C in the sense of Tchebycheff to an analytic function with zeros interior to C by polynomials or by more general analytic functions which are required not to vanish interior to C?*

For instance if we choose C as the unit circle and $f(z)$ as the function z, then it can be shown that we have $R_{0n}(z) \equiv 0$, $\mu_{0n} = 1$,

$$\lim_{n \to \infty} R_{0n}(z) = 0, \quad \text{for all values of } z.$$

4. Thus far we have considered the rational functions $R_{mn}(z)$ merely as the *analogues* of Padé's rational functions $\varrho_{mn}(z)$ for approximation to the same original function $f(z)$. The connection is, however, surely closer than this. We indicate a relation for the case $n = 0$, that is, that the rational functions $R_{mn}(z)$ and $\varrho_{mn}(z)$ are polynomials.

If C is an interval (variable) of the axis of reals, and if $f(z)$ is real for real z and analytic at the origin, then as C approaches the origin the function $R_{m0}(z)$ approaches the function $\varrho_{m0}(z)$, uniformly for z on any limited point set.

The function $f(z) - R_{m0}(z)$ vanishes at least $m+1$ times in the interval C [10]) — provided merely that C lies in a circle whose center is the origin within which $f(z)$ is analytic. When the interval C approaches the origin, these points $\beta_1, \beta_2, \ldots, \beta_{m+1}$ of interpolation of $R_{m0}(z)$ to $f(z)$ necessarily approach the origin.

[10]) Borel, Fonctions de variables réelles, Paris 1905. S. 88.

If γ denotes a curve containing in its interior the origin and all the points β_i but no singularity of $f(z)$, we have

$$f(z) - R_{m0}(z) = \frac{1}{2\pi i}\int_\gamma \frac{(z-\beta_1)\dots(z-\beta_{m+1})f(t)\,dt}{(t-\beta_1)\dots(t-\beta_{m+1})(t-z)}, \quad z \text{ interior to } \gamma,$$

$$f(z) - \varrho_{m0}(z) = \frac{1}{2\pi i}\int_\gamma \frac{z^{m+1}f(t)\,dt}{t^{m+1}(t-z)}, \qquad z \text{ interior to } \gamma.$$

There follows the equation

$$R_{m0}(z) - \varrho_{m0}(z) = \frac{1}{2\pi i}\int_\gamma \left[\frac{z^{m+1}}{t^{m+1}} - \frac{(z-\beta_1)\dots(z-\beta_{m+1})}{(t-\beta_1)\dots(t-\beta_{m+1})}\right]\frac{f(t)\,dt}{t-z}.$$

Since this equation is valid for z interior to γ it is also valid for z exterior to γ, for the integrand has no singularity in z on γ. Our proposition now follows at once by inspection.

The actual analyticity of the function $f(z)$ at the origin need not be assumed; the existence and continuity of certain derivatives is in fact sufficient.

The following similar result is also true:

If C is a (variable) circle whose center is the origin and if $f(z)$ is analytic at the origin, then as the radius of C approaches zero the function $R_{m0}(z)$ approaches the function $\varrho_{m0}(z)$, uniformly for z on any limited point set.

There undoubtedly exist other ways in which (perhaps the most general of) the functions $\varrho_{mn}(z)$ can be considered as limiting cases of the functions $R_{mn}(z)$.

5. The problem we have mentioned in § 1 can be fruitfully generalized: (a) by approximating to more general functions, say functions meromorphic interior to C, or merely continuous or uniformly limited on C; (b) by allowing C to be a Jordan curve or other point set instead of a Jordan region; (c) by using other measures of approximation, such as a line or surface integral, or the Tchebycheff measure of approximation with a norm function.

In (a), if the function approximated is meromorphic on and within the Jordan curve bounding C, the situation is quite analogous to that studied in § 3, with m and n interchanged. In particular the results of § 5 can be easily extended.

(Eingegangen am 2. Januar 1933.)

ON THE DEGREE OF CONVERGENCE OF SEQUENCES OF RATIONAL FUNCTIONS*

BY

J. L. WALSH

The writer has recently studied† the convergence of certain sequences of rational functions of the complex variable, under the hypothesis that the poles of these functions are prescribed and satisfy certain asymptotic conditions. The rational functions are determined either by interpolation to a given analytic function, or by some extremal property of best approximation to such a function. Degree of convergence and regions of uniform convergence of the sequence of rational functions are then obtained (op. cit.). It is the object of the present paper to go more deeply than hitherto into properties of degree of convergence of sequences of rational functions, to make more precise the previous results, and especially to introduce and study the concept of maximal convergence of a sequence of rational functions with preassigned poles; this is a generalization of the corresponding concept for sequences of polynomials. The analogy between convergence properties of sequences of polynomials and convergence properties of sequences of more general rational functions is strong, but has hitherto not been sufficiently strong to justify the use of the term maximal convergence in the latter case (compare op. cit., p. 258). We show now that maximal convergence is characteristic of various sequences of rational functions determined by interpolation and by extremal properties. The present results would seem to be more or less definitive in form.

1. **Introductory results.** We choose as point of departure the following relatively simple but typical formulation (op. cit., §8.3):

THEOREM 1. *Let R be an annular region bounded by two Jordan curves C_1 and C_2, with C_2 interior to C_1. Let the points $\alpha_{n1}, \alpha_{n2}, \cdots, \alpha_{nn}$ lie exterior to C_1, and let the points $\beta_{n1}, \beta_{n2}, \cdots, \beta_{n,n+1}$ lie on or interior to C_2.‡ Let $\overline{R}$ denote the*

* Presented to the Society, December 30, 1938, under the title *Maximal convergence of sequences of rational functions*; received by the editors September 22, 1939.

† *Interpolation and Approximation by Rational Functions in the Complex Domain*, American Mathematical Society Colloquium Publications, vol. 20, New York, 1935. See especially Chapters VIII and IX. Unless otherwise indicated, all references in this paper are to this work, to which the reader should refer also for terminology.

‡ It is a matter of taste whether or not to allow points α_{nk} to lie on C_1 and points β_{nk} to lie on C_2, and whether or not to require that (1) should hold in R or on suitably restricted closed sets in R. There are a variety of allowable choices here. The one we have adopted seems to the writer the most

closure of R. Let the relation

$$(1)\qquad \lim_{n\to\infty}\left| A_n \frac{(z-\beta_{n1})(z-\beta_{n2})\cdots(z-\beta_{n,n+1})}{(z-\alpha_{n1})(z-\alpha_{n2})\cdots(z-\alpha_{nn})}\right|^{1/n} = \Phi(z)$$

hold uniformly on any closed set in $\overline{R}$ *containing no point of* C_2. *Let the function* $\Phi(z)$ *be continuous in the closed region* $\overline{R}$, *and take constant values* γ_1 *and* $\gamma_2<\gamma_1$ *on* C_1 *and* C_2 *respectively. Denote generically* by* C_γ *the curve* $\Phi(z)=\gamma$ *in* $\overline{R}$, $\gamma_2\leqq\gamma\leqq\gamma_1$.

If the function $f(z)$ *is analytic throughout the interior of* C_γ *but is not analytic throughout the interior of any* $C_{\gamma'}$, *with* $\gamma'>\gamma$, *then we have* $(\gamma_2<\lambda<\gamma)$

$$(2)\qquad \limsup_{n\to\infty}\,[\max|f(z)-r_n(z)|,\ z \text{ on } C_\lambda]^{1/n}\leqq\lambda/\gamma,$$

and we have also the limiting case of (2):

$$(3)\qquad \limsup_{n\to\infty}\,[\max|f(z)-r_n(z)|,\ z \text{ on } C_2]^{1/n}\leqq\gamma_2/\gamma,$$

where $r_n(z)$ *is the rational function of degree n whose poles lie in the points* $\alpha_{n1},\alpha_{n2},\cdots,\alpha_{nn}$, *and which interpolates to* $f(z)$ *in the points* $\beta_{n1},\beta_{n2},\cdots,\beta_{n,n+1}$.

Inequality (3) is a direct consequence of (2), by means of the relation

$$[\max|f(z)-r_n(z)|,\ z \text{ on } C_2]\leqq[\max|f(z)-r_n(z)|,\ z\text{ on } C_\lambda],$$

and by allowing λ in (2) to approach γ_2.

The form (1) obviously breaks down whenever a point α_{nk} is infinite, a highly important case which we do not intend to exclude. We therefore use the convention (op. cit., §§8.1, 8.2, 8.5) that in such an expression as the left-hand member of (1) one or more of the points α_{nk} may be infinite; under

convenient in view of the applications. If other choices are made, the conclusions corresponding to Theorem 1 can be read off at once from the present Theorem 1. In later parts of the present paper other choices seem more desirable. There is an obvious asymmetry in Theorem 1 relative to α_{nk} and C_1 on the one hand, and β_{nk} and C_2 on the other hand. This is due to the fact that we assume $f(z)$ analytic on C_2 but not on C_1, and desire to study degree of convergence of $r_n(z)$ to $f(z)$ on C_2. It is then desirable to allow the points β_{nk} to lie on (not merely within) C_2; on occasion the points β_{nk} are to be chosen uniformly distributed on C_2. We need, however, a curve C_1 or C_μ, $\gamma_1\geqq\mu>\gamma$, in such a relation as (4); it is desirable for explicitness to allow μ to be γ_1, hence desirable to assume (1) valid on C_1 and undesirable to allow points α_{nk} to lie on C_1. That is to say, points β_{nk} are readily and conveniently admitted to $\overline{R}$, but not points α_{nk}.

In Theorem 1 the demands on the location of the α_{nk} and β_{nk} may without change in proof be replaced by the demands that no more than a finite number of the α_{nk} shall lie on or within C_1, and that the β_{nk} shall have no limit point exterior to C_2. But if this new hypothesis is used, it may occur that for small n some of the β_{nk} lie outside of the domain of definition of $f(z)$; thus $r_n(z)$ need not be defined for sufficiently small n, but nevertheless is defined for n sufficiently large.

* The notation C_1 and C_2 is exceptional to this, but no confusion should arise; there is double notation for both C_1 and C_2.

those conditions the corresponding factors $z-\alpha_{nk}$ are simply to be omitted; and all of our formulas and conclusions remain correct without other modification.

We have hitherto (op. cit.) employed systematically condition (1) with the restriction $A_n=1$. Nevertheless the present more general form has appeared on occasion (for example, op. cit., pp. 206, 235, 261, 266, 274–275) and requires in the proof of Theorem 1 no change in method over the simpler form with $A_n=1$. Throughout the present paper we shall adopt (1) as standard.*

Even some elementary situations that are of interest are included in Theorem 1 but are not included if we require $A_n=1$. For instance we may take C_1 and C_2 as the circles $|z|=r_1$ and $|z|=r_2<r_1<r_0$, the numbers $\alpha_{n1}, \cdots, \alpha_{nn}$ as the nth roots of an arbitrary a_n whose modulus is not less than r_0^n, and the numbers $\beta_{n1}, \cdots, \beta_{n,n+1}$ as the $(n+1)$st roots of an arbitrary b_n whose modulus is not greater than r_2^{n+1}. Equation (1) is valid with $A_n=a_n$, independently of the behavior of a_n and b_n satisfying the conditions given; but equation (1) is not valid with $A_n=1$ unless the numbers $|a_n|^{1/n}$ approach a finite limit.

For the truth of (2) itself we assume $f(z)$ analytic throughout the interior of C_γ, but need not assume $f(z)$ to be analytic throughout the interior of no $C_{\gamma'}$ with $\gamma'>\gamma$. Indeed $f(z)$ may be analytic throughout the closed interior of C_{γ_1}. For our later purposes in the present paper, however, we find it desirable to make the complete assumption of Theorem 1. For appropriate examples illustrating convergence and degree of convergence when $f(z)$ is analytic throughout R, the reader may refer to the book already mentioned, page 239.

An interesting complement to Theorem 1 is

THEOREM 2. *Under the hypothesis of Theorem* 1 *we have for arbitrary* μ, $\gamma<\mu\leqq\gamma_1$,

$$\limsup_{n\to\infty}\,[\max\,|r_n(z)|,\ z \text{ on } C_\mu]^{1/n} \leqq \mu/\gamma. \tag{4}$$

* Contrary to the situation involving asymptotic conditions for poles, there would seem to be no advantage in the study of asymptotic conditions *for points* β_{nk} *of interpolation* in replacing the condition that

$$\lim_{n\to\infty}\,|(z-\beta_{n1})\cdots(\beta-\beta_{n,n+1})|^{1/n} \tag{a}$$

should exist uniformly by the condition that $\lim_{n\to\infty}|B_n(z-\beta_{n1})\cdots(z-\beta_{n,n+1})|^{1/n}$ should exist uniformly and not vanish identically; for it follows from Theorem 8 with $\alpha_{nk}=\infty$ that each of these conditions implies the other—this entire remark is made on the assumption that (a) is studied in the usual geometric situation, exterior to a curve within which the β_{nk} lie. On the other hand, such a relation as (34) is ordinarily considered interior to a curve to which the α_{nk} are exterior.

A formula for $r_n(z)$ is (op. cit., p. 186)

$$r_n(z) = \frac{1}{2\pi i}\int_\Gamma \left[1 - \frac{(z-\beta_{n1})\cdots(z-\beta_{n,n+1})}{(z-\alpha_{n1})\cdots(z-\alpha_{nn})} \cdot \frac{(t-\alpha_{n1})\cdots(t-\alpha_{nn})}{(t-\beta_{n1})\cdots(t-\beta_{n,n+1})}\right]\frac{f(t)dt}{t-z}, \qquad z \neq \alpha_{nk}, \tag{5}$$

where Γ is a Jordan curve on and within which $f(z)$ is analytic, and which contains all the points β_{nk} in its interior. Equation (5) is valid for all finite values of z other than the α_{nk}, even exterior to Γ, provided of course the integrand when not defined for such a value of z is replaced by its limit for that value of z.

Choose the numbers μ_1, μ_2, and μ_3 with $\mu_1>\mu>\gamma>\mu_2>\mu_3>\gamma_2$, and choose Γ in (5) as the locus C_{μ_2}. For z on C_μ and t on C_{μ_2} we have by (1) when n is sufficiently large

$$\left| A_n \frac{(z-\beta_{n1})\cdots(z-\beta_{n,n+1})}{(z-\alpha_{n1})\cdots(z-\alpha_{nn})}\right| \leqq \mu_1^n,$$

$$\left| \frac{(t-\alpha_{n1})\cdots(t-\alpha_{nn})}{A_n(t-\beta_{n1})\cdots(t-\beta_{n,n+1})}\right| \leqq 1/\mu_3^n.$$

From (5) we read off at once

$$\limsup_{n\to\infty} [\max |r_n(z)|, z \text{ on } C_\mu]^{1/n} \leqq \mu_1/\mu_3,$$

and by allowing μ_1 to approach μ and μ_3 to approach γ we obtain (4).

2. **Degree of convergence.** We shall shortly obtain inequalities in opposite senses to (2) and (4), but in order to do this it is important to show how a certain degree of convergence on C_2 of rational functions $F_n(z)$ with poles in the points α_{nk} implies a corresponding degree of convergence on C_λ. The difficulty here lies in using the relation (1) directly, for the function

$$F_n(z) \div A_n \frac{(z-\beta_{n1})\cdots(z-\beta_{n,n+1})}{(z-\alpha_{n1})\cdots(z-\alpha_{nn})}$$

may have poles on C_2, and cannot be used for immediate comparison.* Nevertheless we shall prove

THEOREM 3. *Under the hypothesis of Theorem* 1 *let us suppose*

$$\limsup_{n\to\infty} [\max |F_n(z)|, z \text{ on } C_2]^{1/n} = q, \tag{6}$$

* Condition (44) is a consequence of (1) and applies here directly. Nevertheless condition (1) is more elementary and more natural; it seems desirable to prove Theorem 3 without using as intermediary Theorem 12, whose proof involves a different order of ideas.

where $F_n(z)$ is a rational function of degree n with its poles in the points α_{nk}. Then we have $(\gamma_2 < \lambda \leq \gamma_1)$

$$\limsup_{n\to\infty} [\max |F_n(z)|, z \text{ on } C_\lambda]^{1/n} \leq \lambda q/\gamma_2. \tag{7}$$

Let $q_1 > q$ be arbitrary, so that we have for n sufficiently large $|F_n(z)| \leq q_1^n$, z on C_2. Let us denote by $w = \phi(z)$ a function which maps the exterior of C_2 onto $|w| > 1$ so that the points at infinity correspond to each other, and denote generically by J_N, $N > 1$, the locus $|\phi(z)| = N$ exterior to C_2. All the points α_{nk} lie exterior to a suitably chosen J_A, so for $Z < A$, $Z > 1$ we have (op. cit., p. 250, Lemma I)*

$$|F_n(z)| \leq q_1^n \left[\frac{AZ - 1}{A - Z}\right]^n, \qquad z \text{ on } J_Z.$$

By the principle of maximum for analytic functions we have for z on and exterior to J_Z

$$\left| F_n(z) \div \left\{ A_n \frac{(z - \beta_{n1}) \cdots (z - \beta_{n,n+1})}{(z - \alpha_{n1}) \cdots (z - \alpha_{nn})} \right\} \right|$$
$$\leq q_1^n \left[\frac{AZ - 1}{A - Z}\right]^n \div \left[\min \left| A_n \frac{(z - \beta_{n1}) \cdots (z - \beta_{n,n+1})}{(z - \alpha_{n1}) \cdots (z - \alpha_{nn})} \right|, z \text{ on } J_Z \right],$$

$$\limsup_{n\to\infty} [\max |F_n(z)|, z \text{ on } C_\lambda]^{1/n} \leq \lambda q_1 \frac{AZ - 1}{A - Z} \div [\min \Phi(z), z \text{ on } J_Z].$$

When we allow Z to approach unity, the curve J_Z approaches C_2, so by allowing q_1 to approach q we obtain (7).

Our most important preliminary result is now available:

THEOREM 4. *Under the hypothesis of Theorem 1 there exists no sequence of rational functions $R_n(z)$ of respective degrees n with poles in the points α_{nk} such that we have either of the relations*

$$\limsup_{n\to\infty} [\max |f(z) - R_n(z)|, z \text{ on } C_2]^{1/n} = \kappa/\gamma, \qquad \kappa < \gamma_2, \tag{8}$$

or $(\gamma_2 < \lambda < \gamma)$

$$\limsup_{n\to\infty} [\max |f(z) - R_n(z)|, z \text{ on } C_\lambda]^{1/n} = \kappa/\gamma, \qquad \kappa < \lambda. \tag{9}$$

Consequently, in Theorem 1 the equality sign holds in both (2) *and* (3).

* In the extension (§11) of Theorem 3 to a set C_2 composed of several Jordan curves, this inequality for $F_n(z)$ is established on a level curve J_Z corresponding to each of the Jordan curves; the remainder of the proof holds without change.

If we assume (8) or (9) to hold, we find by means of (3) or (2) for a specific λ, $\gamma_2 \leqq \lambda < \gamma$,

$$\limsup_{n\to\infty} [\max | r_n(z) - R_n(z) |, z \text{ on } C_\lambda]^{1/n} \leqq \lambda/\gamma,$$

whence from Theorem 3 for arbitrary μ, $\gamma < \mu \leqq \gamma_1$,

$$\limsup_{n\to\infty} [\max | r_n(z) - R_n(z) |, z \text{ on } C_\mu]^{1/n} \leqq \mu/\gamma;$$

a change of notation is necessary in this application of Theorem 3 if $\lambda > \gamma_2$. By Theorem 2 we now derive $(\gamma < \mu \leqq \gamma_1)$

$$\limsup_{n\to\infty} [\max | R_n(z) |, z \text{ on } C_\mu] \leqq \mu/\gamma. \tag{10}$$

The function $\log \Phi(z)$ is harmonic at every point of R, as the uniform limit of the sequence of harmonic functions

$$\frac{1}{n} \log \left| A_n \frac{(z-\beta_{n1}) \cdots (z-\beta_{n,n+1})}{(z-\alpha_{n1}) \cdots (z-\alpha_{nn})} \right|, \tag{11}$$

and is continuous in $\overline{R}$, equal to $\log \gamma_1$ and $\log \gamma_2$ on C_1 and C_1 respectively.* In any closed simply connected region interior to R, suitably chosen conjugates of the functions (11) converge uniformly to a suitably chosen conjugate of the function $\log \Phi(z)$. As z traces a curve C_λ in the counterclockwise sense, the conjugate of (11), which is the argument (that is, angle or amplitude) of the function

$$\left[\frac{(z-\beta_{n1}) \cdots (z-\beta_{n,n+1})}{(z-\alpha_{n1}) \cdots (z-\alpha_{nn})}\right]^{1/n},$$

increases by $2\pi(n+1)/n$; so as z traces C_λ the conjugate of $\log \Phi(z)$ increases by 2π. Consequently $\Phi(z)$ is not identically constant, and we have $\gamma_1 \neq \gamma_2$.

Theorem 4 is a consequence of the following theorem, a treatment of which the writer hopes to publish shortly in these Transactions. The theorem may be proved from the two-constant theorem, in a manner similar to that previously used by the present writer.† Theorem 5 is much more general than we need at the moment, and will be applied also later in the present paper.

THEOREM 5. *Let S be a region bounded by two disjoint Jordan curves K_0 and*

* We cannot have $\gamma_2 = 0$, for in that case the analytic function $\exp[\log \Phi(z) + i\Psi(z)]$, where $\Psi(z)$ is conjugate to $\log \Phi(z)$ in R, would approach the boundary value zero everywhere on C_2, and would be locally single-valued and analytic in R, hence (op. cit., §1.9) would vanish identically.

† Proceedings of the National Academy of Sciences, vol. 24 (1938), pp. 477–486; these Transactions, vol. 46 (1939), pp. 46–65.

K_{-1}, with K_{-1} interior to K_0. Let the function $u(x, y)$ be harmonic in S, continuous in the corresponding closed region $\bar{S}$, equal to zero and -1 on K_0 and K_{-1} respectively. Denote generically by K_σ the locus $u(x, y)=\sigma$, $0>\sigma>-1$, and by S_σ the open set $\sigma>u(x, y)>-1$ in S bounded by K_σ and K_{-1}; denote by $\bar{S}_\sigma$ the closure of S_σ.

Let the function $f(z)$ be analytic throughout S_ρ but not analytic throughout any $S_{\rho'}$, $\rho'>\rho$, and let $f(z)$ be continuous in the two-dimensional sense on K_{-1} with respect to the domain $\bar{S}$. Let the function $f_n(z)$ be analytic in S, continuous in $\bar{S}$, with the relations

$$\limsup_{n\to\infty} [\max |f_n(z)|, z \text{ on } K_0]^{1/n} \leqq e^\alpha > 1, \tag{12}$$

$$\limsup_{n\to\infty} [\max |f(z) - f_n(z)|, z \text{ on } K_{-1}]^{1/n} \leqq e^\beta < 1. \tag{13}$$

Then we must have

$$\alpha + \alpha\rho - \beta\rho \geqq 0; \tag{14}$$

if the equality sign holds here we have

$$\limsup_{n\to\infty} [\max |f_n(z)|, z \text{ on } K_\mu]^{1/n} = e^{(\alpha-\beta)(\mu-\rho)}, \qquad 0 \geqq \mu \geqq \rho, \tag{15}$$

$$\limsup_{n\to\infty} [\max |f(z) - f_n(z)|, z \text{ on } K_\sigma]^{1/n} = e^{(\alpha-\beta)(\sigma-\rho)}, \qquad \rho > \sigma \geqq -1. \tag{16}$$

If Q is an arbitrary continuum in $\bar{S}$ not a single point, and if the equality sign holds in (14), we have

$$\limsup_{n\to\infty} [\max |f_{n+1}(z) - f_n(z)|, z \text{ on } Q]^{1/n} = e^{(\alpha-\beta)(U-\rho)}, \tag{17}$$

where $U=\max\ [u(x, y)$ on $Q]$. Consequently if the second member of (17) is greater than unity, the first member is equal to

$$\limsup_{n\to\infty} [\max |f_n(z)|, z \text{ on } Q]^{1/n}; \tag{18}$$

if the second member of (17) is less than unity, the first member is equal to

$$\limsup_{n\to\infty} [\max |f(z) - f_n(z)|, z \text{ on } Q]^{1/n}. \tag{19}$$

It is a consequence of (17) and (18) that the sequence $f_n(z)$ converges throughout no region containing in its interior a point of K_μ, $0>\mu\geqq\rho$.

We apply Theorem 5 to the situation of Theorem 4 by identifying C_1 and C_λ $(\gamma_2\leqq\lambda<\gamma)$ with K_0 and K_{-1} respectively. Inequalities (10) for $\mu=\gamma_1$ and (8) or (9) are identified with (12) and (13), so we have

$$\alpha = \log \gamma_1 - \log \gamma, \qquad \beta = \log \kappa - \log \gamma.$$

We also set

$$u(x, y) = -\frac{\log \Phi(z) - \log \gamma_1}{\log \lambda - \log \gamma_1}, \qquad \rho = -\frac{\log \gamma - \log \gamma_1}{\log \lambda - \log \gamma_1}.$$

Direct substitution yields for (14) the inequality

$$\frac{\log \gamma - \log \gamma_1}{\log \lambda - \log \gamma_1}(\log \kappa - \log \lambda) \geqq 0,$$

whence $\kappa \geqq \lambda$, and Theorem 4 is established.

Let the equality sign hold in (14), under the conditions of Theorem 5. If the sequence $f_n(z)$ converges throughout some region containing in its interior a point of K_μ, $0 \geqq \mu \geqq \rho$, it follows, from Osgood's theorem to the effect that in a region of convergence subregions of uniform convergence are everywhere dense, that the sequence $f_n(z)$ converges uniformly in some closed region Q containing in its interior a point of some $K_{\mu'}$, $0 > \mu' > \rho$, contrary to the equality of (18) with the second member of (17).

The sequence $f_n(z)$ can converge like a convergent geometric series on no continuum in $\overline{S}$ exterior to K_ρ and consisting of more than one point.

3. **Maximal convergence.** Theorem 4 is our chief justification for the

DEFINITION. *Under the hypothesis of Theorem* 1, *any sequence of rational functions* $R_n(z)$ *of respective degrees* n *with poles in the points* $\alpha_{n1}, \alpha_{n2}, \cdots, \alpha_{nn}$ *is said to converge maximally to* $f(z)$ *on the set* C *constituting the closed interior of* C_2 *provided we have*

$$\limsup_{n\to\infty} [\max |f(z) - R_n(z)|, z \text{ on } C]^{1/n} = \gamma_2/\gamma. \tag{20}$$

We mention explicitly that maximal convergence is not defined if $f(z)$ is analytic throughout the interior of C_1.

As an immediate consequence of the definition we have from Theorems 1 and 4

THEOREM 6. *Under the hypothesis of Theorem* 1, *the sequence* $r_n(z)$ *converges maximally to* $f(z)$ *on* C.

Of course the condition

$$\limsup_{n\to\infty} [\max |f(z) - R_n(z)|, z \text{ on } C_\lambda]^{1/n} \leqq \lambda/\gamma,$$

holding for all λ greater than but sufficiently near γ_2, is sufficient for maximal convergence, for we have

$$[\max |f(z) - R_n(z)|, z \text{ on } C_2] \leqq [\max |f(z) - R_n(z)|, z \text{ on } C_\lambda],$$

and after making the corresponding substitution we may allow λ to approach γ_2.

As an alternative in the definition of maximal convergence we may replace (20) by

$$\limsup_{n\to\infty} [\max |R_{n+1}(z) - R_n(z)|, z \text{ on } C]^{1/n} \leqq \gamma_2/\gamma, \tag{21}$$

provided that the sequence $R_n(z)$ is assumed to converge to $f(z)$ on C; under this assumption we may also replace (20) by (21) with the sign $\leqq$ replaced by the equality sign. This remark is an immediate consequence of the easily proved relation

$$\limsup_{n\to\infty} [\max |R_{n+1}(z) - R_n(z)|, z \text{ on } C]^{1/n} = \limsup_{n\to\infty} [\max |f(z) - R_n(z)|, z \text{ on } C]^{1/n},$$

provided either of these expressions is less than unity.

THEOREM 7. *Under the hypothesis of Theorem* 1, *let* $R_n(z)$ *be a sequence of rational functions of respective degrees n whose poles lie in the points* α_{nk}, *and which converges maximally to* $f(z)$ *on* C. *Then we have* ($\gamma_2 \leqq \lambda < \gamma$, $\gamma \leqq \mu \leqq \gamma_1$)

$$\limsup_{n\to\infty} [\max |f(z) - R_n(z)|, z \text{ on } C_\lambda]^{1/n} = \lambda/\gamma, \tag{22}$$

$$\limsup_{n\to\infty} [\max |R_n(z)|, z \text{ on } C_\mu]^{1/n} = \mu/\gamma. \tag{23}$$

If Q *is an arbitrary continuum in* $\overline{R}$ *not a single point, we have*

$$\limsup_{n\to\infty} [\max |R_{n+1}(z) - R_n(z)|, z \text{ on } Q]^{1/n} = [\max \Phi(z), z \text{ on } Q]/\gamma. \tag{24}$$

Consequently if the second member of (24) *is less than unity we have*

$$\limsup_{n\to\infty} [\max |f(z) - R_n(z)|, z \text{ on } Q]^{1/n} = [\max \Phi(z), z \text{ on } Q]/\gamma; \tag{25}$$

if the second member of (24) *is greater than or equal to unity we have*

$$\limsup_{n\to\infty} [\max |R_n(z)|, z \text{ on } Q]^{1/n} = [\max \Phi(z), z \text{ on } Q]/\gamma. \tag{26}$$

It is a consequence of (26) *that the sequence* $R_n(z)$ *converges throughout no region containing in its interior a point of* C_μ, $\gamma_1 > \mu \geqq \gamma$.

From the assumed maximal convergence of $R_n(z)$ and from the maximal convergence of $r_n(z)$ (Theorem 6) we have

$$\limsup_{n\to\infty} [\max | r_n(z) - R_n(z) |, z \text{ on } C_2]^{1/n} \leqq \gamma_2/\gamma,$$

whence from Theorem 3

$$\limsup_{n\to\infty} [\max | r_n(z) - R_n(z) |, z \text{ on } C_1]^{1/n} \leqq \gamma_1/\gamma.$$

By application of Theorem 2 we may now write

$$\limsup_{n\to\infty} [\max | R_n(z) |, z \text{ on } C_1]^{1/n} \leqq \gamma_1/\gamma. \tag{27}$$

Inequalities (27) and (20) with C replaced by C_2 place us in a position to apply Theorem 5 again. We identify C_1 and C_2 with K_0 and K_{-1} respectively, and we set from (27) and (20)

$$\alpha = \log \gamma_1 - \log \gamma, \qquad \beta = \log \gamma_2 - \log \gamma.$$

Moreover we have

$$u(x, y) = -\frac{\log \Phi(z) - \log \gamma_1}{\log \gamma_2 - \log \gamma_1}, \qquad \rho = -\frac{\log \gamma - \log \gamma_1}{\log \gamma_2 - \log \gamma_1}.$$

Direct computation shows that the equality sign is valid in (14), and hence the conclusion of Theorem 7 follows.

Theorem 7 is a generalization and a sharpening of the corresponding result (op. cit., §§4.7 and 4.8) for the case of approximation by polynomials; relations (24), (25), (26) are new even in the latter case. For approximation by polynomials we have $\alpha_{nk} = \infty$; according to the usual convention (op. cit., §§8.1, 8.2, 8.5) the corresponding factors $z - \alpha_{nk}$ in (1) are simply to be omitted. With $\alpha_{nk} = \infty$ there exist (op. cit., chap. 7) various sets of points β_{nk} for which the relation (1) obtains; for instance we may take $\Phi(z) = \Delta |\phi(z)|$, where $w = \Delta \cdot \phi(z)$ maps the exterior of C_2 onto $|w| > \Delta$ with $\phi'(\infty) = 1$; the first such set β_{nk} was exhibited by Fejér, uniformly distributed on C_2 with respect to a suitably chosen parameter.

In connection with Theorem 7 it is worth remarking that maximal convergence of the sequence $R_n(z)$ to $f(z)$ on the closed interior of C_2 implies the maximal convergence of the sequence $R_n(z)$ to $f(z)$ on the closed interior of every C_λ, $\gamma_2 < \lambda < \gamma$.

4. **Completion of a partial sequence of α_{nk}.** Theorem 1 is valid if the α_{nk} and β_{nk}, and hence also the functions $r_n(z)$, are defined not for every n but merely for an infinite sequence of indices n. But in the proof of Theorem 7 we have employed both $R_n(z)$ and $R_{n+1}(z)$, and thus have made essential use of the fact that the $R_n(z)$ are defined for every n. It is sufficient for our purposes thus far, even in the study of maximal convergence, as the reader may

notice, if the α_{nk} and $R_n(z)$ are defined not for every n but for an infinite sequence of indices n_j, with $n_{j+1}>n_j$ and $n_{j+1}-n_j$ bounded for all j. But if the α_{nk} and $R_n(z)$ are defined only for a sequence of indices n_j, and if $n_{j+1}-n_j$ is not bounded, our fundamental conclusions (22) and (23) may fail even for the specific functions $R_n(z)\equiv r_n(z)$ of Theorem 1, as we now proceed to show by examples. These examples, chosen from Taylor's series, are closely related to gap theorems and to overconvergence in the sense of Ostrowski.

Choose $\alpha_{nk}=\infty$, $\beta_{nk}=0$, so that we have $\Phi(z)\equiv|z|$. Choose C_2 as the circle $|z|=1/2$, C_γ as the circle $|z|=1$, and C_1 as the circle $|z|=2$. Choose the integer n_1 so that $2^{n_1-1}>3$, the integer $n_2>n_1$ so that $2^{(n_2-1)/n_1}>3$, and in general the integer $n_{k+1}>n_k$ so that $2^{(n_{k+1}-1)/n_k}>3$. It is sufficient to choose $n_0=0$, $n_1=3$, $n_{k+1}=2n_k+1$. If we set

$$f(z) = 1 + z^{n_1} + z^{n_2} + z^{n_3} + \cdots, \tag{28}$$

and denote by $r_n(z)$ the sum of the first $n+1$ terms of the corresponding series with all of the powers of z present by indication, we have by Theorem 7

$$\limsup_{n\to\infty} [\max |f(z) - r_n(z)|, \text{ for } |z| = 1/2]^{1/n} = 1/2. \tag{29}$$

But we have for $|z|=1/2$

$$|f(z) - r_{n_k}(z)| \leqq \frac{1}{2^{n_{k+1}}} + \frac{1}{2^{n_{k+2}}} + \cdots < \frac{1}{2^{n_{k+1}-1}},$$

whence

$$\limsup_{k\to\infty} [\max |f(z) - r_{n_k}(z)|, \text{ for } |z| = 1/2]^{1/n_k} \leqq \limsup_{k\to\infty} \frac{1}{2^{(n_{k+1}-1)/n_k}} \leqq \frac{1}{3},$$

in contrast to (29).

With the same choice of α_{nk}, β_{nk}, C_2, C_1, and C_γ, let us now choose the integer n_1 so that $2^{1/(n_1-1)}<3/2$, the integer $n_2>n_1$ so that $2^{(n_1+1)/(n_2-1)}<3/2$, and in general the integer n_{k+1} so that $2^{(n_k+1)/(n_{k+1}-1)}<3/2$; it is sufficient to choose $n_0=0$, $n_{k+1}=2n_k+3$. Again we define $f(z)$ by equation (28), and we denote by $r_n(z)$ the sum of the first $n+1$ terms of the corresponding series with all the powers of z indicated. From Theorem 7 we have

$$\limsup_{n\to\infty} [\max |r_n(z)|, \text{ for } |z| = 2]^{1/n} = 2. \tag{30}$$

But from (28) we may write for $|z|=2$

$$|r_{n_k-1}(z)| \leqq 1 + 2^{n_1} + 2^{n_2} + \cdots + 2^{n_{k-1}} \leqq 2^{n_{k-1}+1},$$

whence

$$\limsup_{k\to\infty} [\max |r_{n_k-1}(z)|, \text{ for } |z| = 2]^{1/(n_k-1)}$$

$$\leqq \limsup_{k\to\infty} 2^{(n_k-1+1)/(n_k-1)} \leqq 3/2,$$

in contrast to (30).

The examples just formulated, in connection with such inequalities as (2), (3), and (4), suggest the following problem. Let C_1, C_2, R, $\Phi(z)$ be given so that the conditions of Theorem 1 are satisfied, including equation (1) but where the α_{nk} and β_{nk} are defined not for all n, only for an infinite sequence of indices; naturally, in equation (1) only those indices are admitted. *To define now new numbers α_{nk} and β_{nk} where necessary so that equation* (1) *shall hold for all n.* We proceed now to the discussion and solution of this problem. In the solution we need the following preliminary result:

THEOREM 8. *Let the relation for some infinite sequence of indices n*

$$\lim_{n\to\infty} \left| A_n \frac{(z-\beta_{n1})\cdots(z-\beta_{n,n+1})}{(z-\alpha_{n1})\cdots(z-\alpha_{nn})} \right|^{1/n} = e^{U(x,y)} \tag{31}$$

be valid uniformly on every closed set interior to an annular region R bounded by Jordan curves C_1 and C_2, with C_2 interior to C_1, where each β_{nk} lies on or interior to C_2 and each α_{nk} lies on or exterior to C_1. Then the function $U(x, y)$ is harmonic in R and not identically constant in R; indeed we have*

$$\int_\Gamma \frac{\partial U}{\partial \nu} ds = 2\pi, \tag{32}$$

where ν indicates the exterior normal, where Γ is an arbitrary analytic Jordan curve separating C_1 and C_2, and where integration is in the counterclockwise sense. At every finite point exterior to C_2 we have

$$\begin{aligned}&\lim_{n\to\infty} |(z-\beta_{n1})\cdots(z-\beta_{n,n+1})|^{1/n} \\ &= \exp\left[\frac{1}{2\pi}\int_{\Gamma_2}\left(U\frac{\partial \log r}{\partial \nu} - \log r \frac{\partial U}{\partial \nu}\right) ds\right], \qquad z = x+iy,\end{aligned} \tag{33}$$

uniformly on any closed finite set exterior to C_2, where Γ_2 is an analytic Jordan curve in R containing C_2 but not (x, y) in its interior; at every point (x, y) interior to C_1 we have

* The left-hand member of (31) if existent uniformly on any closed set in R vanishes at every point of R or at no point of R. For the corresponding analytic functions are locally single-valued, different from zero, and form a normal family in any simply connected subregion of R. By Hurwitz's theorem any limit function of the family vanishes at every point or at no point of R.

$$\lim_{n\to\infty} \left| (z-\alpha_{n1}) \cdots (z-\alpha_{nn})/A_n \right|^{1/n} \tag{34}$$
$$= \exp\left[\frac{-1}{2\pi}\int_{\Gamma_1}\left(U\frac{\partial \log r}{\partial \nu} - \log r\frac{\partial U}{\partial \nu}\right)ds\right], \qquad z = x+iy,$$

uniformly on any closed set interior to C_1, *where* Γ_1 *is an analytic Jordan curve in* R *containing* (x, y) *and* C_2 *but not* C_1 *in its interior. In* (33) *and* (34) *the limits are to be taken over the given sequence of indices for which* (31) *holds. The integrals over* Γ_1 *and* Γ_2 *are to be taken in the counterclockwise and clockwise senses respectively, and* ν *denotes exterior and interior normal respectively.*

Theorem 8 is only a slight modification of a previous result (Theorem 18, op. cit., p. 266), and the proof is therefore left to the reader. In the direction of a converse we have

THEOREM 9. *Let* R *denote an annular region bounded by Jordan curves* C_1 *and* C_2 *with* C_2 *interior to* C_1. *Let the function* $U(x, y)$ *be harmonic interior to* R, *continuous in the corresponding closed region, taking the values* g_1 *and* $g_2 < g_1$ *on* C_1 *and* C_2 *respectively, and satisfying equation* (32), *where* Γ *is an arbitrary analytic Jordan curve separating* C_1 *and* C_2. *Then for every* n *there exist points* α'_{nk} *on* C_1 *and points* β'_{nk} *on* C_2 *and constants* A_n *such that we have*

$$\lim_{n\to\infty}\left| A_n \frac{(z-\beta'_{n1})\cdots(z-\beta'_{n,n+1})}{(z-\alpha'_{n1})\cdots(z-\alpha'_{nn})}\right|^{1/n} = e^{U(x,y)} \tag{35}$$

uniformly on any closed set interior to R.

If we replace A_n in equation (35) by $q^n A_n$, where q is positive, the equation persists with $U(x, y)$ replaced by $U(x, y)+\log q$. In particular, then, it is no loss of generality to assume $g_1 = 0$ in the proof of Theorem 9; we make this assumption.

To exhibit the points α'_{nk} and β'_{nk} desired it is now sufficient to choose the α'_{nk} and β'_{nk} uniformly distributed on C_1 and C_2 respectively with respect to the parameter σ, where

$$d\sigma = \frac{\partial U}{\partial \nu}ds \text{ on } C_1, \qquad d\sigma = -\frac{\partial U}{\partial \nu}ds \text{ on } C_2,$$

for we have the equation for (x, y) in R

$$U(x, y) = \frac{1}{2\pi}\int_{C_2}\log r\, d\sigma - \frac{1}{2\pi}\int_{C_1}\log r\, d\sigma. \tag{36}$$

These equations for $d\sigma$ and $U(x, y)$ still have a meaning in an extended sense even if the Jordan curves C_1 and C_2 are not analytic, and so also does the concept of uniform distribution of points on C_1 and C_2; compare op. cit., §§7.6 and 9.12. Further details of the proof of Theorem 9 are so similar to a proof given elsewhere (Theorem 9, op. cit., pp. 210–211) that they are omitted here. Equation (35) is valid with $A_n=1$ and with $U(x, y)$ replaced by $U(x, y)-g_1$, hence is valid in the original form with $A_n=e^{ng_1}$.

We are now in a position to solve the proposed problem:

THEOREM 10. *Under the hypothesis of Theorem* 8, *where* $U(x, y)$ *takes constant values on* C_1 *and* C_2, *new numbers* α_{nk} *and* β_{nk} *can be defined for those values of* n *for which the original* α_{nk} *and* β_{nk} *are not employed in* (31), *in such a way that* (31) *holds uniformly on any closed set interior to* R *for the entire sequence* $n=1, 2, 3, \cdots$.

The original sequences α_{nk} and β_{nk} used in (31) yield by Theorem 8 a function $U(x, y)$ which assumes constant values g_1 and g_2 on C_1 and C_2, and which satisfies (32), where Γ is an arbitrary analytic Jordan curve separating C_1 and C_2, and where ν indicates exterior normal. It then follows that we have $g_2<g_1$. By virtue of Theorem 9 there exist for every n points α'_{nk} and β'_{nk} which satisfy (35), uniformly on any closed set interior to R. For the values of n that do not appear in the sequence in the relation (31) of our hypothesis we now define $\alpha_{nk}=\alpha'_{nk}$, $\beta_{nk}=\beta'_{nk}$ for those values of n. Then the points α_{nk} and β_{nk} are now defined for every n, and equation (31) is valid uniformly on any closed set interior to R for the complete sequence $n=1, 2, 3, \cdots$.

Theorem 10 applies directly in the situation of Theorem 1 but where (1) is assumed merely for a suitable sequence of indices n. But it is to be noticed that in Theorem 1 equation (1) is assumed to hold uniformly on C_1, whereas in Theorem 10 equation (1) holds uniformly merely on any closed set interior to R.

Proof of Theorem 10 by means of Theorems 8 and 9 is essentially the execution of a program previously outlined (op. cit., p. 268 ff.).

5. **Determination of the β_{nk} when C and the α_{nk} are given.** Our definition of maximal convergence involves the points α_{nk}, the point set C, and the function $\Phi(z)$, but does not involve the points β_{nk} directly; of course the β_{nk} are intimately related to the function $\Phi(z)$. This raises the question of the determination of $\Phi(z)$ and the β_{nk} when C and the α_{nk} are given, a question that we proceed to discuss.

By Theorem 8, condition (34) is a consequence of (31), so (34) or some similar relation is the natural hypothesis for us to use on the points α_{nk}. Such a condition as (34) is fulfilled whenever the points α_{nk} are uniformly distributed on a Jordan curve with respect to a continuous parameter.

THEOREM 11. *Let the point set* C *be the closed interior of a Jordan curve* C_2.

Let C_1' be a Jordan curve containing in its interior the set C but none of the points α_{nk}, and let us suppose

$$\lim_{n\to\infty} \left| (z-\alpha_{n1}) \cdots (z-\alpha_{nn})/A_n \right|^{1/n} = e^{U(x,y)} \tag{37}$$

interior to C_1', uniformly on any closed set interior to C_1'. Then there exists a region R bounded by C_2 and by a Jordan curve C_1 containing C_2 in its interior, and there exists a function $V(x, y)$ harmonic in R, continuous in the corresponding closed region, constant on C_1 and on C_2; for suitably chosen points β_{nk} on C_2 we have interior to R

$$\lim_{n\to\infty} \left| A_n \frac{(z-\beta_{n1}) \cdots (z-\beta_{n,n+1})}{(z-\alpha_{n1}) \cdots (z-\alpha_{nn})} \right|^{1/n} = e^{V(x,y)}, \tag{38}$$

uniformly on any closed set interior to R.

The function $U(x, y)$ can be written as the limit of the sequence of harmonic functions

$$U_n(x, y) \equiv (1/n) \log \left| (z-\alpha_{n1}) \cdots (z-\alpha_{nn})/A_n \right|$$

uniformly on any closed set interior to C_1', so $U(x, y)$ is harmonic interior to C_1'.

Let $G(x, y)$ denote Green's function for the complement K of C with pole at infinity; thus $G(x, y) - \frac{1}{2} \log (x^2+y^2)$ is harmonic at infinity, and $G(x, y)$ vanishes on C_2. Let $W(x, y)$ denote the unique function which is harmonic in K (even at infinity), continuous in the corresponding closed region, and equal to $U(x, y)$ on C_2. Introduce the notation

$$V'(x, y) = W(x, y) + G(x, y) - U(x, y), \tag{39}$$

so that $V'(x, y)$ is harmonic in the annular region R' bounded by C_1' and C_2, and is continuous with the value zero on C_2. If Γ denotes an arbitrary analytic Jordan curve separating C_1' and C_2, and if ν denotes exterior normal, we have

$$\int_\Gamma \frac{\partial V'}{\partial \nu} ds = \int_\Gamma \frac{\partial W}{\partial \nu} ds + \int_\Gamma \frac{\partial G}{\partial \nu} ds - \int_\Gamma \frac{\partial U}{\partial \nu} ds = 2\pi; \tag{40}$$

for the first and third integrals of the second member vanish because $W(x, y)$ and $U(x, y)$ are harmonic respectively in the closed exterior and closed interior of Γ.

We shall now show* that $V'(x, y)$ *is positive at every point of R'*. Consider

* The entire proof of Theorem 11 can be interpreted as the carrying out in detail of a method previously indicated (op. cit., p. 269). But the inequality $V'>0$ is an indispensable condition for the validity of that method, and the present proof of that inequality is the first ever given of wide generality.

the function (compare (39))

$$V_n'(x, y) = W(x, y) + G(x, y) - U_n(x, y),$$

which approaches $V'(x, y)$ uniformly on any closed set in $R'+C_2$. The function $V_n'(x, y)$ is defined and harmonic at every point exterior to C_2, except the points α_{nk} and except at infinity. When (x, y) approaches a finite point $z=\alpha_{nk}$, the functions $-U_n(x, y)$ and $V_n'(x, y)$ become positively infinite. When (x, y) becomes infinite, the functions $W(x, y)$, $G(x, y)-\frac{1}{2}\log(x^2+y^2)$, and $\frac{1}{2}\log(x^2+y^2)-U_n(x, y)$ approach finite limits, except that $\frac{1}{2}\log(x^2+y^2)-U_n(x, y)$ becomes positively infinite if one or more of the points α_{nk} are infinite. Consequently the minimum of the function $V_n'(x, y)$ considered in the entire closed region exterior to C_2 exists and occurs for (x, y) on the curve C_2 itself:

$$[V_n'(x, y), \text{ for } (x, y) \text{ exterior to } C_2] \geqq [\min V_n'(x, y), \text{ on } C_2].$$

By the uniformity of convergence of $V_n'(x, y)$ to $V'(x, y)$ on C_2 as n becomes infinite, this right-hand member approaches the minimum of $V'(x, y)$ on C_2; thus

$$[V'(x, y), \text{ for } (x, y) \text{ in } R'] \geqq [\min V'(x, y), \text{ on } C_2].$$

Consequently for (x, y) in R' we have proved $V'(x, y)\geqq 0$. The function $V'(x, y)$ is zero on C_2, can approach no negative value as (x, y) in R' approaches C_1', and is harmonic in the region bounded by C_1' and C_2. But $V'(x, y)$ is not identically zero in R', by equation (40). It follows from the well known properties of the minima of harmonic functions that the inequality $V'(x, y)>0$ persists throughout R'.

We shall now exhibit the desired points β_{nk}. Let us suppose for the moment that C_2 is an analytic Jordan curve. For (x, y) exterior to C_2 we have (op. cit., p. 266, Lemma IV)

$$\begin{aligned} &W(x, y) + G(x, y) \\ &= \frac{1}{2\pi}\int_{C_2}\left[(W+G)\frac{\partial \log r}{\partial \nu} - \log r\,\frac{\partial(W+G)}{\partial \nu}\right]ds + q, \end{aligned} \tag{41}$$

where ν indicates the interior normal for C_2, and where q is suitably chosen. For (x, y) exterior to C_2 we also have (op. cit., p. 265, Lemma III)

$$0 = \frac{1}{2\pi}\int_{C_2}\left(U\frac{\partial \log r}{\partial \nu} - \log r\,\frac{\partial U}{\partial \nu}\right)ds;$$

this equation is first established with U replaced by U_n, and then use is made of the uniformity of convergence of U_n and $\partial U_n/\partial\nu$ on C_2. It follows by subtraction from (41) for (x, y) exterior to C_2 that

$$
\begin{aligned}
W(x, y) + G(x, y) &= \frac{1}{2\pi}\int_{C_2}\left(V'\,\frac{\partial \log r}{\partial\nu} - \log r\,\frac{\partial V'}{\partial\nu}\right)ds + q \\
&= \frac{-1}{2\pi}\int_{C_2}\log r\,\frac{\partial V'}{\partial\nu}\,ds + q.
\end{aligned} \tag{42}
$$

Thanks to the inequality $V'(x, y)>0$ in R', we have on C_2 the inequality $d\sigma = -(\partial V'/\partial\nu)ds \geqq 0$. If now the points β_{nk} are chosen uniformly distributed on C_2 with respect to the parameter σ, we may write from (42) and from $-\int_{C_2}(\partial V'/\partial\nu)ds = 2\pi$ (a consequence of (40)) the relation

$$
\lim_{n\to\infty} |(z-\beta_{n1})\cdots(z-\beta_{n,n+1})|^{1/n} = e^{W(x,y)+G(x,y)-q} \tag{43}
$$

for z exterior to C_2, uniformly on any closed set exterior to C_2 (compare op. cit., §8.7). It follows now from (37) and (43) by the choice $V(x, y) = V'(x, y) -q$ that equation (38) is valid uniformly on any closed set interior to R'.

In (41) and later equations we have for the sake of convenience assumed C_2 to be an analytic Jordan curve. It is sufficient if C_2 is an arbitrary Jordan curve, provided the integrals are interpreted in an extended sense (compare op. cit., §7.6). Even if C_2 is a still more general set, the integrals may be taken over analytic Jordan curves near but exterior to C_2 on which $V(x, y)$ is constant (which does not alter the validity of (42) or of the other equations), and the points β_{nk} may be chosen on these curves by the method of op. cit., §4.4; but here we must relax the requirement that the β_{nk} shall lie on or interior to C_2, and may replace it by the requirement that the β_{nk} shall have no limit point exterior to C_2.

A region R satisfying the requirements of Theorem 11 is easily defined. Let g_1 denote the least upper bound of $V'(x, y)$ in R', and let g be arbitrary, $0<g<g_1$. Then the locus C_1: $V'(x, y) = g$ in R' is an analytic Jordan curve; the annular region bounded by C_1 and C_2 fulfills all the conditions of Theorem 11. Our fundamental results on approximation apply only to loci $V'(x, y) = \text{const.}$, so it is no great disadvantage to cut R off along such a curve.

It is not uninteresting to note that if in Theorem 11 we assume equation (37) to hold merely for an infinite sequence of indices n, then (as in Theorem 10) new points α_{nk} can be provided, and also the points β_{nk}, so that both the α_{nk} and β_{nk} shall be defined for every n with (38) valid. Indeed, the definition of $U(x, y)$ and the proof of (43) do not assume the α_{nk} defined for every n; Theorem 11 can be used to determine the β_{nk} for the values of n for which the α_{nk} are defined, and then Theorem 10 applies. Of course equation (43) is substantially the same as equation (33) of Theorem 8.

6. **Invariant formulation of results.** The general problem of greatest degree of approximation to a given analytic function $f(z)$ by rational functions with prescribed poles α_{nk}, culminating in Theorem 7, is invariant under linear transformation of the complex variable. But our fundamental condition (1) is not invariant, and has no obvious invariant properties. We proceed to discuss conditions analogous to (1) but having properties of invariance.

THEOREM 12. *Under the conditions of Theorem* 1 (*it is sufficient if* (1) *holds uniformly merely on any closed set in R*) *let the function* $w=\phi(z)$ *map the exterior of* C_2 *onto* $|w|>1$ *so that the points at infinity correspond to each other. Then we have*

$$\lim_{n\to\infty}\left|\frac{[\bar{\phi}(\alpha_{n1})\phi(z)-1]\cdots[\bar{\phi}(\alpha_{nn})\phi(z)-1]}{[\phi(z)-\phi(\alpha_{n1})]\cdots[\phi(z)-\phi(\alpha_{nn})]}\right|^{1/n}=\Phi_1(z), \tag{44}$$

uniformly on any closed set in R, where $\Phi_1(z)$ *is a suitable constant multiple of the function* $\Phi(z)$ *of* (1).

From condition (1) it follows that we have (34) fulfilled; it follows from (34) (method of op. cit., §9.12) that

$$\lim_{n\to\infty}|[\phi(z)-\phi(\alpha_{n1})]\cdots[\phi(z)-\phi(\alpha_{nn})]/A_n|^{1/n}$$

exists uniformly on any closed set interior to R. By another method previously employed (op. cit., §9.4) it follows that (44) is valid uniformly on any closed set in R, where $\Phi_1(z)$ is suitably chosen. It is to be noted that the function whose absolute value appears in (44) can be considered defined even on C_2, and to be continuous there, whence $\Phi_1(z)$ also is continuous on C_2: $\Phi_1(z)\equiv 1$ on C_2. Then the limit in (44) can be considered uniform on any closed set in $\bar{R}$ containing no point of C_1. The function whose absolute value appears in (44) is greater than unity at every point exterior to C_2, so we have $\Phi_1(z)\geqq 1$ in R.

From (1) and (44) we may write

$$\lim_{n\to\infty}\left|A_n\frac{(z-\beta_{n1})\cdots(z-\beta_{n,n+1})}{(z-\alpha_{n1})\cdots(z-\alpha_{nn})}\cdot\frac{[\phi(z)-\phi(\alpha_{n1})]\cdots[\phi(z)-\phi(\alpha_{nn})]}{[\bar{\phi}(\alpha_{n1})\phi(z)-1]\cdots[\bar{\phi}(\alpha_{nn})\phi(z)-1]}\right|^{1/n}=\frac{\Phi(z)}{\Phi_1(z)}, \tag{45}$$

uniformly on any closed set in R. Denote by $\Psi_n(z)(z-\beta_{n1})$ the function whose modulus occurs in (45). Then $\Psi_n(z)$ is analytic and different from zero at every point of the extended plane exterior to C_2, when the function is suitably defined in the points α_{nk}.* Moreover $\lim_{n\to\infty}|z-\beta_{n1}|^{1/n}=1$ uniformly on any closed bounded set exterior to C_2. Then (45), being valid on an arbitrary C_γ, $\gamma_1>\gamma>\gamma_2$ when suitably interpreted, is valid at every finite point in the closed exterior of C_γ; the logarithm of the right-hand member of (45) can be considered defined and harmonic at every point of the extended plane exterior to C_2, is continuous in the corresponding closed region, and has the constant value $\log\gamma_2$ on C_2. Hence throughout R we have $\Phi(z)/\Phi_1(z)\equiv\gamma_2$, and Theorem 12 is established.

Under the conditions of Theorem 12 it may occur that C_1 is a curve $|\phi(z)|=$const., say c; under such circumstances the function

$$\log\Phi(z)-\frac{\log\gamma_1-\log\gamma_2}{\log c}\log|\phi(z)|-\log\gamma_2$$

is harmonic in R, continuous in $\overline{R}$, zero on C_1 and C_2, hence is zero in $\overline{R}$:

$$\log\Phi(z)=\frac{\log\gamma_1-\log\gamma_2}{\log c}\log|\phi(z)|+\log\gamma_2.$$

But by (32) we have

$$\int_\Gamma\frac{\partial\log\Phi(z)}{\partial\nu}ds=2\pi,$$

where Γ is an analytic Jordan curve separating C_1 and C_2; also (op. cit., p. 71)

$$\int_\Gamma\frac{\partial\log|\phi(z)|}{\partial\nu}ds=2\pi,$$

whence we deduce $(\log\gamma_1-\log\gamma_2)/\log c=1$, and for z in $\overline{R}$,

$$\Phi(z)\equiv\gamma_2|\phi(z)|.$$

This equation is satisfied, it may be added, provided the points α_{nk} and β_{nk} are distributed on C_1 and C_2 respectively uniformly with respect to the harmonic function conjugate to $\log|\phi(z)|$; see op. cit., §4.3.

The converse of Theorem 12 can also be established:

THEOREM 13. *Let R be the region between and bounded by two Jordan curves C_1 and C_2, with C_2 interior to C_1. Let $w=\phi(z)$ map the exterior of C_2 onto the*

* When a point α_{nk} lies at infinity, the function $[\phi(z)-\phi(\alpha_{nk})]/[\overline{\phi}(\alpha_{nk})\phi(z)-1]$ in (44) and (45) is to be replaced by its limit as $\alpha_{nk}\to\infty$, namely $1/\phi(z)$. In accordance with the convention already made (§1) concerning factors $z-\alpha_{nk}$, there is here no exception in the behavior of $\Psi_n(z)$ even if points α_{nk} are infinite.

region $|w|>1$ *so that the points at infinity correspond to each other. Let the points* α_{nk} *lie exterior to* C_1, *and suppose* (44) *valid uniformly on any closed set in* $\overline{R}$ *not containing a point of* C_1. *Then for suitably chosen numbers* A_n *and suitably chosen points* β_{nk} *on* C_2, *equation* (1) *is valid uniformly on any closed set in* R, *where* $\Phi(z)$ *is a constant multiple of* $\Phi_1(z)$.

From equation (44) follows (op. cit., p. 274, Corollary 2) an equation of type (37), valid uniformly on any closed set interior to C_1. By Theorem 11 we now derive a condition of form (1), and inspection of the proof of Theorem 11 shows that equation (1) is valid uniformly on any closed set in R. It follows now from Theorem 12 that $\Phi(z)$ is a constant multiple of $\Phi_1(z)$, so Theorem 13 is established.

In Theorem 13 the Jordan curve C_1 naturally need not be a locus obtained by setting $\Phi(z)$ equal to a constant.

We remark (op. cit., p. 274) that condition (44), valid uniformly on any closed set in R, implies the existence of

$$\lim_{n\to\infty}\left|\frac{(z-\alpha_{n1})\cdots(z-\alpha_{nn})}{\phi(\alpha_{n1})\cdots\phi(\alpha_{nn})}\right|^{1/n}$$

uniformly on any closed set in R; if a particular α_{nk} is infinite, the corresponding quotient $(z-\alpha_{nk})/\phi(\alpha_{nk})$ is to be replaced by the limit of that quotient as α_{nk} becomes infinite, namely $-1/\phi'(\infty)$. By virtue of (33), a consequence of Theorems 13 and 8, it now follows that equation (1) is valid uniformly on any closed set in R for a suitably chosen $\Phi(z)$ provided we have

$$A_n = \phi(\alpha_{n1})\cdots\phi(\alpha_{nn}). \tag{46}$$

The proof just given, that (44) implies the existence of

$$\lim_{n\to\infty}\left|\phi(\alpha_{n1})\cdots\phi(\alpha_{nn})\frac{(z-\beta_{n1})\cdots(z-\beta_{n,n+1})}{(z-\alpha_{n1})\cdots(z-\alpha_{nn})}\right|^{1/n},$$

shows the uniform existence of this limit merely on any closed set in R; the limit is different from zero there. But if (1) itself is satisfied under the conditions of Theorem 1, it then follows that for the original A_n

$$\lim_{n\to\infty}\left|\frac{\phi(\alpha_{n1})\cdots\phi(\alpha_{nn})}{A_n}\right|^{1/n}$$

exists and is different from zero. Consequently (1) with the substitution (46) made is valid uniformly on any closed set in $\overline{R}$ containing no point of C_2. It is not essentially more general to consider (1) in its original form than to consider (1) with (46) satisfied.

Let us now consider the function $\Psi_n(z)$ defined in connection with (45), where A_n is given by (46). We see by inspection that $\Psi_n(\infty)=1$, from which it follows that $\Phi(z)/\Phi_1(z)$ is identically unity. Thus we have proved

COROLLARY 1. *Under the conditions of Theorem 1, we have*

$$\lim_{n\to\infty}\left|\phi(\alpha_{n1})\cdots\phi(\alpha_{nn})\frac{(z-\beta_{n1})\cdots(z-\beta_{n,n+1})}{(z-\alpha_{n1})\cdots(z-\alpha_{nn})}\right|^{1/n}=\Phi_1(z), \tag{47}$$

uniformly on any closed set in $\overline{R}$ containing no point of C_2, where $\Phi_1(z)$ is given by (44).

If z_0 is an arbitrary point interior to C_1, the hypothesis of Theorem 8 implies (34) for $z=z_0$, so for the given sequence of indices

$$\lim_{n\to\infty}\left|(z_0-\alpha_{n1})\cdots(z_0-\alpha_{nn})/A_n\right|^{1/n}$$

exists. This limit is different from zero, by the form of (34) itself. It follows now from (31) that for the given sequence of indices

$$\lim_{n\to\infty}\left|\frac{(z_0-\alpha_{n1})\cdots(z_0-\alpha_{nn})(z-\beta_{n1})\cdots(z-\beta_{n,n+1})}{(z-\alpha_{n1})\cdots(z-\alpha_{nn})}\right|^{1/n}$$

exists uniformly on any closed set in R. Consequently whenever equation (1) is satisfied, that equation is also satisfied for some function $\Phi(z)$ with

$$A_n=(z_0-\alpha_{n1})\cdots(z_0-\alpha_{nn}),$$

where z_0 is an arbitrary point interior to C_1.

As a matter of record we formulate

COROLLARY 2. *Under the hypothesis of Theorem 13, let $R_n(z)$ be a sequence of rational functions of respective degrees n, whose poles lie in the points α_{nk}. Denote generically by C_γ the curve $\Phi_1(z)=\gamma$ in R, and let the function $f(z)$ be analytic throughout the interior of C_γ but not throughout the interior of any $C_{\gamma'}$, $\gamma'>\gamma$.*

A sufficient condition for the maximal convergence of the sequence $R_n(z)$ to $f(z)$ on C (closed interior of C_2) is

$$\limsup_{n\to\infty}\left[\max|f(z)-R_n(z)|,\ z\text{ on }C_2\right]^{1/n}\leqq 1/\gamma; \tag{48}$$

if the inequality sign holds here, so also does the equality sign.

Condition (44) is invariant under linear transformation of the complex variable, and in some respects is therefore more advantageous than condition (1). It is obviously immaterial whether the loci C_γ and the right-hand members of (22)–(26) are determined from $\Phi(z)$ or from $\Phi_1(z)$; in the future we shall use these functions interchangeably.

The method of proof of Theorem 12, namely the use of such an equation as (45), yields

COROLLARY 3. *Under the conditions of Theorem* 1, *suppose also that points* β'_{nk} *exist on or within the curve* C_γ, $\gamma_1>\gamma\geqq\gamma_2$, *so that the relation*

$$\lim_{n\to\infty}\left|A'_n\frac{(z-\beta'_{n1})\cdots(z-\beta'_{n,n+1})}{(z-\alpha_{n1})\cdots(z-\alpha_{nn})}\right|^{1/n}=\Phi_2(z)\not\equiv\text{const.}$$

holds uniformly on any closed set in $\overline{R}'$ *containing no point of* C_γ, *where* R' *is the region between and bounded by* C_1 *and* C_γ, *and where* $\Phi_2(z)$ *is constant on* C_γ. *Then the quotient* $\Phi_2(z)/\Phi(z)$ *is identically constant in* R'.

We remark too that this corollary can be established by the use of Theorem 7 itself.

7. **Maximal convergence of extremal sequences.** We are now in a position expeditiously to treat extremal sequences.

THEOREM 14. *Under the conditions of Theorem* 1, *let* C *denote the closed interior of* C_2 *and let* $R_n(z)$ *be the (or a) rational function of degree* n *with poles in the points* α_{nk}, *of best approximation to* $f(z)$ *on* C *in the sense of Tchebycheff, or in the sense of least pth powers* ($p>0$) *over* C_2 *(assumed rectifiable), or in the sense of least pth powers* ($p>0$) *as measured by a surface integral over* C, *or in the sense of least pth powers over* γ: $|w|=1$ *when* K *is mapped onto the exterior of* γ *so that the points at infinity correspond to each other, or in the sense of least pth powers over the circumference* $|w|=1$ *or over the closed region* $|w|\leqq 1$ *when* C *is mapped onto* $|w|\leqq 1$, *in every case with a positive continuous weight function. Then the sequence* $R_n(z)$ *converges maximally to* $f(z)$ *on* C.

Inequality (48) is established for the present sequence $R_n(z)$ precisely as in previous cases (op. cit., p. 264, Theorem 17; p. 254, Theorem 12). The details are left to the reader.

The restriction that the weight function be positive and continuous can be considerably lightened in the various cases; compare op. cit., §5.7.

8. **Necessary conditions on the β_{nk} for uniform convergence.** We have hitherto presented such conditions as (1) and (47) as sufficient conditions on the β_{nk} for maximal convergence. We now consider their necessity as conditions for uniform and maximal convergence.

THEOREM 15. *Let* C_1 *and* C_2 *be Jordan curves, with* C_2 *interior to* C_1, *and let* R *denote the annular region between them. Let the points* α_{nk} *lie exterior to* C_1 *and satisfy the condition*

$$\lim_{n\to\infty}\left|\frac{[\overline{\phi(\alpha_{n1})}\phi(z)-1]\cdots[\overline{\phi(\alpha_{nn})}\phi(z)-1]}{[\phi(z)-\phi(\alpha_{n1})]\cdots[\phi(z)-\phi(\alpha_{nn})]}\right|^{1/n}=\Phi(z) \tag{49}$$

uniformly in $\overline{R}$; *such a condition as* (49) *is satisfied whenever the condition* (37) *is fulfilled uniformly on and within* C_1. *Suppose the points* β_{nk} *are given on or within* C_2, *and suppose the corresponding sequence of interpolating rational functions* $r_n(z)$ *of respective degrees* n *converges uniformly to* $f(z)$ *on and within* C_2 *whenever* $f(z)$ *is analytic on and within* C_2. *Then we have uniformly on any closed set in* R *the equation*

$$\lim_{n\to\infty}\left|\phi(\alpha_{n1})\cdots\phi(\alpha_{nn})\frac{(z-\beta_{n1})\cdots(z-\beta_{n,n+1})}{(z-\alpha_{n1})\cdots(z-\alpha_{nn})}\right|^{1/n}=\Phi(z). \tag{50}$$

Consequently the sequence $r_n(z)$ *converges maximally to* $f(z)$ *on the closed interior of* C_2 *whenever* $f(z)$ *is analytic in that closed region but not analytic throughout* R.

This theorem is due to Kalmár for the case $\alpha_{nk}=\infty$; the present method is due to the present writer (op. cit., §7.3), an extension of the method in the case $\alpha_{nk}=\infty$.

We introduce the notation

$$\Psi_n(z)=\phi(\alpha_{n1})\cdots\phi(\alpha_{nn})\frac{(z-\beta_{n1})\cdots(z-\beta_{n,n+1})}{\phi(z)(z-\alpha_{n1})\cdots(z-\alpha_{nn})};$$

it is obvious that $\lim_{n\to\infty}|\phi(z)|^{1/n}=1$ uniformly on any closed limited set exterior to C_2, so (50) is equivalent to

$$\lim_{n\to\infty}|\Psi_n(z)|^{1/n}=\Phi(z) \tag{51}$$

uniformly on any closed set in R. The relation $\Phi(z)\geqq 1$ in R follows at once from (49).

The sequence

$$\left|\frac{(z-\alpha_{n1})\cdots(z-\alpha_{nn})}{\phi(\alpha_{n1})\cdots\phi(\alpha_{nn})}\right|^{1/n}$$

converges uniformly on any closed set in R to a nonvanishing function (op. cit., p. 274). Hence the sequence $\log|\Psi_n(z)|^{1/n}$ is uniformly bounded on any closed set in R, and forms a normal family of harmonic functions in R;* these

* We are using here easily proved properties of sequences of harmonic and analytic functions, uniformly bounded and hence equicontinuous on any closed set interior to R.

We also use the fact that a uniformly convergent sequence of nonvanishing analytic functions converges to a nonvanishing analytic function unless the limit function vanishes identically.

functions are obviously single-valued and harmonic in R. Denote by $\log|\Psi(z)|$ an arbitrary limit function of the family. We shall prove the

LEMMA. *Either $|\Psi(z)|$ approaches the constant unity as z in R approaches C_2 or for some points of R the function $|\Psi(z)|$ has a value greater than unity and for other points of R this function $|\Psi(z)|$ has a value less than unity.*

A subsequence of the functions $|\Psi_n(z)|^{1/n}$ approaches $|\Psi(z)|$ uniformly on any closed subset of R, so the corresponding sequence

$$\left| \Psi_n(z) \frac{[\phi(z) - \phi(\alpha_{n1})] \cdots [\phi(z) - \phi(\alpha_{nn})]}{[\overline{\phi(\alpha_{n1})}\phi(z) - 1] \cdots [\overline{\phi(\alpha_{nn})}\phi(z) - 1]} \right|^{1/n} \tag{52}$$

approaches $|\Psi(z)|/\Phi(z)$ uniformly on any closed subset of R. The function whose modulus appears in (52) is analytic and different from zero exterior to C_2, even in the points α_{nk} when properly defined there, and has the modulus $1/|\phi'(\infty)|$ at infinity. The sequence (52) therefore converges uniformly on any closed set of the extended plane exterior to C_2, to the modulus of a function analytic exterior to C_2 and which has unit modulus at infinity. Thus $\log|\Psi(z)/\Phi(z)|$ has a meaning and is harmonic in the extended plane exterior to C_2, and has the value zero at infinity, even though $\Phi(z)$ and $\Psi(z)$ are not properly defined exterior to C_1.

Of course each of the quantities in (49) whose limit is $\Phi(z)$ is greater than unity in R; the function $\Phi(z)$ is not constant in R, so we have $\Phi(z)>1$ in R (compare op. cit., p. 229); but $\Phi(z)$ is continuous and equal to unity on C_2. The lemma now follows from the well known properties of the maxima and minima of harmonic functions.

We introduce the notation $\omega_n(z)=\phi(z)\Psi_n(z)$. In the expansion of the function $1/(t-z)$, where t is exterior to C_2, by interpolating functions $r_n(z)$, we have (op. cit., §8.1)

$$f(z) - r_n(z) = \frac{\omega_n(z)}{\omega_n(t)(t - z)}. \tag{53}$$

If we introduce the notation

$$M_n = \max\, [\,|\,\omega_n(z)\,|,\ z \text{ on } C_2], \tag{54}$$

our hypothesis implies

$$\lim_{n\to\infty} M_n/\omega_n(t) = 0. \tag{55}$$

If (51) is not satisfied, the normality of the family $|\Psi_n(z)|^{1/n}$ implies that some limit function $|\Psi(z)|$ of this family is different from $\Phi(z)$. Then for some point t, which may be chosen in R, we have by the lemma for suitably chosen indices n_k

$$\lim_{k\to\infty} |\Psi_{n_k}(t)|^{1/n_k} = \Psi(t) < 1.$$

We may also write

$$\lim_{k\to\infty} |\omega_{n_k}(t)|^{1/n_k} = \Psi(t) < 1,$$

and for suitably large index

$$(56) \qquad |\omega_{n_k}(t)| < 1.$$

Choose this point t as the value in (53). From the definition of $\omega_n(z)$ and from the properties of $\phi(z)$ on C_2 we have

$$M_n = \max [|\omega_n(z)|, z \text{ on } C_2] = \max [|\Psi_n(z)|, z \text{ on } C_2].$$

From the properties at infinity of the function whose modulus appears in (52) we have

$$(57) \qquad \max [|\Psi_n(z)|, z \text{ on } C_2] \geqq 1/|\phi'(\infty)|, \qquad M_n \geqq 1/|\phi'(\infty)|.$$

Inequalities (56) and (57) are in contradiction to (55), so equation (51) is established. The remainder of Theorem 15 is a consequence of Theorem 6.

A result closely related to Theorem 15, an extension of a result due to Fekete (op. cit., p. 163), is

THEOREM 16. *Let C_1, C_2, R, and the α_{nk} satisfy the conditions of Theorem* 15, *including* (49). *Let the points β_{nk} lie on or within C_2, and let M_n be defined by* (54). *Then*

$$(58) \qquad \lim_{n\to\infty} M_n^{1/n} = 1$$

is a necessary and sufficient condition that the sequence of rational functions $r_n(z)$ of respective degrees n with poles in the points α_{nk} and interpolating to $f(z)$ in the points β_{nk}, should converge uniformly to $f(z)$ on and within C_2 whenever $f(z)$ is analytic on and within C_2. Thus (58) *is also a necessary and sufficient condition for the maximal convergence of the sequence $r_n(z)$ to $f(z)$ whenever $f(z)$ is analytic on and within C_2 but is not analytic throughout R.*

If condition (58) is satisfied, the functions

$$(59) \qquad \left[\Psi_n(z)\frac{[\phi(z)-\phi(\alpha_{n1})]\cdots[\phi(z)-\phi(\alpha_{nn})]}{[\overline{\phi(\alpha_{n1})}\phi(z)-1]\cdots[\overline{\phi(\alpha_{nn})}\phi(z)-1]}\right]^{1/n}$$

are locally single-valued and analytic in the extended plane exterior to C_2 and form a normal family in that region in an extended sense (op. cit., p. 162). No limit function of the family can have a modulus in that region greater than unity, by (58). But each function of the family has a value at infinity of modulus $|\phi'(\infty)|^{-1/n}$, and each limit function has a value at infinity of modulus unity. Consequently every limit function of the family is of modulus unity throughout the extended plane exterior to C_2, and the sequence (52) approaches unity uniformly on any closed set exterior to C_2. Then by (49) equation (50) is valid uniformly on any closed set in R, which implies the uniform convergence of $r_n(z)$ to $f(z)$ whenever $f(z)$ is analytic on and within C_2, and the maximal convergence under these conditions if $f(z)$ is not analytic throughout R.

Conversely, let us now assume that the sequence $r_n(z)$ converges uniformly to $f(z)$ on and within C_2 whenever $f(z)$ is analytic on and within C_2. If (58) does not hold, we have from the reasoning used on the functions (59) for a suitable infinite sequence n_k and for a suitable $\Delta > 1$

$$M_{n_k} \geqq \Delta^{n_k} > 1. \tag{60}$$

By Theorem 15 equation (50) is satisfied. Choose t in R with $1 < \Phi(t) < \Delta$. By (51) we have for sufficiently large n

$$|\omega_n(t)| \leqq \Delta^n. \tag{61}$$

For the function $f(z) = 1/(t-z)$ inequalities (60) and (61) contradict our hypothesis (55). This contradiction proves (58), and implies the maximal convergence of $r_n(z)$ to $f(z)$ whenever $f(z)$ is analytic on and within C_2 but not analytic throughout R. Theorem 16 is established.

It should be mentioned that results closely related to Theorems 15 and 16 have been previously established by Shen (op. cit., p. 258); but the hypothesis on the α_{nk} is there geometric, not asymptotic, and the necessary and sufficient conditions derived, such as (50) and (58), are not for uniform convergence, but for an inequality corresponding to (2).

Theorems 15 and 16 are of interest chiefly in connection with maximal convergence. Nevertheless under the conditions of both those theorems (compare the remarks made concerning Theorem 1), whenever $f(z)$ is analytic throughout the interior of $\Phi(z) = T$ in R, we have

$$\limsup_{n\to\infty} [\max |f(z) - r_n(z)|,\ z \text{ on } C_2]^{1/n} \leqq 1/T.$$

In connection with Theorems 15 and 16 we shall prove the

COROLLARY. *Let C_1, C_2, R, and the α_{nk} satisfy the conditions of Theorem* 15, *including* (49). *Let the points β_{nk} lie on C_2. Then a necessary and sufficient con-*

dition that the sequence of rational functions $r_n(z)$ of respective degrees n with poles in the points α_{nk} and interpolating to $f(z)$ in the points β_{nk} should converge uniformly to $f(z)$ on and within C_2 whenever $f(z)$ is analytic on and within C_2, is that the points β_{nk} be uniformly distributed on C_2 with respect to the function conjugate to $\log \Phi(z)$ *in R, continuous in $\overline{R}$.*

The sufficiency of this condition (due to Fejér if $\alpha_{nk}=\infty$) is essentially contained in the proof of Theorem 11, in the light of the proof of Theorem 13. To prove the necessity of the condition, which is due to Kalmár if $\alpha_{nk}=\infty$, we note that (50) is a consequence of Theorem 15 and that (33) then follows; in (33) we may take Γ_2 as identical with C_2 if the integral is interpreted in an extended sense; by the fact that $U(x, y)$ is constant on C_2 we may write (33) in the form

$$\lim_{n\to\infty} \left| (z-\beta_{n1}) \cdots (z-\beta_{n,n+1}) \right|^{1/n} = \exp\left[\frac{-1}{2\pi}\int_{C_2} \log r \frac{\partial U}{\partial \nu} ds\right],$$

uniformly on any closed bounded set exterior to C_2. The conclusion follows at once by methods previously set forth (op. cit., §7.6).

On the topic of uniform distribution of points it is appropriate to establish

THEOREM 17. *Let C_1, C_2, R, and the α_{nk} satisfy the conditions of Theorem* 8, *including* (31), *and let $U(x, y)$ be constant on C_1. Let the points α_{nk} lie on C_1. Then the α_{nk} are uniformly distributed on C_1 with respect to the function conjugate to $U(x, y)$ in R, continuous in $\overline{R}$.*

Let z_0 be an arbitrary fixed point interior to C_1. In (34) we take the integral (in an extended sense) over C_1; by the fact that $U(x, y)$ is constant on C_1 we have

$$\lim_{n\to\infty} \left| \frac{(z-\alpha_{n1}) \cdots (z-\alpha_{nn})}{(z_0-\alpha_{n1}) \cdots (z_0-\alpha_{nn})} \right|^{1/n} = \exp\left[\frac{1}{2\pi}\int_{C_1} \log \frac{r}{r^0}\cdot\frac{\partial U}{\partial \nu} ds\right],$$

uniformly for z on any closed set interior to C_1, where α is a variable on C_1 and $r=|z-\alpha|$, $r^0=|z_0-\alpha|$. If we set

$$du = \frac{1}{2\pi}\frac{\partial U}{\partial \nu} ds,$$

we may then write

$$\lim_{n\to\infty} \frac{1}{n}\sum_{k=1}^{n} \log \frac{r_{nk}}{r_{nk}^0} = \int_{C_1} \log \frac{r}{r^0}\cdot du,$$

$$r_{nk} = |z-\alpha_{nk}|, \qquad r_{nk}^0 = |z_0-\alpha_{nk}|.$$

By virtue of (32), this integral over C_1 can be considered an integral over the interval $0 \leqq u \leqq 1$. An arbitrary function of α continuous on C_1 can be uniformly approximated on C_1 as closely as desired by a suitably chosen constant plus a linear combination of functions

$$\log [|z_k - \alpha| / |z_0 - \alpha|],$$

where the z_k are suitably chosen points interior to C_1; this follows by the transformation $z' = 1/(z_0 - \alpha)$ from a former lemma (op. cit., p. 169, Lemma II).* An arbitrary function $\chi(\alpha)$ constant on C_1 has the trivial property

$$\lim_{n\to\infty} \frac{1}{n} \sum_{k=1}^{n} \chi(\alpha_{nk}) = \int_{C_1} \chi(\alpha) du,$$

so it follows from the possibility of approximation that every function continuous on C_1 has the corresponding property. It follows (compare op. cit., §§7.5 and 7.6) that the set α_{nk} is uniformly distributed on C_1 with respect to the parameter u.

9. **Maximal convergence of sequences interpolating in the Fekete-Shen points.** Certain sets of points, first introduced by Fekete for the case of polynomials, were used by him as points of interpolation to define polynomials converging favorably to a given analytic function on a given point set. The analogous points for the case of rational functions with preassigned poles α_{nk} were introduced and used by Shen (op. cit., §9.7) when the α_{nk} are subject to geometric conditions; Shen's results on degree of convergence, even for the case $\alpha_{nk} = \infty$, are more precise than those of Fekete. These same points were used subsequently by the present writer (op. cit., §9.10) when the α_{nk} satisfy asymptotic conditions. We are now in a position to prove

THEOREM 18. *Let C_1, C_2, R, and the α_{nk} satisfy the conditions of Theorem* 15, *including* (49). *Let the β_{nk} be the Fekete-Shen points for C_2. Then conditions* (50) *and* (58) *are fulfilled. Consequently whenever $f(z)$ is analytic on and within C_2 but not analytic throughout R, the sequence of rational functions $r_n(z)$ of respective degree n with poles α_{nk} defined by interpolation to $f(z)$ in the points β_{nk} converges maximally to $f(z)$ on and within C_2.*

If $f(z)$ is analytic throughout the interior of the curve C_T: $\Phi(z) = T$ in R, we have (op. cit., p. 263, Corollary 2)

$$\limsup_{n\to\infty} [\max |f(z) - r_n(z)|, z \text{ on } C_2]^{1/n} \leqq 1/T.$$

* It is essential to admit an additive constant, for otherwise no nonvanishing constant could itself be approximated; a uniformly convergent sequence of approximating functions converges to a uniform limit in the closed exterior of C_1, and each function $\log |(z_k - \alpha)/(z_0 - \alpha)|$ is harmonic and vanishes at infinity: $\alpha = \infty$.

The maximal convergence of the sequence $r_n(z)$ follows from Corollary 2 to Theorem 13, if $f(z)$ is not analytic throughout the interior of any $C_{T'}$, $T'>T$. The remainder of Theorem 18 follows from Theorems 15 and 16.

It is of interest to note that the Fekete-Shen points β_{nk} lie on C_2 itself, and hence by the Corollary to Theorem 16 these points are uniformly distributed on C_2 with respect to the function conjugate to $\log \Phi(z)$ in R, continuous in $\overline{R}$.

It is of some interest to note that the entire theory of maximal convergence of sequences of rational functions may be developed by taking not (1) but condition (49) as fundamental, using the Fekete-Shen points (as in op. cit., §9.10) to obtain a sequence of rational functions $r_n(z)$ satisfying the above inequality; the analogue of (4) is readily proved, so that Theorem 5 applies. The remainder of the theory can then be built up. It seems simpler and more natural to the writer to make (1) fundamental rather than (49).

10. **A general extremal problem on approximation by functions with preassigned poles.** The problem of approximation by rational functions with preassigned poles can be considered a special case of approximation by more general functions with preassigned poles.

THEOREM 19. *Let C_1, C_2, R, and the points α_{nk} and β_{nk} satisfy the conditions of Theorem* 1, *including condition* (1). *Let R_γ denote generically the annular region bounded by C_2 and C_γ. Suppose the function $F(z)$ to be analytic in R_γ, $\gamma_1>\gamma>\gamma_2$, but not analytic throughout any $R_{\gamma'}$, $\gamma'>\gamma$; suppose $F(z)$ continuous in the two-dimensional sense on C_2.*

Denote by K the region exterior to C_2. Then there exist functions $F_n(z)$ analytic in K except for possible poles in the points α_{nk}, continuous in the two-dimensional sense on C_2, such that we have

$$(62)\qquad \limsup_{n\to\infty} [\max |F(z) - F_n(z)|, z \text{ on } C_2]^{1/n} = \gamma_2/\gamma;$$

but there exists no such sequence $F_n(z)$ for which the left-hand member of (62) *is less than γ_2/γ.*

Any sequence $F_n(z)$ for which (62) *holds possesses also the following properties:*

$$(63)\qquad \limsup_{n\to\infty} [\max |F(z) - F_n(z)|, z \text{ on } C_\lambda]^{1/n} = \lambda/\gamma, \qquad \gamma_2 \leqq \lambda < \gamma,$$

$$(64)\qquad \limsup_{n\to\infty} [\max |F_n(z)|, z \text{ on } C_\mu]^{1/n} = \mu/\gamma, \qquad \gamma \leqq \mu \leqq \gamma_1.$$

If Q is an arbitrary continuum in $\overline{R}$ not a single point, we have

$$(65)\qquad \limsup_{n\to\infty} [\max |F_{n+1}(z) - F_n(z)|, z \text{ on } Q]^{1/n} = [\max \Phi(z), z \text{ on } Q]/\gamma.$$

The function $F(z)$ can be written as the sum of its two components:

$$F(z) \equiv f_1(z) + f_2(z), \qquad z \text{ in } R_\gamma, \tag{66$_1$}$$

$$f_1(z) \equiv \frac{1}{2\pi i}\int_{\Gamma_1} \frac{F(t)dt}{t - z}, \qquad z \text{ interior to } C_\gamma, \tag{66$_2$}$$

$$f_2(z) \equiv \frac{1}{2\pi i}\int_{\Gamma_2} \frac{F(t)dt}{t - z}, \qquad z \text{ exterior to } C_2, \tag{66$_3$}$$

where Γ_1 is a rectifiable Jordan curve in R_γ containing z and C_2 in its interior, and where Γ_2 is a rectifiable Jordan curve in R_γ containing C_2 in its interior but having z exterior to it; the curves Γ_1 and Γ_2 depend then on z, but the functions $f_1(z)$ and $f_2(z)$ are independent of the particular curves chosen. Equation (66_1) is valid for z in R_γ. Nevertheless the integrals in (66_2) and (66_3) define $f_1(z)$ and $f_2(z)$ as functions analytic respectively throughout the interior of C_γ and throughout the exterior of C_2 (even at infinity, by a limiting process). Equation (66_1) can be used to define $f_2(z)$ on C_2; with this additional definition, the function $f_2(z)$ is continuous on C_2. The function $f_1(z)$ is analytic throughout the interior of no $C_{\gamma'}$, $\gamma' > \gamma$.

We are now in a position to identify the function $f(z)$ of Theorem 1 with the present $f_1(z)$. If we write

$$f(z) \equiv f_1(z), \qquad r_n(z) \equiv F_n(z) - f_2(z),$$

equation (62) is a consequence of (indeed identical with) equation (3) (see Theorem 4).

The method of proof of Theorems 3 and 4 applies also in the present case, and shows that there exists no sequence $F_n(z)$ for which the left-hand member of (62) is less than γ_2/γ.

Let us suppose now $F_n(z)$ to be an arbitrary sequence satisfying (62). With the aid of the comparison sequence $r_n(z) + f_2(z)$ just considered, which satisfies (62) in place of $F_n(z)$, we find as in the proof of (10)

$$\limsup_{n\to\infty} \left[\max |F_n(z)|, z \text{ on } C_\mu\right]^{1/n} \leqq \mu/\gamma, \qquad \gamma < \mu \leqq \gamma_1.$$

Theorem 5 now applies, and yields the remaining parts of Theorem 19.

A slight modification of Theorem 19 is of interest. Let us replace the requirement that $F(z)$ be continuous in the two-dimensional sense on C_2 by the requirement that $F(z)$ be bounded in the neighborhood of C_2, with the identical modification in the requirements on $F_n(z)$. The expression

$$\max |F(z) - F_n(z)|, \quad z \text{ on } C_2,$$

which occurs in (62) is then to be interpreted as

(67) $$\limsup_{\lambda\to\gamma_2} [\max |F(z) - F_n(z)|, z \text{ on } C_\lambda],$$

which is necessarily finite. With this understanding, also similarly for (63), (64), and (65), Theorem 5 is valid for the corresponding sequences, and Theorem 19 is valid in the generalized form.

With this new convention, there exists for each n at least one admissible function $F_n(z)$ for which (67) *is least.* By an admissible function $F_n(z)$ we understand here a function analytic in K except for possible poles in the points α_{nk}, and bounded in the neighborhood of C_2. Let N_0 denote the greatest lower bound of all numbers (67) for the class of admissible functions. Let $F_n^{(1)}(z), F_n^{(2)}(z), \cdots$ be admissible functions for which the corresponding numbers $N^{(1)}, N^{(2)}, \cdots$ approach N_0:

(68) $$\lim_{k\to\infty} N^{(k)} = N_0.$$

We have

(69) $$\begin{aligned} &[\max |F_n^{(1)}(z) - F_n^{(k)}(z)|, z \text{ on } C_\lambda] \\ &\le [\max |F(z) - F_n^{(1)}(z)|, z \text{ on } C_\lambda] + [\max |F(z) - F_n^{(k)}(z)|, z \text{ on } C_\lambda], \\ &\limsup_{\lambda\to\gamma_2} [\max |F_n^{(1)}(z) - F_n^{(k)}(z)|, z \text{ on } C_\lambda] \le N^{(1)} + N^{(k)}. \end{aligned}$$

The function $F_n^{(1)}(z) - F_n^{(k)}(z)$ is analytic in K except for poles in the points α_{nk}. The function

(70) $$\frac{[F_n^{(1)}(z) - F_n^{(k)}(z)][\phi(z) - \phi(\alpha_{n1})] \cdots [\phi(z) - \phi(\alpha_{nn})]}{[\overline{\phi(\alpha_{n1})}\phi(z) - 1] \cdots [\overline{\phi(\alpha_{nn})}\phi(z) - 1]},$$

when suitably defined in the points α_{nk}, is analytic throughout K and by (69) is uniformly bounded (independently of k) in K. From any infinite sequence of functions (70) can be extracted a subsequence which converges throughout K, uniformly on any closed set in K, to a bounded function of form (70) with $F_n^{(k)}(z)$ replaced by an admissible function $F_n^{(0)}(z)$. The corresponding subsequence of the $F_n^{(k)}(z)$ then converges to $F_n^{(0)}(z)$ throughout K except in the points α_{nk}, uniformly on any closed set in K containing no point α_{nk}. Let us suppose (this supposition involves merely a change of notation) that the original sequence $F_n^{(k)}(z)$ so converges.

Let μ be chosen arbitrarily but fixed, $\gamma > \mu > \gamma_2$, and let us suppose $\gamma_2 < \lambda < \mu$, where λ will be allowed to approach γ_2. We now apply the two-constant theorem (or Hadamard three-circle theorem), which is valid under the present generalized conditions on the functions involved:

$$\begin{aligned}&[\max |F(z) - F_n^{(k)}(z)|, z \text{ on } C_\lambda] \\ (71) \qquad &\leqq [\max |F(z) - F_n^{(k)}(z)|, z \text{ on } C_2]^{(\mu-\lambda)/(\mu-\gamma_2)} \\ &\qquad \cdot [\max |F(z) - F_n^{(k)}(z)|, z \text{ on } C_\mu]^{(\lambda-\gamma_2)/(\mu-\gamma_2)}.\end{aligned}$$

The second square bracket in the second member of (71) has a bound N independent of k; the first square bracket in the second member of (71) approaches N_0, by (68). Thus we have by letting k become infinite

$$\begin{aligned}&[\max |F(z) - F_n^{(0)}(z)|, z \text{ on } C_\lambda] = \lim_{k\to\infty} [\max |F(z) - F_n^{(k)}(z)|, z \text{ on } C_\lambda] \\ (72) \qquad &\leqq N_0^{(\mu-\lambda)/(\mu-\gamma_2)} \cdot N^{(\lambda-\gamma_2)/(\mu-\gamma_2)};\end{aligned}$$

the first equality sign is a consequence of the uniform convergence of $F_n^{(k)}(z)$ to $F_n^{(0)}(z)$ on C_λ. If we allow λ to approach γ_2 in (72), we have

$$[\max |F(z) - F_n^{(0)}(z)|, z \text{ on } C_2] \leqq N_0;$$

the inequality sign is impossible here by the definition of N_0, so we have established the existence of at least one extremal function $F_n(z)$. The question of the uniqueness of this extremal function is still open.

Naturally the extremal functions $F_n(z)$ whose existence has just been established satisfy condition (62); and for them the conclusion of Theorem 19 is valid.

We remark too that the requirements of continuity of $F(z)$ and $F_n(z)$ on C_2 need not be replaced by the requirement of boundedness in the neighborhood of C_2 but may simply be omitted, without essentially altering our conclusions; the square bracket in the left-hand member of (62) is merely understood as (67); a suitable comparison sequence $F_n(z)$ can be obtained as before from Theorem 1; at least one extreme function $F_n(z)$ exists.

Whether or not this last generalization is allowed, we shall prove

THEOREM 20. *Let the conditions of Theorem 19 be fulfilled, including equation (62), and let the respective components of the function $F_n(z)$ be $f_{n1}(z)$ and $f_{n2}(z)$, defined by equations analogous to (66). Then the function $f_{n1}(z)$ is a rational function of degree n with poles in the points α_{nk}, and the sequence $f_{n1}(z)$ converges maximally to $f_1(z)$ on the closed interior of C_2.*

In the region R_γ we have the equation

$$F_n(z) = f_{n1}(z) + f_{n2}(z),$$

where $f_{n1}(z)$ is analytic throughout the interior of C_γ, and $f_{n2}(z)$ is analytic throughout the exterior of C_2, even at infinity. But $F_n(z)$ is analytic throughout K except in the points α_{nk}, so $f_{n1}(z)$ can be extended analytically from R_γ, and can be considered defined and analytic throughout K except for poles in the points α_{nk}. Consequently $f_{n1}(z)$ is a rational function of degree n whose poles lie in the points α_{nk}.

In the defining equations for $f_1(z)$ and $f_{n1}(z)$ we may choose Γ_1 as the curve C_λ, $\gamma_2<\lambda<\gamma$, whence

$$f_1(z) - f_{n1}(z) = \frac{1}{2\pi i}\int_{C_\lambda} \frac{F(t) - F_n(t)}{t - z}\, dt, \qquad z \text{ on } C_2.$$

Then equation (63) yields

$$\limsup_{n\to\infty} \left[\max |f_1(z) - f_{n1}(z)|, z \text{ on } C_2\right]^{1/n} \leqq \lambda/\gamma.$$

In this inequality we may allow λ to approach γ_2:

$$\limsup_{n\to\infty} \left[\max |f_1(z) - f_{n1}(z)|, z \text{ on } C_2\right]^{1/n} \leqq \gamma_2/\gamma. \tag{73}$$

The function $f_2(z)$ is analytic throughout the region K, and $F(z)$ is analytic throughout the interior of C_γ but not throughout the interior of any $C_{\gamma'}$, $\gamma'>\gamma$. Then $f_1(z)$ also is analytic throughout the interior of C_γ but not throughout the interior of any $C_{\gamma'}$. Inequality (73) thus establishes the conclusion of Theorem 20.

Results analogous to but less explicit than Theorems 19 and 20 lie at hand if we replace the original requirement on $F(z)$ by the requirement that $F(z)$ be analytic throughout R. For instance there exist functions $F_n(z)$ analytic in K except for possible poles in the points α_{nk} such that the first member of (62) is not greater than γ_2/γ_1.

We have several comments to make regarding the general significance of the problem studied in Theorems 19 and 20.

1°. Theorem 19 is a result concerning a problem which is invariant under arbitrary one-to-one conformal transformation of the region K, even under some circumstances (compare §11 below) if K is transformed into a region not bounded by a Jordan curve. But the full import of this invariant property appears best if the hypothesis (1) is replaced by the equivalent hypothesis (49).

2°. The two groups, Theorems 1–4 on the one hand and Theorems 19 and 20 on the other, are intimately related to each other, especially in view of Theorem 5. Theorem 1 can be employed as above to furnish a sequence $F_n(z)$ which converges as rapidly (in the sense indicated by (62)) as does any sequence of admissible functions. Reciprocally, Theorem 20 shows that Theorem 19 can be employed to furnish a sequence of rational functions which converges maximally.

3°. Theorem 1 can be formulated so as to be invariant under arbitrary linear transformation of the complex variable. Condition (49) has the more general property of being essentially invariant under arbitrary one-to-one conformal transformation of the closed exterior of C_2. Remarks 1° and 2° serve to explain this latter invariant property, which has been in the past somewhat mystifying.

4°. Theorem 19 is a companion piece (Gegenstück) to results previously established by the present writer concerning interpolation by functions analytic and bounded in a given simply connected region K.* Prescribed poles on the one hand correspond to prescribed points of interpolation on the other, in each case satisfying asymptotic conditions in an annular region R interior to K, one of whose boundaries is the boundary of K. In the problem of interpolation there are two equivalent conditions on the prescribed points, completely analogous to (1) and (49). The function approximated to in the one problem and the function interpolated to in the other are assumed analytic in respective regions mutually complementary with respect to K, and convergence of approximating and interpolating functions is established throughout those (suitably chosen) regions of analyticity bounded by loci $\Phi=\text{const.}$ in R. Indeed, this entire pair of configurations exhibits a complete duality, in the sense already defined (op. cit., §8.3), except that the present duality is concerned with functions meromorphic or analytic in a region rather than with rational functions as in the former duality. Theorem 5 is sufficiently powerful to apply to both configurations.

11. **Generalizations and extensions.** In connection with the foregoing results, there suggests itself the question of extensions to regions of higher connectivity, and to regions not bounded by Jordan curves. We now proceed to discuss these questions in order.

We have hitherto limited ourselves to a region bounded merely by two Jordan curves C_1 and C_2 instead of two such sets of curves, entirely for the sake of simplicity and ease of exposition. Our methods are applicable with only minor modifications in the more general case. For instance we sketch rapidly the proof of

THEOREM 21. *Let R be a finite region bounded by two disjoint sets C_1 and C_2, each composed of a finite number of disjoint Jordan curves; let one of the Jordan curves C_1^0 of the former set contain in its interior all points of $C_1-C_1^0+C_2$. Let*

* These Transactions, vol. 46 (1939), pp. 46–65.

the points $\alpha_{n1}, \alpha_{n1}, \cdots, \alpha_{nn}$ *be separated from* R *by* C_1, *and let the points* $\beta_{n1}, \beta_{n2}, \cdots, \beta_{n,n+1}$ *not lying on* C_2 *be separated from* R *by* C_2. *Let* $\overline{R}$ *denote the closure of* R.

Let the relation (1) *hold uniformly on any closed set in* $\overline{R}$ *containing no point of* C_2. *Let the function* $\Phi(z)$ *be continuous in the closed region* $\overline{R}$, *and take constant values* γ_1 *and* $\gamma_2<\gamma_1$ *on* C_1 *and* C_2 *respectively. Denote generically by* C_γ *the locus* $\Phi(z)=\gamma$ *in* $\overline{R}$, $\gamma_2\leqq\gamma\leqq\gamma_1$. *For* $\gamma_2<\gamma<\gamma_1$, *the locus* C_γ *consists of a finite number of mutually disjoint analytic Jordan curves, except that for each of a finite number of values of* γ *there are a finite number of multiple points, each common to a finite number of the Jordan curves composing* C_γ. *Denote generically by* R_γ *the open set composed of the interiors of the Jordan curves composing* C_2 *plus the point set in* $\overline{R}$ *on which we have* $\gamma_2\leqq\Phi(z)<\gamma$; *thus* R_γ *need not be connected, but is the sum of a finite number of regions bounded by the entire locus* C_γ.

Let the function $f(z)$ *be single-valued and analytic on* R_γ *but not single-valued and analytic on any* $R_{\gamma'}$ *with* $\gamma'>\gamma$. *Then we have* (2), (3), *and* (4) *fulfilled, where* $r_n(z)$ *is the rational function of degree* n *whose poles lie in the points* α_{nk} *and which interpolates to* $f(z)$ *in the points* β_{nk}.

There exists no sequence of rational functions $R_n(z)$ *of respective degrees* n *with poles in the points* α_{nk} *such that either* (8) *or* (9) *is satisfied.*

Whenever a set of rational functions $R_n(z)$ *of respective degrees* n *with poles in the points* α_{nk} *satisfies* (20), *we shall say that the sequence* $R_n(z)$ *converges maximally to* $f(z)$ *on the set* C *composed of the Jordan curves* C_2 *and their respective interiors. Consequences of maximal convergence are equations* (22) *and* (23), *and if* Q *is an arbitrary continuum in* $\overline{R}$ *not a single point, also equations* (24)–(26).

In Theorem 21 the function $f(z)$ need not be a monogenic analytic function; nevertheless $f(z)$ can be extended analytically from the neighborhood of C_2 by paths in R_γ so as to be single-valued and analytic at every point of R_γ; but this property is shared by no set $R_{\gamma'}$ with $\gamma'>\gamma$. Then (a) for some path of analytic extension in R_γ from C_2 to C_γ the function $f(z)$ has a singularity on C_γ, or (b) the locus C_γ has a multiple point A and the analytic extensions of $f(z)$ from C_2 along paths in the various regions into which C_γ falls of which A is a boundary point fail to be analytic and identical throughout a neighborhood of A; or both (a) and (b) occur.

Theorem 5 extends to a region R bounded by two sets of Jordan curves, and the original proofs already given yield now Theorem 21.

With only minor modifications, which we now mention, the entire discussion of §§1–10 of the present paper is valid for the more general region R introduced in Theorem 21.

Under the hypothesis of Theorem 12 (that is to say, the hypothesis of Theorem 1), equation (44) may be written

$$\lim_{n\to\infty} \frac{1}{n}\sum_{k=1}^{n} G(z, \alpha_{nk}) = \log \Phi_1(z) \tag{74}$$

uniformly on any closed set in R, where

$$G(z, \alpha_{nk}) = \log \left| \frac{\overline{\phi(\alpha_{nk})}\phi(z) - 1}{\phi(z) - \phi(\alpha_{nk})} \right|$$

is Green's function* with pole in the point α_{nk} for the region exterior to the Jordan curves C_2. With this change in the manner of writing (44), the new formulation of Theorem 12 is obviously valid; theorem and proof extend without great difficulty to the situation of Theorem 21. But equation (44) is obviously simpler than (74), and more easily applied. Of course C_1 may consist of several Jordan curves even when C_2 is a single Jordan curve; under such circumstances (44) and similar relations need no revision.

Let us be more explicit about the extension of Theorem 12, using now the hypothesis of Theorem 21 rather than of Theorem 1; it is sufficient if (1) holds uniformly merely on any closed set in R. On an arbitrary closed set C' in R the functions $G(z, \alpha_{nk})$ are uniformly bounded, for the points α_{nk} are bounded from C', and the functions $G(z, \alpha_{nk})$ in R depend continuously on z and α_{nk} except in the neighborhood of α_{nk}. The functions

$$\frac{1}{n}\sum_{k=1}^{n} G(z, \alpha_{nk}) \tag{75}$$

are harmonic in R and uniformly limited on any closed set C' in R, hence form a normal family of harmonic functions in R. Let $\log \Phi_1(z)$ denote an arbitrary limit function of the family (75), the uniform limit on any closed set in R of a suitably chosen subsequence of the family. The functions (75) are continuous in the two-dimensional sense on C_2 and vanish there, so the subsequence converges uniformly on any closed set in $\overline{R}$ containing no point of C_1. The limit function $\log \Phi_1(z)$ is continuous in the two-dimensional sense on C_2 and vanishes there. Consider now the function

$$\begin{aligned}\frac{1}{n}\log\left| A_n \frac{(z-\beta_{n1})\cdots(z-\beta_{n,n+1})}{(z-\alpha_{n1})\cdots(z-\alpha_{nn})}\right| \\ - \frac{1}{n}\log\sum_{k=1}^{n} G(z, \alpha_{nk}) - \frac{1}{n}\log G(z, \infty),\end{aligned} \tag{76}$$

* The term Green's function is sometimes used for the negative of the present function.

which when suitably defined in the points α_{nk} is harmonic without exception in the region exterior to and bounded by the curves C_2. The sequence (76) or any subsequence converges uniformly on any closed set interior to R whenever that is true of the corresponding sequence or subsequence (75). By essentially the reasoning previously used in the proof of Theorem 12 we now have $\Phi(z)/\Phi_1(z)\equiv\gamma_2$ throughout R. Then the function $\Phi_1(z)$ is uniquely determined; from every subsequence of the sequence (75) can be extracted a new subsequence converging uniformly on any closed set in R to the function $\log \Phi_1(z)$, hence the sequence (75) itself converges uniformly on any closed set in R to the function $\log \Phi_1(z)$; the extension of Theorem 12 to the situation of Theorem 21 rather than of Theorem 1 is established.

In the extension of Theorem 13 to a region R bounded by more than two Jordan curves, we naturally replace (46) by

$$\text{og } A_n = \sum_{k=1}^{n} G(\alpha_{nk}, \infty); \tag{77}$$

since only the modulus of A_n is of significance in (1), equation (77) is essentially equivalent to (46) under the hypothesis of Theorem 13. Of course the validity of (1) with the substitution (77) is to be established by the method indicated for the proof of (1) with the substitution (46) (op. cit., p. 274). The indicated extension of the remainder of §§1–10 now presents no difficulty, as the reader may verify; but in §8 a somewhat unusual convention is desirable regarding normal families of analytic functions (compare op. cit., §7.3).

We turn next to the question of replacing by more general sets the Jordan curves in Theorem 21 and in the corresponding results of §§1–10. Theorem 5 extends without difficulty and with only minor modifications to the case that R is bounded by a finite number of continua, none of which is a single point. There is no inherent difficulty in extending almost all of the discussion of the present paper to this more general case. However, with this more general hypothesis, such integrals as appear in (36) no longer have the meaning previously assigned, and points α_{nk} and β_{nk} cannot be taken or proved uniformly distributed on C_1 and C_2 by the previous methods. There are then two possible procedures for determining the points α_{nk} and β_{nk}: (i) consider integrals not over C_1 and C_2 but over sets of curves Γ_1 and Γ_2 in R and approaching C_1 and C_2 respectively, choosing points α_{nk} and β_{nk} on the sets Γ_1 and Γ_2 (as in op. cit., §4.4); but of course these points α_{nk} and β_{nk} then lie interior to R, and (1) may not be valid uniformly on C_1; (ii) if each point of C_1 and C_2 is a boundary point of the complement of $\overline{R}$, the harmonic function $U(x, y)$ in (36) may be uniformly approximated in $\overline{R}$ by functions harmonic throughout $\overline{R}$, by Lebesgue's theory of variable harmonic functions; these approximating functions may be handled by the original method of uniform distribution of points, and yield points α_{nk} and β_{nk} exterior to $\bar{R}$ (compare op. cit., §4.4); this method can be used also for the appropriate points α_{nk} or β_{nk} if only one of the sets C_1 and C_2 has the property that each of its points is a boundary point of the complement of $\bar{R}$. It will be noticed that method (i) enables us to use for C, the point set on which the given function is approximated, any closed bounded set not separating the plane having only a finite number of components of which none is a single point. By an extension of Theorem 5 we may here take C an arbitrary closed bounded set whose complement is connected and possesses a Green's function.

Neither (i) nor (ii) enables us to establish uniform distribution on C_1 or C_2 of given points, and thus the necessity of the uniform distribution in the Corollary to Theorem 16, in Theorem 17, and in §9, is still unproved. In this connection there suggests itself another method: (iii) consider the integrals in a still more general sense, say in the sense of Stieltjes with respect to a variable defined on the boundary elements of the region; in the generalization of (36) this variable would be the conjugate of U, locally single-valued and continuous in $\bar{R}$. It seems to the writer probable that this method would be completely satisfactory both in determining points α_{nk} and β_{nk} from such an equation as (36), and in establishing the uniform distribution of given points as in the Corollary to Theorem 16; but this method has never been completely carried out.

We have hitherto considered merely a single region R, but it is possible (op. cit., §§8.7 and 8.8) to have such a relation as (1) valid simultaneously interior to two mutually disjoint regions of the kind considered in Theorem 1. Indeed if these regions R are preassigned, if the corresponding closed regions are mutually disjoint, and if no region separates another from the point at infinity, suitable harmonic functions can be used as generating functions to define the sequences α_{nk} and β_{nk} so that (1) is valid on any closed set in each region R. Then our entire discussion of §§1–10 carries over to the new situation, with a single exception. The function $\Phi(z)$ of (1) is harmonic in each R, and a locus C_γ occurs in each R. But it may occur that γ as defined from the function $f(z)$ is limited by the behavior of $f(z)$ in one region R, and not by the function $f(z)$ in a second region R; for instance $f(z)$ may be analytic throughout the latter. Theorem 5 then fails to apply to this second region R, and the conclusions (22)–(26) that we have drawn from Theorem 5 apply to the first region R but have not been established for the second region R.*

* But for a series of interpolation (78) the principal results are valid also for the second region R. For various other specific examples the corresponding statement is true; the general question is still open.

An important special case of Theorem 1 is that in which the α_{nk} and β_{nk} are independent of n. Under these circumstances the interpolating functions $r_n(z)$ of Theorem 1 are the partial sums of a series of interpolation

$$f(z) = a_0 + a_1 \frac{z - \beta_1}{z - \alpha_1} + a_2 \frac{(z - \beta_1)(z - \beta_2)}{(z - \alpha_1)(z - \alpha_2)} + \cdots, \tag{78}$$

and some of the results of Theorem 7 so far as it applies to the sequence $r_n(z)$ can be read off by elementary methods, without recourse to Theorem 5. For instance under the hypothesis of Theorem 1 we know (op. cit., §8.3) that the series in (78) converges to $f(z)$ uniformly on any closed set interior to C_γ, but by the definition of γ can converge uniformly throughout the closed interior of no $C_{\gamma'}$, $\gamma' > \gamma$. There follows (op. cit., §3.4, Theorem 5) the relation

$$\limsup_{n \to \infty} |a_n / A_n|^{1/n} = 1/\gamma. \tag{79}$$

For the series (78) we deduce by inspection inequalities (2)–(4). Impossibility of the inequality sign in (2) and (4) follows at once from (1) and (79). Much of Theorem 7 for the series (78) is also a consequence of (1) and (79).

In generalizing the results of the present paper, it is also of interest to notice that points α_{nk} and β_{nk} interior to R may be admitted in restricted numbers, and that the equation (1) need not be supposed to hold uniformly on any closed set in R; under suitable circumstances it is sufficient if (1) holds uniformly on C_1 and on each C_λ for λ everywhere dense in the interval $\gamma_2 < \lambda < \gamma_1$. A sufficiently general instance of this remark has already been elaborated elsewhere (op. cit., §8.5).

HARVARD UNIVERSITY,
CAMBRIDGE, MASS.

OVERCONVERGENCE, DEGREE OF CONVERGENCE, AND ZEROS OF SEQUENCES OF ANALYTIC FUNCTIONS

By J. L. Walsh

In 1904 Porter published the first known example of a Taylor development of which a suitably chosen sequence of partial sums converges in a region containing both points interior and exterior to the circle of convergence. Porter mentioned that this phenomenon has close relations with the accumulation of zeros of partial sums at points of the circle of convergence. In the intervening years, this general topic has been further and extensively studied by Jentzsch, Ostrowski, Szegö, Pólya, Carlson, Bourion, and a number of others; Bourion [1] has recently published a summary of the results obtained, to which the reader may refer for further historical and technical details.

It is the object of the present paper to undertake an analogous study for sequences of *arbitrary* analytic functions, when we assume neither that the functions are polynomials nor that the sequences are defined in terms of a linear series, of the form

$$\sum_{n=0}^{\infty} a_n q_n(z),$$

where the a_n are constants and where the asymptotic properties of the $q_n(z)$ are known.

Although the results of the present study are of considerable generality, involve geometric and analytic configurations which are extremely broad in scope, and yet include all known results on Taylor's series except those of special form which seem to apply only to sequences of polynomials or to sequences of otherwise heavily restricted functions, their inception is not due merely to a search for greater generality. The present results are the outgrowth of, and have immediate application to, the study of maximal convergence of sequences of polynomials, maximal convergence of sequences of rational functions, convergence of analytic functions of prescribed norm of best approximation, convergence of functions of given measure of approximation and minimum norm, and convergence of interpolating functions of minimum norm. The writer hopes on another occasion to consider further application of the present methods, notably to series of harmonic functions and to Dirichlet series.

The methods employed in the present paper involve especially the continued use of a harmonic majorant, a method employed systematically both by Ostrowski and Bourion; we introduce, however, the concept of *exact* harmonic

Received November 26, 1941; revision received February 20, 1946; presented to the American Mathematical Society, April 3, 1942.

majorant for a sequence of analytic functions, a new concept which seems to unify and simplify the treatment, and thereby to have some advantages for the present purposes over the two-constant theorem or three-region theorem as introduced by Ostrowski. This explicit concept of exact harmonic majorant (without the terminology) was used in a recent paper by the present writer [12].

The present paper may thus be interpreted as a chapter in the theory of subharmonic functions and their application. Our study of overconvergence itself applies in detail only to the case where the sequences of functions converge comparably to the partial sums of a geometric series. The methods, however, extend to more general situations.

It is appropriate to mention here in more detail the unpublished work of Mosesson on the general topic which concerns us. In his Harvard doctoral dissertation of 1937, Mosesson investigates this topic for the case of maximally convergent sequences of polynomials, and for such sequences establishes a part of the present results; this seems to be the first investigation of overconvergence for general sequences not arising from a linear series of the special form already mentioned. Mosesson's method involves especially the two-constant theorem in a form in which one of the point sets involved is an arbitrary set of positive capacity; Bourion [2; §10] had previously employed a somewhat similar argument. Of course Mosesson employs consistently the concept of harmonic majorant (but not that of exact harmonic majorant); many of his proofs apply only to sequences of polynomials and thus are essentially less general than those of the present paper; but for the important special case of maximally convergent sequences of polynomials, Mosesson establishes the main results of the present paper, except those of §6; for the case of Taylor's series itself those results had in the main already been established, principally by Ostrowski and Bourion.

In §1 we define the more immediate properties of the exact harmonic majorant. In §2 we determine various sufficient conditions for overconvergence, and in §3 devote some attention to the singularities of the limit function. (Throughout the present paper we use the term *overconvergence* in the sense of Ostrowski; compare Theorem 5 below. The writer has also used that term elsewhere, in a somewhat different connection.) Subsequences which converge with especial rapidity are considered in §4. A critical study of our results with reference to the generality of the point sets involved is given in §5, and finally §6 is devoted to the relation between zeros of approximating functions and both overconvergence and degree of convergence. Some of the results of §6 are new even for the classical case of Taylor's series.

For simplicity we ordinarily deal in the sequel with the plane of finite points, and the regions and other point sets involved are tacitly assumed to be bounded. But the discussion holds with only minor and obvious modifications if the point sets are not bounded, provided we deal with the extended plane or the sphere.

1. The exact harmonic majorant. The following situation is of relatively frequent occurrence: Let the function $V(z)$ be harmonic in the region R of the

z-plane, let the functions $F_n(z)$ be locally single-valued and analytic in R except perhaps for branch points, and let $|F_n(z)|$ be single-valued in R. If for every continuum Q (not a single point) in R we have the relation

$$(1) \qquad \limsup_{n\to\infty} [\max |F_n(z)|, z \text{ on } Q] = [\max e^{V(z)}, z \text{ on } Q],$$

we say that $V(z)$ is *an exact harmonic majorant of the sequence* $F_n(z)$ in R. (It would perhaps be more accurate and descriptive terminology to say that $V(z)$ is an exact harmonic majorant for the sequence $\log |F_n(z)|$ rather than for the sequence $F_n(z)$, but for the present purpose the terminology introduced is simpler and shorter.) If for every continuum Q in R we have the weaker relation

$$(2) \qquad \limsup_{n\to\infty} [\max |F_n(z)|, z \text{ on } Q] \leq [\max e^{V(z)}, z \text{ on } Q],$$

we say merely that $V(z)$ *is a harmonic majorant of the sequence* $F_n(z)$ in R. In this terminology, we have readily ([12; Theorem 2] or compare §5 below):

THEOREM 1. *If $V(z)$ is a harmonic majorant of the sequence $F_n(z)$ in R, and if for a single continuum Q_0 in R (Q_0 may be a single point) we have*

$$(3) \qquad \limsup_{n\to\infty} [\max |F_n(z)|, z \text{ on } Q_0] = [\max e^{V(z)}, z \text{ on } Q_0],$$

then $V(z)$ is an exact harmonic majorant of the sequence $F_n(z)$ in R.

For convenience in reference we state the immediate

COROLLARY. *If $V(z)$ is a harmonic majorant of the sequence $F_n(z)$ in R, and if for a single continuum Q_0 in R consisting of more than one point we have*

$$\limsup_{n\to\infty} [\max |F_n(z)|, z \text{ on } Q_0] < [\max e^{V(z)}, z \text{ on } Q_0],$$

then this strong inequality holds for every continuum Q_0 in R.

It is thus obvious that if the sequence $F_n(z)$ has the exact harmonic majorant $V(z)$ in R, then that sequence has no smaller harmonic or other continuous majorant in the sense of (2) either in R or in any subregion of R; any harmonic majorant of $F_n(z)$ in such a subregion is greater than or equal to $V(z)$. On the other hand, the existence of a harmonic majorant in R does not imply the existence of an exact harmonic majorant in R; this is illustrated by such a sequence as $F_n(z) \equiv [z^n - 1]^{1/n}$, which has the harmonic majorant log 2 in the region R: $|z - 1| < 1$, but has no exact harmonic majorant in that region; in the subregion of R in $|z| < 1$, the sequence $F_n(z)$ has the exact harmonic majorant zero, and in the subregion of R in $|z| > 1$ has the exact harmonic majorant $\log |z|$. Indeed, no subsequence of $F_n(z)$ has an exact harmonic majorant in R or in any subregion of R containing a point of $|z| = 1$.

The usefulness of the concept of exact harmonic majorant as compared with the older concept of harmonic majorant is precisely due to the requirement in

(1) that Q shall be a continuum. It is essential not to allow Q to be an arbitrary closed set, as we shall show in §5. It is also essential not to admit Q as a single point, for that would exclude such an example as $F_n(z) \equiv z^{1/n}$, $\nabla (z) \equiv 0$, R: $|z| < \infty$.

If $V(z)$ is an exact harmonic majorant of the sequence $F_n(z)$ in R, it is clear that for a particular Q in R and for a suitably chosen subsequence $F_{n_k}(z)$, we have

$$\lim_{n_k \to \infty} [\max |F_{n_k}(z)|, z \text{ on } Q] = [\max e^{V(z)}, z \text{ on } Q],$$

and consequently $V(z)$ is an exact harmonic majorant in R of every subsequence of the sequence $F_{n_k}(z)$.

Both harmonic majorant and exact harmonic majorant persist under one-to-one conformal transformation of the independent variable.

A boundary point P of a region is said to be *regular* provided the Dirichlet problem for the region and for arbitrary continuous boundary values has a solution which takes on at P the prescribed values; this property is a local one, depending only on the nature of the boundary in the neighborhood of P (see [5; 328]). The original method of proof of Theorem 1 yields with modifications which we postpone until §5:

THEOREM 2. *If $V(z)$ is an exact harmonic majorant of the sequence $F_n(z)$ in the region R, and if Q is an arbitrary closed set in R which does not separate R and which is such that every point of Q on the boundary of the region $R - Q$ is regular for that region, then* (1) *is valid.*

If we assume here that $V(z)$ is a harmonic majorant of $F_n(z)$ in R but not an exact harmonic majorant, then we conclude (2) instead of (1).

It is a matter of convenience to assume in Theorem 2 that Q does not separate R. If we have given a closed set Q_1 in R which separates R but which separates no pair of boundary points of R, then we adjoin to Q_1 all points of R separated by Q_1 from the boundary of R, and thus obtain a closed set Q to which Theorem 2 may apply. This replacing of Q_1 by Q alters neither of the values

$$[\max |F_n(z)|, z \text{ on } Q_1], \qquad [\max e^{V(z)}, z \text{ on } Q_1].$$

It is an advantage to make this replacement, however, for $R - Q_1$ cannot be a region; also the conclusion of Theorem 2 applies even if some of the boundary points of $R - Q_1$ are irregular, namely if all the irregular boundary points of $R - Q_1$ are separated from the boundary of R by the set Q_1 itself.

If a given set Q does separate boundary points of R, and if Q can be divided into a finite number of closed sets not necessarily disjoint each of which satisfies the conditions of Theorem 2, then the conclusion (1) holds also for the sum of those sets, namely Q itself.

THEOREM 3. *A necessary and sufficient condition that $V(z)$ be a harmonic majorant of $F_n(z)$ in R is that on every closed set S in R we have for an arbitrary positive ϵ*

$$\limsup_{n\to\infty} [\max [\,|F_n(z)| - e^{V(z)+\epsilon}],\ z \text{ on } S] \le 0; \tag{4}$$

or that for n sufficiently large (n depending on S and ϵ*) we have*

$$|F_n(z)| \le e^{V(z)+\epsilon}, \tag{5}$$

z on S.

The equivalence of (4) and (5) is obvious. Moreover (2) for every Q is an obvious consequence of (5) for every S. It remains to show that (2) implies (5). Let $\epsilon > 0$ be given, and an arbitrary S. Each point of S lies in a corresponding circle which with its interior lies in R, such that in the circle we have

$$\max V(z) \le V(z) + \epsilon/2. \tag{6}$$

By the Heine-Borel theorem, the set S is covered by a finite number of these circles, in each of which we have for n sufficiently large by (2) and (6)

$$|F_n(z)| \le \max |F_n(z)| \le \max e^{V(z)+\epsilon/2} \le e^{V(z)+\epsilon}; \tag{7}$$

the extreme inequality is (5).

The proof just given has made use of (2) only for small circles in R, so we may formulate with the aid of the Corollary to Theorem 1:

COROLLARY. *A necessary and sufficient condition that $V(z)$ be a harmonic majorant of the sequence $F_n(z)$ in R is that* (2) *hold for the closed interior Q of every sufficiently small circle in R. A necessary and sufficient condition that $V(z)$ be an exact harmonic majorant of the sequence $F_n(z)$ in R is that* (1) *hold for the closed interior Q of every sufficiently small circle in R.*

The term *sufficiently small* here is intended to indicate that the circles considered may be restricted to those of center z in R and radius less than $r(z)$, where $r(z)$ is positive for every z in R.

THEOREM 4. *Let the region R be bounded by a finite number of mutually disjoint Jordan curves $J_1, J_2, \cdots, J_\nu$. Let each of the functions $V(z)$, $-V(z)$, $|F_n(z)|$ have a finite upper bound in R, let $V(z)$ be harmonic in R, and let $F_n(z)$ be locally single-valued and analytic in R except perhaps for branch points, with $|F_n(z)|$ single-valued in R. Let the set $J_1 + J_2 + \cdots + J_\nu$ be expressed as the sum of Jordan arcs $A_1, A_2, \cdots, A_\mu$ which are mutually disjoint except perhaps for end-points $B_1, B_2, \cdots, B_\lambda$. Suppose for every positive ϵ we have*

$$\limsup_{n\to\infty} \{\text{l.u.b.}\ [|F_n(z)| - e^{V(z)+\epsilon}],\ z \text{ on } A_k\} \le 0 \tag{8}$$

for each $k = 1, 2, \cdots, \mu$, where l.u.b. means the least upper bound of all limit values on A_k with the exception of limit values in the points B_j. Then $V(z)$ is a harmonic majorant of the sequence $F_n(z)$ in R.

Theorem 4 is essentially an application of the Phragmèn-Lindelöf principle. The proof depends on the use of the inequality

(9) $$\log |F_n(z)| \leq V(z) + \epsilon$$

on each A_k except in the points B_j, and on consideration of the inequality ($\delta \downarrow 0$)

(10) $$\log |F_n(z)| \leq V(z) + \epsilon - \delta \sum_{k=1}^{\lambda} \log |z - \beta_k|;$$

we assume that the diameter of R is not greater than unity; inequality (10) holds for n sufficiently large for all z on the A_k, where the B_k are no longer excepted. Details of the proof are left to the reader. (It may be noted that in the previous paper [12; 304] the expression "l.u.b." should read "g.l.b.")

As an application of Theorem 4 we prove

COROLLARY 1. *Let the function $V(z)$ be a harmonic majorant but not an exact harmonic majorant for the sequence $F_n(z)$ in a region R. Then in an arbitrary subregion R_1 of R whose boundary B also lies in R there exists a harmonic majorant $V_1(z)$ for the sequence $F_n(z)$ with $V_1(z) < V(z)$ in R_1.*

Enclose $R_1 + B$ in a region R_2 which is contained in R and which is bounded by a set J consisting of a finite number of mutually disjoint Jordan curves in R. We assume that one of these curves contains an arc A on which $V(z)$ is constant, say $V(z) = a$; this circumstance can always be arranged. By the Corollary to Theorem 1 we may write

$$\limsup_{n\to\infty} [\max |F_n(z)|, z \text{ on } A] \leq e^{a_1} < e^{a}.$$

Denote by $V_1(z)$ the unique function which is harmonic and bounded in R_2, continuous in the corresponding closed region except in the end-points of A, equal to a_1 in the interior points of A and to $V(z)$ in the interior points of $J - A$. Of course we have $V_1(z) < V(z)$ in R_2 and hence in R_1. For every positive ϵ we now have by the Corollary to Theorem 1 for z on A and by Theorem 3 for z on $J - A$.

$$\limsup_{n\to\infty} \{\text{l.u.b.} [|F_n(z)| - e^{V_1(z)+\epsilon}]\} \leq 0;$$

it follows from Theorem 4 that $V_1(z)$ is a harmonic majorant for $F_n(z)$ in R_2 and hence in R_1.

An immediate consequence of Theorem 4 and Theorem 1 is stated for reference in the precise form for numerous applications:

COROLLARY 2. *Let the region R be bounded by loci C_ρ and C_ν, and let the function $V(z)$ be harmonic in R, continuous in the corresponding closed region, and equal to ρ on C_ρ and to ν on C_ν, $\rho \neq \nu$. Let the functions $F_n(z)$ be locally single-valued and analytic in R except perhaps for branch points, with $|F_n(z)|$ single-valued and continuous in the corresponding closed region.*

If we have

$$\limsup_{n\to\infty} [\max |F_n(z)|, z \text{ on } C_\rho] \leq e^\rho,$$

$$\limsup_{n\to\infty} [\max |F_n(z)|, z \text{ on } C_\nu] \leq e^\nu,$$

then $V(z)$ is a harmonic majorant of the sequence $F_n(z)$ in R. If in addition on a single locus $C_{\rho'}$: $V(z) = \rho'$, $\rho < \rho' < \nu$, in R we have

$$\limsup_{n\to\infty} [\max |F_n(z)|, z \text{ on } C_{\rho'}] = e^{\rho'},$$

then $V(z)$ is an exact harmonic majorant of the sequence $F_n(z)$ in R.

2. **Sufficient conditions for overconvergence.** We wish to consider a relatively general situation, which we formulate as follows:

THEOREM 5. *Let R be a region of the z-plane, let the function $V(z)$ be harmonic but not identically constant in R, and let the functions $f_n(z)$ be analytic in R. Let the sets $V(z) = \rho$ and $V(z) < \rho$ in R be denoted generically by C_ρ and R_ρ ; the set R_ρ is an open set. We suppose the set C_0 not to be empty.*

Let $V(z)$ be an exact harmonic majorant of the sequence $[f_{n+1}(z) - f_n(z)]^{1/n}$ in R; then $f_n(z)$ converges throughout R_0, uniformly on any closed set in R_0, to some limit function $f(z)$. The function $V(z)$ is an exact harmonic majorant of the sequence $[f(z) - f_n(z)]^{1/n}$ in R_0 and of the sequence $[f_n(z)]^{1/n}$ in $R - R_0$.

If on a particular continuum Q_0 (not a single point) in R_0 we have the relation

$$(11) \qquad \limsup_{n_k\to\infty} [\max |f(z) - f_{n_k}(z)|^{1/n_k}, z \text{ on } Q_0] < [\max e^{V(z)}, z \text{ on } Q_0],$$

or if on a particular continuum Q_0 (not a single point) in $R - R_0$ we have the relation

$$(12) \qquad \limsup_{n_k\to\infty} [\max |f_{n_k}(z)|^{1/n_k}, z \text{ on } Q_0] < [\max e^{V(z)}, z \text{ on } Q_0],$$

for a specific sequence of indices n_k, then for that same sequence of indices the sequence of functions $f_{n_k}(z)$ converges uniformly throughout some neighborhood of each point α on C_0 at which $f(z)$ is analytic, provided α and Q_0 are not separated in R by a natural boundary of $f(z)$.

Conversely, if a specific sequence $f_{n_k}(z)$ converges uniformly throughout the neighborhood of some point α on C_0, then (11) and (12) are valid in the respective cases that Q_0 lies in R_0 or in $R - R_0 - C_0$, provided that α and Q_0 are not separated in R by a natural boundary of $f(z)$.

In the statement of Theorem 5 we are in reality using a slight generalization of our original definition of exact harmonic majorant, for the set $R - R_0$ is not a region or an open set, but is an open set plus the points of C_0 ; nevertheless the corresponding extension of our original definition presents no difficulty in this situation.

It is a slight advantage here not to assume R to be bounded by Jordan curves, for if by way of illustration the functions $f_n(z)$ are the partial sums of order n of a power series $\sum a_n z^n$ having the unit circle as circle of convergence, we have $\limsup_{n\to\infty} |a_n|^{1/n} = 1$, $\limsup_{n\to\infty} |a_n z^n|^{1/n} = |z|$, and we may take R as the region $0 < |z| < \infty$, with $V(z) = \log |z|$.

The locus C_ρ consists in the neighborhood of each of its points P in R, of an analytic Jordan arc or of a finite number of analytic Jordan arcs whose tangents at P are equally spaced.

On any continuum Q_0 in R_0 we may make use of the inequality

$$\begin{aligned}\limsup_{n\to\infty}\ &[\max |f_{n+1}(z) - f_n(z)|^{1/n},\ z \text{ on } Q_0] \\ &= [\max e^{V(z)},\ z \text{ on } Q_0] < 1,\end{aligned} \tag{13}$$

from which the uniform convergence of the sequence $f_n(z)$ on Q_0 follows. If we denote the second member of (13) by r, we choose an arbitrary r_1, $r < r_1 < 1$, and write for z on Q_0 and for n sufficiently large

$$\begin{gathered}|f_{n+1}(z) - f_n(z)| < r_1^n, \\ f(z) - f_n(z) = [f_{n+1}(z) - f_n(z)] + [f_{n+2}(z) - f_{n+1}(z)] + \cdots, \\ |f(z) - f_n(z)| \le r_1^n/(1 - r_1), \\ \limsup_{n\to\infty}\ [\max |f(z) - f_n(z)|^{1/n},\ z \text{ on } Q_0] \le r_1,\end{gathered} \tag{14}$$

from which it follows that we have

$$\limsup_{n\to\infty}\ [\max |f(z) - f_n(z)|^{1/n},\ z \text{ on } Q_0] \le r = [\max e^{V(z)},\ z \text{ on } Q_0]; \tag{15}$$

thus $V(z)$ is a harmonic majorant of the sequence $[f(z) - f_n(z)]^{1/n}$ in R_0. Elementary inequalities show that the inequality sign in (15) is impossible by virtue of (13), so $V(z)$ is an exact harmonic majorant of $[f(z) - f_n(z)]^{1/n}$ in R_0.

By a method similar to the one just used, setting

$$\begin{aligned}f_n(z) = f_N(z) &+ [f_{N+1}(z) - f_N(z)] + [f_{N+2}(z) - f_{N+1}(z)] \\ &+ \cdots + [f_n(z) - f_{n-1}(z)] \qquad (n \ge N),\end{aligned} \tag{16}$$

it follows also that $V(z)$ is an exact harmonic majorant of the sequence $[f_n(z)]^{1/n}$ in $R - R_0$; details are left to the reader; we are here using the extension of the notion of exact harmonic majorant; the fundamental properties persist; compare Walsh [12; Theorem 3].

The reader will now readily verify the exactness of

COROLLARY 1. *If $V(z)$ is negative in a region, a necessary and sufficient condition that $V(z)$ be a (or an exact) harmonic majorant for the sequence $[f_{n+1}(z) - f_n(z)]^{1/n}$*

is that $V(z)$ be a (or an exact) harmonic majorant for the sequence $[f(z) - f_n(z)]^{1/n}$, where $f(z)$ is the limit of the sequence $f_n(z)$ in the region.

If $V(z)$ is non-negative in a region, a necessary and sufficient condition that $V(z)$ be a (or an exact) harmonic majorant for the sequence $[f_{n+1}(z) - f_n(z)]^{1/n}$ is that $V(z)$ be a (or an exact) harmonic majorant for the sequence $[f_n(z)]^{1/n}$.

In continuing the proof of the latter part of Theorem 5, we find a lemma convenient:

LEMMA 1. *Under the hypothesis of the first part of Theorem 5 (that is to say the hypothesis not involving (11) and (12)), let Q be a closed point set in R which contains points of R_0 and of $R - R_0 - C_0$. Let $f(z)$ be analytic on Q. If we suppose $V(z) \geq -\eta < 0$ on Q, and if $\epsilon > 0$ is arbitrary, then we have for N sufficiently large*

$$\{\max [|f(z) - f_N(z)|^{1/N} - e^{V(z)+\epsilon+\eta}], z \text{ on } Q\} \leq 0. \tag{17}$$

We are not in a position to assume the convergence of $f_n(z)$ to $f(z)$ on Q—indeed, such convergence if Q contains a region in $R - R_0 - C_0$ is impossible—but we may write

$$\begin{aligned} f(z) - f_N(z) = [f(z) - f_k(z)] - [f_{k+1}(z) - f_k(z)] \\ - [f_{k+2}(z) - f_{k+1}(z)] - \cdots - [f_N(z) - f_{N-1}(z)]. \end{aligned} \tag{18}$$

It follows from Theorem 3 that we have on Q for j sufficiently large

$$|f_{j+1}(z) - f_j(z)| \leq e^{j[V(z)+\epsilon]} \leq e^{j[V(z)+\epsilon+\eta]}; \tag{19}$$

it is important to use the last member, a quantity greater than unity on Q.

By virtue of the arbitrariness of ϵ, inequality (17) easily follows from (18) and (19).

A further lemma is to be used frequently:

LEMMA 2. *Under the hypothesis of the first part of Theorem 5, let Q be a closed point set in R, and let $f(z)$ be analytic on Q. If $\epsilon > 0$ is arbitrary we have for n sufficiently large*

$$\{\max [|f(z) - f_n(z)|^{1/n} - e^{V(z)+\epsilon}], z \text{ on } Q\} \leq 0. \tag{20}$$

We divide Q into two closed sets Q_1 and Q_2, which are not necessarily disjoint: Q_1 shall be the intersection of Q and the set $R_{-\epsilon/2} + C_{-\epsilon/2}$; the set Q_2 shall be the intersection of Q and the set $R - R_{-\epsilon/2}$; on Q_2 we have $V(z) \geq -\epsilon/2$. By Theorem 3 we deduce (20) with Q replaced by Q_1; by Lemma 1 we deduce (20) with Q replaced by Q_2. Combination of these two inequalities yields (20) in its present form, for n sufficiently large. It is a consequence of Lemma 2 that $V(z)$ is a harmonic majorant for the sequence $[f(z) - f_n(z)]^{1/n}$ in any subregion of R in which $f(z)$ is analytic.

We proceed now with the proof of the remainder of Theorem 5. If $z = \alpha$ is a multiple point of the locus C_0, our hypothesis of the analyticity of $f(z)$ at

$z = \alpha$ is intended to include the single-valuedness and analyticity of $f(z)$ throughout a neighborhood of that point, that is to say, is intended to include the identity in the neighborhood of $z = \alpha$ of the various analytic extensions of $f(z)$ from the various regions composing R_0 of which α is a boundary point. For the sake of simplicity of exposition we suppose in the sequel that α is a simple point of the locus C_0; the requisite modifications for the contrary case are not difficult and are left to the reader.

Let the positive number η be chosen so small that the following conditions are fulfilled: a closed Jordan arc A_1 of the locus $C_{-\eta}$, a closed Jordan arc A_2 of the locus C_η, and two closed line segments L form a Jordan curve K which together with its interior lies in R, contains $z = \alpha$ in its interior, and contains on and within it only points of analyticity of $f(z)$.

We suppose that (11) is satisfied, where for the present Q_0 and A_1 are not separated in R by C_0; denote by R' the largest subregion of R_0 which contains Q_0 and A_1, and denote by R'' the largest subregion of $R - R_0 - C_0$ which contains A_2. By virtue of our hypothesis (11) and the Corollary to Theorem 1 applied to the region R' we may write

$$\limsup_{n_k\to\infty} [\max |f(z) - f_{n_k}(z)|^{1/n_k}, z \text{ on } A_1]$$
$$< [\max e^{V(z)}, z \text{ on } A_1] = e^{-\eta}.$$

Consequently there exists some $\delta > 0$ such that for n_k sufficiently large we have

$$[\max |f(z) - f_{n_k}(z)|^{1/n_k}, z \text{ on } A_1] \leq e^{-\eta-\delta} = e^{V(z)-\delta}. \tag{21}$$

Let $\epsilon > 0$ be arbitrary; from Lemma 2 we have for n_k sufficiently large

$$\{\max [|f(z) - f_{n_k}(z)|^{1/n_k} - e^{V(z)+\epsilon}], z \text{ on } A_2 + L\} \leq 0. \tag{22}$$

Let us denote by $V_\theta(z)$, where θ is constant, the unique function which is harmonic and bounded interior to K, continuous in the corresponding closed region except in the end-points of A_1, equal to $-\eta-\delta$ in the interior points of A_1 and to $V(z) + \theta$ in the interior points of the Jordan arc $A_2 + L$. The function $V_0(z)$ is less than $V(z)$ in the interior points of A_1 and equal to $V(z)$ in the interior points of $A_2 + L$, hence is less than $V(z)$ at every point interior to K; in particular we have $V_0(\alpha) < 0$. The function $V_\theta(z)$ is represented by Green's formula involving Green's function, so when θ approaches zero the function $V_\theta(z)$ approaches $V_0(z)$ at every point interior to K. We choose $\theta = \epsilon$ sufficiently small but positive so that we have $V_\epsilon(\alpha) < 0$, so throughout some closed neighborhood $N(\alpha)$ of α in K we have also $V_\epsilon(\alpha) < 0$. From inequalities (21) and (22) and from the subharmonic properties of the function $\log |f(z) - f_{n_k}(z)|^{1/n_k}$ we have for n_k sufficiently large

$$\{\max [|f(z) - f_{n_k}(z)|^{1/n_k} - e^{V_\epsilon(z)}], z \text{ on } N(\alpha)\} \leq 0, \tag{23}$$

$$|f(z) - f_{n_k}(z)|^{1/n_k} \leq r < 1 \qquad (z \text{ on } N(\alpha)),$$

from which it follows that the sequence $f_{n_k}(z)$ converges uniformly to $f(z)$ throughout $N(\alpha)$; this is the phenomenon of overconvergence, namely convergence across some arc of C_0, and indeed across every arc of C_0 (not separated from Q_0 by a natural boundary of $f(z)$) on which $f(z)$ is analytic, which we were to establish.

We are in a position to remark now that as a consequence of (23) we have on some continuum Q_1, (not a single point) which lies in R'', namely an arbitrary continuum in $N(\alpha)$ and in R'',

$$\limsup_{n_k \to \infty} [\max |f_{n_k}(z)|^{1/n_k}, z \text{ on } Q_1] \leq 1 \tag{24}$$
$$< [\max e^{V(z)}, z \text{ on } Q_1],$$

so by the Corollary to Theorem 1 the inequality of the extreme members is valid for every continuum Q_1 in R''.

Let us now choose (12) rather than (11) as our hypothesis; we are to establish overconvergence throughout some neighborhood of the point $z = \alpha$; let η and K be defined as before; we assume for the present that Q_0 and A_2 are not separated in R by C_0, but lie in a maximal subregion R'' of $R - R_0 - C_0$. Let R' denote the largest subregion of R_0 which contains A_1. By (12) and the Corollary to Theorem 1 applied to the region R'' we have

$$\limsup_{n_k \to \infty} [\max |f_{n_k}(z)|^{1/n_k}, z \text{ on } A_2] < [\max e^{V(z)}, z \text{ on } A_2] = e^{\eta}. \tag{25}$$

Obvious algebraic inequalities enable us to write from (25) by the known analyticity and therefore boundedness of $f(z)$ on A_2

$$\limsup_{n_k \to \infty} [\max |f(z) - f_{n_k}(z)|^{1/n_k}, z \text{ on } A_2] < [\max e^{V(z)}, z \text{ on } A_2].$$

For a suitably chosen $\delta > 0$ we consequently have for n_k sufficiently large

$$[\max |f(z) - f_{n_k}(z)|^{1/n_k}, z \text{ on } A_2] \leq e^{V(z)-\delta}. \tag{26}$$

If $\epsilon > 0$ is arbitrary we find by Lemma 2 for n_k sufficiently large

$$\{\max [|f(z) - f_{n_k}(z)|^{1/n_k} - e^{V(z)+\epsilon}], z \text{ on } A_1 + L\} \leq 0. \tag{27}$$

As the reader will notice, inequalities (26) and (27) can be employed precisely as were the inequalities (21) and (22) to establish the overconvergence of the sequence $f_{n_k}(z)$ throughout some neighborhood of the point $z = \alpha$. This method shows also by the Corollary to Theorem 1 that we have for every continuum Q_1 in R'

$$\limsup_{n_k \to \infty} [\max |f(z) - f_{n_k}(z)|^{1/n_k}, z \text{ on } Q_1] \tag{28}$$
$$< [\max e^{V(z)}, z \text{ on } Q_1].$$

The reasoning just given is essentially valid if we suppose (12) satisfied on a continuum Q_0 in $R - R_0$, even if Q_0 has points in common with C_0; compare Walsh [12; Theorem 3].

We can now remove the restriction previously made that Q_0 and A_1 (or A_2) are not separated in R by C_0. For definiteness we suppose (11) satisfied, and suppose that Q_0 and α are not separated in R by a natural boundary of $f(z)$. Then Q_0 and α may be joined by a Jordan arc J in R which intersects no natural boundary of $f(z)$; the Jordan arc J may be so chosen that it has only a finite number of points α_1., $\alpha_2, \cdots, \alpha_\mu$ in common with C_0, so that these are points of analyticity of $f(z)$, and so that J crosses C_0 at each of these points except the last one. Let the notation of the points α_k correspond to their ordering on J, with $\alpha_\mu = \alpha$. It follows from our hypothesis (11) and from (23) as already proved that overconvergence occurs throughout a neighborhood of α_1, and it follows from (24) that (12) now holds on every continuum Q_0 in the maximal subregion of the set $R - R_0 - C_0$ having the arc $\alpha_1\alpha_2$ on its boundary. It then follows, by the part of Theorem 5 already established, that (28) holds on every continuum Q_1 in the maximal subregion of the set R_0 having on its boundary the arc $\alpha_2\alpha_3$. We can proceed by this method along J, deriving alternately inequalities of the form (24) and (28); we establish eventually the overconvergence of the sequence $f_{n_k}(z)$ throughout some neighborhood of the point α.

Conversely, if a specific sequence $f_{n_k}(z)$ converges uniformly throughout the neighborhood of some point α of C_0, then on an arbitrary continuum Q_0 of R'' (in the previous notation) in this neighborhood of α the first member of (12) is not greater than unity; consequently inequality (12) is valid for this particular Q_0. Inequality (28) is implied by (12). It follows that (11) and (12) are valid in the respective cases that Q_0 lies in R_0 or in $R - R_0 - C_0$, provided that α and Q_0 are not separated in R by a natural boundary of $f(z)$. This method shows that overconvergence of a specific sequence $f_{n_k}(z)$ in the neighborhood of some point α of C_0 implies that (12) is fulfilled even for any continuum Q_0 in $R - R_0$ (i.e. not necessarily in $R - R_0 - C_0$), provided that α and no point of Q_0 are separated in R by a natural boundary of $f(z)$, and provided that either $f(z)$ is analytic on Q_0 or Q_0 has points which lie in $R - R_0 - C_0$. Theorem 5 is now completely proved.

Obviously inequality (11) is a necessary and sufficient condition that a harmonic majorant $V(z)$ be not an exact harmonic majorant in R_0 for the sequence $[f(z) - f_{n_k}(z)]^{1/n_k}$, and (12) is a necessary and sufficient condition that a harmonic majorant $V(z)$ be not an exact harmonic majorant in $R - R_0 - C_0$ for the sequence $[f_{n_k}(z)]^{1/n_k}$. Moreover, in our proof of the parts of Theorem 5 involving (11) and (12) as hypothesis, we have not made explicit use of $V(z)$ as an *exact* harmonic majorant; a corresponding change in the formal statement of Theorem 5 may be made; but the most interesting case would seem to be the formulated one involving the exact harmonic majorant. If $V(z)$ is a harmonic majorant but not an exact harmonic majorant in R for the sequence $[f_{n+1}(z) - f_n(z)]^{1/n}$, then by Corollary 1 to Theorem 4, in every region which together with its boundary lies in R, that sequence has a smaller harmonic majorant. In particular corresponding to each point α of C_0 there is a neighborhood in which the sequence has a *negative* harmonic majorant; the sequence $f_n(z)$ converges uniformly throughout some neighborhood of α.

The remark just made is at times useful in proving that a harmonic majorant is an exact harmonic majorant—the existence of a single singularity of $f(z)$ on C_0 is sufficient for this conclusion under the conditions specified even if C_0 separates into several disjoint parts; if C_0 has a multiple point α it is sufficient if $f(z)$ cannot be defined so as to be single-valued and analytic throughout a neighborhood of α. But in the situation of the first part of Theorem 5 the function $f(z)$ need not have a singularity on C_0 in R; this is in contrast with the situation for Taylor's series. A trivial example of this fact is found by choosing $f_n(z) = 1 + z + \cdots + z^n$, with $f(z)$ the function $(1 - z)^{-1}$, and R the region $0 < |z| < \infty$ with the point $z = 1$ omitted. A less trivial example is the choice $f_n(z) = z^n + 1$, with $f(z)$ the function unity, and R the region $0 < |z| < \infty$. Still another example is the choice found by expanding in a Jacobi series of interpolation

$$a_0 + a_1(z + 1) + a_2(z^2 - 1) + a_3(z + 1)(z^2 - 1) + \cdots$$

the function identically zero in the right oval of the locus $|z^2 - 1| \leq 1/4$ and a function analytic interior to but having a singularity on the left oval; the functions $f_n(z)$ are the partial sums of the series, with $f(z)$ the function zero, and R may be taken as any region in $0 < |z - 1| < \infty$ containing the right oval and containing no point of the left oval; compare Walsh [11; 82–83].

Of course a sufficient condition for (12) is that the sequence $f_{n_k}(z)$ be uniformly bounded or converge uniformly on the continuum Q_0. We discuss this question further in §5 below.

Theorem 5 is of significance in indicating that overconvergence is not merely a local property; if overconvergence occurs in the neighborhood of a particular point α of C_0, it occurs in the neighborhood of every point β of C_0 at which $f(z)$ is analytic, provided β is not separated in R from α by a natural boundary of $f(z)$. The non-local character of overconvergence for the case of Taylor's series was emphasized by Ostrowski.

Of course inequalities (11) and (12) are not consequences of each other if we omit the requirement that α and Q_0 shall not be separated by a natural boundary of $f(z)$ in R. This is shown by the example of Taylor's series. We set for instance

$$f(z) \equiv 1 + z^{n_1} + z^{n_2} + \cdots \qquad (n_{k+1}/n_k > 1 + \delta > 1),$$

$$f_{n_k}(z) \equiv 1 + z^{n_1} + \cdots + z^{n_k}.$$

We have $f_{n_k}(r) = 1 + r^{n_1} + \cdots + r^{n_k} > r^{n_k}$, whence

$$[\max |f_{n_k}(z)|^{1/n_k}, \text{ for } |z| = r > 1] \geq r,$$

$$\limsup_{n_k \to \infty} [\max |f_{n_k}(z)|^{1/n_k}, \text{ for } |z| = r > 1] \geq r;$$

except that the sign $<$ is replaced by $\geq$, this last inequality is of form (12) with $V(z) \equiv \log |z|$. Nevertheless we have for $|z| < 1$

$$f(z) - f_{n_k}(z) = z^{n_{k+1}} + z^{n_{k+2}} + \cdots,$$

whence we have

$$[\max |f(z) - f_{n_k}(z)|^{1/n_k}, \text{ for } |z| = r < 1] \leq r^{n_{k+1}}/(1-r),$$

$$\limsup_{n_k \to \infty} [\max |f(z) - f_{n_k}(z)|^{1/n_k}, \text{ for } |z| = r < 1]$$

$$\leq \limsup_{n_k \to \infty} r^{n_{k+1}/n_k} \leq r^{1+\delta} < r;$$

the extreme members represent an inequality of form (11). Thus the inequalities given show that (11) does not imply (12). The sequence

$$f_{n_{k+1}-1}(z) \equiv 1 + z^{n_1} + \cdots + z^{n_k}$$

similarly shows that (12) does not imply (11).

Our proof of Theorem 5 yields readily also

Corollary 2. *Under the hypothesis of the first part of Theorem 5, let the function $f(z)$ be continuous on a closed arc Q_0 of C_0, in the sense that the limit values of $f(z)$ as z in R_0 approaches C_0 exist. Suppose the relation* (11) *holds for Q_0:*

(29) $$\limsup_{n_k \to \infty} [\max |f(z) - f_{n_k}(z)|^{1/n_k}, z \text{ on } Q_0] < 1;$$

then the sequence $f_{n_k}(z)$ converges uniformly throughout some neighborhood of each point α on C_0 at which $f(z)$ is analytic, provided α and points in R_0 and in a neighborhood of Q_0 are not separated in R by a natural boundary of $f(z)$.

Conversely, if overconvergence of a sequence $f_{n_k}(z)$ occurs throughout some neighborhood of a point α on C_0, then on each closed arc Q_0 of C_0 on which $f(z)$ is analytic inequality (29) *is valid, provided α and Q_0 are not separated in R by a natural boundary of $f(z)$.*

Under the hypothesis (29), let K be a quadrilateral which consists of the arc Q_0, a Jordan arc A_1 of the locus $C_{-\eta}$ where η is positive, and of two rectilinear segments L in R; we suppose that K is a Jordan curve all of whose interior points lie in R_0; for simplicity we limit our phraseology to apply to the case that Q_0 contains no multiple point of C_0.

The method of proof of Lemma 2 applies here, and gives for arbitrary positive ϵ and for n_k sufficiently large

(30) $$\{\max [|f(z) - f_{n_k}(z)|^{1/n_k} - e^{V(z)+\epsilon}], z \text{ on } L\} \leq 0.$$

By the conditions of Theorem 5 we have for n_k sufficiently large

(31) $$\{\max [|f(z) - f_{n_k}(z)|^{1/n_k} - e^{V(z)+\epsilon}], z \text{ on } A_1\} \leq 0.$$

We are now in a position to use (29) together with (30) and (31) as we have previously used (21) and (22); we obtain an inequality of form (23) for z on an arbitrary continuum interior to K; this is essentially (11), which implies the conclusion of the first part of Corollary 2.

In the proof of Theorem 5 we constructed the quadrilateral K containing a given point α in its interior. Precisely the same method applies for an arbitrary arc of C_0 on which $f(z)$ is analytic. This yields (29) under the conditions specified; the details are left to the reader.

Theorem 5 is closely related to the work of Ostrowski, Bourion, and Mosesson, but is apparently both more specific in characterizing the properties of the sequences involved and more general in hypothesis than any previous result.

Broad sufficient conditions for the hypothesis of Theorem 5 in a form immediately useful for the applications are contained in a recent result, see [12; Theorem 3].

It is an immediate consequence of Theorem 5 together with results previously established [14; Theorems 7, 14, 21] that sequences of rational functions which converge maximally possess exact harmonic majorants and hence are included under the theory developed in the present paper. Thus are included numerous sequences of rational functions (including polynomials) found by interpolation, and also sequences of rational functions (including polynomials) of best approximation in the sense of Tchebycheff, in the sense of least p-th powers ($p > 0$) measured by a line integral, and in the sense of least p-th powers ($p > 0$) measured by a surface integral. There are also included sequences of interpolating functions of minimum norm [13]. The writer hopes to develop on another occasion applications of the present theory to approximating functions of minimum norm.

3. **Isolated singularities of limit function.** Not every function can be approximated in an assigned region by a sequence of analytic functions which exhibits the phenomenon of overconvergence:

THEOREM 6. *Under the hypothesis of the first part of Theorem 5, let a sequence $f_{n_k}(z)$ exhibit the phenomenon of overconvergence on two disjoint open arcs of C_0 which have a common end-point $z = \alpha$. Then $z = \alpha$ cannot be an isolated singularity of $f(z)$.*

Assume α to be an isolated singularity of $f(z)$; we shall reach a contradiction. Construct a circle Γ with center α which together with its interior lies in R, whose radius is less than unity, such that $f(z)$ is analytic in the closed interior of Γ except at α. The function $V(z)$ is a harmonic majorant for the sequence $[f(z) - f_{n_k}(z)]^{1/n_k}$ in an annular region containing Γ in its interior, by Lemma 2. But $V(z)$ is not an exact harmonic majorant for that sequence in that region, by (11) as established in the last part of Theorem 5. Consequently by Corollary 1 to Theorem 4 we have for arbitrary positive ϵ and for n sufficiently large

$$\{\max [| f(z) - f_{n_k}(z)|^{1/n_k} - e^{V_1(z)+\epsilon}], z \text{ on } \Gamma\} \leq 0, \tag{32}$$

where $V_1(z)$ is continuous and less than $V(z)$ on Γ. We denote also by $V_1(z)$ interior to Γ the values of the function harmonic throughout the interior of Γ and continuous in the corresponding closed region; the inequality $V_1(z) < V(z)$

on Γ implies that same inequality throughout the interior of Γ; in particular we have $V(\alpha) = 0$, $V_1(\alpha) < 0$, whence $V_1(z) < 0$ on a suitably chosen circle γ interior to and concentric with Γ. Let Γ_1 be a circle interior to and concentric with γ; we denote the radius of Γ_1 by ρ. Let a (necessarily positive) denote the maximum value of $V(z)$ on γ, from which it follows that $V(z) \leq a$ throughout the interior of γ; in particular that inequality is valid on Γ_1. Again by Lemma 2 we may write

$$\{\max [| f(z) - f_{n_k}(z)|^{1/n_k} - e^{a+\epsilon}], z \text{ on } \Gamma_1\} \leq 0. \tag{33}$$

Denote by b (necessarily negative) the minimum of $V_1(z)$ on γ, from which it follows that the minimum of $V_1(z)$ on Γ_1 is not less than b:

$$V_1(z) \geq b, z \text{ on } \Gamma_1 .$$

The function

$$V_2(z) \equiv V_1(z) + (a - b) \frac{\log | z - \alpha |}{\log \rho}$$

is harmonic in the annular region R_1 bounded by Γ and Γ_1. On Γ we have $V_2(z) > V_1(z)$ and on Γ_1 we have $V_2(z) \geq a$. It follows from (32) and (33) and Theorem 4 that $V_2(z)$ is a harmonic majorant for the sequence $[f(z) - f_{n_k}(z)]^{1/n_k}$ in R_1 and in particular on γ. When ρ approaches zero, $V_2(z)$ approaches $V_1(z)$ (which is negative) on γ. If we choose $\epsilon > 0$ and ρ in such a manner that we have $V_2(z) + \epsilon < 0$ on γ, we have

$$\limsup_{n_k \to \infty} [\max | f(z) - f_{n_k}(z) |^{1/n_k}, z \text{ on } \gamma] < 1;$$

the sequence $f_{n_k}(z)$ converges uniformly on γ, hence converges uniformly throughout the closed interior of γ. The limit function is analytic throughout the closed interior of γ, in particular is analytic in the point α; this contradiction completes the proof.

There exist a number of examples of overconvergent Taylor developments (mostly modeled after Porter's original example) which are elementary in the sense that overconvergence of suitably chosen subsequences is more or less obvious by inspection. There exist no examples, however, which are elementary in the sense that the functions expanded are elementary. Theorem 6 serves to justify the non-elementary nature of the functions in the known illustrations.

For the case of Taylor's series, Theorem 6 is well known. Bourion [1] remarks that more generally there exists no circle with center α in which the singularities of $f(z)$ form a set of capacity zero. The proof already given is readily modified to include the latter conclusion even under the wider hypothesis of Theorem 5; compare the methods of §5 below.

4. **Rapidly convergent subsequences.** For the particular case of Taylor's series, Ostrowski has shown that subsequences which converge with great

rapidity have special and striking properties. These results extend to the more general types of sequences of the present paper, as we now proceed to demonstrate.

THEOREM 7. *Under the hypothesis of Theorem 4 for given $V(z)$, let the relation*

$$\lim_{n\to\infty} [\text{l.u.b.} \mid F_n(z) \mid, z \text{ on } A_1] = 0 \tag{34}$$

be valid. Then on every continuum Q in R there is valid

$$\lim_{n\to\infty} [\max \mid F_n(z) \mid, z \text{ on } Q] = 0. \tag{35}$$

Theorem 7 can be proved as follows. Denote by $V_\sigma(z)$ the unique function which is bounded and harmonic in R, continuous in the corresponding closed region except perhaps in the points B_k, equal to σ in the interior points of A_1 and equal to M in the interior points of the Jordan arcs $A_2 + \cdots + A_\mu$, where M is the least upper bound of $V(z)$ in R. Every $V_\sigma(z)$ is a harmonic majorant for the sequence $\mid F_n(z) \mid$, by Theorem 4. At an arbitrary point z of R we may write

$$V_\sigma(z) = m_1(z) \cdot \sigma + m_2(z) \cdot M,$$

where $m_1(z)$ and $m_2(z)$ are the harmonic measures (see for instance [7; Chapter III]) of A_1 and of $A_2 + \cdots + A_\mu$ with respect to z. The numbers $m_1(z)$ have a positive lower bound for all z on Q, and the numbers $m_2(z)$ have a positive upper bound for all z on Q. It follows that if τ is arbitrary, there exist σ and $\epsilon > 0$ such that we have

$$V_\sigma(z) + \epsilon < \tau, z \text{ on } Q.$$

Consequently, we have by Theorem 4

$$\limsup_{n\to\infty} [\max \mid F_n(z) \mid, z \text{ on } Q] \leq e^\tau,$$

so (35) follows. Of course this method of proof is essentially the use of the two-constant theorem.

THEOREM 8. *If the sequence $F_n(z)$ has a harmonic majorant $V(z)$ in a region R, and if on a single continuum Q_0 (not a single point) in R we have*

$$\lim_{n\to\infty} [\max \mid F_n(z) \mid, z \text{ on } Q_0] = 0, \tag{36}$$

then the corresponding relation (35) *is valid on every continuum Q in R.*

The conclusion of Theorem 8 may be expressed by saying that every function harmonic in R is a harmonic majorant for the sequence $F_n(z)$.

Theorem 8 is a consequence of Theorem 7, by a known method of proof involving a conformal map [12; Theorem 1] if Q does not intersect Q_0, and otherwise by a slight modification of that proof [12; Theorem 2].

THEOREM 9. *Under the conditions of the first part of Theorem 5 suppose we have on a particular continuum Q_0 (not a single point) in R_0 the relation*

$$\lim_{n_k \to \infty} [\max | f(z) - f_{n_k}(z) |^{1/n_k}, z \text{ on } Q_0] = 0. \tag{37}$$

Then on any continuum Q in R which together with Q_0 is contained in a region R' of analyticity of $f(z)$ we have the relation

$$\lim_{n_k \to \infty} [\max | f(z) - f_{n_k}(z) |^{1/n_k}, z \text{ on } Q] = 0. \tag{38}$$

Consequently the sequence $f_{n_k}(z)$ exhibits overconvergence, provided there exist points α of C_0 at which $f(z)$ is analytic, and provided these points α are not separated in R from Q_0 by a natural boundary of $f(z)$; indeed the sequence $f_{n_k}(z)$ converges throughout the largest region R' in R which contains Q_0 and throughout which $f(z)$ is analytic; the sequence $f_{n_k}(z)$ converges uniformly on any closed set interior to R'.

Theorem 9 is an immediate consequence of Lemma 2 and Theorem 8.

It is a consequence of Theorem 9 that R' cannot be multiply connected unless R itself is multiply connected; no Jordan region whose boundary lies in R can contain in its interior a boundary point of R' unless it contains in its interior a boundary point of R. Any singularity of $f(z)$ in R must belong to a continuum of singularities in R; such a continuum must either reach to the boundary of R or separate R' from some boundary points of R. In particular if R_0 consists of a single region and if R is simply connected, the region R' is a simply connected subregion of R; either the boundary of R' separates R_0 from the boundary of R or the boundary of R' consists of one or more continua reaching to the boundary of R.

For the special case of Taylor's series $f(z) = \sum_{k=0}^{\infty} a_k z^k$, $a_k = 0$ for $n_n < k \leq m_n$, with radius of convergence unity, we set

$$f_n(z) = \sum_{k=0}^{n} a_k z^k.$$

We have

$$\limsup_{n \to \infty} [\max | f(z) - f_n(z) |, \text{ for } | z | = r < 1]^{1/n} = r,$$

which for the subsequence m_k by virtue of $f_{n_k}(z) = f_{m_k}(z)$ yields

$$\limsup_{m_k \to \infty} \{[\max | f(z) - f_{n_k}(z) |, \text{ for } | z | = r < 1]^{1/n_k}\}^{n_k/m_k} \leq r.$$

Provided that we now assume $n_k/m_k \to 0$, we have

$$\limsup_{n_k \to \infty} [\max | f(z) - f_{n_k}(z) |, \text{ for } | z | = r < 1]^{1/n_k} = 0.$$

The conditions of Theorem 9 are satisfied. Consequently—this result is due to Ostrowski—the sequence $f_{n_k}(z)$ converges throughout the largest region within which $f(z)$ is analytic; the region of existence of $f(z)$ is a simply connected region of the z-plane (point at infinity not adjoined).

Theorem 5 deals with both (11) and (12); Theorem 9 is concerned with (11) and has an analogue which is concerned with (12);

THEOREM 10. *If the functions $f(z)$ and $f_{n_k}(z)$ are analytic in a region R', if the function zero is a harmonic majorant for the sequence $[f_{n_k}(z)]^{1/n_k}$ in R', and if on a single continuum Q_0 in R' consisting of more than one point we have the relation*

$$\limsup_{n_k \to \infty} [\max | f(z) - f_{n_k}(z) |^{1/n_k}, z \text{ on } Q_0] < 1,$$

then the sequence $f_{n_k}(z)$ converges uniformly to the function $f(z)$ throughout R', uniformly on any closed set in R'. Indeed, the sequence $[f(z) - f_{n_k}(z)]^{1/n_k}$ has a negative harmonic majorant in every subregion of R' whose boundary lies interior to R'.

Under the conditions of the first part of Theorem 5, suppose R' is a region in R within which $f(z)$ is analytic and which contains points of R_0. Suppose on every continuum Q (not a single point) which lies in R' and in $R - R_0 - C_0$ we have

$$\limsup_{n_k \to \infty} [\max | f_{n_k}(z) |^{1/n_k}, z \text{ on } Q] \leq 1. \tag{39}$$

Then the sequence $f_{n_k}(z)$ converges to $f(z)$ throughout R', uniformly on any closed set interior to R'. In particular if (39) is valid on every continuum Q in $R - R_0 - C_0$, the region R' may be chosen as an arbitrary subregion of R which contains points of R_0 and within which $f(z)$ is analytic.

More explicitly, under the conditions of the first part of Theorem 5, let R'' be a region (whose closure lies in R) bounded by a finite number of mutually disjoint Jordan curves intersecting C_0 in a finite number of points, R'' having points in common both with R_0 and $R - R_0 - C_0$. Let $f(z)$ be analytic throughout the closure of R'', let $V_1(z) \geq 0$ be continuous on the part of the boundary of R'' in $R - R_0$, and let $V_1(z)$ be a majorant for the sequence $[f_{n_k}(z)]^{1/n_k}$ on every arc B which belongs to the boundary of R'' and lies in $R - R_0 - C_0$, in the sense that for every positive ϵ we have

$$\limsup_{n_k \to \infty} \{\max [| f_{n_k}(z)|^{1/n_k} - e^{V_1(z)+\epsilon}], z \text{ on } B\} \leq 0.$$

Then a harmonic majorant in R'' for the sequence $[f(z) - f_{n_k}(z)]^{1/n_k}$ is the function $U(z)$ which is harmonic and bounded in R'', continuous in the corresponding closed region except in the boundary points of R'' which lie on C_0, equal to $V_1(z)$ in the boundary points of R'' which lie in $R - R_0 - C_0$ and to $V(z)$ in the boundary points of R'' which lie in R_0.

Under the hypothesis of the first part of Theorem 10, the function zero is a harmonic majorant also for the sequence $[f(z) - f_{n_k}(z)]^{1/n_k}$ in R', but is not an exact harmonic majorant; our conclusion follows from Corollary 1 to Theorem 4.

The remaining parts of Theorem 10 are slightly more complicated, due to our ignorance of a harmonic majorant for $[f_{n_k}(z)]^{1/n_k}$ on C_0 itself; we indicate a more general result:

THEOREM 11. *Let the sequence of functions $F_n(z)$ locally analytic except perhaps for branch points in a region R have a harmonic majorant $V(z)$ in R. Let R' be a region in R bounded by a finite number of Jordan arcs A in R, which fall into closed sets B_1 and B_2, where B_1 and B_2 have no common points except end-points of A. Let the function $V_1(z)$ be continuous on B_1, and on any closed arc B of B_1 not containing an end-point of an arc A let $V_1(z)$ be a majorant for the sequence $F_n(z)$, in the sense that corresponding to an arbitrary positive ϵ we have*

$$\limsup_{n\to\infty} \{\max [|F_n(z)| - e^{V_1(z)+\epsilon}], z \text{ on } B\} \leq 0.$$

Let the function $V_2(z)$ be continuous on B_2, and on any closed arc B of B_2 not containing an end-point of an arc A let $V_2(z)$ be a majorant for the sequence $F_n(z)$, in this same sense.

Let $U(z)$ denote the function harmonic and bounded in R', continuous in the corresponding closed region except in the end-points of A, equal to $V_1(z)$ on B_1 and to $V_2(z)$ on B_2 except in those end-points. Then $U(z)$ is a harmonic majorant for the sequence $F_n(z)$ in R'.

The proof of Theorem 11 is so similar to the proofs of Theorems 4 and 5 that the details are left to the reader.

It may be noticed that the theorem is formulated to apply to only *two* different values $V_1(z)$ and $V_2(z)$ on the boundary of R'; an arbitrary finite number may obviously be used.

We return to the proof of the second and third parts of Theorem 10, which follow now from Theorem 11 thanks to Lemma 2.

In (39) we have employed the sign $\leq$ instead of the equality sign, but obviously the inequality can be valid only if $f(z)$ vanishes identically on Q and hence throughout R'.

In Theorem 9 and the second part of Theorem 10 we may consider R' as the *largest* region with certain properties within which $f(z)$ is analytic. It may readily occur under the conditions of the first part of Theorem 5 that R_0 consists of several separated subregions; under such circumstances there may be a region R' (of Theorem 9 or 10) corresponding to each of these subregions of R_0. Theorems 9 and 10 apply, by a suitable new definition of R if necessary, to each such region R'. No two distinct regions R' of this sort can overlap (i.e. for a given sequence n_k) unless the limit functions $f(z)$ in them coincide, for the $f_n(z)$ are single-valued in R. If several such regions R' overlap, they may be considered as a single R'; in any case the remarks on the singularities of $f(z)$ made in connection with Theorem 9 apply also here.

5. Convergence on more general point sets. In our treatment hitherto we have emphasized convergence and degree of convergence on *continua*. We discuss now in more detail the necessity and sufficiency of other properties of point sets, notably regularity and capacity.

We proceed to give in some detail the proof of Theorem 2, a modification of the proof already given [12; Theorem 2] of Theorem 1. We assume that Theorem 2 is false and write

$$\limsup_{n\to\infty} [\max |F_n(z)|, z \text{ on } Q] = M$$

$$< [\max e^{V(z)}, z \text{ on } Q];$$

we shall reach a contradiction; the reverse inequality cannot hold here, by Theorem 3.

Let $V(z)$ assume its maximum value on Q in the point z_1 of Q, and let the circle γ whose center is z_1 have a radius so small that γ and its interior lie in R and that the inequality $V(z) > \log M$ persists throughout the closed interior of γ. Denote by Q_1 the subset of Q in the closed interior of γ.

Either (i) there is a continuum belonging to Q_1 which contains both z_1 and points of γ, or (ii) all the points of Q_1 on γ can be separated from z_1 by a Jordan curve J which lies in R; this follows at once by transforming z_1 to infinity by a linear transformation, then expressing (compare for instance Walsh [11; §1.3]) the transform of $R - Q$ as the sum of an infinite number of nested regions, each contained in the preceding and each bounded by a finite number of Jordan curves.

In case (i) we denote by Q_2 the continuum having the property mentioned; in case (ii) we denote by Q_2 the points of Q_1 not separated from z_1 by J; in either case there exists a Jordan curve J_1 in R not intersecting Q_2, such that the subregion R_1 of R bounded by J_1 contains Q_2; the region $R_1 - Q_2$ is a region R_2 for which the boundary points of R_2 belonging to Q_2 are regular; this last remark follows from the fact that regularity is a local property.

Let $V_1(z)$ denote the function harmonic in R_2, continuous in the corresponding closed region, equal to $V(z)$ on J_1 and equal to $\log M < V(z)$ on Q_2; there follows the inequality $V_1(z) < V(z)$ throughout R_2. For arbitrary positive ϵ and for n sufficiently large we have

$$\log |F_n(z)| \leq V_1(z) + \epsilon$$

for z on J_1 and also for z on Q_2; consequently, by the subharmonic property of the first member, this inequality is valid throughout R_2. On an arbitrary continuum Q_3 in R_2 we therefore have

$$\limsup_{n\to\infty} [\max |F_n(z)|, z \text{ on } Q_3]$$

$$\leq \limsup_{n\to\infty} [\max e^{V_1(z)}, z \text{ on } Q_3]$$

$$< \limsup_{n\to\infty} [\max e^{V(z)}, z \text{ on } Q_3].$$

But Q_3 is also in R, and this inequality contradicts our hypothesis that $V(z)$ is an exact harmonic majorant in R for the sequence $F_n(z)$. Theorem 2 is established.

If $V(z)$ is an exact harmonic majorant of the sequence $F_n(z)$ in a region R, the fundamental relation (1) holds for every continuum Q (not a single point) in R; this relation (1) also holds (Theorem 2) provided merely that Q is a closed set in R which does not separate R, and if each boundary point on Q of the region $R - Q$ is regular for that region. However, the relation (1) need not hold if Q is merely required to be a closed set in R, or a closed set of positive capacity. For instance we may choose $F_n(z) \equiv z(z-1)^{1/n}$, with R the region $0 < |z| < \infty$, with $V(z) \equiv \log |z|$, and Q as the set $|z - \frac{1}{2}| \leq 1/4$ plus the point $z = 1$. However, our method of proof of Theorem 2 also yields the

COROLLARY TO THEOREM 2. *If $V(z)$ is an exact harmonic majorant of the sequence $F_n(z)$ in a region R, and if Q is a closed set in R of positive capacity, then we have not merely* (2), *as for an arbitrary closed set, but have also*

$$\limsup_{n\to\infty} [\max |F_n(z)|, z \text{ on } Q] \geq [\text{g.l.b. } e^{V(z)}, z \text{ on } Q'], \tag{40}$$

where Q' is the set of points of Q regular for the region $R - Q$.

(It follows from the Kellogg-Evans-Vasilesco Lemma [4] that Q' is not empty.)

We assume (40) false:

$$\limsup_{n\to\infty} [\max |F_n(z)|, z \text{ on } Q] = M < [\text{g.l.b. } e^{V(z)}, z \text{ on } Q'],$$

and shall reach a contradiction. Denote by z_1 a point of the closure of Q' at which $V(z)$ takes on the greatest lower bound of the values of $V(z)$ on Q'. If z_1 (which must belong to Q) belongs to a continuum of points of Q, or is a limit point of continua of points of Q, the method of proof already used is adequate. In the contrary case, there exists (compare the proof of Theorem 2) a Jordan curve J_1 in R bounding a subregion R_1 of R which contains z_1, which (i.e., J_1) does not intersect Q, and with the property $V(z) > \log M$ at every point of J_1. Let R_2 denote the region R_1 minus the points of Q in R_1; this set is of course connected, and its boundary contains points of Q that are regular for the region $R - Q$.

Let $V_1(z)$ denote the sequence solution of the Dirichlet problem for the region R_2, with the boundary values $V(z)$ on J_1 and $\log M$ in the remaining boundary points of R_2. By virtue of the relation $V(z) > \log M$ on J_1 it follows that we have $V_1(z) \geq \log M$ throughout R_2, because all the assigned boundary values for $V_1(z)$ are greater than or equal to $\log M$. This last inequality, together with the equation $V_1(z) = V(z)$ on J_1, shows that $V_1(z)$ is a harmonic majorant in R_2 for the sequence $F_n(z)$. Moreover all the assigned boundary values for $V_1(z)$ are less than or equal to those of $V(z)$, so we have $V_1(z) \leq V(z)$ throughout R_2. The boundary of R_2-interior to R_1 must contain at least one point which is regular for the region $R - Q$, hence also regular for the region R_2, so we have

$V_1(z) < V(z)$ throughout R_2. The proof of the corollary follows henceforth the proof of the theorem itself.

It is of course a consequence of the Corollary to Theorem 2 that if $V(z)$ is a harmonic majorant for the sequence $F_n(z)$ in R, then the relation

$$\limsup_{n\to\infty} [\max |F_n(z)|, z \text{ on } Q] < [\min e^{V(z)}, z \text{ on } Q],$$

where Q is a set of positive capacity in R, implies that $V(z)$ is not an exact harmonic majorant for the sequence $F_n(z)$ in R.

Bourion [1; §10] and Mosesson use a special form of the two-constant theorem involving arbitrary closed sets of positive capacity, in connection with the study of degree of convergence of partial sums of Taylor's series and of sequences of polynomials; their results even for polynomials are slightly less precise than the Corollary to Theorem 2.

In Theorem 5, inequalities (11) and (12), it is now not essential to assume Q_0 a continuum in R_0 (or $R - R_0 - C_0$) not a single point, by virtue of Theorem 2; it is sufficient if Q_0 is a closed set in R_0 (or $R - R_0 - C_0$) which does not separate R, and with the property that every boundary point of $R - Q_0$ which lies on Q_0 is regular for $R - Q_0$. Moreover, it is clear by the Corollary to Theorem 2 that (11) and (12) may be replaced by the negation of (40) for the specific sequences $[f(z) - f_{n_k}(z)]^{1/n_k}$ and $[f_{n_k}(z)]^{1/n_k}$.

Let us investigate in more detail inequality (12) of Theorem 5, and its consequences for the behavior of the sequence $f_n(z)$ and subsequences in $R - R_0 - C_0$.

The sequence $f_n(z)$ can converge throughout no subregion of $R - R_0 - C_0$; any sequence $f_{n_k}(z)$ which converges throughout a subregion of $R - R_0 - C_0$ satisfies (12) *on some continuum Q_0 in $R - R_0 - C_0$.* We use the theorem of Osgood which states that when a sequence of analytic functions converges throughout a region, subregions of uniform convergence are everywhere dense. If the sequence $f_n(z)$ converges uniformly throughout a subregion of $R - R_0 - C_0$, that sequence is uniformly bounded in a subregion of $R - R_0 - C_0$; in such a subregion the sequence $[f_{n+1}(z) - f_n(z)]^{1/n}$ has zero as a harmonic majorant; this contradicts the conditions of Theorem 5 concerning $V(z)$. Similarly, let the sequence $f_{n_k}(z)$ converge throughout a subregion of $R - R_0 - C_0$, and therefore converge uniformly throughout a subregion of $R - R_0 - C_0$; the sequence $[f_{n_k}(z)]^{1/n_k}$ is uniformly bounded in such a subregion Q_0; the first member of (12) is unity, so (12) is satisfied. Of course this proof establishes also that $f_n(z)$ can be bounded in no subregion of $R - R_0 - C_0$, and that $f_{n_k}(z)$ if bounded in such a subregion satisfies (12).

A more general result than that just established is related to the Corollary to Theorem 1:

THEOREM 12. *Suppose $V(z)$ is a harmonic majorant of the sequence $F_n(z)$ in the region R, and suppose E is a closed point set in R of positive capacity or of positive linear or superficial measure. Then the relation*

$$(41) \qquad \limsup_{n\to\infty} | F_n(z) | < e^{V(z)}$$

in each point z of E implies that V(z) is not an exact harmonic majorant of $F_n(z)$ in R.

Denote by E_0 the subset of E on which

$$\limsup_{n\to\infty} | F_n(z) | < e^{V(z)} - 1,$$

and denote by E_k for $k = 1, 2, 3, \cdots$ the subset of E on which

$$e^{V(z)} - 2^{-k+1} \leq \limsup_{n\to\infty} | F_n(z) | < e^{V(z)} - 2^{-k}.$$

Each of the sets E_k is measurable; these sets are mutually disjoint; their sum is E. Consequently at least one E_k, say for definiteness E_m, must be of positive capacity or measure. (So far as concerns capacity, we are using here and below the theorem that if a bounded set is the sum of a countable infinity of sets each of capacity zero, then the original set is also of capacity zero. See for instance [10]. The corresponding proposition for measure is of course well known.)

Let F_1 denote the subset of E_m on which we have

$$| F_n(z) | < e^{V(z)} - 2^{-m} \qquad (n = 1, 2, \cdots);$$

let F_2 denote the subset of $E_m - F_1$ on which we have

$$| F_n(z) | < e^{V(z)} - 2^{-m} \qquad (n = 2, 3, \cdots);$$

and let F_j denote the subset of $E_m - F_1 - F_2 - \cdots - F_{j-1}$ on which we have

$$(42) \qquad | F_n(z) | < e^{V(z)} - 2^{-m} \qquad (n = j, j + 1, \cdots).$$

Then $E_m = F_1 + F_2 + \cdots$; the sets F_j are measurable and mutually disjoint. At least one of the sets F_j is known to be of positive capacity or is of positive measure and hence (see [7; 145]) also of positive capacity. For this particular j, inequality (42) is incompatible with (40), because F_j can be separated into a finite number of subsets, at least one of which is of positive capacity, and on each of which the oscillation of $e^{V(z)}$ is less than 2^{-m-1}. On the closure of such a subset the inequality

$$| F_n(z) | \leq [\min e^{V(z)}] - 2^{-m-1}$$

is valid for all sufficiently large n. Theorem 12 is established.

It is an obvious consequence of Theorem 12 that under the conditions of Theorem 5 the relation

$$\limsup_{n_k\to\infty} | f(z) - f_{n_k}(z) |^{1/n_k} < e^{V(z)}$$

at every point of a set E of positive linear or superficial measure in a subregion of R_0 implies (11) on every continuum Q_0 in that subregion; the relation

$$(43) \qquad \limsup_{n_k\to\infty} | f_{n_k}(z) |^{1/n_k} < e^{V(z)}$$

at every point of a set E of positive linear or superficial measure in a subregion of $R - R_0 - C_0$ implies (12) on every continuum Q_0 in that subregion; in particular the condition (43) is satisfied on a set E of $R - R_0 - C_0$ if the sequence $f_{n_k}(z)$ converges or is bounded on E. Thus the sequence $f_n(z)$ itself can converge or be bounded on no set E of positive linear or superficial measure in $R - R_0 - C_0$.

It is appropriate to mention that sets of positive capacity are intimately related to the discussion of §4. Theorem 8 persists if Q_0 is now assumed a particular set of positive capacity in R; the proof (similar to that of Theorem 7) requires no modification except that we use the sharper form of Ostrowski's two-constant theorem used by Bourion and Mosesson; compare the proof of the Corollary to Theorem 2. Theorems 9 and 10 admit a corresponding extension.

In our study of harmonic majorants and the Dirichlet problem or of a corresponding inequality involving subharmonic functions, we have had occasion to treat: (i) regions bounded by a finite number of Jordan curves, with discontinuities in the assigned boundary values at a finite number of points, as in Theorem 4; (ii) regions bounded by a finite number of Jordan curves, assuming a specific harmonic majorant merely on more or less arbitrary closed subarcs of the boundary, as in Theorem 11; (iii) regions bounded in part by a set known merely to be of positive capacity, as in the Corollary to Theorem 2; (iv) regions with all boundary points regular, as in Theorem 2. We have set forth methods for the study of these various cases; an all-including general theorem can be formulated by the reader; the results useful for our present purposes are those already stated explicitly.

In connection with (40) and Theorem 12 we have emphasized sets of positive capacity. A justification of this emphasis is to be found in Theorems 13 and 14 now to be established, which are in the nature of counter-examples, and which refer respectively to sets of zero capacity in $R - R_0 - C_0$ and in R_0 in the notation of Theorem 5. Theorems 13 and 14 themselves employ a notation (see Walsh [11]) customary in the study of approximation by polynomials, but different from that of Theorem 5.

THEOREM 13. *Let C be a closed bounded set whose complement is connected and regular. Let the function $f(z)$ be single-valued and analytic on C and throughout the interior of a given C_ρ, but not throughout the interior of any $C_{\rho'}$, $\rho' > \rho$. Let D be a closed bounded set of zero capacity exterior to C_ρ, and let $g(z)$ be analytic on D. Then there exists a sequence of polynomials $p_n(z)$ of respective degrees n converging maximally to $f(z)$ on C and converging to $g(z)$ on D.*

We consider approximation by polynomials on the closed limited set $C' = C + D$, whose complement is connected but not regular. Nevertheless [11; §4.9] it is appropriate to study maximal convergence on C'. Green's function for the complement of C is the same as Green's function (in the extended sense) for the complement of C', except perhaps on D; this follows from the fact that regularity of a point is a local property. (Compare Nevanlinna [7; 132, Satz 2].) Consequently the loci C_ρ and $C_{\rho'}$ coincide except perhaps on D itself.

There exists [11; §4.9] a sequence of polynomials $p_n(z)$ converging maximally on C' to the function

$$f(z),\ z \text{ on } C; \qquad g(z),\ z \text{ on } D;$$

this sequence converges maximally to $f(z)$ on C and converges uniformly to $g(z)$ on D, as we were to prove.

For the sequence $p_n(z)$ we have

$$\limsup_{n\to\infty} [\max |p_n(z)|^{1/n},\ z \text{ on } D] \leq 1;$$

we use the sign $\leq$ here because it is possible for $g(z)$ to vanish identically on D. Nevertheless the sequence $[p_n(z)]^{1/n}$ has (see Walsh [14]) in the region exterior to C_ρ (notation of Theorem 13) the exact harmonic majorant $G(z)$, Green's function for the exterior of C_ρ with pole at infinity, and the inequality $G(z) > 1$ holds at every point exterior to C_ρ. Indeed, the sequence $p_n(z)$ satisfies the conditions of Theorem 5 for the sequence $f_n(z)$, where R of Theorem 5 is the complement of C and we have $V(z) \equiv G(z)$.

Precisely as Theorem 13 is a justification of the use of sets of positive capacity in connection with Theorem 12, Theorem 14 is a justification of the use of sets of positive capacity in connection with the Corollary to Theorem 2:

THEOREM 14. *Let C be a closed limited set whose complement is connected and regular. Let $\rho > 1$ be given, and let the function $f(z)$ be single-valued and analytic throughout the interior of C_ρ but not single-valued and analytic throughout the interior of any $C_{\rho'}$, $\rho' > \rho$. Let D be a closed point set of capacity zero interior to C_ρ but disjoint to C. Then there exists a sequence of polynomials $p_n(z)$ converging maximally to $f(z)$ on C and satisfying the relation*

$$\limsup_{n\to\infty} [\max |f(z) - p_n(z)|,\ z \text{ on } D]^{1/n} \leq 1/\rho. \tag{44}$$

Theorem 14 can be established by precisely the method used in Theorem 13. Of course an inequality of the form (44) is impossible for a set D of positive capacity interior to C_ρ but disjoint to C.

6. Zeros of approximating functions.

The intimate relation between overconvergence and the zeros of functions of the approximating sequence was recognized by Porter and studied in more detail by Jentzsch, Ostrowski, and later writers. We proceed to establish certain properties concerning this topic, under our general conditions. For reference we state

LEMMA 3. *Under the conditions $|\alpha| \leq r < 1$, $r < |z| < 1$, we have the inequalities*

$$\frac{|z| - r}{1 - r|z|} \leq \left|\frac{z - \alpha}{1 - \bar{\alpha}z}\right| \leq \frac{|z| + r}{1 + r|z|} < 1.$$

Of course the function $w = (z - \alpha)/(1 - \bar{\alpha}z)$ transforms $|z| < 1$ into $|w| < 1$ and $|z| = 1$ into $|w| = 1$. The proof of Lemma 3 is not difficult, and is omitted here. See for instance Walsh [11; 290, 229].

For the sake of simplicity we have assumed in Lemma 3 that α lies in the circle $|z| \leq r$ *concentric with the unit circle.* Inequalities similar to those of Lemma 3 can be established at once if α is restricted to lie in any circle γ interior to the unit circle, for there exists a linear transformation which leaves the unit circle unchanged and transforms γ into a circle whose center is the origin; the function $|(z - \alpha)/(1 - \bar{\alpha}z)|$ is invariant under such a transformation simultaneously operating on α and z. This remark applies to a number of the following results.

THEOREM 15. *Let C be the circle $|z| = 1$, and γ the circle $|z| = r < 1$. Let the functions $f_n(z)$ be analytic interior to C and have N_n zeros $\alpha_{n1}, \alpha_{n2}, \cdots, \alpha_{nN_n}$ interior to γ, where we have*

$$\lim_{n\to\infty} N_n/n = 0. \tag{45}$$

If we set

$$\varphi_n(z) = \prod_{k=1}^{N_n} \frac{z - \alpha_{nk}}{1 - \bar{\alpha}_{nk} z}, \quad F_n(z) = f_n(z)/\varphi_n(z),$$

where the function $F_n(z)$ is defined to be continuous in the points α_{nk}, then we have on any closed set Q in C but exterior to γ

$$\limsup_{n\to\infty} [\max |f_n(z)|^{1/n}, z \text{ on } Q]$$
$$= \limsup_{n\to\infty} [\max |F_n(z)|^{1/n}, z \text{ on } Q].$$

The sequence $f_n(z)$ need not be defined for every n; it is sufficient for it to be defined for an infinite sequence of indices n.

If for any particular n the number N_n is zero, we set $\varphi_n(z) \equiv 1$ for that index. By Lemma 3 we write for $r < |z| < 1$

$$N_n \log \frac{|z| - r}{1 - r|z|} \leq \log |\varphi_n(z)| \leq N_n \log \frac{|z| + r}{1 + r|z|};$$

these inequalities are satisfied uniformly on Q; it follows that we have uniformly on Q

$$\lim_{n\to\infty} |\varphi_n(z)|^{1/n} = 1,$$

so the theorem follows.

THEOREM 16. *Let $V(z)$ be an exact harmonic majorant in a region R for the sequence of functions $[f_n(z)]^{1/n}$, and let $V(z)$ also be an exact harmonic majorant in R for every subsequence, where the functions $f_n(z)$ are analytic in R. If γ is an* *arbitrary circle which together with its interior lies in R, and if N_n is the number of zeros of $f_n(z)$ in γ, then equation (45) is satisfied. The sequence $f_n(z)$ need not be defined for every n, but merely for an infinite sequence of indices n, provided $V(z)$ is an exact harmonic majorant in R for the corresponding sequence $[f_n(z)]^{1/n}$ and for every subsequence.*

We construct a circle C concentric with γ and which together with its interior also lies in R. We assume, as we may do without loss of generality, that C is the circle $|z| = 1$. As in Theorem 15 we define the functions $\varphi_n(z)$ and $F_n(z)$. A consequence of Lemma 3 is for z in C but not in γ

$$(46)\qquad \begin{aligned} \frac{1}{n}\log|f_n(z)| - \frac{N_n}{n}\log\frac{|z|+r}{1+r|z|} &\leq \frac{1}{n}\log|F_n(z)| \\ &\leq \frac{1}{n}\log|f_n(z)| - \frac{N_n}{n}\log\frac{|z|-r}{1-r|z|}. \end{aligned}$$

The circumference C lies in R, whence by Theorem 3 for arbitrary positive ϵ

$$\limsup_{n\to\infty}\{\max[|f_n(z)|^{1/n} - e^{V(z)+\epsilon}], z \text{ on } C\} \leq 0.$$

On C we have $|f_n(z)| = |F_n(z)|$, whence

$$\limsup_{n\to\infty}\{\max[|F_n(z)|^{1/n} - e^{V(z)+\epsilon}, z \text{ on } C\} \leq 0.$$

It then follows by Theorem 4 that $V(z)$ is a harmonic majorant for the sequence $[F_n(z)]^{1/n}$ interior to C, so if Q is a continuum in C but exterior to γ we may write

$$\limsup_{n\to\infty}[\max|F_n(z)|^{1/n}, z \text{ on } Q] \leq [\max e^{V(z)}, z \text{ on } Q].$$

We obviously have by our hypothesis

$$(47)\qquad \limsup_{n\to\infty}[\max|f_n(z)|^{1/n}, z \text{ on } Q] = [\max e^{V(z)}, z \text{ on } Q];$$

the superior limit may be replaced by the limit itself. We then find from the first of inequalities (46) by choosing $[\max|f_n(z)|^{1/n}, z \text{ on } Q]$,

$$(48)\qquad \limsup_{n\to\infty}\left\{\max\left[\frac{|z|+r}{1+r|z|}\right]^{N_n/n}, z \text{ on } Q\right\} = 1,$$

from which our conclusion follows.

Under the hypothesis of Theorem 16, equation (45) represents a limitation on the asymptotic number of zeros possible in γ, —a limitation characteristic of any sequence $f_{n_k}(z)$ if $[f_{n_k}(z)]^{1/n_k}$ and every subsequence thereof has in R an exact harmonic majorant. Naturally $V(z) \equiv -\infty$ is here excluded as a majorant.

Corollary. *Under the conditions of Theorem 16, but without the assumption that $V(z)$ is an exact harmonic majorant for every subsequence of the $[f_n(z)]^{1/n}$, the zeros of $f_n(z)$ in γ may be suppressed in the sense that we may replace $f_n(z)$ by $F_n(z)$*

without altering the exact harmonic majorant of the sequence $[f_n(z)]^{1/n}$ *in the circle* C, *an arbitrary circle which contains* γ *and which together with its interior lies in* R.

Here we do not require that (45) be satisfied. This Corollary follows from the relation $|f_n(z)| = |F_n(z)|$ on C, whence $V(z)$ is a harmonic majorant for the sequence $[F_n(z)]^{1/n}$ interior to C, and from the relation $|f_n(z)| \leq |F_n(z)|$ interior to C, by the Corollary to Theorem 1 for Q in C exterior to γ.

If in the Corollary the function $V(z)$ is an exact harmonic majorant for any particular subsequence of $[f_n(z)]^{1/n}$, it is also an exact harmonic majorant for the corresponding subsequence of $[F_n(z)]^{1/n}$ interior to C.

It is a matter of indifference both in Theorem 16 and this Corollary whether we suppress the zeros of $f_n(z)$ merely interior to γ or in the closed interior of γ.

Of course Theorem 16 is not valid if we omit the requirement that $V(z)$ shall be an exact harmonic majorant in R for every subsequence of the sequence $[f_n(z)]^{1/n}$. This is shown by such an example as

$$f_{2m}(z) \equiv z^{N_{2m}}, \qquad f_{2m+1}(z) \equiv 1, \qquad R: |z| < 1;$$

we take m here as an arbitrary integer. The function $V(z) \equiv 0$ is an exact harmonic majorant for the sequence $[f_n(z)]^{1/n}$, but is not an exact harmonic majorant for the sequence $[f_{2m}(z)]^{1/2m}$ if $\lim_{m\to\infty} N_{2m}/2m$ exists and is positive. Indeed, equation (45) is a necessary and sufficient condition that $V(z)$ should be an exact harmonic majorant for every subsequence of the given sequence.

The reader will easily establish the following, by the theory of normal families of functions:

Remark 1. *Let* $V(z)$ *be an exact harmonic majorant in a region* R *for the sequence of functions* $[f_n(z)]^{1/n}$, *where the functions* $f_n(z)$ *are analytic and different from zero in* R. *Then every infinite subsequence of the sequence* $[f_n(z)]^{1/n}$ *has a new subsequence which converges throughout* R, *uniformly on every closed set in* R, *to a function whose modulus in* R *is not greater than* $e^{V(z)}$. *There exists a subsequence of the sequence* $[f_n(z)]^{1/n}$ *such that* $V(z)$ *is an exact harmonic majorant for every subsequence of this subsequence. If we assume that* $V(z)$ *is an exact harmonic majorant for every subsequence of the original sequence* $[f_n(z)]^{1/n}$, *it follows that every limit function of that family is analytic in* R *and has a modulus in* R *of precisely* $e^{V(z)}$; *indeed if the n-th roots are suitably chosen, the original sequence converges to such a limit function.*

The last clause can be proved as follows. Let $z = \alpha$ be a point of R, and choose the n-th roots in such a way that we have

$$|\arg\{[f_n(\alpha)]^{1/n}\}| \leq \frac{\pi}{n}.$$

Each limit function $f(z)$ of the family fails to be identically zero, and hence must be different from zero throughout R and in the point α, by Hurwitz's theorem; thus all limit functions have a positive value at $z = \alpha$; the quotient of any two limit functions is analytic and of modulus unity in R, and has the value unity in the point α, hence is unity throughout R; any two limit functions are identical; the remark is established.

We proceed now to an extension of Theorem 15; we do not assume as in the Corollary to Theorem 16 the existence of an exact harmonic majorant:

THEOREM 17. *Let the functions $f_{n_k}(z)$ be analytic throughout the interior of the unit circle C, and let the functions $[f_{n_k}(z)]^{1/n_k}$ be uniformly bounded there. Suppose N_{n_k} denotes the number of zeros of $f_{n_k}(z)$ interior to C, with the relation*

$$\lim_{n_k \to \infty} N_{n_k}/n_k = 0. \tag{49}$$

Let the functions $\varphi_{n_k}(z)$ and $F_{n_k}(z)$ be defined as in Theorem 15, where the $\alpha_{n_k i}$ are the zeros of $f_{n_k}(z)$ interior to a second circle γ interior to and concentric with C. Then on every continuum Q interior to C we have

$$\begin{aligned} &\limsup_{n_k \to \infty} [\max |f_{n_k}(z)|^{1/n_k}, z \text{ on } Q] \\ &= \limsup_{n_k \to \infty} [\max |F_{n_k}(z)|^{1/n_k}, z \text{ on } Q]. \end{aligned} \tag{50}$$

This result is contained in Theorem 15 if Q is exterior to γ; we treat next the case that Q is interior to γ. The inequality $|\varphi_{n_k}(z)| \leq 1$ interior to C assures us that the first member of (50) is not greater than the second member. Let us choose the particular sequence of indices m_k in such manner that we have for the particular Q considered

$$\begin{aligned} &\lim_{m_k \to \infty} [\max |F_{m_k}(z)|^{1/m_k}, z \text{ on } Q] \\ &= \limsup_{n_k \to \infty} [\max |F_{n_k}(z)|^{1/n_k}, z \text{ on } Q]. \end{aligned} \tag{51}$$

Let J be a Jordan curve interior to γ containing Q in its interior, and let K be a region consisting of a finite number of circles interior to γ covering J but having no point of Q in the interior or on the boundary of K. It follows from Theorem 15 that suppression of the zeros of the sequence $f_{m_k}(z)$ in K does not alter the first member of (50). From the sequence $f_{m_k}^{(1)}(z)$ of thus modified functions can be extracted a sequence $f_{p_k}^{(1)}(z)$ such that $|f_{p_k}^{(1)}(z)|^{1/p_k}$ converges throughout K, uniformly on any closed set in K, say to a function of modulus $e^{U(z)}$ in K. For the sequence $[f_{p_k}^{(1)}(z)]^{1/p_k}$ is bounded and hence forms a normal family in each of the circles of K; to be sure, the function $[f_{p_k}^{(1)}(z)]^{1/p_k}$ need not be single-valued in the totality of circles K, but its modulus is single-valued; the function $U(z)$ is harmonic in K. Further suppression of the zeros on the point set K_1 interior to J but not in K of the sequence $f_{p_k}^{(1)}(z)$ yields a sequence $f_{p_k}^{(2)}(z)$ with no zeros on or within J; this suppression of zeros is accomplished by covering K_1 with a finite number of circles interior to J; we suppose that these circles leave uncovered one or more regions K_2 simultaneously in K and in the interior of J. Replacement of the sequence $f_{p_k}^{(1)}(z)$ by the sequence $f_{p_k}^{(2)}(z)$ does not alter the exact harmonic majorant of the sequence in the regions K_2. In such regions

K_2 the exact harmonic majorant is $U(z)$ for $[f_{p_k}^{(1)}(z)]^{1/p_k}$ and for every subsequence thereof, hence is $U(z)$ for $[f_{p_k}^{(2)}(z)]^{1/p_k}$ and for every subsequence. The sequence $f_{p_k}^{(2)}(z)$ has no zeros interior to J, hence there exists a subsequence of $[f_{p_k}^{(2)}(z)]^{1/p_k}$ converging throughout that region, uniformly in any closed subregion. Any limit function of any subsequence has the modulus $e^{U(z)}$ in K_2 ; hence $U(z)$ if suitably defined is harmonic throughout the interior of J and is throughout that interior an exact harmonic majorant for the entire sequence $[f_{p_k}^{(2)}(z)]^{1/p_k}$; in particular we have

$$\limsup_{p_k\to\infty} [\max | f_{p_k}^{(2)}(z) |^{1/p_k}, z \text{ on } Q] = [\max e^{U(z)}, z \text{ on } Q].$$

Since $U(z)$ is a harmonic majorant in the region K containing J for the sequence $[f_{p_k}^{(1)}(z)]^{1/p_k}$, it follows from Theorem 4 that $U(z)$ is a harmonic majorant interior to J for that sequence, whence

$$\limsup_{p_k\to\infty} [\max | f_{p_k}^{(1)}(z) |^{1/p_k}, z \text{ on } Q] \leq [\max e^{U(z)}, z \text{ on } Q].$$

If the strong inequality sign held here, the function $U(z)$ would not be an exact harmonic majorant for the sequence $[f_{p_k}^{(1)}(z)]^{1/p_k}$ in K_2, contrary to our definition of $U(z)$; thus by the Corollary to Theorem 1 we have

$$\begin{aligned}\limsup_{p_k\to\infty} [\max | f_{p_k}^{(1)}(z) |^{1/p_k}, z \text{ on } Q] &= [\max e^{U(z)}, z \text{ on } Q] \\ &= \limsup_{p_k\to\infty} [\max | f_{p_k}^{(2)}(z) |^{1/p_k}, z \text{ on } Q].\end{aligned} \tag{52}$$

The first term of this equation is also equal to

$$\limsup_{p_k\to\infty} [\max | f_{p_k}(z) |^{1/p_k}, z \text{ on } Q];$$

the last term is equal to

$$\limsup_{p_k\to\infty} [\max | F_{p_k}(z) |^{1/p_k}, z \text{ on } Q],$$

for further removal of zeros of $f_{p_k}^{(2)}(z)$ interior to γ but exterior to J does not alter the last member of (52); since the first member of (50) is not greater than the second member, Theorem 17 is established for the case that Q lies in γ.

The proof as given requires formal modification if both members of (51) vanish, but under these conditions equation (50) follows from the inequality $| F_{n_k}(z) | \geq | f_{n_k}(z) |$ interior to C.

If now Q is not assumed to lie in γ, but to lie interior to C, we consider the auxiliary circle $\gamma^{(1)}$ interior to C containing both Q and γ in its interior. If $F_{n_k}^{(1)}(z)$ denotes the modified form of $f_{n_k}(z)$ found by suppressing the zeros of $f_{n_k}(z)$ in $\gamma^{(1)}$, we obviously have for z in C

$$| F_{n_k}^{(1)}(z) | \geq | F_{n_k}(z) | \geq | f_{n_k}(z) |.$$

From the facts already proved we have

$$\limsup_{n_k\to\infty} [\max | F^{(1)}_{n_k}(z) |^{1/n_k}, z \text{ on } Q]$$
$$= \limsup_{n_k\to\infty} [\max | f_{n_k}(z) |^{1/n_k}, z \text{ on } Q];$$

it follows that each of these members is equal to the intermediate value

$$\limsup_{n_k\to\infty} [\max | F_{n_k}(z) |^{1/n_k}, z \text{ on } Q],$$

and Theorem 17 is completely proved.

Theorem 17 yields

Remark 2. *Let* $V_1(z)$ *be a harmonic majorant in a region* R *for the sequence of functions* $[f_{n_k}(z)]^{1/n_k}$, *where the functions* $f_{n_k}(z)$ *are analytic in* R. *Suppose for every circle* γ *in* R *the relation* (49) *is valid, where* N_{n_k} *denotes the number of zeros of* $f_{n_k}(z)$ *interior to* γ. *Then there exists a subsequence of the sequence* $[f_{n_k}(z)]^{1/n_k}$ *such that a suitably chosen* $V(z)$ *is an exact harmonic majorant in* R *for every subsequence of this subsequence. If* $V_1(z)$ *is an exact harmonic majorant for the given sequence, we may choose* $V(z)$ *as* $V_1(z)$.

We cover R by a countable infinity of circles $\gamma_1, \gamma_2, \cdots$, which together with their interiors lie in R, in such a way that an arbitrary closed set in R is covered by a finite number of these circles. Let circles $C_1, C_2, \cdots$ together with their interiors also lie in R and contain respectively the circles $\gamma_1, \gamma_2, \cdots$. We can remove the zeros of $f_{n_k}(z)$ in γ_1 without altering the properties of the harmonic majorant in C_1; with these zeros removed, the sequence $[f_{n_k}(z)]^{1/n_k}$ forms a normal family in γ_1; a suitable subsequence converges uniformly on every closed set interior to γ_1, and thus possesses an exact harmonic majorant in γ_1; this exact harmonic majorant is the same for every further subsequence; this property of the exact harmonic majorant persists if the zeros of the original functions are restored, by Theorem 17. We operate on the subsequence possessing an exact harmonic majorant in γ_1 in a similar way with the circle γ_2; by continuing in this manner using the Cantor diagonal process we arrive at a subsequence of the original sequence which has an exact harmonic majorant $V(z)$ in each of the circles γ_j, $j = 1, 2, 3, \cdots$. This majorant is harmonic throughout R, by the monodromy theorem. The conclusion now follows from the Corollary to Theorem 3; if we assume that $V_1(z)$ is an exact harmonic majorant for the given sequence, we may choose $V_1(z)$ as the exact harmonic majorant in γ_1 for the first subsequence chosen and for every new subsequence thereof. The proof carried through as before then yields $V(z)$ equal to $V_1(z)$ in γ_1, hence throughout R.

In Remark 2 it may occur exceptionally that the chosen subsequence of the $[f_{n_k}(z)]^{1/n_k}$ converges uniformly in γ_1 *to the function zero*; then (compare Theorem 8) *every* function harmonic in R is a harmonic majorant for this subsequence and for every further subsequence; there is properly no exact harmonic majorant in R for the subsequence; but in a sense we may (by a temporary modification of our usual convention) consider $V(z) \equiv -\infty$ as an exact harmonic majorant; Remark 2 is to be understood with the revised convention as a possibility.

The following theorem may be considered a converse of Theorem 16, relative to the existence of a harmonic majorant throughout the whole of a given region; together the two theorems indicate the influence of zeros of functions on the exact harmonic majorant:

THEOREM 18. *Let the functions $f_n(z)$ be analytic in a region R, and let $[f_n(z)]^{1/n}$ be bounded in R. Let subregions R_1 and R_2 of R exist such that a particular function $V_1(z)$ is an exact harmonic majorant in R_1 for the sequence $[f_{n_k}(z)]^{1/n_k}$ and for every subsequence, and a function $V_2(z)$ not the harmonic extension of $V_1(z)$ from R_1 is an exact harmonic majorant in R_2 for the sequence $[f_{n_k}(z)]^{1/n_k}$ and for every subsequence.*

Let γ be a circle which together with its interior lies in R and which contains points of both R_1 and R_2. Then if N_n denotes the number of zeros of $f_n(z)$ in γ we have

$$\liminf_{n_k \to \infty} N_{n_k}/n_k > 0. \tag{53}$$

Suppose now we have (49) valid, where if necessary the notation may refer to a subsequence of the original subsequence; we shall reach a contradiction. Modify the functions $f_{n_k}(z)$ by removing the zeros interior to γ; by Theorem 17 this does not affect the first member of (50) nor an exact harmonic majorant, either for $f_{n_k}(z)$ or for a subsequence. The modified sequence $[f_{n_k}(z)]^{1/n_k}$ forms a normal family in γ; a suitable subsequence converges throughout the interior of γ, uniformly on any closed set interior to γ, and has a certain exact harmonic majorant throughout that interior; this contradicts our hypothesis of different harmonic majorants in the subregions interior to γ and in R_1 and R_2 respectively. The proof is complete. We have at once also the

COROLLARY. *If we modify the hypothesis of Theorem 18 so that $V_2(z)$ is not required to be an exact harmonic majorant in R_2 for every subsequence, but is still required to be an exact harmonic majorant for the original sequence, then the conclusion (53) is to be modified by writing*

$$\limsup_{n_k \to \infty} N_{n_k}/n_k > 0. \tag{54}$$

We cannot assume in this Corollary merely that $V_1(z)$ and $V_2(z)$ are exact harmonic majorants in R_1 and R_2, as is shown by the example $f_{2m}(z) \equiv z^{2m}$, $f_{2m+1}(z) \equiv 1$, R: $|z - 1| < 1$. In the subregion R_1 of R exterior to $|z| = 1$, we have the exact harmonic majorant $V_1(z) \equiv \log |z|$, and in the subregion R_2 of R interior to $|z| = 1$ we have the exact harmonic majorant $V_2(z) \equiv 0 \not\equiv V_1(z)$. Yet the functions $f_n(z)$ have no zeros in R.

As a simple specific example related to Theorem 18 and the Corollary, it is instructive to consider $f_{n_k}(z) \equiv z^{Mn_k} - 1$, so that $[f_{n_k}(z)]^{1/n_k}$ with every subsequence has in $|z| < 1$ the exact harmonic majorant $V_1(z) \equiv 0$, and the sequence $[f_{n_k}(z)]^{1/n_k}$ has in $|z| > 1$ the exact harmonic majorant $V_2(z) \equiv \log|z| \cdot \lim\sup (M_{n_k}/n_k)$; this last expression is assumed finite. If γ is a circle with center $z = 1$, the conditions (54) and $\lim\sup (M_{n_k}/n_k) > 0$ are equivalent, and are necessary and sufficient that $V_2(z)$ not be the harmonic extension of $V_1(z)$; the conditions (53) and $\lim\inf (M_{n_k}/n_k) > 0$ are equivalent, and are necessary and sufficient that each subsequence of $[f_{n_k}(z)]^{1/n_k}$ should have a new subsequence whose exact harmonic majorant in $|z| > 1$ exists but is not the harmonic extension of $V_1(z)$.

The scope of Theorem 18 is considerably greater than the situation of Theorem 5, at least if we restrict ourselves to the sequence $f_n(z)$ in Theorem 5, because under the conditions of Theorem 5 the regions R_0 (or a subregion) and $R - R_0 - C_0$ (or a subregion) have whatever common boundary points may exist located on the locus C_0, composed (in any closed region interior to R) of a finite number of analytic Jordan arcs. On the other hand, if C is an arbitrary closed limited set whose complement K is connected and regular (in the sense that K possesses a Green's function $G(z)$ with pole at infinity), then there exists a sequence of polynomials $\omega_n(z)$ whose roots have no limit point except on the boundary of K, and the functions $[\omega_n(z)]^{1/n}$ have in K as exact harmonic majorant a constant plus $G(z)$, and have in the interior points of C as exact harmonic majorant a constant; the boundary between these two sets need not be an analytic curve or even a Jordan curve; compare a discussion by the present writer [11; §4.4].

Although Theorem 18 is stated for γ a circle, it is obvious by means of a conformal map that (53) is valid when N_n denotes the number of zeros of $f_n(z)$ in an arbitrary Jordan region γ in R, provided γ contains points of both R_1 and R_2.

The primary importance of Theorems 17 and 18 in our present investigation arises from the fact that under the conditions of Theorem 5, the function $V(z) \equiv 0$ is an exact harmonic majorant for the sequence $[f_n(z)]^{1/n}$ or for any subsequence in any subregion of R_0 in which $f(z)$ fails to vanish identically; thus the Corollary to Theorem 18 contains and makes more explicit Jentzsch's theorem for Taylor's series, that every point of the circle of convergence is a limit point of zeros of partial sums. Moreover, Theorem 17 may apply also under circumstances other than those of Theorem 5:

THEOREM 19. *Let the functions $f_{n_k}(z)$ be analytic in a region R, and let the sequence $[f_{n_k}(z)]^{1/n_k}$ have there a harmonic majorant. Let the sequence $f_{n_k}(z)$ converge in a subregion R' to an analytic function $f(z)$ in such a way that the sequence $[f(z) - f_{n_k}(z)]^{1/n_k}$ has in R' a negative harmonic majorant.*

Let the point $z = \alpha$ in R be a boundary point of R', let $f(z)$ be analytic but not identically zero throughout a neighborhood of α, and suppose no subsequence of $f_{n_k}(z)$ converges uniformly throughout a neighborhood of α. Then if γ is an arbitrary circle in R whose center is α we have (53) *satisfied, where N_{n_k} indicate the number of zeros of $f_{n_k}(z)$ interior to γ.*

In particular it follows that if α is a boundary point in R of R' and if $f(z)$ is analytic but not identically zero throughout the neighborhood of α, then either the sequence $f_{n_k}(z)$ converges uniformly throughout such a neighborhood or α is a limit point of the zeros of the $f_{n_k}(z)$.

Theorems 9, 10, and 11 are obviously of significance in connection with Theorem 5 in establishing a negative harmonic majorant for the sequence $[f(z) - f_{n_k}(z)]^{1/n_k}$ in R', but of course under the hypothesis of Theorem 5 there may exist a subregion R' of $R - R_0 - C_0$ and a function $f(z)$ analytic in R' yet not the analytic extension of $f(z)$ from any point of R_0, while the sequence $[f(z) - f_{n_k}(z)]^{1/n_k}$ has a negative harmonic majorant in R'. Theorem 19 and its Corollary below apply to a boundary point α of any such region R'; the point α may or may not lie on C_0.

If (53) is not satisfied, there exists a γ in R and a sequence of indices m_k with

$$\lim_{m_k \to \infty} N_{m_k}/m_k = 0.$$

By Remark 2 a subsequence of the sequence m_k exists such that the subsequence of functions $[f_{m_k}(z)]^{1/m_k}$ corresponding and every subsequence thereof have an exact harmonic majorant interior to γ; this majorant is zero in the subregion of R' in γ, hence is zero throughout γ. It follows from Theorem 10 that the subsequence converges uniformly on any closed set interior to γ; this contradiction of our hypothesis completes the proof of Theorem 19. Inspection of this proof yields also the

COROLLARY. *If in Theorem 19 we omit the requirement that no subsequence of $f_{n_k}(z)$ shall converge throughout a neighborhood of $z = \alpha$, but require instead that the original sequence $f_{n_k}(z)$ shall not converge throughout such a neighborhood, then the conclusion is to be modified by writing (54) instead of (53); it is still true that α is a limit point of the zeros of the functions $f_{n_k}(z)$.*

If the conditions of Theorem 19 are fulfilled under the hypothesis of the first part of Theorem 5 (whether or not $V(z)$ is assumed an *exact* harmonic majorant), where now R' is a region contained in R_0, it will be noticed that the condition that no subsequence of $f_{n_k}(z)$ converge uniformly throughout a neighborhood of α is not a local condition, but if fulfilled for a specific point α in R on the boundary of R' at which $f(z)$ is analytic is also fulfilled for every such point α. Under these circumstances the property expressed by (53) is thus not a local property, but if fulfilled for a single point α is fulfilled for every α in R on the boundary of R' at which $f(z)$ is analytic and not identically zero.

The conclusion (53) under the hypothesis corresponding to Theorem 19 has been stated by Ostrowski [8] for the special case of Taylor's series; see Bourion [1]. A less specific conclusion under a weaker hypothesis is also due to Ostrowski [9].

It may be noticed that if the conditions of Theorem 19 on R, R', $f(z)$, $f_{n_k}(z)$ are fulfilled, if R' is a maximal region in R with the required properties, and if

α' in R is a boundary point of R' with the property that every neighborhood of α' contains boundary points of R' at which $f(z)$ (not identically zero) is analytic, then each such neighborhood of α' contains a point α (a boundary point of R') and a neighborhood of α in which $f(z)$ is analytic but in which a suitably chosen subsequence of the $f_{n_k}(z)$ fails to converge uniformly; consequently (54) is valid, where N_{n_k} denotes either the number of zeros of $f_{n_k}(z)$ in a neighborhood of α or the number of zeros in a neighborhood of α'; if R' is a maximal region of uniform convergence not merely for $f_{n_k}(z)$ but for every subsequence, we have (53) in either notation. In particular, as Ostrowski indicated in his special case, if α' in R is a singular point of $f(z)$ not contained in a continuum (more than a single point) of singular points of $f(z)$, then each neighborhood of α' contains a Jordan curve J in R containing in its interior α' and only points of R, with $f(z)$ analytic at every point of J and with some points of J (but not all) in R'; thus if an arbitrary neighborhood of α' is chosen, containing a suitable J, then from any sequence of the $f_{n_k}(z)$ can be chosen a subsequence such that on J lies a boundary point α of the corresponding maximal region of uniform convergence (by Remark 3 necessarily existent if (53) is not satisfied for α'); consequently (53) is valid, where N_{n_k} denotes the number of zeros of $f_{n_k}(z)$ in an arbitrary neighborhood of α'.

In Theorem 19 and its Corollary, the condition $f(z) \not\equiv 0$ cannot be omitted, as is illustrated by the example $f_n(z) \equiv z^n$; R: $|z - 1| < \frac{1}{2}$; R': $|z - 1| < \frac{1}{2}$, $|z| < 1$; α: $z = 1$; the point α is not a limit point of zeros of $f_n(z)$.

Under the hypothesis of Theorem 5, Theorems 5, 10, 18, and 19 furnish a fairly complete picture of the total possibilities relative to the number of zeros of the sequences $f_n(z)$ and $f_{n_k}(z)$ in the neighborhood of a point α on C_0, provided $f(z) \not\equiv 0$ in a neighborhood of α. The sequence $[f_n(z)]^{1/n}$ has $V(z)$ as exact harmonic majorant in $R - R_0 - C_0$, and has zero as exact harmonic majorant in R_0. A suitably chosen subsequence $[f_{n_k}(z)]^{1/n_k}$ and every subsequence thereof have $V(z)$ as exact harmonic majorant in a neighborhood of α and interior to $R - R_0 - C_0$, and $f_{n_k}(z)$ converges to $f(z)$ interior to R_0, hence $[f_{n_k}(z)]^{1/n_k}$ with every new subsequence has zero as exact harmonic majorant in the neighborhood of α interior to R_0; Theorem 18 applies, and inequality (53) is valid. The point α is a limit point of zeros of the sequence $f_{n_k}(z)$. Our results are, however, more explicit in application to particular sequences. If $f(z)$ is analytic throughout a neighborhood of α, the sequence $f_n(z)$ fails to converge uniformly throughout such a neighborhood, so the Corollary to Theorem 19 applies and (54) is valid for $n_k = k$; if $f_{n_k}(z)$ is any subsequence of which no new subsequence converges uniformly throughout a neighborhood of α, Theorem 19 applies and (53) is valid. If $f(z)$ is not analytic throughout a neighborhood of α, we have already indicated the validity of (53) for a suitably chosen subsequence $f_{n_k}(z)$; the reasoning which establishes (53) is valid for a given sequence or any subsequence thereof unless zero is an exact harmonic majorant in the neighborhood of α (even in $R - R_0 - C_0$) for the n_k-th roots of the given sequence and every subsequence. Under the latter conditions we draw no conclusions regarding the number of zeros of the functions in the neighborhood of α; this situation can actually occur even for Taylor's series, as was indicated by Jentzsch and studied in more detail by Bourion (although without the concept of exact harmonic majorant). Indeed we shall demonstrate that if $f(z)$ is not analytic throughout a neighborhood of α and if for a sequence $f_{n_k}(z)$ inequality (54) is not satisfied, then zero is an exact harmonic majorant for the sequence $[f_{n_k}(z)]^{1/n_k}$ throughout some neighborhood of α. By the method of Theorem 17, the zeros of the functions $f_{n_k}(z)$ can be removed without altering the exact harmonic majorant (if existent). With the zeros removed, the modified functions $[f_{n_k}(z)]^{1/n_k}$ form a normal family in some neighborhood $N(\alpha)$ of α; from every subsequence of $[f_{n_k}(z)]^{1/n_k}$ a suitable new subsequence converges uniformly on every closed subset of $N(\alpha)$; the limit of this sequence is of modulus unity in R_0, since $f(z) \equiv 0$ is impossible by the non-analyticity of $f(z)$ in the neighborhood of α; hence the limit of this sequence is of modulus unity throughout $N(\alpha)$; thus zero is an exact harmonic majorant in $N(\alpha)$ for the given sequence of n_k-th roots and for every subsequence.

The following theorem is a direct consequence of Theorem 10, but is so closely related to our present discussion that it seems to deserve explicit statement here:

THEOREM 20. *Under the conditions of the first part of Theorem* 5, *let $f(z)$, $f_{n_k}(z)$, and R' satisfy the hypothesis of Theorem* 19. *Suppose $f(z)$ not analytic throughout the neighborhood $N(\alpha)$ of the point $z = \alpha$. Suppose zero is an exact harmonic majorant in $N(\alpha)$ for the sequence $[f_{n_k}(z)]^{1/n_k}$. If $z = \beta$ lies in $N(\alpha)$ but is not a singular point of $f(z)$, and if the sequence $f_{n_k}(z)$ does not converge in a neighborhood of β, then β cannot be joined to points of R_0 by a Jordan arc on which $f(z)$ is analytic and which remains in a subregion of R on which zero is an exact harmonic majorant for the sequence $[f_{n_k}(z)]^{1/n_k}$.*

Let now $f_{n_k}(z)$ and R' satisfy the conditions of Theorem 20; we remark that if the sequence $f_{n_k}(z)$ omits the value zero in $N(\alpha)$ that sequence cannot omit a second value there; if a second value were omitted, the sequence $f_{n_k}(z)$ would be normal in $N(\alpha)$, and being uniformly convergent on any closed subset of $R' \cdot N(\alpha)$ would be uniformly convergent on any closed subset of $N(\alpha)$, so $f(z)$ would be analytic throughout $N(\alpha)$, contrary to hypothesis.

It is fairly obvious that the foregoing theorems have application to the study of zeros not merely of the functions $f_n(z)$ but also of other functions. To be explicit in a very elementary case, let the hypothesis of the first part of Theorem 5 be satisfied, let α be a point of C_0 in R, and let $f(z)$ be analytic or not throughout the neighborhood $N(\alpha)$ of α. Let a new function $\Phi(z)$ be analytic throughout $N(\alpha)$, and $f(z) + \Phi(z) \not\equiv 0$ in $N(\alpha)$. Then the sequence $[f_n(z) + \Phi(z)]^{1/n}$ has zero as an exact harmonic majorant for the sequence and for every subsequence in $R_0 \cdot N(\alpha)$, and has $V(z)$ as an exact harmonic majorant for the sequence $[f_n(z) + \Phi(z)]^{1/n}$ in $(R - R_0 - C_0) \cdot N(\alpha)$. It follows from the Corollary to Theorem 18 that (54) is satisfied with $n_k = n$, where γ is an arbitrary circle whose center is α, and N_n now denotes the number of zeros of $[f_n(z) + \Phi(z)]$ in γ.

We add the remark that Theorem 17 and its method of proof have fairly obvious applications in establishing the existence of an exact harmonic majorant. For instance let the functions $q_n(z)$ be analytic in a region R; let R_1 be an arbitrary simply connected subregion of R; we set

$$Q_n(z) = q_n(z)/\varphi_n(z),$$

where $\varphi_n(z)$ is defined as in Theorem 15, after R_1 is mapped conformally into the interior of the unit circle, and where we suppose (49) satisfied. If we have for every such R_1

$$\lim_{n\to\infty} |Q_n(z)|^{1/n} = e^{U(z)}$$

uniformly on any closed set interior to R_1, where $U(z)$ is harmonic throughout R, we consider the series $\sum a_n q_n(z)$, and we introduce the notation

$$a = \limsup_{n\to\infty} |a_n|^{1/n}.$$

The function $U(z) + \log a$ is an exact harmonic majorant for the sequence of n-th roots of the partial sums of order n of this series in any subregion of R in which that function is positive, and is an exact harmonic majorant for the n-th root of the remainder after n terms of this series in any subregion of R in which that function is negative; compare Corollary 1 to Theorem 5.

The reverse reasoning is not without interest, for it follows for instance also from Corollary 1 to Theorem 5 that the polynomials $q_n(z)$ normal and orthogonal on an arbitrary rectifiable Jordan curve C have the property that Green's function for the exterior of C with pole at infinity is an exact harmonic majorant exterior to C for the sequence $[q_n(z)]^{1/n}$. This is a consequence of the maximal convergence on C of the sequence corresponding to the series $\sum a^n q_n(z)$, where $a < 1$ is arbitrary. This result persists if the set C is even more general, if a suitable weight function is employed, and if orthogonality is measured in any one of a number of ways; compare Walsh [11; §6.6]; asymptotic formulas for such general orthogonal polynomials have never been established. (Compare also [6].)

We turn now to the behavior in general of the harmonic majorant under differentiation and integration of functions. If the functions $f_n(z)$ are analytic in a region R, and if the sequence $[f_n(z)]^{1/n}$ has the harmonic majorant $V(z)$ in R, we choose an arbitrary continuum Q in R. If $\epsilon > 0$ is arbitrary, we have by Theorem 3 in a suitable region R_1 in R containing Q and for n sufficiently large

$$|f_n(z)|^{1/n} \le \exp\{\max[V(z) + \epsilon],\ z \text{ on } Q\}.$$

By differentiation under the integral sign of Cauchy's integral formula for the function $f_n(z)$ taken over a contour in R_1 which contains Q in its interior, we have for z on Q

$$|f_n'(z)|^{1/n} \le M^{1/n} \cdot \exp\{\max[V(z) + \epsilon],\ z \text{ on } Q\},$$

where M is independent of n and z. It follows that $V(z)$ is a harmonic majorant for the sequence $[f_n'(z)]^{1/n}$ in R. We cannot show that $V(z)$ is an exact harmonic majorant in R for the sequence $[f_n'(z)]^{1/n}$ if $V(z)$ is an exact harmonic majorant in R for $[f_n(z)]^{1/n}$, for that need not be the case, as is illustrated by $f_n(z) \equiv 1 - z^n$, R: $|z| < 1$, $V(z) = 0$.

Similarly if the functions $f_n(z)$ are analytic in a region R, and if the sequence $[f_n(z)]^{1/n}$ has the harmonic majorant $V(z)$ in R, it may be shown readily that the functions

$$\left[\int_{z_0}^{z} f_n(z)\, dz \right]^{1/n} \tag{55}$$

have in an arbitrary closed subregion R_1 of R the harmonic majorant $V(z)$ provided z_0 lies in R_1 and $V(z_0) = [\min V(z),\ z \text{ in } R_1]$, and provided for each z in R_1 a path exists in R_1 joining z_0 with z on which $V(z)$ varies monotonically; we assume these paths of uniformly bounded lengths. Some condition here on z_0 is essential, as may be seen from the example $f_n(z) \equiv nz^{n-1}$, $z_0 = 1$, $V(z) \equiv \log |z|$, R: $|z - 1| < \frac{1}{2}$; the integrals are the functions $z^n - 1$. Under the general conditions laid down, an exact harmonic majorant for $[f_n(z)]^{1/n}$ in R is an exact harmonic majorant in R_1 for the functions (55).

One further remark seems appropriate. Let the hypothesis of the first part of Theorem 5 be satisfied, and let the sequence $n_1 < n_2 < n_3 \cdots$ have the property that $\lim_{k\to\infty} n_k/k = \tau$ exists. We necessarily have $\tau \geq 1$. If we set $\psi_k(z) \equiv f_{n_k}(z)$, what can be said of the harmonic majorant (exact or not) of the sequence $[\psi_k(z)]^{1/k}$? Merely by writing

$$\begin{aligned}\psi_{k+1}(z) - \psi_k(z) \equiv{} & [f_{n_k+1}(z) - f_{n_k}(z)] + [f_{n_k+2}(z) - f_{n_k+1}(z)] \\ & + \cdots + [f_{n_{k+1}}(z) - f_{n_{k+1}-1}(z)],\end{aligned}$$

and by using obvious algebraic inequalities on geometric sums, we observe that the sequence $[\psi_{k+1}(z) - \psi_k(z)]^{1/k}$ has the harmonic majorant $\tau V(z)$ in both R_0 and $R - R_0 - C_0$, hence by Theorem 11 that sequence has the harmonic majorant $\tau V(z)$ in R. It may occur, as in the case of Taylor's series, that a singularity on C_0 of the limit function of the sequence $f_n(z)$ (or other information) enables us to conclude that $\tau V(z)$ is an *exact* harmonic majorant for the sequence $[\psi_{k+1}(z) - \psi_k(z)]^{1/k}$ in R. Under these conditions the discussion of §6 yields many new results on the zeros of the sequence $\psi_k(z)$, results related to the theorems of Jentzsch, Szegö, Ostrowski, and Bourion for Taylor's series.

Bibliography

1. G. Bourion, *L'Ultraconvergence dans les séries de Taylor*, Actualités scientifiques et industrielles, no. 472, Paris, 1937.
2. G. Bourion, *Recherches sur l'ultraconvergence*, Annales Scientifiques de l'École Normale Supérieure, (3), vol. 50(1933), pp. 245–318.
3. G. Bourion, *Sur les zéros des polynomes-sections d'une série de Taylor*, Compositio Mathematica, vol. 1(1934), pp. 163–176.

4. G. C. Evans, *On potentials of positive mass*. II, Transactions of the American Mathematical Society, vol. 38(1935), pp. 201–236.
5. O. D. Kellogg, *Potential Theory*, Berlin, 1929.
6. P. Korovkin, *Expression asymptotique des polynômes orthogonaux sur un contour rectifiable*, Comptes Rendus de l'Académie des Sciences de L'URSS, vol. 27(1940), pp. 531–534.
7. R. Nevanlinna, *Eindeutige analytische Funktionen*, Berlin, 1936.
8. A. Ostrowski, *On representation of analytical functions by power series*, Journal of the London Mathematical Society, vol. 1(1926), pp. 251–263.
9. A. Ostrowski, *Über vollständige Gebiete gleichmässiger Konvergenz von Folgen analytischer Funktionen*, Abhandlungen aus dem mathematischen Seminar der Hamburgischen Universität, vol. 1(1922), pp. 327–350.
10. F. Vasilesco, *La notion de capacité*, Actualités scientifiques et industrielles, no. 571, Paris, 1937.
11. J. L. Walsh, *Interpolation and approximation by rational functions in the complex domain*, vol. 20, American Mathematical Society, Colloquium Publications, New York, 1935.
12. J. L. Walsh, *Note on the degree of convergence of sequences of analytic functions*, Transactions of the American Mathematical Society, vol. 47(1940), pp. 293–304.
13. J. L. Walsh, *On interpolation by functions analytic and bounded in a given region*, Transactions of the American Mathematical Society, vol. 46(1939), pp. 46–65.
14. J. L. Walsh, *On the degree of convergence of sequences of rational functions*, Transactions of the American Mathematical Society, vol. 47(1940), pp. 254–292.

Harvard University.

*Padé Approximants as Limits of Rational Functions of Best Approximation**

J. L. WALSH

Communicated by D. GILBARG

We shall call a rational function *of type* (n, ν) provided it can be written in the form

$$\frac{s_0 + s_1 z + \cdots + s_n z^n}{t_0 + t_1 z + \cdots + t_\nu z^\nu}, \qquad \sum |t_k| \neq 0.$$

In a thesis [1] that has become classical, Padé studied for each pair (n, ν) the rational function $P_{n\nu}(z)$ of type (n, ν) which has contact of highest order at the origin with a given analytic function

$$(1) \qquad f(z) \equiv a_0 + a_1 z + a_2 z^2 + \cdots, \qquad a_0 \neq 0.$$

Later Montessus de Ballore [2] studied the convergence of the $P_{n\nu}(z)$ and showed under suitable mild conditions that if $f(z)$ is a function of z meromorphic with precisely ν poles in the disk $D_0 : |z| < \rho \ (\leqq \infty)$ and if D denotes that disk with the ν poles deleted, then the sequence $P_{n\nu}(z)$ of Padé functions approaches $(n \to \infty)$ $f(z)$ uniformly on any closed bounded set in D, and the finite poles of $P_{n\nu}(z)$ approach respectively these ν poles of $f(z)$. With this same condition on $f(z)$, the present writer has shown [3] (as a special case of a much more general result) that if $f(z)$ is also analytic in the closed disk δ: $|z| \leqq \epsilon < \rho$, $\epsilon > 0$, then the rational functions $R_{n\nu}(\epsilon, z)$ of type (n, ν) of best Tchebycheff approximation to $f(z)$ on δ converge uniformly $(n \to \infty)$ to $f(z)$ on any closed bounded set in D and the finite poles of $R_{n\nu}(z)$ approach respectively these ν poles of $f(z)$; the method of proof is by study of degree of convergence in δ and D of the sequence $R_{n\nu}(\epsilon, z)$.

The object of the present note is to show that there is more than an analogy between these two sets of rational functions, by showing (i) that under suitable conditions the rational functions $R_{n\nu}(\epsilon, z)$ for fixed n and ν approach $(\epsilon \to 0)$ the Padé function $P_{n\nu}(z)$, and (ii) that for fixed ϵ the geometric degrees of convergence $(n \to \infty)$ to $f(z)$ of the sequences $P_{n\nu}(z)$ and $R_{n\nu}(\epsilon, z)$ in that part of D exterior to δ are the same. In this sense the Padé functions are limit cases

* This research was supported (in part) by the Air Force Office of Scientific Research. Abstract published in *Amer. Math. Soc. Notices*, 11 (1964) 132.

Journal of Mathematics and Mechanics, Vol. 13, No. 2 (1964).

of the $R_{n\nu}(\epsilon, z)$, and the results of Montessus de Ballore are limiting cases of the results [3] on convergence of the rational functions of best approximation.

The results of Montessus de Ballore show that the finite poles of the $P_{n\nu}(z)$ approach ($n \to \infty$) respectively the finite poles of $f(z)$, but nothing seems to be known hitherto of the behavior of the poles of $R_{n\nu}(\epsilon, z)$ for fixed n and ν as $\epsilon \to 0$, not even whether the origin can be a limit point of such poles, and therein lie the difficulties of the proofs of (i) and (ii).

1. We give for reference a brief résumé of Padé's results. With $f(z)$ defined by (1), we need to determine

$$P_{n\nu}(z) \equiv \frac{s_0 + s_1 z + \cdots + s_n z^n}{t_0 + t_1 z + \cdots + t_\nu z^\nu} \equiv \sum_{k=0}^{n+\nu} a_k z^k + \sum_{n+\nu+1}^{\infty} c_k z^k. \tag{2}$$

Padé shows (as can be deduced directly) that the determination of the s_i and t_i is equivalent to the determination of $t_0, t_1, \cdots, t_\nu$ and the d_i, where we set

$$\sum_{j=0}^{\infty} a_j z^j \cdot \sum_{k=0}^{\nu} t_k z^k \equiv \sum_{i=0}^{\infty} d_i z^i, \tag{3}$$

and where $d_{n+1} = d_{n+2} = \cdots = d_{n+\nu} = 0$. This determination is equivalent in turn to the solution for the numbers $t_0, t_1, \cdots, t_\nu$ from the two sets of equations

$$\begin{aligned} a_0 t_0 &= d_0 = s_0, \\ a_1 t_0 + a_0 t_1 &= d_1 = s_1, \\ &\cdots\cdots\cdots\cdots \\ a_n t_0 + a_{n-1} t_1 + \cdots + a_{n-\nu} t_\nu &= d_n = s_n; \end{aligned} \tag{4}$$

$$\begin{aligned} a_{n+1} t_0 + a_n t_1 + \cdots + a_{n-\nu+1} t_\nu &= d_{n+1} = 0, \\ a_{n+2} t_0 + a_{n+1} t_1 + \cdots + a_{n-\nu+2} t_\nu &= d_{n+2} = 0, \\ &\cdots\cdots\cdots\cdots \\ a_{n+\nu} t_0 + a_{n+\nu-1} t_1 + \cdots + a_n t_\nu &= d_{n+\nu} = 0. \end{aligned} \tag{5}$$

Equations (4) and (5) are written for the case $n \geqq \nu$; in the contrary case the numbers a_i with negative subscripts are to be taken as zero.

We treat $R_{n\nu}(\epsilon, z)$ by equations precisely similar to (3), (4), and (5), where $f(z)$ is still given by (1), except that $R_{n\nu}(\epsilon, z)$ of type (n, ν) is now determined by its property of best approximation to $f(z)$ on the disk δ; we have

$$R_{n\nu}(\epsilon, z) \equiv \frac{u_0 + \cdots + u_n z^n}{v_0 + \cdots + v_\nu z^\nu} \equiv \sum_0^\infty b_k z^k, \tag{6}$$

where the coefficients depend on ϵ.

These coefficients $b_0, b_1, \cdots, b_{n+\nu}$ are related to the u_j and v_j by the sets of equations

$$\sum_{j=0}^{\infty} b_j z^j \cdot \sum_{k=0}^{\nu} v_k z^k \equiv \sum_{i=0}^{n} u_i z^i, \tag{7}$$

$$b_0v_0 = u_0\ ,$$

(8)
$$b_1v_0 + b_0v_1 = u_1\ ,$$
$$\cdots\cdots\cdots\cdots\cdots\cdots$$
$$b_nv_0 + b_{n-1}v_1 + \cdots + b_{n-\nu}v_\nu = u_n\ ,$$

$$b_{n+1}v_0 + b_nv_1 + \cdots + b_{n-\nu+1}v_\nu = u_n\ ,$$

(9)
$$b_{n+2}v_0 + b_{n+1}v_1 + \cdots + b_{n-\nu+2}v_\nu = 0,$$
$$\cdots\cdots\cdots\cdots\cdots\cdots\cdots\cdots\cdots$$
$$b_{n+\nu}v_0 + b_{n+\nu-1}v_1 + \cdots + b_nv_\nu = 0;$$

these equations too are written for $n \geqq \nu$, but for $\nu > n$ we consider all b_i with negative subscripts to be zero. Of course equations (9) can be continued indefinitely, but that is not necessary for our present purposes.

2. We are now in a position to establish a result:

Theorem 1. *Let the function $f(z)$ defined by (1) be analytic at $z = 0$, and for $\epsilon(> 0)$ sufficiently small and fixed n and ν let $R_{n\nu}(\epsilon, z)$ denote the function of type $(n\ \nu)$ of best approximation to $f(z)$ in the sense of Tchebycheff on the disk δ: $|z| \leqq \epsilon$. Suppose we have*

(10)
$$\Delta_{n-1,\nu-1} \equiv \begin{vmatrix} a_n & a_{n-1} & \cdots & a_{n-\nu+1} \\ a_{n+1} & a_n & \cdots & a_{n-\nu+2} \\ \cdots & \cdots & \cdots & \cdots \\ a_{n+\nu-1} & a_{n+\nu} & \cdots & a_n \end{vmatrix} \neq 0;$$

then as ϵ approaches zero $R_{n\nu}(\epsilon, z)$ approaches the Padé function $P_{n\nu}(z)$ of (2) uniformly on any closed bounded set containing no zero of $\sum_0^\nu t_kz^k$.

Proof. Both $P_{n\nu}(z)$ and $R_{n\nu}(\epsilon, z)$ are of type (n, ν), so by the extremal property of $R_{n\nu}(\epsilon, z)$ we have

(11)
$$[\max |f(z) - R_{n\nu}(\epsilon, z)|, z \text{ on } \delta] \leqq [\max |f(z) - P_{n\nu}(z)|, z \text{ on } \delta],$$

and since the functions $f(z)$ and $P_{n\nu}(z)$ agree for $n + \nu + 1$ terms of their respective Maclaurin developments, we have for ϵ sufficiently small

$$[\max |f(z) - R_{n\nu}(\epsilon, z)|, z \text{ on } \delta] = O(\epsilon^{n+\nu+1}).$$

We write for the Maclaurin coefficients (where b_k depends on ϵ)

(12)
$$a_k - b_k = \frac{1}{2\pi i}\int_{|z|=\epsilon} \frac{f(z) - R_{n\nu}(\epsilon, z)}{z^{k+1}}\, dz,$$

(13)
$$a_k - b_k = O(\epsilon^{n+\nu-k+1}),$$

so the first member of (13) approaches zero for $k = 0, 1, \cdots, n + \nu$.

The conclusion of Theorem 1 now follows, from the fact that these $n + \nu + 1$ coefficients b_k are "near" the corresponding a_k, the equations (9) and (8) are "near" equations (5) and (4) respectively, and hence their unique solutions s_i, t_i and u_i, v_i are "near". To be more explicit, let us adjoin to the system (9) the equation $v_0 = v$, where v is a multiplicative parameter. We now have $\nu + 1$ equations in the $\nu + 1$ unknowns $v_0, v_1, \cdots, v_\nu$; for ϵ sufficiently small the determinant of the system does not vanish, by (10) and (13). The numbers $v_1, v_2, \cdots, v_\nu$ and $u_0, u_1, \cdots, u_n$ are then uniquely determined by (8) from $b_0, b_1, \cdots, b_{n+\nu}$ in terms of the parameter v. Of course equation (6) determines the u_i and v_i from the b_k merely to within a multiplicative constant; we shall consider such determination as determining the u_i and v_i uniquely. We adjoin similarly the equation $t_0 = v$ to the system (5), so (5) determines $t_0, t_1, \cdots, t_\nu$, and (4) determines the numbers $s_0, s_1, \cdots, s_n$ uniquely in terms of the multiplicative parameter v. The coefficients u_i and v_i in (6) can be made to differ by as small an amount as we please from the corresponding coefficients s_i and t_i in (2), merely by choosing ϵ sufficiently small, and we may choose $v_0 = t_0 = v = 1$; the conclusion of Theorem 1 follows.

3. Corollary. *Let Theorem 1 be modified so that $R_{n\nu}(\epsilon, z)$ now denotes the function of type (n, ν) which has no poles on δ and minimizes for some fixed $p(\geqq 0)$ the pth root of*

$$\int_{|z|=\epsilon} |f(z) - R_{n\nu}(\epsilon, z)|^p \, |dz|;$$

then the conclusion of Theorem 1 persists.

Proof. Here we replace (11) by writing

$$\int_{|z|=\epsilon} |f(z) - R_{n\nu}(\epsilon, z)|^p \, |dz| \leqq \int_{|z|=\epsilon} |f(z) - P_{n\nu}(z)|^p \, |dz|, \tag{14}$$

which compares the extremal norm on $|z| = \epsilon$ of $f(z) - R_{n\nu}(\epsilon, z)$ with that of $f(z) - P_{n\nu}(z)$, a function of type (n, ν) which likewise is analytic on δ for ϵ sufficiently small. The second member of (14) is $O(\epsilon^{(n+\nu+1)p+1})$, so for $p > 1$ the Hölder inequality applied to (12) yields (13), and the proof can be concluded as before.

This proof for $p \geqq 1$ is direct, but does not include $0 < p < 1$; for the latter case a further revision of the proof is necessary; this new proof includes also the case $p > 1$. It is convenient to use the following lemma, a slight sharpening of [4, p. 101, Lemma]:

Lemma. *If the function $P_0(z)$ is analytic and uniformly bounded interior to γ: $|z| = \epsilon$, then the inequality*

$$\int_\gamma |P_0(z)|^p \, |dz| \leqq L^p, \qquad p > 0, \qquad L \geqq 0,$$

implies for z on an arbitrary closed set E_0 interior to γ

$$|P_0(z)| \leqq L'L, \tag{15}$$

where L' may be chosen as $(2\pi h)^{-1/p}$, and h is the distance from γ to E_0.

We apply this lemma to (14), where E_0 is the set $|z| \leqq \epsilon/2$, and use the fact that the second member of (14) is $O(\epsilon^{(n+\nu+1)p+1})$; we obtain

(16) $$|f(z) - R_{n\nu}(\epsilon, z)| = O(\epsilon^{n+\nu+1}), \qquad |z| \leqq \epsilon/2.$$

Equation (12) can be modified by integrating over $|z| = \epsilon/2$ instead of $|z| = \epsilon$, whence (16) yields (13) for $k = 0, 1, \cdots, n + \nu$, and the conclusion of the Corollary follows.

Both Theorem 1 and the Corollary can be broadly generalized:

Theorem 2. *Let $E = E(\epsilon)$ denote a variable point set contained in δ and containing the disk $|z| \leqq \lambda\epsilon$, $0 < \lambda \leqq 1$, where λ is independent of ϵ. Let Theorem 1 be modified so that $R_{n\nu}(\epsilon, z)$ is now the function of type (n, ν) of best approximation to $f(z)$ on E in the sense of Tchebycheff. Then the conclusion of Theorem 1 persists.*

Proof. As before, we deduce (11) modified by taking z on E in both members, and in (12) we integrate over $|z| = \lambda\epsilon$ to deduce (13). The conclusion follows.

The Corollary to Theorem 1 can be generalized in this same spirit, where we require $E(\epsilon)$ to be a Jordan region in δ whose boundary γ is of length not greater than $\Lambda\epsilon$, and which contains the disk $|z| \leqq \lambda\epsilon$, $0 < \lambda \leqq 1$, where Λ and λ are independent of ϵ. A modification of the Lemma already used, proved for instance by use of a conformal map of $E(\epsilon)$ and where γ need no longer be a circle, shows that (15) persists for $|z| \leqq \lambda\epsilon/2$ with $L' = (\pi\lambda\epsilon)^{-1/p}$, and enables us to establish Theorem 2 modified so that $R_{n\nu}(\epsilon, z)$ is the function of type (n, ν) of least p^{th} power approximation to $f(z)$ on γ, $p > 0$.

Theorem 1, previously [5] mentioned in the literature for $\nu = 0$, is related to a discussion by Motzkin and Walsh [6] of the zeros near the origin of $f(z) - R_{n0}(\epsilon, z)$ as ϵ approaches zero, where the rational function is a polynomial. In this connection, it is of interest to note [4, §9.1] that under the conditions of the corollary to Theorem 1 with $p = 2$, the function $f(z) - R_{n\nu}(\epsilon, z)$ has a zero in the origin of order $n + 1$, and a zero in each inverse in the circle $|z| = \epsilon$ of a finite pole of $R_{n\nu}(\epsilon, z)$, so for ϵ sufficiently small that function has at least $n + \nu + 1$ zeros in δ.

With the hypothesis of Theorem 1, its Corollary, or Theorem 2, the function $f(z) - P_{n\nu}(z)$ has a zero of order at least $n + \nu + 1$ in $z = 0$, so it follows from Hurwitz's theorem that as $\epsilon \to 0$ at least $n + \nu + 1$ zeros of $f(z) - R_{n\nu}(\epsilon, z)$ approach $z = 0$.

Theorem 1, its Corollary, and Theorem 2 can be generalized so as to include norms defined by surface integration over δ or $E(\epsilon)$ of $|f(z) - R_{n\nu}(z)|^p$, $p > 0$. In addition, suitable weight functions may be introduced throughout.

4. The significance of inequality (10) in the hypothesis of Theorem 1 should be discussed briefly. Without the condition (10), it may readily occur that equations (5) do not determine the ratios of the t_i uniquely, and hence that $P_{n\nu}(z)$ is not uniquely determined. Without such uniqueness the relation $R_{n\nu}(\epsilon, z) \to P_{n\nu}(z)$ would require a special interpretation and deeper study. However, the assumption (10) in Theorem 1 may be replaced by the requirement of the non-vanishing of any determinant of order ν from the matrix of coefficients a_i in (5).

In Padé's general study of the complete table of fractions $P_{n\nu}(z)$, he derives [1, p. 34] a necessary and sufficient condition that a given fraction $P_{n\nu}(z)$ in its lowest terms should occur but once in that table, namely that all the determinants

$$\Delta_{n\nu}, \quad \Delta_{n-1,\nu-1}, \quad \Delta_{n,\nu-1}, \quad \Delta_{n-1,\nu}$$

should be different from zero. Thus (10) is by no means an unusual condition in the theory of Padé approximation.

Montessus de Ballore in proving [2] his results already mentioned, makes the assumption that the determinant Δ_{jk} shall be different from zero for all values of j and k.

5. Hitherto we have considered primarily the behavior of $R_{n\nu}(\epsilon, z)$ as $\epsilon \to 0$. We now discuss, with the hypothesis chosen by Montessus de Ballore, degree of convergence of the sequence $P_{n\nu}(z)$ as $n \to \infty$ in the disk D_0, for he establishes uniform convergence on every closed set in D_0 containing no pole of $f(z)$ without considering degree of convergence. If the circle $\Gamma_0 : |z| = r_0$ lies in D_0 and contains all the ν poles of $f(z)$ in D_0 in its interior, we write

$$(17) \qquad |f(z) - P_{n\nu}(z)| \leq \epsilon_n, \quad z \text{ on } \Gamma_0, \quad \epsilon_n \to 0.$$

We shall prove

Theorem 3. *Let $f(z)$ satisfy the hypothesis of Theorem 1 and be meromorphic with precisely ν poles in the disk $D_0 : |z| < \rho$ ($\leq \infty$), let $P_{n\nu}(z)$ denote the Padé function in (2), where we suppose $\Delta_{jk} \neq 0$ for all j and k, and let D denote D_0 with the deletion of the ν poles of $f(z)$ interior to D_0. Then for any closed bounded set T in D we have*

$$(18) \qquad \limsup_{n\to\infty} [\max |f(z) - P_{n\nu}(z)|, z \text{ on } T]^{1/n} \leq [\max |z| \text{ on } T]/\rho.$$

Proof. Obvious consequences of (17) are

$$|P_{n+1,\nu}(z) - P_{n\nu}(z)| \leq \epsilon_n + \epsilon_{n+1}, \quad z \text{ on } \Gamma_0,$$

and with the notation

$$(19) \qquad S_{n\nu}(z) \equiv \frac{P_{n+1,\nu}(z) - P_{n\nu}(z)}{z^{n+\nu+1}},$$

$$(20) \qquad |S_{n\nu}(z)| \leq \frac{\epsilon_n + \epsilon_{n+1}}{r_0^{n+\nu+1}}, \quad z \text{ on } \Gamma_0.$$

The function $f(z) - P_{jk}(z)$ has generically a zero of order at least $j + k + 1$ at $z = 0$, so the function $S_{n\nu}(z)$ defined by (19) is a rational function with no pole at $z = 0$ and clearly with no pole at $z = \infty$ even if $\nu = 0$. Thus $S_{n\nu}(z)$ is a rational function of degree $2\nu + 1$ whose only poles in the extended plane lie in the finite poles of $P_{n\nu}(z)$ and of $P_{n+1,\nu}(z)$, poles which approach the ν poles of $f(z)$ in D_0 as $n \to \infty$.

By Cauchy's Integral Formula

$$(21) \qquad S_{n\nu}(z) - S_{n\nu}(\infty) = \frac{1}{2\pi i} \int_{\Gamma_0} \frac{S_{n\nu}(t)}{t - z}\, dt, \qquad z \text{ exterior to } \Gamma_0 ,$$

and by (20) used to estimate both the integral in (21) and $S_{n\nu}(\infty)$, since $S_{n\nu}(z)$ has no poles on or exterior to Γ_0, we have in an arbitrary closed bounded region exterior to Γ_0

$$(22) \qquad |S_{n\nu}(z)| \leqq A(\epsilon_n + \epsilon_{n+1})/r_0^{n+\nu+1}, \qquad n \text{ sufficiently large},$$

where the constant A (independent of n and z) is suitably chosen. By a convenient lemma [3, Theorem 4] it follows whether the second member of (22) approaches zero or not, that (22) remains valid with mere modification of the constant A, on any point set of the extended plane containing no limit point of the poles of $S_{n\nu}(z)$, hence remains valid on T. We choose r_0 ($< \rho$) greater than [max $|z|$ on T], as we may do, whence by (22) for z on T and by (19)

$$\limsup_{n\to\infty} [\max |P_{n+1,\nu}(z) - P_{n\nu}(z)|, z \text{ on } T]^{1/n} \leqq [\max |z| \text{ on } T]/r_0 ;$$

Here we may allow r_0 to approach ρ, and (18) follows.

If $f(z)$, δ, and $R_{n\nu}(\epsilon, z)$ are as in Theorem 3, it is proved in [3] that for fixed ϵ (> 0) and for T in D but not wholly interior to δ (T may coincide with δ) we have

$$(23) \qquad \limsup_{n\to\infty} [\max |f(z) - R_{n\nu}(\epsilon, z)|, z \text{ on } T]^{1/n} \leqq [\max |z| \text{ on } T]/\rho,$$

a comparison of which with (18) shows that indeed (18) is a limiting case of (23).

It is shown in [3] that if ρ is the largest number such that $f(z)$ is meromorphic with precisely ν poles in $|z| < \rho$, and if T is $|z| = r$, $\epsilon \leqq r < \rho$, and passes through no pole of $f(z)$, then the equality sign holds in (23). Under these same conditions, and if T is any continuum (not a single point) in D, the equality sign holds also in (18), as follows by the same methods of [3] taken together with [7, Corollary to Theorem 1].

References

[1] H. Padé, Sur la représentation approchée d'une fonction par des fractions rationnelles. Thèse, Paris, 1892.

[2] R. de Montessus de Ballore, Sur les fractions continues algébriques, *Bull. Soc. Math. de France*, **30** (1902) 28–36.

[3] J. L. Walsh, *The convergence of sequences of rational functions of best approximation*. To appear.

[4] J. L. Walsh, Interpolation and approximation, *Amer. Math. Soc. Colloquium Pubs.*, **20** (1935).

[5] J. L. Walsh, On approximation to an analytic function by rational functions of best approximation, *Math. Zeit.*, **38** (1934) 163–176.
[6] T. S. Motzkin & J. L. Walsh, Zeros of the error function for Tchebycheff approximation in a small region, *Proc. London Math. Soc.*, **13** (3) (1963) 90–98.
[7] J. L. Walsh, Overconvergence, degree of convergence, and zeros of sequences of analytic functions, *Duke Math. Jour.*, **13** (1946) 195–234.

Harvard University

Reprinted from
Henry L. Garabedian (editor)
Approximation of Functions
Elsevier Publishing Company,
Amsterdam 1965
Printed in Northern Ireland

THE CONVERGENCE OF SEQUENCES OF RATIONAL FUNCTIONS OF BEST APPROXIMATION WITH SOME FREE POLES[1]

J. L. WALSH

Harvard University, Cambridge, Massachusetts

There have recently been published a number of papers [**6**, **7**, **8**, **9**] concerning the convergence of sequences of rational functions of best approximation with some poles prescribed and others free, with reference both to degree of approximation and regions of convergence. In the present paper various previous results are brought together and unified, new ones are added, and open problems are particularly emphasized.

In §1 we discuss degree of convergence, in §2 regions of convergence, and in §3 exact degree of convergence. In §4 we treat degree of convergence of poles, in §5 norms involving pth powers, and in §6 properties of the Padé functions. Sharper results on degree of convergence are treated in §7, further sequences of rational functions in §8, and the Γ-function in §9.

1. **Degree of convergence.** Let the function $f(z)$ be continuous on a closed bounded set E of the z-plane, and let j and k be non-negative integers. Let $R_{jk}(z)$ denote the (or a) rational function of type (j, k), namely of form

$$R_{jk}(z) \equiv \frac{a_0 z^j + a_1 z^{j-1} + \cdots + a_j}{b_0 z^k + b_1 z^{k-1} + \cdots + b_k}, \qquad \sum |b_i| \neq 0, \tag{1}$$

of best approximation to $f(z)$ on E; such a rational function exists (if E is dense in itself) but need not be unique [**5**, Chapter 12]. Best approximation here means minimizing the Tchebycheff norm

$$\|f(z) - r_{jk}(z)\| = [\max |f(z) - r_{jk}(z)|, \ z \text{ on } E] \tag{2}$$

of $f(z) - r_{jk}(z)$ for all functions of type (j, k). The $R_{jk}(z)$ thus form a table of double entry [**4**], analogous to that of Padé [**1**], who bases his table of rational functions $P_{jk}(z)$ on a given function $\phi(z) = c_0 + c_1 z + c_2 z^2 + \cdots$ analytic at the origin, and requires that for each pair (j, k) the function $P_{jk}(z)$ of form (1) shall be chosen so that $\phi(z) - P_{jk}(z)$ has a zero of the highest possible order at $z = 0$.

For each (j, k) we set (with $R_{jk}(z)$ extremal)

$$\delta_{jk} = \|f(z) - R_{jk}(z)\|, \tag{3}$$

and there follows the inequality

$$\delta_{mn} \leqq \delta_{jk}, \qquad m \geqq j, \qquad n \geqq k. \tag{4}$$

[1] Research sponsored (in part) by Air Force Office of Scientific Research.

We shall proceed to establish [6]:

THEOREM 1. *Let E be a closed bounded point set whose complement is connected, and is regular in the sense that it possesses a Green's function $G(z)$ with pole at infinity. Let Γ_σ denote generically the locus $G(z) = \log \sigma$ (>0) and E_σ the interior of Γ_σ. Let $f(z)$ be analytic on E, meromorphic with precisely ν poles (that is, poles of total multiplicity ν) in E_ρ, $1 < \rho \leqq \infty$. Then we have*

$$\limsup_{n\to\infty} \delta_{n\nu}^{1/n} \leqq 1/\rho. \tag{5}$$

Theorem 1 is analogous to, and is to be proved by means of, the first part of a theorem on approximation by polynomials [5, §4.6]:

THEOREM 2. *With the notation of Theorem 1, let the function $\phi(z)$ be analytic throughout E_ρ, $1 < \rho \leqq \infty$; then there exist polynomials $p_n(z)$ of respective degrees n, namely of form $a_{nn}z^n + a_{n,n-1}z^{n-1} + \cdots + a_{n0}$, such that for the Tchebycheff norm on E we have*

$$\limsup_{n\to\infty} \|\phi(z) - p_n(z)\|^{1/n} \leqq 1/\rho. \tag{6}$$

Conversely, if $\phi(z)$ is defined on E and if (6) is valid for suitably chosen polynomials $p_n(z)$ of respective degrees n, defined for every n, then the sequence $p_n(z)$ converges uniformly on any closed bounded set in E_ρ, so $\phi(z)$ can be analytically extended from E to be analytic throughout E_ρ.

The first part of Theorem 2 is precisely the case of Theorem 1 for $\nu = 0$.

To prove Theorem 1 we denote by $r_0(z)$ the sum of the principal parts of the poles of $f(z)$ interior to E_ρ, so that $f(z) - r_0(z)$ is analytic interior to E_ρ. By the first part of Theorem 2 we have for suitably chosen polynomials $p_n(z)$ of respective degrees n

$$\limsup_{n\to\infty} \|f(z) - [p_n(z) + r_0(z)]\|^{1/n} \leqq 1/\rho\,.$$

The function $p_n(z) + r_0(z)$ is a rational function $r_{n+\nu,\nu}(z)$ of type $(n + \nu, \nu)$, so we may write

$$\limsup_{n\to\infty} \|f(z) - r_{n+\nu,\nu}(z)\|^{1/(n+\nu)} \leqq 1/\rho\,,$$

and there follows by the extremal properties of the $R_{n+\nu,\nu}(z)$

$$\limsup_{n\to\infty} \|f(z) - R_{n+\nu,\nu}(z)\|^{1/(n+\nu)} \leqq 1/\rho,$$

which is (5).

We mention explicitly that E may consist of several—indeed of an infinite number of—components, and $f(z)$ need not be a monogenic function on E if the locus Γ_ρ separates the plane into more than two regions. In the latter case we consider E_ρ to be the union of the interiors of the Jordan curves composing Γ_ρ.

2. **Regions of convergence.** Regions of convergence of the sequence $R_{n\nu}(z)$ under the hypothesis of Theorem 1 depend more on the degree of convergence than on extremal properties [6, 7]:

THEOREM 3. *Let E and $f(z)$ satisfy the conditions of Theorem 1, and let the $R_{n\nu}(z)$ of respective types (n, ν) but not necessarily extremal satisfy*

$$\limsup_{n\to\infty} \|f(z) - R_{n\nu}(z)\|^{1/n} \leqq 1/\rho. \tag{7}$$

Let D denote E_ρ with the ν poles of $f(z)$ deleted. Then for n sufficiently large the function $R_{n\nu}(z)$ has precisely ν finite poles, which approach respectively the ν poles of $f(z)$ in E_ρ. The functions $R_{n\nu}(z)$ approach $f(z)$ throughout D. For any closed bounded set S in D and in the closed interior of E_σ, $1 < \sigma < \rho$, we have

$$\limsup_{n\to\infty} [\max |f(z) - R_{n\nu}(z)|, z \text{ on } S]^{1/n} \leqq \sigma/\rho. \tag{8}$$

Proof of the second part of Theorem 2 is based primarily on the generalized Bernstein lemma [5, §4.6], that if $P_n(z)$ is a polynomial in z of degree n, then we have

$$|P_n(z)| \leqq \sigma^n \|P_n(z)\|, \qquad z \text{ on } E_\sigma + \Gamma_\sigma. \tag{9}$$

The proof of Theorem 3 is correspondingly based (compare [6] and [7]) on

LEMMA 1. *With the notation and conditions on E of Theorem 1, let rational functions $r_{n\nu}(z)$ of respective types (n, ν) satisfy the inequality*

$$\limsup_{n\to\infty} \|r_{n\nu}(z)\|^{1/n} \leqq 1/\rho_1, \qquad 1 < \rho_1 \leqq \infty, \tag{10}$$

where ν is constant. Suppose the finite poles of the $r_{n\nu}(z)$ are uniformly bounded. Let S be a closed set in the closed interior of E_σ, $1 < \sigma < \rho_1$, and containing no limit point of the poles of the $r_{n\nu}(z)$. Then the sequence $r_{n\nu}(z)$ converges uniformly to zero on S, and we have

$$\limsup_{n\to\infty} [\max |r_{n\nu}(z)|, z \text{ on } S]^{1/n} \leqq \sigma/\rho_1. \tag{11}$$

The $r_{n\nu}(z)$ need not be defined for every n.

The function $r_{n\nu}(z)$ can be written as the quotient of a polynomial $P_n(z)$ of degree n by a polynomial of degree ν having unity as the coefficient of its highest power of z. For n sufficiently large and for z on S, the latter polynomial has a positive lower bound m_1 independent of n, whence

$$|r_{n\nu}(z)| \leqq |P_n(z)|/m_1, \qquad z \text{ on } S. \tag{12}$$

For z on E the denominator polynomial of $r_{n\nu}(z)$ has a bound M_1 independent of n, whence

$$|P_n(z)|/M_1 \leqq |r_{n\nu}(z)|, \qquad z \text{ on } E, \tag{13}$$

$$\|P_n(z)\|/M_1 \leqq \|r_{n\nu}(z)\|. \tag{14}$$

Inequality (9) is useful here in the form

$$[\max |P_n(z)|,\ z \text{ on } S] \leqq \sigma^n \|P_n(z)\|. \tag{15}$$

Inequality (11) now follows by successive application of (12), (15), (14), and (10).

We proceed to establish Theorem 3. Let $f_0(z)$ denote $f(z)$ minus the sum $r_0(z)$ of the principal parts of the ν poles of $f(z)$ in E_ρ, and let $r_n(z)$ denote the sum of the principal parts of the finite poles of $R_{n\nu}(z)$. By Theorem 2 there exist polynomials $p_n(z)$, taken now of respective degrees $n - \nu$, satisfying

$$\limsup_{n\to\infty} \|f_0(z) - p_n(z)\|^{1/n} \leqq 1/\rho. \tag{16}$$

We denote the polynomial $R_{n\nu}(z) - r_n(z)$ by $p_n(z) + q_n(z)$, whence by (7) and (16)

$$\limsup_{n\to\infty} \|r_0(z) - r_n(z) - q_n(z)\|^{1/n} \leqq 1/\rho, \tag{17}$$

an inequality of form (10), where n and ν in (10) are to be replaced by $n + \nu$ and 2ν of Theorem 3. If $R_{n\nu}(z)$ has effectively ν finite poles, it may be written as a rational function $r_n(z)$ of type $(\nu - 1, \nu)$ plus a polynomial $p_n(z) + q_n(z)$ of degree $n - \nu$; if $R_{n\nu}(z)$ has fewer than ν finite poles, the degree of the remaining polynomial is correspondingly increased, but not beyond n. We assume temporarily that the finite poles of the $R_{n\nu}(z)$ are uniformly bounded; this restriction will be removed later.

The function $r_0(z)$ in (17) is independent of n, and we now discuss the finite poles of $r_n(z)$, namely of the function $r_n(z) + q_n(z)$ which approaches $r_0(z)$. If α is a typical pole of $r_0(z)$ interior to E_ρ, we construct $\nu + 1$ mutually disjoint open annuli A_j interior to E_ρ, each with center α, so that each A_j separates α from all zeros of $r_0(z)$ and from all poles of $r_0(z)$ other than α. *If any subsequence of the $r_n(z)$ is chosen, there exists a new subsequence having no limit point of poles in at least one annulus A_j.* If the original subsequence has no limit point of poles in A_1, this conclusion is established; if it has a limit point of poles in A_1, a new subsequence has for each term at least one pole in A_1 and that same limit point of poles. If the new subsequence has no limit point of poles in A_2, the conclusion is established, and if it has a limit point of poles in A_2 we continue the former procedure. We must eventually reach an annulus $A_{\nu+1}$ which for n sufficiently large contains no pole of some subsequence of the original subsequence. If C is a circle which lies in $A_{\nu+1}$ and whose center is α, the subsequence of the $r_n(z) + q_n(z)$ converges uniformly to $r_0(z)$ on C, by Lemma 1. Under such conditions it is readily shown [**6**, Lemma 1] that for n sufficiently large the subsequence of the $r_n(z)$ must have at least as many poles interior to C as does $r_0(z)$. This discussion applies to each of the ν poles α of $r_0(z)$ in E_ρ, so every subsequence of the $r_n(z)$ (which has at most ν finite poles) admits a new subsequence having for n sufficiently large in a suitable neighborhood of each α at least as many poles as does $r_0(z)$, so the new subsequence has in such a neighborhood precisely as many poles as does $r_0(z)$, namely a totality of ν. It follows that the original sequence $r_n(z)$ has this same property, of having for n sufficiently large in the neighborhood of each α precisely as many poles as does $r_0(z)$, and has no other poles. That is to say, for n sufficiently large, $R_{n\nu}(z)$ has precisely ν finite poles which approach respectively the ν poles of $f(z)$ interior to E_ρ. Inequality (8) follows from (17) by Lemma 1 and from (16) = (6) and the corollary to Theorem 2 proved in [5, §4.7]:

$$\limsup_{n\to\infty} [\max|\phi(z) - p_n(z)|, z \text{ on } S]^{1/n} \leqq \sigma/\rho; \tag{18}$$

inequality (18) is indeed a consequence of the generalized Bernstein lemma, and of (6).

To complete the proof of Theorem 3 it remains to be proved that the finite poles of the $R_{n\nu}(z)$ are uniformly bounded. If these poles are not uniformly bounded, as we now suppose, Lemma 1 cannot be applied to the study of the sequence $r_0(z) - r_n(z) - q_n(z)$. There exists a subsequence of the latter functions denoted by $\psi_n(z)$, such that a certain number μ of the finite poles α_j of each $\psi_n(z)$ become infinite as n becomes infinite, while the remaining poles ($2\nu - \mu$ or fewer in number) are uniformly in modulus less than some A, where E_ρ lies in the circle whose center is the origin and radius A. If, say, μ' poles β_j of $\psi_n(z)$ are in modulus greater than A, we replace $\psi_n(z)$ in the sequence by

$$\phi_n(z) \equiv \psi_n(z)\cdot\prod_{j=1}^{\mu'}\frac{z-\beta_j}{-\beta_j}, \qquad 1 \leqq \mu' \leqq \mu, \tag{19}$$

where the β_j depend on n, and where $\phi_n(z)$ has no more than $2\nu - \mu$ finite poles. This replacement alters neither the limit of $\psi_n(z)$ interior to the circle nor such a relation as (17). Our discussion of the $\psi_n(z)$ as already given, commencing with (17), applies now to the modified sequence, and shows that the modified functions $r_n(z)$ have for n sufficiently large at least ν finite poles approaching the respective poles of $r_0(z)$, which is impossible because the modified $r_n(z)$ have fewer than ν finite poles. This contradiction shows that the finite poles of the $r_n(z)$ are bounded, and completes the proof of Theorem 3.

3. **Exact degree of convergence.** Theorem 3 thus far has dealt merely with an upper bound to the degree of convergence of the $R_{n\nu}(z)$, but more specific results exist:

THEOREM 4. *With the hypothesis of Theorem* 3, *let the* $R_{n\nu}(z)$ *be defined for every* n *and let* ρ *be the largest number such that* $f(z)$ *is meromorphic with precisely* ν *poles in* E_ρ; *then the equality sign holds in* (7) *and* (8), *provided* S *is a locus* Γ_σ *not passing through a pole of* $f(z)$.

Let us suppose the first member of (7) to be $1/\rho_1$ ($<1/\rho$); then for the norm on E we have

$$\limsup_{n\to\infty} \|R_{n+1,\nu}(z) - R_{n\nu}(z)\|^{1/n} \leqq 1/\rho_1,$$

so by Lemma 1 the sequence $R_{n\nu}(z)$ converges uniformly, necessarily to $f(z)$ or its analytic extension, throughout some annular region (or set of annular regions) containing E_ρ in its interior, which contradicts the definition of ρ. We have used here the new hypothesis that the $R_{n\nu}(z)$ are defined for every n.

If $S = \Gamma_\sigma$ contains precisely μ poles of $f(z)$ in its interior, we have $(\Gamma_\sigma)_{\rho/\sigma} = \Gamma_\rho$, so by (8) the hypothesis of Theorem 4 is satisfied where now ρ/σ, $\nu - \mu$, and E_σ take the roles of ρ, ν, and E in Theorem 4; to be sure there is the distinction that both $f(z)$ and $R_{n\nu}(z)$ have μ poles in E_σ, but that is unessential. By the part of Theorem 4 already proved, it now follows that the equality sign holds in (8), so Theorem 4 is established.

In Theorems 2–4 we have not assumed the $R_{n\nu}(z)$ to be rational functions of type (n, ν) of best approximation to $f(z)$ on E, but it is clear from Theorem 1 that they may be so chosen:

THEOREM 5. *If the $R_{n\nu}(z)$ are the rational functions of respective types (n, ν) of best (Tchebycheff) approximation to $f(z)$ on E, the hypotheses of Theorems 2–4 are satisfied; in particular, (7) and (8) follow if the $R_{n\nu}(z)$ are defined for every n and if ρ is the largest number such that E_ρ contains precisely ν poles of $f(z)$.*

We add a further result [7; **10, 11**] without proof here, based in part on results due to Ostrowski:

THEOREM 6. *With the hypothesis of Theorem 4, let S in D be a continuum not a single point, and let S lie in the closed interior of E_σ but not in the closed interior of any $E_{\sigma'}$, $\sigma' < \sigma$. Then we have*

$$\limsup_{n\to\infty} [\max |f(z) - R_{n\nu}(z)|,\ z \text{ on } S]^{1/n} = \sigma/\rho. \tag{20}$$

The sequence $R_{n\nu}(z)$ converges uniformly in no region containing a point of Γ_ρ.

Theorem 5 deals with the convergence of various rows of the table of the $R_{n\nu}(z)$ mentioned in connection with (2), namely the convergence of those rows $R_{n\nu}(z)$ where some E_ρ contains precisely ν poles of $f(z)$. If $f(z)$ is meromorphic in each finite point of the plane; if no more than one pole lies on each E_σ, Theorem 5 applies in turn to each row of the array; but if more than one pole lies on some E_σ, the convergence properties of certain rows of the array are not included in Theorem 5.

It is of some interest to note that if $f(z)$ is given merely continuous on E (not known to satisfy the conditions of Theorem 1), and if the $R_{n\nu}(z)$ (extremal or not) are given for every n satisfying (7) with or without the equality sign, then little is known about the further properties of $f(z)$. To be sure, we can deduce

$$\limsup_{n\to\infty} [\max |R_{n+1,\nu}(z) - R_{n\nu}(z)|,\ z \text{ on } E]^{1/n} \leqq 1/\rho,$$

where the function whose modulus occurs is rational of type $(n + \nu + 1, 2\nu)$, but its poles may conceivably be everywhere dense in the plane. We can deduce that

$f(z)$ is not meromorphic with precisely μ poles in E_ρ, $\nu < \mu < \infty$, for then (7) could be used as in the proof of Theorem 3 to show that each $R_{n\nu}$ for n sufficiently large has at least μ finite poles, approaching the μ poles of $f(z)$ in E_ρ. We cannot disprove the possibility that $f(z)$ may be meromorphic with fewer than ν poles in E_ρ; compare [7, Theorem 5].

4. **Degree of convergence of poles.** Since we have considered (§§1–3) degree of convergence of functions $R_{n\nu}(z)$, it is of some interest to study the degree of convergence of the ν finite poles of the $R_{n\nu}(z)$ to the poles of $f(z)$ interior to Γ_ρ, a subject not treated in any of the references yet highly important in numerical analysis. Indeed, the results to be proved apply in far more general circumstances than §§1–3.

THEOREM 7. *Let the function $f(z)$ have a simple pole in the point $z = \alpha$, but otherwise be analytic and different from zero in the closed disk δ: $|z - \alpha| \leqq 3b/2$, $b > 0$, let the function $f_n(z)$ have a simple pole α_n interior to δ but otherwise be analytic and different from zero in δ, and suppose*

$$(21) \qquad |f(z) - f_n(z)| \leqq \varepsilon_n \quad \text{in} \quad \delta_1\colon b/2 \leqq |z - \alpha| \leqq 3b/2,$$

where $\varepsilon_n \to 0$. Then we have

$$(22) \qquad \alpha - \alpha_n = O(\varepsilon_n).$$

We write for z on γ: $|z - \alpha| = b$

$$f'(z) - f_n'(z) \equiv \frac{1}{2\pi i}\int \frac{f(t) - f_n(t)}{(t - z)^2}\, dt,$$

where the integral is taken over the boundary of δ_1; there follows

$$(23) \qquad [\max |f'(z) - f_n'(z)|,\ z \text{ on } \gamma] = O(\varepsilon_n).$$

We have further

$$(24) \qquad \alpha_n - \alpha = \frac{1}{2\pi i}\int_\gamma \frac{zf'(z)f_n(z) - zf(z)f_n'(z)}{f(z)f_n(z)}\, dz,$$

where the numerator in the integrand can be written $zf'(f_n - f) + zf(f' - f_n')$. The relation (22) now follows by (21) and (23).

We continue by studying the slightly more general situation where $f(z)$ has a pole of order k in $\alpha = 0$ but is otherwise analytic and different from zero in δ, and where $f_n(z)$ has a totality of k poles $\alpha_1, \alpha_2, \cdots, \alpha_k$ (depending on n) interior to δ but is otherwise analytic and different from zero in δ, and where (21) is valid. The relation (23) follows as before, and we add to (24) the equations

$$(25) \qquad -k\alpha^m + \sum_{j=1}^{k} \alpha_j^m = \frac{1}{2\pi i}\int_\gamma \frac{z^m f'(z)f_n(z) - z^m f(z)f_n'(z)}{f(z)f_n(z)}\, dz, \qquad m \leqq k,$$

where for convenience we take $\alpha = 0$. For fixed n the α_j are the zeros of a polynomial $Q(z) \equiv z^k + p_1 z^{k-1} + \cdots + p_k$, where we have $s_m = \sum_{j=1}^{k} \alpha_j^m$,

$$\begin{aligned} &s_1 + p_1 = 0, \\ &s_2 + p_1 s_1 + 2p_2 = 0, \\ &\cdots\cdots\cdots, \\ &s_k + p_1 s_{k-1} + \cdots + p_{k-1}s_1 + kp_k = 0. \end{aligned} \tag{26}$$

By (25) we have $s_m = O(\varepsilon_n)$, and by (26) we then have

$$p_m = O(\varepsilon_n), \qquad m \leqq k. \tag{27}$$

By a theorem due to R. D. Carmichael **[14]** an upper bound to the moduli of the zeros of $Q(z)$ is $\sum_{j=1}^{k} |p_j|^{1/j} = O(\varepsilon_n^{1/k})$, so we have established

$$\alpha - \alpha_j = O(\varepsilon_n^{1/k}), \tag{28}$$

of which (22) is a special case.

Inequality (28) is sharp, when considered to be a consequence of (21), as is shown by the example $f(z) \equiv z^{-k}$, $f_n(z) \equiv (z^k - \varepsilon_n)^{-1}$, and on the unit circumference γ we have

$$f(z) - f_n(z) \equiv \frac{-\varepsilon_n}{z^k(z^k - \varepsilon_n)} \sim \varepsilon_n.$$

The poles of $f_n(z)$ have the common modulus $\varepsilon_n^{1/k}$, whence for each j

$$|\alpha - \alpha_j| \sim \varepsilon_n^{1/k}.$$

Our conclusions (22) and (28) clearly apply with the hypothesis of Theorem 6, for we may choose the set S of Theorem 6 as the annulus δ_1 of Theorem 7, with σ appropriately chosen. It is to be noticed, however, that Theorem 6 gives an exact degree of convergence of the $R_{n\nu}(z)$, while (22) and (28) give merely upper bounds on $\alpha - \alpha_j$. With the hypothesis (7) and using the equality sign or not, we may have $\alpha_j = \alpha$ for all j and n, as in the proof of Theorem 1, so a reverse inequality to (22) and (28) does not exist.

These general results on degree of convergence of poles might well be compared with a result on degree of convergence of the principal parts of the poles in Theorem 3:

THEOREM 8. *With the hypothesis of Theorem* 3, *let* $R_0(z)$ *and* $R_n(z)$ *denote respectively the principal parts of a pole* α *of* $f(z)$ *in* E_ρ *and of the totality of poles of* $R_{n\nu}(z)$ *which approach* α. *Then we have*

$$\limsup_{n\to\infty} [\max |R_0(z) - R_n(z)|, z \text{ on } T]^{1/n} \leqq \sigma/\rho,$$

where T *is any closed bounded set not containing* α *and* α *lies on* Γ_σ.

Let γ be a circle exterior to T with center α and lying in E_λ, $\sigma < \lambda < \rho$, containing on or within it no pole of $f(z)$ other than α. For z in T we have for n sufficiently large

$$R_0(z) - R_n(z) \equiv \frac{1}{2\pi i}\int_\gamma \frac{R_0(t) - R_n(t)}{t - z}\,dt \equiv \frac{1}{2\pi i}\int_\gamma \frac{r_0(t) - r_n(t) - q_n(t)}{t - z}\,dt.$$

By (17) and Lemma 1 we have

$$\limsup_{n\to\infty}\,[\max |r_0(z) - r_n(z) - q_n(z)|,\ z \text{ on } \gamma]^{1/n} \leqq \lambda/\rho,$$

so the conclusion of Theorem 8 follows with the second member replaced by λ/ρ, and we can allow λ to approach σ.

It may be noted that here as in Theorem 7 we have merely an upper bound on degree of convergence; this is the most that one can expect from the hypothesis of degree of convergence of the $R_{n\nu}(z)$, for in the auxiliary functions used to prove Theorem 1 and to obtain the degree of convergence of the $R_{n\nu}(z)$ we have $R_n(z) \equiv R_0(z)$.

Still with the hypothesis of Theorem 3, it seems worth while to consider the degree of convergence of $q_n(z)$ to zero. Let $\alpha_1, \alpha_2, \cdots, \alpha_\nu$ be the poles of $f(z)$ in E_ρ, where $\Phi(\alpha_1) \leqq \Phi(\alpha_2) \leqq \cdots \leqq \Phi(\alpha_\nu)$, $\Phi(z) \equiv \exp G(z)$ in the notation of Theorem 1. Suppose S is a closed bounded set in D; if S is contained in E we set $\sigma_0 = 1$, otherwise $\sigma_0 = [\max \Phi(\alpha_\nu), z \text{ on } S]$. Choose circles $\gamma_1, \gamma_2, \cdots, \gamma_\nu$ (mutually exterior when distinct) whose centers are the respective α_k, which lie exterior to S but in E_σ where $\sigma > \sigma_0$, $\rho > \sigma > \Phi(\alpha_\nu)$; for n sufficiently large and z on S we integrate over $\Gamma_\sigma + \gamma_1 + \gamma_2 + \cdots + \gamma_\nu$:

$$r_0(z) - r_n(z) - q_n(z) \equiv \frac{1}{2\pi i}\int \frac{r_0(t) - r_n(t) - q_n(t)}{t - z}\,dt.$$

By Lemma 1 and (17) we have

$$\limsup_{n\to\infty}\,[\max |r_0(z) - r_n(z) - q_n(z)|,\ z \text{ on } S]^{1/n} \leqq \sigma/\rho,$$

and by allowing σ to approach $\sigma_1 = \max\,[\sigma_0, \Phi(\alpha_\nu)]$,

$$\limsup_{n\to\infty}\,[\max |r_0(z) - r_n(z) - q_n(z)|,\ z \text{ on } S]^{1/n} \leqq \sigma_1/\rho.$$

Combination of this inequality with Theorem 8 applied to each pole α_k yields the desired result:

$$\limsup_{n\to\infty}\,[\max |q_n(z)|,\ z \text{ on } S]^{1/n} \leqq \sigma_1/\rho.$$

If S here lies in the closed interior of the locus $\Phi(z) = \Phi(\alpha_\nu)$, and in particular if $S = E$, we have $\sigma_1 = \Phi(\alpha_\nu)$.

5. **Norms involving *p*th powers.** Hitherto we have considered only the Tchebycheff norm as a measure of approximation of $R_{n\nu}(z)$ to $f(z)$ on E, but it is clear that

other norms may be used in defining the table of functions $R_{n\nu}(z)$ and the norms $\delta_{n\nu}$. For instance, suppose E is a closed rectifiable Jordan arc C or a closed Jordan region bounded by a rectifiable curve C and $f(z)$ is continuous on E; then

$$\delta_{n\nu}^{(p)} = \left[\int_C |f(z) - R_{n\nu}(z)|^p \, |dz|\right]^{1/p}, \qquad p > 0, \tag{29}$$

is a suitable norm to measure the deviation of $R_{n\nu}(z)$ from $f(z)$ on C. Use of $\delta_{n\nu}^{(p)}$ as norm leads to a new double-entry table of extremal functions $R_{n\nu}(z)$ of respective types (n, ν) and a new table of deviations. With the notation (2) and (3) it is clear from (29) that we have for all n and ν

$$\delta_{n\nu}^{(p)} \leqq l^{1/p} \cdot \delta_{n\nu}$$

where l is the length of C, so with E as just described and with the hypothesis of Theorem 1 on $f(z)$ there follows the analogue of (5):

$$\limsup_{n\to\infty} \delta_{n\nu}^{(p)1/n} \leqq 1/\rho. \tag{30}$$

Under these same conditions on E and $f(z)$, the analogue of Theorem 3 is valid with the hypothesis (30) instead of (7). The analogue of Lemma 1 is true if the norm in (29) is used in (10), thanks to a previously known lemma concerning the norm of (29) as used in approximation by polynomials, and (7) in its present form follows [7, Corollary to Theorem 1] for the $R_{n\nu}(z)$ minimizing the pth power norm, as do the consequences of (7) already set forth.

If E consists of a finite number of mutually exterior closed rectifiable Jordan arcs or closed Jordan regions each bounded by a rectifiable Jordan curve, or of the union of a finite number of mutually disjoint such arcs and curves, we may use as norm on E the sum of the norms on the respective components. It is still true [7, Theorem 8] that (7) for the Tchebycheff norm on E follows for the $R_{n\nu}(z)$ minimizing the pth power norm, with numerous consequences.

If E consists of one or several mutually exterior Jordan regions, a suitable norm is (29) where the integral over C is now replaced by the double integral over E. If $f(z)$ satisfies the conditions of Theorem 1, again by a suitable modification of Lemma 1, inequality (7) for the new norm and for the Tchebycheff norm on E follows [7]; there are again numerous consequences.

6. **Properties of Padé functions.** The Padé functions $P_{jk}(z)$ defined in §1 have been widely studied as approximating functions, especially in their relation to continued fractions. An important result, proved by Montessus de Ballore [2] by use of Hadamard's theory of the polar singularities of a function represented by Taylor's series, is closely related to that part of Theorem 3 which describes the convergence of $R_{n\nu}(z)$ (but not degree of convergence) and its poles, and indeed gave the impetus for the proof of Theorem 3:

THEOREM 9. (Montessus de Ballore) *Let the function* $f(z) \equiv c_0 + c_1 z + c_2 z^2 + \cdots$ *be analytic in* $z = 0$, *and meromorphic with precisely* ν *poles in* $|z| < \rho$. *Let the*

Padé table be normal, in the sense that certain determinants formed from the c_k *are different from zero so that the* P_{jk} *are uniquely determined. Then the sequence* $P_{n\nu}(z)$ *which forms the* $(\nu + 1)$*st row of the Padé table converges to* $f(z)$ *throughout the region D formed by* $|z| < \rho$ *with the poles of* $f(z)$ *deleted, uniformly on any closed bounded set in D. The* ν *finite poles of* $P_{n\nu}(z)$ *approach respectively the* ν *poles of* $f(z)$ *in* $|z| < \rho$.

Of course the methods used by Montessus de Ballore are so closely identified with the use of the Taylor series that they are not useful in the proof of Theorem 3, even when D is a circular disk. Nevertheless there is far more than an analogy between Theorem 9 and Theorem 3, for the $P_{jk}(z)$ are the limits of $R_{jk}(z)$, extremal on E, as the suitably chosen point set E tends to 0:

THEOREM 10. *Let* $f(z)$ *satisfy the conditions of Theorem 9, and let* $R_{jk}(\varepsilon, z)$ *be the function of type* (j, k) *of best Tchebycheff approximation to* $f(z)$ *on the set* E: $|z| \leqq \varepsilon$ (>0). *Then we have uniformly on any closed bounded set containing no pole of* $P_{jk}(z)$

$$\lim_{\varepsilon \to 0} R_{jk}(\varepsilon, z) = P_{jk}(z).$$

Theorem 10 can be proved **[9]** by studying explicitly the formulas for the determination of $P_{jk}(z)$ and $R_{jk}(\varepsilon, z)$ in terms of the Taylor coefficients of $f(z)$.

Montessus de Ballore did not consider degree of convergence of the $P_{jk}(z)$ to $f(z)$, but on this topic the analogy with Theorems 3 and 4 is still striking **[9]**:

THEOREM 11. *Under the conditions of Theorem 9 let* ρ *be the largest number such that* $f(z)$ *is meromorphic with precisely* ν *poles interior to* $|z| < \rho$. *If S is any closed bounded continuum (not a single point) in D, and if* $\sigma = [\max |z|, z$ *on* $S]$, *then we have*

$$\limsup_{n \to \infty} [\max |f(z) - P_{n\nu}(z)|, z \text{ on } S]^{1/n} = \sigma/\rho. \tag{31}$$

Of course (31) is well known in the case that $f(z)$ has no poles in $|z| < \rho$, namely the case that $\nu = 0$ and $P_{n0}(z)$ is a section of the Taylor development of $f(z)$.

7. **Sharper results on degree of convergence.** There exist various refinements of such degrees of convergence as are indicated by (5), (7), and (8), which have not hitherto received attention in the literature, and which we shall consider very briefly.

Let E be the closed interior of an analytic Jordan curve, and let $f(z)$ be analytic on E, meromorphic with precisely ν poles on E_ρ, and **[13]** on Γ_ρ of class $L(k, \alpha)$ with $0 < \alpha < 1$ or of class $Z(k)$ with $\alpha = 1$, where k may be negative. We set $f(z) \equiv f_0(z) + r_0(z)$, z in E_ρ, where $r_0(z)$ is the sum of the principal parts of the ν poles of $f(z)$ in E_ρ, and $f_0(z)$ is analytic in E_ρ and on Γ_ρ of class $L(k, \alpha)$ with $0 < \alpha < 1$ or of class $Z(k)$ with $\alpha = 1$. There exist **[13]** polynomials $p_n(z)$ of respective degrees n such that we have

$$|f_0(z) - p_n(z)| \leqq \frac{A}{\rho^n n^{k+\alpha}}, \qquad z \text{ on } E; \tag{32}$$

here and below the letter A with or without a subscript denotes a constant independent of n and z, not necessarily the same constant with repeated occurrences. We can write (32) as

$$|f(z) - r_{n+\nu,\nu}(z)| \leqq \frac{A}{\rho^n n^{k+\alpha}}, \qquad z \text{ on } E,$$

with $r_{n+\nu,\nu}(z) \equiv r_0(z) + p_n(z)$, a function of type $(n + \nu, \nu)$, or we can write

$$|f(z) - r_{n\nu}(z)| \leqq \frac{A_1}{\rho^n n^{k+\alpha}}, \qquad z \text{ on } E. \tag{33}$$

That is to say, we have established the existence of the $r_{n\nu}(z)$ of type (n, ν) satisfying (33). The functions $R_{n\nu}(z)$ of best Tchebycheff approximation to $f(z)$ on E satisfy (33) a fortiori. Henceforth we suppose the $r_{n\nu}(z)$ of type (n, ν) to satisfy (33), whether extremal or not. It follows by Theorem 3 that as $n \to \infty$ the poles of $r_{n\nu}(z)$ approach those of $f(z)$ in E_ρ.

Let the $r_{n\nu}(z)$ satisfying (33) be given, and let $r_n(z)$ denote the sum of the principal parts of the finite poles of $r_{n\nu}(z)$, where $f(z)$ is as before. With (33) we combine (32), where $p_n(z)$ is now a polynomial of degree $n - \nu$, and obtain

$$|r_0(z) - r_n(z) - q_n(z)| \leqq \frac{A_1}{\rho^n n^{k+\alpha}}, \qquad z \text{ on } E, \tag{34}$$

where $q_n(z)$ is a polynomial of degree $n - \nu$. The function whose modulus appears in (34) is of type $(n + \nu, 2\nu)$, whose finite poles lie in the poles of $r_0(z)$ and in the poles of $r_n(z)$, which approach the poles of $r_0(z)$. By the method of proof of Lemma 1, we have for an arbitrary closed set S containing no pole of $r_0(z)$ and lying in the closed interior of E_σ, $1 < \sigma < \rho$,

$$|r_0(z) - r_n(z) - q_n(z)| \leqq \frac{A_2\sigma^n}{\rho^n n^{k+\alpha}}, \qquad z \text{ on } S. \tag{35}$$

It follows [13] from (32) that we have

$$|f_0(z) - p_n(z)| \leqq \frac{A_3\sigma^n}{\rho^n n^{k+\alpha}}, \qquad z \text{ on } S, \tag{36}$$

so by (35) there follows

$$|f(z) - r_{n\nu}(z)| \leqq \frac{A_4\sigma^n}{\rho^n n^{k+\alpha}}, \qquad z \text{ on } S; \tag{37}$$

it is to be noted that (37) is a consequence merely of (33) and of the hypothesis on $f(z)$; inequality (37) is considerably sharper than (8).

For this same $f(z)$, inequality (37) holds for the same functions $r_{n\nu}(z)$ on Γ_ρ, provided we have $k + \alpha > 0$, for in this case (35) holds with $\sigma = \rho$ on Γ_ρ and (36) holds with $\sigma = \rho$ if the $p_n(z)$ are suitably chosen, so we have

$$|f(z) - r_{n\nu}(z)| \leqq \frac{A_5}{n^{k+\alpha}}, \qquad z \text{ on } \Gamma_\rho. \tag{38}$$

This result is valid and seems to be new even for the case $\nu = 0$.

However, if (33) is given for functions $r_{n\nu}(z)$ defined for every n without knowledge of the properties of $f(z)$ except its continuity on E, we deduce

$$|r_{n+1,\nu}(z) - r_{n\nu}(z)| \leqq \frac{A}{\rho^n n^{k+\alpha}}, \qquad z \text{ on } E,$$

so by the method of proof of Lemma 1 we have merely

$$|r_{n+1,\nu}(z) - r_{n\nu}(z)| \leqq \frac{A_1\sigma^n}{\rho^n n^{k+\alpha}}, \qquad z \text{ on } S;$$

where S lies in the closed interior of E_σ, $1 < \sigma < \rho$, and contains no limit point of the poles of the $r_{n\nu}(z)$; if the $r_{n\nu}(z)$ have finite poles that are not bounded, we proceed as in the proof of Theorem 3. Conceivably these poles are everywhere dense interior to E_ρ, so our knowledge of the properties of $f(z)$ remains slight. But if now we assume also that $f(z)$ is meromorphic with precisely ν poles in E_ρ, our previous discussion shows that the poles of $r_{n\nu}(z)$ approach those of $f(z)$. We then have

$$|r_{n+1,\nu}(z) - r_{n\nu}(z)| \leqq \frac{A_1}{n^{k+\alpha}}, \qquad z \text{ on } \Gamma_\rho,$$

whence it follows [12] that $f(z)$ is of class $L(k - 1, \alpha)$ on Γ_ρ provided $k + \alpha > 1$ and provided Γ_ρ has no multiple points.

Further results analogous to (33), (37), and (38) can be established, as the writer plans to show on another occasion.

8. **Further sequences of rational functions.** The reader will have noticed that under certain conditions the degree of approximation to a given function $f(z)$ on a set E by rational functions may be readily obtainable, whereas the location of free poles and regions of convergence may be far less accessible; indeed, the free poles may lie everywhere dense in the entire plane or a portion of it. This observation may be carried further, as we shall now indicate.

In the preceding theorems we have considered convergence of rows of the table of functions $R_{n\nu}(z)$ of best approximation; ν is fixed while n becomes infinite. Both degree of approximation and regions of convergence depend heavily on the location of the poles of $f(z)$, where we suppose $f(z)$ to be analytic on E, meromorphic in a region containing E. We proceed now to consider convergence of columns of the table $R_{\nu n}(z)$ of functions of best approximation; ν is fixed while n becomes infinite. Here both degree of approximation and regions of convergence depend essentially on the location of the zeros of $f(z)$, where we suppose $f(z)$ to be analytic on E, meromorphic in a region containing E. The following theorem has been established [8]:

THEOREM 12. *Let E and E_σ be as in Theorem 1. Let the function $F(z)$ be analytic and different from zero on E, meromorphic with precisely ν zeros interior to E_ρ,*

$1 < \rho \leqq \infty$. *Let the rational functions* $R_{\nu n}(z)$ *of respective types* (ν, n) *satisfy for the Tchebycheff norm on* E

(39) $$\limsup_{n\to\infty} \|F(z) - R_{\nu n}(z)\|^{1/n} \leqq 1/\rho.$$

Then the poles of the $R_{\nu n}(z)$ *interior to* E_ρ *approach either* E_ρ *or the respective poles of* $F(z)$ *interior to* E_ρ, *and every pole of* $F(z)$ *interior to* E_ρ *is approached by precisely an equal multiplicity of poles of* $R_{\nu n}(z)$. *If* D *denotes the interior of* E_ρ *with the poles of* $F(z)$ *deleted, the functions* $R_{\nu n}(z)$ *approach* $F(z)$ *throughout* D, *and for any closed bounded set* S *in* D *and in the closed interior of* E_σ *we have*

(40) $$\limsup_{n\to\infty} [\max |F(z) - R_{\nu n}(z)|, z \text{ on } S]^{1/n} \leqq \sigma/\rho.$$

If the $R_{\nu n}(z)$ *are the rational functions of type* (ν, n) *of best Tchebycheff approximation to* $F(z)$ *on* E, *then* (39) *is satisfied.*

Thus far the $R_{\nu n}(z)$ *need not be defined for every* n, *but below they shall be so defined.*

Whether the $R_{\nu n}(z)$ *are extremal or not, let* ρ *be the largest number such that* $F(z)$ *is meromorphic with precisely* ν *zeros interior to* E_ρ, $1 < \rho \leqq \infty$. *If* (39) *holds, then* (39) *holds with the equality sign, as does* (40) *with* $S = E_\sigma$, *provided* $1 < \sigma < \rho$, *and provided no pole of* $F(z)$ *lies on* E_σ.

Theorem 12 may be regarded as the dual of previous theorems, where the roles of zeros and poles of both $f(z)$ and the approximating functions are interchanged, that is, where $f(z)$ is replaced by its reciprocal; this not to imply that the extremals $R_{n\nu}(z)$ and $R_{\nu n}(z)$ are reciprocals of each other, but nevertheless in suitable comparison sequences $(n \to \infty)$ rational functions of types (n, ν) and (ν, n) approximating $f(z)$ and $1/f(z)$ respectively may be chosen as mutually reciprocal.

The treatment given in [8], like Theorem 12, is restricted to use of the Tchebycheff norm. Other norms may be treated at once provided E consists of a finite number of mutually disjoint closed Jordan regions, by the method used in [6].

Theorem 12 is clearly complementary to Theorems 1, 3, 4, and 5. Together these theorems give considerable information concerning the convergence of rows and columns of the array $R_{jk}(z)$ of rational functions of best approximation. These theorems apply also to the degree of convergence of diagonal (i.e., not horizontal or vertical) sequences formed from that array. As an illustration, suppose $f(z)$ to be meromorphic for all finite values of z, and suppose the hypothesis of Theorem 1 fulfilled for the pairs of parameters of the sequence $\nu = \nu_1, \nu_2, \nu_3, \cdots \to \infty$; $\rho = \rho_1, \rho_2, \rho_3, \cdots \to \infty$. With the notation (3) we have by (4) for n sufficiently large and for fixed ν_j

(41) $$\lim_{n\to\infty} \delta_{nn}^{1/n} = 0, \qquad \delta_{nn} \leqq \delta_{n\nu_j}, \qquad \limsup_{n\to\infty} \delta_{nn}^{1/n} \leqq \limsup_{n\to\infty} \delta_{n\nu_j}^{1/n} \leqq 1/\rho_j.$$

Inequality (41) follows also by a similar consideration of the functions $R_{\nu n}(z)$ if $f(z)$ has no zeros on E.

If $f(z)$ is no longer meromorphic for all finite values of z, but is meromorphic at every point interior to E_{ρ_0}, and if $f(z)$ has infinitely many poles interior to E_{ρ_0}, we can consider appropriate pairs of parameters of the sequences $\nu = \nu_1, \nu_2, \cdots \to \infty$; $\rho = \rho_1, \rho_2, \cdots \to \rho_0$, so chosen that for each pair the hypothesis of Theorem 1 is fulfilled. There follows by the method just used

$$\limsup_{n\to\infty} \delta_{nn}^{1/n} \leqq 1/\rho_0.$$

Further results on the degree of convergence of the $R_{nn}(z)$ are established in [3], namely (41) remains valid if $f(z)$ is analytic on E and if the singularities of $f(z)$ in the extended plane form a reducible set, that is to say, a set one of whose derivatives is a null set. However, the general investigation of the sequence δ_{nn} for an arbitrary $f(z)$ (studied also in [15]), and more particularly the determination of the regions of convergence of the sequence $R_{nn}(z)$, must be regarded as an open problem.

9. **The Γ-function.** The theory outlined in §§1–8 above and some computational work on rational approximations to the Γ-function [16] by Dr. John R. Rice were commenced and well developed quite independently of each other. Rice chooses E as the segment $2 \leqq z \leqq 3$; the function $w = w(z)$ which maps the plane slit along E onto $|w| > 1$ with $w(\infty) = \infty$ is given by

$$z = \frac{1}{4}\left(w + \frac{1}{w}\right) + \frac{5}{2}, \qquad w = 2z - 5 \pm 2(z^2 - 5z + 6)^{1/2}.$$

Thus the locus E_ρ passes through the point $z(\leqq 0)$ if and only if

$$\rho = 5 - 2z + 2(z^2 - 5z + 6)^{1/2}.$$

The poles of $\Gamma(z)$ lie in the points $0, -1, -2, -3, \cdots$, and the corresponding values of ρ are $5 + 2 \cdot 6^{1/2}$, $7 + 4 \cdot 3^{1/2}$, $9 + 4 \cdot 5^{1/2}$, $11 + 2 \cdot 30^{1/2}, \cdots$. Since the Γ-function is meromorphic at every point of the plane, this is precisely the case considered in §8 with the (greatest) values of ρ just enumerated, and respectively $\nu_1 = 0$, $\nu_2 = 1$, $\nu_3 = 2, \cdots$. Rice has computed $R_{n\nu}(z)$ for suitable values of n and ν, and strangely enough, even for small values of those indices the law (7) with the equality sign can be verified quite exactly; compare [16].

References

1. H. Padé, *Sur la représentation approchée d'une fonction par des fractions rationelles*, Thèse, Paris, 1892.

2. R. de Montessus de Ballore, *Sur les fractions continues algébriques*, Bull. Soc. Math. de France **30** (1902), 28–36.

3. J. L. Walsh, *On the overconvergence of certain sequences of rational functions of best approximation*, Acta Math. **57** (1931), 411–435.

4. ———, *On approximation to an analytic function by rational functions of best approximation*, Math. Zeit. **38** (1934), 163–176.

5. ———, *Interpolation and approximation by rational functions in the complex domain*, Amer. Math. Soc. Coll. Pubs. **20** (1935).

6. ———, *The convergence of sequences of rational functions of best approximation*, Math. Annalen **155** (1964), 252–264.

7. ———, *The convergence of sequences of rational functions of best approximation.* II.[1]

8. ———, *Note on the convergence of approximating rational functions of prescribed type*, Proc. Nat. Acad. Sci. **50** (1963), 791–794.

9. ———, *Padé approximants as limits of rational functions of best approximation*, J. Math. and Mech. **13** (1964), 305–312.

10. ———, *The analogue for maximally convergent polynomials of Jentzsch's theorem*, Duke Math. J. **26** (1959), 605–616.

11. ———, *Overconvergence, degree of convergence, and zeros of sequences of analytic functions*, Duke Math. J. **13** (1946), 195–234.

12. ———, *Note on approximation by bounded analytic functions*, Proc. Nat. Acad. of Sciences **37** (1951), 821–826.

13. J. L. Walsh and H. M. Elliott, *Polynomial approximation to harmonic and analytic functions: generalized continuity conditions*, Trans. Amer. Math. Soc. **68** (1950), 183–203.

14. R. D. Carmichael, *Elementary inequalities for the roots of an algebraic equation*, Bull. Amer. Math. Soc. **24** (1917–18), 286–296.

15. V. Erohin, *On the best approximation of analytic functions by rational functions with free poles*, Doklady Akad. Nauk **128** (1959), 29–32.

16. J. R. Rice, *On the* L_∞ *Walsh Arrays for* $\Gamma(x)$ *and* Erf c(x), Math. Comp. **18** (1964).

[1] To be published in Trans. Amer. Math. Soc.

COLLOQUIUM MATHEMATICUM

XVI DÉDIÉ À M. FRANCISZEK LEJA 1967

*AN EXTENSION OF THE GENERALIZED BERNSTEIN LEMMA**

BY

JOSEPH L. WALSH (CAMBRIDGE, MASS., U. S. A.)

This note is dedicated to Professor Franciszek Leja on the occasion of his 80th anniversary, in recognition of his admirable contributions to the theory of extremal point sets and approximation by polynomials.

The following lemma has shown itself useful [1, 2] in the study of sequences of rational functions in case the finite poles are uniformly bounded. The proposition without that qualification is also true, and its proof is the object of the present note.

We say that a rational function of the form

$$r_{jk}(z) \equiv \frac{a_0 z^j + a_1 z^{j-1} \ldots + a_j}{b_0 z^k + b_1 z^{k-1} \ldots + b_k}, \qquad \sum |b_i| \neq 0,$$

is *of type* (j, k).

LEMMA. *Let E be a closed bounded point set whose complement is connected, and regular in the sense that it possesses a Green's function $G(z)$ with pole at infinity. We denote generically by Γ_σ the locus $G(z) = \log\sigma(>0)$ and its interior by E_σ. Let rational functions $r_{n\nu}(z)$ of respective types (n, ν), where ν is constant, satisfy the inequality*

$$(1) \qquad \limsup_{n\to\infty} [\max |r_{n\nu}(z)|,\ z \text{ on } E]^{1/n} \leqslant 1/\varrho_1, \qquad 1 < \varrho_1 \leqslant \infty.$$

Let S be a closed set in the closed interior of E_σ, $1 < \sigma < \varrho_1$, and containing no limit point of the poles of the $r_{n\nu}(z)$. Then the sequence $r_{n\nu}(z)$ converges uniformly to zero on S, and we have

$$(2) \qquad \limsup_{n\to\infty} [\max |r_{n\nu}(z)|,\ z \text{ on } S]^{1/n} \leqslant \sigma/\varrho_1.$$

The $r_{n\nu}(z)$ need not be defined for every n.

This lemma has already been proved [1, 2] for the case that the finite poles of the $r_{n\nu}(z)$ are uniformly bounded, so we here consider the

* Research sponsored by Air Force Office for Scientific Research. Abstract published in Notices of the American Mathematical Society 12 (1965) p. 714.

contrary case. Choose a sequence of the $r_{n\nu}(z)$ such that

$$\lim_{n\to\infty}[\max|r_{n\nu}(z)|, z \text{ on } S]^{1/n}$$

equals the first member of (2). Choose a subsequence of that sequence such that at least one finite pole of the successive terms becomes infinite, then (if possible) a subsequence of that subsequence such that at least two finite poles of the successive terms become infinite, and so on as long as possible. Eventually we reach a subsequence such that precisely μ finite poles of the terms become infinite, $1 \leqslant \mu \leqslant \nu$, and the remaining $\nu-\mu$ (or fewer) finite poles are uniformly bounded. For the elements $\psi_n(z)$ of that last sequence we denote by β_j (depending on n) the μ poles of the successive $\psi_n(z)$ that become infinite. Let the point set E_σ and the uniformly bounded poles of the sequence lie in some disk $|z| \leqslant A$, and for each sufficiently large n define

$$\varphi_n(z) \equiv \psi_n(z) \prod_{j=1}^{\mu} \frac{z-\beta_j}{-\beta_j}$$

when the corresponding β_j are in modulus greater than A. The rational function $\varphi_n(z)$ is of type $(n, \nu-\mu)$. It is clear that replacement in the sequence $\psi_n(z)$ [$\equiv r_{n\nu}(z)$]) of $\psi_n(z)$ by $\varphi_n(z)$ does not alter the first member of (1) nor of (2). The replacement sequence $\varphi_n(z)$ has all its finite poles uniformly bounded, so the original proof of the original lemma implies (2) for the subsequence, and hence for the $r_{n\nu}(z)$.

The phraseology of this proof assumes that the $\psi_n(z)$ have effectively ν finite poles, but the only necessary modification for the contrary case is the choice of a suitable ν', $0 \leqslant \nu' < \nu$, and of a suitable subsequence of the $\psi_n(z)$ each of type (n, ν') with effectively ν' finite poles; the reasoning as given is then applicable.

With $\nu = 0$, the Lemma is essentially the generalized Bernstein lemma for polynomials ([3], § 4.6) which has many applications to polynomial approximation. The new lemma has many applications to the study of approximation by rational functions, as the writer plans to indicate elsewhere.

REFERENCES

[1] J. L. Walsh, *The convergence of rational functions of best approximation*, Mathematische Annalen 155 (1964), p. 252-264.

[2] — *The convergence of sequences of rational functions of best approximation with some free poles*, Proceedings of General Motors Symposium of Approximation, September 1964, edited by H. L. Garabedian, published by Elsevier (Amsterdam), 1965, p. 1-16.

[3] — *Interpolation and approximation*, American Mathematical Society Colloquium Publications 20 (1935).

Reçu par la Rédaction le 30. 8. 1965

DEGREE OF APPROXIMATION BY RATIONAL FUNCTIONS AND POLYNOMIALS

J. L. Walsh

In a recent paper [1], Newman proved the striking result that the function $|x|$ can be uniformly approximated on the interval $[-1, 1]$ by rational functions of degree n, with an error $O(e^{-\sqrt{n}})$. This represents much more rapid convergence than the error $O(1/n)$ for best approximation by polynomials that is given (for the same function and the same interval) by the theory of Jackson, Bernstein, Montel, and de La Vallée Poussin.

The special rational functions used by Newman have the origin as a limit point of their poles. This can be shown without reference to the precise formulas involved; indeed, we now formulate a general theorem relevant to this topic.

THEOREM. *If the function* $f(z)$ *is approximable on a closed Jordan arc* C *to the order* $n^{-\alpha}$ $(\alpha > 0)$ *by rational functions* $Q_n(z)$ *(of degree* n*) whose poles have no limit point on* C, *then* $f(z)$ *is also approximable on* C *to the order* $n^{-\alpha}$ *by polynomials* $p_n(z)$ *of degree* n.

In Newman's case, this theorem applies to each closed subinterval of $[-1, 1]$ containing the origin in its interior, and it shows that rational functions of respective degrees n having no limit point of poles on the subinterval can not converge to $|x|$ with approximation of order $n^{-\alpha}$ $(\alpha > 1)$, on the subinterval. Thus Newman's rational functions must have at least one limit point of poles on each such subinterval; hence the origin must be a limit point of poles.

The theorem is an immediate consequence of the following two propositions (see [2, Section 9.7, Lemma 1] and [3, Theorem 1]).

LEMMA 1. *Let* C *be a nondegenerate, bounded continuum whose complement* K *is simply connected. Let* $w = \phi(z)$ *map* K *conformally onto the domain* $|w| > 1$, *with* $\phi(\infty) = \infty$, *and for* $R > 1$ *let* $C(R)$ *denote the preimage of the circle* $|w| = R$. *If* $Q(z)$ *is a rational function of degree* n *whose poles lie on the curve* $C(R_0)$ *or in its exterior, for some* $R_0 > 1$, *and if* $|Q(z)| \leq M$ *on* C, *then*

$$|Q(z)| \leq M\left(\frac{R_0 R - 1}{R_0 - R}\right)^n$$

on the curve $C(R)$, *for* $1 < R < R_0$.

LEMMA 2. *Let* $\{\varepsilon_n\}$ *be a sequence of positive numbers such that*

$$\varepsilon_{[\mu n]} = O(\varepsilon_n) \quad \textit{and} \quad r^n = O(\varepsilon_n)$$

for each positive μ *and each* r $(0 < r < 1)$. *Let* D *be a domain,* C *a Jordan arc in* D, $f(z)$ *a function defined on* C, *and* $\{f_n(z)\}$ *a sequence of functions that are analytic in* D. *If*

Revised April 22, 1966.
This research was supported in part by the U. S. Air Force Office of Scientific Research, Air Research and Development Command.

$$|f(z) - f_n(z)| \leq A_1 \varepsilon_n \qquad (z \in C)$$

and

$$|f_n(z)| \leq A_2 M_0^n \qquad (z \in D)$$

for some constants A_1, A_2, M_0 *and for* $n = 1, 2, \cdots$, *then there exist polynomials* $p_n(z)$ *of degree* n $(n = 1, 2, \cdots)$ *such that for some constant* A_3

$$|f(z) - p_n(z)| \leq A_3 \varepsilon_n \qquad (z \in C).$$

We note that every sequence $\{\varepsilon_n\} = \{n^{-\alpha}\}$ $(\alpha > 0)$ satisfies the restriction in Lemma 2.

REFERENCES

1. D. J. Newman, *Rational approximation to* $|x|$, Michigan Math. J. 11 (1964), 11-14.

2. J. L. Walsh, *Interpolation and approximation by rational functions in the complex domain*, Amer. Math. Soc. Colloquium Publications 20 (1935).

3. ———, *Note on polynomial approximation on a Jordan arc*, Proc. Nat. Acad. Sci. U.S.A. 46 (1960), 981-983.

University of Maryland
College Park, Maryland 20740

TRANSACTIONS OF THE
AMERICAN MATHEMATICAL SOCIETY
Volume 159, September 1971

SOME EXAMPLES IN DEGREE OF APPROXIMATION BY RATIONAL FUNCTIONS(1)

BY
D. AHARONOV AND J. L. WALSH

Abstract. We exhibit examples of (1) series that converge more rapidly than any geometric series where the function represented has a natural boundary, (2) the convergence of a series with maximum geometric degree of convergence yet having limit points of poles of the series everywhere dense on a circumference in the complement of E, (3) a Padé table for an entire function whose diagonal has poles everywhere dense in the plane and (4) a corresponding example for the table of rational functions of best approximation of prescribed type.

Specific examples are frequently of service in a developing theory, both to suggest new propositions and as counterexamples to refute possible conjectures. In the present paper we present some examples of this nature concerning degree of approximation by rational functions.

To be more explicit, we study (§1) examples of series that converge more rapidly than any geometric series where the function represented has a natural boundary, in §2 the convergence of a series with maximum geometric degree of convergence yet having limit points of poles everywhere dense on a circumference in the complement of E. In §3 we give an illustration of rational functions of degree n in the Padé table for an entire function where the diagonal has poles everywhere dense in the plane, and in §4 a corresponding example for the diagonal sequence in the analogous table of rational functions of best approximation.

1. Rapidly converging series.

An infinite series

$$f(z) \sim u_1(z)+u_2(z)+\cdots \tag{1}$$

of rational functions of respective degrees n, is said to *converge to* $f(z)$ *like a geometric series on a set* E provided on that set for the sum $S_n(z)$ of the first n terms of the second member of (1) we have

$$\limsup_{n\to\infty} [\max |f(z)-S_n(z)|, z \text{ on } E]^{1/n} < 1. \tag{2}$$

Received by the editors November 9, 1970.
AMS 1969 *subject classifications.* Primary 3070; Secondary 4140, 4141.
Key words and phrases. Approximation, rational functions, Padé table, best approximation, overconvergence.
(1) Research sponsored (in part) by USAFOSR Grant 69-1690.

The series (1) is said *to converge on E more rapidly than any (nontrivial) geometric series* if we have

$$\lim_{n\to\infty} [\max |f(z)-S_n(z)|,\ z \text{ on } E]^{1/n} = 0. \tag{3}$$

If $u_n(z)$ is a rational function of degree n, if (2) is satisfied, and if there is no limit point of poles of the $u_n(z)$ on the closed region E, then the series (1) has important properties relative to convergence, notably that the series (1) *overconverges*, namely converges to $f(z)$ uniformly in a larger region containing E in its interior. If under these same conditions (3) is satisfied, then [3] the series (1) converges, uniformly on compact sets in any region E_1 containing E and containing no limit points of the poles of the $u_n(z)$. Moreover (3) is satisfied if E is replaced by any closed subregion of E_1. If the function $f(z)$ is meromorphic at every finite point of the plane, and if the closed region E is bounded, it is known [3] that there exists a sequence of rational functions of respective degrees n whose poles lie in the poles of $f(z)$ and which satisfies (3). This raises the question as to whether (3) always implies that $f(z)$ has no natural boundary, a question which we proceed to answer here in the negative, by counterexamples.

THEOREM 1. *Let E be the closed unit disk $|z|\leqq 1$, and the points $\alpha_1, \alpha_2, \ldots$ chosen in the disk E': $|z|<3$ so that α_1 lies on the circle $|z|=5/2$, the next two α_k are equally spaced on $|z|=8/3$, the next three α_k are equally spaced on $|z|=11/4$, and so on for the circles $|z|=3-1/N$. We choose $A_n=1/n^n$. Then the function*

$$f(z) \equiv \sum_{1}^{\infty} \frac{A_k}{z-\alpha_k} \tag{4}$$

exists throughout E', is meromorphic there, is analytic on E and satisfies

$$\lim_{n\to\infty} \left[\max \left| f(z)-\sum_{k=1}^{n} \frac{A_k}{z-\alpha_k}\right|,\ z \text{ on } E\right]^{1/n} = 0. \tag{5}$$

We note the relations

$$\sum_{k=n+1}^{\infty} A_k = \sum_{k=n+1}^{\infty} \frac{1}{k^k} \leq \sum_{k=n+1}^{\infty} \frac{1}{(n+1)^k} = \frac{1}{n(n+1)^n} < A_n, \tag{6}$$

$$\lim_{n\to\infty} \left(\sum_{n+1}^{\infty} A_k\right)^{1/n} = 0. \tag{7}$$

If z_0 is a point in E' not an α_n, we have for all n $|z_0-\alpha_n|\geqq 2\delta$ (>0), where δ depends on z_0, and for z in the neighborhood $|z-z_0|<\delta$ we have $|z-\alpha_n|>\delta$. Thus the series in (4) is dominated by the series $\sum A_k/\delta$, and by the Weierstrass M-test the series (4) converges uniformly in that neighborhood, so $f(z)$ is analytic in the neighborhood, in particular is analytic on E. The same proof shows that if α_n is arbitrary in E', then $f(z)-A_n/(z-\alpha_n)$ is analytic in a neighborhood of α_n, so

$f(z)$ is meromorphic in E'. For z on E we have for every n the inequality $|z-\alpha_n|>1$, whence

$$f(z)-\sum_{k=1}^{n}\frac{A_k}{z-\alpha_k} \equiv \sum_{k=n+1}^{\infty}\frac{A_k}{z-\alpha_k};$$

this last series is dominated by the first member of (6), and (7) implies (5). The rational function in (5) is of degree n, so (5) is of form (3). The analog of (5) holds for degree of approximation to $f(z)$ on E by the rational functions of degrees n of best approximation.

The function $f(z)$ clearly has the circle $|z|=3$ as a natural boundary.

A somewhat similar function but no longer with poles in E' is presented in

THEOREM 2. *Let E be the closed unit disk, E' the disk $|z|<3$, and the points $\alpha_1, \alpha_2, \ldots$ chosen all distinct but everywhere dense on the circle $|z|=3$ (e.g. $\alpha_n=3e^{in}$). We choose $A_n=1/n^n$. Then the function*

$$f(z) \equiv \sum_{1}^{\infty}\frac{A_n\alpha_n}{3(z-\alpha_n)} \tag{8}$$

exists throughout E': $|z|<3$, is analytic there, and satisfies (where $\arg z=\arg\alpha_n$, $|z|<3$)

$$\lim_{z\to\alpha_n}|f(z)| = \infty. \tag{9}$$

Consequently, each α_n is a singularity of $f(z)$ and the circle $|z|=3$ is a natural boundary for $f(z)$.

It follows from the Weierstrass M-test that the series in (8), being dominated by the series $\sum A_n/(3-r)$ on the disk $|z|\leqq r<3$, is analytic on that disk and throughout E'. We note too by (6) the inequality

$$\sum_{k=N+1}^{\infty} A_k = A' < A_N. \tag{10}$$

Since the α_n are distinct, the function

$$S_{N-1}(z) = \sum_{k=1}^{N-1}\frac{A_k\alpha_k}{3(z-\alpha_k)}$$

is continuous at the point $z=\alpha_N$ and takes there some (finite) value A. Moreover, for $0<|z|<3$, $\arg z=\arg\alpha_N$, we have

$$|z-\alpha_N| < |z-\alpha_k| \quad \text{for } k>N \text{ as } z\to\alpha_N,$$

whence

$$f(z) \equiv S_{N-1}(z)+\frac{A_N\alpha_N}{3(z-\alpha_N)}+\sum_{k=N+1}^{\infty}\frac{A_k\alpha_k}{3(z-\alpha_k)}.$$

The first term in the second member approaches A, the second term is real and negative, of absolute value $A_N/|z-\alpha_N|$, and the third term is by (10) not greater in absolute value than

$$\sum_{k=N+1}^{\infty} \frac{A_k}{|z-\alpha_N|} = \frac{A'}{|z-\alpha_N|} < \frac{A_N}{|z-\alpha_N|}.$$

Consequently we have ($\arg z = \arg \alpha_N$)

$$\lim_{z\to\alpha_N} \mathscr{R}[f(z)] = -\infty, \tag{11}$$

so (9) is established, and α_N is a singularity of $f(z)$. Thus the circle $|z|=3$ is a natural boundary for $f(z)$ in E'. We note incidentally that the proof of (11) is valid too for $z\to\alpha_N$, $|z|>3$, so the circle $|z|=3$ is also a natural boundary for $f(z)$ in $|z|>3$.

The function $f(z)$ in Theorem 2 is of significance for overconvergence, for we have for z on E

$$|f(z)-S_n(z)| \equiv \sum_{n+1}^{\infty} \left| \frac{A_k\alpha_k}{3(z-\alpha_k)} \right| < \sum_{n+1}^{\infty} \frac{A_k}{2},$$

and by (7) we have (5). Of course (5) is of form (3). The analog of (5) holds for approximation to $f(z)$ on E by the rational functions of degrees n of best approximation.

A rational function $R_{mn}(z)$ of the form

$$R_{mn}(z) \equiv \frac{a_0z^m+a_1z^{m-1}+\cdots+a_m}{b_0z^n+b_1z^{n-1}+\cdots+b_n}, \qquad \sum |b_i| \neq 0,$$

is said to be *of type* (m, n). If the function $f(z)$ is continuous on a closed bounded set E, with no isolated points, for each type (m, n) there exists a function of that type of best approximation to $f(z)$ on E; these functions can then be arranged [2] in a table, analogous to that of Padé,

$$\begin{array}{l} R_{00}, R_{10}, R_{20}, \ldots, \\ R_{01}, R_{11}, R_{21}, \ldots, \\ R_{02}, R_{12}, R_{22}, \ldots, \\ \ldots \;\; \ldots \;\; \ldots \;\; \ldots. \end{array} \tag{12}$$

Various sequences from this table have been studied [1], [2], [3]. The functions $R_{nn}(z)$ of the diagonal are precisely the rational functions of respective degrees n of best approximation to $f(z)$ on E.

Perron has pointed out [4, p. 466] that there exists a function $f(z)$ such that the second row of the Padé table consists of rational functions whose poles are everywhere dense in the circle of convergence of $f(z)$. We prove now an analogous theorem concerning rational function of degree n of best approximation to $f(z)$, the diagonal in the analog (12) of the Padé table.

2. Comparison series.

THEOREM 3. *Let $f(z)$ be a function regular in a closed bounded region E and let $\{R_n(z)\}$ be the functions of degree n of best approximation to $f(z)$ on E. Also assume*

$$\limsup_{n\to\infty} \|f(z)-R_n(z)\|^{1/n} = A. \tag{13}$$

(The norm is any of the ordinary norms.) Then a comparison series $T_n(z)$ may be found such that

(i) $\limsup_{n\to\infty} \|f(z)-T_n(z)\|^{1/n}=A$,

(ii) *the $T_n(z)$ are rational functions of degree n,*

(iii) *the set of poles of $\{T_n(z)\}$ form a dense subset in the complement $C(E)$.*

Proof. Let $\{1/\alpha_0^{(n)}\}_{n=1,2,\dots}$ be a dense subset of $C(E)$ such that the $1/\alpha_0^{(n)}$ are different from the poles of the $\{R_n(z)\}_{n=1,2,\dots}$.

Let $A_0^{(n)}$ be so small that for $z \in E$ we have

$$\lim_{n\to\infty} \left\| \frac{A_0^{(n)}}{1-\alpha_0^{(n)}z} \right\|^{1/(n+1)} = 0.$$

We define $T_{n+1}(z)=R_n(z)+A_0^{(n)}/(1-\alpha_0^{(n)}z)$, whence we have

$$\begin{aligned} \|f-R_{n+1}(z)\|^{1/(n+1)} &\leqq \|f-T_{n+1}(z)\|^{1/(n+1)} \\ &= \|f-R_n(z)-A_0^{(n)}/(1-\alpha_0^{(n)}z)\|^{1/(n+1)} \\ &\leqq \|f-R_n\|^{1/(n+1)}+\|A_0^{(n)}/(1-\alpha_0^{(n)}z)\|^{1/(n+1)}. \end{aligned}$$

But

$$\limsup_{n\to\infty} \|f-R_{n+1}\|^{1/(n+1)} = A, \qquad \limsup_{n\to\infty} \|f-R_n\|^{1/(n+1)} = A,$$

$$\lim_{n\to\infty} \left\| \frac{A_0^{(n)}}{1-\alpha_0^{(n)}z} \right\|^{1/(n+1)} = 0.$$

So we have

$$\limsup_{n\to\infty} \|f-T_{n+1}(z)\|^{1/(n+1)} = A, \tag{14}$$

and the theorem follows.

We may replace the hypothesis (13) by

$$\limsup_{n\to\infty} \|f(z)-R_n(z)\|^{1/n} \leqq A,$$

and deduce the conclusion

$$\limsup_{n\to\infty} \|f(z)-T_n(z)\|^{1/n} \leqq A.$$

The rational functions $R_n(z)$ of best approximation are not necessarily unique, but we do not assert that the $T_n(z)$ are best approximating. The degree of convergence used in (13) and (14) does not distinguish between the two sequences, but presumably a more refined measure of degree of convergence would do so.

In the study of equation (3), one might conjecture that (3) is possible only when the poles of the $S_n(z)$ lie in the poles of $f(z)$, or at least approach the singularities of $f(z)$. Theorem 3 of §2 effectively shows that the conjecture is false.

3. Examples on Padé table.

LEMMA 1 [5, p. 377]. *Let $f(z)$ be a function analytic in a circle containing the origin with the expansion $f(z)=\sum_{k=0}^{\infty} a_k z^k$. Let $R_{nm}(z)$ be a rational function of type (n, m) (i.e. a polynomial of order n divided by a polynomial of order m) with the expansion $R_{nm}(z)\equiv\sum_{k=0}^{\infty} b_k z^k$ near the origin. Then if $a_k=b_k$, $k=0, 1, \ldots, n+m$, $R_{nm}(z)$ is the Padé approximant of type (n, m) to the function $f(z)$.*

Proof. Denote $R_{nm}=P_n/Q_n$,

$$f(z)-\frac{P_n(z)}{Q_m(z)} = \sum_{k=n+m+1}^{\infty} C_k z^k. \tag{15}$$

If $\tilde{P}_n(z)/\tilde{Q}_m(z)$ is a Padé approximant to $f(z)$, then by definition of the Padé approximant and (15) we get

$$f(z)-\frac{\tilde{P}_n(z)}{\tilde{Q}_m(z)} = \sum_{k=n+m+1}^{\infty} d_k z^k. \tag{16}$$

From (15) and (16) there follows

$$\frac{P_n(z)}{Q_m(z)}-\frac{\tilde{P}_n(z)}{\tilde{Q}_m(z)} = \sum_{k=n+m+1}^{\infty} t_k z^k. \tag{17}$$

So we have from (17)

$$P_n(z)\tilde{Q}_m(z)-\tilde{P}_n(z)Q_m(z) = \sum_{k=n+m+1}^{\infty} V_k z^k. \tag{18}$$

Since on the left side of (18) we have a polynomial of degree $n+m$ at most it follows that $V_k=0$, $k=n+m+1, \ldots,$

$$P_n(z)/Q_m(z) \equiv \tilde{P}_n(z)/\tilde{Q}_m(z), \tag{19}$$

which shows that R_{nm} is the unique Padé approximant of type (n, m).

EXAMPLE 1. Let

$$f(z) = \frac{A_0}{1-\alpha_0 z}+\frac{A_1}{1-\alpha_1 z}+\sum_{v=0}^{\infty} h_v z^v = \sum_{k=0}^{\infty} t_k z^k,$$

such that $\sum_{v=0}^{\infty} h_v z^v$ is an entire function, $0<|\alpha_0|<\infty$, $0<|\alpha_1|<\infty$. Let $g_{m1}(z) =b/(1-\beta z)+\sum_{v=0}^{m-1} C_v z^v=\sum_{k=0}^{\infty} d_k z^k$. In order that g_{m1} should be the Padé approximant of type $(m, 1)$ of $f(z)$ it is sufficient (by Lemma 1) that $d_k=t_k$,

$k=0, 1, 2, \ldots, m+1$. So we have the set of equations

$$
\begin{aligned}
A_0+A_1+h_0 &= b+C_0,\\
A_0\alpha_0+A_1\alpha_1+h_1 &= b\beta+C_1,\\
\vdots\qquad & \qquad \vdots\\
A_0\alpha_0^{m-1}+A_1\alpha_1^{m-1}+h_{m-1} &= b\beta^{m-1}+C_{m-1};\\
A_0\alpha_0^{m}+A_1\alpha_1^{m}+h_{m} &= b\beta^{m},\\
A_0\alpha_0^{m+1}+A_1\alpha_1^{m+1}+h_{m+1} &= b\beta^{m+1}.
\end{aligned} \tag{20}
$$

From the last two equations we get

$$\beta = (A_0\alpha_0^{m+1}+A_1\alpha_1^{m+1}+h_{m+1})/(A_0\alpha_0^{m}+A_1\alpha_1^{m}+h_m). \tag{21}$$

We now consider two cases:

$$|\alpha_0| > |\alpha_1|, \qquad |\alpha_0| = |\alpha_1|.$$

Denote $\alpha_1=\tau\alpha_0$. Then

$$\beta = \frac{\alpha_0^{m+1}}{\alpha_0^m}\,\frac{A_0+A_1\tau^{m+1}+h_{m+1}/\alpha_0^{m+1}}{A_0+A_1\tau^m+h_m/\alpha_0^m}.$$

If $|\alpha_1|<|\alpha_0|$ then $|\tau|<1$. So $\beta\to\alpha_0$ as $m\to\infty$. If $|\alpha_1|=|\alpha_0|$ denote $\tau=e^{i\phi}$. Then

$$\beta = \alpha_0\,\frac{A_0+(A_1\tau)\tau^m+h_{m+1}/\alpha_0^{m+1}}{A_0+A_1\tau^m+h_m/\alpha_0^m}.$$

For $\phi=2\pi/l$ where l is an integer we get for β the l limit points

$$\alpha_0\,\frac{A_0+(A_1\tau)e^{2\pi ik/l}}{A_0+A_1e^{2\pi ik/l}}, \qquad k = 0, 1, \ldots, l-1.$$

If $\phi/2\pi$ is an irrational number we get the whole circle (straight line)

$$\alpha_0\,\frac{A_0+(A_1\tau)w}{A_0+A_1w}, \qquad |w| = 1,$$

as limit points of β, as $m\to\infty$.

Example 2. Our aim is now to construct an example of an entire function $H(z)=\sum_{v=0}^{\infty} h_v z^v$ such that the diagonal in the Padé table will diverge in a dense subset of the plane. For convenience we consider the sequence $R_{n-1,n}(z)$ instead of the sequence $R_{nn}(z)$. The difference is only an additional additive constant.

Let α_0, $0<|\alpha_0|<\infty$, be arbitrary. Consider the function $A_0/(1-\alpha_0 z)$, where A_0 is a constant that will be fixed later.

In order that $A_0/(1-\alpha_0 z)$ will be $R_{01}(z)$ of $H(z)=\sum_{v=0}^{\infty} h_v z^v$, it is sufficient (Lemma 1) that

$$A_0 = h_0, \qquad A_0\alpha_0 = h_1. \tag{22}$$

We now fix A_0 so that $|h_0|, |h_1| < 1/1!$. Next we consider the rational function

$$\frac{b_0}{1-\beta_0 z}+\frac{b_1}{1-\beta_1 z}+\frac{b_2}{1-\beta_2 z}$$

where β_0, $0<|\beta_0|<\infty$, is arbitrary, $\beta_1=q\bar{\beta}_1$, $\beta_2=q\bar{\beta}_2$, for $1>q>0$, and q will be fixed later; $\bar{\beta}_1\neq\bar{\beta}_2$ are arbitrary (but $\bar{\beta}_1, \bar{\beta}_2\neq 0, \infty$). We now look for the conditions that this rational function will be $R_{23}(z)$ of $H(z)=\sum_{v=0}^{\infty} h_v z^v$ (only the coefficients h_0, h_1 are determined for the time being). Again, by Lemma 1, we have the set of equations:

$$\begin{aligned} b_0 + b_1 + b_2 &= h_0,\\ b_0\beta_0+b_1\beta_1+b_2\beta_2 &= h_1,\\ b_0\beta_0^2+b_1\beta_1^2+b_2\beta_2^2 &= h_2,\\ \vdots \qquad \vdots \qquad &\quad \vdots\\ b_0\beta_0^5+b_1\beta_1^5+b_2\beta_2^5 &= h_5. \end{aligned} \tag{23}$$

First b_0 is determined so that

$$|b_0\beta_0^j| < 1/2\cdot 5!, \qquad j = 2, 3, 4, 5. \tag{24}$$

We next denote

$$h_0-b_0 = \lambda_0, \qquad h_1-b_0\beta_0 = \lambda_1. \tag{25}$$

From (23) and (25) we get

$$b_1+b_2 = \lambda_0, \qquad b_1\beta_1+b_2\beta_2 = \lambda_1. \tag{26}$$

Using the given conditions $\beta_1=q\bar{\beta}_1$, $\beta_2=q\bar{\beta}_2$ we have

$$\begin{vmatrix} \lambda_0 & 1 \\ \lambda_1 & \bar{\beta}_2 q \end{vmatrix} \Big/ \begin{vmatrix} 1 & 1 \\ \bar{\beta}_1 q & \bar{\beta}_2 q \end{vmatrix} = \frac{\lambda_0\bar{\beta}_2-\lambda_1/q}{\bar{\beta}_2-\bar{\beta}_1}. \tag{27}$$

We have a similar formula for b_2, so

$$|b_1|, |b_2| < M/q, \quad M \text{ depends only on } \lambda_0, \lambda_1, \bar{\beta}_1, \bar{\beta}_2. \tag{28}$$

From (28) we get

$$\begin{aligned} |b_1\beta_1^j+b_2\beta_2^j| &= |b_1\bar{\beta}_1^j q^j+b_2\bar{\beta}^j q^j| \\ &< 2(M/q)q^j \operatorname{Max}(|\bar{\beta}_1|^j, |\bar{\beta}_2|^j), \qquad j = 2, 3, 4, 5. \end{aligned} \tag{29}$$

Denote

$$L = \operatorname*{Max}_{2\leq j\leq 5} \operatorname{Max}(|\bar{\beta}_1|^j, |\bar{\beta}_2|^j). \tag{30}$$

From (29) and (30) we get

$$|b_1\beta_1^j+b_2\beta_2^j| < 2MqL, \qquad j = 2, 3, 4, 5. \tag{31}$$

If q is sufficiently small we have

$$|b_1\beta_1^j+b_2\beta_2^j| < 1/2\cdot 5!, \qquad j = 2, 3, 4, 5. \tag{32}$$

From (23), (24) and (32) we obtain

$$|h_j| < 1/5!, \qquad j = 2, 3, 4, 5.$$

Up to now the coefficients $h_0, h_1, \ldots, h_5$ have been determined; also we know that R_{01}, R_{23} of the Padé table each has one arbitrary pole. In order to continue the procedure we have to construct R_{67} (one arbitrary pole and six other poles because we have six coefficients of $H(z)$ already determined). So let $\sum_{k=0}^{6} C_k/(1-\gamma_k z)$ be a rational function such that γ_0, $0<|\gamma_0|<\infty$, is arbitrary. $\gamma_k=q_1\tilde{\gamma}_k$, $1>q_1>0$; $k=1, 2, \ldots, 6$; $\tilde{\gamma}_j\neq\tilde{\gamma}_l, j\neq l, j, l\geqq 2$. In order that this function should be R_{67} of $H(z)$ the following set of equations has to be fulfilled:

$$\begin{aligned} C_0+C_1+\cdots+C_6 &= h_0, \\ C_0\gamma_0+\cdots+C_6\gamma_6 &= h_1, \\ \vdots \qquad\quad \vdots \quad & \quad \vdots \\ C_0\gamma_0^5+\cdots+C_6\gamma_6^5 &= h_5; \\ C_0\gamma_0^6+\cdots+C_6\gamma_6^6 &= h_6, \\ \vdots \qquad\quad \vdots \quad & \quad \vdots \\ C_0\gamma_0^{13}+\cdots+C_6\gamma_6^{13} &= h_{13}. \end{aligned} \tag{33}$$

Exactly as we have done before we now choose C_0 so small that

$$|C_0\gamma_0^j| < 1/2\cdot 13!, \qquad j = 6, \ldots, 13. \tag{34}$$

Again, we define $\lambda_{j1}=h_j-C_0\gamma_0^j$, $j=0, \ldots, 5$. From the first six equations in (33) we have

$$\begin{aligned} C_1+\cdots+C_6 &= \lambda_{01}, \\ C_1\gamma_1+\cdots+C_6\gamma_6 &= \lambda_{11}, \\ \vdots \qquad\quad \vdots \quad & \quad \vdots \\ C_1\gamma_1^5+\cdots+C_6\gamma_6^5 &= \lambda_{51}. \end{aligned} \tag{35}$$

Using now the conditions $\gamma_k=q_1\tilde{\gamma}_k$ we get (in the same way as we obtained (28)),

$$|C_k| < M_1/q_1, \qquad k = 1, 2, \ldots, 6, \tag{36}$$

M_1 depends only on $\{\lambda_{j1}\}\cup\{\gamma_l\}$, $0\leqq j\leqq 5$, $1\leqq l\leqq 6$. Exactly in the same manner as before we obtain

$$|C_1\gamma_1^j+\cdots+C_6\gamma_6^j| < 6M_1L_1q_1, \qquad j = 6, \ldots, \tag{37}$$

where L_1 is defined by

$$L_1 = \max_{6\leqq j\leqq 13}\ \max_{1\leqq k\leqq 6} (|\tilde{\gamma}_k|^j). \tag{38}$$

For q_1 sufficiently small we get from (37)

$$(39)\qquad |C_1\gamma_1^j + C_2\gamma_2^j + C_2\gamma_2^j + \cdots + C_6\gamma_6^j| < 1/2\cdot 13!, \qquad j = 6, \ldots, 13.$$

From (33), (34), and (39) we obtain

$$(40)\qquad |h_j| < 1/13!, \qquad j = 6, \ldots, 13.$$

In this way we continue to construct the subsequence of the diagonal of Padé table. Each of the terms of this subsequence has one arbitrary pole. It is clear that the $H(z)$ constructed is majorized by the exponential function and so is an entire function.

REMARK. By a slight change in the argument we could construct a subsequence of the diagonal such that each member of this subsequence would have more than one arbitrary pole. In fact the number of the arbitrary poles may increase monotonically to ∞. This is, of course, not needed here for the construction of the desired example.

4. Examples on table (12).

LEMMA 1. *Suppose $S_p(z)$ is a rational function of order p having poles at $1/\alpha_j$, $1 \leqq j \leqq p$, $1/|\alpha_j| > 1$, $1/\alpha_j \neq 1/\alpha_k$ if $k \neq j$, $|1/\alpha_j - 1/\alpha_1| \geqq 2r > 0$ for $2 \leqq j \leqq p$. Then there exists $\varepsilon = \varepsilon(\alpha_1, r, p)$ such that for any rational function $R_p(z)$ of order p satisfying*

$$(4.1)\qquad \sup_{|z|\leqq 1} |R_p(z) - S_p(z)| < \varepsilon |A_1|, \qquad S_p(z) = \frac{A_1}{1-\alpha_1 z} + \cdots + \frac{A_p}{1-\alpha_p z},$$

$R_p(z)$ has at least one pole in the circle $|z - 1/\alpha_1| \leqq 2r$.

Proof. Suppose the lemma is false. Then for a particular rational function $S_p(z)$ and $r > 0$ there exist a sequence $\{\varepsilon_n\}$, such that $\varepsilon_1 \geqq \varepsilon_2 \geqq \cdots$, $\varepsilon_n \to 0$ and a sequence of rational functions $\{R_p^{(n)}(z)\}_{n=1}^{\infty}$ of order p such that

$$(4.2)\qquad \sup_{|z|\leqq 1} |R_p^{(n)}(z) - S_p(z)| < \varepsilon_n |A_1|,$$

and all the poles of $R_p^{(n)}(z)$ for any n lie outside the disc $|z - 1/\alpha_1| \leqq 2r$.

Denote the poles of $R_p^{(n)}(z)$ by $1/\beta_{kn}$, $k = 1, 2, \ldots, p$. (It may occur that $1/\beta_{kn} = 1/\beta_{ln}$ for $l \neq k$.) Then by (4.2)

$$(4.3)\quad \sup_{|z|=1} |R_p^{(n)}(z) - S_p(z)| \left| \frac{(1-\alpha_1 z)\cdots(1-\alpha_p z)(1-\beta_{1n} z)\cdots(1-\beta_{pn} z)}{(z-\bar\alpha_1)\cdots(z-\bar\alpha_p)(z-\bar\beta_{1n})\cdots(z-\bar\beta_{pn})} \right| < \varepsilon_n |A_1|.$$

But the function on the left side of (4.3) is regular on $|z| \geqq 1$; so [1, §9.4]

$$(4.4)\qquad \sup_{|z|\geqq 1} |R_p^{(n)}(z) - S_p(z)| < \varepsilon_n |A_1| \frac{|(z-\bar\alpha_1)\cdots(z-\bar\beta_{pn})|}{|(1-\alpha_1 z)\cdots(1-\beta_{pn} z)|}.$$

Now there is no loss of generality in assuming that $2r < 1/|\alpha_1| - 1$. So we have

$$(4.5)\qquad \sup_{|z-1/\alpha_1| = r} |R_p^{(n)}(z) - S_p(z)| < \varepsilon_n |A_1| \frac{|(z-\bar\alpha_1)\cdots(z-\bar\beta_{pn})|}{|(1-\alpha_1 z)\cdots(1-\beta_{pn} z)|}.$$

Let $R=1/|\alpha_1|+r$. If $1/|\alpha_j|\geqq 2R$ we have for $|z-1/\alpha_1|=r$

$$1/|1-\alpha_j z| \leqq 2. \tag{4.6}$$

Indeed, for $|z|\leqq R$ we have $|\alpha_j|\leqq 1/2R$ so $|1-\alpha_j z|\geqq 1/2$. We have for $|z-1/\alpha_1|=r$, $|1/\alpha_j|\geqq 2R$,

$$|(z-\bar{\alpha}_j)/(1-\alpha_j z)| \leqq 2(R+1/2R). \tag{4.7}$$

A similar result is obtained for $1/|\beta_j|\geqq 2R$.

Now assume that $1/|\alpha_j|<2R$. Then

$$\left|\frac{z-\bar{\alpha}_j}{1-\alpha_j z}\right| = \left|\frac{\bar{\alpha}_j(1-z/\alpha_j)}{\alpha_j(z-1/\alpha_j)}\right| < \frac{1+2R^2}{|z-1/\alpha_j|}, \qquad 1\leqq j\leqq p.$$

But by our assumption $|1/\alpha_j-1/\alpha_1|\geqq 2r$. So on the circle $|z-1/\alpha_1|=r$ we have $|z-1/\alpha_j|\geqq r$ for $2\leqq j\leqq p$, and for $|z-1/\alpha_1|=r_1$ we get $|1/\alpha_j|<2R$,

$$|(z-\bar{\alpha}_j)/(1-\alpha_j z)| < (1+2R^2)/r, \qquad 1\leqq j\leqq p. \tag{4.8}$$

By our assumption on $R_p^{(n)}(z)$ we know that $|1/\beta_{jn}-1/\alpha_1|\geqq 2r$ for $1\leqq j\leqq p$. So we get a similar result for $1/|\beta_{jn}|<2R$. Combining (4.7) and (4.8) and the similar results for the $1/\beta_j$ we get from (4.5)

$$\sup_{|z-1/\alpha_1|=r} |R_p^{(n)}(z)-S_p(z)| < \varepsilon_n|A_1|[2(R+1/2R)]^{2p}[(1+2R^2)/r]^{2p}. \tag{4.9}$$

Now, the function $R_p^{(n)}(z)-S_p(z)$ has only the pole $1/\alpha_1$ on the disc $|z-1/\alpha_1|\leqq r$. So by integration of this function along the circle $|z-1/\alpha_1|=r$ we have from (4.9)

$$|A_1|/|\alpha_1| < \varepsilon_n|A_1|[2(R+1/2R)]^{2p}[(1+2R^2)/r]^{2p} = \varepsilon_n M(R,r,p)|A_1|. \tag{4.10}$$

But now we let $n\to\infty$ and then $\varepsilon_n\to 0$ and we get the desired contradiction. (In fact we get the contradiction for $\varepsilon_n<1/M(R,r,p)|\alpha_1|$. So ε in the theorem can be taken as $\varepsilon=1/2M(R,r,p)|\alpha_1|=\varepsilon(\alpha_1,r,p)$.)

Lemma 1, so far as concerns the existence of $\varepsilon=\varepsilon(\alpha_1,r,p)$ but not the specific formula (4.10), can be proved qualitatively by the properties of sequences of rational functions.

The existence of $R_p^{(n)}(z)$ satisfying (4.2) shows that the rational function $R_p^{(n)}(z)-S_p(z)$ of degree $2p$ approaches zero uniformly on $|z|<1$, hence [1, §12.1] admits a subsequence which approaches zero in the extended plane with the omission of at most $2p$ points. Since $S_p(z)$ has a pole in $1/\alpha_1$ of fixed principal part independent of n, every such subsequence has at least one pole (necessarily belonging to $R_p^{(n)}(z)$) which approaches $1/\alpha_1$; we see this by integrating $R_p^{(n)}(z)-S_p(z)$ over a circle $|z-1/\alpha_1|=r_1<r$.

LEMMA 2. *Let $\eta>0$, $\varepsilon>0$, $r>0$, and $n+1$ complex numbers $C_0, C_1, \ldots, C_n$ be given. Also let $0<|\alpha_1|<1$. Then there exists $m>n$ and a rational function $S_p(z)$ of* *order $p=n+2$ with a pole at $1/\alpha_1$ and also at the points $\{1/\alpha_j\}_{j=2}^{p}$, $\alpha_j \neq \alpha_k$ for $j\neq k$, $|1/\alpha_j|>1$ such that for*

$$S_p(z) = \frac{A_1}{1-\alpha_1 z}+\frac{A_2}{1-\alpha_2 z}+\cdots+\frac{A_p}{1-\alpha_p z} = \sum_{k=0}^{\infty} \mu_k z^k \tag{4.11}$$

we have

(a) $\mu_k=C_k$, $k=0, 1, \ldots, n$,
(b) $|\mu_{n+1}|+|\mu_{n+2}|+\cdots<\eta$,
(c) $|\mu_{n+1}|, |\mu_{n+2}|, \ldots, |\mu_m|<1/m!$,
(d) $|\mu_{m+1}|+|\mu_{m+2}|+\cdots<\varepsilon|A_1|/4$,
(e) $|1/\alpha_j-1/\alpha_1|\geqq 2r$, $2\leqq j\leqq p$.

Proof. Let $\{\alpha_j'\}$, $2\leqq j\leqq p$, be arbitrary complex nonvanishing numbers, $\alpha_j'\neq\alpha_k'$ for $j\neq k$. Let $0<q<1$.

The exact value of q will be determined later. Let $\alpha_j=q\alpha_j'$ where $1/\alpha_j$ are the poles of $S_p(z)$. We have by (a)

$$\begin{aligned} A_1 + A_2 + \cdots + A_p &= \mu_0 = C_0,\\ A_1\alpha_1+A_2\alpha_2+\cdots+A_p\alpha_p &= \mu_1 = C_1,\\ &\vdots\\ A_1\alpha_1^n+A_2\alpha_2^n+\cdots+A_p\alpha_p^n &= \mu_n = C_n. \end{aligned}\tag{4.12}$$

Also

$$\begin{aligned} A_1\alpha_1^{n+1}+A_2\alpha_2^{n+1}+\cdots+A_p\alpha_p^{n+1} &= \mu_{n+1},\\ A_1\alpha_1^{n+2}+ \cdots + A_p\alpha_p^{n+2} &= \mu_{n+2},\\ &\vdots \end{aligned}\tag{4.13}$$

First we choose $m>n$ such that

$$|\alpha_1|^{m+1}+|\alpha_1|^{m+2}+\cdots = |\alpha_1|^{m+1}/(1-|\alpha_1|) < \varepsilon/8. \tag{4.14}$$

Now we determine $A_1\neq 0$ so that

$$|A_1|\,|\alpha_1|^{n+1},\quad |A_1|\,|\alpha_1|^{n+2},\quad \ldots,\quad |A_1|\,|\alpha_1|^m < 1/2\cdot m!, \tag{4.15}$$

$$|A_1|\,|\alpha_1|^{n+1}+|A_1|\,|\alpha_1|^{n+2}+\cdots < \eta/2. \tag{4.16}$$

From (4.12) we get

$$\begin{aligned} A_2 + \cdots + A_p &= C_0-A_1 = C_0',\\ A_2\alpha_2+\cdots+A_p\alpha_p &= C_1-A_1\alpha_1 = C_1',\\ &\vdots\\ A_2\alpha_2^n+\cdots+A_p\alpha_p^n &= C_n-A_1\alpha_1^n = C_n'. \end{aligned}\tag{4.17}$$

But $\alpha_j = q\alpha_j'$ for $\alpha_j' \neq \alpha_l'$ if $j \neq l$. So $\alpha_j \neq \alpha_l$ for $j \neq l$. Also $p = n+2$. So we get from (4.17)

$$A_j = \frac{\begin{vmatrix} 1 & \cdots & C_0' & \cdots & 1 \\ \alpha_2 & \cdots & & \cdots & \alpha_p \\ \vdots & & \vdots & & \vdots \\ \alpha_2^n & \cdots & C_n' & \cdots & \alpha_p^n \end{vmatrix}}{\begin{vmatrix} 1 & \cdots & 1 & \cdots & 1 \\ \alpha_2 & \cdots & \alpha_j & \cdots & \alpha_p \\ \vdots & & \vdots & & \vdots \\ \alpha_2^n & \cdots & \alpha_j^n & \cdots & \alpha_p^n \end{vmatrix}}$$

$$= \frac{\begin{vmatrix} 1 & \cdots & C_0' & \cdots & 1 \\ \alpha_2' q & \cdots & & \cdots & \alpha_p' q \\ \vdots & & \vdots & & \vdots \\ (\alpha_2')q^n & \cdots & C_n' & \cdots & (\alpha_p')^n q^n \end{vmatrix}}{\begin{vmatrix} 1 & \cdots & 1 & \cdots & 1 \\ \alpha_2' q & \cdots & \alpha_j' q & \cdots & \alpha_p' q \\ \vdots & & \vdots & & \vdots \\ (\alpha_2')^n q^n & \cdots & (\alpha_j')^n q^n & \cdots & (\alpha_p')^n q^n \end{vmatrix}}$$

$$= \frac{\begin{vmatrix} 1 & \cdots & C_0' & \cdots & 1 \\ \alpha_2' & \cdots & C_1'/q & \cdots & \alpha_p' \\ \vdots & & \vdots & & \vdots \\ (\alpha_2')^n & \cdots & C_n'/q^n & \cdots & (\alpha_p')^n \end{vmatrix}}{\begin{vmatrix} 1 & \cdots & 1 & \cdots & 1 \\ \alpha_2' & \cdots & \alpha_j' & \cdots & \alpha_p' \\ \vdots & & \vdots & & \vdots \\ (\alpha_2')^n & \cdots & (\alpha_j')^n & \cdots & (\alpha_p')^n \end{vmatrix}}.$$

So we get (note that $0 < q < 1$)

$$|A_j| < M/q^n, \qquad j = 2, 3, \ldots, p, \tag{4.18}$$

where M depends on $(C_0', \ldots, C_n')$ and $(\alpha_2', \ldots, \alpha_p')$ but not on q. Denote, now, $N = \max_{2 \leq j \leq p} |\alpha_j'|$. Then from (4.18)

$$\sum_{k=2}^{p} |A_k|\, |\alpha_k|^{n+j} < (p-1)MN^{n+j}q^j, \qquad j = 1, 2, \ldots. \tag{4.19}$$

We now restrict q to the interval $(0, 1/N(1+2r|\alpha_1|)]$. Then

$$|\alpha_j' q| = |\alpha_j| \leq Nq \leq \frac{|\alpha_1|}{1+2r|\alpha_1|}, \qquad 2 \leq j \leq p.$$

So

$$\left|\frac{1}{\alpha_j}-\frac{1}{\alpha_1}\right| \geqq \left|\frac{1}{\alpha_j}\right| - \left|\frac{1}{\alpha_1}\right| \geqq \frac{1+2r|\alpha_1|}{|\alpha_1|}-\frac{1}{|\alpha_1|} = 2r$$

for $2 \leqq j \leqq p$, and (e) is fulfilled.

We now restrict q to a smaller interval, if needed, so that

$$\frac{(p-1)MN^{n+1}q}{1-qN} < \min\left(\frac{1}{2\cdot m!}, \frac{\eta}{2}, \frac{\varepsilon}{8}|A_1|\right). \tag{4.20}$$

From (4.19) we get, summing on j,

$$\sum_{j=1}^{\infty}\sum_{k=2}^{p}|A_k|\,|\alpha_k|^{n+j} < \frac{(p-1)MN^{n+1}q}{1-qN}. \tag{4.21}$$

From (4.20) and (4.21) we clearly have

$$\begin{aligned} &\sum_{j=1}^{\infty}\sum_{k=2}^{p}|A_k|\,|\alpha_k|^{n+j} < \frac{\eta}{2}, \\ &\sum_{k=2}^{p}|A_k|\,|\alpha_k|^{n+j} < \frac{1}{2\cdot m!}, \qquad j = 1, 2, \ldots, m-n, \\ &\sum_{j=1}^{\infty}\sum_{k=2}^{p}|A_k|\,|\alpha_k|^{m+j} < \frac{\varepsilon}{8}|A_1|. \end{aligned} \tag{4.22}$$

(b) follows from (4.13), (4.16) and (4.22), (c) follows from (4.13), (4.15) and (4.22), (d) follows from (4.13), (4.14) and (4.22), and the proof of Lemma 2 is complete.

LEMMA 3. *Let* $\{1/\alpha_{1k}\}_1^\infty$ *be a sequence of points outside the unit disc, dense in* $|z| \geqq 1$. *Let* $r_1 \geqq r_2 \geqq \cdots r_n \to 0$. *Consider the sequence of discs* $\{U_k\colon |z-1/\alpha_{1k}| \leqq 2r_k\}$. *Suppose* $\{\tau_k\}_1^\infty$ *is an arbitrary sequence of points such that* $\tau_k \in U_k$, $k=1, 2, \ldots$. *Then* $\{\tau_k\}_1^\infty$ *form a dense subset of* $|z| \geqq 1$.

Proof. Let z_0 be any point in $|z| \geqq 1$. Since $\{1/\alpha_k\}$ forms a dense subset of $|z| \geqq 1$ we can construct a subsequence $\{1/\alpha_{1k_l}\}_{l=1}^\infty$ such that $1/\alpha_{1k_l} \to z_0$. Consider the sequence $\{\tau_{k_l}\}_{l=1}^\infty$. Then $|\tau_{k_l}-1/\alpha_{1k_l}| \leqq 2r_{k_l} \to 0$ as $l \to \infty$. So $|\tau_{k_l}-z_0| \to 0$ as $l \to \infty$ and the proof is complete.

EXAMPLE. (Compare example in §3.) *We now construct an example of an entire function such that the diagonal in the table* (12) *for the unit disc and the sup norm diverges at a dense subset of* $|z| \geqq 1$.

Let $\{1/\alpha_{1k}\}_1^\infty$ be a sequence of points outside the closed unit disc, dense in $|z| > 1$. Let $r_1 \geqq r_2 \geqq \cdots$ such that $r_n \to 0$. Our aim is first to construct a sequence of rational functions $\{S_{p_l}(z)\}_{l=1}^\infty$ of orders $\{p_l\}_{l=1}^\infty$ respectively. These rational functions will have a simple pole at $\{1/\alpha_{1l}\}_{l=1}^\infty$ respectively. Then an entire function is constructed with the aid of these rational functions. The subsequence $\{R_{p_l}(z)\}_{l=1}^\infty$ in table (12) will have simple poles in the discs $|z-1/\alpha_{1l}| \leqq 2r_l$, $l=1, 2, \ldots$, and so our assertion will follow by Lemma 3. The construction of the $\{S_{p_l}(z)\}_{l=1}^\infty$ makes use of Lemma 2, while for the proof that the poles of $\{R_{p_l}(z)\}_{l=1}^{\infty}$ lie near poles of $\{S_{p_l}(z)\}_{l=1}^{\infty}$ we need Lemma 1.

So our first aim is to construct the sequence $\{S_{p_l}(z)\}_{l=1}^{\infty}$. For this aim we need also to construct sequences $\{m_l\}_{l=1}^{\infty}$, $\{\eta_l\}_{l=2}^{\infty}$, $\{\varepsilon_l\}_{l=1}^{\infty}$. Also we will use the given sequences $\{r_l\}_{l=1}^{\infty}$ and $\{1/\alpha_{1l}\}_{l=1}^{\infty}$ for the construction of $\{S_{p_l}(z)\}_{l=1}^{\infty}$.

We first define $S_{p_1}(z)$ and then successively the sequence $\{S_{p_l}(z)\}$. Let $p_1=1$; $S_{p_1}(z)=A_{11}/(1-\alpha_{11}z)=\sum_{k=0}^{\infty}\mu_k^{(1)}z^k$; A_{11} will be fixed later. Now ε_1 is chosen as $\varepsilon_1=\varepsilon(\alpha_{11}, r_1, p_1)$ where $\varepsilon(\alpha_{11}, r_1, p_1)$ is the function appearing in Lemma 1; m_1 is defined as a solution of the inequality

$$|\alpha_{11}|^{m_1+1}/(1-|\alpha_{11}|) < \varepsilon_1/4. \tag{4.23}$$

$A_{11}\neq 0$ is now chosen small enough to give

$$|A_{11}|\,|\alpha_{11}|^k = |\mu_k^{(1)}| < 1/m_1!, \qquad k = 0, 1, 2, \ldots, m_1. \tag{4.24}$$

(In fact, since $|\alpha_{11}|<1$ it is sufficient to take $|A_{11}|<1/m_1!$.) Let $l\geqq 2$. Suppose now that $\{S_{p_k}(z)\}_{k=1}^{l-1}$, $\{m_k\}_{k=1}^{l-1}$, $\{\varepsilon_k\}_{k=1}^{l-1}$ have been defined. For reference we denote

$$S_{p_j}(z) = \frac{A_{1j}}{1-\alpha_{1j}z}+\frac{A_{2j}}{1-\alpha_{2j}z}+\cdots+\frac{A_{p_jj}}{1-\alpha_{p_jj}z} = \sum_{k=0}^{\infty}\mu_k^{(j)}z^k. \tag{4.25}$$

We now use Lemma 2 for $n_l=m_{l-1}$ (n_l instead of n in Lemma 2), $p_l=n_{l+2}$ (p_l instead of p in Lemma 2), α_{1l} (instead of α_1), r_l (instead of r),

$$\eta_l = \min_{1\leq j\leq l-1} |A_{1j}\varepsilon_j|/2^{l+2-j}$$

(η_l instead of η), $\mu_k^{(l)}$ (instead of μ_k). The constants $C_0, C_1, \ldots, C_n$ appearing in Lemma 2 we take as the first n (i.e., n_l or m_{l-1} in our notation) coefficients of the function $S_{p_{l-1}}$. Thus, $C_k=\mu_k^{(l-1)}$, $k=0, 1, \ldots, m_{l-1}$. The ε appearing in Lemma 2 will be taken as $\varepsilon_l=\varepsilon(\alpha_{1l}, r_l, p_l)$ where $\varepsilon(\alpha_{1l}, r_l, p_l)$ is the function appearing in Lemma 1.

After the new notation in Lemma 2 we use the result of Lemma 2 to get the function $S_{p_l}(z)$ and the integer m_l (m in Lemma 2) such that for

$$S_{p_l}(z) = \frac{A_{1l}}{1-\alpha_{1l}z}+\cdots+\frac{A_{p_ll}}{1-\alpha_{p_ll}z} = \sum_{k=0}^{\infty}\mu_k^{(l)}z^k \tag{4.25'}$$

and for m_l we have

$$\begin{aligned}
&\text{(a)}\quad \mu_k^{(l)} = \mu_k^{(l-1)}, \qquad k = 0, 1, \ldots, m_{l-1},\\
&\text{(b)}\quad |\mu_{m_{l-1}+1}^{(l)}|+|\mu_{m_{l-1}+2}^{(l)}|+\cdots < \eta_l = \min_{1\leq j\leq l-1}\frac{|A_{1j}\varepsilon_j|}{2^{l+2-j}},\\
&\text{(c)}\quad |\mu_{m_{l-1}+1}^{(l)}|, |\mu_{m_{l-1}+2}^{(l)}|, \ldots, |\mu_{m_l}^{(l)}| < 1/m_l!,\\
&\text{(d)}\quad |\mu_{m_l+1}^{(l)}|+|\mu_{m_l+2}^{(l)}|+\cdots < (\varepsilon_l/4)|A_{1l}|,\\
&\text{(e)}\quad |1/\alpha_{jl}-1/\alpha_{1l}| \geqq 2r_l, \qquad 2\leqq j\leqq p_l.
\end{aligned} \tag{4.26}$$

In this way we have defined the sequences $\{S_{p_l}(z)\}_{l=1}^{\infty}$, $\{m_l\}_{l=1}^{\infty}$. Notice the special definitions of $S_{p_1}(z)$, m_1; the fact that η_1 was not defined, and that (4.26(c)) is complementary to (4.24) (in (4.26) we have $l \geqq 2$). Also, it is worth noting that $S_{p_l}(z)$ agrees with the first $m_{l-1}+1$ coefficients of $S_{p_{l-1}}(z)$, the first $m_{l-2}+1$ coefficients of $S_{p_{l-2}}(z)$ and so on (as follows from (4.26(a)) after substituting $l-1$, $l-2$, instead of l).

We now are in a position to define our function $f(z)$. It will be convenient to set

$$f(z) = \sum_{k=0}^{\infty} d_k z^k. \tag{4.27}$$

We now simply demand

$$d_k = \mu_k^{(l)}, \qquad k = 0, 1, \ldots, m_l,\ l = 1, 2, \ldots. \tag{4.28}$$

We recall that because of (4.26(a)) the function is well defined. It follows from (4.28) that, because of (4.26(c)) and (4.24), $f(z)$ is an entire function. We have (recall that the norm is the sup norm on $|z| \leqq 1$) $\|f(z)-S_{p_1}(z)\| = \|\sum_{k=m_1+1}^{\infty} d_k z^k - \sum_{k=m_1+1}^{\infty} \mu_k^{(1)} z^k\|$; this is because of (4.28) for $l=1$. So

$$\|f(z)-S_{p_1}(z)\| \leqq \sum_{k=m_1+1}^{\infty} |d_k| + \sum_{k=m_1+1}^{\infty} |\mu_k^{(1)}|. \tag{4.29}$$

From (4.23) and the definition of $S_{p_1}(z)$ we have

$$\sum_{k=m_1+1}^{\infty} |\mu_k^{(1)}| = A_{11} \frac{|\alpha_{11}|^{m_1+1}}{1-|\alpha_{11}|} < \frac{e_1|A_{11}|}{4}. \tag{4.30}$$

From (4.26(b)) we get, summing on l and taking $j=1$,

$$\begin{aligned} \sum_{l=2}^{\infty} \sum_{k=1}^{\infty} |\mu_{m_{l-1}+k}^{(l)}| &< \sum_{l=2}^{\infty} \eta_l \\ &\leqq \frac{|e_1|\,|A_{11}|}{2}\left(\frac{1}{4}+\frac{1}{8}+\cdots\right) = \frac{|e_1|\,|A_{11}|}{4}. \end{aligned} \tag{4.31}$$

We now deduce from (4.28)

$$d_k = \mu_k^{(l)}, \qquad m_{l-1}+1 \leqq k \leqq m_l,\ l = 2, 3, \ldots. \tag{4.32}$$

From (4.31) and (4.32) we clearly have

$$\sum_{k=m_1+1}^{\infty} |d_k| < \frac{|e_1|\,|A_{11}|}{4}. \tag{4.33}$$

From (4.29), (4.30) and (4.33) we get

$$\|f(z)-S_{p_1}(z)\| < \frac{|e_1|\,|A_{11}|}{2}. \tag{4.34}$$

We now obtain a similar result for $\|f(z)-S_{p_l}(z)\|$, $2 \leqq l$. The calculation is very similar to the above calculation. Indeed, from (4.28) we have

$$\|f(z)-S_{p_l}(z)\| < \sum_{k=m_l+1}^{\infty} |d_k| + \sum_{k=m_l+1}^{\infty} |\mu_k^{(l)}|. \tag{4.35}$$

From (4.26(d)) we have

$$\sum_{k=m_l+1}^{\infty} |\mu_k^{(l)}| < \frac{\varepsilon_l}{4}|A_{1l}|. \tag{4.36}$$

Now a similar formula to (4.31) has to be obtained. For this objective, we change l to t in (4.26(b)) and sum on t from $l+1$ to ∞, taking $j=l$. So we get

$$\sum_{t=l+1}^{\infty} \sum_{k=1}^{\infty} |\mu_{m_{t-1}+k}^{(t)}| < \sum_{t=l+1}^{\infty} \eta_t \leqq \frac{|\varepsilon_l|\,|A_{1l}|}{4}. \tag{4.37}$$

From (4.32) and (4.37) we get

$$\sum_{k=m_l+1}^{\infty} |d_k| < \frac{\varepsilon_l|A_{1l}|}{4}, \qquad l = 2, 3, \ldots. \tag{4.38}$$

So from (4.35), (4.36) and (4.38) we deduce

$$\|f(z)-S_{p_l}(z)\| < \varepsilon_l|A_{1l}|/2, \qquad l = 2, 3, \ldots. \tag{4.39}$$

We now come to the last step, namely, showing that the functions of best approximation have poles "near" $\{1/\alpha_{1l}\}$. Indeed, let $\{R_p(z)\}_{p=1}^{\infty}$ be the diagonal in the table (12). Consider the subsequence $\{R_{p_l}(z)\}$. Since these are functions of best approximation we have from (4.34) and (4.39)

$$\|f(z)-R_{p_l}(z)\| < \varepsilon_l|A_{1l}|/2, \qquad l = 1, 2, \ldots. \tag{4.40}$$

From (4.34), (4.39) and (4.40) we get

$$\|S_{p_l}(z)-R_{p_l}(z)\| < \varepsilon_l|A_{1l}|, \qquad l = 1, 2, \ldots. \tag{4.41}$$

Recalling now (4.26(e)) and the way the $\{\varepsilon_l\}_{l=1}^{\infty}$ were chosen we deduce from Lemma 1 that $S_{p_l}(z)$ has a pole (at least one) in the disc $|z-1/\alpha_{1l}| \leqq 2r_l$. So our assertion follows now from Lemma 3.

REMARK. There are possibilities for generalization. By minor changes one gets a similar generalization to that mentioned in the remark following the corresponding example in §3. Also, one may consider other norms rather than the sup norms. For this aim we may use the fact that for a regular function $g(z)$ in $|z| \geqq 1$, the integral $\int_0^{2\pi} |g(re^{i\theta})|\,d\theta$, $p>0$, is a monotonic decreasing function of r in the interval $1 \leqq r \leqq \infty$. This enables us to change Lemma 1 and to get the above example for the L_p norm.

REFERENCES

1. J. L. Walsh, *Interpolation and approximation by rational functions in the complex domain*, Amer. Math. Soc. Colloq. Publ., vol. 20, Amer. Math. Soc., Providence, R. I., 1935.

2. ———, *On approximation to an analytic function by rational functions of best approximation*, Math. Z. **38** (1934), 163–176.

3. ———, *On the overconvergence of certain sequences of rational functions of best approximation*, Acta Math. **57** (1931), 411–435.

4. O. Perron, *Die Lehre von den Kettenbrüchen*, Chelsea, New York, 1929.

5. H. S. Wall, *Analytic theory of continued fractions*, Chelsea, New York, 1967.

UNIVERSITY OF MARYLAND,
COLLEGE PARK, MARYLAND 20742

Commentary by E.B. Saff[1]

Comments on [31-c*]

It follows immediately from Walsh's result [9, Section 8.7] relating to interpolation by rational functions with fixed poles that, if f is a single-valued analytic function whose set of singularities has zero capacity, then

$$\lim_{n\to\infty} \rho_n^{1/n} = 0,$$

where ρ_n is the error in best approximation of f in the uniform metric on a compact set $E \subset \mathbf{C}$ by rational functions of degree at most n. In particular, for entire and meromorphic functions the corresponding limit exists and is equal to zero. Karlsson [3] proved that

$$\lim_{n\to\infty} \frac{\log \rho_n}{n \log n} \leq -\frac{1}{\sigma}$$

for meromorphic functions of finite order $\sigma \geq 0$. Prokhorov [7] strengthened this result by proving that if f is a meromorphic function of finite order $\sigma \geq 0$, then

$$\limsup_{n\to\infty} \frac{\log(\rho_1 \cdots \rho_n)}{n^2 \log n} \leq -\frac{1}{\sigma}$$

[1]The editors are grateful to Prof. V. Prokhorov for providing many of the references to the Russian literature.

and

$$\liminf_{n\to\infty} \frac{\log \rho_n}{n \log n} \le -\frac{2}{\sigma}.$$

For further results on the degree of best rational approximation to entire functions, see [8], [4].

Concerning the analogous problem for Padé approximants, Nuttall [5] and Pommerenke [6] obtained the results on the convergence in measure and in capacity of diagonal sequences of Padé approximants of meromorphic functions. We also mention Gonchar's papers [1], [2] relating to the convergence of the classical and multipoint Padé approximants of meromorphic functions.

References

[1] A. A. Gonchar, *The convergence of Padé approximations*, Mat. Sb. **92(134)** (1973) 152–164; English transl. in Math. USSR Sb. **21** (1973).

[2] A. A. Gonchar, *On the convergence of generalized Padé approximants of meromorphic functions*, Mat. Sb. **98(140)** (1975) 564–577; English transl. in Math. USSR Sb. **27** (1975).

[3] Johan Karlsson, *Rational interpolation and best rational approximation*, J. Math. Anal. Appl. **53** (1976), 38–52.

[4] A. V. Krot, V. A. Prokhorov, E. B. Saff, *On ray sequences of the best rational approximations of entire functions*, Approximation Theory IX (eds. Charles K. Chui and Larry L. Schumaker), Vanderbilt University Press, Nashville, 1998.

[5] J. Nuttall, *The convergence of Padé approximants of meromorphic functions*, J. Math. Anal. Appl. **31** (1970), 147–153.

[6] Ch. Pommerenke, *Padé approximants and convergence in capacity*, J. Math. Anal. Appl. **41** (1973), 775–780.

[7] V. A. Prokhorov, *On the degree of rational approximation of meromorphic functions*, Mat. Sb. **185** (1994), 3–26; English transl. in Russian Acad. Sci. Sb. Math. **81** (1995).

[8] V. A. Prokhorov, *Rational approximation of analytic functions*, Mat. Sb. **184** (1993), 3–32; English transl. in Russian Acad. Sci. Sb. Math. **78** (1994).

[9] J. L. Walsh, *Interpolation and Approximation by Rational Functions in the Complex Domain*, 5th ed., Amer. Math. Soc., Providence, RI, 1969.

Comments on [34-b*]

This paper is of fundamental importance. Here Walsh first introduced the table $\{R_{mn}\}_{n,m=0}^{\infty}$ of best uniform rational approximations, now called the *Walsh table*, and studied the convergence of the sequences formed from the table. In particular, he investigated the convergence of reciprocals of polynomials. In this direction, Jackson type theorems were later obtained by Levin and Saff [3], [4], and Leviatan,

Levin and Saff [5]. Walsh also gave an example of a continuous complex-valued function on a compact set in the plane whose best uniform complex rational approximation of order 1 is not unique. It is well known that for a real function f on the real interval $E = [a, b]$, the best uniform approximation of f by real rational functions is unique (see, for example, [1]). Nonuniqueness of best *complex* rational approximation can even hold in approximating a real function on a closed interval. We refer the reader to the papers by Lungu [6], Saff and Varga [8], [9]. The case when E is the unit disk was investigated by Gutknecht and Trefethen [2]. For results that give conditions for uniqueness of best complex rationals, see [7], [10].

References

[1] N. I. Achieser, *Theory of Approximation*, Frederick Ungar, New-York, 1956.
[2] M. H. Gutknecht and L. N. Trefethen, *Nonuniqueness of best rational Chebyshev approximations on the unit disk*, J. Approx. Theory **39** (1983), 275–288.
[3] A. L. Levin and E. B. Saff, *Degree of approximation of real functions by reciprocals of real and complex polynomials*, SIAM J. Math. Anal. **19** (1988), 233–245.
[4] A. L. Levin and E. B. Saff, *Jackson type theorems in approximation by reciprocals of polynomials*, Rocky Mountain J. **19** (1989), 243–249.
[5] D. Leviatan, A. L. Levin and E. B. Saff, *On approximation in the L^p-norm by reciprocals of polynomials*, J. Approx. Theory **57** (1989), 322–331.
[6] K. N. Lungu, *The best approximations by rational functions*, Mat. Zametki **10** (1971), 11–15 (Russian).
[7] A. Ruttan, *A characterization of best complex rational approximants in a fundamental case*, Constr. Approx. **1** (1985), 287–296.
[8] E. B. Saff and R. S. Varga, *Nonuniqueness of best approximating complex rational functions*, Bull. Amer. Math. Soc. **83** (1977), 375–377.
[9] E. B. Saff and R. S. Varga, *Nonuniqueness of best complex rational approximations to real functions on real intervals*, J. Approx. Theory **23** (1978), 78–85.
[10] J. P. Thiran and M. P. Istace, *Optimality and uniqueness conditions in complex rational Chebyshev approximation with examples*, Constr. Approx. **9** (1993), 83–103.

Comments on [40-b*]

In this paper Walsh investigated questions related to interpolation by rational functions with fixed poles. The corresponding methods allowed him [10, Section 8.7] to obtain the fundamental results in the theory of approximation of analytic functions characterizing the degree of rational approximation. Let E be an arbitrary compact set in the complex plane $\mathbf{C}$. Consider a function f holomorphic on $\bar{\mathbf{C}} \backslash F$, where F is a compact set in the extended complex plane $\bar{\mathbf{C}}$ such that $E \cap F = \emptyset$.

The following inequality of Walsh holds:

$$\limsup_{n\to\infty} \rho_n^{1/n} \le \frac{1}{\rho}, \qquad \rho = \exp(1/C(E,F)),$$

where ρ_n is the distance of f in the uniform metric on E from the class of rational functions of degree at most n and $C(E, F)$ denotes the capacity of the condenser (E, F). Walsh's inequality is sharp in the class of all functions that are holomorphic on $\bar{\mathbb{C}}\setminus F$. However, there is always a subsequence of integers for which this rate can be improved. Using Hankel operator methods, Parfenov [6] for the case when E is the unit disk and Prokhorov [7] in the general case proved the inequalities

$$\limsup_{n\to\infty} (\rho_1\rho_2\cdots\rho_n)^{1/n^2} \le \frac{1}{\rho}$$

and

$$\liminf_{n\to\infty} \rho_n^{1/n} \le \frac{1}{\rho^2}, \tag{1}$$

where the last estimate was conjectured by Gonchar [2]. In [8] Prokhorov and Saff obtained similar estimates for the rate of convergence of ray sequences from the Walsh table of the best rational approximations of holomorphic functions.

We remark that the ordinary limit exists and equality holds in (1) for functions that can be represented in the form

$$f(z) = \int_F \frac{d\mu(\xi)}{\xi - z},$$

where F and (in the general case) complex measure $d\mu$ satisfy special conditions (see [1], [3], [4], [5], [9]). In particular, for analytic functions having finitely many branch points outside E it has been proved that

$$\lim_{n\to\infty} \rho_n^{1/n} = \frac{1}{\rho^2}, \qquad \rho = \exp(1/C(E,F)),$$

where the compact set F is uniquely determined by the compact set E order the branch points of f, as the solution to a minimum condenser capacity problem.

References

[1] W. Barrett, *On the convergence of sequences rational approximations to analytic functions of a certain class*, J. Inst. Math. Appl. **7** (1971), 308–323.

[2] A. A. Gonchar, *Rational approximation of analytic functions*, Linear and Complex Analysis Problem Book (V. P. Havin [Khavin] et al., editors) Lecture Notes in Math., vol. 1043, Springer-Verlag, Berlin, 1984, pp. 471–474.

[3] A. A. Gonchar, *The rate of rational approximation of analytic functions*, Trudy Mat. Inst. Steklov. **166** (1984), 52–60; English transl. in Proc. Steklov Inst. Math. (1986) **1 (166)**.

[4] A. A. Gonchar, *Rational approximations of analytic functions*, Proceedings of International Congress of Mathematicians (Berkeley, CA, 1986), vol. 1, 2,

Amer. Math. Soc., Providence, RI (1987), 739–748; English transl. in Amer. Math. Soc. Transl. (2) **147** (1990).
[5] A. A. Gonchar, *The rate of rational approximation of certain analytic functions*, Mat. Sb. **105 (147)** (1978), 147–163; English transl. in Math. USSR Sb. **34** (1978).
[6] O. G. Parfenov, *Estimates for the singular numbers of the Carleson embedding operator*, Mat. Sb. **131(173)** (1986), 501–518; English transl. in Math. USSR Sb. **59** (1988).
[7] V. A. Prokhorov, *Rational approximation of analytic function*, Mat. Sb. **184** (1993), 3–32; English transl. in Russian Acad. Sci. Sb. Math. **78** (1994).
[8] V. A. Prokhorov and E. B. Saff, *Rates of best uniform rational approximation of analytic functions by ray sequences of rational functions*, Constr. Approx. **15** (1999), 155–173.
[9] H. Stahl, *Orthogonal polynomials with complex-valued weight functions*, I, II, Constr. Approx. **2** (1986), 225–240, 240–251.
[10] J. L. Walsh, *Interpolation and Approximation by Rational Functions in the Complex Domain*, 5th ed., Amer. Math. Soc., Providence, RI, 1969.

Comments on [46-c*]

This benchmark paper develops the concept of exact harmonic majorants which provides a powerful tool for the analysis of the overconvergence phenomenon for sequences of analytic functions. The concept (without the terminology) was introduced by Walsh in [40-c]. Its main application in [46-c*] is stated as Theorem 5, which unifies many of the then previously known results on overconvergence.

Several authors (see the partial list below) have utilized the concept of exact harmonic majorants in their applications of potential theory to problems in approximation.

References

[1] H.-P. Blatt, E. B. Saff, *Behavior of zeros of polynomials of near best approximation*, J. Approx. Theory, **46** (1986), 323–344.
[2] H.-P. Blatt, E. B. Saff, *Distribution of zeros of polynomial sequences, especially best approximations*, International Series of Numerical Mathematics, **74** (1985), 71–82.
[3] H.-P. Blatt, E. B. Saff, M. Simkani, *Jentzsch-Szegö type theorems for the zeros of best approximants*, J. London Math. Soc., **(2) 38** (1988), 307–316.
[4] R. Grothmann, *Ostrowski gaps, overconvergence and zeros of polynomials*, In: Approximation Theory VI, Vol. I (College Station, TX, 1989), 303–306, Academic Press, Boston, 1989.
[5] R. Grothmann, *Distribution of interpolation points*, Ark. Mat., **34** (1996), 103–117.

[6] R. Kovacheva, *Diagonal rational Chebyshev approximants and holomorphic continuation of functions*, Analysis, **10** (1990), 147–161.
[7] R. Kovacheva, *On uniform convergence of rational functions of best L_p-approximation*, Resultate der Mathematik (to appear).

Comments on [64-a*]

In this well known paper Walsh compares the Padé approximants $P_{n\nu}(z)$ (which are best approximants to f in the sense of matching power series coefficients) and the best rational approximants $R_{n\nu}(\varepsilon, z)$ to f in the uniform metric on the closed disk $E = \{z : |z| \le \varepsilon\}$. It is proved that if a certain determinant is not zero then for fixed n and ν, $R_{n\nu}(\varepsilon, z)$ converges as $\varepsilon \to 0$ uniformly to $P_{n\nu}(z)$ on any closed set with the poles of $P_{n\nu}(z)$ deleted. In [74-b] Walsh proved a similar result for the case when $E = [0, \varepsilon]$ and $R_{n\nu}(\varepsilon, z)$ is the best rational approximant to $f \in C^{(n+\nu+1)}[0, 1]$ in the uniform metric on E.

Comments on [65-i*]

This paper continues the study of the convergence of the rows $\{R_{n\nu}\}_{n=0}^{\infty}$ (with fixed denominator degree ν) of the Walsh table $\{R_{jk}\}_{j,k=0}^{\infty}$ of the best rational approximants to a meromorphic function f started in [64-d], [64-h], [65-e]. Theorem 3 is an analog of the classical Montessus de Ballore theorem [2] related to Padé approximants. Assuming the function f has exactly ν poles in E_ρ, Walsh proved that the sequence $\{R_{n\nu}\}_{n=0}^{\infty}$ converges as $n \to \infty$ uniformly on compact sets to f in E_ρ with the poles of f deleted and that the poles of $R_{n\nu}$ tend with the rate of a geometric progression to the poles of f in E_ρ. When f has at most ν poles in E_ρ, Gonchar [6] showed that the sequence $\{R_{n\nu}\}_{n=0}^{\infty}$ converges in capacity (or, more precisely, m_1-almost uniformly on compact subsets of E_ρ) to the function f. Furthermore, Saff [13], and Liu and Saff [11] have studied the case when locally uniform convergence in E_ρ of the best rational approximants $\{R_{n\nu}\}_{n=0}^{\infty}$ is possible for a mermorphic function f having more than ν poles in a closed region $\bar{E}_\rho$.

Saff [14] and Gonchar [6] determined a connection between the rate of convergence to zero of the best approximation error $\delta_{n\nu}$ and the domain of meromorphicity of a given function. It was proved that

$$\limsup_{n\to\infty} \delta_{n\nu}^{1/n} = \frac{1}{\rho_\nu},$$

where ρ_ν is the index of the largest region of type of E_ρ in which f can be continued as a meromorphic function with no more that ν poles (counting multiplicities). We also mention Walsh's papers [66-e], [67-b], [68-d], [69-b] dealing with the rows of the Walsh table.

For the Padé table, the convergence properties and the behavior of the poles in a row of this table were investigated by Gonchar [7], Baker [1], Baker and

Graves-Morris [3], Saff [15], and Graves-Morris [9]. Several authors have studied the converse theorems, in which from the given asymptotic behavior of the poles in a row of the Padé table of a power series, we can draw conclusions concerning the possible meromorphic extension of the function and the location of its poles. The most important results were proved by Gonchar [8], Suetin [16] (see also [4], [5], [10], [17], [18], [19]). In this direction, for rows of the Walsh table, we mention the paper by Prokhorov [12].

References

[1] G. A. Baker, *Essentials of Padé Approximants*, New-York, Academic Press, 1975.

[2] G. A. Baker and P. R. Graves–Morris, *Padé Approximants*, 2nd edition, Encyclopedia of Mathematics, vol. 59, Cambridge University Press, Cambridge, 1996.

[3] G. A. Baker and P. R. Graves–Morris, *Convergence of rows of the Padé table*, J. Math. Anal. Appl. **57** (1977), 323–339.

[4] V. I. Buslaev, *Relations for the coefficients and the singular points of a function*, Mat. Sb. **131(173)** (1986), 357–384 (Russian).

[5] V. I. Buslaev, *On the poles of the m-th line of a Padé table*, Mat. Sb. **117(159)** (1982), 435–441 (Russian).

[6] A. A. Gonchar, *On a theorem of Saff*, Mat. Sb. **94(136)** (1975), 152–157 (Russian).

[7] A. A. Gonchar, *The convergence of Padé approximants of meromorphic functions*, Mat. Sb. **98(140)** (1975), 564–577: English transl. in Math. USSR Sb. **27** (1975).

[8] A. A. Gonchar, *Poles of rows of the Padé table and mermorphic continuation of functions*, Mat. Sb. **115(157)** (1981), 590–613; English transl. in Math. USSR Sb. **43** (1982).

[9] P. R. Graves–Morris, *Convergence of rows of the Padé table*, Lecture Notes in Physics **57** (1977), 55–68.

[10] G. L. López, V. V. Vavilov, *Survey on recent advances in inverse problems of Padé approximation theory*, Rational approximation and interpolation (Tampa, Fla., 1983), 11–26, Lecture Notes in Math., 1105, Springer, Berlin-New York, 1984.

[11] X. Liu and E. B. Saff, *Intermediate rows of the Walsh array of best rational approximants to meromorphic functions*, Methods and Applications of Analysis **2(3)** (1995), 269–284.

[12] V. A. Prokhorov, *Poles of rational Chebyshev approximations and meromorphic continuation of functions*, Mat. Sb. **181** (1990), 354–366; English transl. in Math. USSR Sb. **69** (1991).

[13] E. B. Saff, *On the row convergence of the Walsh array for meromorphic functions*, Trans. Amer. Math. Soc. **146** (1969), 241–257.

[14] E. B. Saff, *Regions of meromorphy determined by the degree of best rational approximation*, Proc. Amer. Math. Soc. **29** (1971), 30–38.

[15] E. B. Saff, *An extension of Montessus de Ballore's theorem on the convergence of interpolating rational functions*, J. Approx. Theory **6** (1972), 63–67.
[16] S. P. Suetin, *An inverse problem for the m-th row of a Padé table*, Mat. Sb. **122(164)** (1984), 238–250 (Russian).
[17] S. P. Suetin, *Poles of the m-th row of a Padé table*, Mat. Sb. **120(162)** (1983), 500–504 (Russian).
[18] V. V. Vavilov, *The convergence of Padé approximants of meromorphic functions*, Mat. Sb. **101(143)** (1976), 44-56; English transl. in Math USSR Sb. **30** (1976).
[19] V. V. Vavilov, V. A. Prokhorov, S. P. Suetin, *Poles of the m-th line of a Padé table and singular points of a function*, Mat. Sb. **122(164)** (1983), no. 4, 475–480 (Russian).

Comments on [67-c*]

Walsh [3, Section 4.6] generalized a classical result of S. N. Bernstein in several directions. These results, now called *Bernstein-Walsh lemmas*, concern the growth of a polynomial in the complex plane given an estimate for the polynomial on a compact subset of the plane. The lemma stated in the paper printed above is yet another generalization, namely to the case of rational functions with a fixed number of poles (see also Gonchar [1]). Walsh used this lemma to study the convergence of the rows of the Walsh table of best rational approximants (see [64-d], [65-e], [65-i]).

For the case of weighted polynomials of the form $w(z)^n P_n(z)$, with deg $P_n \leq n$, a generalization of the Bernstein-Walsh lemma appears in [2, Section III.2].

References

[1] A. A. Gonchar, *Local conditions for the single-valuedness of analytic functions*, Mat. Sb. **89(131)** (1972), 148–164 (Russian).
[2] E. B. Saff and V. Totik, *Logarithmic Potentials with External Fields*, Springer-Verlag, Heidelberg, 1997.
[3] J. L. Walsh, *Interpolation and Approximation by Rational Functions in the Complex Domain*, 5th ed., Amer. Math Soc., Providence, RI, 1969.

Comments on [68-a*]

For $f(x) = |x|$ on $[-1, 1]$, it was shown by Blatt, Iserles, and Saff [1] that the best uniform rational approximants to f on $[-1, 1]$ have all their poles on the imaginary axis $[-1, 1]$. Furthermore, Saff and Stahl [3], [4] proved that the normalized counting measure of these poles converges weak-star to the unit point mass at zero.

Let $\rho_n(|x|)$ be the error in best approximation to $|x|$ in the class of all rational functions of degree at most n. Newman [2] showed that

$$\frac{1}{2}e^{-9\sqrt{n}} \leq \rho_n(|x|) \leq 3e^{-\sqrt{n}}, \qquad n = 4, 5, \ldots.$$

Stahl [5] proved that the conjecture of Varga, Ruttan, and Carpenter [7] is true; namely, that

$$\lim_{n\to\infty} e^{\pi\sqrt{n}} \rho_n(|x|) = 8.$$

More generally Stahl [6] showed that

$$\lim_{n\to\infty} e^{\pi\sqrt{\alpha n}} \rho_n(|x|^\alpha) = 4^{1+\alpha/2}|\sin(\alpha\pi/2)|.$$

References

[1] H.-P. Blatt, A. Iserles and E. B. Saff, *Remark on the behavior of zeros and poles of best approximating polynomials and rational functions*. In: Algorithms for Approximation, (eds. J. C. Mason and M. G. Cox), Inst. Math. Appl. Conf. Ser. New Ser. **10**, Claredon Press, Oxford, 1987, 437–445.

[2] D. J. Newman, *Rational approximation to* $|x|$, Michigan Math. J. **11** (1964), 11–14.

[3] E. B. Saff and H. Stahl, *Sequences in the Walsh table for* x^α. In: Constructive Theory of Functions (eds. K. Ivanov, P. Petrushev and Bl. Sendov), Bulgarian Academy of Science, Sofia, 1992, 246–259.

[4] E. B. Saff and H. Stahl, *Ray sequences of best rational approximants for* $|x|^\alpha$, Can. J. Math. **49(5)** (1997), 1034–1065.

[5] H. Stahl, *Best uniform rational approximation of* $|x|$ *on* $[-1, 1]$, Mat. Sb. **183** (1992), 85–118.

[6] H. Stahl, *Best uniform rational approximation of* x^α *on* $[0, 1]$, Bull. Amer. Math. Soc. **28** (1993), 116–122.

[7] R. S. Varga, A. Ruttan and A. J. Carpenter, *Numerical results on best uniform rational approximation of* $|x|$ *on* $[-1, 1]$, Mat. Sb. **182** (1991), 1523–1541 (Russian).

Comments on [71-b*]

In connection with Theorem 2, we remark that Gonchar [1], [2], [3] used Borel series

$$\sum_{n=1}^{\infty} \frac{A_n}{z - \alpha_n}$$

to give a simple method for constructing a function with a given domain of analyticity. Let $\bar{G}$ be a closed region and G be the set of interior points of $\bar{G}$. Denote by $\{\alpha_n\}_{n=1}^{\infty}$ a sequence of points lying in the complement of $\bar{G}$, such that the set of their limit points coincides with $\partial G = \bar{G} \backslash G$. Let

$$f(z) = \sum_{n=1}^{\infty} \frac{A_n}{z - \alpha_n}, \qquad z \in G.$$

If

$$\limsup_{n\to\infty} |A_n|^{1/n} < 1$$

and

$$A_n \neq 0, \qquad n = 1, 2, \ldots,$$

then every point of ∂G is a singular point of the analytic function $f(z)$, $z \in G$.

Gonchar [4] also defined the class R^0 of functions f analytic at the origin such that

$$\lim_{n\to\infty} \rho_n^{1/n} = 0,$$

where ρ_n is the error in best uniform rational approximation of f on a disk centered at the origin. For functions from the class R^0 he investigated the convergence of best rational approximants. In this direction we also mention Walsh's papers [72-a] and [72-c]. In particular, it was proved by Gonchar [4] that if the function $f \in R^0$, then the diagonal sequence of Padé approximants converges to f in capacity inside of its Weierstrass natural domain of existence W_f. We also single out the following result of Gonchar [2]:

If

$$\liminf_{n\to\infty} \rho_n^{1/n} = 0,$$

then f is a single-valued analytic function in its natural domain of existence.

An example similar to that of Example 2 in the above reprinted paper was obtained by Wallin [5].

References

[1] A. A. Gonchar, *Properties of functions related to their rate of approximability by rational fractions*, Proc. Internat. Congr. Math. (Moscow, 1966) 329–356, Mir, Moscow; Amer. Math. Soc. Transl. **91** (1970).

[2] A. A. Gonchar, *Local condition of single-valuedness of analytic functions*, Mat. Sb. **89(131)** (1972) 148–164; English transl. in Math. USSR Sb. **18** (1972).

[3] A. A. Gonchar, *Generalized analytic continuation*, Mat. Sb. **76(118)** (1968) 135–146; English transl. in Math. USSR Sb. **5** (1968).

[4] A. A. Gonchar, *The convergence of Padé approximants*, Mat. Sb. **92(134)** (1973); English transl. Math. USSR Sb. **21** (1973), 155–166.

[5] H. Wallin, *On the convergence theory of Padé approximants*. In: Linear operators and approximation (Proceedings of the Conference Oberwolfach, 1971), International Series Numer. Mathematics, Vol. 20, 461–469, Birkhauser, Basel, 1972.

Chapter 7

Spline Functions

Reprinted from the PROCEEDINGS OF THE NATIONAL ACADEMY OF SCIENCES
Vol. 52, No. 6, pp. 1412–1419. December, 1964.

FUNDAMENTAL PROPERTIES OF GENERALIZED SPLINES

BY J. H. AHLBERG,* E. N. NILSON,† AND J. L. WALSH

DEPARTMENT OF MATHEMATICS, HARVARD UNIVERSITY

Communicated October 15, 1964

1. *Introduction.*—A spline function of a single variable (more simply, a spline), defined on an interval $[a,b]$ of the real line, is composed of segments of polynomial functions of degree $2n - 1$ so joined that the resulting composite function is of class C^{2n-2} $[a,b]$. The cubic spline function ($n = 2$) represents the analytic counterpart of the draftsman's spline in consequence of the small deflection property of beams.

In 1946, Schoenberg[1] studied the use of splines in the smoothing of equidistant data. Holladay[2] in 1957 demonstrated the minimum curvature property for certain cubic splines of interpolation, which corresponds to the minimum strain energy property of beams. In 1962–1963 we[3] extended this property to periodic cubic splines and exhibited the best-approximation and other characteristic properties together with proofs of convergence for cubic splines of interpolation.

In 1963, we[4] first announced the extension of the extremal properties to periodic splines of odd degree. Subsequently, DeBoor[5] and later Schoenberg[6] exhibited these properties for nonperiodic splines of odd degree. In addition, we recently announced[7] the orthogonality of splines and expansion of functions in series of orthogonal splines, including a consideration of complex-valued splines and the approximation of analytic functions of a complex variable.

In 1960, Birkhoff and Garabedian[8] considered the problem of approximating surfaces by specialized piecewise bicubic surfaces; in 1962, DeBoor[9] demonstrated the existence and uniqueness of spline surfaces of interpolation, restricting his attention to bicubic splines with particular boundary conditions. We[10] have now extended these results on existence and uniqueness to a large variety of multidimensional splines. In addition, we derive best approximation, minimum norm, orthogonality, and convergence properties.

All of the preceding work has thus been based upon approximation by piecewise-polynomial functions. The present generalization of this to piecewise solutions of a general linear differential equation constitutes a major extension in the study of splines and represents an objective of first importance. As a step in this direction, Schoenberg[11] introduced the so-called "trigonometric splines," relating these to a linear differential operator with constant coefficients, and suggested that the splines associated with a general linear differential operator of finite order could be similarly handled. In 1964, Greville[12] considered such generalized splines and exhibited smoothness and best-approximation results for the nonperiodic case. Both Schoenberg and Greville employed the method of Lagrange's multipliers to obtain these extremal properties. It is the relationship, however, between the linear differential operator and its adjoint, considered to some extent by Greville, which plays a fundamental role not previously presented in the literature.

We have developed, in our series of papers, a simple integral relation which was first expressed by Holladay[2] for cubic splines. This relationship is a fundamental and powerful tool in the demonstration of the extremal properties of generalized splines, but yields the existence and uniqueness theorems as well. It is the purpose of the present communication to derive this basic integral relation, to present the resulting best-approximation and minimum norm together with existence, uniqueness, and orthogonality properties for both periodic and nonperiodic generalized splines. Convergence properties are announced which are closely related to the Birkhoff-de Boor[13] recent results for ordinary cubic splines, and extensions to multidimensional generalized splines are indicated.

2. *Generalized Splines and the Integral Relation.*—Let L be a linear differential operator of order n,

$$L = p_0(x)D^n + p_1(x)D^{n-1} + \ldots + p_{n-1}(x)D + p_n(x)\cdot, \qquad (1)$$

where it is assumed that $p_k(x)$ is of class $C^n[0,1]$, $0 \leq k \leq n$. Let L^* be the formal adjoint,

$$L^* = (-1)^n D^n(p_0\cdot) + (-1)^{n-1}D^{n-1}(p_1\cdot) + \ldots - D(p_{n-1}\cdot) + p_n\cdot, \qquad (2)$$

and $P(u,v)$ the bilinear concomitant,[14]

$$P(u,v) \equiv \sum_{j=0}^{n-1}\sum_{k=0}^{n-j-1}(-1)^k u^{(n-j-k-1)}(p_j v)^{(k)} \equiv \sum_{i=0}^{n-1}\sum_{k=0}^{i} u^{(n-i-1)}(-1)^k(p_{i-k}v)^{(k)}. \qquad (3)$$

Choose a subdivision, $\Delta: 0 = x_0 < x_1 < \ldots < x_N = 1$, of the unit interval. A spline associated with the linear differential operator L on the subdivision Δ is a $C^{2n-2}[0,1]$ function S_Δ which $\epsilon\ C^{2n}$ on each interval of this subdivision and there satisfies

$$L^*LS_\Delta = 0. \qquad (4)$$

The spline S_Δ is said to be *a spline of interpolation* to a function f defined on $[0,1]$ if $S_\Delta(x_j) = f(x_j), j = 0,1, \ldots, N$. When f is of class $C^{n-1}[0,1]$, with $f^{(n-1)}$ absolutely continuous and $f^{(n)}$ of Lebesgue class $L^2[0,1]$, and S_Δ is any spline associated with L on the mesh Δ, we may write

$$\int_0^1 (Lf)^2dx = \int_0^1 (L(f - S_\Delta))^2dx + 2\int_0^1 L(f - S_\Delta)LS_\Delta dx + \int_0^1 (LS_\Delta)^2dx, \qquad (5)$$

and the integral relation results when the middle integral on the right vanishes.

Now from the relation between L and L^*, $vLu = uL^*v + (d/dx)P(u,v)$, we obtain

$$\int_0^1 L(f - S_\Delta)LS_\Delta dx = \int_0^1 \frac{d}{dx}P(f - S_\Delta, LS_\Delta)dx + \int_0^1 (f - S_\Delta)L^*LS_\Delta dx, \qquad (6)$$

the last term of which vanishes by (4). To handle the discontinuities in $S_\Delta^{(2n-1)}$ at the junctions x_j, the first integral on the right is written

$$\sum_{i=0}^{n-2}\sum_{k=0}^{i}(-1)^k(f - S_\Delta)^{(n-i-1)}[p_{i-k}LS_\Delta]^{(k)}\Big|_{x=0}^{x=1} + \sum_{k=0}^{n-2}(-1)^k(f - S_\Delta)[p_{n-k-1}LS_\Delta]^{(k)}\Big|_0^1 + (-1)^{n-1}\sum_{j=1}^{N}(f - S_\Delta)[p_0LS_\Delta]^{(n-1)}\Big|_{x_{j-1}}^{x_j}.$$

If S_Δ interpolates to f at the junctions x_j, the last two terms vanish. We thus arrive at the integral relation,

$$||f||^2 = ||f - S_\Delta||^2 + ||S_\Delta||^2, \qquad ||f||^2 \equiv \int_0^1 (Lf)^2dx, \qquad (7)$$

if, for $k = 1,2, \ldots, n-1$, $S_\Delta^{(k)}(x_i) = f^{(k)}(x_i)$ or $\sum_{j=0}^{k-1}(-1)^j\{p_{k-1-j}\cdot LS_\Delta\}^{(j)}(x_i) = 0$, $i = 0$ and N;

or if f and S_Δ are of class $C^{2n-2}(-\infty, \infty)$ and have period 1.

These requirements lead to a natural selection of three types of splines; we require that S_Δ take on prescribed values at the subdivision points, $S_\Delta(x_j) = f_j$ $(j = 0,1, \ldots, N)$, and further that one of the following end conditions is met:

Type I. $S_\Delta^{(k)}(x_i) = A_{k,i}$, $i = 0$ and N, $k = 1,2, \ldots, n-1$, $A_{k,i}$ prescribed.
Type II. $(LS_\Delta)^{(k)}(x_i) = B_{k,i}$, $i = 0$ and N, $k = 0,1, \ldots, n-2$, $B_{k,i}$ prescribed.
Type III. S_Δ of class $C^{2n-2}(-\infty, \infty)$ and of period 1.

When the prescribed values in Type I and Type II are all zero, these classes are referred to as Type I′ and Type II′. In general, a real-valued function of class $C^{2n-2}[0,1]$ satisfying one of these three sets of conditions is referred to as being of the corresponding type. We have proved the following theorem:

THEOREM 1. *Let f be of class* $C^{n-1}[0,1]$ *with* $f^{(n-1)}$ *absolutely continuous and* $f^{(n)}$ *of class* $L^2[0,1]$. *Let* Δ: $0 = x_0 < x_1 < \ldots < x_N = 1$ *be a subdivision of* $[0,1]$ *and* S_Δ *a spline on* Δ *associated with the operator* (1) *for which* $p_k \in C^n[0,1]$ $(k = 0,1, \ldots, n)$, *interpolating to f at the points* x_j $(j = 0,1, \ldots, N)$. *If* $f - S_\Delta$ *is of Type I′, if* S_Δ *is of Type II′, or if f and* S_Δ *are of Type III, then*

$$\|f\|^2 = \|f - S_\Delta\|^2 + \|S_\Delta\|^2.$$

It should be noted that although the preceding remarks have been made under the assumption that the functions involved are real, the extension to complex-valued functions is immediate.

3. *Extremal Properties.*—With precisely the methods of proof employed by the authors for piecewise-polynomial splines the extremal properties for generalized splines are obtained. The following theorem displays the best-approximation property.

THEOREM 2. *Let f,* Δ, *and L satisfy the conditions of Theorem* 1. *If* $f - S_\Delta$ *is of Type I′, or if* S_Δ *is of Type II, or if f and* S_Δ *are of Type III, then* $\|f - S_\Delta\|$ *is minimum for all such splines* S_Δ *when* S_Δ *is a spline of interpolation, provided the latter exists.*

Proof: The existence and uniqueness of the spline of interpolation is considered in the next section. Here let $S_{\Delta,f}$ represent a spline of interpolation and in (7) replace f by $f - S_\Delta$, S_Δ by $S_{\Delta,f} - S_\Delta$. There results

$$\|f - S_\Delta\|^2 = \|S_{\Delta,f} - S_\Delta\|^2 + \|f - S_{\Delta,f}\|^2,$$

and the conclusion of the theorem follows. It is further seen that $\|f - S_\Delta\| > \|f - S_{\Delta,f}\|$ unless $\|S_\Delta - S_{\Delta,f}\| = 0$: that is, unless S_Δ differs from $S_{\Delta,f}$ by at most a solution of $Lu = 0$.

The minimum-norm property is given in the next theorem.

THEOREM 3. *Let f,* Δ, *and L satisfy the conditions of Theorem* 1. *Let g satisfy the conditions on f and interpolate to f on* Δ. *Then if* $S_{\Delta,f}$ *exists and* $g - S_{\Delta,f} \in$ *Type I′, or* $S_{\Delta,f} \in$ *Type II′, or f,g and* $S_{\Delta,f} \in$ *Type III,*

$$\|g\|^2 = \|g - S_{\Delta,f}\|^2 + \|S_{\Delta,f}\|^2.$$

If $g^{(k)}(x_i)$ *is prescribed,* $i = 0$ *and N* $(k = 1, \ldots, n-1)$, *then* $\|g\|$ *is minimum when g coincides with the* $S_{\Delta,f}$ *for which* $g - S_{\Delta,f} \in$ *Type I′. If no added restriction is placed upon g, then* $\|g\|$ *is minimum when* $g = S_{\Delta,f}$ *where* $S_{\Delta,f} \in$ *Type II′. In the periodic case,* $\|g\|$ *is minimum when* $g = S_{\Delta,f}$.

Proof: In (7) replace f by g and S_Δ by $S_{\Delta,f}$. Again $||g||$ minimum implies $L(g - S_{\Delta,f}) = 0$ a.e. If $g - S_{\Delta,f} \epsilon$ Type I′, however, then $g = S_{\Delta,f}$: the spline of minimum norm is unique in this case.

COROLLARY. *Under the conditions of Theorem 3, if* $g - S_{\Delta,f} \epsilon$ *Type I′, then* $g = S_{\Delta,f}$.

These are the basic extremal properties. Other related extremal properties carry over to generalized splines in a similar fashion.

4. *Existence and Uniqueness.*—It will be assumed that the differential equation (4) has $2n$ independent solutions of class $C^{2n}[0,1]$: $v_i(x)$, $i = 1,2, \ldots, 2n$. In the interval $x_{j-1} \leq x \leq x_j$, express the spline S_Δ as

$$S_\Delta = \sum_{i=1}^{2n} a_{j,i}v_i.$$

Continuity conditions on S_Δ require, for $j = 1,2, \ldots, N - 1$ (and also N in the periodic case), that

$$\sum_{i=1}^{2n} a_{j,i}v_i^{(p)}(x_j) = \sum_{i=1}^{2n} a_{j+1,i}v_i^{(p)}(x_j),\ p = 0,1, \ldots, 2n - 2.$$

In addition, if S_Δ interpolates to f at the points of Δ, we have

$$\sum_{i=1}^{2n} a_{j,i}v_i(x_j) = f(x_j),\ j = 0,1, \ldots, N.$$

In the periodic case, there result $2nN$ relations on the $2nN$ quantities $a_{j,i}$. If in the nonperiodic cases it is assumed that $f - S_\Delta \epsilon$ Type I′, or $S_\Delta \epsilon$ Type II, there are $2n - 2$ additional relations in addition to the $(2n - 1)(N - 1) + N + 1$ expressed above, yielding in all again $2nN$. It is to be shown that under suitable restrictions the solution of each such system exists and is unique.

Let f be the null function. There exists a spline of interpolation to f, the null spline, which gives zero measure of approximation and zero norm. If $f - S_\Delta \epsilon$ Type I′, S_Δ is unique by the corollary to Theorem 3. For $S_\Delta \epsilon$ Type II, if $\tilde{S}_\Delta$ were any other spline of interpolation with $\tilde{S}_\Delta - S_\Delta \epsilon$ Type II′, its measure of approximation would be zero and so $L(\tilde{S}_\Delta - S_\Delta) = 0$. There is a maximum number of zeros N_L which a nontrivial solution to $L(f) = 0$ can have in $[0,1]$ in view of the assumptions made upon the differential operator. If $N > N_L$, it must follow that $\tilde{S}_\Delta = S_\Delta$. In the periodic case, it again follows that $\tilde{S}_\Delta = S_\Delta$ if $N > N_L$. Alternatively, it may be noted that when the intervals of the subdivision are all sufficiently short, on n adjacent ones the equation possesses what Pólya[15] calls "Property W" and on these $\tilde{S}_\Delta - S_\Delta$ can have at most $n - 1$ zeros without vanishing identically. We are again led to the uniqueness of S_Δ.

In all three of these situations, the only solution of the homogeneous system of $2nN$ simultaneous equations in the quantities $a_{j,i}$, resulting when $f \equiv 0$, is the null solution. Thus for general f, the spline of interpolation exists and is unique.

THEOREM 4. *Let f, Δ, and L satisfy the conditions of Theorem 1. The spline associated with the operator L interpolating to f on the points of Δ exists and is unique if $f - S_\Delta \epsilon$ Type I′, if $S_\Delta \epsilon$ Type II and $N > N_L$ (the maximum number of zeros of a nontrivial solution of $Lf = 0$ on $[0,1]$), or if f and S_Δ are of Type III and $N > N_L$.*

5. *Orthogonality.*—The orthogonality properties enunciated for the simple

spline extend in a natural manner to generalized splines. We introduce the pseudo-inner product

$$(u,v) = \int_0^1 LuLvdx,$$

and describe u and v as "orthogonal" if $(u,v) = 0$. We obtain directly the following theorem.

THEOREM 5. *Let Δ and $\bar{\Delta}$ be two subdivisions of* [0,1] *and let $\Delta \subset \bar{\Delta}$. Let S_Δ and $S_{\bar{\Delta}}$ be splines associated with L on Δ and $\bar{\Delta}$, respectively, with $S_{\bar{\Delta}} = 0$ on Δ. If $S_{\bar{\Delta}}$ is of Type I′ or S_Δ is of Type II′, or if S_Δ and $S_{\bar{\Delta}}$ are of Type III, then S_Δ and $S_{\bar{\Delta}}$ are orthogonal.*

Proof: The integral

$$\int_0^1 L(S_\Delta)L(S_{\bar{\Delta}})dx$$

is shown to be zero in precisely the same way that the integral of the cross product in (5) was shown to vanish, with the additional consideration that integrals over [0,1] be expressed as sums of integrals over intervals of Δ.

Consider now a monotone sequence of subdivisions $\{\Delta_k\}$ of [0,1]: $\Delta_k \subset \Delta_{k+1}$. For a given f satisfying the conditions of Theorem 1, let $\{S_{\Delta_k}\}$ be the associated sequence of splines of interpolation to f satisfying one of the following conditions: $f - S_{\Delta_k} \in$ Type I′, $S_{\Delta_k} \in$ Type II′, or f and $S_{\Delta_k} \in$ Type III. (In the latter two situations, it is assumed that the number of mesh points on Δ exceeds N_L, the maximum number of zeros of a nontrivial solution of $Lf = 0$ on [0,1].) We obtain

$$\begin{aligned}\|f - S_{\Delta_k}\|^2 &= \|f - S_{\Delta_{k+1}}\|^2 + \|S_{\Delta_k} - S_{\Delta_{k+1}}\|^2 \\ &= \lim_{p\to\infty} \|f - S_{\Delta_p}\|^2 + \sum_{j=k}^{\infty} \|S_{\Delta_{j+1}} - S_{\Delta_j}\|^2,\end{aligned}$$

since $\{\|f - S_{\Delta_j}\|\}$ is monotone decreasing. Also

$$\begin{aligned}\|S_{\Delta_k}\|^2 &= \|S_{\Delta_{k+1}}\|^2 - \|S_{\Delta_{k+1}} - S_{\Delta_k}\|^2 \\ &= \lim_{p\to\infty} \|S_{\Delta_p}\|^2 - \sum_{j=k}^{\infty} \|S_{\Delta_{j+1}} - S_{\Delta_j}\|^2,\end{aligned}$$

since $\|f\| \geq \|S_{\Delta_p}\|$ and $\{\|S_{\Delta_p}\|\}$ is monotone increasing.

Sequences of orthogonal splines enter in the following natural way. For Type I′ situations, take a sequence of distinct points on [0,1]: $P_0(x = 0)$, $P_1(x = 1)$, $P_2, P_3, \ldots$. Set $\Delta_1 = \{P_0, P_1\}$, $\Delta_m = \{P_0, P_1, \ldots, P_m\}$, $m > 1$. Introduce the spline S_{Δ_1} such that $f - S_{\Delta_1} \in$ Type *I′*. Next introduce the splines V_k of Type *I′* such that $V_k = 1$ at P_k and $V_k = 0$ on Δ_{k-1}. Then

$$(V_k, V_{k'}) = 0, \qquad k \neq k'.$$

Also, the spline S_{Δ_k} of interpolation to f on Δ_k is given by

$$S_{\Delta_k} = S_{\Delta_1} + \sum_{i=2}^{k} a_i V_i,$$

where

$$a_i = (S_{\Delta_k}, V_i)/(V_i, V_i) = (f, V_i)/(V_i, V_i).$$

Thus

$$||f - S_{\Delta_k}||^2 = ||f||^2 - ||S_{\Delta_1}||^2 - \sum_{i=2}^{k} a_i^2||V_i||^2.$$

For Type II′ and Type III situations, orthogonal splines can be introduced in a similar manner, provided Δ_1 contains more than N_L distinct points in order to ensure the existence (and uniqueness) of the splines involved.

6. *Convergence.*—The convergence properties of generalized splines together with the possibility of expansion of functions in series of orthogonal generalized splines will be presented in a subsequent communication now in preparation. Results obtained previously for ordinary splines carry over essentially to the generalized splines. It is possible, however, to obtain these results by simpler and more direct methods than those employed previously. These more efficient methods are a consequence of the manner in which the integral identity for generalized splines was derived.

We shall restrict ourselves here to a statement of two of the principal results.

THEOREM 6. *Let* $f \in C^{n-1}[0,1]$ *with* $f^{(n-1)}$ *absolutely continuous and* $f^{(n)}$ *Lebesgue square integrable on* $[0,1]$. *Let* $\{\Delta_k\}$ *be a sequence of subdivisions of* $[0,1]$, Δ_k: $0 = x_{k,0} < x_{k,1} < \ldots < x_{k,n_k} = 1$ *with* $|\Delta_k| = \max_j |x_{k,j+1} - x_{k,j}| \rightarrow 0$. *Let* S_{Δ_k} *be a spline of interpolation to* f *on* Δ_k *associated with the linear differential operator* $L = D^n + p_1D^{n-1} + \ldots + p_{n-1}D + p_n$, $p_i \in C^n[0,1]$. *Then if* $f - S_{\Delta_k} \in$ *Type I′, if* $S_{\Delta_k} \in$ *Type II′, or if* f *and* S_{Δ_k} *are of Type III, it follows that* $(f - S_{\Delta_k})^{(p)}$ *is* $O(|\Delta_k|^{n-p-1/2})$, $p = 0,1, \ldots, n-1$, *and* $S_{\Delta_k}^{(n)} \rightarrow f^{(n)}$ *in the mean.*

THEOREM 7. *Let* L, $\{\Delta_k\}$, *and* $\{S_{\Delta_k}\}$ *satisfy the conditions of Theorem 6 with* $f - S_{\Delta_k}$ *of Type I′ or Type II′, or* f *and* S_{Δ_k} *of Type III. Let* f *be of class* $C^{2n-1}[0,1]$ *and* $f^{(2n-1)}$ *be absolutely continuous on* $[0,1]$. *Then* $(f - S_{\Delta_k})^{(p)}$ *is* $O(|\Delta_k|^{2n-p-1})$, $p = 0,1,\ldots, n-1$. *If, further,*

$$\max_{i,j} \frac{x_{k,j+1} - x_{k,j}}{x_{k,i+1} - x_{k,i}} < b < \infty,$$

then $(f - S_{\Delta_k})^{(p)}$ *is* $O(|\Delta_k|^{2n-p-1})$, $p = n, n+1, \ldots, 2n-1$.

7. *Multidimensional Splines.*—The method of deriving the generalized integral relation also affords, through the relation of a linear partial differential operator and its adjoint, the method for making the fundamental extension to multidimensional generalized splines. For the case in which this operator is of the form L_xL_y and the generalized splines S_Δ satisfy, on each rectangle of the mesh, both $L_x{}^*L_xS = 0$ and $L_y{}^*L_yS = 0$, the extension is immediate. For the case of the bicubic this operator is simply $D_x{}^2D_y{}^2$ and the integral identity is

$$\iint [D_x{}^2D_y{}^2f]^2dxdy = \iint [D_x{}^2D_y{}^2(f - S_\Delta)]^2dxdy + \iint [D_x{}^2D_y{}^2S_\Delta]^2dxdy.$$

The authors will present this extension in detail in another paper.

Note added in proof: There has recently appeared an abstract of a paper on the same subject as the present paper [de Boor, C., and R. E. Lynch, *Notices Amer. Math. Soc.*, **11**, 681 (1964)]. An abstract of the present paper also appeared *loc. cit.*, p. 680.

* With United Aircraft Corporation Research Laboratories.

† Chief, Scientific Staff, Pratt and Whitney Aircraft.

[1] Schoenberg, I. J., *Quart. Appl. Math.*, **4**, 45–99 and 112–141 (1946).

[2] Holladay, J. C., *Math. Tables Aids Comput.*, **11**, 233–243 (1947).

[3] Walsh, J. L., J. H. Ahlberg, and E. N. Nilson, *J. Math. Mech.*, **11**, 225–234 (1962); Ahlberg, J. H., and E. N. Nilson, *J. Soc. Ind. Appl. Math.*, **11**, 95–104 (1963), presented at the International Congress of Mathematicians in Stockholm, August, 1962.

[4] Ahlberg, J. H., E. N. Nilson, and J. L. Walsh, "Best approximation and convergence properties of higher order spline approximations," *J. Math. Mech.*, in press; abstract in *Notices Amer. Math. Soc.*, **10**, 202 (1963).

[5] deBoor, C., *J. Math. Mech.*, **12**, 747–749 (1963).

[6] Schoenberg, I. J., these PROCEEDINGS, **51**, 24–28 (1964), and *Bull. Amer. Math. Soc.*, **70**, 143–148 (1964).

[7] Ahlberg, J. H., E. N. Nilson, and J. L. Walsh, "Orthogonality properties of spline functions," *J. Math. Anal. Appl.*, in press; abstract in *Notices Amer. Math. Soc.*, **11**, 468 (1964).

[8] Birkhoff, G., and H. Garabedian, *J. Math. Phys.*, **39**, 258–268 (1960).

[9] deBoor, C., *J. Math. Phys.*, **41**, 212–218 (1962).

[10] Ahlberg, J. H., E. N. Nilson, and J. L. Walsh, "Extremal, orthogonality and convergence properties of multidimensional splines," *J. Math. Anal. Appl.*, in press; abstract in *Notices Amer. Math. Soc.*, **11**, 468 (1964).

[11] Schoenberg, I. J., *J. Math. Mech.*, **13**, 795–825 (1964).

[12] Greville, T. N. E., "Interpolation by generalized spline functions," *MRC Tech. Rep.* 476, Univ. of Wisconsin, May 1964.

[13] Birkhoff, G., and C. de Boor, *J. Math. Mech.*, **13**, 827–835 (1964).

[14] Ince, E. L., *Ordinary Differential Equations* (New York: Dover, 1956), p. 124.

[15] Pólya, G., *Trans. Amer. Math. Soc.*, **24**, 233–243 (1922).

COMPLEX CUBIC SPLINES

BY
J. H. AHLBERG, E. N. NILSON AND J. L. WALSH

Introduction. The past two decades have witnessed an increasing intensity of investigation, [1]–[26], into the properties of spline functions. These functions, which in their simplest form yield the analytic counterpart of the draftsman's tool for drawing a smooth curve through a number of prescribed points, play an important and fundamental role in many parts of numerical analysis.

Many of the properties which these functions possess, such as the minimum curvature property [3], [8], [13], [14], are associated with obvious attributes of the elastic beam. Some relate to the best approximation of linear operators [15], [20], and have rather profound meanings in approximation theory. Others, such as orthogonality [24], rate of convergence [12], [17], [23], [25], and completeness [25] of certain bases of splines, are at first rather surprising. A few (cf. [22, p. 241]) are still quite puzzling.

The application of spline theory to the approximation of a function analytic interior to a rectifiable Jordan curve and continuous in the corresponding closed region has been considered to a limited extent in an earlier paper, [24], by the authors. There splines were treated which are piecewise polynomial in the arc length s on the curve. The present development is concerned with cubic splines in the complex variable z and provides some insight into the structure of the spline approximation generally. In particular, it serves to establish a connection with the classical theory of approximation to an analytic function.

As part of this development, proofs are given for the convergence of the complex cubic spline and its derivatives for the situations in which the approximated function is of class C^α ($\alpha=0, 1, 2, 3$, or 4) on the boundary. These may be modified in an obvious manner for the standard real cubic spline and for the convergence properties of second and third derivatives. There result noteworthy simplifications over proofs already existing in the literature [12], [17], for $\alpha=2, 3, 4$.

The convergence properties of cubic splines for cases in which the approximated function is assumed merely to be continuous or to have continuous first derivative constitute significant new developments in spline theory. In addition, a curious spline property is here presented relating to the approximation of the fourth derivative.

The complex spline approximation. Let K be a rectifiable Jordan curve in the complex plane. Let $t_1, t_2, \ldots, t_N$ be points on K arranged in counterclockwise order, separating K into arcs K_j ($j=1, \ldots, N$) with K_j the arc from t_{j-1} to t_j ($t_{N+1}=t_1$).

Presented to the Society, November 8, 1965; received by the editors November 15, 1965.

For this subdivision Δ of K we form a spline $q_\Delta(t)$ on K composed of complex cubics: for t on K_j,

$$q_\Delta(t) = \frac{M_{j-1}}{6h_j}(t_j-t)^3 + \frac{M_j}{6h_j}(t-t_{j-1})^3 + \left(\frac{f_{j-1}}{h_j} - \frac{M_{j-1}h_j}{6}\right)(t_j-t) + \left(\frac{f_j}{h_j} - \frac{M_jh_j}{6}\right)(t-t_{j-1}), \tag{1}$$

where $h_j = t_j - t_{j-1}$, $f_j = q_\Delta(t_j)$, and $M_j = q''_\Delta(t_j)$. The quantities M_j are to be so chosen that the limiting values of $q'_\Delta(t_j)$, obtained by approach to t_j on K_{j+1} and K_j, are equal: for $j = 1, 2, \ldots, N$,

$$\frac{h_j}{6}M_{j-1} + \frac{h_j+h_{j+1}}{3}M_j + \frac{h_{j+1}}{6}M_{j+1} = \frac{f_{j+1}-f_j}{h_{j+1}} - \frac{f_j - f_{j-1}}{h_j}. \tag{2}$$

If we set

$$\lambda_j = \frac{h_{j+1}}{h_j + h_{j+1}}, \quad \mu_j = 1 - \lambda_j,$$

we obtain

$$\mu_j M_{j-1} + 2M_j + \lambda_j M_{j+1} = 6f[t_{j-1}, t_j, t_{j+1}]. \tag{3}$$

Here $f[t_{j-1}, t_j, t_{j+1}]$ is the second divided difference [27] involving functional values f_{j-1}, f_j, f_{j+1} at the points t_{j-1}, t_j, t_{j+1}.

The existence of the spline $q_\Delta(t)$ on K, assuming prescribed values f_j at the mesh locations t_j, rests upon the possibility of solving the N simultaneous equations (3) for the quantities M_j. Now the coefficient matrix A has dominant main diagonal provided that, for each j ($j = 1, 2, \ldots, N$),

$$2|h_j + h_{j+1}| > |h_j| + |h_{j+1}|.$$

This condition is equivalent to requiring, for each j, that t_j lies within the ellipse with foci t_{j-1} and t_{j+1} and with eccentricity 1/2. If K is a smooth Jordan curve (i.e., has a continuously turning tangent), then this condition is satisfied for sufficiently small mesh norm

$$\|\Delta\| = \max_{1 \leq j \leq N} |h_j|. \tag{4}$$

We assume throughout the remainder of the paper that K is smooth and that $\|\Delta\|$ is sufficiently small that A be nonsingular. The piecewise cubic function $q_\Delta(t)$ then exists and is unique for arbitrary complex values $f_1, f_2, \ldots, f_N$.

Alternatively we may represent the spline $q_\Delta(t)$ in terms of its first derivative at the mesh points, $m_j = q'_\Delta(t_j)$. On K_j we have

$$q_\Delta(t) = \frac{m_{j-1}}{h_j^2}(t_j-t)^2(t-t_{j-1}) - \frac{m_j}{h_j^2}(t-t_{j-1})^2(t_j-t) + \frac{f_{j-1}}{h_j^3}(t_j-t)^2[2(t-t_{j-1})+h_j] + \frac{f_j}{h_j^3}(t-t_{j-1})^2[2(t_j-t)+h_j], \tag{5}$$

and the quantities m_j satisfy the requirement that $q''_\Delta(t_j-)=q''_\Delta(t_j+)$ for $j=1, 2, \ldots, N$:

$$\frac{m_{j-1}}{h_j}+2\left(\frac{1}{h_j}+\frac{1}{h_{j+1}}\right)m_j+\frac{m_{j+1}}{h_{j+1}} = 3\frac{f_{j+1}-f_j}{h_{j+1}^2}+3\frac{f_j-f_{j-1}}{h_j^2}, \tag{6}$$

which may be rewritten in the form

$$\lambda_j m_{j-1}+2m_j+\mu_j m_{j+1} = 3\left[\mu_j\frac{f_{j+1}-f_j}{h_{j+1}}+\lambda_j\frac{f_j-f_{j-1}}{h_j}\right]. \tag{7}$$

The complex spline $S_\Delta(z)$ is now defined interior to K by the Cauchy integral

$$S_\Delta(z) = \frac{1}{2\pi i}\int_K \frac{q_\Delta(t)\,dt}{t-z}, \tag{8}$$

with $S_\Delta(t)$, t on K, as the limiting value (cf. [29]) for approach from within K,

$$S_\Delta(t) = \tfrac{1}{2}q_\Delta(t)+\frac{1}{2\pi i}\int_K \frac{q_\Delta(\tau)\,d\tau}{\tau-t}. \tag{9}$$

Convergence on the boundary. We consider first the convergence of $\{q_{\Delta_k}(t)\}$ and the corresponding sequences of derivatives for a sequence of meshes $\{\Delta_k\}$ with $\|\Delta_k\|\to 0$ as $k\to\infty$. Let the mesh points of Δ_k be $t_{k,1}, t_{k,2}, \ldots, t_{k,N_k}$. The row-max norm of the inverse of the coefficient matrix A_k in (3), or B_k in (7) does not exceed (cf. [10, p. 97]) $\{\min_j [2-|\lambda_{k,j}|-|\mu_{k,j}|]\}^{-1}$ and for arbitrary $\eta>0$ this bound can be made not to exceed $1+\eta$ by taking $\|\Delta_k\|$ sufficiently small. Let $b_{k;i,j}^{-1}$ represent the general element of the inverse matrix B_k^{-1}. For convenience in notation, we will frequently drop the index k on the mesh Δ_k as in the following proof.

Consider the interval K_j. On K_j set $t=(t_{j-1}+t_j)/2+\varepsilon$. Let $f(t)$ be continuous on K. Then from (5) we have

$$\begin{aligned} q_\Delta(t)-f(t) &= (\tfrac{1}{2}h_j-\varepsilon)^2(\tfrac{1}{2}h_j+\varepsilon)\frac{m_{j-1}}{h_j^2}-(\tfrac{1}{2}h_j+\varepsilon)^2(\tfrac{1}{2}h_j-\varepsilon)\frac{m_j}{h_j^2} \\ &\quad +\frac{2f_{j-1}}{h_j^3}(\tfrac{1}{2}h_j-\varepsilon)^2(h_j+\varepsilon)+\frac{2f_j}{h_j^3}(\tfrac{1}{2}h_j+\varepsilon)^2(h_j-\varepsilon)-f(t). \end{aligned}$$

Thus

$$\begin{aligned} |q_\Delta(t)-f(t)| &\leq (\tfrac{1}{2}|h_j|+\varepsilon)^3\frac{3(1+\eta)\cdot 2}{|h_j|^2}\max_i\left[|\mu_i|\frac{|f_{i+1}-f_i|}{|h_{i+1}|}+|\lambda_i|\frac{|f_i-f_{i-1}|}{|h_i|}\right] \\ &\quad +\left|\frac{f_j+f_{j+1}}{2}-f(t)\right|+\left|\frac{3}{2}\frac{\varepsilon}{h_j}-\frac{2\varepsilon^3}{h_j^3}\right||f_j-f_{j-1}|, \end{aligned}$$

since

$$m_j = 3\sum_{j=1}^{N} b_{j,i}^{-1}\left(\mu_i\frac{f_{i+1}-f_i}{h_{i+1}}+\lambda_i\frac{f_i-f_{i-1}}{h_i}\right) \tag{10}$$

and $\sum_k |b_{j,k}^{-1}|\leq 1+\eta$.

Now $|\varepsilon|/|h_j| \leqq 1/2$, $|\mu_j| < 1$, and $|\lambda_j| < 1$ for $\|\Delta\|$ sufficiently small. If we assume that the meshes Δ possess the property that

$$\left[\max_j \|\Delta\|/|h_j|\right] \leqq C_1 < \infty \tag{11}$$

and if we designate by $\mu(\delta; f)$ the modulus of continuity of the function $f(t)$, then on K_j

$$|q_\Delta(t) - f(t)| \leqq [12(1+\eta)C_1 + 2]\mu(\|\Delta\|, f).$$

Since the constants are independent of j, we have proved

THEOREM 1. *Let $f(t)$ be continuous on K. Let $\{\Delta_k\}$ be a sequence of subdivisions of K with* $\lim_{k\to\infty} \|\Delta_k\| = 0$ *and with* $[\max_j \|\Delta_k\|/|h_{k,j}|] \leqq C_1 < \infty$. *Let $q_{\Delta_k}(t)$ be the complex cubic spline on K of interpolation to $f(t)$ on Δ_k. Then $\{q_{\Delta_k}(t)\} \to f(t)$ uniformly as $\|\Delta_k\| \to 0$. If $f(t)$ satisfies a Hölder condition of order β on K $(0 < \beta \leqq 1)$, then* $|q_{\Delta_k}(t) - f(t)| = O(\|\Delta_k\|^\beta)$.

In order to obtain the convergence properties of the complex spline defined in equation (8), it is necessary to show that $q_\Delta(t) - f(t)$ or its derivatives satisfy suitable Hölder conditions. These conditions become involved in the definition of the Cauchy Principal Value of the integral in (9).

For t and τ on K_j we obtain from (5) $(t^* = (t_{j-1} + t_j)/2)$

$$\begin{aligned} q_\Delta(t) - q_\Delta(\tau) &\\ = (t-\tau)\Big\{&\left[m_{j-1} + m_j - 2\frac{f_j - f_{j-1}}{h_j}\right]\left[\frac{(t-t^*)^2 + (t-t^*)(\tau - t^*) + (\tau - t^*)^2}{h_j^2} - \frac{1}{2}\right] \\ &+ (m_j - m_{j-1})\frac{t + \tau - 2t^*}{2h_j} + \frac{(f_j - f_{j-1})}{h_j}\Big\}. \end{aligned} \tag{12}$$

If $f(t)$ satisfies a Hölder condition of order β $(0 < \beta \leqq 1)$ and if $0 \leqq \delta \leqq \beta$, it follows when $\|\Delta\|$ is sufficiently small that for t and τ on K_j,

$$\begin{aligned} |[q_\Delta(t) - f(t)] - [q_\Delta(\tau) - f(\tau)]| &= |t-\tau|^\delta \|\Delta\|^{\beta-\delta} \cdot \frac{|t-\tau|^{\beta-\delta}}{\|\Delta\|^{\beta-\delta}} \\ \cdot\Big\{\Big[|m_{j-1} + m_j| \cdot |t-\tau|^{1-\beta} &+ 2\cdot\frac{|f_j - f_{j-1}|}{|h_j|^\beta}\left|\frac{t-\tau}{h_j}\right|^{1-\beta}\Big]\cdot\frac{1}{4} \\ + \frac{1}{2}|m_j - m_{j-1}|\,|t-\tau|^{1-\beta} &+ \left|\frac{t-\tau}{h_j}\right|^{1-\beta}\cdot\frac{|f_j - f_{j-1}|}{|h_j|^\beta} + \frac{f(\tau) - f(t)}{|\tau - t|^\beta}\Big\}. \end{aligned}$$

Thus, by means of (10), it is seen that $[q_\Delta(t) - f(t)]/\|\Delta\|^{\beta-\delta}$ satisfies a uniform Hölder condition of order δ.

COROLLARY. *Under the conditions of Theorem 1 with $f(t)$ satisfying a Hölder condition of order β $(0 < \beta \leqq 1)$, the function $[q_{\Delta_k}(t) - f(t)]/\|\Delta_k\|^{\beta-\delta}$ satisfies a Hölder condition of order δ, $0 < \delta \leqq \beta$, uniformly with respect to k.*

We turn to a consideration of the case in which $f(t)$ has a continuous derivative on K.

THEOREM 2. *Let $f(t)$ be of class C^1 on K. Let Δ_k: $t_{k,1}, t_{k,2}, \ldots, t_{k,N_k}$ $(k=1,2,\ldots)$ represent a sequence of subdivisions of K with $\lim_{k\to\infty} \|\Delta_k\|=0$. Let $q_{\Delta_k}(t)$ be the complex cubic spline on K of interpolation to $f(t)$ at the points of Δ_k. Then $\{q'_{\Delta_k}(t)\}$ converges uniformly on K to $f'(t)$ and $[q_{\Delta_k}(t)-f(t)]=o(\|\Delta_k\|)$.*

Proof. If we write (7) as

$$B_k\mathbf{m}_k = 3\mathbf{e}_k$$

where $\mathbf{e}_k$ is the N_k-vector $(e_{k,1}, e_{k,2}, \ldots, e_{k,N_k})^{\mathsf{T}}$ with

$$e_{k,j} = \mu_{k,j}[f(t_{k,j+1})-f(t_{k,j})]/h_{k,j+1}+\lambda_{k,j}[f(t_{k,j})-f(t_{k,j-1})]/h_{k,j}$$

and $\mathbf{m}_k$ the corresponding vector of spline first derivatives, $(m_{k,1}, m_{k,2}, \ldots, m_{k,N_k})^{\mathsf{T}}$, then

$$B_k(\mathbf{m}_k-\mathbf{e}_k) = (3I_k-B_k)\mathbf{e}_k. \tag{13}$$

Here I_k is the $N_k\times N_k$ identity matrix. The jth row in $3I_k-B_k$ has three nonzero elements centered on the main diagonal position: $\lambda_{k,j}$, -1, $\mu_{k,j}$. Thus the right-hand member of (13) has the jth element

$$\lambda_{k,j}e_{k,j-1}-e_{k,j}+\mu_{k,j}e_{k,j+1} = \lambda_{k,j}(e_{k,j-1}-e_{k,j})+\mu_{k,j}(e_{k,j+1}-e_{k,j}).$$

Now $f'(t)$ is continuous on K so that $\max_j |f'(t_{k,j})-e_{k,j}|$ can be made arbitrarily small, uniformly with respect to k, by taking $\|\Delta_k\|$ sufficiently small. Thus

$$\|(3I_k-B_k)\mathbf{e}_k\| = 3[\max_j|\lambda_{k,j}(e_{k,j-1}-e_{k,j})+\mu_{k,j}(e_{k,j+1}-e_{k,j})|]$$

can be made uniformly (in k) arbitrarily small. Hence $\max_j |m_{k,j}-f'(t_{k,j})| \to 0$ as $\|\Delta_k\| \to 0$. For t on $K_{k,j}$ with $t=(t_{k,j-1}+t_{k,j})/2+\varepsilon$, we have

$$\begin{aligned} &q'_{\Delta_k}(t)-\frac{f(t_{k,j})-f(t_{k,j-1})}{h_{k,j}} \\ &= \left(\frac{3\varepsilon^2}{h_{k,j}^2}-\frac{1}{4}\right)\left(m_{k,j-1}+m_{k,j}-2\,\frac{f(t_{k,j})-f(t_{k,j-1})}{h_{k,j}}\right)+\frac{\varepsilon}{h_{k,j}}(m_{k,j}-m_{k,j-1}). \end{aligned} \tag{14}$$

For $\|\Delta_k\|$ sufficiently small, $|\varepsilon/h_j|\leqq 1$. The uniform convergence of $\{q'_{\Delta_k}(t)\}$ to $f'(t)$ is a direct consequence. Since for t on $K_{k,j}$ we have

$$q_{\Delta_k}(t)-f(t) = \int_{t_{k,j-1}}^{t} [q'_{\Delta_k}(t)-f'(t)]\, dt,$$

it is evident that $[q_{\Delta_k}(t)-f(t)]=o(\|\Delta_k\|)$.

Since we may write, for t on $K_{k,j}$,

$$f'(t)-\frac{f(t_{k,j})-f(t_{k,j-1})}{h_{k,j}} = \frac{1}{h_{k,j}}\int_{t_{k,j-1}}^{t_{k,j}} [f'(t)-f'(\tau)]\, d\tau,$$

it follows from (14) that $[q'_{\Delta_k}(t)-f'(t)]=O(\|\Delta_k\|^{\beta})$ if $f'(t)$ satisfies a Hölder condition of order β on K, $0<\beta\leq 1$. Thus we have

COROLLARY 1. *Under the conditions of Theorem 2 on $f(t)$, $\{\Delta_k\}$, and $\{q_{\Delta_k}(t)\}$, if $f'(t)$ satisfies a Hölder condition of order β on K $(0<\beta\leq 1)$, then for $p=0, 1$ we have $f^{(p)}(t)-q^{(p)}_{\Delta_k}(t)=O(\|\Delta_k\|^{1+\beta-p})$.*

We can also prove

COROLLARY 2. *Under the conditions of Corollary 1, $[q'_{\Delta_k}(t)-f'(t)]/\|\Delta_k\|^{\beta-\delta}$ satisfies a uniform Hölder condition of order δ, $0\leq\delta\leq\beta$, provided*

$$\left[\max_j \|\Delta_k\|/|h_{kj}|\right] \leq C_1 < \infty.$$

Proof. We obtain from (5) (dropping the index k; again $t^*=(t_{j-1}+t_j)/2$)

$$q'_\Delta(t)-q'_\Delta(\tau) = (t-\tau)\left\{\left[\frac{m_{j-1}-(f_j-f_{j-1})/h_j}{h_j}+\frac{m_j-(f_j-f_{j-1})/h_j}{h_j}\right]\cdot\frac{3(t+\tau-2t^*)}{h_j}+\frac{m_j-m_{j-1}}{h_j}\right\}. \tag{15}$$

It is seen from (13) and the properties of $f'(t)$ that the quantities

$$\frac{m_j-(f_j-f_{j-1})/h_j}{\|\Delta\|^{\beta}}, \qquad \frac{m_{j-1}-(f_j-f_{j-1})/h_j}{\|\Delta\|^{\beta}}$$

are uniformly bounded as is $(m_j-m_{j-1})/\|\Delta\|^{\beta}$. Hence for $\|\Delta\|$ sufficiently small,

$$\begin{aligned}&|[q'_\Delta(t)-f'(t)]-[q'_\Delta(\tau)-f'(\tau)]|\\ &\leq |t-\tau|^{\delta}\|\Delta\|^{\beta-\delta}\cdot\left\{\left[3\,\frac{|m_{j-1}-(f_j-f_{j-1})/h_j|}{\|\Delta\|^{\beta}}+3\,\frac{|m_j-(f_j-f_{j-1})/h_j|}{\|\Delta\|^{\beta}}\right.\right.\\ &\qquad\left.\left.+\frac{|m_j-m_{j-1}|}{\|\Delta\|^{\beta}}\right]\left|\frac{t-\tau}{h_j}\right|^{1-\delta}\cdot\frac{\|\Delta\|^{\delta}}{|h_j|^{\delta}}+\left|\frac{f'(\tau)-f'(t)}{|\tau-t|^{\beta}}\right|\right\}.\end{aligned}$$

The conclusion of the corollary follows.

It is to be noted that the mesh restriction (11) appearing in Theorem 1 is not required in Theorem 2 or in the next theorem. We remark also that by (12) we have, without this mesh restriction,

COROLLARY 3. *Under the conditions of Theorem 2, for arbitrary δ, $0<\delta\leq 1$, the quantity $[q_{\Delta_k}(t)-f(t)]/\|\Delta_k\|^{1-\delta}$ satisfies a Hölder condition of order δ on K uniformly with respect to k.*

THEOREM 3. *Let $f(t)$, $\{\Delta_k\}$, and $\{q_{\Delta_k}(t)\}$ satisfy the requirements of Theorem 2 but with $f(t)$ of class C^2 on K. Then $\{q''_{\Delta_k}(t)\}\to f''(t)$ uniformly on K as $k\to\infty$.*

Proof. Write (3) as

$$A_k\mathbf{M}_k = 3\mathbf{d}_k,$$

where $\mathbf{M}_k=(M_{k,1}, M_{k,2}, \ldots, M_{k,N_k})^{\mathrm{T}}$ and d_k is the N_k-vector whose jth component is $2f[t_{k,j-1}, t_{k,j}, t_{k,j+1}]$. Then

$$A_k(\mathbf{M}_k-\mathbf{d}_k) = (3I_k-A_k)\mathbf{d}_k.$$

The vector $(3I_k-A_k)\mathbf{d}_k$ has norm equal to

$$3 \max_j |\mu_{k,j}(d_{k,j-1}-d_{k,j})+\lambda_{k,j}(d_{k,j+1}-d_{k,j})|.$$

If $f''(t)$ is continuous, then $3d_{k,j}$ (the right-hand member of (3)) can be made uniformly close to $3f''(t_{k,j})$ by taking $\|\Delta_k\|$ sufficiently small. In fact, we have from the Taylor theorem with integral remainder,

$$\begin{aligned} &f[t_{k,j-1}, t_{k,j}, t_{k,j+1}]-\tfrac{1}{2}f''(t_{k,j}) \\ (16)\quad &= \frac{1}{h_{k,j}+h_{k,j+1}}\left\{\frac{1}{h_{k,j+1}}\int_{t_{k,j}}^{t_{k,j+1}} (t_{k,j+1}-\tau)[f''(\tau)-f''(t_{k,j})]\,d\tau \right.\\ &\qquad \left. +\frac{1}{h_{k,j}}\int_{t_{k,j-1}}^{t_{k,j}} (\tau-t_{k,j-1})[f''(\tau)-f''(t_{k,j})]\,d\tau\right\}. \end{aligned}$$

The left-hand member of (16), therefore, does not exceed in absolute value the quantity

$$\begin{aligned} &\left[\max_{t \text{ on } K_{k,j}} |f''(t)-f''(t_{k,j})|\right]\cdot\frac{|h_{k,j+1}|}{|h_{k,j}+h_{k,j+1}|} \\ (16')\quad &+\left[\max_{t \text{ on } K_{k,j-1}} |f''(t)-f''(t_{k,j})|\right]\cdot\frac{|h_{k,j}|}{|h_{k,j}+h_{k,j+1}|} \leqq 2\mu(|t-t_{k,j}|, f'') \end{aligned}$$

for $\|\Delta_k\|$ sufficiently small.

From the boundedness of $\|A_k^{-1}\|$ (uniform with respect to k) it follows that $\|\mathbf{M}_k-\mathbf{d}_k\| \to 0$ as $k\to\infty$. The linearity of $q''_{\Delta_k}(t)$ between junctions $t_{k,j}$ and the uniform continuity of $f''(t)$ now give the conclusion of the theorem.

If $f''(t)$ satisfies a Hölder condition of order β on K, $0<\beta\leqq 1$, then it follows from (16) that $\|\mathbf{M}_k-\mathbf{d}_k\|=O(\|\Delta_k\|^\beta)$. Moreover, for t on $K_{k,j}$ we have

$$\frac{f(t)-f(t_{k,j-1})}{t-t_{k,j-1}}-f'(t_{k,j-1}) = \frac{1}{t-t_{k,j-1}}\int_{t_{k,j-1}}^{t} (t-\tau)f''(\tau)\,d\tau,$$

and so from the interpolation property

$$q'_{\Delta_k}(t_{k,j-1})-f'(t_{k,j-1}) = \frac{1}{h_{k,j}}\int_{t_{k,j-1}}^{t_{k,j}} (t_{k,j}-\tau)[q''_{\Delta_k}(\tau)-f''(\tau)]\,d\tau.$$

Thus $[q'_{\Delta_k}(t_{k,j})-f'(t_{k,j})]=O(\|\Delta_k\|^{1+\beta})$. By using, for t on $K_{k,j}$, the relation

$$q'_{\Delta_k}(t)-f'(t) = q'_{\Delta_k}(t_{k,j})-f'(t_{k,j})+\int_{t_{k,j}}^{t} [q''_{\Delta_k}(\tau)-f''(\tau)]\,d\tau$$

and by a repetition of this argument for $q_{\Delta_k}(t)-f(t)$ we obtain

COROLLARY 1. *Under the conditions of Theorem 2, if $f''(t)$ satisfies a Hölder condition of order β on K $(0<\beta\leqq 1)$, then* $[q^{(p)}_{\Delta_k}(t)-f^{(p)}(t)]=O(\|\Delta_k\|^{2+\beta-p})$ $(p=0,1,2)$. *If $f''(t)$ is continuous on K, then* $[q^{(p)}_{\Delta_k}(t)-f^{(p)}(t)]=o(\|\Delta_k\|^{2-p})$.

We have also

COROLLARY 2. *If $f''(t)$ satisfies a Hölder condition of order β $(0<\beta\leqq 1)$, then* $[q''_{\Delta_k}(t)-f''(t)]/\|\Delta_k\|^{\beta-\delta}$ *satisfies a Hölder condition of order δ on K $(0<\delta\leqq\beta)$ provided* $[\max_j \|\Delta_k\|/|h_{k,j}|]\leqq C_1<\infty$.

Proof. We have, for t and τ on K_j (dropping mesh index k),

$$(17)\qquad |[q''_\Delta(t)-f''(t)]-[q''_\Delta(\tau)-f''(\tau)]| = |t-\tau|\left|\frac{M_j-M_{j-1}}{h_j}-\frac{f''(t)-f''(\tau)}{t-\tau}\right|.$$

Since $M_j-M_{j-1}=O(\|\Delta\|^\beta)$, it is seen that the left-hand member of (17) does not exceed

$$|t-\tau|^\delta\|\Delta\|^{\beta-\delta}\left\{\frac{|t-\tau|^{1-\delta}\|\Delta\|^\delta}{|h_j|}\cdot\left[\frac{M_j-M_{j-1}}{\|\Delta\|^\beta}+\frac{f''(t)-f''(\tau)}{|t-\tau|^\beta}\cdot\frac{|h_j|}{\|\Delta\|^\beta|t-\tau|^{1-\beta}}\right]\right\},$$

and that the term within braces is uniformly bounded with respect to the meshes Δ.

Without the mesh restriction (11) we have

COROLLARY 3. *If $f(z)$ is of class C^1 on K, then* $[q'_{\Delta_k}(t)-f'(t)]/\|\Delta_k\|^{1-\delta}$ *satisfies a Hölder condition of order δ on K, $0<\delta\leqq 1$.*

Proof. Since by (1) we have for t and τ on $K_{k,j}$

$$(18)\qquad q'_{\Delta_k}(t)-q'_{\Delta_k}(\tau) = (t-\tau)\left\{M_{k,j-1}\frac{2t_{k,j}-t-\tau}{2h_{k,j}}+M_j\frac{t+\tau-2t_{k,j-1}}{2h_{k,j}}\right\},$$

it follows that, for $\|\Delta_k\|$ sufficiently small,

$$|[q'_{\Delta_k}(t)-f'(t)]-[q'_{\Delta_k}(\tau)-f'(\tau)]| = |t-\tau|^\delta\|\Delta_k\|^{1-\delta}\cdot\left\{|M_{k,j-1}|+|M_{k,j}|+\left|\frac{f'(\tau)-f'(t)}{\tau-t}\right|\right\}.$$

Sharma and Meir [28] have announced the results of Theorem 3 for the case of a real nonperiodic spline with special end conditions. The proof given above is readily adapted to a wide class of splines including that of Sharma and Meir.

The third derivatives of the spline have jump discontinuities at the junctions t_j of the mesh Δ; nevertheless they possess important convergence properties relative to the function approximated. Some of these have been demonstrated by Birkhoff and deBoor [17] for the special spline later considered by Sharma and Meir. Their proofs are rather involved. We study the problem here for the complex cubic splines (periodic) $q_\Delta(t)$, giving a simple proof which is readily adapted to real splines of general type.

A somewhat more surprising result is the relation of the jumps in $q'''_\Delta(t)$ to $f^{iv}(t)$ when the latter is continuous. This is presented in Theorem 5.

THEOREM 4. *Let $f'''(t)$ be continuous on K. Let $\{\Delta_k\}$ be a sequence of meshes on K with* $\lim_{k\to\infty} \|\Delta_k\| = 0$. *Let $q_{\Delta_k}(t)$ be the complex cubic spline of interpolation to $f(t)$ on Δ_k. If* $[\max_j \|\Delta_k\|/|h_{k,j}|]$ *is uniformly bounded with respect to k, then* $\{q'''_{\Delta_k}(t)\} \to f'''(t)$ *uniformly on K as $k \to \infty$.*

Proof. Set $\sigma_j = (M_j - M_{j-1})/h_j$. In equations (3) we subtract the $(j-1)$th equation from the jth. Noting that

$$\frac{h_{j-1}}{h_{j-1}+h_j+h_{j+1}} = \frac{\mu_{j-1}\mu_j}{1-\mu_{j-1}\lambda_j}, \qquad \frac{h_j}{h_{j-1}+h_j+h_{j+1}} = \frac{\lambda_{j-1}\mu_j}{1-\mu_{j-1}\lambda_j},$$

$$\frac{h_{j+1}}{h_{j-1}+h_j+h_{j+1}} = \frac{\lambda_{j-1}\lambda_j}{1-\mu_{j-1}\lambda_j},$$

we set

$$C = \begin{bmatrix} \frac{(1+\mu_N+\lambda_1)\lambda_N\mu_1}{1-\mu_N\lambda_1} & \frac{\lambda_N\lambda_1^2}{1-\mu_N\lambda_1} & 0 & \cdots & \frac{\mu_N^2\mu_1}{1-\mu_N\lambda_1} \\ \frac{\mu_1^2\mu_2}{1-\mu_1\lambda_2} & \frac{(1+\mu_1+\lambda_2)\lambda_1\mu_2}{1-\mu_1\lambda_2} & \frac{\lambda_1\lambda_2^2}{1-\mu_1\lambda_2} & \cdots & 0 \\ \vdots & & & & \\ \frac{\lambda_{N-1}\lambda_N^2}{1-\mu_{N-1}\mu_N} & 0 & \cdots & \frac{\mu_{N-1}^2\mu_N}{1-\mu_{N-1}\lambda_N} & \frac{(1+\mu_{N-1}+\lambda_N)\lambda_{N-1}\mu_N}{1-\mu_{N-1}\lambda_N} \end{bmatrix}.$$

Then

$$C\sigma = 6\mathbf{r}, \tag{19}$$

where $\sigma = (\sigma_1, \sigma_2, \ldots, \sigma_N)^T$ and $\mathbf{r} = (r_1, r_2, \ldots, r_N)^T$, with $r_j = f[t_{j-2}, t_{j-1}, t_j, t_{j+1}]$. We may write

$$GCH = E, \tag{20}$$

where G and H are the diagonal matrices

$$G = \begin{bmatrix} \frac{1-\mu_N\lambda_1}{\mu_1\lambda_N} & & & & 0 \\ & \frac{1-\mu_1\lambda_2}{\mu_1\mu_2} & & & \\ & & \frac{(1-\mu_2\lambda_3)\lambda_1}{\mu_1\mu_2\mu_3} & & \\ 0 & & & & \frac{(1-\mu_{N-1}\lambda_N)\lambda_1\lambda_2\cdots\lambda_{N-2}}{\mu_1\mu_2\cdots\mu_N} \end{bmatrix},$$

$$H = \begin{bmatrix} 1 & & & & 0 \\ & \frac{\mu_1}{\lambda_1} & & & \\ & & \frac{\mu_1\mu_2}{\lambda_1\lambda_2} & & \\ & & & \ddots & \\ 0 & & & & \frac{\mu_1\mu_2\cdots\mu_{N-1}}{\lambda_1\lambda_2\cdots\lambda_{N-1}} \end{bmatrix},$$

and, since $\lambda_1\lambda_2\cdots\lambda_N=\mu_1\mu_2\cdots\mu_N$,

$$E=\begin{bmatrix} 1+\mu_N+\lambda_1 & \lambda_1 & 0 & \cdots & \mu_N \\ \mu_1 & 1+\mu_1+\lambda_2 & \lambda_2 & \cdots & 0 \\ \vdots & & & & \\ 0 & \cdots & \mu_{N-2} & 1+\mu_{N-2}+\lambda_{N-1} & \lambda_{N-1} \\ \lambda_N & \cdots & 0 & \mu_{N-1} & 1+\mu_{N-1}+\lambda_N \end{bmatrix}.$$

Now $\|E^{-1}\|\leqq 1+\eta$ (arbitrary $\eta>0$) provided $\|\Delta\|$ is sufficiently small. We assume, for all meshes Δ, that we have $\|\Delta\|/\min_j|h_j|\leqq C_1<\infty$. We find, for arbitrary positive integers i and p, that

$$C_1^{-1}\leqq\frac{\lambda_i}{\mu_i}\cdot\frac{\lambda_{i+1}}{\mu_{i+1}}\cdots\frac{\lambda_{i+p}}{\mu_{i+p}}=\left|\frac{h_{i+p+1}}{h_i}\right|\leqq C_1,$$

and that for $\|\Delta\|$ sufficiently small,

$$|\mu_i|=\frac{|h_i|}{|h_i+h_{i+1}|}\leqq\frac{C_1+\eta}{1+C_1},\qquad|\lambda_i|\leqq\frac{C_1+\eta}{1+C_1},$$

from the smoothness of the curve K. We require $\eta<1$. Then

$$\frac{1}{|1-\mu_{i-1}\lambda_i|}\leqq\frac{1}{1-(C_1+\eta)^2/(1+C_1)^2}=\frac{(1+C_1)^2}{(1-\eta)(2C_1+\eta+1)}.$$

It follows that $\|C^{-1}\|\leqq(1+C_1)^2(1+\eta)C_1^2/[(1-\eta)(2C_1+\eta+1)]$.

Write (19) as

$$C(\boldsymbol{\sigma}-6\mathbf{r})=6(I-C)\mathbf{r}.$$

The sum of the elements of each row of $I-C$ is zero. Thus the jth component of $(I-C)\mathbf{r}$ is

$$-\frac{\lambda_{j-1}\lambda_j^2}{1-\mu_{j-1}\lambda_j}(r_{j+1}-r_j)+\frac{\mu_{j-1}^2\mu_j}{1-\mu_{j-1}\lambda_j}(r_j-r_{j-1}). \tag{21}$$

Taylor's Theorem with integral remainder gives

$$f(t)=f(t_j)+(t-t_j)f'(t_j)+(t-t_j)^2f''(t_j)/2!+\frac{1}{2!}\int_{t_j}^{t}(t-\tau)^2f'''(\tau)\,d\tau.$$

Thus we obtain

$$f[t_{j-1},t_j,t_{j+1}]$$
$$=\frac{f_j''}{2}+\frac{(1/h_{j+1})\int_{t_j}^{t_{j+1}}(t_{j+1}-\tau)^2f'''(\tau)\,d\tau-(1/h_j)\int_{t_{j-1}}^{t_j}(t_{j-1}-\tau)^2f'''(\tau)\,d\tau}{2(h_j+h_{j+1})}$$

and

$$f[t_{j-2},t_{j-1},t_j,t_{j+1}]-\frac{(2h_{j-1}+h_j)f'''_{j-1}+(h_j+2h_{j+1})f'''_j}{12(h_{j-1}+h_j+h_{j+1})}$$

$$\begin{aligned}(22)\quad &= \frac{1}{2(h_{j-1}+h_j+h_{j+1})}\left\{\int_{t_{j-1}}^{t_j}\left[f'''(\tau)-\frac{f'''_{j-1}+f'''_j}{2}\right]d\tau\right.\\ &+\frac{1}{h_j+h_{j+1}}\left[\frac{1}{h_{j+1}}\int_{t_j}^{t_{j+1}}(t_{j+1}-\tau)^2[f'''(\tau)-f'''_j]\,d\tau\right.\\ &\qquad\left.-\frac{1}{h_j}\int_{t_{j-1}}^{t_j}(t_{j-1}-\tau)^2[f'''(\tau)-f'''_j]\,d\tau\right]\\ &+\frac{1}{h_{j-1}+h_j}\left[-\frac{1}{h_j}\int_{t_{j-1}}^{t_j}(t_j-\tau)^2[f'''(\tau)-f'''_{j-1}]\,d\tau\right.\\ &\qquad\left.\left.+\frac{1}{h_{j-1}}\int_{t_{j-2}}^{t_{j-1}}(t_{j-2}-\tau)^2[f'''(\tau)-f'''_{j-1}]\,d\tau\right]\right\}.\end{aligned}$$

Now if we write $t_{j+1}-\tau=\rho e^{i\phi}$, then $|d\tau|=|1+\rho i\,d\phi/d\rho|\,|d\rho|$ and $|d\phi/d\rho|\leqq H_1$ on K for some constant H_1. Thus

$$(1/|h_{j+1}|)\int_{t_j}^{t_{j+1}}|t_{j+1}-\tau|^2|d\tau| \leqq |h_{j+1}|^2(1/3+H_1|h_{j+1}|/4).$$

It follows that the left-hand member of (22) is in absolute value not greater than

$$\frac{\mu(\|\Delta\|,f''')}{2|h_{j-1}+h_j+h_{j+1}|}\left\{1+\frac{|h_j|^2+|h_{j+1}|^2}{|h_j+h_{j+1}|}+\frac{|h_{j-1}|^2+|h_j|^2}{|h_{j-1}+h_j|}\right\}\cdot\left(\frac{1}{3}+\frac{H_1\|\Delta\|}{4}\right).$$

Here the coefficient of $\mu(\|\Delta\|, f''')$ is seen to be dominated by a function of H_1 alone provided $\|\Delta\|$ is sufficiently small.

The left-hand member of (22) is the difference between r_j and a weighted mean of f'''_{j-1} and f'''_j. We may conclude that $\|(I-C)\mathbf{r}\|\to 0$ as $\|\Delta\|\to 0$. From the uniform boundedness of $\|C^{-1}\|$ we have $\|\boldsymbol{\sigma}-6\mathbf{r}\|\to 0$ by (21) and $\{q'''(t)\}\to f'''(t)$ uniformly on K as $\|\Delta\|\to 0$ ($q'''(t)$ is constant on each K_j and $f'''(t)$ is uniformly continuous on K). This completes the proof of the theorem.

If $f'''(t)$ satisfies a Hölder condition of order β on K $(0<\beta\leqq 1)$, then the quantities (21) may be shown to be less than $b\cdot\|\Delta\|^\beta$ for some constant b independent of Δ. Thus $f'''(t)-q'''_\Delta(t)=O(\|\Delta\|^\beta)$.

To obtain the degree of convergence of $\{q''_{\Delta_k}(t)\}$ we must circumvent the difficulty that $q'''_{\Delta_k}(t)$ is not continuous. We have

$$\frac{(f_{j+1}-f_j)/h_{j+1}-(f_j-f_{j-1})/h_j}{h_j+h_{j+1}}$$

$$=\frac{f''_j}{2}+\frac{(1/h_{j+1})\int_{t_j}^{t_{j+1}}(t_{j+1}-\tau)^2f'''(\tau)\,d\tau-(1/h_j)\int_{t_{j-1}}^{t_j}(t_{j-1}-\tau)^2f'''(\tau)\,d\tau}{2(h_j+h_{j+1})}.$$

Using (3) and the interpolation property of $q_\Delta(t)$, we obtain

$$f_j'' - M_j$$
$$= \frac{(1/h_{j+1})\int_{t_j}^{t_{j+1}}(t_{j+1}-\tau)^2[f'''(\tau)-\sigma_{j+1}]\,d\tau - (1/h_j)\int_{t_{j-1}}^{t_j}(t_{j-1}-\tau)^2[f'''(\tau)-\sigma_j]\,d\tau}{h_j+h_{j+1}}.$$

Thus $f_j''-M_j = O(\|\Delta\|^{1+\beta})$ and, for t on K_j,

$$f''(t)-q_\Delta''(t) = f_j''-M_j+\int_{t_j}^{t}[f'''(\tau)-\sigma_j]\,d\tau,$$

so that $[f''(t)-q_\Delta''(t)] = O(\|\Delta\|^{1+\beta})$.

The corresponding convergence rates for $[f'(t)-q_\Delta'(r)]$ and $[f(t)-q_\Delta(t)]$ now result by integration as before and we have

COROLLARY 1. *Under the conditions of Theorem 4, let $f'''(t)$ satisfy a Hölder condition of order β $(0<\beta\leq 1)$ on K. Then $q_{\Delta_k}^{(p)}(t)-f^{(p)}(t) = O(\|\Delta_k\|^{3+\beta-p})$, $p=0, 1, 2, 3$. If $f'''(t)$ is continuous on K, then $[q_{\Delta_k}^{(p)}(t)-f^{(p)}(t)] = o(\|\Delta_k\|^{3-p})$.*

The function $q'''(t)-f'''(t)$ does not, of course, satisfy a Hölder condition on K. We do, however, have

COROLLARY 2. *If $f'''(z)$ satisfies a Hölder condition of order β $(0<\beta\leq 1)$, then for arbitrary δ $(0<\delta\leq\beta)$ the functions $[q_{\Delta_k}''(t)-f''(t)]/\|\Delta_k\|^{1+\beta-\delta}$ satisfy a Hölder condition of order δ uniformly with respect to k.*

Proof. We have on $K_{k,j}$

$$(23)\qquad |[q_{\Delta_k}''(t)-f''(t)]-[q_{\Delta_k}''(\tau)-f''(\tau)]| = |t-\tau|\left|\sigma_{k,j}-\frac{f''(t)-f''(\tau)}{t-\tau}\right|.$$

Now

$$\left|f'''(t)-\frac{f''(t)-f''(\tau)}{t-\tau}\right| = \frac{1}{|t-\tau|}\left|\int_\tau^t [f'''(t)-f'''(\tau)]\,d\tau\right|$$

is $O(\|\Delta_k\|^\beta)$ and so is $|\sigma_{k,j}-f'''(t)|$ by Corollary 1. Thus the left-hand member of (23) does not exceed

$$|t-\tau|\cdot\|\Delta_k\|^{1+\beta-\delta}\cdot\frac{|t-\tau|^{1-\delta}}{\|\Delta_k\|^{1-\delta}}\cdot\left|\frac{\sigma_{k,j}-f'''(t)}{\|\Delta_k\|^\beta}+\frac{f'''(t)-[f''(t)-f''(\tau)]/(t-\tau)}{\|\Delta_k\|^\beta}\right|.$$

The corollary follows.

COROLLARY 3. *Under the conditions of Theorem 4 but without the additional mesh restriction* (11), *for arbitrary δ $(0<\delta\leq 1)$ the functions*

$$[q_{\Delta_k}''(t)-f''(t)]/\|\Delta_k\|^{1-\delta}$$

satisfy a Hölder condition of order δ on K uniformly with respect to k.

Proof. The right-hand member of (23) does not exceed

$$|t-\tau|^{\delta}\|\Delta_k\|^{1-\delta}\left[\sup_k \max_j |\sigma_{k,j}| + \sup_K \left|\frac{f''(t)-f''(\tau)}{t-\tau}\right|\right]$$

and the bracketed expression is finite. This completes the proof.

Consider next the jump in $q_\Delta'''(t)$ at the junctions t_j of Δ. Define

$$\delta_j = \frac{\sigma_{j+1}-\sigma_j}{h_j+h_{j+1}}.$$

Subtract each of the component equations of (19) from its successor and obtain ($j=1, 2, \ldots, N$) the equation

$$\frac{\mu_{j-1}^2\mu_j}{1-\mu_{j-1}\lambda_j}(\sigma_j-\sigma_{j-1})+\mu_j\lambda_j\left(\frac{1}{1-\mu_{j-1}\lambda_j}+\frac{1}{1-\mu_j\lambda_{j+1}}\right)(\sigma_{j+1}-\sigma_j)$$
$$+\frac{\lambda_j\lambda_{j+1}^2}{1-\mu_j\lambda_{j+1}}(\sigma_{j+2}-\sigma_{j+1}) = 6(r_{j+1}-r_j).$$

Note that

$$\frac{\lambda_{j-1}\lambda_j+\mu_j\mu_{j+1}}{h_{j-1}+h_j+h_{j+1}+h_{j+2}} = \frac{\mu_j\mu_{j+1}}{h_{j-1}+h_j} = \frac{\lambda_{j-1}\mu_{j+1}}{h_j+h_{j+1}} = \frac{\lambda_j\lambda_{j-1}}{h_{j+1}+h_{j+2}}.$$

Set $\theta_j=(1-\mu_{j-1}\lambda_j)^{-1}$ and

$$D=\begin{bmatrix} \frac{\lambda_N\lambda_1\mu_1\mu_2(\theta_1+\theta_2)}{\lambda_N\lambda_1+\mu_1\mu_2} & \frac{\lambda_N\lambda_1^2\lambda_2^2\theta_2}{\lambda_N\lambda_1+\mu_1\mu_2} & 0 & \cdots & \frac{\mu_N^2\mu_1^2\mu_2\theta_1}{\lambda_N\lambda_1+\mu_1\mu_2} \\ \frac{\mu_1^2\mu_2^2\mu_3\theta_2}{\lambda_1\lambda_2+\mu_2\mu_3} & \frac{\lambda_1\lambda_2\mu_2\mu_3(\theta_2+\theta_3)}{\lambda_1\lambda_2+\mu_2\mu_3} & \frac{\lambda_1\lambda_2^2\lambda_3^2\theta_3}{\lambda_1\lambda_2+\mu_2\mu_3} & \cdots & 0 \\ \vdots & & & & \\ \frac{\lambda_{N-1}\lambda_N^2\lambda_1^2\theta_1}{\lambda_{N-1}\lambda_N+\mu_N\mu_1} & 0 & \cdots & \frac{\mu_{N-1}^2\mu_N^2\mu_1\theta_N}{\lambda_{N-1}\lambda_N+\mu_N\mu_1} & \frac{\lambda_{N-1}\lambda_N\mu_N\mu_1(\theta_N+\theta_1)}{\lambda_{N-1}\lambda_N+\mu_N\mu_1} \end{bmatrix};$$

also let $\boldsymbol{\delta}=(\delta_1, \delta_2, \ldots, \delta_N)^T$ and $\mathbf{d}=(d_1, d_2, \ldots, d_N)^T$, where

$$d_j = 6f[t_{j-2}, t_{j-1}, t_j, t_{j+1}, t_{j+2}].$$

Then

$$D\boldsymbol{\delta} = \mathbf{d}. \tag{24}$$

As in the proof of Theorem 4 we transform D into a matrix with dominant main diagonal by left and right multiplication by suitable diagonal matrices. We multiply the rows of D by

$$(\lambda_N\lambda_1+\mu_1\mu_2)\lambda_N/(\mu_1^2\mu_2),\ (\lambda_1\lambda_2+\mu_2\mu_3)\lambda_N^2\lambda_1/((\mu_1\mu_2)^2\mu_3), \ldots,$$
$$(\lambda_{N-2}\lambda_{N-1}+\mu_{N-1}\mu_N)(\lambda_N\lambda_1\cdots\lambda_{N-3})^2\lambda_{N-2}/((\mu_1\cdots\mu_{N-1})^2\mu_N)$$
$$= (\lambda_{N-2}\lambda_{N-1}+\mu_{N-1}\mu_N)\mu_N/(\lambda_{N-1}^2\lambda_{N-2}),\ (\lambda_{N-1}\lambda_N+\mu_N\mu_1)/(\lambda_{N-1}\mu_1),$$

and the columns by

$$\mu_1/(\lambda_N^2\lambda_1),\ \mu_1^2\mu_2/((\lambda_N\lambda_1)^2\lambda_2), \ldots, (\mu_1\cdots\mu_{N-2})^2\mu_{N-1}/((\lambda_N\lambda_1\cdots\lambda_{N-2})^2\lambda_{N-1})$$
$$= \lambda_{N-1}/(\mu_{N-1}\mu_N^2),\ 1/(\lambda_N\mu_N).$$

We obtain as a result the matrix

$$(25)\qquad \begin{bmatrix} \theta_1+\theta_2 & \lambda_2\theta_2 & 0 & \dots & 0 & 0 & \mu_N\theta_1 \\ \mu_1\theta_2 & \theta_2+\theta_3 & \lambda_3\theta_3 & \dots & 0 & 0 & 0 \\ \vdots & & & & & & \\ 0 & 0 & \dots & \dots & \mu_{N-2}\theta_{N-1} & \theta_{N-1}+\theta_N & \lambda_N\theta_N \\ \lambda_1\theta_1 & 0 & \dots & \dots & 0 & \mu_{N-1}\theta_N & \theta_N+\theta_1 \end{bmatrix}.$$

If we assume for the meshes Δ under consideration that $\|\Delta\|/\min_j |h_j| \leqq C^* < \infty$, we find by the methods used previously that $\|D^{-1}\|$ is uniformly bounded (the matrix (25) has dominant main diagonal for $\|\Delta\|$ sufficiently small).

Now consider the *sum* of elements in the jth row of D,

$$\begin{aligned} G_j &= \frac{\mu_{j-1}^2\mu_j^2\mu_{j+1}\theta_j+\lambda_{j-1}\lambda_j\mu_j\mu_{j+1}(\theta_j+\theta_{j+1})+\lambda_{j-1}\lambda_j^2\lambda_{j+1}^2\theta_{j+1}}{\lambda_{j-1}\lambda_j+\mu_j\mu_{j+1}} \\ &= \frac{(\mu_{j-1}^2\mu_j/\lambda_{j-1}\lambda_j)\theta_j+(\theta_j+\theta_{j+1})+(\lambda_j\lambda_{j+1}^2/\mu_j\mu_{j+1})\theta_{j+1}}{1/\mu_j\mu_{j+1}+1/\lambda_{j-1}\lambda_j}. \end{aligned}$$

If the meshes Δ become asymptotically uniform as $k \to \infty$; that is, if

$$\max_j |\lambda_j - 1/2| \to 0,$$

then $\max_j |G_j - 1/2| \to 0$. In general, let G be the diagonal matrix with G_j the diagonal element in the jth row. Then for $G^{-1}D-I$ the sum of elements in each row is zero ($G_j \neq 0$ for $\|\Delta\|$ sufficiently small).

We write (24) as

$$D\boldsymbol{\delta} - G^{-1}D\mathbf{d} = (I-G^{-1}D)\mathbf{d}.$$

Let $f^{\text{iv}}(t)$ be continuous on K. The quantities $4d_j$ differ from a weighted mean of $f_{j-1}^{\text{iv}}, f_j^{\text{iv}}, f_{j-1}^{\text{iv}}$ by an amount which is $O(\|\Delta\|)$. This follows from the uniform continuity of f^{iv} and properties of the d_j. Thus $\|(I-G^{-1}D)\mathbf{d}\| \to 0$ as $\|\Delta\| \to 0$. Since $\|D^{-1}\|$ is uniformly bounded, it is evident that $\|\boldsymbol{\delta} - D^{-1}G^{-1}D\mathbf{d}\| \to 0$. If the meshes become asymptotically uniform as $\|\Delta\| \to 0$, then $\|\boldsymbol{\delta} - 2\mathbf{d}\| \to 0$. Thus we have

THEOREM 5. *If $f^{\text{iv}}(t)$ is continuous on K and if, as $\|\Delta_k\| \to 0$, we have*

$$\left[\|\Delta_k\|/\min_j |h_{k,j}|\right] \leqq C < \infty,$$

then $\|\boldsymbol{\delta}_k - D_k^{-1}G_k^{-1}D_k\mathbf{d}_k\| \to 0$. If the meshes become asymptotically uniform as $\|\Delta_k\| \to 0$, then

$$\lim_{\|\Delta_k\|\to 0}\left[\max_j\left|\frac{q_{\Delta_k}'''(t_{k,j}+)-q_{\Delta_k}'''(t_{k,j}-)}{h_j+h_{j+1}} - \frac{f^{\text{iv}}(t_{k,j})}{2}\right|\right] = 0.$$

Finally we show that the convergence rate of $q_{\Delta_k}^{(p)}(t)-f^{(p)}(t)$ to zero can be no higher than $O(\|\Delta_k\|^{4-p})$. Let us assume $f^{\text{iv}}(t)$ to be continuous on K and that

$|q^{(p)}_{\Delta_k}(t)-f^{(p)}(t)| \leqq B_1\|\Delta_k\|^{4-p+\mu}$ for some $\mu>0$ and some p ($p=0, 1, 2, 3$, or 4). If $p=4$, then $f^{\text{iv}}(t)\equiv 0$. If $p=3$, then as in the proof of Corollary I of Theorem 4 we have $q''_{\Delta_k}(t)-f''(t)=O(\|\Delta_k\|^{2+\mu})$; hence, as in the proof of Corollary 1 of Theorem 3, $q^{(p)}_{\Delta_k}(t)-f^{(p)}(t)=O(\|\Delta_k\|^{4+\mu-p})$ for $p=0, 1$. In all cases, therefore, we are led to the situation in which $|q_{\Delta k}(t)-f(t)| \leqq B\|\Delta_k\|^{4+\mu}$. Subdivide each interval $[t_{k,j-1}, t_{k,j}]$ into four subintervals of equal length by points $\xi_{j,0}=t_{j-1}$, $\xi_{j,1}$, $\xi_{j,2}$, $\xi_{j,3}$, $\xi_{j,4}=t_j$ (dropping mesh index k); set $h_{j,n}=\xi_{j,n}-\xi_{j,n-1}$ ($n=1, 2, 3, 4$). Form the difference of fourth divided differences

$$q_\Delta[\xi_{j,0}, \xi_{j,1}, \ldots, \xi_{j,4}]-f[\xi_{j,0}, \xi_{j,1}, \ldots, \xi_{j,4}] = (q_\Delta-f)[\xi_{j,0}, \xi_{j,1}, \ldots, \xi_{j,4}].$$

The right-hand member of this equation does not exceed

$$(\|\Delta\|/|h_j|)^{4+\mu}\cdot(|h_j|/\min_n |h_{j,n}|)^4\cdot|h_j|^{\mu}/6.$$

If $\|\Delta\|/|h_j| \leqq C_1<\infty$, then the right-hand member is $O(\|\Delta\|^{\mu})$. Since $q_\Delta(t)$ is cubic in t over the interval t_{j-1}, t_j, and since $f[\xi_{j,0}, \xi_{j,1}, \ldots, \xi_{j,4}]-f^{\text{iv}}(t_{j-1})\to 0$, we have $f^{\text{iv}}(t)\equiv 0$ on K. Hence

THEOREM 6. *Let $f^{\text{iv}}(t)$ be continuous on K. Let $\{\Delta_k\}$ be a sequence of meshes on K with* $\lim_k \to \|\Delta_k\|=0$ *and* $[\max_j \|\Delta_k\|/|h_{k,j}|]\leqq C_1<\infty$. *If*

$$|q^{(p)}_{\Delta_k}(t)-f^{(p)}(t)| = O(\|\Delta_k\|^{4+\mu-p})$$

for any p ($p=0, 1, 2, 3, 4$) and $\mu>0$, then $f^{\text{iv}}(t)\equiv 0$ on K.

The analytic spline. Let $f(t)$ be continuous on K. Let Δ be a mesh on K and $q_\Delta(t)$ the complex cubic spline of interpolation to $f(t)$ on Δ. If R represents the region interior to K, then the complex spline $S_\Delta(z)$, z in R, is defined by (8).

We note that if for z in R we let $z\to t$ on K, then [29] $S_\Delta(z)\to S_\Delta(t)$, where

$$S_\Delta(t) = \frac{1}{2}q_\Delta(t)+\frac{1}{2\pi i}\int_K \frac{q_\Delta(\tau)}{\tau-t}\,d\tau. \tag{31}$$

Here the Cauchy principal value of the integral on the right is intended. On K, if $q_\Delta(t)-f(t)$ satisfies a Hölder condition of order $\delta>0$, we have

$$S_\Delta(t)-f(t) = \frac{1}{2}[q_\Delta(t)-f(t)]+\frac{1}{2\pi i}\int_K \frac{q_\Delta(\tau)-f(\tau)}{\tau-t}\,d\tau. \tag{32}$$

From the Corollary to Theorem 1 and Corollary 3 of Theorem 2, together with an application of the Principle of Maximum for analytic functions, we obtain

THEOREM 7. *Let $f(z)$ be analytic in R and continuous in $\bar{R}=R+K$. Let $\{\Delta_k\}$ be a sequence of meshes on K with* $\lim_{k\to\infty}\|\Delta_k\|=0$ *and* $[\max_j \|\Delta_k\|/|h_{k,j}|]\leqq C_1<\infty$. *If $f(t)$ satisfies a Hölder condition on K of order β ($0<\beta\leqq 1$), then for z in $\bar{R}$ and any β', $0<\beta'<\beta$, we have*

$$S_{\Delta_k}(z)-f(z) = O(\|\Delta_k\|^{\beta'}). \tag{33}$$

If $f'(t)$ exists and is continuous on K, then (33) *is valid for arbitrary* β', $0<\beta'<1$, *with only the restriction* $\lim_{k\to\infty}\|\Delta_k\|=0$ *on the meshes* Δ_k.

Let $f'(t)$ satisfy a Hölder condition on K. Then we have

$$S'_\Delta(t)-f'(t) = \frac{1}{2}[q'_\Delta(t)-f'(t)]+\frac{1}{2\pi i}\int_K \frac{q'_\Delta(\tau)-f'(\tau)}{\tau-t}\,d\tau.$$

We obtain from Corollary 2 of Theorem 2 and Corollary 3 of Theorem 3 the following:

THEOREM 8. *Let $f(z)$ be analytic in R and of Class C^1 in $\bar{R}$. Let $\{\Delta_k\}$ be a sequence of meshes on K with $\lim_{k\to\infty}\|\Delta_k\|=0$ and $\max_j\|\Delta_k\|/|h_{k,j}|\leqq C<\infty$. If $f'(t)$ satisfies a Hölder condition of order β on K $(0<\beta\leqq 1)$, then for z in $\bar{R}$ and any β', $0<\beta'<\beta$, we have*

$$S^{(p)}_{\Delta_k}(z)-f^{(p)}(z) = O(\|\Delta_k\|^{1+\beta'-p}) \qquad (p=0,1). \tag{34}$$

If $f''(t)$ exists and is continuous on K, then (34) *is valid for arbitrary β', $0<\beta'<1$, with only the restriction $\lim_{k\to\infty}\|\Delta_k\|=0$ on the meshes Δ_k. In this case also $[S^{(p)}_{\Delta_k}(z)-f^{(p)}(z)]=O(\|\Delta_k\|^{1+\beta'})$ for $p=0,1,\ldots,$ on any closed subset of R.*

The last conclusion of the theorem follows from the Cauchy Integral Formula. By Corollary 2 of Theorem 3 and Corollary 3 of Theorem 4 we are led to conclude the following theorem:

THEOREM 9. *Let $f(z)$ be analytic in R and of Class C^2 in $\bar{R}$. Let $\{\Delta_k\}$ be a sequence of meshes on K with $\lim_{k\to\infty}\|\Delta_k\|=0$ and $\max_j\|\Delta_k\|/h_{k,j}|\leqq C<\infty$. If $f''(t)$ satisfies a Hölder condition of order β on K $(0<\beta\leqq 1)$, then for z in $\bar{R}$ and any β', $0<\beta'<\beta$, we have $(p=0,1,2)$*

$$S^{(p)}_{\Delta_k}(z)-f^{(p)}(z) = O(\|\Delta_k\|^{2+\beta'-p}). \tag{35}$$

If $f'''(t)$ exists and is continuous on K, then (35) *is valid for arbitrary β', $0<\beta'<1$, without the additional mesh restriction. Here also $S^{(p)}_{\Delta_k}(z)-f^{(p)}(z)$ is $O(\|\Delta_k\|^{2+\beta'})$ $(p=0,1,2)$ on any closed subset of R.*

If $f(z)$ is analytic in R and of Class C^3 in $\bar{R}$, we wish to examine the convergence of the sequence $\{S'''_{\Delta_k}(z)\}$. The method employed in the preceding three theorems does not apply without modification since $q'''_k(t)$ is discontinuous at the mesh points $t_{k,j}$. However, using Corollary 2 of Theorem 4 we obtain

THEOREM 10. *Let $f(z)$ be analytic in R and of Class C^3 in $\bar{R}$. Let $\{\Delta_k\}$ be a sequence of meshes on K with $\lim_{k\to\infty}\|\Delta_k\|=0$, and $[\max_j\|\Delta_k\|/|h_{k,j}|]\leqq C<\infty$. If $f'''(t)$ satisfies a Hölder condition of order β on K, $0<\beta\leqq 1$, then for z in $\bar{R}$ and any β', $0<\beta'<\beta$ we have $(p=0,1,2)$*

$$S^{(p)}_{\Delta_k}(z)-f^{(p)}(z) = O(\|\Delta_k\|^{3+\beta'-p}). \tag{36}$$

On any closed subset of R, $S^{(p)}_{\Delta_k}(z)-f^{(p)}(z)=O(\|\Delta_k\|^{3+\beta'})$ $(p=0,1,2,3)$.

Structure of the analytic spline. The integration indicated in (8) may be carried out explicitly. We have

$$2\pi i S_\Delta(z) = \sum_{j=1}^{N} \int_{K_j} \frac{q_{\Delta,j}(\tau)}{\tau - z}\, d\tau,$$

when $q_{\Delta,j}(t)$ is the cubic given by (1) or (5) which coincides with $q_\Delta(t)$ on the arc K_j. Employing (1), rearrange $q_{\Delta,j}(\tau)$ as follows:

$$\begin{aligned} q_{\Delta,j}(\tau) = {} & \frac{M_{j-1}}{6h_j}[(t_j-z)^3 - 3(t_j-z)^2(\tau-z) + 3(t_j-z)(\tau-z)^2 - (\tau-z)^3] \\ & + \frac{M_j}{6h_j}[(z-t_{j-1})^3 + 3(z-t_{j-1})^2(\tau-z) + 3(z-t_{j-1})(\tau-z)^2 + (\tau-z)^3] \\ & + (f_{j-1}/h_j - M_{j-1}h_j/6)[(t_j-z) - (\tau-z)] \\ & + (f_j/h_j - M_jh_j/6)[(z-t_{j-1}) + (\tau-z)]. \end{aligned}$$

We obtain, using $\sum_{j=1}^{N} M_j(h_j+h_{j+1}) = 0$, the relation

$$(36) \quad 2\pi i S_\Delta(z) = \sum_{j=1}^{N} q_{\Delta,j}(z) \log\frac{t_j - z}{t_{j-1}-z} - \frac{5}{36}\sum_{j=1}^{N} M_j(h_j+h_{j+1})(t_{j-1}+t_j+t_{j+1}).$$

The quantity z is, of course, generally outside the domain of definition of $q_{\Delta,j}(t)$. Here $q_{\Delta,j}(z)$ represents the result of substituting z for t in (1). Thus

$$2\pi i S'_\Delta(z) = \sum_{j=1}^{N} q'_{\Delta,j}(z)\log\frac{t_j-z}{t_{j-1}-z} - \sum_{j=1}^{N}\frac{(t_j-z)^2}{6}\left[\frac{M_{j+1}-M_j}{h_{j+1}} - \frac{M_j - M_{j-1}}{h_j}\right],$$

$$2\pi i S''_\Delta(z) = \sum_{j=1}^{N} q''_{\Delta,j}(z)\log\frac{t_j-z}{t_{j-1}-z} + \sum_{j=1}^{N}\frac{5(t_j-z)}{6}\left[\frac{M_{j+1}-M_j}{h_{j+1}} - \frac{M_j - M_{j-1}}{h_j}\right],$$

$$2\pi i S'''_\Delta(z) = \sum_{j=1}^{N} q'''_{\Delta,j}(z)\log\frac{t_j-z}{t_{j-1}-z} = \sum_{j=1}^{N}\sigma_j \log\frac{t_j-z}{t_{j-1}-z},$$

$$2\pi i S^{\mathrm{iv}}_\Delta(z) = \sum_{j=1}^{N}\frac{\sigma_{j+1}-\sigma_j}{t_j - z}.$$

We have previously established the fact that, for a sequence of meshes $\{\Delta_k\}$ on K with $\|\Delta_k\| \to 0$, $[\max_j \|\Delta_k\|/|h_{k,j}|] \leqq C < \infty$ and $[\max_j |\lambda_j - 1/2|] \to 0$ as $k \to \infty$, and a function $f(t)$ with $f^{\mathrm{iv}}(t)$ continuous on K, we have

$$\lim_{k\to\infty}\left[\max_j \left|\frac{\sigma_{k,j+1}-\sigma_{k,j}}{h_{k,j}+h_{k,j+1}} - \frac{1}{2}f^{\mathrm{iv}}(t_j)\right|\right] = 0.$$

If in addition $f(t)$ represents the boundary values of a function $f(z)$ analytic in R, then we know from the convergence properties of $\{S_{\Delta k}(z)\}$ that

$$(37) \qquad \lim_{k\to\infty}\frac{1}{2\pi i}\sum_{j=1}^{N_k}\frac{\sigma_{k,j+1}-\sigma_{k,j}}{(h_{k,j}+h_{k,j+1})/2}\cdot\frac{1}{t_j - z}\cdot\frac{h_{k,j}+h_{k,j+1}}{2} = \frac{1}{2\pi i}\int_K \frac{f^{\mathrm{iv}}(t)}{t-z}\,dt.$$

Furthermore, for z on any closed subset of R the rate of convergence is $O(\|\Delta_k\|^4)$.

This relationship, however, sheds some light on the nature of the spline approximation. The sum in the left-hand member of (37) may be interpreted as a "trapezoidal" approximation of the integral appearing on the right, and more generally relates to the classical theorem by Runge concerning the approximation of a function analytic in R by rational functions with poles on K [30].

The last term in (36) can be further simplified. If we multiply equation (2) by t_j and sum over j, the term in M_j is

$$M_j\left(\frac{h_j}{6}t_{j-1}+\frac{h_j+h_{j+1}}{3}t_j+\frac{h_{j+1}}{6}t_{j+1}\right) = \frac{M_j}{6}(h_j+h_{j+1})(t_{j-1}+t_j+t_{j+1}).$$

Thus we have

$$\begin{aligned}\frac{1}{6}\sum_{j=1}^{N} M_j(h_j+h_{j+1})(t_{j-1}+t_j+t_{j+1}) &= \sum_{j=1}^{N} t_j\left(\frac{f_{j+1}-f_j}{h_{j+1}}-\frac{f_j-f_{j-1}}{h_j}\right)\\ &= \sum_{j=1}^{N}(f_j-f_{j-1}),\end{aligned}$$

which is equal to 0. Thus (36) becomes

$$S_\Delta(z) = \frac{1}{2\pi i}\sum_{j=1}^{N} q_{\Delta,j}(z)\log\frac{t_j-z}{t_{j-1}-z}.$$

The unit circle. The inverse of the coefficient matrix for (3) or (7) when the curve K is the unit circle and the intervals t_{k-1}, t_k are of equal length can be easily obtained. We have $t_{k-1}=e^{ik\alpha}$ $(\alpha=2\pi/N)$ and

$$\lambda_k = \frac{t_{k+1}-t_k}{t_{k+1}-t_{k-1}} = \frac{e^{i\alpha}}{e^{i\alpha}+1} = \lambda.$$

The coefficient matrix for (3) in this situation is the $N\times N$ circulant

$$C[2,\lambda,0,\ldots,0,1-\lambda],$$

$$\begin{bmatrix} 2 & \lambda & 0 & \cdots & 1-\lambda \\ 1-\lambda & 2 & \lambda & \cdots & 0 \\ 0 & 1-\lambda & 2 & \cdots & 0 \\ \vdots & & & & \\ \lambda & 0 & 0 & \cdots & 2 \end{bmatrix}.$$

Representing the inverse matrix by the circulant $C[a_1^{-1}, a_2^{-1},\ldots,a_N^{-1}]$, we have

$$\begin{aligned}2a_1^{-1}+\lambda a_N^{-1}+(1-\lambda)a_2^{-1} &= 1,\\ 2a_2^{-1}+\lambda a_1^{-1}+(1-\lambda)a_3^{-1} &= 0,\\ &\vdots\\ 2a_N^{-1}+\lambda a_{N-1}^{-1}+(1-\lambda)a_1^{-1} &= 0.\end{aligned} \tag{38}$$

Thus the quantities a_k^{-1} satisfy the difference equation $\lambda a_{k-1}^{-1}+2a_k^{-1}+(1-\lambda)a_{k+1}^{-1}=0$ $(k=2, 3, \ldots, N-1)$, subject to the two conditions represented by the first and last of equations (38).

The roots of the characteristic equation are

$$(-1\pm(1-\lambda(1-\lambda))^{1/2})/\lambda = e^{-i\alpha/2}[-2\cos\alpha/2\pm(4\cos^2\alpha/2-1)^{1/2}].$$

Here the radicand is positive if $\alpha<2\pi/3$; i.e., if $N>3$. Designate these roots by r and s. Then

$$a_k^{-1} = Ar^k+Bs^k,$$

and

$$A[\lambda r^{N-1}+2r^N+(1-\lambda)r]+B[\lambda s^{N-1}+2s^N+(1-\lambda)s] = 0,$$
$$A[\lambda r^N+2r+(1-\lambda)r^2]+B[\lambda s^N+2s+(1-\lambda)s^2] = 1.$$

Hence

$$A = \frac{-s}{\lambda(r-s)(r^N-1)}, \qquad B = \frac{r}{\lambda(r-s)(s^N-1)},$$

and, since $rs=\lambda/(1-\lambda)=e^{i\alpha}$,

$$a_k^{-1} = -\frac{\cos\alpha/2}{(4\cos^2\alpha/2-1)^{1/2}}\cdot\frac{r^{N-k+1}+r^{k-1}}{r^N-1}.$$

Alternative type of complex spline. Closely related to the same problem of approximating a function analytic interior to the curve K are the complex-valued splines of the real variable s (arc length) on K. These have been investigated to a limited extent previously [24]. Here we can sharpen and extend the earlier results.

Equations (3) become

$$\mu_j\tilde{M}_{j-1}+2\tilde{M}_j+\lambda_j\tilde{M}_{j+1} = 6f[s_{j-1}, s_j, s_{j+1}], \tag{39}$$

with $\tilde{M}_j=d^2f/ds^2$, but now the quantities $h_j=s_j-s_{j-1}$ and $\mu_j=1-\lambda_j=h_j/(h_j+h_{j+1})$ are real and positive. The coefficient matrix has dominant main diagonal for all meshes on K and the inverse matrix has row-max norm not greater than unity. Similarly, equations (7) become

$$\lambda_j\tilde{m}_{j-1}+2\tilde{m}_j+\mu_j\tilde{m}_{j+1} = 3(\mu_jf[s_j, s_{j+1}]+\lambda_jf[s_{j-1}, s_j]) \tag{40}$$

with $\tilde{m}_j=df/ds$.

The complex spline within the region R is defined as

$$\tilde{S}_\Delta(z) = \frac{1}{2\pi i}\int_K \frac{\tilde{q}_\Delta(s)}{t(s)-z}\,dt(s),$$

where $t=t(s)$ is the complex representation of the curve K and $\tilde{q}_\Delta(s)$ is the complex-valued spline given in the interval $s_{j-1}\leqq s\leqq s_j$ by

$$\tilde{q}_\Delta(s) = \frac{\tilde{M}_{j-1}}{6h_j}(s_j-s)^3+\frac{\tilde{M}_j}{6h_j}(s-s_{j-1})^3$$
$$+(f_{j-1}/h_j-M_{j-1}h_j/6)(s_j-s)+(f_j/h_j-M_jh_j/6)(s-s_{j-1}),$$

or, equivalently, by

$$q_\Delta(s) = \frac{\tilde{m}_{j-1}}{h_j^2}(s_j-s)^2(s-s_{j-1})+\frac{\tilde{m}_j}{h_j^2}(s-s_{j-1})^2(s_j-s)$$

$$+\frac{f_{j-1}}{h_j^3}(s_j-s)^2[2(s-s_{j-1})+h_j]+\frac{f_j}{h_j^3}(s-s_{j-1})^2[2(s_j-s)+h_j].$$

As z in R approaches $t(s)$ on K, $\tilde{S}_\Delta(z)$ approaches

$$\tilde{S}_\Delta(t) = \frac{1}{2}q_\Delta(s)+\frac{1}{2\pi i}\int_K \frac{\tilde{q}_\Delta(s)}{\tau(s)-t}\,d\tau(s).$$

Similarly $\tilde{S}'_\Delta(z)$ and $\tilde{S}''_\Delta(z)$ approach, respectively,

$$\tilde{S}'_\Delta(t) = \frac{1}{2}\frac{d\tilde{q}_\Delta}{ds}\cdot\frac{ds}{dt}+\frac{1}{2\pi i}\int_K \frac{(d\tilde{q}_\Delta/ds)\cdot(ds/d\tau)}{\tau(s)-t}\,d\tau(s),$$

and

$$S''_\Delta(t) = \frac{1}{2}\left[\frac{d^2\tilde{q}_\Delta}{ds^2}\left(\frac{ds}{dt}\right)^2+\frac{d\tilde{q}_\Delta}{ds}\cdot\frac{d^2s}{dt^2}\right]+\frac{1}{2\pi i}\int_K \frac{\dfrac{d^2\tilde{q}_\Delta}{ds^2}\left(\dfrac{ds}{d\tau}\right)^2+\dfrac{d\tilde{q}_\Delta}{ds}\cdot\dfrac{d^2s}{d\tau^2}}{\tau(s)-t}\,d\tau(s).$$

The properties corresponding to Theorems 1–6 and their corollaries carry over to the present situation. Some simplification in the proofs results from the fact that we are concerned here with splines in the real variable s. The convergence properties set forth in Theorem 7–10 are valid in this situation and follow as in the preceding section.

The cubic splines in the real variable s have the important property that they satisfy two fundamental integral relations. We have

$$\int_K \left|\frac{d^2f}{ds^2}\right|^2 ds = \int_K \left|\frac{d^2f}{ds^2}-\frac{d^2\tilde{q}_\Delta}{ds^2}\right|^2 ds+\int_K \left|\frac{d^2\tilde{q}_\Delta}{ds^2}\right|^2 ds+2\mathcal{R}\int_K \left(\frac{d^2f}{ds^2}-\frac{d^2\tilde{q}_\Delta}{ds^2}\right)\overline{\left(\frac{d^2\tilde{q}_\Delta}{ds^2}\right)}\,ds,$$

and integration by parts gives

$$\int_K \left(\frac{d^2f}{ds^2}-\frac{d^2\tilde{q}_\Delta}{ds^2}\right)\overline{\left(\frac{d^2q}{ds^2}\right)}\,ds = \sum_{j=1}^N \left(\frac{df}{ds}-\frac{d\tilde{q}_\Delta}{ds}\right)\overline{\left(\frac{d^2\tilde{q}_\Delta}{ds^2}\right)}\Bigg|_{t_{j-1}}^{t_j} - \sum_{j=1}^N (f-\tilde{q}_\Delta)\overline{\left(\frac{d^3q}{ds^3}\right)}\Bigg|_{t_{j-1}}^{t_j} = 0.$$

The First Integral Identity which results since $\tilde{q}_\Delta$ is a spline of interpolation to f on Δ is

$$\int_K \left|\frac{d^2f}{ds^2}\right|^2 ds = \int_K \left|\frac{d^2f}{ds^2}-\frac{d^2\tilde{q}_\Delta}{ds^2}\right|^2 ds+\int_K \left|\frac{d^2\tilde{q}_\Delta}{ds^2}\right|^2 ds.$$

Actually we require only that df/ds be absolutely continuous on K and that d^2f/ds^2 be Lebesque square integrable.

As an immediate consequence we have [24] the following extremal properties:

(i) The quantity $\int_K |d^2(f-h_\Delta)/ds^2|^2\,ds$ for arbitrary cubic splines $h_\Delta(s)$ on Δ is minimum when $h_\Delta(s)=\tilde{q}_\Delta(s)$.

(ii) Of all the C^2 functions $g(s)$ on interpolating to $f(s)$ on Δ, the quantity $\int_K |q''(s)|^2\,ds$ is minimum when $g(s)=\tilde{q}_\Delta(s)$.

We define a pseudo inner product

$$(u, v) = \int_K u''(s)\overline{v''(s)}\,ds \tag{41}$$

when $u(s)$ and $v(s)$ are absolutely continuous on K (necessarily periodic in s) with L^2-integrable second derivatives. We say that u and v are orthogonal if $(u, v)=0$. An immediate consequence [21] is that if Δ and Δ' are two meshes on K with $\Delta\subset\Delta'$, if h_Δ and $h_{\Delta'}$ are splines on Δ and Δ' such that $h_{\Delta'}=0$ on Δ, then h_Δ and $h_{\Delta'}$ are orthogonal.

Using a sequence of imbedded meshes $\{\Delta_k\}$ with $\|\Delta_k\|\to 0$ and with Δ_{k+1} consisting of the points of Δ_k together with a point s_{k+1} not in Δ_k, and taking $h_{\Delta_{k+1}}$ to be the complex cubic spline on Δ_{k+1} which is 1 at s_{k+1} and 0 on Δ_k, we obtain a complete sequence of orthogonal splines which is a dense subset of the space of continuous functions on K. Many other orthogonal sequences may be constructed, of course.

We obtain further that if $\{\Delta_k\}$ is any sequence of subdivisions of K with $\|\Delta_k\|\to 0$ as $k\to\infty$, and if $f'(s)$ is absolutely continuous with $f''(s)$ L^2-integrable, with $q_{\Delta_k}(s)$ the corresponding cubic splines on K of interpolation to $f(s)$ on Δ_k, then $q''_{\Delta_k}(s)$ converges in the mean to $f''(s)$:

$$\lim_{k\to\infty}\int_K |(f-q_{\Delta_k})''|^2\,ds = 0. \tag{42}$$

The set of all functions $f(s)$ with absolutely continuous first derivatives and L^2-integrable second derivatives form a Hilbert space with respect to the pseudo-scalar product (41). A basis for this space may be taken, for example, to be the sequence of orthogonal splines introduced above.

We note for the sake of completeness the so-called *second integral relation* [23], valid if $f(s)$ is Lebesgue integrable and $q_\Delta(s)$ is the complex cubic spline of interpolation to $f(s)$ on K:

$$\int_K |(f-q_\Delta)''|^2\,ds = \int_K (f-q_\Delta)\overline{f^{\mathrm{iv}}(s)}\,ds. \tag{43}$$

REFERENCES

1. I. J. Schoenberg, *Contributions to the problem of approximation of equidistant data by analytic functions*, Quart. Appl. Math. 4 (1946), 45–99, 112–141.

2. I. J. Schoenberg and Anne Whitney, *On Pólya frequency functions*. III. *The positivity of translation determinants with application to the interpolation problem by spline curves*, Trans. Amer. Math. Soc. 74 (1953), 246–259.

3. J. C. Holladay, *Smoothest curve approximation*, Math. Comp. 11 (1957), 233–243.

4. I. J. Schoenberg, *Spline functions, convex curves, and mechanical quadrature*, Bull. Amer. Math. Soc. **64** (1958), 352–357.

5. R. S. Johnson, *On monosplines of least deviation*, Trans. Amer. Math. Soc. **96** (1960), 458–477.

6. G. Birkhoff and H. Garabedian, *Smooth surface interpolation*, J. Math. Phys. **39** (1960), 258–268.

7. F. Theilheimer and W. Starkweather, *The fairing of ship lines on a high-speed computer*, Math. Comp. **15** (1961), 338–355.

8. J. L. Walsh, J. H. Ahlberg and E. N. Nilson, *Best approximation properties of the spline fit*, J. Math. Mech. **11** (1962), 225–234.

9. C. deBoor, *Bicubic spline interpolation*, J. Math. Phys. **41** (1962), 212–218.

10. B. Asker, *The spline curve, a smooth interpolating function used in numerical design of ship-lines*, Nordisk Tidskr. Informations-Behandling **2** (1962), 76–82.

11. F. Landis and E. N. Nilson, "The determination of therodynamic properties by direct differentiation techniques," *Progress in International Research on Thermodynamics and Transport Properties*, Amer. Soc. of Mech. Engineers, 1962.

12. J. H. Ahlberg and E. N. Nilson, *Convergence properties of the spline fit*, J. Soc. Indust. Appl. Math. **11** (1963), 95–104.

13. C. deBoor, *Best approximation properties of spline functions of odd degree*, J. Math. Mech. **12** (1963), 747–749.

14. I. J. Schoenberg, *Spline interpolation and the higher derivatives*, Proc. Nat. Acad. Sci. U.S.A. **51** (1964), 24–28.

15. ———, *Spline interpolation and best quadrature formulae*, Bull. Amer. Math. Soc. **70** (1964), 143–148.

16. ———, *On trigonometric spline interpolation*, J. Math. Mech. **13** (1964), 795–826.

17. G. Birkhoff and C. deBoor, *Error bounds for spline interpolation*, J. Math. Mech. **13** (1964), 827–836.

18. T. N. E. Greville, *Numerical procedures for interpolation by spline functions*, J. SIAM Ser. B Numer. Anal. **1** (1964), 53–68.

19. ———, *Interpolation by generalized splines*, MRC Rep. 476, Univ. of Wisconsin, Madison, 1964.

20. I. J. Schoenberg, *On best approximation of linear operators*, Nederl. Akad. Wetensch. Proc. Ser. A **67** = Indag. Math. **26** (1964), 155–163.

21. J. H. Ahlberg, E. N. Nilson and J. L. Walsh, *Fundamental properties of generalized splines*, Proc. Nat. Acad. Sci. U.S.A. **52** (1964), 1412–1419.

22. ———, *Best approximation properties of higher-order spline approximations*, J. Math. Mech. **14** (1965), 231–244.

23. ———, *Convergence properties of generalized splines*, Proc. Nat. Acad. Sci. U.S.A. **54** (1965), 344–350.

24. J. H. Ahlberg and E. N. Nilson, *Orthogonality properties of spline functions*, J. Math. Anal. Appl. **11** (1965), 321–337.

25. J. H. Ahlberg, E. N. Nilson and J. L. Walsh, *Extremal, orthogonality, and convergence properties of multidimensional splines*, J. Math. Anal. Appl. **11** (1965), 27–48.

26. J. H. Ahlberg and E. N. Nilson, *Solution of differential equations by the method of cardinal splines*, J. SIAM Numer. Anal. **3** (1966), 173–182, Walsh Jubilee Volume.

27. F. B. Hildebrand, *Introduction to numerical analysis*, McGraw-Hill, New York, 1956, p. 38.

28. A. Sharma and A. Meir, *Convergence of spline functions*, Abstract 64T–496, Notices Amer. Math. Soc. **11** (1964), 768.

29. N. I. Muskhelishvili, *Some basic problems of the mathematical theory of elasticity*, Noordhoff, Groningen, 1953, pp. 57ff.

30. J. L. Walsh, *Interpolation and approximation by rational functions in the complex domain*, Colloq. Publ. Vol. 20, Amer. Math. Soc., Providence, R. I., 1935, p. 10.

United Aircraft Research Laboratories,
East Hartford, Connecticut
Pratt & Whitney Aircraft,
East Hartford, Connecticut
University of Maryland,
College Park, Maryland

JOURNAL OF APPROXIMATION THEORY 1, 5–10 (1968)

Cubic Splines on the Real Line[1]

J. H. AHLBERG

United Aircraft Research Laboratories, East Hartford, Connecticut 06108

E. N. NILSON

Pratt and Whitney Aircraft Company, East Hartford, Connecticut 06108

J. L. WALSH[2]

Department of Mathematics, University of Maryland, College Park, Maryland 20740

We shall limit ourselves to the consideration of cubic splines since we wish to take full advantage of the special properties of the system of linear equations which arises in this situation. For the same reason we restrict our considerations to *simple splines* of interpolation. These splines are required by definition to interpolate only to $f(x)$ at mesh points and not to both $f(x)$ and $f'(x)$; they possess continuous second derivatives throughout their domain of definition.

For most purposes, the restriction to simple cubic splines is innocuous since at a mesh point where interpolation to both $f(x)$ and $f'(x)$ is required, a cubic spline of interpolation is separated into two independent splines. For instance, if there are two double points of interpolation x_{j_0} and x_{j_1} such that $x_{j_0} < x_{j_1}$, the spline reduces on the closed interval $[x_{j_0}, x_{j_1}]$ to a type I spline of interpolation ([*1*], p. 75) to $f(x)$; the theory of such splines is well known. If, however, there is no point of double interpolation x_{j_1} to the right of x_{j_0}, we must consider splines $S_\Delta(f;x)$ on the infinite interval $[x_{j_0}, \infty)$ which interpolate to $f(x)$ at the mesh points of a mesh Δ on (x_{j_0}, ∞). We shall consider this situation in detail, and also indicate the rather minor modifications required when the interval (x_{j_0}, ∞) is replaced by the interval $(-\infty, \infty)$.

In the interval $x_{i-1} \le x \le x_i$ a cubic spline $S_\Delta(f;x)$ which interpolates to the values $f_j = f(x_j)$, $j = 0, 1, \ldots$, at the points x_j of the mesh Δ: $x_0 < x_1 < \ldots$ is given ([*1*], p. 10) by

---

[1] Abstract published in *Notices, Am. Math. Soc.* **15** (1968), 68T-10.

[2] The research of this author was sponsored, in part, by the U.S. Air Force Office of Scientific Research.

$$S_\Delta(f;x) = [M_{i-1}(x_i - x)^3 + M_i(x - x_{i-1})^3]\frac{1}{6l_i} + \left[\frac{f_i}{l_i} - M_i\frac{l_i}{6}\right](x - x_{i-1}) + \left[\frac{f_{i-1}}{l_i} - \frac{M_{i-1}l_i}{6}\right](x_i - x). \tag{1}$$

Here $l_j = x_j - x_{j-1}$ and $M_j = S''_\Delta(f;x_j)$. As a consequence of (1) we have on $x_{i-1} \leqslant x \leqslant x_i$

$$S'_\Delta(f;x) = -\frac{M_{i-1}(x_i - x)^2}{2l_i} + \frac{M_i(x - x_{i-1})^2}{2l_i} + \frac{f_i - f_{i-1}}{l_i} - \frac{M_i - M_{i-1}}{6}l_i, \tag{2}$$

$$S''_\Delta(f;x) = \frac{M_i(x - x_{i-1})}{l_i} + \frac{M_{i-1}(x_i - x)}{l_i}. \tag{3}$$

It is immediate from (1) that $S_\Delta(f;x_i) = f_i$ and from (3) that $S''_\Delta(f;x_i) = M_i$ ($i = 0, 1, \ldots$). Consequently, $S_\Delta(f;x)$ and $S''_\Delta(f;x)$ are in $C(x_0, \infty)$. In order to insure that $S'_\Delta(f;x)$ is also in $C(x_0, \infty)$, the following infinite system of equations must be satisfied:

$$\tfrac{1}{6}l_i M_{i-1} + \tfrac{1}{3}(l_i + l_{i+1})M_i + \tfrac{1}{6}l_{i+1}M_{i+1} = \frac{f_{i+1} - f_i}{l_{i+1}} - \frac{f_i - f_{i-1}}{l_i} \equiv d_i \qquad (i = 1, 2, \ldots). \tag{4}$$

In order that $S'_\Delta(f;x_0) = f_0' \equiv f'(x_0)$ we have to satisfy the additional equation

$$\frac{1}{3}l_1 M_0 + \frac{1}{6}l_1 M_1 = \frac{f_1 - f_0}{l_1} - f_0' \equiv d_0. \tag{5}$$

In matrix form we must solve the equation

$$AM = d, \tag{6}$$

where

$$A = \begin{pmatrix} \frac{1}{3}l_1 & \frac{1}{6}l_1 & 0 & 0 & \cdot & \cdot & \cdot \\ \frac{1}{6}l_1 & \frac{1}{3}(l_1 + l_2) & \frac{1}{6}l_2 & 0 & \cdot & \cdot & \cdot \\ 0 & \frac{1}{6}l_2 & \frac{1}{3}(l_2 + l_3) & \frac{1}{6}l_3 & \cdot & \cdot & \cdot \\ \cdot & \cdot & \cdot & \cdot & \cdot & \cdot & \cdot \\ \cdot & \cdot & \cdot & \cdot & \cdot & \cdot & \cdot \\ \cdot & \cdot & \cdot & \cdot & \cdot & \cdot & \cdot \end{pmatrix}, \tag{7}$$

$$M = (M_0, M_1, \ldots)^T, \tag{8}$$

$$d = (d_0, d_1, \ldots)^T. \tag{9}$$

When the interval (x_0, ∞) is replaced by the real line $(-\infty, \infty)$, Eqs. (4) apply except that now $i = 0, \pm 1, \pm 2, \ldots$, and Eq. (5) is omitted. In this case

$$A = \begin{pmatrix} \cdot & \cdot & \cdot & \cdot & \cdot & \cdot & \cdot & \cdot \\ \cdot & \cdot & \cdot & \cdot & \cdot & \cdot & \cdot & \cdot \\ \cdot & \cdot & \frac{1}{6}l_{-1} & \frac{1}{3}(l_{-1}+l_0) & \frac{1}{6}l_0 & 0 & 0 & \cdot \\ \cdot & \cdot & 0 & \frac{1}{6}l_0 & \frac{1}{3}(l_0+l_1) & \frac{1}{6}l_1 & 0 & \cdot \\ \cdot & \cdot & 0 & 0 & \frac{1}{6}l_1 & \frac{1}{3}(l_1+l_2) & \frac{1}{6}l_2 & \cdot \\ \cdot & \cdot & \cdot & \cdot & \cdot & \cdot & \cdot & \cdot \\ \cdot & \cdot & \cdot & \cdot & \cdot & \cdot & \cdot & \cdot \end{pmatrix}, \tag{7'}$$

$$M = (\ldots, M_{-1}, M_0, M_1, \ldots)^T, \tag{8'}$$

$$d = (\ldots, d_{-1}, d_0, d_1, \ldots)^T. \tag{9'}$$

Moreover, if we reorder the components of M (and d) such that in rearranged form

$$M = (M_0, M_{-1}, M_1, M_{-2}, M_2, \ldots),$$

then A will be a singly-infinite matrix such as (6) which is symmetric and has the same diagonal elements as those of A, although they will be reordered.

Let us introduce the notation

$$\|\Delta\| = \sup_i l_i,$$
$$\delta_\Delta = \inf_i l_i, \tag{10}$$
$$\mathcal{R}_\Delta = \|\Delta\|/\delta_\Delta, \ \delta_\Delta \neq 0.$$

If we require $\delta_\Delta > 0$, then $A = (a_{ij})$, as given by (7) or (7′), is a symmetric matrix, which is strongly diagonally dominant in the sense that

$$\inf_i \left\{ |a_{ii}| - \sum_{j \neq 1} |a_{ij}| \right\} > \frac{1}{3}\delta_\Delta > 0. \tag{11}$$

Moreover, $S_\Delta(f; x)$ exists and is unique if and only if Eq. (6) has a unique solution. In addition, the standard convergence results ([1], Chapter II) can be obtained for the intervals $[x_0, \infty)$ and $(-\infty, \infty)$ if

$$\|A^{-1}\|_\infty = \sup_j \sum_j |b_{ij}| < K/\delta_\Delta, \tag{12}$$

where K is a constant independent of Δ and

$$A^{-1} = (b_{ij}). \tag{13}$$

By l^∞ we shall mean the collection of all vectors

$$v = (v_0, v_1, \ldots)^T, \tag{14}$$

where v_j $(j = 0, 1, \ldots)$ is a complex number and $\|v\|_\infty \equiv \sup_j |v_j| < \infty$. The set l^∞ is a Banach space under this norm. If $A = (a_{ij})$, $(i, j = 0, 1, 2, \ldots)$ is a matrix for which

$$\sup_i \sum_{j=0}^{\infty} |a_{ij}| < \infty,$$

then A defines a bounded linear transformation on l^∞; and if $\|A\|_\infty$ is the infimum of all positive numbers C for which

$$\|Av\|_\infty < C\|v\|_\infty, \tag{15}$$

then

$$\|A\|_\infty = \sup_i \sum_{j=0}^{\infty} |a_{ij}|. \tag{16}$$

We shall denote by l^2 the set of all vectors (14) for which

$$\|v\|^2 = \sum_{j=0}^{\infty} |v_j|^2 < \infty.$$

The space l^2 is a Hilbert space under the inner product

$$(v, u) = \sum_{j=0}^{\infty} v_j \bar{u}_j.$$

A matrix $A = (a_{ij})$ defines a linear transformation of l^2 into l^2 if Av is in l^2 whenever v is in l^2. We shall let $\|A\|$ denote the infimum of all positive numbers C for which

$$\|Av\| < C\|v\|. \tag{17}$$

If $\|A\|$ is finite, A defines a bounded linear transformation of l^2 into l^2. Although we lack a convenient expression for $\|A\|$, Schur's theorem ([2], p. 328) asserts

$$\|A\| < \left\{\left(\sup_i \sum_{j=0}^{\infty} |a_{ij}|\right)\left(\sup_j \sum_{i=0}^{\infty} |a_{ij}|\right)\right\}^{1/2}. \tag{18}$$

If A is either symmetric or Hermitian, (18) becomes

$$\|A\| < \|A\|_\infty. \tag{19}$$

The following theorem establishes the existence of $S_\Delta(f; x)$ if $\|\Delta\| < \infty$, $\delta_\Delta > 0$, and shows that (12) is satisfied so that convergence results can be derived.

THEOREM. *Let $A = (a_{ij})$ $(i, j = 0, 1, \ldots)$ be a real symmetric matrix for which*

$$\begin{aligned} &\inf_i \left\{|a_{ii}| - \sum_{j \neq i} |a_{ij}|\right\} \equiv \delta > 0, \\ &\sup_i \sum_{j=0}^{\infty} |a_{ij}| \equiv \|A\|_\infty < \infty. \end{aligned} \tag{20}$$

Under these conditions, A^{-1} exists and, with the notation $A^{-1} = (b_{ij})$, we have

$$\|A^{-1}\|_\infty \equiv \sup_i \sum_{j=0}^{\infty} |b_{ij}| < \frac{1}{\delta}. \tag{21}$$

Proof. Consider, momentarily, A as a linear transformation on l^2. Since (19) holds, A is a bounded linear transformation, and since A is Hermitian, its spectrum is real; thus for $\lambda > 0$, $A_\lambda^{-1} \equiv (A - i\lambda I)^{-1}$ exists. Moreover, for positive λ we have

$$\inf_i \left\{ |a_{ii} - i\lambda| - \sum_{j \neq i} |a_{ij}| \right\} \equiv \delta_\lambda > 0. \tag{22}$$

In addition $\delta_\lambda \to \delta$ as $\lambda \to 0$. Now, A_λ^{-1} is a bounded linear operator on l^2 and can be represented by an infinite matrix $(a_{ij}^{(-\lambda)})$ which is symmetric since A_λ, while not Hermitian, is symmetric. Let v be in l^2 and choose $\epsilon > 0$. There exists i_ϵ such that $\|v\|_\infty - |v_{i_\epsilon}| < \epsilon$. Consequently, if $A_\lambda = (a_{ij}^\lambda)$,

$$\begin{aligned} \|A_\lambda v\|_\infty &\geq |a_{i_\epsilon i_\epsilon}^\lambda| |v_{i_\epsilon}| - \|v\|_\infty \sum_{j \neq i_\epsilon} |a_{i_\epsilon j}^\lambda| \\ &\geq \|v\|_\infty \left\{ |a_{i_\epsilon i_\epsilon}^\lambda| - \sum_{j \neq i_\epsilon} |a_{i_\epsilon j}^\lambda| \right\} - |a_{i_\epsilon i_\epsilon}^\lambda| \epsilon. \end{aligned} \tag{23}$$

Since (23) holds for all $\epsilon > 0$ and since $|a_{i_\epsilon i_\epsilon}^\lambda| < \|A_\lambda\|_\infty$, we have

$$\|A_\lambda v\|_\infty \geq \delta_\lambda \|v\|_\infty. \tag{24}$$

Indeed the inequality (24) holds not only for v in l^2 but for any v in l^∞. However, for ω in l^2 we can choose v such that $v = A_\lambda^{-1} \omega$, and obtain

$$\|A_\lambda^{-1} \omega\|_\infty \leq \frac{1}{\delta_\lambda} \|\omega\|_\infty. \tag{25}$$

Now let

$$Y_k^N = (\exp(-i \text{ any } a_{k,0}^{(-\lambda)}), \exp(-i \text{ any } a_{k,1}^{(-\lambda)}), \ldots, \exp(-i \text{ any } a_{k,n}^{(-\lambda)}), 0, 0, \ldots)^T. \tag{26}$$

Since Y_k^N is in l^2 and $\|Y_k^N\|_\infty = 1$, it follows that

$$|a_{k,0}^{(-\lambda)}| + |a_{k,1}^{(-\lambda)}| + \ldots + |a_{k,N}^{(-\lambda)}| \leq \frac{1}{\delta_\lambda}. \tag{27}$$

But this is true for $N = 0, 1, \ldots,$ and for each k. Hence,

$$\sup_k \sum_{j=0}^{\infty} |a_{kj}^{(-\lambda)}| \leq \frac{1}{\delta_\lambda} < \infty,$$

and A_λ^{-1} is actually a bounded linear transformation on l^∞ which is inverse to A_λ. Moreover, we can choose λ such that

$$\|A - A_\lambda\|_\infty = |\lambda| < \delta_\lambda \leq \frac{1}{\|A_\lambda^{-1}\|_\infty};$$

hence ([2], p. 164) A^{-1} exists. Furthermore, we have for every $\lambda > 0$

$$\|A^{-1}\|_\infty < \frac{\|A_\lambda^{-1}\|_\infty}{1 - \|A_\lambda^{-1}\|_\infty \lambda}.$$

If we let $\lambda \to 0$, then

$$\|A^{-1}\|_\infty < \frac{1}{\delta},$$

which proves the theorem.

A simpler proof can be given in the case $\mathscr{R}_\Delta < 2$. Let D be the diagonal matrix whose diagonal is that of A. Then D^{-1} exists and

$$\|D^{-1}\|_\infty < \frac{3}{2\delta_\Delta}.$$

In addition, we have

$$\|A - D\|_\infty < \frac{\|\Delta\|}{3} = \frac{\mathscr{R}_\Delta}{3}\delta_\Delta < \frac{2\delta_\Delta}{3} < \frac{1}{\|D^{-1}\|_\infty}$$

under the condition $\mathscr{R}_\Delta < 2$. The theorem now follows ([2], p. 164).

We have demonstrated that if a sequence of $\{y_i\}$ $(i = 0, 1, \ldots)$ is prescribed such that for the associated vector d we have $\|d\|_\infty < \infty$, then there is a unique cubic spline $S_\Delta(x)$ interpolating to y_i at x_i, having a prescribed value y_0' for $S'_\Delta(x)$ at x_0, and having $\|M\|_\infty < \infty$. If we do not require $\|d\|_\infty < \infty$, then there is a one-parameter family of splines on $[x_0, \infty]$ having these properties. This s also true even if $\|d\|_\infty < \infty$; but in this case, $\|M\|_\infty = \infty$ except for one value of the parameter. In support of these assertions we observe that there is a unique cubic $C(x)$ on the interval $[x_0, x_1]$ with $C(x_0)$, $C'(x_0)$, $C''(x_0)$, $C(x_1)$ arbitrarily prescribed. Using the values of $C'(x)$ and $C''(x)$ at x_1 we now (by repeating the construction) can extend the domain of definition of $C(x)$ to $[x_0, x_2]$ and have $C(x)$ in $C^2[x_0, x_2]$. Further repetition of this construction gives the desired spline function. Similarly, $C(x)$ can be extended to the interval $(-\infty, \infty)$. In this case there are two more degrees of freedom ($C'(x_0)$, $C''(x_0)$) than when we require $\|d\|_\infty < \infty$, $\|M\|_\infty < \infty$. When $\|d\|_\infty < \infty$ in order to avoid a contradiction to our earlier existence theorem, $\|M\|_\infty = \infty$ with one exception.

The following inequalities typify the behavior of $C(x)$: If $C''(x) < Q$ for $x > x_0 = 0$, then $C(x) < y_0 + y_0'x + \frac{1}{2}Qx^2$ and $C'(x) < y_0' + Qx$ for $x > 0$. Similarly, if $C''(x) > Q_1$ for $x > x_0 = 0$, then $C(x) > y_0 + y_0'x + \frac{1}{2}Q_1x^2$ and $C'(x) > y_0' + Q_1x$ for $x > 0$. These inequalities are the best possible since they become equalities if $C(x)$ is a quadratic. Of course $C''(x_i) \leqq Q$ for $i \geqq 0$ implies $C''(x) \leqq Q$ for $x \geqq x_0$; $C''(x_i) \geqq Q_1$ for $i \geqq 0$ implies $C''(x) \geqq Q_1$ for $x \geqq x_0$.

References

1. J. H. Ahlberg, E. N. Nilson, and J. L. Walsh, "The Theory of Splines and Their Applications". Academic Press, New York, 1967.
2. A. E. Taylor, "Introduction to Functional Analysis". Wiley, New York, 1958.

Commentary by Walter Schempp

The smoothing of functions on an interval I of the real line $\mathbf{R}$ by polynomials of a fixed degree yields a process in numerical analysis that is not flexible enough to perform a close fit of the polynomials on the prescribed data set. According to the initial concept of a smooth interpolating function used as a draftman's tool for the numerical design of ship lines, a spline function represents locally a polynomial function. This means that pieces of polynomials are joined together so that the resulting function retains a sufficiently high order of differentiability at the mesh points of I. A related localization problem in the field of signal processing has recently motivated the introduction of the concept of wavelet to Fourier analysis. Wavelets allow for expansions in terms of simple building blocks that are reasonably localized in space and frequency. Thus the tendency of technology at the end of this century of moving from adapted analysis in mechanical engineering to electrical engineering also finds its counterpart in the recent developments of mathematicians. These developments which use splines to generate quadrature filters will have far reaching technological applications to subband coding.

In 1946, I. J. Schoenberg started a systematic study of the use of spline functions in the smoothing of equidistant data with the cubic splines as prototype of this exciting new class of approximating functions. This interpolation problem by cubic splines on the real line leads to a system of linear equations which admits an interesting structure. Due to the fact that the matrix associated to the linear system is symmetric and has a dominant diagonal, it allows for a functional analytic treatment of the existence and uniqueness problem via spectral theory. ([8]).

As discovered by J. C. Holladay in 1957, there exists a mechanical engineering fundament of the theory of cubic splines which is deeper and less obvious. As suggested by the initial concept of spline function, the link to mechanical engineering is given by the minimum strain energy property of beams. This extremal property, which corresponds to the minimum curvature property of interpolating cubic splines, forms the key to the development of quadratic variational principles in the theory of spline functions. It suggests to estimate the deviation of the interpolated function in terms of energy norms and to characterize the proximum in terms of Sobolev type norms. The standard L^2 Sobolev inequality for the distributional derivatives suggests to abandon the approximation by piecewise polynomial functions and to replace it by the concept of generalized splines which form piecewise solutions of linear differential equations. In analogy to the Sturm-Liouville theory of ordinary differential equations, the key idea is to consider a variable coefficient linear differential operator L of order n given by

$$L\cdot = \sum_{0\leq j\leq n} p_j(x)\mathrm{D}^{n-j}.$$

with coefficients $(p_j)_{0\leq j\leq n}$ of class C^n. The formal adjoint $L^\star$ of L takes the form

$$L^\star\cdot = \sum_{0\leq j\leq n} (-1)^{n-j}\mathrm{D}^{n-j}(p_j(x)\cdot)$$

and allows one to formulate in an elegant way the basic integral relations for a subdivision of the unit interval in terms of the operator product

$$LL^\star$$

on the Sobolev space. Quadratic variational arguments then allow one to establish proximum and minimum norm properties together with existence, uniqueness, and orthogonality properties for both periodic, non-periodic, and multivariate generalized splines in a unified manner. In this way, a window was opened up for functional analysis on spline theory to make the structure of spline functions transparent and to initiate a functional analytic treatment ([1], [2], [3], [4], [5]) of the various classes of spline functions. A summary of the functional analytic approach has been collected in the monograph ([7]) which presents an impressive approach to a comprehensive theory of generalized spline approximation based on quadratic variational principles.

Another aspect of the approximation by cubic splines, independent of the application to extremal energy problems in mechanical engineering, is the extension to complex spline functions. The complex spline integral is defined by the Cauchy integral along a rectifiable Jordan curve C in the complex plane $\mathbf{C}$ which is approximated by interpolating complex cubic splines. Provided the function f is holomorphic inside the domain D bounded by the curve C with parameter interval I, a careful analysis of the boundary behavior in terms of Hölder conditions for the function f and its derivatives $(\mathrm{D}^j f)_{1\leq j\leq 3}$ allows for a study of the convergence properties of the Cauchy integrals for sequences of successively refined meshes on C. An application of the maximum principle yields an estimate of the pointwise deviation in the domain $D \subset \mathbf{C}$ in terms of the mesh widths and the order of the Hölder condition on the boundary $C = \partial D$ of D ([6], [9], [10]). This elegant procedure permits the determination of structural properties of the holomorphic function f being approximated on D.

References

[1] J. L. Walsh, J. H. Ahlberg, and E. N. Nilson, Best approximation properties of the spline fit. *J. Math. Mech.* **11**, 225–234 (1962)
[2] J. H. Ahlberg, E. N. Nilson, and J. L. Walsh, Fundamental properties of generalized splines. *Proc. Nat. Acad. Sci. USA* **52**, 1412–1419 (1964)
[3] J. H. Ahlberg, E. N. Nilson, and J. L. Walsh, Best approximation properties of higher-order spline approximations. *J. Math. Mech.* **14**, 231–244 (1965)
[4] J. H. Ahlberg, E. N. Nilson, and J. L. Walsh, Convergence properties of generalized splines. *Proc. Nat. Acad. Sci. USA* **54**, 344–350 (1965)
[5] J. H. Ahlberg, E. N. Nilson, and J. L. Walsh, Extremal, orthogonality and convergence properties of multidimensional splines. *J. Math. Anal. Appl.* **11**, 27–48 (1965)
[6] J. H. Ahlberg, E. N. Nilson, and J. L. Walsh, Complex cubic splines. *Trans. Amer. Math. Soc.* **129**, 391–413 (1967)
[7] J. H. Ahlberg, E. N. Nilson, and J. L. Walsh, *The theory of splines and their applications*. Academic Press, New York 1967
[8] J. H. Ahlberg, E. N. Nilson, and J. L. Walsh, Cubic splines on the real line. *J. Approx. Theory* **1**, 5–10 (1968)
[9] J. H. Ahlberg, E. N. Nilson, and J. L. Walsh, Properties of analytic splines (I): Complex polynomial splines. *J. Math. Anal. Appl.* **27**, 262–278 (1969)
[10] J. H. Ahlberg, E. N. Nilson, and J. L. Walsh, Complex polynomial splines on the unit circle. *J. Math. Anal. Appl.* **33**, 234–257 (1971)

Lehrstuhl fuer Mathematik I, University of Siegen, D-57068 Siegen, Germany
schempp@mathematik.uni-siegen.d400.de